Advances in Rock Mechanics and Geotechnical Engineering—Volume I

Advances in Rock Mechanics and Geotechnical Engineering—Volume I

Editors

Mahdi Hasanipanah
Danial Jahed Armaghani
Jian Zhou

MDPI

Basel • Beijing • Wuhan • Barcelona • Belgrade • Novi Sad • Cluj • Manchester

Editors
Mahdi Hasanipanah
Institute of Research and Development
Duy Tan University
Da Nang, Vietnam

Danial Jahed Armaghani
School of Civil and Environmental Engineering
University of Technology Sydney
Sydney, Australia

Jian Zhou
School of Resources and Safety Engineering
Central South University
Changsha, China

Editorial Office
MDPI
St. Alban-Anlage 66
4052 Basel, Switzerland

This is a reprint of articles from the Special Issue published online in the open access journal *Sustainability* (ISSN 2071-1050) (available at: https://www.mdpi.com/journal/sustainability/special_issues/advances_rock_mechanics_geotechnical_engineering).

For citation purposes, cite each article independently as indicated on the article page online and as indicated below:

Lastname, A.A.; Lastname, B.B. Article Title. *Journal Name* **Year**, *Volume Number*, Page Range.

Volume I
ISBN 978-3-0365-9772-0 (Hbk)
ISBN 978-3-0365-9773-7 (PDF)
doi.org/10.3390/books978-3-0365-9773-7

Set
ISBN 978-3-0365-9762-1 (Hbk)
ISBN 978-3-0365-9763-8 (PDF)

Contents

About the Editors

Mahdi Hasanipanah

Dr. Mahdi Hasanipanah obtained his BSc degree in Mining Engineering from the University of Kashan, Iran, in 2008; his MSc degree in Mining Engineering from the Islamic Azad University—South Tehran Branch, Iran, in 2010; and his PhD in Mining Engineering from University of Kashan, Iran, in 2020. His research interests include rock blasting, rock mechanics, machine learning methods, and optimization algorithms. He has cooperated in a large number of projects in Iran and Malaysia. He is also the chief Guest Editor of *Sustainability* (MDPI)'s Special Issue "Advances in Rock Mechanics and Geotechnical Engineering". His citation and H-index are 5600 and 49, respectively, which indicate the significant impact and influence of his research in the academic and scientific communities. He was among the top 2% of scientists for two consecutive years, including 2022 and 2023, according to Stanford University.

Danial Jahed Armaghani

Dr. Danial Jahed Armaghani is a prominent researcher in the field of civil and geotechnical engineering. His work has made significant contributions to the understanding and mitigation of geotechnical and geological hazards, earning him a reputation as an excellent researcher in his field. Dr. Danial's research focuses on a wide range of topics, including slope stability analysis, rock mechanics, tunnel construction, surface and deep excavations, and applying machine learning models and optimization algorithms to solve various geotechnical problems. Dr. Danial obtained his MEng (2012) and PhD (2015) in the area of Civil-Geotechnics, from Universiti of Teknologi Malaysia (UTM), Malaysia. He is currently working as a Postdoctoral Research Fellow at the School of Civil and Environmental Engineering, University of Technology Sydney (UTS), Australia. He published more than 300 articles in well-established ISI and Scopus journals and at national and international conferences. His published works have received more than 19,000 citations, which indicate the significant impact and influence of his research in the academic and scientific communities. He was among the top 2% of scientists for four consecutive years from 2020 to 2023, according to Stanford University.

Jian Zhou

Assoc. Prof. Jian Zhou obtained his BSc degree (2008) and PhD (2015) from Central South University (CSU), China, and was a visiting scholar with Mine Design Laboratory at McGill University from 2013 to 2014. Currently, he is an associate professor in the School of Resources and Safety Engineering at CSU, China. His current research interests include geological and geotechnical hazards prediction and mitigation, applying predictive models in rock mechanics, and mining engineering. Dr. Zhou is the Highly Cited Researcher in the field of Cross-Field (Clarivate); a Highly Cited Chinese Researcher in the field of Mining Engineering (Elsevier); one of the World's Top 2% Scientists according to Stanford University; and a recipient of the Distinguished Young Scholars Fund of Hunan Province, China. He has published more than 180 papers in international journals on mining and geotechnical issues and has received China's 100 Most Influential International Academic Papers Award, the *Journal of Rock Mechanics and Geotechnical Engineering's* Best Paper Award, and the *Journal of Central South University's* Best Paper Award. His citation and H-index are 9600 and 55, respectively.

sustainability

MDPI

Article

Optimized Data-Driven Models for Prediction of Flyrock due to Blasting in Surface Mines

Xiaohua Ding [1,2], Mehdi Jamei [3,4], Mahdi Hasanipanah [5,6,*], Rini Asnida Abdullah [6] and Binh Nguyen Le [5,7]

1 School of Mines, China University of Mining and Technology, Xuzhou 221116, China
2 State Key Laboratory of Coal Resources and Safe Mining, China University of Mining and Technology, Xuzhou 221116, China
3 Faculty of Engineering, Shohadaye Hoveizeh Campus of Technology, Shahid Chamran University of Ahvaz, Dashte Azadegan 78986, Iran
4 New Era and Development in Civil Engineering Research Group, Scientific Research Center, Al-Ayen University, Thi-Qar, Nasiriyah 64001, Iraq
5 Institute of Research and Development, Duy Tan University, Da Nang 550000, Vietnam
6 Department of Geotechnics and Transportation, Faculty of Civil Engineering, Universiti Teknologi Malaysia, Johor Bahru 81310, Malaysia
7 School of Engineering and Technology, Duy Tan University, Da Nang 550000, Vietnam
* Correspondence: hasanipanahmahdi@duytan.edu.vn

Abstract: Using explosive material to fragment rock masses is a common and economical method in surface mines. Nevertheless, this method can lead to some environmental problems in the surrounding regions. Flyrock is one of the most dangerous effects induced by blasting which needs to be estimated to reduce the potential risk of damage. In other words, the minimization of flyrock can lead to sustainability of surroundings environment in blasting sites. To this aim, the present study develops several new hybrid models for predicting flyrock. The proposed models were based on a cascaded forward neural network (CFNN) trained by the Levenberg–Marquardt algorithm (LMA), and also the combination of least squares support vector machine (LSSVM) and three optimization algorithms, i.e., gravitational search algorithm (GSA), whale optimization algorithm (WOA), and artificial bee colony (ABC). To construct the models, a database collected from three granite quarry sites, located in Malaysia, was applied. The prediction values were then checked and evaluated using some statistical criteria. The results revealed that all proposed models were acceptable in predicting the flyrock. Among them, the LSSVM-WOA was a more robust model than the others and predicted the flyrock values with a high degree of accuracy.

Keywords: blast-induced flyrock; LSSVM; optimization; prediction models

Citation: Ding, X.; Jamei, M.; Hasanipanah, M.; Abdullah, R.A.; Le, B.N. Optimized Data-Driven Models for Prediction of Flyrock due to Blasting in Surface Mines. *Sustainability* **2023**, *15*, 8424. https://doi.org/10.3390/su15108424

Academic Editor: Rajesh Kumar Jyothi

Received: 6 April 2023
Revised: 13 May 2023
Accepted: 18 May 2023
Published: 22 May 2023

1. Introduction

Drilling and blasting is an indispensable technique for breakage and displacement of rock masses in open-pit mines. Nevertheless, some undesirable phenomena, such as ground vibration, airblast, flyrock (FR), and backbreak are produced by blasting operations [1–5]. Any blasting event produces a sudden ejection of rock pieces, which are referred to as "FR". This phenomenon is one of the most hazardous environmental issues induced by blasting which may lead to various problems for humans, including fatalities [6–9]. As mentioned in previous studies, some blast design factors, such as burden (*B*), spacing (*S*), stemming (*ST*), weight charge (*WC*), and powder factor (*PF*), are the effective factors in the intensity of *FR* [8–10]. Aside from the aforementioned factors, the properties of the rock mass, such as rock density and uniaxial compressive strength, are considered the effective factors on the *FR*, called uncontrollable factors [9,10]. The *FR* can occur based on three different mechanisms, i.e., rifling, face burst, and cratering [11,12]. The poor *ST* material, the small ratio of *ST* to blast-hole diameter, and inadequate *B* are the most important causes for the rifling, cratering, and face burst mechanisms [9,13].

In the literature, numerical simulations are considered to be the common methods to study the blasting mechanism in rock masses. According to Kutter and Fairhurst [14], three main zones, i.e., the crushed, cracked, and elastic vibration zones, can be formed after each blasting event. Additionally, the in-situ stress has an important effect on the propagation of cracks produced by blasting. To study the failure responses of rocks, the distinct element method (DEM), finite difference method (FDM), and finite element method (FEM) have been extended in recent years by many scholars [15–20]. As an example, a two-dimensional discrete element method was used to numerically simulate the mechanism of rock fragmentation produced by blasting in the study conducted by Hajibagherpour et al. [19]. They showed that the proposed numerical model can be effectively employed to simulate the crack propagation process around a blast-hole. Aside from numerical modelling, several empirical models have been employed to predict flyrock [20]. These empirical models have been formulated based on considering only one or two of the effective factors of flyrock. For this reason, the accuracy of the mentioned empirical models is not good enough. Therefore, the use of artificial intelligence methods can be a good solution to predict *FR* with a high degree of performance. Additionally, the use of artificial intelligence methods in different fields of mingling and civil engineering indicates the effectiveness of these methods for predicting and optimizing aims [21–31].

An artificial neural network (ANN) model was employed to predict *FR* in the study conducted by Monjezi et al. [32], and its performance was compared with statistical models. Their results indicated the performance of ANN was better than statistical models in predicting *FR*. For the same purpose, Ghasemi et al. [33] employed ANN and fuzzy system (FS) and showed better prediction capability of FS over ANN. Moreover, the ANN model was compared with the adaptive neuro-fuzzy inference system (ANFIS) for the prediction of *FR* by Trivedi et al. [34]. They revealed higher performance in respect to accuracy of ANFIS compared with the ANN model. In another study, a genetic programming (GP) model was employed by Faradonbeh et al. [35] to predict *FR*. In their study, the non-linear regression models were also used for comparison aims. They concluded that the GP predicted *FR* with a higher performance in comparison to non-linear regression models. The GP model was also employed by Ye et al. [36] and its results was compared with a random forest model. According to their results, the performance of GP was better than the random forest model.

The present study attempts to propose several efficient hybrid models through the cascaded forward neural network (CFNN) and also the least squares support vector machine (LSSVM) in combination with three optimization algorithms, including artificial bee colony (ABC), gravitational search algorithm (GSA), and whale optimization algorithm (WOA), for the prediction of *FR*.

The rest of this article includes the following sections. More details about the source of the database and the developed models are explained in the second section. Then, the setting parameters in the modelling processes are explained in the third section. The results and discussions are provided in the fourth section; finally, the last section presents the conclusions of the study.

2. Research Significance

FR is considered as an environmental and hazardous problem in mine blasting, which may result in human injuries, fatalities, property damage, and instability of slopes. Hence, a valid and reliable prediction of *FR* has critical implications in mitigating and controlling the adverse effects along with sustainable development and responsible mining. In other words, the control and minimization of *FR* can lead to sustainability of surroundings environment in blasting sites. For the aforementioned aims, the present study attempts to propose several efficient hybrid models through the CFNN and also LSSVM in combination with three optimization algorithms. To the best of our knowledge, this is the first work that predicts the *FR* by using the proposed models.

3. Materials and Methods

3.1. Materials

The database used in this study was collected from three granite quarry sites located in Malaysia, including the Ulu Tiram, Pengerang, and the Masai quarry sites. The values of the rock quality designation (RQD) of the aforementioned quarry sites ranged from 45 to 80, 50 to 70, and 40 to 75, respectively. Additionally, the values of the rock strength ranged from 30 to 110 MPa, respectively. In total, 80 datasets including some effective parameters on the *FR* were used in constructing the predictive models. In this regard, the *S*, *B*, *ST*, *PF*, and density were used as the input parameters, and the *FR* was used as the output parameter. More details about the statistical properties of datasets will be provided in Section 5.

3.2. Methods

In this study, the LSSVM is combined with the ABC, GSA, and WOA to predict *FR*. Additionally, the CFNN model is also used for comparison aims. In this section, the mentioned models are briefly explained.

3.2.1. LSSVM Model

LSSVM is a robust machine learning technique. This method was proposed as an upgraded form of the SVM, which suffered from some drawbacks in its learning stage, namely the demand in calculability and the limitation in dealing with inequality constraints. Therefore, LSSVM becomes an efficient ML technique after fixing the aforesaid issues [37,38]. For a regression task which aims at finding a suitable correlation that emulates the behaviour of a system defined by a set of data having inputs $x = \{x_1, x_2, \ldots, x_N\}$ that $x_j \in R^D$ and N is the number of samples in the set, and targets t defined on R as $y = \{y_1, y_2, \ldots, y_N\}$, the first step in the LSSVM method consists of formulating the following minimization problem [39]:

$$\begin{gathered} \text{minimize } \tfrac{1}{2} w^T w + \tfrac{1}{2}\gamma \sum_{j=1}^{N} \left(e_j^2\right) \\ s.t.\ y_j = w^T \varphi(x_j) + b + e_j,\ i = 1, 2, \ldots, N \end{gathered} \tag{1}$$

where e_j denotes the regression error, γ represents the regularization parameter, ^{T} points out the transpose operator, w and b are the weight and bias parameters, respectively, and φ is a nonlinear mapping function.

The learning phase of LSSVM passes through finding the proper values of w and b. To this end, the formulated minimization problem is transformed into a Lagrangian function using the formula shown below [40]:

$$L(w, b, \alpha, e) = \frac{1}{2} w^T w + \frac{1}{2}\gamma \sum_{j=1}^{N} \left(e_j^2\right) - \sum_{j=1}^{N} \alpha_j \left(w^T \varphi(x_j) + b + e_j - y_j\right) \tag{2}$$

where the coefficients α_i are called Lagrangian multipliers. The solution of L is obtained by solving the following system of equations:

$$\begin{cases} \frac{\partial L(w,b,\alpha,e)}{\partial w} = 0 \Rightarrow w = \sum_{j=1}^{N} \alpha_j \varphi_j(x_j) \\ \frac{\partial L(w,b,\alpha,e)}{\partial b} = 0 \Rightarrow \sum_{j=1}^{N} \alpha_j = 0 \\ \frac{\partial L(w,b,\alpha,e)}{\partial e_j} = 0 \Rightarrow \alpha_j = \gamma e_j, \quad j = 1, 2, \ldots, N \\ \frac{\partial L(w,b,\alpha,e)}{\partial \alpha_j} = 0 \Rightarrow w^T \varphi(x_j) + b + e_j - y_j = 0, \quad j = 1, 2, \ldots, N \end{cases} \tag{3}$$

The above system which defines the vanishing of the partial derivatives of L with regard to w, b, e, and α can be arranged in the following matrix scheme:

$$\begin{bmatrix} 0 & 1_N^T \\ 1_N & \Omega + \gamma^{-1} I_N \end{bmatrix} \begin{bmatrix} b \\ \alpha \end{bmatrix} = \begin{bmatrix} 0 \\ y \end{bmatrix} \tag{4}$$

In the above equation, I_N points out $N \times N$ size identity matrix, $y = [y_1, y_2, \ldots, y_N]^T$, $\alpha = [\alpha_1, \alpha_2, \ldots, \alpha_N]^T$, $1_N = [1, 1, \ldots, 1]^T$, and Ω is the kernel matrix. The elements of this latter term are expressed as follows:

$$\Omega_{j,l} = \varphi(x_j)\varphi(x_l) = K(x_j, x_l) \tag{5}$$

where K is the kernel function. Gaussian radial basis function (RBF) is among the frequently considered kernel functions in LSSVM.

In the last step, the gained LSSVM paradigm can predict the investigated target using the following expression:

$$f(x) = \sum_{j=1}^{N} \alpha_j K(x_j, x_l) + b \tag{6}$$

where (α_j, b) are determined from Equation (4).

It is worth mentioning that the robustness of LSSVM is related to the proper selection of its hyper-parameters, viz.,

σ^2 and γ. To do so, in the present work, three rigorous metaheuristic algorithms were suggested to tune these control parameters.

In this study, three optimization algorithms, including the GSA, WOA, and ABC, are used to improve the LSSVM performance. The aforementioned algorithms are briefly explained in this part.

(A) Gravitational search algorithm (GSA)

The GSA is a metaheuristic algorithm developed by Rashedi et al. [41] based on Newton's law of gravity [37]. The GSA is a population-based algorithm, and this means that a population of possible solutions is considered during the optimization process. The particles of the population are subjected to positions updating using the main governing equations of the algorithms. In this regard, the position of each particle is denoted by a vector x, while the force between two elements i and j at iteration g, is expressed as follows [41]:

$$F_{ij}^g = G^g \frac{M_i^g \, M_j^g}{R_{ij}^g + \varepsilon} \left(x_i^g - x_j^g \right) \tag{7}$$

where ε points out a constant with a small value, R denotes the Euclidian distance between the two particles, while G represents the gravitational constant defined as:

$$G^g = G^{g_0} \frac{g_0^\chi}{g} \quad \chi < 1 \tag{8}$$

where G^{g_0} is the initial value of the gravitational constant. The overall force resulted from the particles of the population on each particle i is determined using the following formula [41]:

$$F_i^g = \sum_{j \in J_{best},\, j \neq i}^{N} r_{1j} F_{ij}^g \tag{9}$$

where J_{best} represents a set of best particles in the population and r_{1j} is a random determined uniformly over iterations from [0, 1].

In another step, the motion law is considered as per following formulas [41]:

$$a_i = \frac{F_i^g}{M_i^g} \tag{10}$$

where a_i points out the acceleration of mass and M_i is the inertia mass which is determined using the equation below:

$$M_i^g = \frac{m_i^g}{\sum_{j=1}^{N} m_j^g} \tag{11}$$

and

$$m_i^g = \frac{f_i^g - w^g}{b^g - w^g} \tag{12}$$

where f is the fitness value of the element i, and w and b represent the worst and best fitness values in the population, respectively. Finally, the velocity and the position of the elements are:

$$v_i^{g+1} = r_{2i} v_i^g + a_i^g \tag{13}$$

$$x_i^{g+1} = v_i^{g+1} + x_i^g \tag{14}$$

where r_{2i} is a random generated uniformly from [0, 1], and v and x point out the velocity and position of elements, respectively.

The steps of GSA based on the stated equations are repeated until a stopping criterion is fulfilled.

(B) Whale optimization algorithm (WOA)

The WOA is another population-based algorithm introduced by Mirjalili and Lewis [42]. The main steps and the governing equations of WOA mimic the hunting process of back whales [37]. Initially, an initial population of whales is created randomly. The positions of the whales represent possible solutions of the optimization problem. In order to evaluate the quality of these positions, a fitness function that emulates the objective function to optimize is applied. Based on the evaluation step, the whales are subjected to update their positions at a given generation $(g+1)$. To do so, the associated shapes of the position are changed to spiral or circular forms with respect to a probability p using the following equation [42]:

$$X^{g+1} = \begin{cases} D' e^{bl} \cos(2\pi t) + X^{*g} & if \;\; p \geq 0.5 \\ X^{*g} - AD & if \;\; p < 0.5 \end{cases} \tag{15}$$

In the above equation, X^* points out the best whale which is located nearby the prey. $D' = |X_i - X^*|$ represents the distance between the whale i and the prey, b is a constant for specifying the spiral shape, and l is a random number from $[-1, 1]$. The other terms of the above equation are defined as per following equations [42]:

$$A = 2ar - a \tag{16}$$

$$C = 2r \tag{17}$$

$$D = |CX^* - X_i| \tag{18}$$

where a is a number decreasing linearly from 2 to 0 over the distance, and r is a random from [0, 1]. According to the value of A, if it is mainly outside the interval $[-1, 1]$, a circular form is considered for upgrading the positions based on a randomly selected whale X_r^g. The following equation shows this process:

$$X^{g+1} = X_r^g - AD \tag{19}$$

The new positions obtained after carrying out the update of the shapes are assessed using the fitness function. Lastly, if the best whale shows an enhancement in its fitness values, it will change its position to this newest one, otherwise, the best existing position will be conserved.

The described steps of this optimization algorithm are repeated until a stopping criterion is achieved.

(C) Artificial Bee Colony (ABC)

The ABC is an intelligent swarm-based metaheuristic algorithm proposed by Karaboga [43]. As indicated in its appellation, ABC emulates the steps performed by honeybees as they search for nectar sources. In this regard, three groups, including employed, onlookers, and scout bees are considered in this process. The role of employed group is to amass the information and expose it to the onlooker group. The scout group consists of changing the positions once no improvement is noticed in some sources of bees.

The main steps of ABC are given below [44]:

- An initial population of bees is generated randomly. Each bee has its own position x. The numbers of employed and onlookers are the same in the population. A fitness function is considered for assessing the quality of the bees.
- Employed bees: this step consists of updating the positions of bees at the generation $(g + 1)$ using the following equation:

$$x_i^{g+1} = x_i^g + \vartheta_i \left(x_i^g - x_\omega^g \right) \quad (20)$$

where ϑ_i is a random from [0, 1] and $\omega \in \{1, 2, \ldots, colony\ size\}_{\omega \neq i}$.

Afterward, the quality of each new position is examined using the fitness function, and if an improvement is obtained, this new position is conserved, otherwise, it is abandoned.

- Onlooker bees: by carrying out the previous step which emulates the exploitation phase, the gained information by the employed bees is exposed to the onlookers, which select the proper ones by applying the following equation bases on the probability P:

$$P_i = \frac{ft_i}{\sum_{i=1}^{E} ft_i} \quad (21)$$

where ft points out the fitness value and E represents the number of employed bees. As in the case of employed bees, if an improvement is obtained, this new position is conserved, otherwise, it is abandoned.

- Scout bees: this step consists of randomly changing the position of a given employed bee after a defined number of generations if it does not show any improvements in its fitness quality.

The optimum solution of the problem is represented by the fittest bee. The above-described steps are recurring until a stopping condition is reached.

3.2.2. CFNN Model

The CFNN is a type of ANN which is recognized by its flexible-based structure [45]. This advantage allows CFNN to generate accurate predictive models for many systems with different degrees of complexity. The structure of CFNN neurons is distributed into three kinds of layers, including input, hidden, and output layers [46]. The input layer receives the data, then this latter is transformed and processed in one or more hidden layers using the so-called activation functions (such as *tansig* and *logsig*), while the results of the paradigm are obtained from the output layer. The number of hidden layers and their involved neurons depend on the complexity of the system as one hidden layer is generally enough for low to medium complicated cases, while more than one hidden layer is requested for highly complex cases. CFNN is characterized by its specific cascaded scheme for linking the neurons to the others [46]. This scheme is ensured by linking each neuron from a preceding layer to the nodes of the subsequent layers [46].

As the other kinds of ANN, the learning phase of CFNN aims at achieving the suitable weight and bias values of its architecture. Backpropagation-based algorithms, such as the

Levenberg–Marquardt algorithm (LMA), are known to be highly efficient for this kind of optimization. In this investigation, LMA algorithms were considered in the optimization of the bias and weights of CFNN. More information about LMA can be found in published literature [47,48].

It is worth mentioning that the considered soft computing approaches in this study, namely CFNN and LSSVM, differ from each other mainly on the learning strategy where in LSSVM, the learning process is done after the formulation of the minimization problem (shown in Equation (1)) and then the problem is resolved by finding the control parameters of the model, while in CFNN, the learning approach is gained by finding the suitable topology and the appropriate weights linking between the neurons of different layers until reaching the low function error (such as root mean square error) value.

3.2.3. Model Performance Evaluation

As stated in the previous section, the modelling task of *FR* using CFNN and LSSVM was extended in this study by investigating the suitable input parameters that can give the most accurate predictions of this vital factor. Before showing the main finding of the modelling tasks using the aforesaid ML models, it is worth mentioning that during the performance evaluation of the models, the following statistical indexes were calculated using the equations shown below [1,4,5,49–52]:

Average Absolute Relative Error (*AARE*):

$$AARE\% = \frac{1}{N}\sum_{i=1}^{N}\left|\frac{FR_{imea} - FR_{ipred}}{FR_{imea}}\right| \times 100 \tag{22}$$

Coefficient of Determination (R^2):

$$R^2 = 1 - \frac{\sum_{i=1}^{N}\left(FR_{imea} - FR_{ipred}\right)^2}{\sum_{i=1}^{N}\left(FR_{ipred} - \overline{FR}\right)^2} \tag{23}$$

Root Mean Square Error (*RMSE*):

$$RMSE = \sqrt{\frac{1}{N}\sum_{i=1}^{N}\left(FR_{imea} - FR_{ipred}\right)^2} \tag{24}$$

In the above equations, the subscripts *mea* and *pred* denote the measured and estimated *FR*, respectively, $\overline{FR}$ mean of *FR* values, and *N* points out the number of samples.

4. Development of Predictive Models

The main purpose of the suggested ML learning methods in this study was to deliver robust models that can estimate the *FR* under different circumstances. For a better investigation, several input parameters were considered in each of the proposed paradigms. Accordingly, six different schemes were involved in the development of these ML-based predictive models. These schemes are summarized in the following equations:

$$(M_1) \qquad FR = f(S, B, ST, PF,\ Density) \tag{25}$$

$$(M_2) \qquad FR = f(S, B, ST, PF) \tag{26}$$

$$(M_3) \qquad FR = f(S, B, ST,\ Density) \tag{27}$$

$$(M_4) \qquad FR = f(S, B, PF,\ Density) \tag{28}$$

$$(M_5) \qquad FR = f(S, ST, PF,\ Density) \tag{29}$$

$$(M_6) \qquad FR = f(B, ST, PF, \ Density) \tag{30}$$

In order to properly implement CFNN-LMA and the proposed hybridization LSSVM-metaheuristic algorithms, including the LSSVM-WOA, LSSVM-GSA, and LSSVM-ABC models, some necessary steps were carried out as shown in the two flowcharts of Figures 1 and 2, respectively. According to these flowcharts, the first step consisted of normalizing the database. This step significantly improves the performance of the considered ML techniques. The normalization procedure of the database is given in the following equation:

$$x_n = \frac{2(x_i - x_{min})}{(x_{max} - x_{min})} - 1 \tag{31}$$

where x and x_n point out the variable and the normalized value, respectively, while x_{max} and x_{min} represent the maximum and minimum values of the variable, respectively.

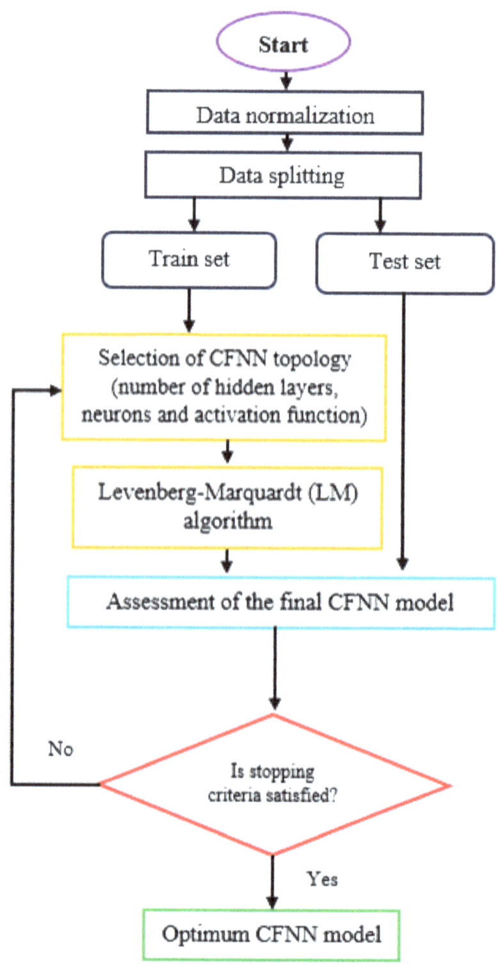

Figure 1. The workflow of the suggested CFNN paradigm.

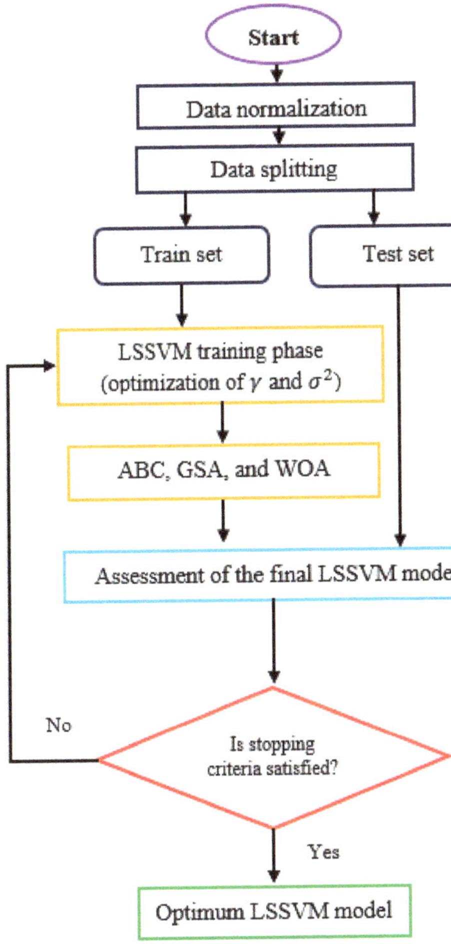

Figure 2. The workflow of the suggested LSSVM paradigms.

After performing the normalization of the data, the latter was divided into train and test sets. The aim of these two groups was to train the models (train set) and certify their robustness of unseen measurements (test set). These two sets covered 80% and 20% of the whole measurements, respectively. During the learning phase of the CFNN- and LSSVM-based models, their control parameters were investigated using the above-discussed algorithms and some other techniques. In this regard, the trial and error method was considered for selecting the best topology of CFNN, while LMA was applied for optimizing the weights and bias values of the network. For the LSSVM model, three meta-heuristic algorithms including ABC, GSA, and WOA were implemented in the optimization of the two impacting LSSVM control parameters, namely γ and σ^2. It is necessary to add that it was proven in some previous works that it is more suitable to consider some specific trust region algorithms such as Levenberg–Marquardt algorithms rather than applying metaheuristic algorithms in the training phase of some feedforward networks such as MLP and CFNN [53]. However, for many other soft-computing approaches such as LSSVM and SVM, metaheuristic algorithms are much more appropriate for finding the control parameters of these techniques [54]. Table 1 reports the main setting of these metaheuristic algorithms. In our suggested workflows (Figures 1 and 2), the constraints and/or evaluation of overfitting were considered by checking the accuracy of the paradigms during both training and testing phases. It is clear from the statistical evaluation of the models that if the prediction accuracy of the latter is very satisfactory during the training and testing phases, the overfitting issue is avoided.

Table 1. The considered control parameters of the three employed metaheuristic algorithms.

Algorithm	Parameter	Value
ABC	Number of employer bees	20
	Number of onlooker bees	20
	Number of generations	30
	Number of generations to scout bees	4
GSA	r_{1j} and r_{2j}	[0, 1]
	Number of generations	30
	Number of individuals	40
WOA	a	2 to 0
	r	[0, 1]
	Number of generations	30
	Number of whales	40

5. Results and Discussion

5.1. Exploratory Analysis

The optimal strategy for solving the high non-linear problems significantly depend on the behavior of the dataset used in the simulation process. Thus, the statistical data description is one of the most crucial tasks of the pre-processing stage in ML-based modelling in engineering problems. Table 2 lists the statistical properties of datasets implemented in the *FR* predicting procedure. The low values of skewness and kurtosis confirmed that all inputs and targets are categorized as a pseudo-normal distribution. Figure 3 demonstrates the Pearson correlation coefficients in form of a correlogram. It can be concluded that the "*S*" parameter with respect to the largest correlation coefficient (0.65) has the most significance in prediction of the *FR* value.

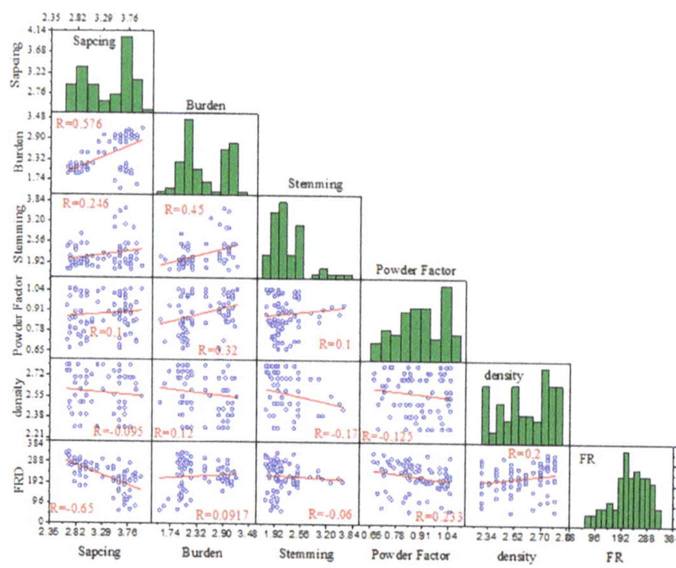

Figure 3. Correlogram for assessing the Pearson correlation coefficient between input and target parameters. Green: Histogram, Red: Regression line, and Blue: Distribution points.

Table 2. Descriptive statistics of all features used in modelling the *FR*.

Metric/Feature	*S* (m)	*B* (m)	*ST* (m)	*PF* (kg/m^3)	Density (gr/cm^3)	*FR* (m)
Minimum	2.65	1.5	1.7	0.67	2.3	61
Maximum	4	3.2	3.6	1.05	2.8	334
Mean	3.324	2.415	2.171	0.8908	2.579	223.5
Std. Deviation	0.4228	0.4776	0.4022	0.113	0.1684	64.61
CV	12.72%	19.78%	18.53%	12.69%	6.529%	28.91%
Skewness	−0.2032	0.2017	1.608	−0.2148	−0.2942	−0.5848
Kurtosis	−1.556	−1.518	2.82	−1.058	−1.247	−0.1065

As stated before, the database considered in this study was divided into training and testing sets. The first set is applied in the models' development, while the test set is devoted to the validation and investigation of the accuracy behavior of the established models when dealing with unseen measurements.

5.2. Modelling Results

By performing the steps described in the previous sections, it was found that 3 hidden layers with *tansig* as an activation function, and 12, 11, and 9 neurons in each of them, respectively, represented the proper CFNN topology in all of the six schemes. For LSSVM models, the achieved σ^2 and γ values using the ABC, GSA, and WOA ranged between 403.15 to 1847.43 and 35,479,174.56 to 67,688,321.04, respectively.

The statistical evaluation of the performance of the obtained ML-based models with respect to the stated six schemes in the previous sections is shown in Table 3. In this table, statistical criteria, namely *AARE*, R^2, and *RMSE* are reported for the training set, the test set, and the whole dataset. According to this table, and based on the schemes, it can be seen that M_1 is the best one, followed by M_2. In the combination of M_1 including all inputs for the testing phase, the LSSVM-WOA in terms of (R^2 = 0.999, *RMSE* = 3.4209 m, and *AARE* = 1.3017) was identified as the superior predictive model, followed by CFNN-LMA (R^2 = 0.9347, *RMSE* = 16.5215 m, and *AARE* = 7.512), LSSVM-GSA (R^2 = 0.904, *RMSE* = 16.1775 m, and *AARE* = 5.5193), and LSSVM-ABC (R^2 = 0.9049, *RMSE* = 19.439 m, and *AARE* = 8.0032), respectively. In the M_2 (in testing phase), as the second-best combination, the LSSVM-WOA with respect to the highest R^2 (0.9896) and smallest *RMSE* (10.4268) outperformed the other models. The result assessment demonstrated that M_6 on account of poorest performance (R^2 = 0.3616 and *RMSE* = 58.5744 for the LSSVM-WOA) was recognized as the worst scheme regardless of the ML type. Typically, it can be understood from this remark that *S* is the most impacting input parameter on *FR* as its exclusion from the input variables (M_6) caused the worst prediction performance regardless of the type of ML techniques, while density has a small effect on *FR* since its elimination from the input parameters (M_2) did not significantly affect the degree of prediction accuracy. In addition, it can also be deduced that for each of the six schemes, the LSSVM-WOA yielded more accurate predictions compared with the other LSSVM-metaheuristic algorithms and the CFNN-LMA. According to Table 3, it was found that M_1 outperformed other combinations followed by M_2, M_4, M_3, M_5, and M_6, respectively. Additionally, it can be said that the LSSVM-WOA in all input combinations was the best predictive model developed in this study for prediction of *FR*. For better comparison between the predictive performances of the provided models in all combinations, the probability density function violin plots are exhibited in Figure 4. According to this figure, considering the best agreement between measured and predicted values of *FR*, it can be clearly implied that the LSSVM-WOA was the superior model for accurately estimating *FR*, the CFNN-LMA was identified as the second-best model, and LSSVM-ABC yielded the worst results in all combinations. Regarding the mentioned analysis, the combination of M_1 was kept for further performance investigation and validations. It is necessary to add that in order to confirm that the ANN scheme did not suffer from the overfitting issue, a 4-fold cross-validation was performed on our best ANN paradigm (the case of M_1) to assess the generalization of the model when dealing with new sets of data. To do so, the database was randomly divided into 4 folds, then, the modelling was done by considering a sole fold as the test sub-data and devoting the rest for the training phase. In order to swap between the folds involved in the training and testing phases, the aforesaid step was repeated 4 times. The results gained from the 4-fold cross-validation are reported in Table 4. As can be seen, the consistency of the model is confirmed for all the folds, thus, the overfitting issue is avoided.

Table 3. Performance of the suggested ML-based models with respect to the six schemes.

Scheme	Model	Statistical Criteria	Train Data	Test Data	All Data
M_1	CFNN-LMA	R^2	0.9875	0.9347	0.977
		AARE	2.5163	7.512	3.5154
		RMSE	7.2292	16.5215	9.8184
	LSSVM-ABC	R^2	0.9867	0.9049	0.971
		AARE	3.0419	8.0032	4.0341
		RMSE	7.4089	19.439	10.9311
	LSSVM-GSA	R^2	0.9828	0.904	0.967
		AARE	3.4215	5.5193	3.8411
		RMSE	8.7308	16.1775	10.6453
	LSSVM-WOA	R^2	0.9926	0.9991	0.9943
		AARE	1.6473	1.3017	1.5782
		RMSE	7.4847	3.4209	6.8671
M_2	CFNN-LMA	R^2	0.9812	0.9366	0.972
		AARE	2.9261	8.5748	4.0558
		RMSE	8.4645	20.2612	11.8077
	LSSVM-ABC	R^2	0.9769	0.9235	0.9675
		AARE	3.7643	7.083	4.428
		RMSE	9.8655	16.7483	11.5743
	LSSVM-GSA	R^2	0.9754	0.9054	0.9614
		AARE	4.254	6.5356	4.7103
		RMSE	10.3962	18.0443	12.312
	LSSVM-WOA	R^2	0.9871	0.9896	0.9875
		AARE	2.6662	2.8723	2.7074
		RMSE	10.105	10.4268	10.1702
M_3	CFNN-LMA	R^2	0.883	0.9172	0.89
		AARE	7.2573	8.2441	7.4546
		RMSE	21.9097	18.6501	21.2977
	LSSVM-ABC	R^2	0.9191	0.7622	0.8811
		AARE	6.7642	12.4379	7.8989
		RMSE	17.7366	34.5376	22.1413
	LSSVM-GSA	R^2	0.8874	0.7979	0.8714
		AARE	7.8757	13.615	9.0236
		RMSE	21.7035	27.6908	23.0259
	LSSVM-WOA	R^2	0.9625	0.9398	0.9305
		AARE	6.6657	21.8618	9.7049
		RMSE	17.5076	43.8141	25.0828

Table 3. *Cont.*

Scheme	Model	Statistical Criteria	Train Data	Test Data	All Data
M_4	CFNN-LMA	R^2	0.8878	0.8706	0.8856
		AARE	6.9643	10.5219	7.6758
		RMSE	21.6209	22.0891	21.7153
	LSSVM-ABC	R^2	0.892	0.8572	0.8849
		AARE	7.0333	9.8644	7.5995
		RMSE	20.9307	24.9094	21.7846
	LSSVM-GSA	R^2	0.8689	0.8748	0.8707
		AARE	8.1285	11.6169	8.8262
		RMSE	22.7463	24.3938	23.0852
	LSSVM-WOA	R^2	0.9921	0.9201	0.9777
		AARE	2.7384	14.8704	5.1648
		RMSE	8.0666	31.6042	15.8689
M_5	CFNN-LMA	R^2	0.8879	0.8456	0.8791
		AARE	7.6983	11.2721	8.413
		RMSE	21.3743	25.7841	22.326
	LSSVM-ABC	R^2	0.8891	0.8376	0.8823
		AARE	8.3536	7.5541	8.1937
		RMSE	22.1859	21.3895	22.0289
	LSSVM-GSA	R^2	0.8983	0.831	0.8825
		AARE	6.6676	11.6499	7.6641
		RMSE	19.9291	28.846	22.0035
	LSSVM-WOA	R^2	0.8979	0.7985	0.8859
		AARE	12.374	13.7799	12.6551
		RMSE	29.5911	30.7061	29.8175
M_6	CFNN-LMA	R^2	0.4324	0.3195	0.4186
		AARE	17.817	30.9799	20.4496
		RMSE	46.3091	58.3475	48.9542
	LSSVM-ABC	R^2	0.4856	0.1575	0.4327
		AARE	17.9507	30.1723	20.395
		RMSE	44.6322	61.0195	48.356
	LSSVM-GSA	R^2	0.5574	0.4669	0.4329
		AARE	18.587	24.6491	19.7994
		RMSE	44.7295	60.6898	48.3449
	LSSVM-WOA	R^2	0.806	0.3616	0.7228
		AARE	15.5962	22.1002	16.897
		RMSE	40.172	58.5744	44.466

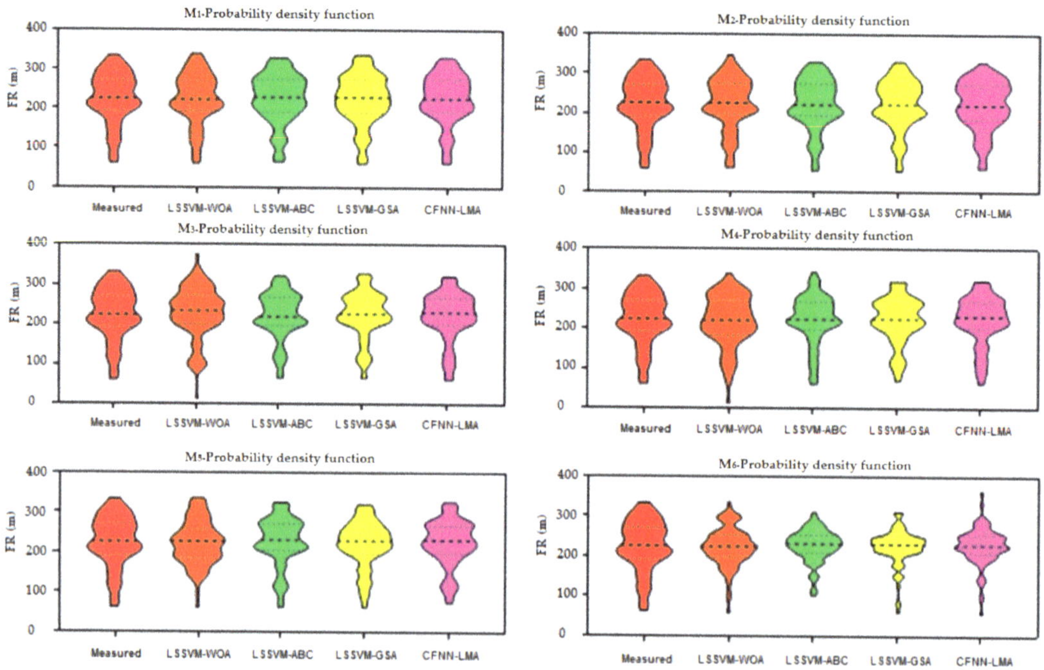

Figure 4. Probability density function of the predictive models of M_1 for prediction of *FR* for all datasets used in the simulation process.

Table 4. Results of the 4-fold cross-validation for the CFNN-LMA model (M_1 scheme).

	Overall R^2	Overall *RMSE*
Fold 1	0.9759	9.8664
Fold 2	0.9761	9.8601
Fold 3	0.9754	9.9361
Fold 4	0.9762	9.8597

In order to extend the examination of the accuracy of the best implemented models, some graphical evaluation techniques were considered. Figure 5 shows the physical trend variation and cross plots related to the M_1 which illustrate a comparison of the measured and predicted *FR* values during both the training and testing phases. Based on the scatter plots, a tight cloud of points is located nearby the line $Y = X$ for all datasets. This means that the LSSVM-WOA can predict *FR* values with a great degree of accuracy as its predictions are very close to the perfect case shown by the unit-slop line. The left side of Figure 6 shows the ability of the predictive models to capture the non-linearity behaviour of the datasets. The LSSVM-WOA yields the best agreement with the measured *FR* compared with other LSSVM models and the CFNN-LMA model.

Another visual tool for inspecting the reliability of the LSSVM-WOA paradigm is dealing with the relative deviation (RD%) distribution diagram that is exhibited in Figure 6. The quartiles of RD for 25% and 75% of datasets are listed in Table 5. The LSSVM-WOA model, owing to having the least values of Q25% = −0.5566, Q75% = 1.093, and IQR = 1.650, yielded more reliable outcomes in comparison with the CFNN-LMA (Q25% = −2.976, Q75% = 2.105, and IQR = 5.081), LSSVM-ABC (Q25% = −2.778, Q75% = 2.504, and IQR = 5.282), and LSSVM-GSA (Q25% = −2.712, Q75% = 2.716, and IQR = 5.428), respectively. Obviously, the above diagnostic analysis reveals that the LSSVM-WOA model has very low prediction errors regardless of the considered conditions.

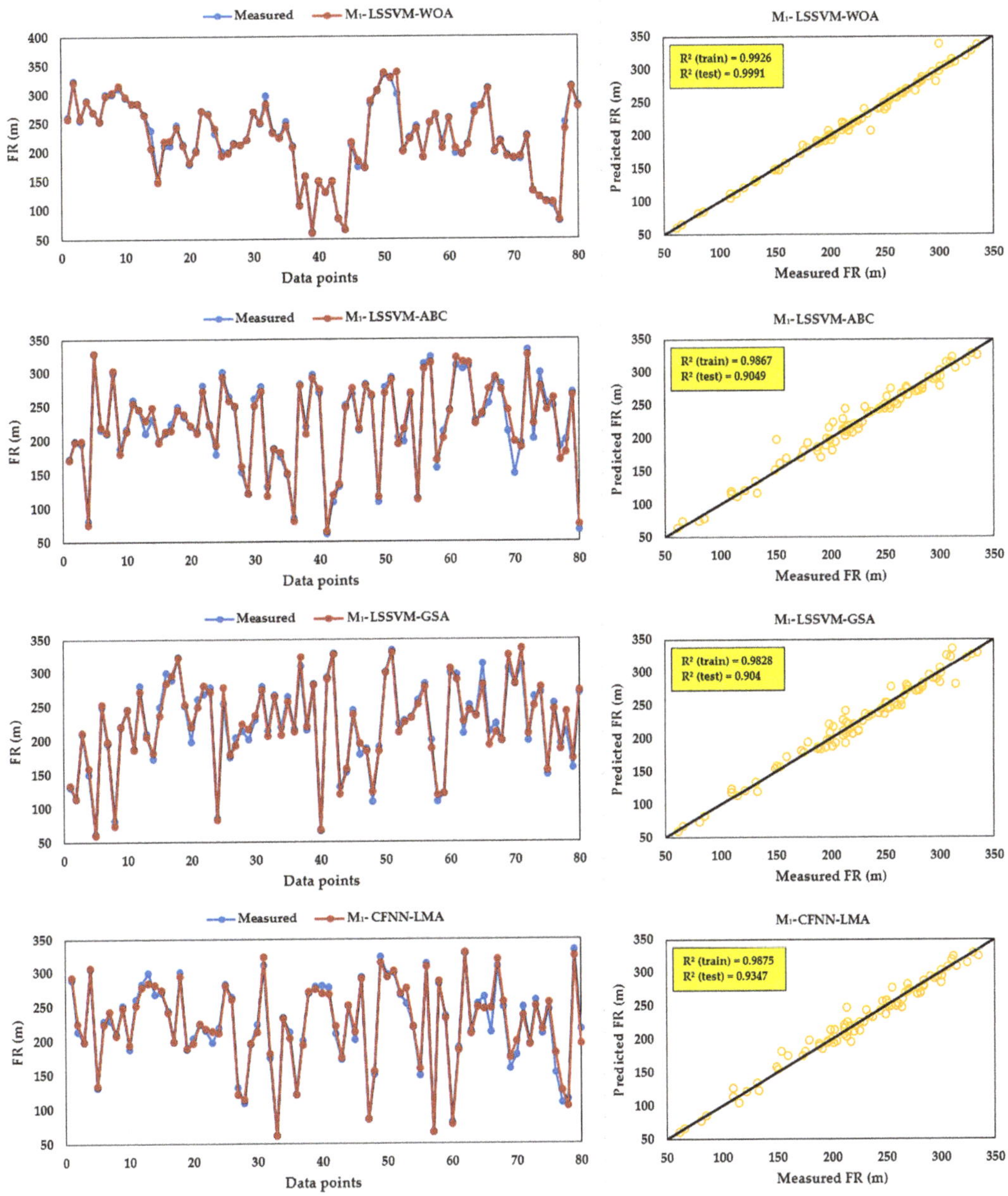

Figure 5. Comparing the predicted and measured *FR* values in the form of the cross plot and trend variation plot for all datasets.

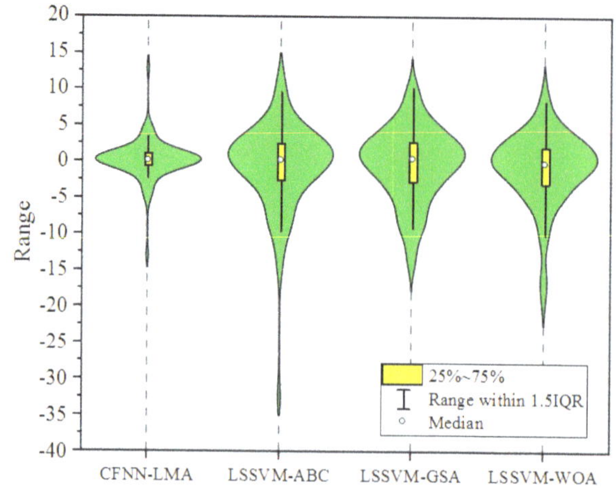

Figure 6. Relative errors of prediction models versus the measured *FR* values.

Table 5. Quartile of RD for 25% and 75% of datasets.

Model	CFNN-LMA	LSSVM-ABC	LSSVM-GSA	LSSVM-WOA
$Q_{25\%}$-RD%	−0.5566	−2.778	−2.712	−2.976
$Q_{75\%}$-RD%	1.093	2.504	2.716	2.105
IQR-RD%	1.650	5.282	5.428	5.081

Lastly, Figure 7 demonstrates the 3D surface of RD variation concerning the LSSVM-WOA versus the *S* and *PF*, as the two most significant variables, to assess the reliability seeking zone for estimating the *FR*. The RD variation reveals that the most reliable and accurate results are obtained in the ranges of $3 \leq S \leq 3.5$ and $0.8 \leq PF \leq 1.1$.

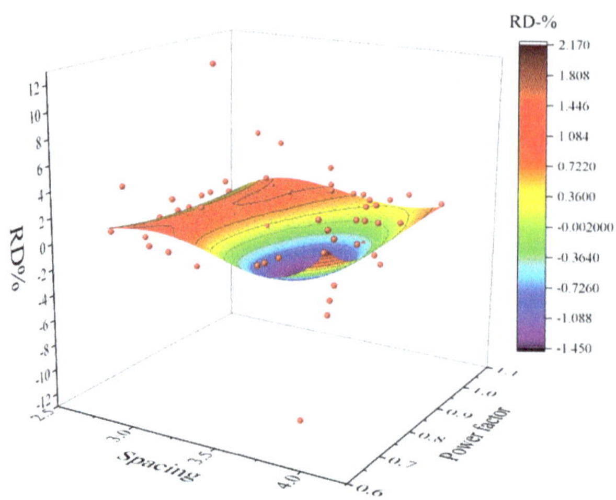

Figure 7. RD variation versus S and PF for LSSVM-WOA model. (Red points: Samples in distribution).

6. Conclusions

FR is one of the most adverse effects induced by blasting in surface mines. In this study, an attempt was made to predict blast-induced *FR* through hybridizing three optimization algorithms including the ABC, GSA, and WOA with the LSSVM model. In addition, CFNN-LMA, as a powerful tool for the prediction aims, was developed. For developing the models, six different schemes based on different combinations of input parameters were employed and in total, twenty-four different models were constructed. After that, three statistical indexes, i.e., *AARE*, R^2, and *RMSE*, were used to check the performance of the models and to compare their results.

Some conclusions are drawn as follows: the results indicated that among the total constructed models, the LSSVM-WOA model was the most accurate model in all six schemes compared with the CFNN-LMA, LSSVM-GSA, and LSSVM-ABC models. The most accurate results of the LSSVM-WOA (*AARE* = 1.3017, R^2 = 0.9991 and *RMSE* = 3.4209), LSSVM-GSA (*AARE* = 5.5193, R^2 = 0.904 and *RMSE* = 16.1775), and CFNN-LMA (*AARE* = 7.512, R^2 = 0.9347 and *RMSE* = 16.5215) were obtained from the first scheme, while that for the LSSVM-ABC model (*AARE* = 7.083, R^2 = 0.9235 and *RMSE* = 16.7483) was obtained from the second scheme. It is important to note that the above results were related to the testing phase. On the other hand, the sixth scheme had the worst performance. In this scheme, the *S* parameter was removed from the modelling. Therefore, it can be suggested that the *S* was an effective parameter in the modelling. Additionally, according to RD%, the LSSVM-WOA model had very low prediction errors regardless of the considered conditions. The presented results in this study cannot be compared with results of the previous studies because different fields investigation as well as different range of input parameters were used in the previous studies. Nevertheless, for a comparison with the literature, the LSSVM-WOA presented in this study predicted the *FR* with a very good R^2, while Koopialipoor et al. [55], Faradonbeh et al. [56], Zhou et al. [57], Nguyen et al. [58], and Marto et al. [59] predicted the *FR* with an R^2 of 0.959, 0.924, 0.944, 0.986, and 0.981, respectively. The aforementioned results indicate the effectiveness of the LSSVM-WOA model in predicting the *FR*.

It is worth mentioning that the proposed models herein are specific to studied cases and the use of these models in other surface mines requires some modification based on blasting and mining conditions. However, our best implemented model can predict the *FR* with high accuracy and the paradigm can be applied for cases filling the applicability conditions considered in this work, namely by respecting the maximum and minimum values of the involved input parameters.

For future works, it can be recommended to use other optimization algorithms, such as seagull optimization algorithm (SOA), sparrow search algorithm (SSA), loin swarm optimization (LSO), and moth-flame optimization (MFO) algorithms, in combination with CFNN and LSSVM models.

Author Contributions: Conceptualization, M.H.; Methodology, X.D. and M.J.; Validation, X.D., M.J. and M.H.; Investigation, M.H. and B.N.L.; Data curation, M.H.; writing—original draft preparation, X.D., M.J., M.H., R.A.A. and B.N.L.; writing—review and editing, X.D., M.J., M.H., R.A.A. and B.N.L.; Supervision, M.H. and R.A.A. All authors have read and agreed to the published version of the manuscript.

Funding: This paper is supported by the National Natural Science Foundation of China (Grant No. 52174131).

Institutional Review Board Statement: Not applicable.

Informed Consent Statement: Not applicable.

Data Availability Statement: The data used in this study may be available on request from the corresponding author.

Acknowledgments: The authors would like to thank Danial Jahed Armaghani for providing the information and facilities required for conducting this research.

Conflicts of Interest: The authors have no conflict of interest to declare that are relevant to the content of this article.

References

1. Raina, A.K.; Chakraborty, A.K.; Choudhury, P.B.; Sinha, A. Flyrock danger zone demarcation in opencast mines: A risk based approach. *Bull. Eng. Geol. Environ.* **2011**, *70*, 163–172. [CrossRef]
2. Stojadinovic, S.; Pantovic, R.; Zikic, M. Prediction of flyrock trajectories for forensic applications using ballistic flight equations. *Int. J. Rock Mech. Min. Sci.* **2011**, *48*, 1086–1094. [CrossRef]
3. Rad, H.N.; Hasanipanah, M.; Rezaei, M.; Eghlim, A.L. Developing a least squares support vector machine for estimating the blast-induced flyrock. *Eng. Comput.* **2018**, *34*, 709–717. [CrossRef]
4. Jamei, M.; Hasanipanah, M.; Karbasi, M.; Ahmadianfar, I.; Taherifar, S. Prediction of flyrock induced by mine blasting using a novel kernel-based extreme learning machine. *J. Rock Mech. Geotech. Eng.* **2021**, *13*, 1438–1451. [CrossRef]
5. Yari, M.; Armaghani, D.J.; Maraveas, C.; Ejlali, A.N.; Mohamad, E.T.; Asteris, P.G. Several Tree-Based Solutions for Predicting Flyrock Distance Due to Mine Blasting. *Appl. Sci.* **2023**, *13*, 1345. [CrossRef]
6. Raina, A.K.; Chakraborty, A.K.; More, R.; Choudhury, P.B. Design of factor of safety based criterion for control of flyrock/throw and optimum fragmentation. *J. Inst. Eng. Ser. A* **2007**, *87*, 13–17.
7. Ghasemi, E.; Sari, M.; Ataei, M. Development of an empirical model for predicting the effects of controllable blasting parameters on flyrock distance in surface mines. *Int. J. Rock Mech. Min. Sci.* **2012**, *52*, 163–170. [CrossRef]
8. Armaghani, D.J.; Mahdiyar, A.; Hasanipanah, M.; Shirani Faradonbeh, R.; Khandelwal, M.; Bakhshandeh Amnieh, H. Risk assessment and prediction of flyrock distance by combined multiple regression analysis and monte-carlo simulation of quarry blasting. *Rock Mech. Rock Eng.* **2016**, *49*, 3631–3641. [CrossRef]
9. Hasanipanah, M.; Jahed Armaghani, D.; Bakhshandeh Amnieh, H.; Abd Majid, M.Z.; Tahir, M. Application of PSO to develop a powerful equation for prediction of flyrock due to blasting. *Neural Comput. Appl.* **2017**, *28*, 1043–1050. [CrossRef]
10. Hasanipanah, M.; Armaghani, D.J.; Amnieh, H.B.; Koopialipoor, M.; Arab, H. A risk based technique to analyze flyrock results through rock engineering system. *Geotech. Geol. Eng.* **2018**, *36*, 2247–2260. [CrossRef]
11. Raina, A.K.; Murthy, V.; Soni, A.K. Flyrock in bench blasting: A comprehensive review. *Bull. Eng. Geol. Environ.* **2014**, *73*, 1199–1209. [CrossRef]
12. Jahed Armaghani, D.; Hajihassani, M.; Monjezi, M.; Edy Tonnizam, M.; Marto, A.; Moghaddam, M.R. Application of two intelligent systems in predicting environmental impacts of quarry blasting. *Arab. J. Geosci.* **2015**, *8*, 9647–9665. [CrossRef]
13. Little, T.N.; Blair, D.P. Mechanistic Monte Carlo models for analysis of flyrock risk. *Rock Fragm. Blasting* **2010**, *9*, 641–647.
14. Kutter, H.K.; Fairhurst, C. On the fracture process in blasting. *Int. J. Rock Mech. Min. Sci. Geomech. Abstr.* **1971**, *8*, 181–202. [CrossRef]
15. Liu, K.; Li, X.; Hao, H.; Li, X.; Sha, Y.; Wang, W.; Liu, X. Study on the raising technique using one blast based on the combination of long-hole presplitting and vertical crater retreat multiple-deck shots. *Int. J. Rock Mech. Min. Sci.* **2019**, *113*, 41–58. [CrossRef]
16. Chao, Z.; Ma, G.; Wang, M. Experimental and numerical modelling of the mechanical behaviour of low-permeability sandstone considering hydromechanics. *Mech. Mater.* **2020**, *148*, 103454. [CrossRef]
17. Wang, M.; Wang, F.; Zhu, Z.; Dong, Y.; Mousavi Nezhad, M.; Zhou, L. Modelling of crack propagation in rocks under SHPB impacts using a damage method. *Fatigue Fract. Eng. Mater. Struct.* **2019**, *42*, 1699–1710. [CrossRef]
18. Wang, M.; Ma, G.; Wang, F. Numerically investigation on blast-induced wave propagation in catastrophic large-scale bedding rockslide. *Landslides* **2021**, *18*, 785–797. [CrossRef]
19. Hajibagherpour, A.R.; Mansouri, H.; Bahaaddini, M. Numerical modeling of the fractured zones around a blasthole. *Comput. Geotech.* **2020**, *123*, 103535. [CrossRef]
20. Trivedi, R.; Singh, T.N.; Raina, A.K. Prediction of blast-induced flyrock in Indian limestone mines using neural networks. *J. Rock Mech. Geotech. Eng.* **2014**, *6*, 447–454. [CrossRef]
21. Yang, H.; Koopialipoor, M.; Armaghani, D.J.; Gordan, B.; Khorami, M.; Tahir, M.M. Intelligent design of retaining wall structures under dynamic conditions. *Steel Compos. Struct.* **2019**, *31*, 629–640.
22. Yang, H.; Liu, X.; Song, K. A novel gradient boosting regression tree technique optimized by improved sparrow search algorithm for predicting TBM penetration rate. *Arab. J. Geosci.* **2022**, *15*, 461. [CrossRef]
23. Yang, H.; Song, K.; Zhou, J. Automated Recognition Model of Geomechanical Information Based on Operational Data of Tunneling Boring Machines. *Rock Mech. Rock Eng.* **2020**, *55*, 1499–1516. [CrossRef]
24. Yang, H.; Wang, Z.; Kanglei, S. A new hybrid grey wolf optimizer-feature weighted-multiple kernel-support vector regression technique to predict TBM performance. *Eng. Comput.* **2020**, *38*, 2469–2485. [CrossRef]
25. Zhang, X.; He, B.; Sabri, M.M.S.; Al-Bahrani, M.; Ulrikh, D.V. Soil Liquefaction Prediction Based on Bayesian Optimization and Support Vector Machines. *Sustainability* **2022**, *14*, 11944. [CrossRef]
26. Khan, N.M.; Cao, K.; Yuan, Q.; Bin Mohd Hashim, M.H.; Rehman, H.; Hussain, S.; Emad, M.Z.; Ullah, B.; Shah, K.S.; Khan, S. Application of Machine Learning and Multivariate Statistics to Predict Uniaxial Compressive Strength and Static Young's Modulus Using Physical Properties under Different Thermal Conditions. *Sustainability* **2022**, *14*, 9901. [CrossRef]
27. Shahani, N.M.; Zheng, X.; Guo, X.; Wei, X. Machine Learning-Based Intelligent Prediction of Elastic Modulus of Rocks at Thar Coalfield. *Sustainability* **2022**, *14*, 3689. [CrossRef]
28. Yu, Q.; Monjezi, M.; Mohammed, A.S.; Dehghani, H.; Armaghani, D.J.; Ulrikh, D.V. Optimized Support Vector Machines Combined with Evolutionary Random Forest for Prediction of Back-Break Caused by Blasting Operation. *Sustainability* **2021**, *13*, 12797. [CrossRef]

29. Hasanipanah, M.; Bakhshandeh Amnieh, H. A fuzzy rule based approach to address uncertainty in risk assessment and prediction of blast-induced flyrock in a quarry. *Nat. Resour. Res.* **2020**, *29*, 669–689. [CrossRef]
30. Huat, C.Y.; Moosavi, S.M.H.; Mohammed, A.S.; Armaghani, D.J.; Ulrikh, D.V.; Monjezi, M.; Hin Lai, S. Factors Influencing Pile Friction Bearing Capacity: Proposing a Novel Procedure Based on Gradient Boosted Tree Technique. *Sustainability* **2021**, *13*, 11862. [CrossRef]
31. Hosseini, S.; Pourmirzaee, R.; Armaghani, D.J.; Sabri Sabri, M.M. Prediction of ground vibration due to mine blasting in a surface lead–zinc mine using machine learning ensemble techniques. *Sci. Rep.* **2023**, *13*, 6591. [CrossRef] [PubMed]
32. Monjezi, M.; Mehrdanesh, A.; Malek, A.; Khandelwal, M. Evaluation of effect of blast design parameters on flyrock using artificial neural networks. *Neural Comput. Appl.* **2013**, *23*, 349–356. [CrossRef]
33. Ghasemi, E.; Amini, H.; Ataei, M.; Khalokakaei, R. Application of artificial intelligence techniques for predicting the flyrock distance caused by blasting operation. *Arab. J. Geosci.* **2014**, *7*, 193–202. [CrossRef]
34. Trivedi, R.; Singh, T.N.; Gupta, N. Prediction of blast-induced flyrock in opencast mines using ANN and ANFIS. *Geotech. Geol. Eng.* **2015**, *33*, 875–891. [CrossRef]
35. Faradonbeh, R.S.; Armaghani, D.J.; Monjezi, M. Development of a new model for predicting flyrock distance in quarry blasting: A genetic programming technique. *Bull. Eng. Geol. Environ.* **2016**, *75*, 993–1006. [CrossRef]
36. Ye, J.; Koopialipoor, M.; Zhou, J.; Jahed Armaghani, D.; He, X. A Novel Combination of Tree-Based Modeling and Monte Carlo Simulation for Assessing Risk Levels of Flyrock Induced by Mine Blasting. *Nat. Resour. Res.* **2021**, *30*, 225–243. [CrossRef]
37. Hemmati Sarapardeh, A.; Larestani, A.; Nait Amar, M.; Hajirezaie, S. *Applications of Artificial Intelligence Techniques in the Petroleum Industry*; Gulf Professional Publishing: Houston, TX, USA, 2020. [CrossRef]
38. Cai, M.; Hocine, O.; Salih Mohammed, A.; Chen, X.; Nait Amar, M.; Hasanipanah, M. Integrating the LSSVM and RBFNN models with three optimization algorithms to predict the soil liquefaction potential. *Eng. Comput.* **2021**, *38*, 3611–3623. [CrossRef]
39. Suykens, J.A.K.; Vandewalle, J. Least squares support vector machine classifiers. *Neural Process. Lett.* **1999**, *9*, 293–300. [CrossRef]
40. Forrester, A.I.J.; Sóbester, A.; Keane, A.J. *Engineering Design via Surrogate Modelling: A Practical Guide*; John Wiley & Sons: Hoboken, NJ, USA, 2008.
41. Rashedi, E.; Nezamabadi-Pour, H.; Saryazdi, S. GSA: A gravitational search algorithm. *Inf. Sci.* **2009**, *179*, 2232–2248. [CrossRef]
42. Mirjalili, S.; Lewis, A. The Whale Optimization Algorithm. *Adv. Eng. Softw.* **2016**, *95*, 51–67. [CrossRef]
43. Karaboga, D. *An Idea Based on Honey Bee Swarm for Numerical Optimization*; Technical Report-TR06; Erciyes University, Engineering Faculty, Computer Engineering Department: Kayseri, Turkey, 2005.
44. Karaboga, D.; Basturk, B. A powerful and efficient algorithm for numerical function optimization: Artificial bee colony (ABC) algorithm. *J. Glob. Optim.* **2007**, *39*, 459–471. [CrossRef]
45. Nait Amar, M. Modelling solubility of sulfur in pure hydrogen sulfide and sour gas mixtures using rigorous machine learning methods. *Int. J. Hydrogen Energy* **2020**, *45*, 33274–33287. [CrossRef]
46. Abujazar, M.S.S.; Fatihah, S.; Ibrahim, I.A.; Kabeel, A.E.; Sharil, S. Productivity modelling of a developed inclined stepped solar still system based on actual performance and using a cascaded forward neural network model. *J. Clean. Prod.* **2018**, *170*, 147–159. [CrossRef]
47. Hemmati-Sarapardeh, A.; Varamesh, A.; Husein, M.M.; Karan, K. On the evaluation of the viscosity of nanofluid systems: Modeling and data assessment. *Renew. Sustain. Energy Rev.* **2018**, *81*, 313–329. [CrossRef]
48. Hemmati-Sarapardeh, A.; Nait Amar, M.; Soltanian, M.R.; Dai, Z.; Zhang, X. Modeling CO_2 Solubility in Water at High Pressure and Temperature Conditions. *Energy Fuels* **2020**, *34*, 4761–4776. [CrossRef]
49. Huang, J.; Zhang, J.; Li, X.; Qiao, Y.; Zhang, R.; Kumar, G.S. Investigating the effects of ensemble and weight optimization approaches on neural networks' performance to estimate the dynamic modulus of asphalt concrete. *Road Mater. Pavement Des.* **2022**, 1–21. [CrossRef]
50. Huang, J.; Zhou, M.; Zhang, J.; Ren, J.; Vatin, N.; Sabri, M. Development of a new stacking model to evaluate the strength parameters of concrete samples in laboratory. *Iran. J. Sci. Technol. Trans. Civ. Eng.* **2022**, *46*, 4355–4370. [CrossRef]
51. Huang, J.; Zhou, M.; Zhang, J.; Ren, J.; Vatin, N.I.; Sabri, M.M.S. The use of GA and PSO in evaluating the shear strength of steel fiber reinforced concrete beams. *KSCE J. Civ. Eng.* **2022**, *26*, 3918–3931. [CrossRef]
52. Huang, J.; Xue, J. Optimization of SVR functions for flyrock evaluation in mine blasting operations. *Environ. Earth Sci.* **2022**, *81*, 434. [CrossRef]
53. Talebkeikhah, M.; Amar, M.N.; Naseri, A.; Humand, M.; Hemmati-Sarapardeh, A.; Dabir, B.; Seghier, M.E.A.B. Experimental measurement and compositional modeling of crude oil viscosity at reservoir conditions. *J. Taiwan Inst. Chem. Eng.* **2020**, *109*, 35–50. [CrossRef]
54. Rostami, A.; Baghban, A.; Shirazian, S. On the evaluation of density of ionic liquids: Towards a comparative study. *Chem. Eng. Res. Des.* **2019**, *147*, 648–663. [CrossRef]
55. Koopialipoor, M.; Fallah, A.; Armaghani, D.J.; Azizi, A.; Mohamad, E.T. Three hybrid intelligent models in estimating flyrock distance resulting from blasting. *Eng. Comput.* **2019**, *35*, 243–256. [CrossRef]
56. Faradonbeh, R.S.; Armaghani, D.J.; Amnieh, H.B.; Mohamad, E.T. Prediction and minimization of blast-induced flyrock using gene expression programming and firefly algorithm. *Neural Comput. Appl.* **2018**, *29*, 269–281. [CrossRef]
57. Zhou, J.; Aghili, N.; Ghaleini, E.N.; Bui, D.T.; Tahir, M.M.; Koopialipoor, M. A Monte Carlo simulation approach for effective assessment of flyrock based on intelligent system of neural network. *Eng. Comput.* **2019**, *36*, 713–723. [CrossRef]

58. Nguyen, H.; Bui, X.-N.; Nguyen-Thoi, T.; Ragam, P.; Moayedi, H. Toward a state-of-the-art of fly-rock prediction technology in open-pit mines using EANNs model. *Appl. Sci.* **2019**, *9*, 4554. [CrossRef]
59. Marto, A.; Hajihassani, M.; Jahed Armaghani, D.; Tonnizam Mohamad, E.; Makhtar, A.M. A novel approach for blast-induced flyrock prediction based on imperialist competitive algorithm and artificial neural network. *Sci. World J.* **2014**, *2014*, 643715. [CrossRef]

sustainability

Article

Simulation of Surface Settlement Induced by Parallel Mechanised Tunnelling

Chia Yu Huat [1], Danial Jahed Armaghani [2,*], Sai Hin Lai [1,3,*], Haleh Rasekh [2] and Xuzhen He [2]

1 Department of Civil Engineering, Faculty of Engineering, University of Malaya, Kuala Lumpur 50603, Malaysia; 17107717@siswa.um.edu.my
2 School of Civil and Environmental Engineering, University of Technology Sydney, Sydney, NSW 2007, Australia; haleh.rasekh@uts.edu.au (H.R.); xuzhen.he@uts.edu.au (X.H.)
3 Department of Civil Engineering, Faculty of Engineering, Universiti Malaysia Sarawak (UNIMAS), Kota Samarahan 94300, Sarawak, Malaysia
* Correspondence: danial.jahedarmaghani@uts.edu.au (D.J.A.); laish@um.edu.my (S.H.L.)

Abstract: Mechanised tunnelling is extensively utilised for twin tunnel construction, particularly in urban areas. A common challenge encountered during this construction method is the occurrence of surface settlement (SS) induced by tunnelling activities. The integrity of nearby structures can be compromised by SS, making it imperative to accurately quantify and mitigate this phenomenon. Several methods for determining SS exist, including empirical formulas and laboratory studies. However, these methods are often constrained by specific soil types and are time-consuming. Moreover, crucial parameters such as tunnel operational factors and construction stages are often omitted from empirical formulas. Given these limitations, this paper aims to address these challenges by employing 3D numerical analysis to simulate tunnelling-induced SS in twin tunnels. This approach takes into account tunnel geometry, construction sequencing, soil properties, and tunnelling operational factors. By incorporating data from in-situ and laboratory tests conducted on the ground, engineering soil parameters are established as inputs for the numerical analysis. The simulated SS results obtained from the 3D numerical analysis are compared with field measurements of SS taken from available ground surface settlement markers. The transverse SS pattern derived from the numerical analysis closely mirrors the field measurements. Additionally, SS values above the first and second tunnels are compared with field measurements, resulting in coefficient of determination (R^2) values of 0.94 and 0.96, respectively. The utilisation of the 3D numerical modelling approach enables the customizable mitigation strategies for managing the SS with project-specific parameters such as tunnel geometry, geotechnical engineering factors, and tunnelling operational variables. This will help plan and construct more sustainable tunnels with minimal effects on the ground and residential areas.

Keywords: twin tunnels; 3D numerical analysis; surface settlement; field measurement; geotechnical and geological conditions

Citation: Yu Huat, C.; Armaghani, D.J.; Lai, S.H.; Rasekh, H.; He, X. Simulation of Surface Settlement Induced by Parallel Mechanised Tunnelling. *Sustainability* **2023**, *15*, 13265. https://doi.org/10.3390/su151713265

Academic Editor: Antonio Caggiano

Received: 18 July 2023
Revised: 17 August 2023
Accepted: 31 August 2023
Published: 4 September 2023

1. Introduction

Tunnelling is one of the most significant transport solutions where the overlying population does not need to be displaced. With the advancement of current technologies, mechanised excavations to construct underground spaces such as tunnels have become popular. Although mechanised tunnel excavations have been widely used all around the world, considerable safety considerations during construction need to be ensured. Several problems can be encountered during tunnel excavations, such as tunnel face instability, excessive wear of the cutter head, and excessive surface settlement (SS) [1,2]. Among them, SS induced by mechanised tunnelling is still one of the common issues in tunnel construction [3,4]. The excessive SS during and after tunnelling projects has an adverse impact on the existing structures. Therefore, it is important to estimate the settlements caused by the tunnelling to minimise the effects on the existing structures. Several methods can be used to estimate the SS due to tunnelling, which is empirical or semi-empirical, laboratory-based, and numerical analysis. One of the pioneer empirical formulas to determine the SS due to tunnelling was proposed by Peck [5], which shows that the volume loss is from the radial deformations along the tunnel perimeter directly linked to the SS. Based on the field observations on the site and the simplification of the equation proposed by Litwiniszyn [6], Peck [5] suggested Equation (1) to predict SS induced by tunnelling for soft clay. The settlement in this equation is calculated based on the pattern of SS caused by loss of ground and approximated by a Gaussian probability curve.

$$S = S_{max} e^{\frac{-x^2}{2i^2}} \tag{1}$$

where S is the surface settlement in the transverse section at a specific distance, x is the distance from the centreline of the tunnel, and i is the point of inflection (settlement through). In this way, the maximum SS can be defined using Equation (2):

$$S_{max} = \frac{V_s}{\sqrt{2 \; x \; \pi \; x \; i}} \tag{2}$$

where V_s is the volume loss of the soil (m^3/m), and can be expressed by Equation (3):

$$V_s = \frac{Volume\ Loss\ (VL)(\%)}{100}\left(\frac{\pi\ D^2}{4}\right) \tag{3}$$

The main inputs for the calculation of the maximum SS are affected by volume loss of the soil (VL) and settlement trough. These volume losses are affected by the type of ground, tunnel geometry, and ground condition [7,8]. Hence, many researchers have carried out their investigations into different types of ground conditions with various ranges of VL, as summarised in Table 1. In addition, several researchers have also proposed various i equations for different ground conditions, as presented in Table 2.

Table 1. Tunnelling methods with various ground conditions of VL.

Author (s)	Ground Condition	VL (%)	Method of Tunnelling
Attewell and Farmer [9]	London clay	1.44	Hand excavation shield tunnelling
O'Reilly and New [10]	London clay	1.0–1.4	Open face shield-driven tunnels
Mair and Taylor [11]	Stiff clay Stiff clay Sand Soft clay	1.0–2.0 0.5–1.5 0.5 1.0–2.0	Open face method NATM Closed face Tunnelling Boring Machine Closed face Tunnelling Boring Machine
Wan et al. [12]	London clay	0.8	Earth Pressure Balance Machine (EPBM)
Amir and Mohammad [13]	Graded gravel to silt/clays	0.2–0.7	EPBM
Le et al. [14]	Sandy	<0.2 to 2.4	EPBM

Table 2. Summary of empirical formulas for estimation of settlement trough width.

Authors	Empirical Formula	Variable Definition	Ground Condition	Tunnelling Excavation Method
Peck [5]	$i = R\prime \left(\frac{Z}{2R\prime}\right)^n$	R' is the radius of the tunnel. Z is the tunnel depth below ground level. n is a constant parameter dependent on soil type (0.8–1).	Various types of soils	Open cutting excavation

Table 2. *Cont.*

Authors	Empirical Formula	Variable Definition	Ground Condition	Tunnelling Excavation Method
O'Reilly and New [10]	$i = kZ$	k is the constant parameter dependent on the soil type. Z is the tunnel depth below ground level.	Various types of soils	Shield tunnelling
Mair et al. [15]	$i = Z[0.175 + 0.325(1 - \frac{Z'}{Z})$	Z' is the depth of the calculated settlement trough from the surface settlement. Z is the tunnel at a depth below ground level.	Clay	Centrifuge model test
Loganathan and Poulos [16]	$i = R'(1.15)(\frac{Z}{2R'})^{0.9}$	R' is the radius of the tunnel. Z is the tunnel at a depth below ground level.	Clay	Tunnelling machine
Wang et al. [17]	$i = m\ (R + Z\tan\left(45 - \frac{\Phi}{2}\right))$	m is the influence coefficient of the width. R' is the radius of the tunnel. Z is the tunnel depth below ground level. Φ is the internal friction angle of the soil.	Sand	Laboratory
Zhu et al. [18]	$i = 0.51Z + 0.48$	Z is the tunnel depth below ground level.	Sand and Clay	Shield machine

According to Table 2, VL and i have a wide range of values, which can be due to different methods of tunnel construction and ground conditions. It can also be seen that most studies are carried out for clay material. In addition, no tunnelling operational parameter is included in the previous empirical formulas. Loganathan and Poulos [16] proposed a semi-empirical equation (i.e., Equation (4)), which is based on the gap parameter proposed by Lee et al. [19] and the exponential function to model the nonlinear ground movement. This formula has been checked with the five case studies. From the findings, four tunnels that were constructed in stiff clay showed good agreement; however, soft clay depicted a much lower estimation of the ground loss. Nevertheless, the method proposed by Loganathan and Poulos [16] is limited to clay and based on an elastic solution.

$$\varepsilon_{xy} = \frac{4gR + g^2}{4R'^2} e^{\frac{1.38x^2}{(Z+R')^2} + \frac{0.69z^2}{Z^2}} \tag{4}$$

where R' is the radius of tunnel, Z is the tunnel depth below ground level, z is the depth of the point, g is the gap parameter, and ε_{xy} is the equivalent ground loss component at the tunnel soil interface due to the ground movements at point x, z.

Aside from the empirical and semi-empirical equations, 2D finite element analysis has been widely used to determine SS induced by tunnelling. Some of these 2D simplified techniques are the contraction method by Vermeer and Brinkgreve [20], the stress reduction method by Addenbrooke et al. [21], and the modified grout pressure method by Surarak [22]. The contraction method of the analysis works with the introduction of the correction factor (which represents the value of reduction and excavated area) as a predefined uniform radial inward strain. However, this assumption may not reflect the actual field displacement because, based on the centrifuge modelling study carried out by Mair [23], the finding shows that little ground displacement happens at the tunnel invert. In addition, the stress reduction method uses an "unloading factor" to stimulate the 3D tunnelling effects in the 2D numerical model, but this method requires trial-and-error on the unloading factor to calculate/determine the SS induced by tunnelling. Additionally, the modified grout pressure method [22] is the method that considers the tunnel operation parameter for the analysis. Although this method can include the operational parameters (face pressure and grout pressure), it does not consider the 3D mechanised tunnelling sequence in the analysis. Every stage of the mechanised tunnelling construction influences the ground SS. The stages of the tunnelling such as excavation using face pressure, advancement of the tunnelling machine, installation of the lining, injection of the tail void grout, and hardening of the grout, affect the surrounding soil area [24,25]. Apart from that, the type of soil model used in the numerical analysis is important; therefore, proper in-situ and laboratory data interpretation is required to stimulate the closest possible ground conditions at the site. In 3D numerical analysis, the tunnelling construction can cause SS at different stages of the construction. Mathew and Lehane [26] deduced the inconsistency of the calculations using volumetric contraction in comparison with field measured settlements due to the complexities of the actual tunnelling procedures. In comparison with a single tunnel, twin tunnels are more complex and more difficult to analyse as more parameters, e.g., the spacing between two tunnel and different tunnelling operational parameters are required to consider in the analyses.

Most studies [3,27] investigated these issues for single tunnels; however, limited studies are available for twin tunnels using earth pressure balance (EPB) shield machines. In the recent numerical tunnelling analysis carried out by Islam and Iskander [28], the effect of geometric parameters and construction sequence on the SS was investigated using MIDAS GTS NX software 2021 (v1.1). However, this case study is not for parallel twin tunnelling and the outputs were not compared with the field measurements. The 3D numerical model is not limited to certain types of ground in comparison with the empirical and semi-empirical methods. In addition, the 3D numerical model requires a shorter period of time to determine the SS than the laboratory works. The 2D numerical methods (contraction and stress reduction methods) do not consider operational parameters in the analysis, whereas modified grout pressure does not consider the construction stages of the tunnelling works such as the advancement of the tunnelling machine. The consideration of important factors affecting the SS due to tunnelling in the analysis is crucial to providing the closest possible results with the actual measurements.

In light of the above discussion, this study considers the holistic approach of 3D modelling shield tunnelling works in the finite element by incorporating the sequence of mechanised tunnel construction. This sequence includes different stages such as excavation, machine advancement, lining installation, grouting, and hardening. Additionally, most previous research, especially empirical formulas, is solely applicable for a single tunnel. Many existing empirical formulas are tailored to single tunnel scenarios and do not account for modern tunnelling machine processes. This work extends the analysis to twin tunnels constructed using mechanised tunnelling, which presents a more complex scenario due to the interactions between adjacent tunnels. The utilisation of the 3D numerical modelling approach provides a sustainable solution for tailoring mitigation strategies to effectively manage SS while accounting for project-specific parameters such as tunnel geometry, geotechnical engineering factors, and tunnelling operational variables. The 3D numerical model approach empowers the customization of SS mitigation strategies. This customization ensures not only the effectiveness of the mitigation strategies but also the adherence to sustainability principles by minimizing adverse impacts on the environment and adjacent structures. By identifying potential SS issues at various construction stages through early estimation, timely adjustments and proactive measures can be implemented, thereby mitigating risks, and reducing the likelihood of detrimental effects on the surroundings.

2. Field Measurement and Input Parameter for Numerical Analysis

The Mass Rapid Transit (MRT) project is located in the Klang Valley, Malaysia, which comprises of total distance track of 52.2 km including the elevated works, underground works and one deport of the MRT system with the purpose of public transport for the citizens. In this project, the total distance of the twin tunnels is 13.5 km. The twin tunnel is two tunnels that can be arranged into four configurations: parallel, stacked, perpendicular crossing, and offset arrangements. The tracked alignment, a combination of parallel and stacked underground tunnels, is from Jalan Ipoh to Bandar Malaysia. Only parallel tunnelling is used for this study, as limited information is only available for stacked underground tunnels. The tunnel is traversed through limestone, granite, alluvium, and the Kenny Hill formation, as shown in Figure 1.

The project used mechanised tunnelling for tunnel excavation, including EPB and variable density tunnelling (VDT) machines with 6.7 m diameter for different ground conditions. VDT is the new mechanised tunnelling machine that was developed to mitigate the risk of blowouts and sinkholes in variable and mixed ground conditions. Whereas EPB was selected for the homogenous and cohesive soils. It uses the excavated material to stabilise the tunnel face with constant pressure. The selected section for the 3D numerical analysis is located between the alluvium and Kenny Hill. Several early geologists [29,30] have used diluvium as a term to describe "alluvium-like" materials comprised of sands, gravels, boulders, and clays on the hard rock in all landscape positions. The alluvium is modern sediments deposited into the ground as a result of recent river activity [31]. Since this soil material is deposited by the river, it can be anticipated that the ground is soft for the alluvium. Furthermore, the Kenny Hill formation in Kuala Lumpur has a thickness of 1200–1500 m of clastic sedimentary rocks and is located in the west and south [32]. It can be stated that this formation is for weathered rock and residual soil [33]; therefore, this formation is anticipated as a firm and hard material. This project encounters two different soil hardness materials: soft (alluvium) and hard (Kenny Hill).

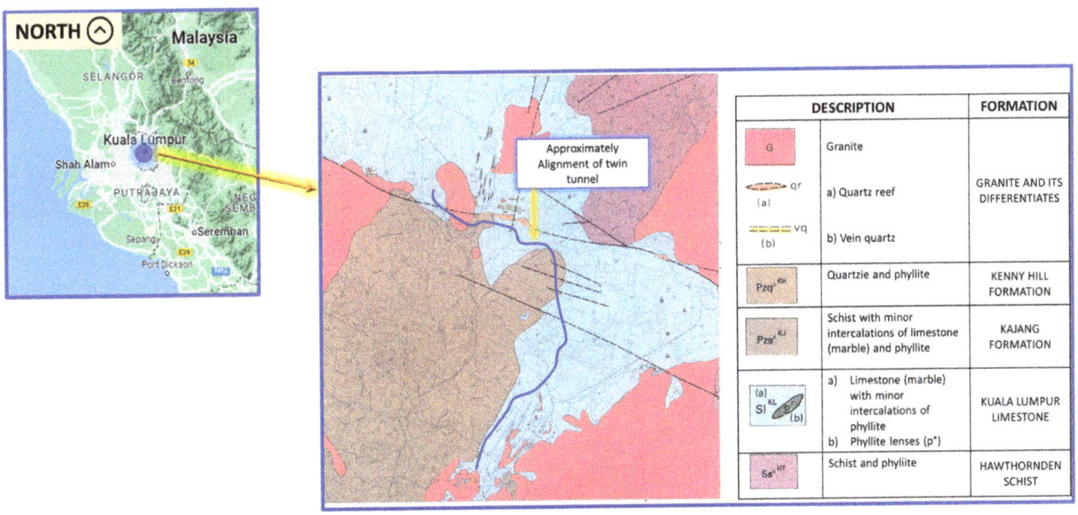

Figure 1. The geological formation and location of the twin tunnels investigated in this research [34].

2.1. Field Measurement of SS

Throughout the chainage of the tunnelling, the overburden of the soil at the tunnel crown is in the range of 6 m to 42 m. For the back analysis of the SS due to tunnelling, five cross sections of the tunnel that traverse through the soil with a sufficient array of ground settlement markers were selected. In this project, the SS values were measured using a total station from the ground settlement marker as shown in Figure 2.

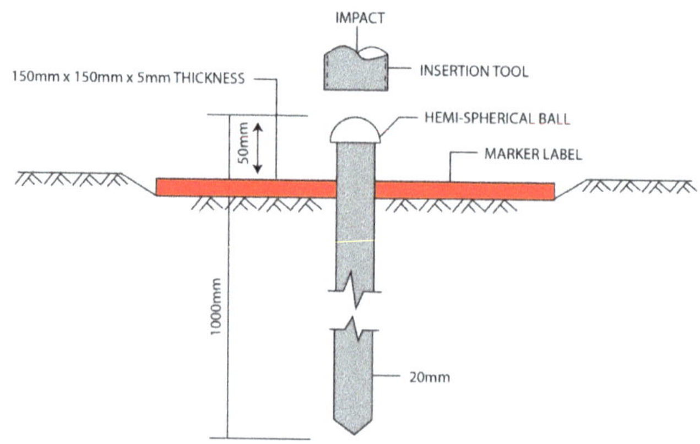

Figure 2. Example of installed ground settlement marker.

The ground settlement marker for this project is an instrument with a steel rod length of 0.95 m embedded in the soil that is covered by a steel plate of 150 mm by 150 mm, which acts as a marker label to ease the identification of the location of the markers. The protrusion of a 0.05 m steel rod above the marker label was used to measure the level of the change. The reduced levels of the tops of the rods of settlement markers based on the nearest benchmark (BM) or temporary BM were measured prior to the commencement of the tunnelling works. The baseline readings were set after taking readings for several days before the commencement of the tunnelling works. The settlement was calculated by taking the change in the reduced level of the top of the rod relative to the base readings and previous readings. In general, the readings of the settlement markers were taken once per day during the construction period of the tunnelling. To evaluate the settlement from the field data, the settlement values were collected when both tunnels had passed by the settlement array, as illustrated in Figure 3, and only the final readings were collected as the SS. There are approximately 304 subsurface investigation (SI) boreholes along the area of the tunnelling chainage. In the numerical analysis, the nearest sections to the SI boreholes were utilised to establish the geological profile of the model.

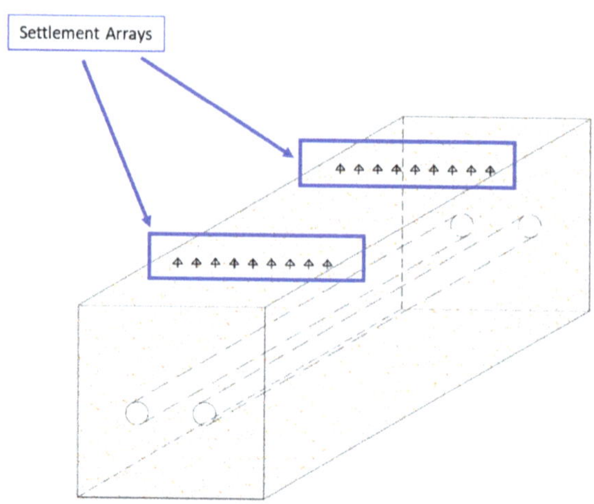

Figure 3. Illustration of the ideal position of settlement marker arrays along the twin tunnels.

2.2. Interpretation of Geotechnical Parameters

To carry out the numerical analysis of any geotechnical application, the accuracy of the soil-structure interaction problems is crucial as the soil is the weakest material compared to others such as concrete and steel. Hence, the analysis is strongly dependent on the soil constitutive model to reflect the actual condition of the ground.

The soil profile for each model comprises two layers of soil and rock. The tunnels are bored through the soil layers (alluvium and Kenny Hill). Hence, the rock layer does not have an impact on the analysis. The range of soil stresses and strains are important in the analysis; hence, this shall be considered in the numerical analysis. For the finite element analysis, two common soil constitutive models were used by many researchers in relevant studies, which are Mohr-Coulomb (MC) [25,35] and Hardening Soil (HS) [36,37]. Gerhard [37] compared the MC and HS of the tunnelling finite element analysis and found that the MC model can provide higher settlement values in comparison with the measurements on the site, while the HS model predicts SS values that are closer to the actual measurements using the 2D grout pressure method. In addition, Hejazi et al. [38] found the SS induced by tunnelling is strongly influenced by the soil constitutive model. From their findings, MC SS appears unrealistic, which is due to the stiffness modulus in the MC model being constant at certain points of strain. Islam and Iskander [28] mentioned that MC is not able to stimulate the behaviour of the soil, especially during unloading and tunnel excavation. However, they stayed with the MC model due to the simplicity of the parameters. Moller [39] carried out both analyses using 2D and 3D finite element analysis for the second Heinenoord slurry shield using the HS model, and the findings showed the simulated SS is close to the site's actual settlement values. His analysis results are aligned with another study conducted by Likitlersuang et al. [36], where they reported very close readings to the measured settlement using the HS model. Choon [40] emphasised that the soil constitutive models play a crucial role in the analysis results and concluded that tunnelling involves unloading and reloading the surrounding soils. Hence, the HS model was recommended by Choon [40] for use in the assessment of SS. In addition, Janin et al. [41] stated that the HS model can distinguish the stiffness modulus for the primary loading and unloading, which makes this numerical analysis closer to the measured SS results.

The soil properties can be classified by the strain levels into three categories, which are very small strain level (stiffness modulus is constant in elastic range), small strain level (stiffness modulus varies non-linearly with the strain), and large strain level (soil is close to the failure, and the soil stiffness is relatively small) where tunnelling works can be considered as large strain level [42]. The illustration of the categories of the strain level can be seen in Figure 4a. The HS model was created with the concept of plasticity, where the total strains of the soil are calculated using a stress-dependent stiffness that is different for both loading and unloading. The MC model indicates that the soil behaviour is linearly elastic and perfectly plastic. In addition, the HS model can consider loading and unloading stages, which can reflect the actual situation of tunnel construction for the removal of soil (unloading), followed by tunnel lining construction (loading). Therefore, the HS model is a more suitable soil model for the tunnel numerical analysis of tunnelling compared to MC [43]. An illustration of an indication of the difference between the MC and HS is shown in Figure 4b.

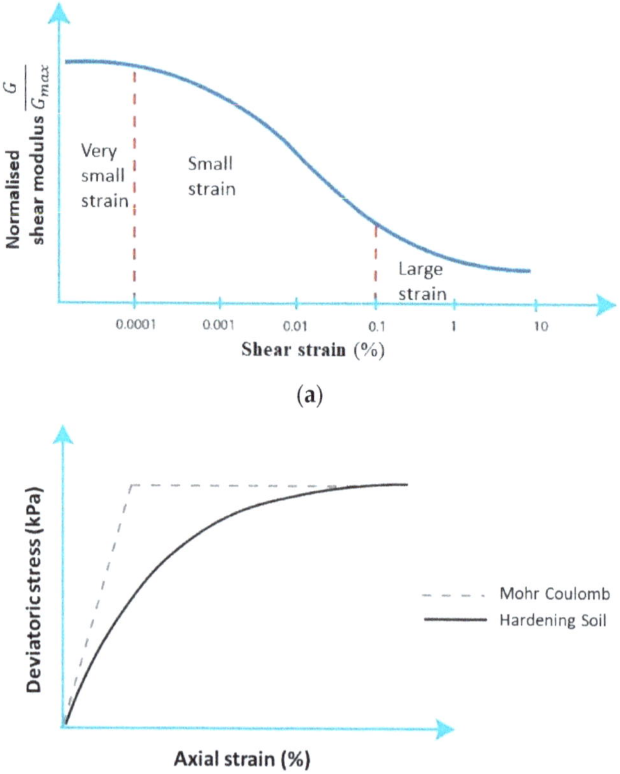

(**b**)

Figure 4. (**a**) Normalised stiffness degradation curve [42], and (**b**) Triaxial drained compression stress of MC and HS models [43].

The soil parameter of effective strength is retrieved by carrying out consolidated isotopically undrained (CIU) tests from the closest boreholes at different depths. In addition, several Menard Pressuremetre Tests (PMTs) have been carried out as per BS 5930 [44] throughout the chainage of the tunnelling. The PMT was placed inside the borehole after the Standard Penetration Test (SPT-N) was taken. The depth of PMT was located at approximately the same depth as SPT-N, as this was conducted to ensure the data from SPT-N could be used as a reference for the comparison of similar lithologies. The corrected SPT-N was correlated with the pressuremetre modulus, E_m. The previous researchers [45,46] carried out the correlation of the corrected SPT-N of N_{60} with the elastic modulus from the PMT. Some of the E_m empirical formulas are presented in Table 3 according to different types of soil. The performance prediction of these formulas is presented based on the coefficient of determination (R^2).

Table 3. Summary of the Em empirical equations to determine the elastic modulus of the soil.

Author(s)	Ground Condition	Method of Tunnelling	R^2
Ohya et al. [45]	$E_m = 19.3(N_{60})^{0.63}$	Clayey soil	0.39
Yagiz et al. [46]	$E_m = 388.67(N_{60}) + 4554$	Silty Clay	0.83
Cheshomi and Ghodrati [47]	$E_m = N_{60} - 2.67$	Silty Clay	0.85
Naseem and Jamil [48]	$E_m = 15.214N_{60} + 89.276$	Sandy soils	0.88

A total of 20 and 41 PMTs were carried out at the alluvium and Kenny Hill formations along the chainage of the tunnel, respectively. After carrying out the removal of outliers using the method of the interquartile range rule for a total of 61 PMTs, the corrected STP-N_{60} is plotted against E_m for 2 different formations, as shown in Figure 5.

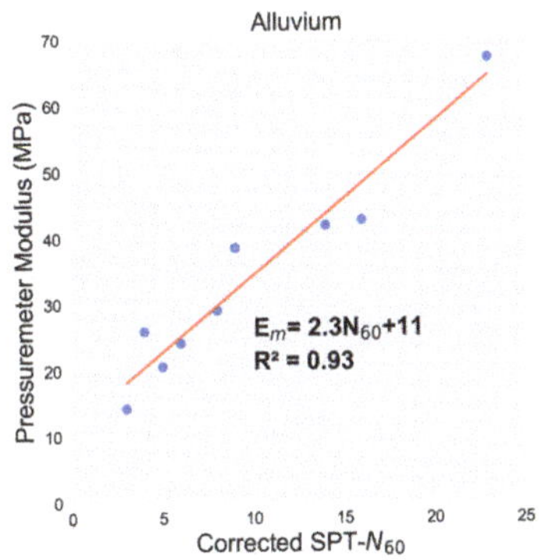

(a)

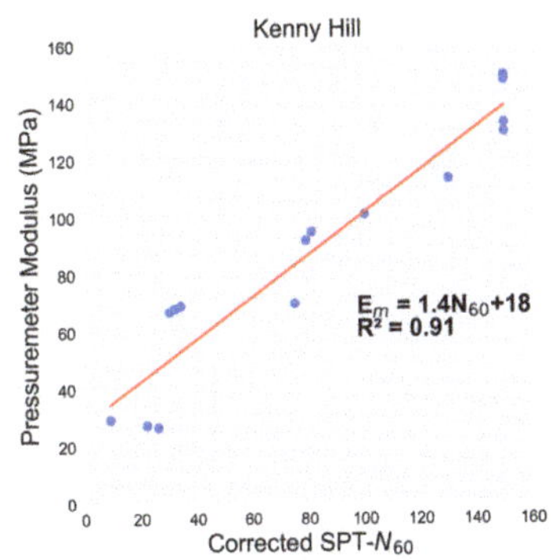

(b)

Figure 5. SPT-N (N_{60} against E_m) for (**a**) Alluvium, (**b**) and Kenny Hill.

From the SPT-N, the selected section for the tunnelling analysis with a borehole is correlated with the elastic modulus according to the type of soil. However, E_m is different from the soil elastic modulus; hence, conversion of the calculation is required. The difference could be due to several reasons, such as the difference in the range of radial, the stress surrounding the borehole wall, the possibility of disturbance during drilling and the installation of probes, and the computation of the modulus with the assumption of cylinder length as infinite [49]. Therefore, correction of the elastic modulus, E, of the soil is introduced to the actual value from E_m [50], which is defined as follows:

$$\alpha = \frac{E_m}{E} \tag{5}$$

where α is a Menard's factor. In the HS model, there are three crucial elastic modulus inputs for the analysis, which are namely: secant stiffness modulus corresponding to 50% of the ultimate deviatoric stress, E_{50}; Oedometer modulus, E_{Oed}; and unloading or reloading modulus, E_{ur}. Secant stiffness from the drained triaxial test at 50% of the ultimate deviatoric stress E_{50} is the confining stress that depends on stiffness modulus for primary loading where the amount of stress is dependent on the power of m.

$$E_{50} = E_{50}^{ref}\left(\frac{c'\cos\Phi' - \sigma_3'\sin\Phi'}{c'\cos\Phi' + p^{ref}\sin\Phi'}\right)^m \quad (6)$$

where p^{ref} is a reference stress of 100 kPa, c' is effective cohesion, and Φ' is effective friction angle. The stiffness is affected by minor effective principal stress. The power m governs the stress dependency. As reported by Janbu [51], the sand content of m is approximately 0.5. Other than this input, the reference oedometer modulus, E_{oed}^{Ref}, acts as the parameter that controls the magnitude of the plastic strains that arise from the yield cap ε_v^{pc}. The oedometer modulus can be defined as follows:

$$E_{Oed} = E_{Oed}^{Ref}\left(\frac{c'\cos\Phi' - \sigma_1'\sin\Phi'}{c'\cos\Phi' + p^{ref}\sin\Phi'}\right)^m \quad (7)$$

where E_{oed}^{Ref} is approximately similar to E_{50}^{Ref} [52] as illustrated in Figure 6. In addition, the stress-dependent stiffening modulus for unloading and reloading can be calculated as follows:

$$E_{ur} = E_{ur}^{Ref}\left(\frac{c'\cos\Phi' - \sigma_3'\sin\Phi'}{c'\cos\Phi' + p^{ref}\sin\Phi'}\right)^m \quad (8)$$

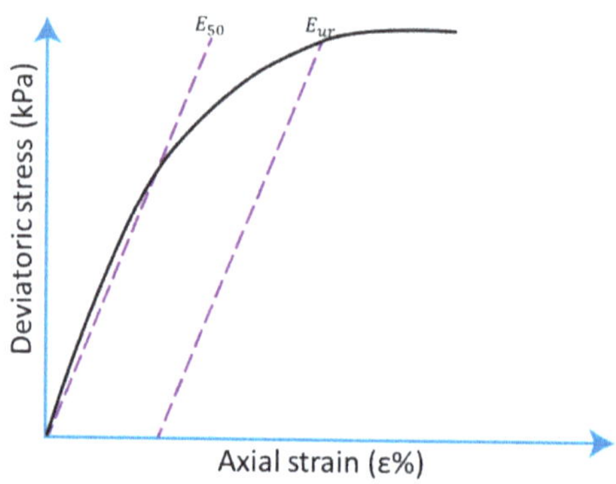

Figure 6. Hyperbolic stress-strain relationship for the E_{50} and E_{ur} [52].

For the ease of numerical computation, several researchers [40,53] have used $E_{50}^{Ref} = 3E_{ur}^{Ref}$ as per recommended by the previous researchers [52,54]. The input of E_{50} is correlated with the range from $0.06E$ to $0.26E$ with $0.5q_f$ (where E is the elastic modulus of the soil and q_f is the deviatoric stress from the secant stiffness modulus reduction curves from static torsional and triaxial shear data on clays and sands) [55]. Based on the interpretation, Table 4 summaries the geotechnical input parameters for these five models.

Table 4. Summary of the stiffness and effective strength of the soil used in the numerical analysis.

Model	Effective Cohesion, c′ (kPa)	Friction Angle (°)	E_{50}^{Ref} (MPa)	E_{oed}^{Ref} (MPa)	E_{ur}^{Ref} (MPa)	Type of Formation
1	2	29	9.3	9.3	27.9	Alluvium
2	3	32	26.0	26.0	78.0	Alluvium
3	2	28	6.1	6.1	18.3	Alluvium
	5	32	11.6	11.6	34.8	Kenny Hill
4	2	30	7.0	7.0	21.0	Alluvium
5	1	25	3.5	3.5	10.5	Alluvium

A total of 36 and 35 bulk density tests are carried out on the alluvium and Kenny Hill formations across the tunnelling alignment at various depths, respectively. The statistical input of bulk density is summarised in Table 5. In addition, the groundwater level recorded from the closest borehole with the readings recorded for morning and evening throughout the SI works will be used as part of the analysis, and these readings are taken as an average of the groundwater level. The groundwater level is one of the important parameters because when the groundwater level is lowered, it can cause a reduction in the pore pressure, which indirectly increases the surface settlement [56]. Figure 7 shows an illustration of how the lowering of groundwater causes the settlement.

Table 5. Summary of the inputs of the bulk density of different ground types.

Type of Ground	Bulk Density ($kN.m^{-3}$)			
	Mean	Maximum	Minimum	Standard Deviation
Alluvium	19	21	16	1.05
Kenny Hill	20	22	15	1.51

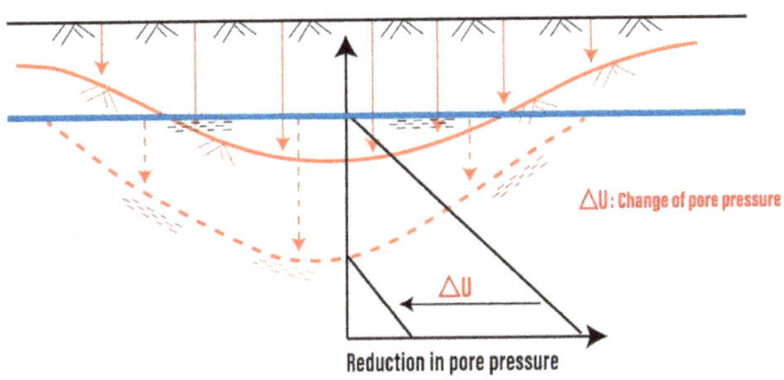

Figure 7. Lowered groundwater causes surface settlement [56].

2.3. Tunnel Boring Machine (TBM) Operational Parameters

Various operational factors of TBMs, including face pressures, rates of penetration, and pressures of grouting the tail void, jointly impact the surface settlement caused by mechanised tunnelling operations. TBMs cause surface settlement mainly by moving soil during excavation, resulting in empty spaces where the neighboring ground eventually subsides. While the TBM constructs the tunnel, it can also shift the ground, prompting settlement. The interplay between the tunnel lining and the nearby soil can trigger certain adaptations and shifts in the ground, indirectly resulting in surface settlement. Numerous studies [3,24] have demonstrated that face pressure and grouting pressure are the two important contributors to the SS caused by mechanised tunnelling. Hence, in this study, these two parameters were considered to be part of the finite element analysis. The EPB shield of the face pressure operates with the cutter head rotating, and cutting tools scrap the ground from the tunnel face while additives are injected into the material. The amount of the excavated soil is controlled and transported from the shield, and the tunnel face can be supported by the soil stored in the chamber. Hence, the face pressure stores in the chamber are crucial to maintaining the stability of the excavation. Different conditions of face pressure result in different SS scenarios. Several studies have found that increasing face pressure causes a decrease in settlement [3,57]. A different ideal scenario for the face pressure that affects the SS is illustrated in Figure 8. Case 1 shows that when the face pressures are equivalent to the overburden pressure, there are no impacts on the ground surface. Case 2 depicts the face pressures being less than the overburden pressures and the ground settling. Lastly, case 3 illustrates when the face pressures are more than the overburden pressures, heaving can be seen. Nevertheless, the amount of face pressure exceeded or lesser than the required pressures is subjected to the soil properties.

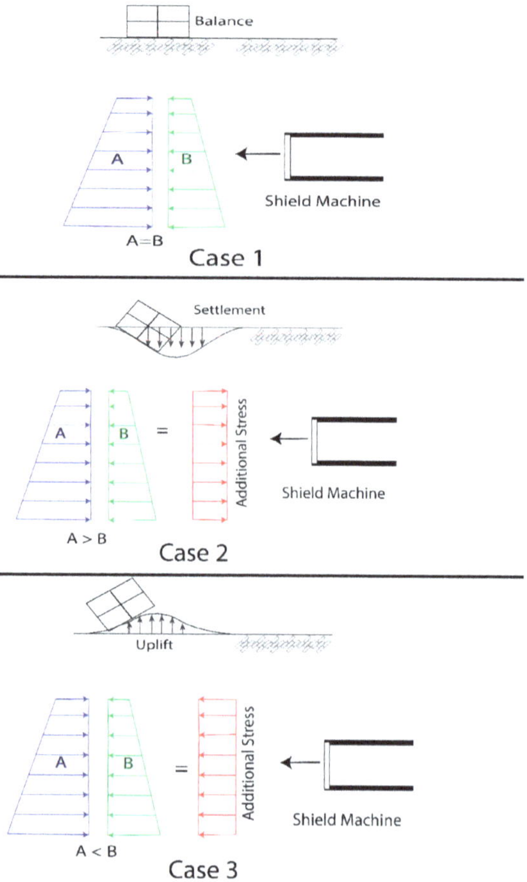

Figure 8. Illustration of the face pressure on the surface settlement in different cases.

The shield diameter is larger than the external diameter of the installed lining, which causes a gap between the segments and the excavated soil mass [58]. This gap difference is known as a tail void or annulus, and it easily occupies the inward deforming soil during the shield and tail skin assembly process. When the tunnelling machine advances, the void can be continually filled with grout in one of two ways: by pumping the grout through pre-cast concrete lining segments or injecting the grout via grout ports at the back of the tail skin. The injection was carried out under high pressure in the tail void formed by the shield tail's back. Thus, this is known as tail void grouting pressure. With high-pressure injection, the void can be filled within a short period of time with grouting material.

Other than these two parameters, another operating parameter that can be modelled into the numerical analysis is jack-force thrust. The distribution of the jack thrust in this study is uniform across the theoretical models. The function of this thrust force is to stimulate the actual condition of the mechanised tunnelling to move forward. The face pressure and grouting pressure, with the average input of the recorded face pressure and grouting pressure, are adopted in the analysis. To avoid the complication of the analysis, the value is rounded off to the nearest two significant values, and these numbers are summarised in Table 6.

Table 6. Summary of the TBM operational parameters.

Model	Tunnel Bound	Average Face Pressure (kPa)	Average Grouting Pressure (kPa)	Average Force Thrust (kPa)
1	1st	140	230	4700
	2nd	180	370	2900
2	1st	290	350	4400
	2nd	230	390	3800
3	1st	190	170	5400
	2nd	200	160	5800
4	1st	190	220	5800
	2nd	210	250	5800
5	1st	220	410	4400
	2nd	250	270	5200

3. Construction of Numerical Model

3.1. General Characteristics

The numerical analysis was carried out using MIDAS GTX NX. In this section, the geometry dimension of the numerical model for the analysis will be further explained, the input for the grouting materials for the analysis will be stated, and the boundary condition of the model will be described.

Before the commencement modelling of the numerical analysis, the extended dimensions of the models affect the analysis required to be identified. In this analysis, the overall geometry of the numerical model was based on the recommendations by Alagha and Chapman [59] and Ruse [60]. Where C is the tunnel cover, D is the tunnel diameter, the recommendation for the minimum numerical model width is $2D$, the depth of the model is $C + 3D/2$, and the length is $3D$. Figure 9 shows a view of the models with various dimensions (x, y, and z directions). The overall adopted numerical model dimension is 115 m (width) × 115 m (depth) × 20 m (thickness). In this numerical model, the diameter of the tunnel is 6.7 m with a segmental lining width of 1.5 m. The tail void grout thickness is 150 mm. Two types of grouting were used in the analysis which are fresh and hardened grouting. The properties of the grouting materials are tabulated in Table 7. In addition, the degree of freedom for the numerical model was constrained in the x-direction, whereas in the y-direction was on both sides. Lastly, only nodes at the bottom were constrained in the z-direction. The top nodal along the ground surface was not constrained to allow the movement of the ground during the tunnelling process.

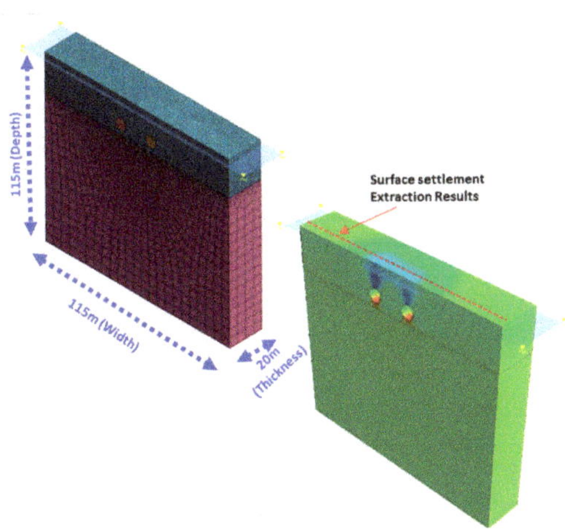

Figure 9. A numerical model for simulating SS induced by twin tunnels.

Table 7. The inputs of the fresh and hardened grouting.

Type of Material	Unit Weight, γ (kN.m^{-3})	Young's Modulus, E (MPa)
Fresh grout	15	7.5
Hardened grout	15	15

3.2. Construction Stages

Several stages have been involved in stimulating tunnel construction. In general, the process of tunnelling includes soil excavation, the installation of segment lining, and the application of external forces. At the beginning of the process, the self-weight and groundwater were activated to simulate the initial stage to establish the geostatic stresses in the soil mass. Subsequently, excavation was carried out by deactivating the soil elements with face pressure on the excavation face and, at the same time, activating the shield shell, which represents the tunnelling machine moving forward (Figure 10a). This process of excavation was repeated until the total length of the shield (equal to six segmental linings) was moved into the soil mass and the force thrust of tunnelling machine was activated to indicate the movement of tunnelling machine, and the segmental lining was installed at this stage (Figure 10b). Subsequently, the tail void grouting pressure was activated with fresh grout material (Figure 10c). Finally, the fresh grout was converted to hardened grout (Figure 10d) to show the later stage of the project. The tunnelling operational procedures of the numerical analysis for the tunnelling machine are shown in Figure 11. The different stages involved in tunnel construction, which encompasses excavation involving face pressure, lining installation, and grouting, exert an influence on SS. During excavation, voids are formed that cause the surrounding soil to settle. The installation of the lining can displace the ground, contributing to settlement as the tunnel's supporting structure interacts with the soil. The grouting process, aimed at filling the gaps between the tunnel and the excavated area, affects the soil's radial distribution, consequently impacting SS. Each of these phases brings about distinct settlement effects, and the overall sequence and coordination of these stages play a pivotal role in effectively managing settlement levels.

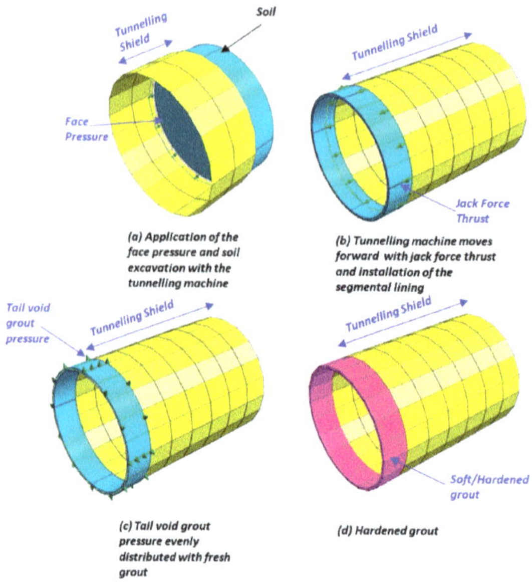

Figure 10. The scenario of the tunnelling boring machine in the numerical analysis.

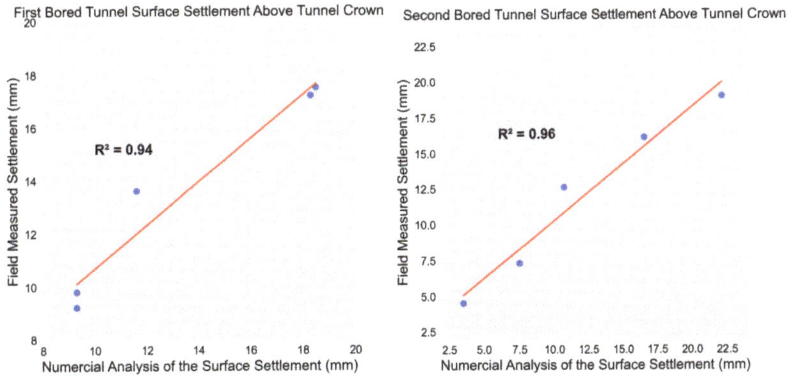

Figure 11. The SS is induced by parallel twin tunnelling for the first and second bored tunnels.

4. Results and Discussion

In this section, the output SS from the numerical analysis is compared with field measurements of SS, and the interpretation of the results will be further discussed. A total of five numerical models were analysed and compared with the measured settlement at the site. A total of five models' geometries can be found in Table 8. One of the tunnel geometries that impacts the SS between twin tunnels is the pillar width. Based on Table 8, it can be observed that Model 1, with the shortest pillar width, exhibits the highest SS between the tunnels. This implies that both the 1st and 2nd bored tunnels contribute to the maximum SS occurring between them.

Table 8. Summary of the tunnel geometry models for the numerical analysis.

Model	Tunnel Depth (m)	Pillar Width (m)	Location of Maximum SS
1	10	6	Between 1st and 2nd tunnel
2	13	10	1st bored tunnel
3	15	11	1st bored tunnel
4	15	19	1st bored tunnel
5	15	12	2nd bored tunnel

The field measurements of SS directly above each bored tunnel closely align with the results obtained from the numerical analysis, as presented in Table 9. A comparison between the numerical analysis and the actual measurements is conducted in terms of percentage differences. The disparities are predominantly below 18% across most models, with the exception of Model 2 of the second bored tunnel, where a notably higher percentage difference of 28.6% is observed. In comparison with the actual measurement, only a minor difference of 1 mm is observed between the actual value and the analysis. In general, the difference between measured and actual measurement is in the range of 0.5 mm to 3.1 mm.

Additionally, the SS values obtained from measurements and numerical analysis for both the first and second bored tunnels are graphically represented in Figure 11, with coefficient of determination (R^2) values of 0.94 and 0.96 for the first and second bored tunnel SS, respectively. These high R^2 values suggest strong correlations between the measured and computed SS values for both tunnels.

The SS output of the numerical analysis for each model is visualised alongside the corresponding field measurement, as depicted in Figure 12. Observing the patterns within Figure 12, it can be seen that the 3D numerical analysis of the tunnel construction process effectively approximates the measured SS.

Table 9. Summary of the SS results through numerical analysis.

Model	Sequence of Boring	SS Obtained by MIDAS (mm)	Actual SS (mm)	Percentage of Difference (%)
1	1st	9.8	9.3	5.4
	2nd	7.3	7.5	2.7
2	1st	9.2	9.3	1.1
	2nd	4.5	3.5	28.6
3	1st	13.6	11.6	17.1
	2nd	12.6	10.7	17.8
4	1st	17.2	18.3	6.0
	2nd	16.1	16.5	2.4
5	1st	17.5	18.5	5.4
	2nd	19.0	22.1	14.0

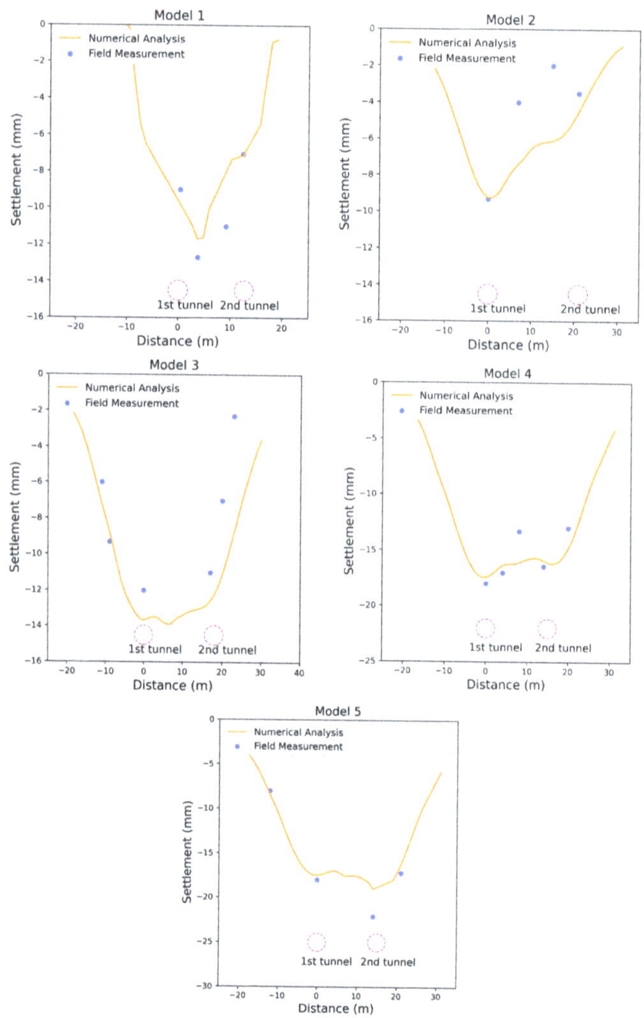

Figure 12. The surface settlement induced by tunnelling analysis graph.

The output of this analysis has shown close similarities between the prediction settlement using MIDAS software and the site measurement. Model 1 has the shortest pillar width (6 m) in comparison with all the models. From this model, it can be seen that the 1st and 2nd bored tunnels do not have much difference (2 mm); however, the maximum SS is found located in between these two tunnels with the shape of a single symmetric settlement trough, which is similar to the findings as per Peck [5] and Likitlersuang et al. [36].

Model 2 depicts the unique trend of SS, where the maximum SS is found at the 1st bored tunnel instead of the 2nd bored tunnel. In the actual scenario of excavation for twin tunnels, the first bored tunnel excavates the undisturbed soil. However, the process of excavation for the first bored tunnel disturbed the soil surrounding it. Hence, this will cause the second tunnel's soil to be disturbed. Terzaghi [61] was the first researcher who published a paper related to the field data of surface settlements above twin tunnels. His findings showed that the settlements above the second tunnel are larger than the first tunnel. However, based on the result of the analysis and the site-recorded SS, the second bore tunnel does not have the highest SS. In the past, most of the tunnelling operations for twin tunnels were carried out using compressed air [62]. In addition, several researchers [63,64] have found that the settlements above the second tunnels were larger than the first tunnels. The finding that the second bored tunnel has the maximum SS is applicable for open-face tunnelling because the ground reaction is influenced by tunnel geometry and geological conditions.

It is obvious that the SS values above the second tunnel could be lower. This is because the tunnelling operational parameters are considered contributing factors to the settlement [65]. Kannangara et al. [66] also found a similar trend in the outputs, where the observed second excavated tunnel has a lower SS when the face and grouting pressure are controlled at the same or higher margin compared with the first excavation. Although Model 3 has an approximately similar pillar width to Model 2 (different by 1 m) with a deeper depth, the 1st and 2nd bored tunnel SS do not have many differences. This could be due to the similarities in the operation parameters between the two tunnels and the deeper tunnel depth. Hence, these two bored tunnel SS of Model 3 have more consistent SS compared to Model 2.

Model 5 exhibits the highest settlement in comparison to the remaining models due to the lowest soil stiffness of the ground. It can be inferred that when twin tunnels have a larger spacing of 3D, there is no interaction between the tunnels [28]. This is evident from Model 4, where the maximum surface settlement at the 1st excavated tunnel is likely due to slightly low face pressure. Consequently, it can be deduced that within this range of pillar width, there is no impact from the first bored tunnel on the area of the second tunnel.

Several researchers have also carried out the analysis of tunnelling using MIDAS software and reported very close findings to the measured settlement for the different scenarios of the analysis as tabulated in Table 10.

Table 10. Summary of tunnelling analysis using MIDAS software for various scenarios.

No	Author (s)	Number of Numerical Models	Comparison of the Numerical Analysis with Site Measurement
1	Zhou et al. [67]	1	The site measured maximum settlement = 9.72 mm at x = 12.25 m. The numerical result of maximum settlement = 9.94 mm at x = 11.25 m. The errors between the numerical results and measured data are small.
2	Ahmed et al. [3]	4	The maximum difference between the measured and numerical analysis is 25%.
3	Ayasrah et al. [68]	5	The highest longitudinal settlement is less than a 3 mm difference.

In previous studies, the most common method to determine the surface settlement induced by tunnelling is the empirical formula. This empirical formula is dependent on the VL, which varies according to the type of ground and installation tunnel. However, as far as the authors know, no literature on VL for the soil in Malaysia has been published. The use of empirical formulas is limited to certain ground conditions and installation methods. Chakeri and Ünver [69] compared the empirical formula with the measured settlement, and from their studies, the calculated and measured settlement diverge by more than two times. In addition, they also carried out the 3D numerical analysis; however, the 3D numerical analysis was carried out in the lithology of the ground for the Gungoren formation, which mainly consists of very stiff clay, hard clay, and dense sand. Furthermore, Choon [40] used the 2D finite element of the contraction method with various VL inputs to retrieve a similar SS as measured at the site. Furthermore, several studies [70–72] have utilised centrifuge laboratory modelling for sandy soil materials. However, these laboratory results have not been compared with field measurements, and they have not taken tunnelling operational parameters into consideration. Therefore, by utilising 3D numerical analysis, it becomes possible to integrate three crucial factors: tunnel geometry, geotechnical soil properties, and tunnelling operational parameters. This integration, combined with the incorporation of tunnelling machine construction processes, allows for the accurate determination of SS. This approach mitigates the constraints associated with alternative methodologies, as discussed previously. Consequently, this demonstrates that this method offers superior advantages compared to other approaches.

5. Limitations and Future Works

In this study, the number of field measurements is limited due to the focused investigation on the alluvium and Kenny Hill formation areas. Therefore, expanding the scope to include more field measurements from various locations, particularly within the soil area, with closer intervals would enhance the development and validation of the model. Additionally, if the budget allows, considering real-time monitoring for tunnelling-induced settlement measurements could be beneficial.

It is important to note that the geological profile is established based on the closest SI boreholes with a diameter of 100 mm, which implies a potential non-homogeneity in the studied geological profile. Furthermore, the input of groundwater levels relies on recorded SI works, which could be influenced by field conditions.

While 3D numerical modelling is time-consuming and unsuitable for real-time analysis during the operation of long tunnels, situations involving geological formations and tunnel operational parameters may arise. Therefore, new computational approaches such as machine learning and intelligent techniques could find applications in this context. Combining these approaches with established theories and empirical equations in relevant domains could yield a more generalised technique for predicting settlement induced by twin tunnels. Moreover, these techniques would offer practicality and relevance to engineers in the field.

6. Conclusions

The findings from the 3D numerical analysis of parallel tunnels using the MIDAS software, incorporating mechanised tunnelling construction stages and a hardening soil model, demonstrate an acceptable level of prediction accuracy for SS when compared with field measurements. Consequently, the industry may consider adopting this approach for SS assessment. Furthermore, the 3D numerical modelling approach enables the customization of mitigation strategies for SS based on specific project parameters, aiming to achieve a more sustainable tunnel construction. The key findings of this investigation can be summarised as follows:

(1) The 3D numerical analysis produced SS above the tunnel crown of twin tunnels has shown R^2 = 0.94 and R^2 = 0.96, respectively for the first and second bored tunnels with the actual field measurements and the largest difference settlement of 3.1 mm.

(2) The second bored tunnel does not consistently exhibit the largest SS because it can be influenced by other factors, such as tunnel geometry, geotechnical soil properties, and tunnel operational parameters. For example, a narrower pillar width can lead to higher SS between tunnels, as demonstrated in Model 1. Similarly, areas with lower soil stiffness could also result in elevated SS. Additionally, insufficient face pressure can contribute to increased settlement.

(3) It can be stated that no effect of the first bored tunnel on the second bored tunnel area at a distance of equal or more than 3D between the tunnels has been observed.

(4) The interpretation of the elastic modulus from the field pressuremetre test and SPT-N can be used as geotechnical soil stiffness parameter inputs in the numerical analysis. The interpreted relationship of the pressuremetre modulus with corrected SPT-N is as follows: $E_m = 2.3\ N_{60} + 11$ (alluvium) and $E_m = 1.4\ N_{60} + 18$ (kenny Hill).

(5) Three main input parameters, namely tunnel geometry, engineering ground parameters, and tunnel operational parameters, considered in the numerical analysis yield results that closely align with the field measurements.

Author Contributions: Conceptualization, C.Y.H., S.H.L., D.J.A., H.R. and X.H.; methodology, C.Y.H., S.H.L. and D.J.A.; software, C.Y.H., S.H.L. and D.J.A.; validation, C.Y.H., S.H.L. and D.J.A.; formal analysis, C.Y.H., S.H.L. and D.J.A.; resources, C.Y.H. and S.H.L.; data curation, C.Y.H. and D.J.A.; writing—original draft preparation, C.Y.H., S.H.L., D.J.A., H.R. and X.H.; writing—review and editing, C.Y.H., S.H.L., D.J.A., H.R. and X.H.; supervision, S.H.L., D.J.A., H.R. and X.H. All authors have read and agreed to the published version of the manuscript.

Funding: This research received no external funding.

Institutional Review Board Statement: Not applicable.

Informed Consent Statement: Not applicable.

Data Availability Statement: The data will be available upon a reasonable request.

Acknowledgments: The authors would like to extend their sincere gratitude to Mass Rapid Transit Corporation Sdn Bhd, the contractor (MMC Gamuda KVMRT Sdn Bhd) and the consultant (G&P Geotechnics Sdn Bhd) for facilitating this study. The authors would also like to express appreciation to Universiti Malaya for supporting this research.

Conflicts of Interest: The authors declare no conflict of interest.

References

1. Chen, R.; Li, J.; Kong, L.; Tang, L. Experimental study on face instability of shield tunnel in sand. *Tunn. Undergr. Sp. Technol.* **2013**, *33*, 12–21. [CrossRef]
2. Wan, C.; Jin, Z. Adaptability of the cutter-head of the earth pressure balance (epb) shield machine in water-rich sandy and cobble strata: A case study. *Adv. Civ. Eng.* **2020**, *2020*, 8847982. [CrossRef]
3. Ahmed, N.Z.; El-Shourbagy, M.; Akl, A.; Metwally, K. Field monitoring and numerical analysis of ground deformation induced by tunnelling beneath an existing tunnel. *Cogent Eng.* **2021**, *8*, 1861731. [CrossRef]
4. Hasanipanah, M.; Noorian-Bidgoli, M.; Armaghani, D.J.; Khamesi, H. Feasibility of PSO-ANN model for predicting surface settlement caused by tunnelling. *Eng. Comput.* **2016**, *32*, 705–715. [CrossRef]
5. Peck, R.B. Deep Excavations and Tunneling in Soft Ground. In Proceedings of the 7th International Conference on Soil Mechanics and Foundation Engineering, Mexico City, Mexico, 1969; pp. 225–290. Available online: http://scholar.google.com/scholar?hl=en&btnG=Search&q=intitle:Deep+excavations+and+tunneling+in+soft+ground#0%5Cnhttp://ci.nii.ac.jp/naid/10007809489/ (accessed on 30 August 2023).
6. Litwiniszyn, J. Application of the equation of stochastic processes to mechanics of loose bodies. *Arch. Mech. Stos.* **1956**, *8*, 393–411.
7. Clough, G.W.; Leca, E. EPB shield tunneling in mixed face conditions. *J. Geotech. Eng.* **1993**, *119*, 1640–1656. [CrossRef]
8. Suwansawat, S.; Einstein, H.H. Artificial neural networks for predicting the maximum surface settlement caused by EPB shield tunnelling. *Tunn. Undergr. Sp. Technol.* **2006**, *21*, 133–150. [CrossRef]
9. Attewell, P.B.; Farmer, I.W. Ground disturbance caused by shield tunnelling in a stiff, overconsolidated clay. *Eng. Geol.* **1974**, *8*, 361–381. [CrossRef]
10. O'Reilly, M.P.; New, B.M. *Settlements above Tunnels in the United Kingdom—Their Magnitude and Prediction*; Institution of Mining & Metallurgy: Carlton, Australia, 1982.

11. Mair, R.J.; Taylor, R.N. Bored tunnelling in the urban environments. In Proceedings of the Fourteenth International Conference on Soil Mechanics and Foundation Engineering, Hamburg, Germany, 6–12 September 1997.
12. Wan, M.S.P.; Standing, J.R.; Potts, D.M.; Burland, J.B. Measured short-term ground surface response to EPBM tunnelling in London Clay. *Geotechnique* **2017**, *67*, 420–445. [CrossRef]
13. Rezaei, A.H.; Ahmadi-adli, M. The volume loss: Real estimation and its effect on surface settlements due to excavation of Tabriz Metro tunnel. *Geotech. Geol. Eng.* **2020**, *38*, 2663–2684. [CrossRef]
14. Le, B.T.; Nguyen, T.T.T.; Divall, S.; Davies, M.C.R. A study on large volume losses induced by EBPM tunnelling in sandy soils. *Tunn. Undergr. Sp. Technol.* **2023**, *132*, 104847. [CrossRef]
15. Mair, R.J.; Taylor, R.N.; Bracegirdle, A. Subsurface settlement profiles above tunnels in clays. *Geotechnique* **1993**, *43*, 315–320. [CrossRef]
16. Loganathan, N.; Poulos, H.G. Analytical prediction for tunneling-induced ground movements in clays. *J. Geotech. Geoenvironmental Eng.* **1998**, *124*, 846–856. [CrossRef]
17. Wang, H.; Sun, H.; Jin, H. The influence and calculation method of ground settlement induced by tunnel construction in sand. *J. Chin. Inst. Eng.* **2020**, *43*, 681–693. [CrossRef]
18. Zhu, B.; Zhang, P.; Lei, M.; Wang, L.; Gong, L.; Gong, C.; Chen, F. Improved analytical solution for ground movements induced by circular tunnel excavation based on ground loss correction. *Tunn. Undergr. Sp. Technol.* **2023**, *131*, 104811. [CrossRef]
19. Lee, K.M.; Rowe, R.K.; Lo, K.Y. Subsidence owing to tunnelling. I. Estimating the gap parameter. *Can. Geotech. J.* **1992**, *29*, 929–940. [CrossRef]
20. Vermeer, P.A.; Brinkgreve, R. *Plaxis Version 5 Manual*; Balkema, A.A., Ed.; PLAXIS MoDeTo: Rotterdam, The Netherlands, 1993.
21. Addenbrooke, T.I.; Potts, D.M.; Puzrin, A.M. The influence of pre-failure soil stiffness on the numerical analysis of tunnel construction. *Géotechnique* **1997**, *47*, 693–712. [CrossRef]
22. Surarak, C. *Geotechnical Aspects of the Bangkok MRT Blue Line Project*; Griffith University: Nathan, Australia, 2011.
23. Mair, R.J. Centrifugal Modelling of Tunnel Construction in Soft Clay. Ph.D. Thesis, University of Cambridge, Cambridge, UK, 1980.
24. Wongsaroj, J.; Borghi, F.X.; Soga, K.; Mair, R.J.; Sugiyama, T.; Hagiwara, T.; Bowers, K.H. Effect of TBM driving parameters on ground surface movements: Channel Tunnel Rail Link Contract 220. *Geotech. Asp. Undergr. Constr. Soft Gr.* **2005**, 335–341.
25. Lai, J.; Zhou, H.; Wang, K.; Qiu, J.; Wang, L.; Wang, J.; Feng, Z. Shield-driven induced ground surface and Ming Dynasty city wall settlement of Xi'an metro. *Tunn. Undergr. Sp. Technol.* **2020**, *97*, 103220. [CrossRef]
26. Mathew, G.V.; Lehane, B.M. Numerical back-analyses of greenfield settlement during tunnel boring. *Can. Geotech. J.* **2013**, *50*, 145–152. [CrossRef]
27. Lu, D.; Li, X.; Du, X.; Lin, Q.; Gong, Q. Numerical simulation and analysis on the mechanical responses of the urban existing subway tunnel during the rising groundwater. *Tunn. Undergr. Sp. Technol.* **2020**, *98*, 103297. [CrossRef]
28. Islam, M.S.; Iskander, M. Effect of Geometric Parameters and Construction Sequence on Ground Settlement of Offset Arrangement Twin Tunnels. *Geosciences* **2022**, *12*, 41. [CrossRef]
29. Cuvier, G.; Knight, D. *Essay on the Theory of the Earth*; Routledge: Oxfordshire, UK, 2018.
30. Taylor, R. *A key to the Knowledge of Nature; or an Exposition of the Mechanical; Chemical, and Physical Laws Imposed on Matter by the Wisdom of the Almighty*; Baldwin, Cradock, and Joy: London, UK, 1825.
31. Woodward, J. *The Ice Age: A Very Short Introduction*; Oxford University Press: Oxford, UK, 2014.
32. Lee, C.P. Scarcity of fossils in the Kenny Hill formation and its implications. *Gondwana Res.* **2001**, *4*, 675–677. [CrossRef]
33. Law, K.H.; Othman, S.Z.; Hashim, R.; Ismail, Z. Determination of soil stiffness parameters at a deep excavation construction site in Kenny Hill Formation. *Measurement* **2014**, *47*, 645–650. [CrossRef]
34. Malaysia, G.S. *Bedrock Geology of Kuala Lumpur, Scale 1:25,000, Wilayah Persekutuan L8010 Part of Sheet 94k*; Jabatan Mineral dan Geosains: Kuala Lumpur, Malaysia, 1993.
35. Xie, X.; Yang, Y.; Ji, M. Analysis of ground surface settlement induced by the construction of a large-diameter shield-driven tunnel in Shanghai. China. *Tunn. Undergr. Sp. Technol.* **2016**, *51*, 120–132. [CrossRef]
36. Likitlersuang, S.; Surarak, C.; Suwansawat, S.; Wanatowski, D.; Oh, E.; Balasubramaniam, A. Simplified finite-element modelling for tunnelling-induced settlements. *Geotech. Res.* **2014**, *1*, 133–152. [CrossRef]
37. Gerhard, V.v.S. Numerical Analysis of Surface Settlements Induced by a Fluid-Supported Tunnelling Machine. 2012. Available online: https://diglib.tugraz.at/download.php?id=576a77b147200&location=browse (accessed on 30 August 2023).
38. Hejazi, Y.; Dias, D.; Kastner, R. Impact of constitutive models on the numerical analysis of underground constructions. *Acta Geotech.* **2008**, *3*, 251–258. [CrossRef]
39. Moller, S. Tunnel Induced Settlements and Structural Forces in Linings. 2006. Available online: http://www.uni-s.de/igs/content/publications/Docotral_Thesis_Sven_Moeller.pdf (accessed on 30 August 2023).
40. Choon, C.H. Field Measurements and Numerical Analysis of Interaction between Closely Spaced Bored Tunnels. Ph.D. Thesis, Nanyang Technological University, Singapore, 2016.
41. Janin, J.P.; Dias, D.; Emeriault, F.; Kastner, R.; Le Bissonnais, H.; Guilloux, A. Numerical back-analysis of the southern Toulon tunnel measurements: A comparison of 3D and 2D approaches. *Eng. Geol.* **2015**, *195*, 42–52. [CrossRef]
42. Atkinson, J.H.; Sallfors, G. Experimental determination of soil properties. General Report to Session 1. In Proceedings of the 10th European Conference on Soil Mechanics and Foundation Engineering, Florence, Italy, 26–30 May 1991; pp. 915–956.

43. Vitali, O.P.M.; Celestino, T.B.; Bobet, A. Construction strategies for a NATM tunnel in São Paulo, Brazil, in residual soil. *Undergr. Sp.* **2022**, *7*, 1–18. [CrossRef]
44. *BS 5930*; Code of Practice for Site Investigations. Br. Stand. Inst. (BSI): London, UK, 1981.
45. Ohya, S.; Imai, T.; Matsubara, M. Relationships between N value by SPT and LLT pressuremeter results. In Proceedings of the 2nd European Symposium on Penetration Testing, Amsterdam, The Netherlands, 24–27 May 1982; pp. 125–130.
46. Yagiz, S.; Akyol, E.; Sen, G. Relationship between the standard penetration test and the pressuremeter test on sandy silty clays: A case study from Denizli. *Bull. Eng. Geol. Environ.* **2008**, *67*, 405–410. [CrossRef]
47. Cheshomi, A.; Ghodrati, M. Estimating Menard pressuremeter modulus and limit pressure from SPT in silty sand and silty clay soils. A case study in Mashhad, Iran. *Geomech. Geoengin.* **2015**, *10*, 194–202. [CrossRef]
48. Naseem, A.; Jamil, S.M. Development of correlation between standard penetration test and pressuremeter test for clayey sand and sandy soil. *Soil Mech. Found. Eng.* **2016**, *53*, 98–103. [CrossRef]
49. Briaud, J.L. *The Pressuremeter*; AA Balkema: Rotterdam, The Netherlands, 1992.
50. Gambim, M.; Rosseau, J. Interpretation and application of pressuremeter test results to foundation design. *Gen. Memo. Rev. Sols Soils* **1975**, *26*.
51. Janbu, N. Soil compressibility as determined by oedometer and triaxial tests Wiesbaden. In Proceedings of the European Conference SMFE, Wiesbaden, Germany, 1963.
52. Schanz, T.; Vermeer, P.A. On the stiffness of sands. In *Pre-Failure Deform. Behav. Geomaterials*; Thomas Telford Publishing: London, UK, 1998; pp. 383–387.
53. Surarak, C.; Likitlersuang, S.; Wanatowski, D.; Balasubramaniam, A.; Oh, E.; Guan, H. Stiffness and strength parameters for hardening soil model of soft and stiff Bangkok clays. *Soils Found.* **2012**, *52*, 682–697. [CrossRef]
54. Schanz, T.; Vermeer, P.A.; Bonnier, P.G. The hardening soil model: Formulation and verification. In *Beyond 2000 Comput. Geotech.*; Routledge: Oxfordshire, UK, 2019; pp. 281–296.
55. Mayne, P.W. Integrated ground behavior: In-situ and lab tests. *Deform. Charact. Geomater.* **2005**, *2*, 155–177.
56. Yoo, C.; Lee, Y.; Kim, S.-H.; Kim, H.-T. Tunnelling-induced ground settlements in a groundwater drawdown environment—A case history. *Tunn. Undergr. Sp. Technol.* **2012**, *29*, 69–77. [CrossRef]
57. Shirlaw, J.N.; Ong, J.C.W.; Rosser, H.B.; Tan, C.G.; Osborne, N.H.; Heslop, P.E. Local settlements and sinkholes due to EPB tunnelling. *Proc. Inst. Civ. Eng. Eng.* **2003**, *156*, 193–211. [CrossRef]
58. Greenwood, J. 3D Analysis of Surface Settlement in Soft Ground Tunnelling. 2002. Available online: http://hdl.handle.net/1721.1/29558 (accessed on 30 August 2023).
59. Alagha, A.S.N.; Chapman, D.N. Numerical modelling of tunnel face stability in homogeneous and layered soft ground. *Tunn. Undergr. Sp. Technol.* **2019**, *94*, 103096. [CrossRef]
60. Ruse, N.M. *Räumliche Betrachtung der Standsicherheit der Ortsbrust beim Tunnelvortrieb*; Institut für Geotechnik: Hanover, Germany, 2004.
61. Terzaghi, K. *Shield Tunnels of the Chicago Subway*; Harvard University, Graduate School of Engineering: Cambridge, MA, USA, 1942.
62. Islam, M.S.; Iskander, M. Twin tunnelling induced ground settlements: A review. *Tunn. Undergr. Sp. Technol.* **2021**, *110*, 103614. [CrossRef]
63. Bartlett, J.V.; Bubbers, B.L. Surface movements caused by bored tunnelling. In Proceedings of the Conference on Subway Construction, Budapest, Hungary, 1970; pp. 513–539.
64. Moretto, O. Discussion on "Deep excavations and tunnelling in soft ground". In Proceedings of the 7th International Conference on Soil Mechanics and Foundation Engineering, Mexico City, Mexico, 1969; pp. 311–315.
65. Suwansawat, S.; Einstein, H.H. Describing settlement troughs over twin tunnels using a superposition technique. *J. Geotech. Geoenviron. Eng.* **2007**, *133*, 445–468. [CrossRef]
66. Kannangara, K.K.P.M.; Ding, Z.; Zhou, W.-H. Surface settlements induced by twin tunneling in silty sand. *Undergr. Sp.* **2022**, *7*, 58–75. [CrossRef]
67. Zhou, Z.; Ding, H.; Miao, L.; Gong, C. Predictive model for the surface settlement caused by the excavation of twin tunnels. *Tunn. Undergr. Sp. Technol.* **2021**, *114*, 104014. [CrossRef]
68. Ayasrah, M.; Qiu, H.; Zhang, X. Influence of cairo metro tunnel excavation on pile deep foundation of the adjacent underground structures: Numerical study. *Symmetry* **2021**, *13*, 426. [CrossRef]
69. Chakeri, H.; Ünver, B. A new equation for estimating the maximum surface settlement above tunnels excavated in soft ground. *Environ. Earth Sci.* **2014**, *71*, 3195–3210. [CrossRef]
70. Lu, H.; Shi, J.; Wang, Y.; Wang, R. Centrifuge modeling of tunneling-induced ground surface settlement in sand. *Undergr. Sp.* **2019**, *4*, 302–309. [CrossRef]
71. Moussaei, N.; Khosravi, M.H.; Hossaini, M.F. Physical modeling of tunnel induced displacement in sandy grounds. *Tunn. Undergr. Sp. Technol.* **2019**, *90*, 19–27. [CrossRef]
72. Athar, M.F.; Sadique, M.R.; Alsabhan, A.H.; Alam, S. Ground Settlement Due to Tunneling in Cohesionless Soil. *Appl. Sci.* **2022**, *12*, 3672. [CrossRef]

sustainability

MDPI

Article

Study on the Impact of Different Parameters on Prediction of Crown Deformations in Underground Caverns

Kadiyala Sudhakar [1,*], Rabindra Kumar Sinha [1] and Sripad Ramachandra Naik [2]

1 Department of Mining Engineering, Indian Institute of Technology (Indian School of Mines), Dhanbad 826004, India; rksinha@iitism.ac.in
2 National Institute of Rock Mechanics, Bengaluru 560070, India; nmd.nirm@gmail.com
* Correspondence: kadiyala.17dp000198@me.iitism.ac.in

Abstract: Crown deformation is a major concern in the design and construction of underground caverns. It can lead to damage to the cavern structure and surrounding infrastructure and can also pose a safety hazard to workers. This paper studies the factors affecting crown deformation in underground caverns. A parametric study was conducted to investigate the effects of seven parameters on crown deformation: *rock mass rating (RMR), uniaxial compressive strength (UCS), Young's modulus of intact rock (Ei), Poisson's ratio (v), tensile strength (σt), angle of internal friction (φ), and cohesion (C)*. The results of the parametric study showed that the following parameters significantly affected crown deformation: RMR, UCS, Ei, and Φ. A multiple linear regression analysis was conducted to develop a regression equation to predict crown deformation. The coefficient of determination (R^2) for the regression equation is 92.92%, which indicates that the equation is a good predictor of crown deformation. The parametric study results and the regression analysis can be used to improve the design and construction of underground caverns. By considering the factors that affect crown deformation, engineers can design more stable caverns that are less likely to experience deformations. The results of the study can be used to improve the design and construction of caverns, making them safer and more sustainable.

Keywords: crown deformation; underground caverns; parametric study; multiple linear regression

Citation: Sudhakar, K.; Sinha, R.K.; Naik, S.R. Study on the Impact of Different Parameters on Prediction of Crown Deformations in Underground Caverns. *Sustainability* **2023**, *15*, 12851. https://doi.org/10.3390/su151712851

Academic Editors: Mahdi Hasanipanah, Danial Jahed Armaghani and Jian Zhou

Received: 5 July 2023
Revised: 17 August 2023
Accepted: 21 August 2023
Published: 25 August 2023

1. Introduction

Underground caverns are widely used for various purposes, such as housing hydroelectric machinery, storing crude oil, gas, and radioactive waste, conducting scientific research, and providing leisure amenities. Due to limited space availability and environmental concerns, construction of underground structures rapidly increased worldwide between the late 20th and early 21st century. Underground caverns have several advantages over surface structures, such as reduced land occupation, higher security, lower cost, and increased environmental benefits. From a sustainability perspective, underground caverns offer several advantages over surface structures: they can help to reduce land occupation, which can help to preserve natural habitats and reduce the impact of development on the environment; they can also help to reduce noise and air pollution; and they can provide a more secure environment for storing hazardous materials. As the world population continues to grow and the demand for resources increases, underground caverns are likely to play an increasingly important role in sustainable development.

However, the stability of these caverns is a critical concern, as most of them are constructed in complex geology and stress conditions, which influences the deformation pattern of the caverns, which, in turn, can lead to structural failure and safety hazards. Sometimes, larger crown failures in caverns hamper the project time, escalate the cost, and cause overruns. Hence, the study of the behavior of the crown prior to the excavation of caverns is essential. The stability of the crown of the cavern is an important consideration as the deformation is influenced by several parameters, such as the size and shape of the cavern, the rock cover, the rock type, the in situ stress field, the geology and rock mass properties, and the construction practices. Therefore, it is essential to understand how these parameters affect crown deformation and to develop reliable methods for predicting and controlling it.

Previous studies have investigated some aspects of crown deformation in underground caverns, such as the effect of cavern geometry, rock cover, rock type, in situ stress, geology and rock mass properties, and construction practices [1–4]. However, most of these studies have focused on specific cases or locations and have not provided a comprehensive and systematic analysis of the impact of different parameters on crown deformation. Moreover, most of these studies have used empirical or analytical methods that have limitations in capturing the complex behavior of rock masses and caverns.

In this paper, we present a unique study on the impact of different parameters on the prediction of crown deformation in underground caverns. We use a numerical modeling approach based on the finite element method (FEM), which can account for the nonlinear and heterogeneous characteristics of rock masses and caverns. We consider a wide range of parameters that cover various aspects of cavern design and site conditions. We conduct a series of parametric analyses to examine how each parameter affects the crown deformation and to identify the most influential ones. A comprehensive statistical analysis with multiple linear regression models was carried out to bring out the most important factors affecting crown deformation.

The advantage of this research is that the empirical equation developed for predicting crown deformation may become a handy tool for practicing engineers in the estimation of the behavior of crowns prior to excavation. The crown deformations predicted in this study can assist in selecting the optimal range of the instruments. The predicted deformations are helpful in setting the warning limits in deformation monitoring and reviewing the planned support systems. Overall, it can be beneficial in terms of project cost savings and the safety of manpower and machinery during construction time. Hence, study of the cavern crown areas is essential. The results of the study can be used to improve the design and construction of caverns, making them safer and more sustainable.

2. Overview of Factors Affecting Crown Deformation

The size and shape of an underground cavern are important factors that affect its crown deformation. Arch-shaped crowns with straight high walls are the most common shape for underground caverns [2]; this shape provides good stability and minimizes stress concentration. However, larger hydro caverns are being constructed as the demand for power increases. These larger caverns may be more susceptible to crown deformation due to their increased span, height, and length.

Another factor affecting crown deformation is the rock cover over the cavern. Caverns at shallow depths with reasonably good-strength rock mass evidence instabilities observed in the form of small wedges [3]. These wedges are formed by intersecting joints or fractures in the rock mass and may detach from the roof under gravity or water pressure. Caverns at deeper depths with low-strength rock mass may experience large-scale failure or collapse due to high stress and low confinement.

Rock mass properties, such as the rock mass rating (RMR), density, strength parameters, and friction angle, are also important factors affecting an underground cavern's crown deformation.

The in situ stress field is another important factor affecting an underground cavern's crown deformation. It is influenced by tectonic forces, gravity, topography, and geological structures. Generally, the long axis of the caverns is aligned in the direction of the major principal stress to reduce stress concentration and optimize stability. However, the magnitude and direction of in situ stresses can vary depending on the location of the cavern. The maximum stress concentration factor in the pillar owing to vertical stress is unaffected by cavern orientation [5].

The geology and rock type of the area where the cavern is being constructed are also important factors that affect its crown deformation. Strong, well-bonded rock is generally more stable than weak, fractured rock; for example, an underground cavern built at India's Peninsular Gneissic complex has a rock type of granitic gneiss, which is strong to extremely strong and traversed by three–four sets of notable discontinuities, including a sub-horizontal joint set of dolerite dykes. The dolerite dykes are weak to medium strong, closely joined with several sets of joints, and brittle. This may significantly modify stresses and displacements in underground openings [3].

The underground powerhouse located in Deccan basalt flows intruded by dolerite dykes and sill seems to be good for tunneling. However, during the construction stage, geotechnical problems such as wedge-type roof falls and vertical and horizontal cracks were observed on sides of the roof and walls [6].

Methodologies Used for the Prediction of Displacements

The prediction of crown deformations in large underground caverns is a complex problem that can be addressed using a variety of methodologies. These methodologies can be broadly categorized into three main groups: analytical, numerical, and empirical.

Analytical methods are based on theoretical models of rock mechanics that can estimate the magnitude and distribution of deformations under idealized conditions. However, these methods have limitations in accounting for the complex geological and geotechnical conditions encountered in real-world projects. One of the most common analytical methods for predicting crown deformation is the convergence–confinement method. This method assumes that the rock mass behaves elastically and that the deformations are governed by the in situ stresses and the geometry of the opening. The convergence–confinement method can estimate the overall convergence of an underground opening but it cannot be used to predict the distribution of deformations within the opening.

Numerical methods are based on computational models that can simulate the excavation of underground caverns and the subsequent deformation of the surrounding rock mass. Numerical methods can account for a wide range of factors, such as the geometry of the cavern, the properties of the rock mass, and the in situ stresses. One of the most widely used numerical methods for predicting crown deformation is the finite element method (FEM). The FEM is a powerful numerical method that can capture complex phenomena and interactions in underground structures. FEM models can be used to account for a wide range of factors, including the geometry of the cavern, the properties of the rock mass, and the in situ stresses. However, FEM models can be computationally expensive to develop and run and can be sensitive to input parameters.

The patterns of deformation, stress status, and distribution of plastic areas for the stability of the huge underground caverns were analyzed and assessed using Rocscience RS2-FEM software [7]. In Greece, the convergence of shallow tunnels (30–120 m overburden thickness) built in various rock bodies has been evaluated as a function of the geological strength index (GSI classification). Maximum vertical and horizontal convergence predictions were made using the FEM and the 'characteristic line' theory during or shortly after tunnel excavation. They were found to be in good accord, with geodetic observations of convergence taken about two months following the excavation [8]. The cavern behavior was investigated using 3D numerical modeling (FEM analysis) and geotechnical instrumentation monitoring to study the sensitive parameters for prediction of convergence [5]. The horseshoe cavern stability was studied using 3DEC under various scenarios in terms of deformations at various points. The dynamic nature of underground construction and rock mass behavior, as well as timing and interpretation (installation of monitoring target and recorded data), are critical for understanding the stability of excavated structures [9].

The effect of the lateral-stress-to-vertical-stress ratio, cavern height, overburden depth, and deformation modulus was investigated and produced as an elastoplastic displacement prediction equation for the cavern's high side wall. This method was used to determine the deformation key sites for a cavern [10].

Empirical methods for predicting the crown deformation of underground caverns are based on observational data and empirical relationships that can predict deformations based on previous experiences and cases. Empirical methods can provide practical and simple solutions for predicting crown deformation, but they may not be applicable to all situations and conditions. This approach is often used in conjunction with analytical or numerical methods to improve the accuracy of predictions.

One of the most common empirical methods for predicting crown deformation is regression analysis. Regression analysis can be used to fit a mathematical equation to a set of data points. The equation can then be used to predict crown deformations for new projects with similar characteristics. Another empirical method for predicting crown deformation is the neural network. Neural networks are machine learning algorithms that can learn complex relationships between variables. Neural networks have been used to predict crown deformations by training them on datasets that contain information about crown deformation and the factors that affect it. Once a neural network is trained, it can be used to predict crown deformation for new projects.

Alongside the above-mentioned methodologies, other methods have been used to predict crown deformations in underground caverns. These methods include the following:

- Support vector machines (SVMs): SVMs are machine learning algorithms that can classify data or predict continuous values. SVMs have been used to predict crown deformations by developing models that relate deformation to various parameters, such as rock mass properties, in situ stresses, and excavation methods.
- Artificial neural networks (ANNs): ANNs are another type of machine learning algorithm that can be used to predict crown deformations. ANNs are trained on datasets that contain information about crown deformation and the factors that affect it. Once an ANN is trained, it can be used to predict crown deformation for new projects.
- Monte Carlo simulation: Monte Carlo simulation is a statistical method that can be used to estimate the probability of various outcomes. Monte Carlo simulation has been used to predict crown deformations by simulating the excavation of an underground cavern multiple times with different random values for the input parameters. The results of the simulations can then be used to estimate the probability of different levels of crown deformation.

Some examples of empirical methods applied to specific cases are as follows.

An evolutionary SVM approach was devised using a combination of SVM and the genetic algorithm for generating a time series analysis of nonlinear slope deformation. The results demonstrate that the established SVMs can accurately describe the evolutionary law of geomaterial deformation at depth and predict the following 6–10 time steps with sufficient accuracy and confidence [11]. ANN approaches were used to thoroughly combine information from monitoring observations and investigations to create two dimensionless indices. The first indicates the predisposition of a portion of the rock mass to destruction, dislocation, and deformation—$C^{P}{}_{RF}$—coefficient of roof fall—predisposition, and the second represents the predisposition and possibility of maintaining the working—$C^{M}{}_{RF}$ coefficient of roof fall—maintenance [12].

The surrounding rock stability of an underground cavern group is an essential issue in the process of cavern excavation, which has the characteristics of large displacement, discontinuity, and uneven deformation, making computation and analysis difficult. By observing the destruction process of a jointed rock mass, the discontinuous deformation analysis for the rock failure approach was used to analyze the stability of the surrounding rock [13]. A support vector machine (SVM) is used to forecast the allowed deformation of surrounding rock. One-hundred sets of multi-factor and multi-level orthogonal experiments are constructed and simulated using ABAQUS-based two-dimensional numerical models. Three parameters are considered: rock mass categorization, cavern buried depth, and cavern size. The mapping association between permitted deformation and the three influencing factors described above is established [13].

The choice of methodology for crown deformation prediction will depend on several factors, including the size and complexity of the cavern, the availability of data, and the desired level of accuracy. A combination of analytical, numerical, and empirical methods is often used to achieve the most reliable results.

3. Study Methodology

The current study examines the stability analysis of underground structures based on factors for rock mass utilizing various prediction techniques. The study methodology is shown in Figure 1.

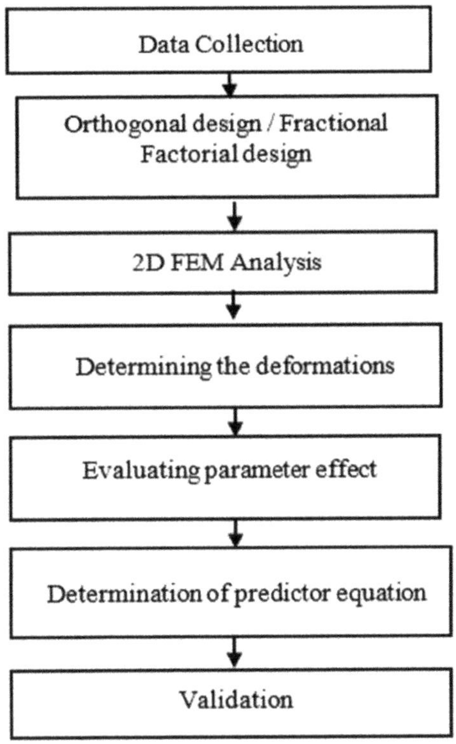

Figure 1. Workflow of the current study.

3.1. Data Collection

The first step in the study methodology is to collect data from case studies of underground caverns in various geological contexts. The data collected include the following:

- Geometrical data include the size and shape of the cavern, the rock cover, and the in situ stress conditions;
- Rock mass properties include the rock mass rating, the strength parameters for compression, tension, cohesion, friction angle, and Poisson's ratio;
- Mechanical characteristics of intact rock samples, such as the uniaxial compressive strength and the tensile strength;
- Instrumentation data; in particular, the measurements of crown displacement made with extensometers.

The data for this study were collected from a variety of sources, including previous research papers, government reports, and industry databases.

The next step in the study methodology is to develop an empirical relationship for the prediction of crown displacements. This relationship will be based on the data collected from the case studies. The relationship will be developed using numerical and statistical methods such as design of experiments (DOE) and regression analysis. Once the empirical relationship has been developed, it was validated using data from additional case studies. The validation process will involve comparing the predicted crown displacements to the actual crown displacements that were measured in the case studies. The final step in the study methodology is to draw conclusions about the effectiveness of the empirical relationship. The conclusions will be based on the results of the validation process.

The study methodology outlined above is a rigorous and systematic approach to develop and validate empirical relationships for predicting crown displacements in underground caverns. The methodology can be used to develop empirical relationships for a variety of underground caverns, regardless of the geological context.

About three–four instrumented sections of data were collected at each cavern. Figure 2 depicts the precise location information for the caverns considered for this investigation. Table 1 displays the parameter's possible range used in this investigation and to study the deformations.

Figure 2. Location of the caverns (Ref. www.mapsofindia.com, accessed on 15 May 2023).

Table 1. List of caverns with geometry and rock properties.

Sr No	Cavern Name	Width, m	Height, m	Rock Cover, m	Rock Mass Rating, RMR	Stress Ratios (K_H, K_h)	UCS, MPa	Young's Modulus, Ei, GPa	Tensile Strength, σt MPa	Poisson's Ratio, υ	Friction Angle, Φ	Cohesion, C MPa	Rock Types	Issues Encountered during Construction	References
1	THP-MH *	20.4	44.5	500	56–72	1.30, 0.87	60–70	21–29.5	0.64–1.7	0.35–0.365	32.82–35.73	1.81–3.09	Phyllites, Philitic quartzite, and quartzite bands	The roof collapse occurred during construction; vertical and horizontal cracks occurred on shotcrete at high walls	[5,14]
2	THP-DC *	13.9	18.5	140	47–53	2.10, 1.30	34–40	20–30	0.03–0.05	0.19–0.20	46.35–50.08	0.65–0.78	Predominantly muscovite-biotite genesis with miner bands of biotite schist, quartzite, and clac silicate rocks; general rock types are good to moderately good	Major cracks occurred in the crown area during bench excavations	[15]
3	MHP-MH *	23	43	180	43–65	1.75, 1.17	182–250	40–72	0.18–1.3	0.165–0.172	58.4–63.23	1.71–5.16	Biotite schist, quartzite, pegmatite with micaceous schist, granite intrusions		[16]
4	NJHP-MH *	20	45	300	70–84	1.34, 0.67	50–52	21.5–29	0.16–0.70	0.24–0.28	51.27–53.95	1.9–2.77	Major rock types encountered are augen gneiss and gneiss	No failures on a rock; minor cracks occurred on the shotcrete while benching	[17]
5	NJHP-DC *	16.3	27.5	450	63–68	1.04, 0.35	51–55	31–35	0.17–0.32	0.35–0.38	41.51–43.38	1.89–2.01	Predominantly massive augen gneiss and gneiss with minor bands of biotite schist	Rockfall occurred in chambers 3 and 4	[18]
6	Tehri-MH *	22	47	370	59–61	0.53, 0.31	56–60	45–46.5	0.17–0.21	0.20–0.21	41.0–42.08	1.48–1.61	Massive thinly bedded phyllitic quartzite	No failures	[19]
7	SSP-PH *	23	57	47.5	54–59	2.5, 1.25	60–80	23–45	0.06–0.13	0.20–0.22	58.9–61.53	0.5–0.75	Deccan trap group-porphyritic basalt, amygdaloidal basalt, and agglomerate	Vertical cracks occurred in the crown and walls during construction	[20,21]
8	SLBHP-MH *	25.7	53	233	57–68	1.55, 0.62	220–255	52–73	0.74–1.95	0.23–0.25	51.3–53.38	2.51–4.95	Hard siltstones; pinkish, fine-grained, and hard crystalline rock	No major problems associated with construction	[22]
9	KLIPP8-PH *	25	34	42	40–60	1.62, 1.08	120–160	32–64	0.03–0.14	0.25–0.265	64.26–68.41	0.48–1.03	Granite	A wedge fall occurred in one place	[23]
10	PYKARA-MH *	20.2	37	532	45–50	1.75, 0.88	72–180	95–100	0.03–0.11	0.26–0.27	49.92–57.51	1.2–1.88	Massive charnokite and migmatite gneiss	No problems noticed during construction	[24,25]

* THP-MH—Tala Hydroelectric Project Machine Hall; THP-DC—Tala Hydroelectric Project Desilting Chambers; MHP-MH—Mangdhechu Hydroelectric Project Machine Hall; NJPC-MH—Nathpa Jhakri Hydroelectric Project Machine Hall; NJPC-DC—Nathpa Jhakri Hydroelectric Desilting Chambers; Tehri MH—Tehri Machine Hall; SSP-PH—Sardar Sarovar Project Power House; SLBHP-MH—Srisailam Left Bank Hydroelectric Project Machine Hall; KLIPP8-PH—Kaleswaram Lift Irrigation Project Package8 Pump House; PYKARA MH—Pykara Hydroelectric Project Machine hall.

3.2. Description and Overview of the Data Collected

The data collected for this study include the following:

- Geometry: The dimensions of the caverns range from 100 to 525 m in length, 14 to 26 m in width, and 18.55 to 57 m in height.
- Overburden depth: The rock cover of the caverns ranges from 42 to 532 m.
- Rock type: The surrounding rock of the caverns ranges from soft rock to medium-hard rocks, such as phyllites and charnockite. The amygdaloidal and porphyritic basalt flows that make up the rocks of the Sardar Sarovar Project underground powerhouse cavern are separated by pockets of agglomerate [6]. The Tala Powerhouse Cavern is located on the southern slopes of the eastern Himalayas, not far from the MCT (Main Central Thrust). This major thrust zone is the boundary between the Lesser and Higher Himalayas. Most of the rocks in the powerhouse are phyllitic quartzite, quartzite, phyllites, and amphibolite schist.
- Physico-mechanical properties: The specimens were prepared from intact rock samples from the project sites. Following ISRM guidelines, specimens were examined at the NIRM laboratory to determine their physical and mechanical characteristics during cavern design and construction [26]. These are needed as input data for numerical modelling.
- The following are the physico-mechanical properties of the intact rock samples:
- Uniaxial compressive strength (UCS): The UCS ranges from 34 to 250 MPa for undamaged rock samples;
- Young's modulus (Ei): The intact rock's Young's modulus ranges from 20 to 100 GPa;
- Poisson's ratio (υ) and tensile strength (σt): These parameters range from 0.16 to 0.38 and 0.03 to 1.7 MPa, respectively;
- Internal friction has an angle (φ) ranging from 32.84° to 68.41°;
- Cohesion (c): The samples' cohesion values range from 0.48 to 3.09 MPa.

4. Instrumentation and Monitoring

Instrumentation and monitoring play a vital role in the construction and operation of underground caverns. They help to ensure the safety and stability of the structure by providing data on rock movement, pore water pressure, and the load on support elements.

The instrumentation layout for large underground caverns typically includes the following:

- Extensometers: These devices measure deformations in the surrounding rock mass;
- Anchor load cells: These devices measure the load on support elements, such as bolts and anchors;
- Piezometers: These devices measure pore water pressure in the rock mass;
- Convergence targets: These devices measure the convergence of the roof and side walls of the cavern;
- Rock bolt stress meters: These devices measure the stress buildup along the length of rock bolts.

The type of instrumentation used can vary depending on the project's specific needs. Simple mechanical instruments are often used for initial monitoring, while more sophisticated remote-type instruments may be used for long-term monitoring.

The effectiveness of the support systems and rock mass stability is verified by monitoring the support elements using loadcells and deformations using the MPBX and identifying the crucial zones of the powerhouse and transformer caverns [15,27]. The sophistication and remoteness of instruments increase with a decrease in the stability of the structure [28].

An extensive geotechnical instrumentation and monitoring program was carried out at the Ingula hydropower caverns to validate design assumptions and monitor long-term creep effects. The results of this monitoring program showed that most of the time-dependent deformation in the power caverns was expected to occur within 6 months to a year following excavation down to the operating floor level [29].

In the last four decades, major hydropower caverns built in India and Bhutan, oil storage caverns, and pump house caverns built in India were instrumented and monitored during construction for their safety and stability. Normally, about four–five sections or more of instrumentation arrays cover the full length of the cavern. Figure 3 explains the typical instrumentation layout of a large cavern section.

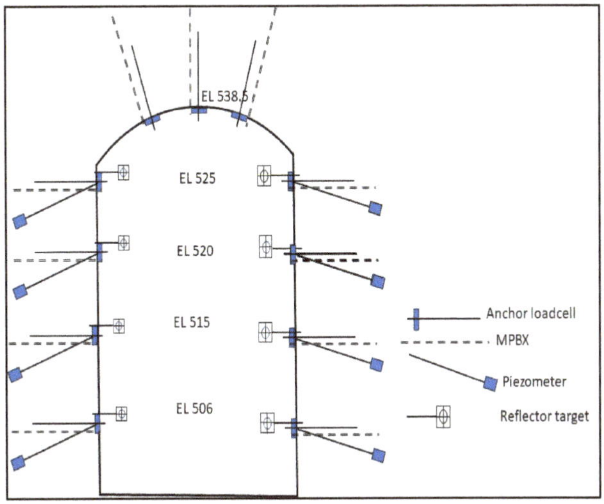

Figure 3. Typical instrumentation layout of an underground cavern.

5. Data Analysis

Initially, data were collected from all ten caverns selected for the present study. The data were checked and sorted, with 31 sections of crown displacements data and the surrounding rock mass conditions. Later, we checked for correlations between measured displacements and variables from geometry, overburden height, and rock mass properties. Finally, we plotted the correlation effect of crown deformations with the selected parameters in Figure 4a–h. The correlation coefficient matrix between the selected variables is shown in Table 2.

Table 2. Correlation coefficient matrix between the selected variables.

	Width, m	Height, m	Rock Cover, m	K_H	K_h	RMR	UCS, MPa	Ei, GPa	Poisson's Ratio υ	Tensile Strength, σt	Friction Angle, Φ	Cohesion (C), MPa	Deformation (mm)
Width, m	1.0												
Height, m	0.8	1.0											
Rock Cover, m	−0.3	0.0	1.0										
KH	0.0	−0.1	−0.6	1.0									
Kh	0.0	−0.2	−0.6	0.9	1.0								
RMR	0.0	0.3	0.2	−0.4	−0.5	1.0							
UCS, MPa	0.7	0.4	−0.3	0.2	0.1	−0.1	1.0						
Ei, GPa	0.4	0.1	0.2	0.1	0.0	−0.3	0.6	1.0					
Poisson's Ratio υ	−0.2	−0.1	0.7	−0.5	−0.5	0.3	−0.3	−0.2	1.0				
Tensile strength, σt	0.4	0.5	0.2	−0.2	−0.2	0.5	0.5	0.1	0.2	1.0			
Friction angle (Φ)	0.5	0.1	−0.7	0.5	0.4	−0.3	0.5	0.3	−0.6	−0.2	1.0		
C, MPa	0.3	0.4	0.2	−0.3	−0.4	0.6	0.6	0.3	0.1	0.9	−0.1	1.0	
Deformation (mm)	−0.3	−0.1	0.6	−0.3	−0.1	0.0	−0.4	−0.3	0.6	0.0	−0.7	−0.1	1.0

Conditional formatting was used to color code the correlations: dark blue color indicates a strong positive correlation, light blue color indicates fair positive correlation, dark red indicates a strong negative correlation, light red indicates fair negative correlation and white indicates low or no correlation among the variables.

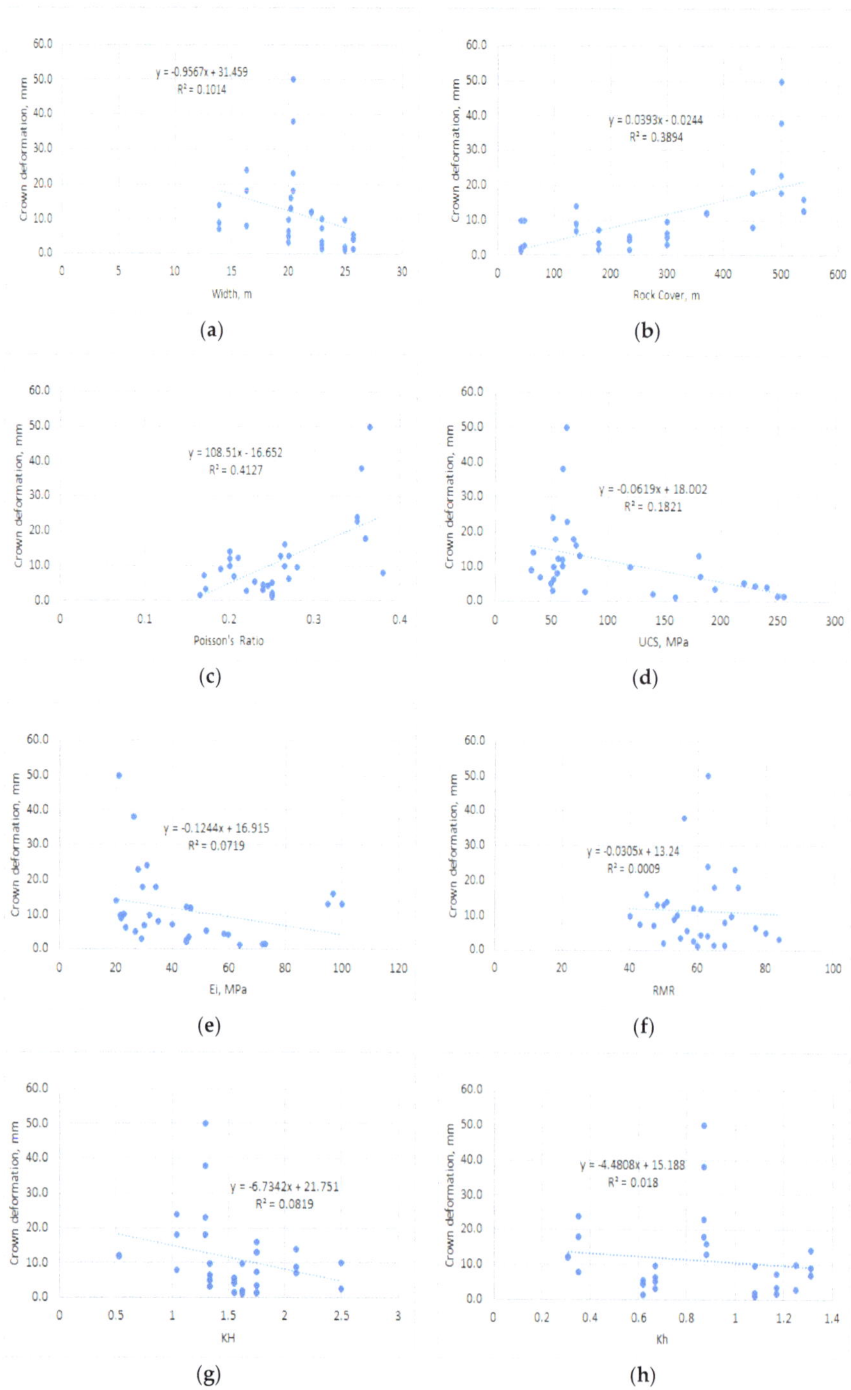

Figure 4. Effect of crown deformation with different variables: (**a**) width; (**b**) rock cover; (**c**) Poisson's ratio; (**d**) UCS; (**e**) Ei; (**f**) RMR; (**g**) K_H; (**h**) K_h.

The following are the findings from the correlation matrix:

- There is a strong positive correlation between crown displacement rock cover and Poisson's ratio;
- There is a weak negative correlation between crown displacement and UCS (Young's modulus);
- There is no significant correlation between crown displacement and other rock mass properties such as RMR, cohesion, and tensile strength; however, there is some nonlinear relation between these parameters and crown displacements.

5.1. Numerical Modeling of Underground Caverns

In this study, the displacements were found using the RS2 (Phase2, Rocscience Inc., Toronto, ON, Canada) 2D finite element method (FEM) program for soil and rock applications. This program can be used to create complex, multi-stage models that can be quickly analyzed, such as tunnels and caverns in weak or jointed rock. The analysis type used in this study was plane strain.

The caverns studied in this paper were all excavated in 10–15 stages. The dimensions of each cavern were 200–250 m in length, 20–25 m in width, and 18–53 m in height. Figure 5 shows the dimensions of one of the caverns studied. The excavation of the caverns was modeled in stages, starting with the top heading (TH) and then moving on to bench 1 (B1) and bench 2 (B2). The bench height in each stage was 2.5 to 4 m. Figure 5 shows the sectional view of the excavation sequence considered for numerical modeling of case 1 of the MHP-MH cavern.

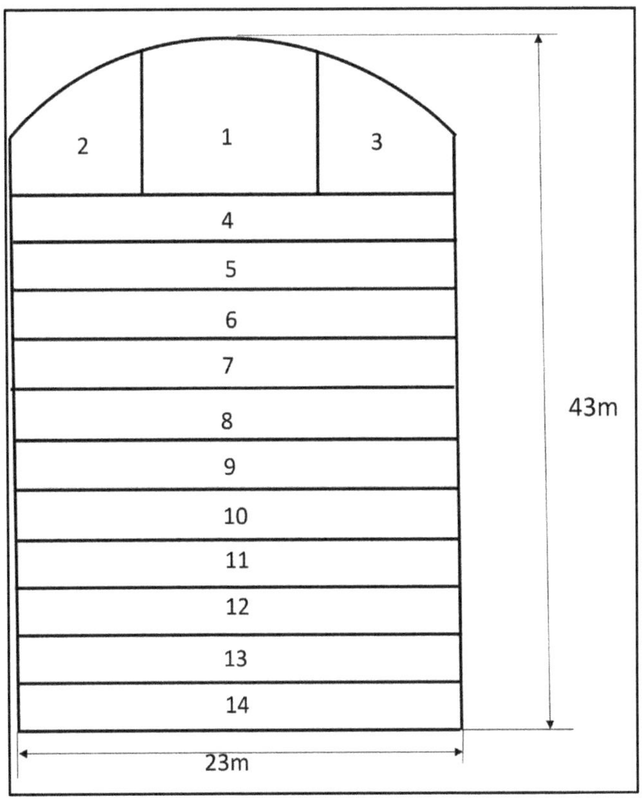

Figure 5. Geometric model of cavern studied parametric study (Case: MHP-MH).

Figure 6 shows the mesh and boundary conditions of the model. The boundary conditions were set such that the sides of the model were restrained in the X direction and, at the bottom, restrained in the XY direction. Figure 7 shows the total displacement contours of the model.

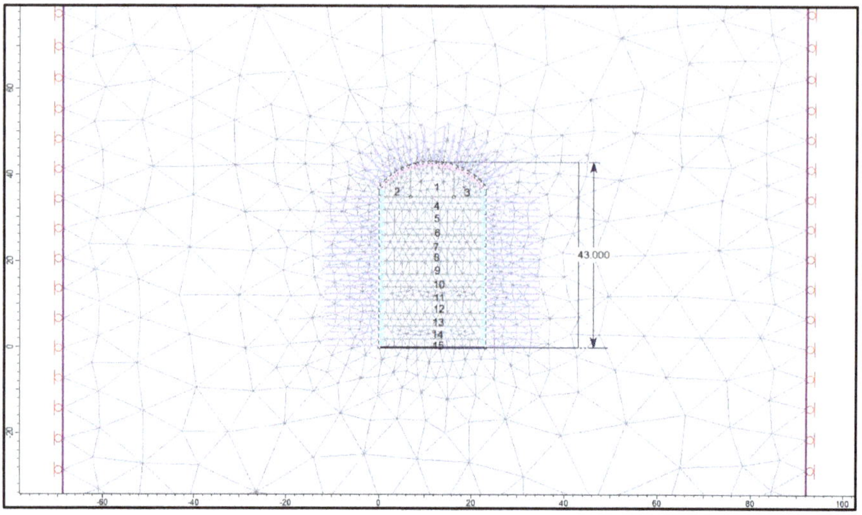

Figure 6. Mesh and boundary condition of the model (MHP-MH, Model 1/64).

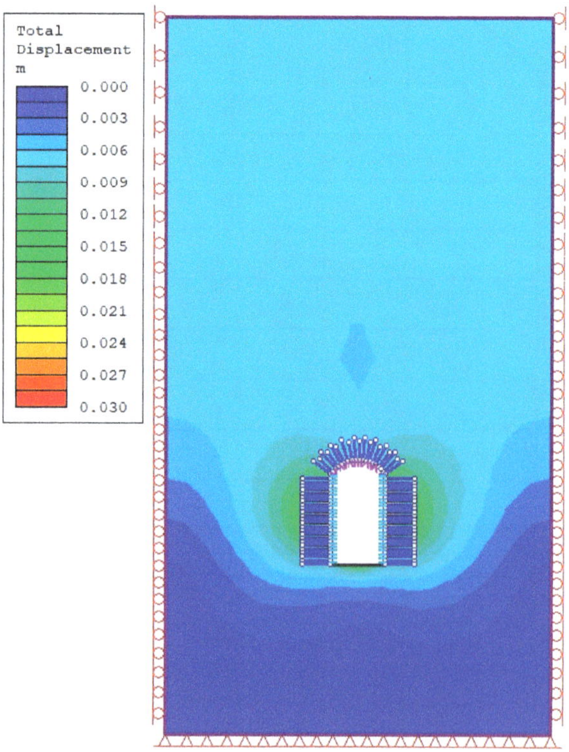

Figure 7. Total displacement contours of the model (MHP-MH, Model 1/64).

The crown displacements for all 64 samples of the machine hall cavern of the MHP project for different values of the parameters are given in Table 3.

Table 3. Predicted crown displacements from the 64 samples of the MHP machine hall cavern for different combinations of the input parameters.

RMR	UCS, MPa	Ei, GPa	Poisson's Ratio	Tensile Strength, MPa	Friction Angle, Degree	C, MPa	Deformation, mm
40	32	100	0.165	0.03	68.41	5.16	11.2
40	255	20	0.165	1.95	32.82	5.16	32.6
84	32	100	0.38	1.95	32.82	5.16	1.72
40	32	100	0.38	1.95	68.41	5.16	16.9
84	255	100	0.165	0.03	68.41	5.16	0.976
40	255	20	0.38	1.95	32.82	0.48	58.7
84	32	100	0.165	0.03	32.82	5.16	0.979
84	32	20	0.165	1.95	32.82	5.16	4.8
84	32	100	0.38	0.03	32.82	0.48	1.72
84	32	100	0.38	1.95	68.41	0.48	1.72
84	32	20	0.38	0.03	32.82	5.16	8.38
84	32	20	0.38	0.03	68.41	0.48	8.39
84	255	100	0.165	0.03	32.82	0.48	0.976
40	32	100	0.38	0.03	32.82	5.16	16.2
40	32	20	0.165	0.03	68.41	0.48	51.1
40	32	20	0.38	0.03	68.41	5.16	72.4
40	32	100	0.165	1.95	68.41	0.48	11.2
40	255	20	0.38	0.03	68.41	0.48	58.8
40	255	100	0.165	1.95	32.82	0.48	7.27
40	32	100	0.38	1.95	32.82	0.48	31.8
40	255	20	0.165	0.03	68.41	5.16	32.7
40	255	100	0.38	1.95	68.41	0.48	12.7
84	255	100	0.165	1.95	32.82	5.16	0.977
40	255	100	0.165	0.03	68.41	0.48	7.27
40	255	20	0.38	1.95	68.41	5.16	58.8
84	255	20	0.165	0.03	68.41	0.48	4.79
40	32	20	0.165	1.95	32.82	0.48	51.2
84	255	20	0.165	1.95	32.82	0.48	4.8
40	255	100	0.38	0.03	68.41	5.16	12.7
40	255	100	0.165	0.03	32.82	5.16	7.27
40	255	100	0.38	0.03	32.82	0.48	12.7
84	32	20	0.165	1.95	68.41	0.48	4.8
40	32	20	0.38	1.95	68.41	0.48	9.39
84	255	20	0.165	0.03	32.82	5.16	4.79
84	32	20	0.38	1.95	68.41	5.16	8.39
84	32	100	0.165	1.95	32.82	0.48	0.979
84	255	20	0.165	1.95	68.41	5.16	4.79
84	32	20	0.165	0.03	68.41	5.16	4.8
40	255	100	0.165	1.95	68.41	5.16	7.27

Table 3. *Cont.*

RMR	UCS, MPa	Ei, GPa	Poisson's Ratio	Tensile Strength, MPa	Friction Angle, Degree	C, MPa	Deformation, mm
40	255	20	0.165	1.95	68.41	0.48	32.7
40	32	100	0.165	0.03	32.82	0.48	7.27
40	255	20	0.165	0.03	32.82	0.48	32.7
84	255	20	0.38	0.03	68.41	5.16	8.27
40	32	100	0.38	0.03	68.41	0.48	13.6
84	32	100	0.38	0.03	68.41	5.16	1.72
84	32	20	0.165	0.03	32.82	0.48	4.8
40	32	20	0.38	0.03	32.82	0.48	72.5
84	255	20	0.38	1.95	32.82	5.16	8.28
84	255	100	0.38	0.03	32.82	5.16	1.69
84	255	100	0.165	1.95	68.41	0.48	0.975
84	255	100	0.38	0.03	68.41	0.48	1.69
84	255	20	0.38	0.03	32.82	0.48	8.27
84	32	100	0.165	1.95	68.41	5.16	0.943
40	32	20	0.38	1.95	32.82	5.16	67.9
40	32	20	0.165	1.95	68.41	5.16	48
84	32	20	0.38	1.95	32.82	0.48	8.08
40	255	20	0.38	0.03	32.82	5.16	56.6
84	255	20	0.38	1.95	68.41	0.48	7.98
40	32	100	0.165	1.95	32.82	5.16	10.7
84	255	100	0.38	1.95	68.41	5.16	1.63
84	32	100	0.165	0.03	68.41	0.48	0.942
40	255	100	0.38	1.95	32.82	5.16	12.2
84	255	100	0.38	1.95	32.82	0.48	1.63
40	32	20	0.165	0.03	32.82	5.16	48

The results of the numerical modeling show that the crown displacements increase with increasing cavern height, decreasing rock mass rating, and increasing overburden depth.

The results of the numerical modeling can be used to improve the design and construction of underground caverns. By taking into account the factors that affect crown displacement, engineers can design caverns that are more stable and less likely to experience deformations.

5.2. Parametric Study on Crown Deformations of a Cavern (Case Study MHP-MH)

5.2.1. Design of Experiments and Fractional Factorial Work

Design of experiments (DOE) is a tool that allows you to obtain information about how factors (X's), alone and in combination, affect the process and its output (Y). DOE allows you to test more than one factor at a time, as well as different settings for each factor. DOE is more cost-effective than trial-and-error methods. Using DOE techniques, you can find the individual and interactive effects of various factors that can influence the output results of your measurements.

One way to do a parametric study is by changing one of the parameters within a range, keeping all other parameters constant, and then studying the effect. This process is then repeated for the remaining parameters one after another. However, this can be time-consuming and resource-intensive, especially if there are many parameters to consider. A more efficient way to conduct a parametric study is to use a factorial design of experiments. In a factorial design, experimental trials (or runs) are performed at all combinations of factor levels. A factorial experiment with k factors, each factor having two levels of value, will require 2^k number of runs. This is known as a 2^k factorial design, and it is read as a two-level k factor design [30–33].

In this study, we used a half-factorial design, which means that we only ran half of the possible combinations of factor levels. This was undertaken because we were only interested in the main effects of the parameters, not the interaction effects.

The values of the parameters used in the parametric study for the MHP machine hall cavern were as follows:

- Rock mass rating of the cavern crown (RMR) :43–65;
- Uniaxial compressive strength (UCS) :182–250 MPa;
- Young's modulus of intact rock (Ei) :40–72 GPa;
- Poisson's ratio (υ) :0.16–0.17;
- Tensile strength (σt) :0.18–1.3 MPa;
- Internal friction angle (Φ) :58.36–63.23 degrees;
- Cohesion (C) :1.42–5.16 MPa.

5.2.2. Main and Interaction Effects

The effect of a factor is defined as the change in response produced by a change in the level of the factor. It is called the main effect because it refers to the study's primary factors. The main effect of all seven parameters has been shown for the cases of the MHP-MH and the Pykara-MH cavern in Figures 8 and 9, respectively.

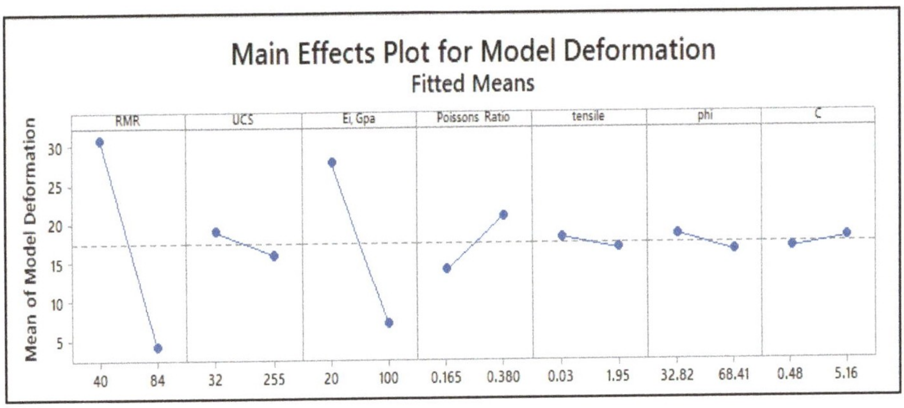

Figure 8. Main effects plot for crown deformation in the MHP-MH cavern.

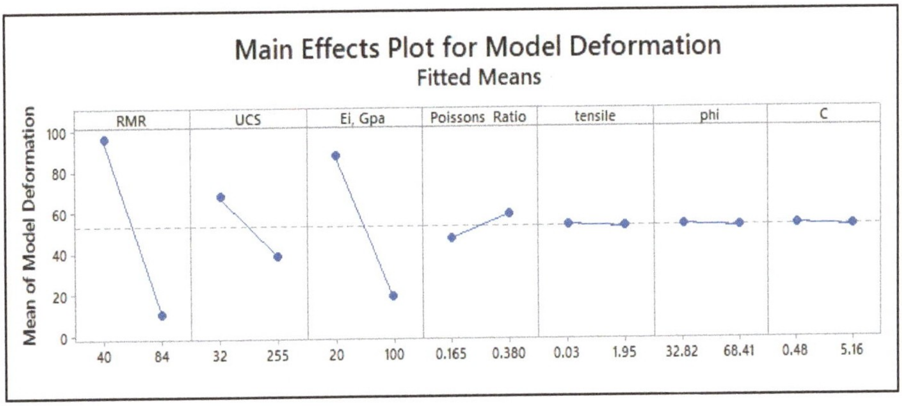

Figure 9. Main effects plot for crown deformation in the Pykara-MH cavern.

In this study, the main effects of the following parameters on crown deformation were investigated:

- Rock mass rating (RMR);
- Uniaxial compressive strength (UCS);
- Young's modulus of intact rock (Ei);
- Poisson's ratio (υ);
- Tensile strength (σt);
- Internal friction angle (φ);
- Cohesion (C).

The results of the study showed that the following main effects were significant:

- RMR: As RMR increases, crown deformation decreases;
- UCS: As UCS increases, crown deformation slightly decreases or does not change significantly;
- Ei: As Ei increases, crown deformation decreases;
- υ: As PR increases, crown deformation increases;
- σt: σt has no significant effect on crown deformation;
- φ: As φ increases, crown deformation slightly decreases;
- C: C has no significant effect on crown deformation.

The interaction effects between the parameters were also investigated. The results showed that there were no significant interaction effects between any of the parameters. This means that the effect of each parameter on crown deformation can be explained via its main effect alone. The parallel and near-to-parallel plots of crown deformation versus different parameters shown in Figure 10 indicate a very rare chance of interaction among the parameters. The same was observed for all other cases; thus, the influence of each parameter can be explained by its main effect alone. In conclusion, the results of this study show that the main effects of RMR, UCS, Ei, and υ are highly significant, and σt, φ, and C are of low significance to crown deformation. The interaction effects between the parameters are not significant.

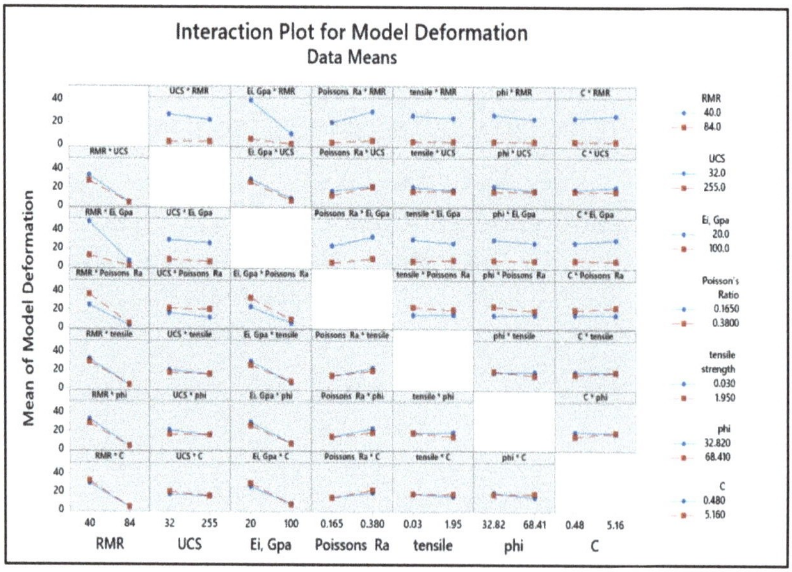

Figure 10. Interaction effects plot of crown deformation between different variables of MHP-MH cavern.

In the case study MHP-MH, it is appropriate to consider the effect of Poisson's ratio on displacement, which is 6.84. This indicates that as the Poisson's ratio value increases from 0.165 to 0.38, the displacement will increase by 6.84 mm. A negative value shows the reduction in displacements. The ranking of each parameter is given based on its absolute value. It is observed that the rock mass rating, uniaxial compressive strength, Young's modulus of intact rock, Poisson's ratio, tensile strength, angle of internal friction, and cohesion have a decisive role to play in influencing the displacements. The main effect of different parameters with respect to other studied cases is given in Table 4. The effect of all parameters cannot be neglected unless the data are analyzed through regression analysis.

Table 4. The main effect of different parameters on displacement for all cases.

Cavern Name		RMR	UCS	Ei	P-Ratio	Tensile Strength	Phi	C
MHP-MH	Effect	−26.77	−3.03	−20.91	6.84	−1.39	−2.15	1.29
	Ranking	1st	3rd	2nd	7th	5th	4th	6th
THP-MH	Effect	−255.6	−187.4	−218.0	−6.6	−33.7	−35.9	1.1
	Ranking	1st	3rd	2nd	6th	5th	4th	7th
THP-DC	Effect	−9.23	−0.64	−7.71	2.62	0.029	−0.12	0.06
	Ranking	1st	3rd	2nd	7th	6th	4th	5th
NJHP-MH	Effect	−51.72	−17.78	−41.98	4.74	0.94	−0.34	−0.39
	Ranking	1st	3rd	2nd	7th	6th	5th	4th
NJHP-DC	Effect	−260.9	−198.0	−186.2	−14.5	0.6	−0.8	0.6
	Ranking	1st	2nd	3rd	4th	6th	5th	7th
Tehri-PH	Effect	−99.9	−42.15	−76.98	−2.33	1.32	−1.24	1.06
	Ranking	1st	3rd	2nd	4th	7th	5th	6th
SSP-PH	Effect	−8.49	−0.041	−7.66	4.71	−0.63	−0.84	0.68
	Ranking	1st	5th	2nd	7th	4th	3rd	6th
SLBHP-MH	Effect	−60.2	−24.82	−46.7	2.59	0.09	0.02	0.10
	Ranking	1st	3rd	2nd	7th	4th	5th	6th
KLIP-P8-PH	Effect	−58.93	−0.03	−4.83	1.40	−0.001	−0.08	0.08
	Ranking	1st	4th	2nd	3rd	5th	6th	7th
PYKARA-MH	Effect	−85.86	−28.94	−69.32	11.80	−1.27	−1.20	−1.21
	Ranking	1st	3rd	2nd	7th	4th	6th	5th

5.3. Development of Regression Equation

A multiple linear regression analysis was conducted to develop a regression equation for predicting crown deformation. The twelve variables considered in the regression analysis were as follows: width and height of the cavern, rock cover, horizontal stress coefficient (KH), vertical stress coefficient (Kh_1), rock mass rating (RMR), uniaxial compressive strength (UCS), Young's modulus of intact rock (Ei), Poisson's ratio (υ), tensile strength (σt), angle of internal friction (φ), and cohesion (C).

A log-linear form of the regression is considered here for analysis. With the twelve variables, in logarithmic form, the linear equation in general can be written in the following form:

$$\begin{aligned} \text{Log (Deformation)} &= \text{a1} + \text{a2} \times \text{Log (Width)} + \text{a3} \times \text{Log (Height)} + \text{a4} \times \text{Log (Rock Cover)} + \text{a5} \times \text{Log (KH)} - \text{a6} \times \text{Log (Kh)} \\ &- \text{a7} \times \text{Log (RMR)} - \text{a8} \times \text{Log (UCS)} - \text{a9} \times \text{Log (Ei)} + \text{a10} \times \text{Log (Poisson's ratio)} - \text{a11} \times \text{Log (Sigma t)} \\ &- \text{a12} \times \text{Log (Phi)} + \text{a13 Log (C)}. \end{aligned} \quad (1)$$

This equation can be simplified as follows:

$$\text{Log (Deformation)} = \log 10^{a1} + \text{Log Width}^{a2} + \text{Log Height}^{a3} + \text{Log Rock Cover}^{a4} + \text{Log } K_H{}^{a5} - \text{Log } K_h{}^{a6} - \text{Log RMR}^{a7} - \text{Log UCS}^{a8} - \text{Log Ei}^{a9} + \text{Log Poisson's ratio}^{a10} - \text{Log Sigmat}^{a11} - \text{Log Phi}^{a12} + \text{Log C}^{a13}, \quad (2)$$

where a1 to a13 are the constants and are obtained in linear regression analysis. The constants are substituted in Equation (2):

$$\text{Log (Deformation)} = 5.448 + 0.091\,\text{Log(Width)} + 0.535\,\text{Log(Height)} + 1.0084\,\text{Log(Rock Cover)} + 0.180\,\text{Log}(K_H) - 0.4060\,\text{Log(Kh)} - 3.1223\,\text{Log(RMR)} - 0.1888\,\text{Log(UCS)} - 0.9818\,\text{Log(Ei)} + 0.3619\,\text{Log(Poisson's ratio)} - 0.00282\,\text{Log(Sigma t)} - 0.0704\,\text{Log(Phi)} + 0.0145\,\text{Log(C)}. \quad (3)$$

Taking the antilog on both sides in Equation (3), the resulting equation for deformation will have the following form:

$$\text{Deformation(mm)} = 10^{5.448}\,(\text{Width}^{0.091}\ \text{Height}^{0.535}\ \text{Rock Cover}^{1.0084}\ \text{KH}^{0.18}\ \text{Poisson's ratio}^{0.3619}\ \text{C}^{0.0145})/(\text{Kh\_1}^{0.406}\ \text{RMR}^{3.1223}\ \text{UCS0.1888 Sigma t}^{0.00282}\ \text{Ei}^{0.9818}\text{Phi}^{0.0704});\ R^2 = 92.92\% \quad (4)$$

The coefficient of determination (R^2) for the regression equation is 92.92%, which indicates that the equation is a good predictor of crown deformation.

The regression equation shows that the most significant factors affecting crown deformation are RMR, UCS, Ei, and Poisson's ratio. The width and height of the cavern, rock cover, and the stress ratios also have a significant effect on crown deformation. The results of the parametric study and the regression analysis can be used to improve the design and construction of underground caverns. By considering the factors that affect crown deformation, engineers can design caverns that are more stable and less likely to experience deformations.

5.4. Validation of the Equation

Validating the equation is necessary for the validation of any empirical relationship developed. One way of validating the equation is to plot the deformation values observed through monitoring the crown deformations and predicted deformation obtained from the empirical equation. The exercise was conducted, and we found that the correlation coefficient (R^2) is 0.84, indicating the level of confidence in the prediction of the crown deformations. Figure 11 depicts the correlation of deformation.

A correlation coefficient of 0.84 indicates a strong positive correlation between the observed and predicted deformations. This means that the predicted deformations are very close to the observed deformations. The high correlation coefficient suggests that the developed equation predicts crown deformation well.

The results of the validation exercise show that the developed equation can be used to predict crown deformation with a high degree of accuracy. This can be used to improve the design and construction of underground caverns. By considering the factors that affect crown deformation, engineers can design more stable caverns that are less likely to experience deformations.

Here are some additional details about the validation exercise:

- The exercise was conducted for ten caverns;
- The observed deformations were measured using monitoring instruments;
- The predicted deformations were obtained using the developed equation;
- The correlation coefficient (R^2) of deformation was calculated for each cavern;
- The average correlation coefficient for all ten caverns was 0.84.

The high correlation coefficient of 0.84 suggests that the developed equation is a good predictor of crown deformation for many caverns. This can be used to improve the design and construction of underground caverns.

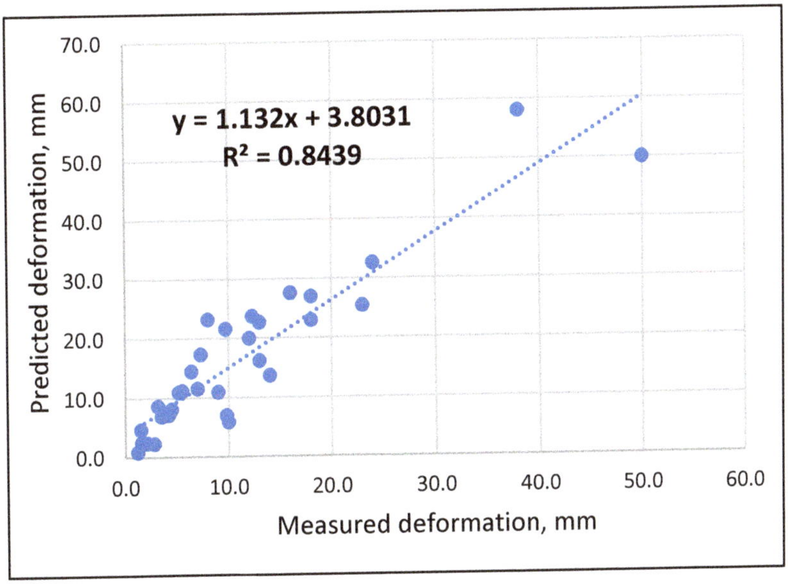

Figure 11. Correlation of deformation.

6. Discussion and Conclusions

In this study, a parametric study using DOE and a multiple linear regression analysis were conducted to investigate the factors affecting crown deformation in underground caverns. The results of the study showed that the following factors have a significant effect on crown deformation:

- Rock mass rating (RMR);
- Uniaxial compressive strength (UCS);
- Young's modulus of intact rock (Ei);
- Poisson's ratio;
- Angle of internal friction (Φ);
- Width of the cavern;
- Height of the cavern;
- Rock cover;
- KH (horizontal stress coefficient);
- Kh_1 (vertical stress coefficient).

A regression equation was developed to predict crown deformation, given below:

$$\text{Deformation(mm)} = 10^{5.448}\,(\text{Width}^{0.091}\ \text{Height}^{0.535}\ \text{Rock Cover}^{1.0084}\ \text{KH}^{0.18}\ \text{Poisson's ratio}^{0.3619}\ \text{C}^{0.0145})/(\text{Kh\_1}^{0.406}\ \text{RMR}^{3.1223}\ \text{UCS0.1888 Sigma t}^{0.00282}\ \text{Ei}^{0.9818}\text{Phi}^{0.0704}).$$

The equation was validated using monitoring data from ten caverns. The results of the validation exercise showed that the developed equation can be used to predict crown deformation with a high degree of accuracy. The results of this study can be used to improve the design and construction of underground caverns. By considering the factors that affect crown deformation, engineers can design caverns that are more stable and less likely to experience deformations.

Here are some additional conclusions that can be drawn from the study:

- The most significant factors affecting crown deformation are RMR, UCS, Ei, and Poisson's ratio;

- The width and height of the cavern, rock cover, and stress ratios also significantly affect crown deformation;
- The developed equation can be used to predict crown deformation with high accuracy for a wide range of caverns;
- The fractional factorial of designs in design of experiments (DOE) is very useful in building the models and no complication calculations are needed;
- The rock mass parameters' combinations were taken in models using fractional factorial analysis to cover all types of rock mass conditions;
- The results of this study can be used to improve the design and construction of underground caverns;
- The prediction values agreed well with the measured values, and the largest relative error was 15–20%;
- Since the behavior of the rock mass conditions are site-specific and all parameters are available by the time of construction, these are to be included in deformation prediction;
- Based on the study, the main influential parameters in crown deformations are rock mass rating, Young's modulus, uniaxial compressive strength of the intact rock samples, and the shear strength parameters;
- The predicted deformations can be used for optimal range selection for the extensometers and other deformation measurement instruments;
- The predicted deformations can be used for Setting the warning/alert signals;
- Finally, the prediction capability of the regression model may further improve with the addition of a greater number of case studies.

Author Contributions: Conceptualization, K.S.; methodology, K.S.; software, K.S. and R.K.S.; validation, K.S.; formal analysis, K.S.; investigation, K.S. and R.K.S.; resources, K.S., R.K.S. and S.R.N.; data curation, K.S.; writing—original draft preparation, K.S.; writing—review and editing, R.K.S. and S.R.N.; visualization, K.S.; supervision, R.K.S. and S.R.N. All authors have read and agreed to the published version of the manuscript.

Funding: This research received no external funding.

Institutional Review Board Statement: Not applicable.

Informed Consent Statement: Not applicable.

Data Availability Statement: The data availability presented in this study are available on request from the corresponding author. The data are not publicly available due to the reason that they pertain to mega engineering projects.

Acknowledgments: The authors thank the project authorities and the director of the National Institute of Rock Mechanics (NIRM) for allowing them to conduct their studies. The support provided by the Department of Mining Engineering, Indian Institute of Technology (Indian School of Mines), Dhanbad, India, and BH Vijaysekar, BNV Siva Prasad, Rabi Bhusan, and Praveen Das Jennifer, Scientists, NIRM, are duly acknowledged. This forms part of the first author's doctoral research at IIT (ISM).

Conflicts of Interest: The authors declare no conflict of interest.

References

1. Zhu, W.S.; Sui, B.; Li, X.J.; Li, S.C.; Wang, W.T. A methodology for studying the high wall displacement of large scale underground cavern complexes and it's applications. *Tunn. Undergr. Space Technol.* **2008**, *23*, 651–664. [CrossRef]
2. Saure, E.; Marcher, T.; John, M. *Decisive Design Basis and Parameters for Power Plant Caverns*; CRC Press: Geneva, Switzerland, 2013.
3. Usmani, A.; Pal, S.; Nanda, A. Continuum and Discontinuum Analysis of Rock Caverns. *Indian Geotech. J.* **2018**, *48*, 663–676. [CrossRef]
4. Dwivedi, R.D.; Singh, M.; Viladkar, M.N.; Goel, R.K. Parametric Analysis of an Empirical Correlation Predicting Deformation of Squeezing Tunnels. In Proceedings of the National Seminar on Innovative Practices in Rock Mechanics, Bengaluru, India, 6–7 February 2014.
5. Naik, S.R.; Sastry, V.R.; Mishra, A. Effect of Orientation on Stability of Caverns—A Numerical Modelling Study. In Proceedings of the 10th Asian Rock Mechanics Symposium (ARMS10), Singapore, 29 October 2018.

6. Prakash, I. Application of observational method in successful construction of underground structures, Sardar Sarovar (Narmada) project. In Proceedings of the Seventh International Conference on Case Histories in Geotechnical Engineering, Chicago, IL, USA, 1 May 2013.
7. Van Kien, D.; Anh, D.N.; Thai, D.N. Numerical Simulation of the Stability of Rock Mass around Large Underground Cavern. *Civ. Eng. J.* **2022**, *8*, 81–91. [CrossRef]
8. Kontogianni, V.A.; Stiros, S.C. Predictions and observations of convergence in shallow tunnels: Case histories in Greece. *Eng. Geol.* **2002**, *63*, 333–345. [CrossRef]
9. Kuili, S.; Sastry, V.R. Prediction and Assessment of Lateral Displacement of Rock Mass Along the Length of the Horseshoe Cavern-A Numerical Modelling Approach Stability analysis of cavern under static and dynamic loads: A numerical modelling approach View project. *Int. J. Geol. Geotech. Eng.* **2017**, *3*, 17–28.
10. Abdollahipour, A.; Rahmannejad, R. Investigating the effects of lateral stress to vertical stress ratios and caverns shape on the cavern stability and sidewall displacements. *Arab. J. Geosci.* **2013**, *6*, 4811–4819. [CrossRef]
11. Feng, X.T.; Zhao, H.; Li, S. Modeling non-linear displacement time series of geo-materials using evolutionary support vector machines. *Int. J. Rock Mech. Min. Sci.* **2004**, *41*, 1087–1107. [CrossRef]
12. Małkowski, P.; Juszyński, D. Roof fall hazard assessment with the use of artificial neural network. *Int. J. Rock Mech. Min. Sci.* **2021**, *143*, 104701. [CrossRef]
13. Ma, H.P.; Daud, N.N.N.; Yusof, Z.M.; Yaacon, W.Z.; He, H.J. Stability Analysis of Surrounding Rock of an Underground Cavern Group and Excavation Scheme Optimization: Based on an Optimized DDARF Method. *Appl. Sci.* **2023**, *13*, 2152. [CrossRef]
14. Basin, R.; Pabst, T.; Li, C. Rock Characterisation, Modelling and Engineering Design Methods. In Proceedings of the 3rd ISRM SINOROCK 2013 Symposium, Shanghai, China, 18–20 June 2013.
15. Gupta, A.K.; Goyal, D.P.; Mishra, R.K.; Khazanchi, R.N. Instability Problems in Desilting Caverns of Tala hydroelectric project, Bhutan. In Proceedings of the SRM India Symp. on Advancing Rock Mechanics Frontiers to Meet the Challenges of 21st Century, New Delhi, India, 24–27 September 2002.
16. Mishra, A.K.; Chaudhary, R.K.; Punetha, P.; Ahmed, I. Geological and Geotechnical Challenges Faced during Excavation of Powerhouse and Head Race Tunnel for Mangdechhu Project, Bhutan, India. *Hydro Nepal J. Water Energy Environ.* **2018**, *23*, 30–41. [CrossRef]
17. Bhasin, R.K.; Barton, N.; Grimstad, E.; Chryssanthakis, P.; Shende, F.P. Comparison of predicted and measured performance of a large cavern in the Himalayas. *Int. J. Rock Mech. Min. Sci. Geomech.* **1996**, *33*, 607–613. [CrossRef]
18. Dasgupta, B.; Singh, R.; Sharma, V.M. Numerical modelling of desilting chambers for Nathpa Jhakri Hydroelectric Project. In Proceedings of the 9th ISRM Congress 99, Paris, France, 25 August 1999.
19. Dasgupta, B.; Sharma, M.K.; Verman, M.; Sharma, V.M. Design of underground caverns for Tehri Hydro Power project, India by numerical modelling. In Proceedings of the 9th ISRM Congress 99, Paris, France, 25 August 1999.
20. Dasgupta, B.; Dham, R.; Lorig, L.J. Three dimensional discontinuum analysis of the underground powerhouse for SardarSarovar Project, India. In Proceedings of the 8th ISRM Congress, Tokyo, Japan, 25 September 1995.
21. Varman, M.; Jethwa, J.L.; Singh, B. Monitoring of a large underground power house cavity. In Proceedings of the 9th ISRM Congress, Aachen, Germany, 16 September 1991.
22. Venkateswarlu, V.; Srinivasulu, T.; Nagaraj, C.; Sudhakar, K. Instrumentation and Strata Monitoring at Srisailam Left Bank Hydro electric scheme. Project Report, NIRM, Kolar Gold Fields, 30 November 1996.
23. Naitani, A.K.; Singh, L.G.; Rawat, D.S.; Jain, P. Engineering Geological Investigations of the Extension of Surge Pool and Pump House Caverns, Kaleswaram Lift Irrigation Project-Package8, Telangana, India, Project Report, NIRM, India. 31 January 2017.
24. Purushothaman, C. Planning design and construction of transformer cavern for an underground hydroelectric project. In Proceedings of the ISRM International Symposium 2010 and 6th Asian Rock Mechanics Symposium—Advances in Rock Engineering, Newdelhi, India, 23 October 2010.
25. Venkateswarlu, V.; Nagaraj, C. Instrumentation for monitoring the underground caverns at Pykara Ultimate Stage Hydro Electric Project, Project Report, NIRM, Kolar Gold Fields, India, 31 March 2000.
26. Jayanthu, S.; Rao, G.M.N.; Udaykumar, S. Physico mechanical properties of Rocks—A data bank. *Kolar Gold Fields.* 2006.
27. Song, S.; Feng, X.; Liao, C.; Cai, D.; Liu, Z.; Yang, Y. Measures for controlling large deformations of underground caverns under high in-situ stress condition—A case study of Jinping I hydropower station. *J. Rock Mech. Geotech. Eng.* **2016**, *8*, 605–618. [CrossRef]
28. Mandal, P.K.; Singh, A.K.; Kumar, R.; Singh, R. Instrumentation and monitoring strategies for study of ground movement during underground extraction of coal. *Min. Eng. J.* **2006**, *8*, 15–27.
29. Kellaway, M.; Taylor, D.; Keyter, G.J. The use of Geotechnical Instrumentation to Monitor Ground Displacements during Excavation of Ingula power caverns, For Model Calibration and Design Purposes. In Proceedings of the Southern African Institute of Mining and Metallurgy South African Tunnelling 2012-Lessons Learnt on Major Projects, Ladysmith, South Africa, 3 December 2012.
30. Montgomery, D.C. *Engineering Statistics*; RGHN; John Wiley & Sons: Hoboken, NJ, USA, 2011.
31. Lye, L.M. Some Applications of Statistical Design of Experiment Methodology in Civil Engineering. In Proceedings of the Annual Conference of the Canadian Society for Civil Engineering, Moncton, NB, Canada, 4–7 June 2003.

32. Sinha, R.K.; Jawed, M.; Sengupta, S. An approach for support design in depillaring panels of coal mines. *Arab. J. Geosci.* **2019**, *12*, 666. [CrossRef]
33. Ruan, J.-K.; Zhu, W.-W. Sensitivity Analysis of Influencing Factors of Building Slope Stability Based on Orthogonal Design and Finite Element Calculation. *E3S Web Conf.* **2018**, *53*, 03076.

sustainability

MDPI

Article

Molecular Dynamics Simulation of Forsterite and Magnesite Mechanical Properties: Does Mineral Carbonation Reduce Comminution Energy?

Akash Talapatra * and Bahareh Nojabaei

Department of Mining and Minerals Engineering, Virginia Tech, Blacksburg, VA 24061, USA; baharehn@vt.edu
* Correspondence: takash@vt.edu

Abstract: This work compares the mechanical properties of two geomaterials: forsterite and magnesite. Various physical conditions are considered to investigate the evolution of stress–strain relationships for these two polycrystals. A molecular-scale study is performed on three-dimensional models of forsterite and magnesite. Three different temperatures (300 K, 500 K, and 700 K) and strain rates (0.001, 0.01, and 0.05 ps^{-1}) are considered to initiate deformation in the polycrystals under tensile and compressive forces. The polycrystalline structures face deformation at lower peaks at high temperatures. The Young's modulus values of forsterite and magnesite are found to be approximately 154.7451 GPa and 92.84 GPa under tensile forces and these values are found to be around 120.457 GPa (forsterite) and 77.04 GPa (magnesite) for compressive forces. Increasing temperature reduces the maximum strength of the polycrystalline structures, but forsterite shows higher ductility compared to magnesite. Strain rate sensitivity and the effect of grain size are also studied. The yield strengths of the forsterite and magnesite drop by 7.89% and 9.09% when the grain size is reduced by 20% and 15%, respectively. This study also focuses on the changes in elastic properties for different pressures and temperatures. In addition, from the radial distribution function (RDF) results, it was observed that the peak intensity of pairwise interaction of Si–O is higher than that of Mg–O. Finally, it is found that the formation of magnesite, which is the product of mineral carbonation of forsterite, is favorable in terms of mechanical properties for the comminution process.

Keywords: mineral carbonation; comminution energy; stress–strain relationship; elastic properties; radial distribution function

Citation: Talapatra, A.; Nojabaei, B. Molecular Dynamics Simulation of Forsterite and Magnesite Mechanical Properties: Does Mineral Carbonation Reduce Comminution Energy? *Sustainability* **2023**, *15*, 12156. https://doi.org/10.3390/su151612156

Academic Editors: Jian Zhou, Mahdi Hasanipanah and Danial Jahed Armaghani

Received: 10 June 2023
Revised: 3 August 2023
Accepted: 6 August 2023
Published: 9 August 2023

1. Introduction

Carbon mineralization is an emerging approach to storing CO_2 in the form of carbonate minerals, particularly in calcium, magnesium, and silicate-rich rocks/geomaterials such as olivine, wollastonite, and serpentine. This process occurs naturally during the weathering of these rocks/geomaterials [1,2]. The major pathways and kinetics of storing enriched CO_2 in carbonate minerals have been described frequently in the literature to discuss the potential and required cost of this process. Researchers have not only focused on the potential of carbon mineralization in minimizing greenhouse gas emissions and assuring storage is non-toxic and permanent, but they have also been searching for emulating and accelerating the spontaneity as well as the balance of this process within the Earth's deep interior [3–5].

Mineral carbonation is a process of reacting carbon dioxide (CO_2) with alkaline and alkaline earth-bearing (magnesium and silicate-rich) minerals to form stable carbonate minerals [6]. A variety of silicate mineral groups containing Ca^{2+}, Mg^{2+}, and Fe^{2+} ions present in nature for targeting mineral carbonation are olivine, serpentine, pyroxene, mica group, and clay minerals. However, past work has claimed that olivine-group minerals, particularly forsterite (Mg_2SiO_4), are the best potential feedstock for the carbon mineralization process and form stable carbonate minerals ($MgCO_3$). Forsterite is abundant in the Earth's crust and comprises the upper mantle. It is a common mineral in ultramafic rocks and formed as a result of the cooling and solidification of magma and it is much more stable at high temperatures and pressure. Furthermore, forsterite has a high surface area-to-volume ratio, making it highly reactive with CO_2 [7,8]. It has a CO_2 sequestration potential of around 2014.7–1896.3 kg/m^3 (Table 1). Magnesite, on the other hand, is a carbonate-rich mineral found in tectonically active regions and comprises the Earth's lower mantle and is less stable at higher temperatures and pressure. Figure 1 shows a schematic view of the 3D atomic and crystalline structures of forsterite and magnesite. Both forsterite and magnesite are polycrystals, with forsterite crystalizing in orthorhombic systems and magnesite crystalizing in cubic systems. A consideration of crystal orientation and lattice parameters is provided in the next section.

Table 1. CO_2 sequestration potential of major rock-forming minerals [9].

Mineral Name	Formula	Potential CO_2 Fixed, kg/m^3 Mineral
Olivine group (forsterite)	Mg_2SiO_4-Fe_2SiO_4	2014.7–1896.3
Pyroxene group (enstatite)	$(Mg, Fe)_2\ Si_2O_6$	1404
Serpentine	$Mg_3Si_2O_5(OH)_4$	1232
Wollastonite	$CaSiO_3$	1097.1
Amphibole group (hornblende)	$Ca_2Na_{0\text{-}1}(Mg,Fe(II)_{3\text{-}5}(Al,Fe(III)_{2\text{-}0}[Si_{6\text{-}8}Al_{2\text{-}0}O_{22}](O,OH)_2$	1000.4
Mica group (biotite)	$K_2(Mg,Fe(II))_{6\text{-}4}(Fe(III),Al)_{0\text{-}2}[Si_{6\text{-}5}Al_{2\text{-}3}O_{20}](OH)_{4\text{-}2}$	671
Plagioclase (anorthite)	$Ca[Al_2Si_2O_8]$	436.4
Clay Minerals (smectite)	$(1/2Ca,Na)_{0.7}(Al,Mg,Fe)_4(Si,Al)_8O_{20}(OH)_4.nH_2O$	161.2

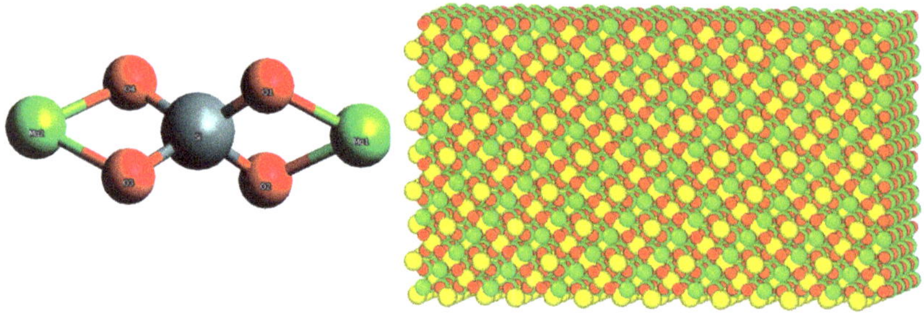

(**a**)

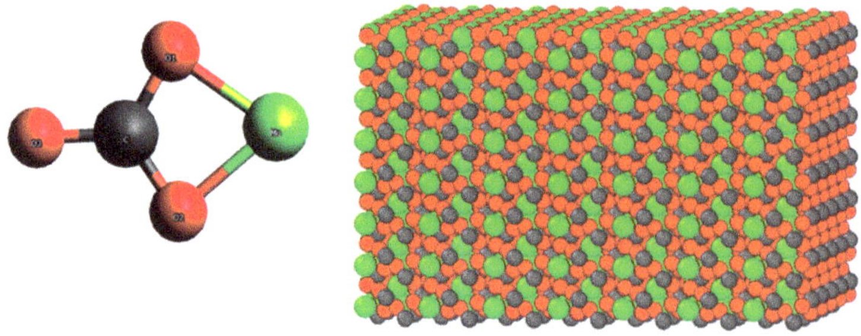

(**b**)

Figure 1. 3D atomic and crystal structure of forsterite (**a**) and magnesite (**b**).

The mineral carbonation reaction between forsterite and CO_2 is expressed as follows:

$$Mg_2SiO_4 + CO_2 \rightarrow MgCO_3 + SiO_2 \quad (1)$$

As mentioned before, mineral carbonation occurs in naturally occurring silicate rocks (alkaline/alkaline earth minerals) when a high concentration of CO_2 is brought into contact with them. This carbonation process can happen through both in situ and ex situ carbonation, which differ based on the time scale and contact of CO_2 [10]. In situ carbonation involves the injection of CO_2 into the geological formation containing silicate minerals (forsterite) to form solid carbonate minerals (magnesite) [11,12], while ex situ carbonation involves reacting silicate minerals with CO_2 in a controlled environment above ground. If the silicate minerals are brought from industrial wastes in the form of fly ash, cement kiln dust, steel slag, etc., they can sequestrate around 200–300 Mt of CO_2 annually [3,13,14]. This process sequestrates a large amount of CO_2 and produces reusable, valuable, and stable carbonate minerals for different industrial purposes [10]. However, both in situ and ex situ mineral carbonation processes are part of carbon-negative solutions and have a great potential to reduce the effects of CO_2 on the environment (Figure 2).

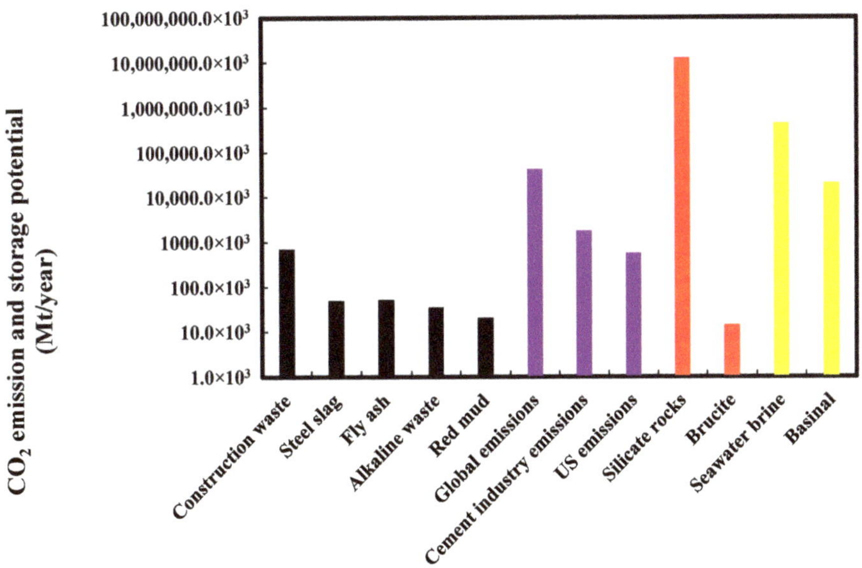

Figure 2. Global capacity to sequester CO_2 via mineralization [15].

The comminution process is a physical pre-treatment method in mineral processing that involves reducing the size of ore particles through crushing, grinding, and breaking the minerals into smaller/finer particles. It is a critical component for separating valuable minerals from waste rocks and requires significant energy input [16]. During the comminution process, factors such as the formation of minerals, mineral structure, presence of defects, mechanical strength, and grain size play a great role. The comminution process requires significant energy input to break rocks during crushing and grinding. CO_2 mineralization can reduce the comminution energy, as silicate-rich minerals require more energy for size reduction than carbonate [17]. Mineral carbonation can also change the surface chemistry of minerals, affecting the efficiency of both froth flotation and separation processes [18]. However, this study does not focus on explaining the reactive phenomena in the flotation process from a chemical point of view.

Comminution, froth flotation, and other mineral processes are affected by changes in the hardness and mechanical strength properties of minerals, which depend on crystalline structure, 3D arrangement of atoms, and interatomic behavior. Mineral carbonation alters these crystalline and interatomic structures. Few studies have been performed on the strength properties of forsterite and magnesite individually. Holyoke et al. [19] performed experiments on two different types of magnesite aggregation (coarse and fine grains) to determine the triaxial deformation over a wide range of temperatures (400–1000 °C) and strain rates (2×10^{-7} s^{-1} to 2×10^{-4} s^{-1}). In both aggregation types, the strengths of the magnesites at higher temperatures were reduced. Both fine- and coarse-grained magnesites showed little chance of recovery at the plastic stage. The study also mentioned that magnesite is more stable at low temperatures [19]. In another study, Liu et al. [20] carried out an atomic simulation (using a transferable empirical interatomic potential) to investigate the structural and elastic properties of magnesite over a wide range of pressure (based on the Earth's mantle's conditions). The simulation work found that magnesite shows anisotropic behavior at lower mantle depths but shows significant change with increasing depth. The percentage anisotropy in the shear and compressibility were calculated for a pressure range from 0 to 150 GPa and it was observed that, at higher pressure ($\geq$120 Gpa), both shear and compressibility values were close to 1 (0 means isotropic and 1 means anisotropic). This result means that magnesite is less stable at higher temperatures [20]. Yao et al. [21] worked on the impacts of pressure and temperature on magnesite using local-density approximation. All of the pressure and temperature values were in lower-mantle conditions. They found that the elastic and thermodynamic properties of magnesite were influenced by zero-point motion (the motion or vibration of the atoms at absolute zero temperature) and increasing temperatures. They also noticed a change (around 4.0%) in the shear and bulk modulus from static to ambient conditions (300 K and 0 Gpa). The authors considered 0 Gpa to indicate that the experiment was performed for Earth surface conditions, without any external pressure [21]. Gonzalez et al. [22] investigated the structural, dielectric, and vibrational spectroscopic properties of the amorphous form of forsterite. The work was conducted using two approaches: classical molecular dynamics (MD) for structural evolution using the empirical charge-based rigid ionic model and density functional theory (DFT) for measuring electronic structure using quantum mechanics. The radial distribution function (RDF) calculations for Mg–O and Si–O show broader profiles in increasing temperatures, which indicates the loss of crystallinity of forsterite. This work also analyzed the degrees of freedom for disordering in the crystal structure at higher temperatures (400 K to 2000 K). The increasing temperatures accelerated the dynamics of melting of forsterite and at some point (above 1800 K), the system became unstable for the disorganization of the interatomic positions [22]. In 2019, Gouriet et al. investigated the mechanical deformation of the orthorhombic structure of forsterite under applied strains. They determined the energy–strain curves and linear elastic regimes to ascertain the ultimate instability of the crystal structure. The maximum stress values from stress–strain curves were found to be approximately 15.9, 12.1, and 29.3 Gpa along the different [001], [010], and [100] directions, respectively. They also observed a similar change in features in ideal shear stress evolution in each direction. The shear stress increased initially, then decreased for a percentage of strain (around 10%) due to the bond divergence between Mg–O. But with the reduction of bond distance between Mg–O, the shear stress as well as stiffness increased again with increasing strain [23]. In addition, Choudhary et al. (2020) studied the mechanical stability and degradation of forsterite and noted that increasing content of forsterite increased the mechanical strength of a composite compared to calcium phosphates and calcium silicates [24]. Several studies also worked on generalizing the relationship between the compressive strength of materials and the grain size of the materials and stated that there is a growing trend of mechanical strength with increasing grain size [25,26]. These studies investigated the petrophysical and mechanical properties of carbonate minerals for different grain sizes and claimed that decreasing the average grain size reduces the strength properties of materials.

Until now, no studies have investigated and compared the mechanical properties of forsterite and magnesite in various conditions (temperature, pressure, and applied forces). This present work focuses on the evolution of the stress–strain relationship according to changes in physical properties. The effects of temperatures, loading rates, and grain sizes on the maximum stress values of two different polycrystals were studied. In addition, polycrystals highly sensitive to the applied forces are also measured. Changes in pressure and thermal effects induced on the elastic properties are observed for the two polycrystals. Later, the radial distribution function is also evaluated for measuring the pairwise interaction between Si–O and C–O bonds. This study compares the strength properties and microstructure of the two polycrystals under the same conditions, to assess the effect of mineral carbonation on comminution energy for mineral processing.

2. Molecular Modeling and Simulation Method

Molecular dynamics (MD) simulation provides a basic representation and interpretation of molecular interaction modeling of any material with detailed information and underlying governing mechanisms. It is a powerful tool for studying particle movement, mechanical behavior, and other properties according to the materials' physical characteristics and chemical reaction kinetics [27]. Using MD, deformation, elasticity, diffusion, yield, and other physical behaviors of materials at the atomic level can be measured. MD can predict the dynamics of materials' behavior under different temperatures and loading conditions by performing computational experiments. The benefits of using MD are flexibility in working with different sizes of materials and visualization of the molecular interaction at the atomic scale which might not be possible in experiments. The fundamental theory of molecular dynamics simulation is to observe the dynamic trajectory of an atomic system and analyze the atomic interactions among the respective atoms by solving Newton's equation of motion. Prior works on different materials using molecular dynamics simulation have offered reliable outputs for illustrating their mechanical properties.

Here in this paper, a molecular dynamics simulation study on forsterite and magnesite polycrystals is performed using the Large-scale Atomic/Molecular Massively Parallel Simulator (LAMMPS) (9 October 2020) software package [28].

Better performance of molecular dynamics simulation depends on the implementation of a suitable and successive potential for polycrystals. Since the model's reliability and validation directly depend on this selected potential, using the right forcefield parameters to obtain accurate results and configuration is important. The accessibility of the interatomic potential to the size of the system is a governing factor for the numerical integration of the above model. Here, in this study, an empirical potential model named the 'Buckingham' potential is used for simulating the atoms of both polycrystals [29]. This potential is freely available with the LAMMPS library package. It considers long-range electrostatic terms with classical Coulombic energy, a short-range repulsive term, and a three-body harmonic term. This potential is a bond-order and short-range interaction-based potential which is attributed to the partial charge of the atoms. Several prior studies have used this potential to reproduce the structural properties of both forsterite and magnesite polycrystals [30,31]. The Buckingham potential can be expressed as follows:

$$E = A_{ij}.e^{-\frac{r_{ij}}{\rho_{ij}}} - \frac{C_{ij}}{r_{ij}^6} \tag{2}$$

$$E = \frac{Cq_iq_j}{\epsilon r_{ij}} \tag{3}$$

where E is the potential energy, A is energy units, e is the elementary charge, ρ_{ij} is the hardness parameter of repulsive energy (indicates the characteristic length scale that determines the range over which the potential decays exponentially), C is the coefficient of dispersive energy, r_{ij} is the interatomic distance between two atoms i and j (i and j being Mg, Si, C, and O), and q_i and q_j are the atoms' partial charges. Though the Buckingham potential seems simple, this potential considers the pair's ion charges, thermal expansion, heat capacity, and long-range coulombic interaction to illustrate the short-range interaction between the paired atoms. For this reason, this potential works slowly compared to other interatomic potentials used mainly for silicate and carbonate molecules. The performance of this potential depends on some forcefield parameters of the forsterite and magnesite. The crystal structures for forsterite and magnesite are considered orthorhombic and trigonal, respectively, and the crystal orientations are made with specified lattice constants. The volume of the simulation models for forsterite and magnesite are 125 nm^3 and 117.65 nm^3. The initial configuration and forcefield parameters are shown in Tables 2 and 3, respectively.

Table 2. Selected materials and atomic model properties.

Material Type	Dimension (Å)	Potential	Number of Atoms	Atomic Bond Type	Lattice Constant (Å)	Crystal System
Forsterite (Mg_2SiO_4)	50 × 50 × 50	Buckingham	4480	Ionic–Covalent	a = 4.787 b = 10.272 c = 6.023	Orthorhombic
Magnesite ($MgCO_3$)	49 × 49 × 49	Buckingham	4250	Ionic–Covalent	a = 4.64 b = 4.64 c = 14.93	Trigonal

Table 3. Forcefield parameters used in this study.

Buckingham Potential			
Ion Pairs	A(eV)	B(Å)	C(eVÅ^6)
Mg-O	1428.5	0.2945	0
Si-O	473.2	0.4157	0
O-O	22,764.30	0.149	60.08
C-O (Morse)	4.71	3.8	1.18
Harmonic 3-Body Term			
	k(eV rad^{-2})	Θ_0 (degrees)	
O-Si-O	2.09	109.47	
O-C-O	1.69	120	
Charges			
Mg	+2.00		
Si	+4.00		
C	+1.135		
O (for forsterite)	+0.84819		
O (for magnesite)	−1.632		

In this study, Moltemplate was used to generate all-atom molecular models for forsterite and magnesite [32]. Moltemplate is a cross-platform text-based molecular builder (for both all-atom and coarse-grained molecular models) made for LAMMPS. The simulation models of these two polycrystals were made with random orientations of the polycrystals as well as by using the Voronoi method including random seeds. The conjugate gradient (CG) algorithm was used to optimize the initial structure and position of the atoms. This minimization algorithm adds the force gradient to the previous iteration's information to compute the new direction perpendicular to the previous search iteration. The Verlet algorithm was used with a timestep of 1 *fs* to calculate the dynamic trajectories of the particles. The long-range Coulombic interaction was calculated using the Ewald summation method. Two different systems approached a specified temperature (300 K) using an NPT ensemble under zero pressure. A Nose–Hoover thermostat and barostat were used successfully to control the temperature and pressure. Periodic boundary conditions were applied in each direction of the simulation box. The tensile force was applied to the polycrystals by using incremental homogeneous strain, which indicates the displacement of the atomic layers along the tensile direction. The components of stress tensors were brought to zero at each deformation state so that the atoms received enough time for relaxation. Before the measurement of the stress–strain properties, in all cases, the polycrystal structures were optimized for equilibrium purposes.

3. Results and Discussion

Stress–strain behavior of the forsterite and magnesite

Multiple multiscale approaches have been developed to determine the strength properties of materials, but the determination of the stress–strain relationship for different thermodynamic conditions is one of the most reliable options among relevant approaches. These material properties should be determined by allowing deformation in the designed microscale models under uniaxial tensile or compression tests as these properties are considerably changed by different strain rates and temperatures. The definition and idea of continuum Cauchy stress in atomistic simulation are slightly different but equivalent to the definition of Virial stress [33]. For stress calculation, the per-atom pressure tensor is computed for each atom in the group.

Here, both uniaxial tensile and compression tests are performed to study and compare the stress–strain behavior of these two polycrystals for different strain rates and temperatures. Three different temperatures (300 K, 500 K, and 700 K) for a constant strain rate of 0.01 ps^{-1} and three different strain rates (0.01 ps^{-1}, 0.03 ps^{-1}, and 0.05 ps^{-1}) for a constant temperature at 300 K are considered during deformation under uniaxial tensile and compression tests. For both polycrystals, the evolution of the stress values as a function of strain values is studied along the [100] plane in the x direction.

3.1. Stress–Strain Behavior of the Geomaterials for Different Temperatures

Figure 3a shows a typical example of a Young's modulus-based stress–strain curve used to study the mechanical properties of the two geomaterials. The curve depicts that the ultimate strength/stress point is the maximum stress for any material before failure occurs under tensile or compression load. The elongation of the elasticity of materials depends on the values of this maximum stress point. Before reaching this point, the loading stress increases with increasing strain, but after the peak point, the material faces deformation. In this region, the stress values usually decrease with increasing strain values. Figure 4a,b represents such types of stress–strain relationships of the studied polycrystals (forsterite and magnesite, respectively) under a constant tensile load for temperatures 300 K, 500 K, and 700 K, at a strain rate of 0.01 ps^{-1}. As expected, a parabolic evolution of the stress values corresponding to the initial linear portion of the stress–strain curves is observed for both polycrystals. For forsterite (Figure 4a), the ultimate strength point is near 26 GPa at 300 K, and the Young's modulus value is around 154.7451 GPa, and this value was validated using the result of the Young's modulus (153.2 GPa) calculated by Gouriet et al. (shown in Figure 3b) [23]. Again, at 500 K and 700 K, the corresponding ultimate strength values are found to be near 21.79 GPa and 17.92 GPa, respectively. The increasing temperatures from 300 K to 500 K and 700 K initiated considerable drops in the maximum stress of forsterite by 16.15% and 31.07%. More importantly, increasing temperature also increased the maximum strain values needed to obtain the maximum stress point. For instance, at 300 K, the maximum stress point was achieved for a strain value of 0.168, whereas at 500 K and 700 K, maximum stress points were obtained at the strain values of 0.181 and 0.199, respectively.

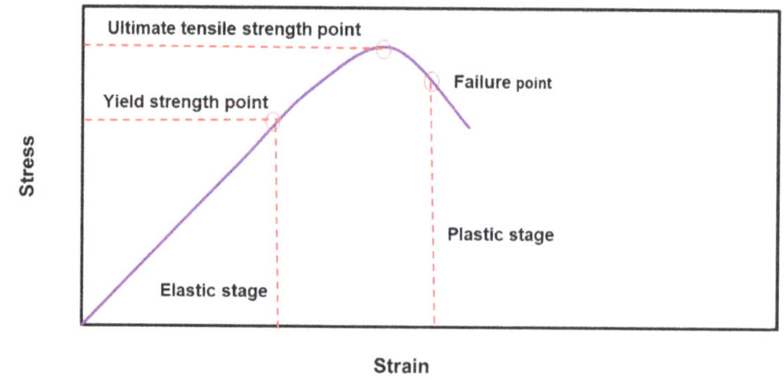

(a)

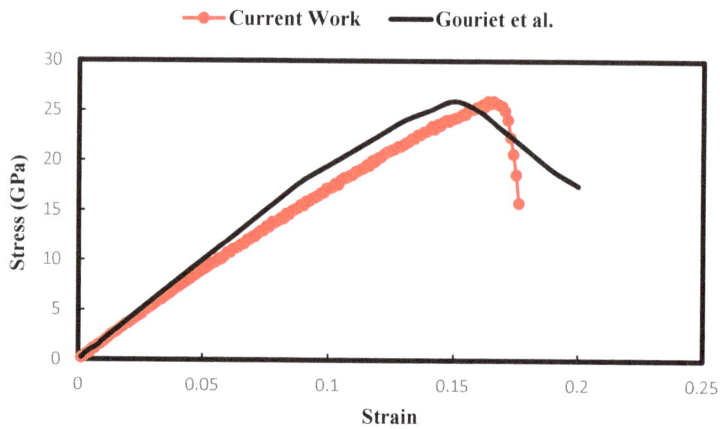

(b)

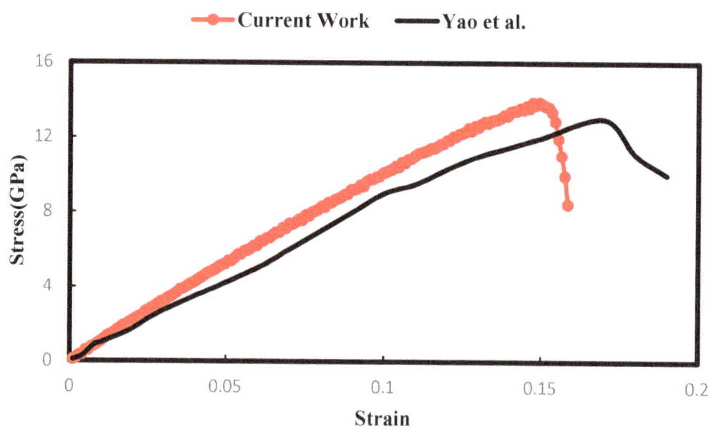

(c)

Figure 3. (**a**) A generalized physical model of the stress–strain relationship for any material. (**b**) The validation of the work for forsterite at 300 K [23]. (**c**) The validation of the work for magnesite at 300 K [21].

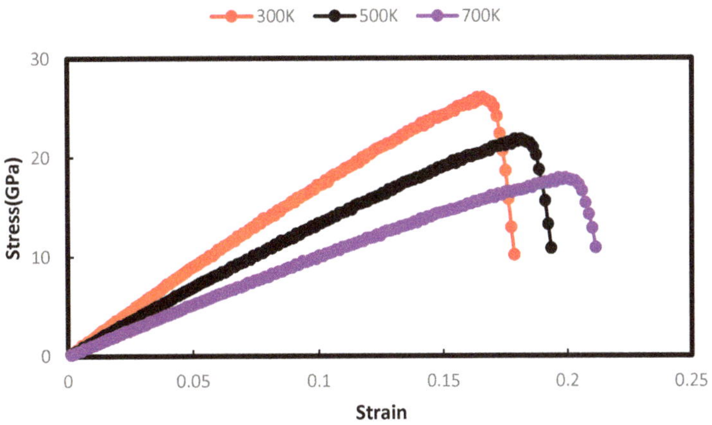

(**a**)

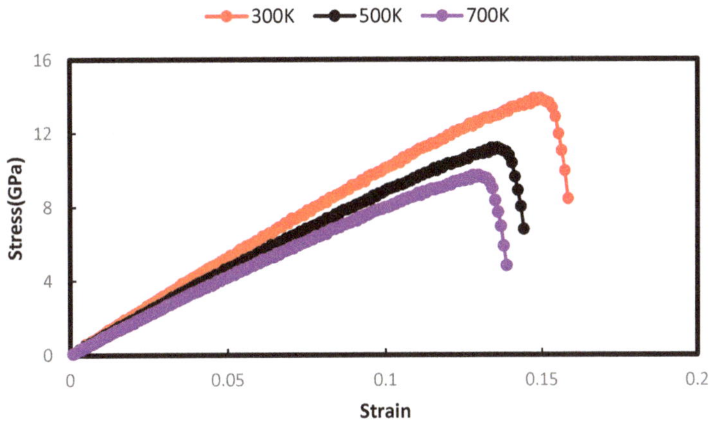

(**b**)

Figure 4. (**a**) The stress–strain curve of forsterite for different temperatures under tensile test (at a strain rate of 0.01 ps^{-1}). (**b**) The stress–strain curve of magnesite for different temperatures under tensile test (at a strain rate of 0.01 ps^{-1}).

From Figure 3c, the ultimate strength point for magnesite is found near 13.90 GPa (at 300 K) with a corresponding value of Young's modulus of 92.84 GPa. This value was validated using the calculated result (ultimate strength of 13.70 GPa at 300 K) of Yao et al. (shown in Figure 3c) [21]. Then, at 500 K and 700 K, the ultimate strength values are decreased by 19.42% and 34.46%, respectively. In addition to that, as opposed to those of forsterite, the maximum strain values for magnesite are decreased by 10.06% (for 500 K) and 13.42% (for 700 K) as the temperature is changed from 300 K. This behavior indicates that at elevated temperatures, forsterite tends to show more ductility than magnesite. Table 4 provides information on the Young's modulus values of forsterite and magnesite for different temperatures.

Table 4. Ultimate strength and Young's modulus of forsterite and magnesite under tensile force.

	Ultimate Strength (Gpa)			Young's Modulus (Gpa)		
	300 K	500 K	700 K	300 K	500 K	700 K
Forsterite	26.00 Gpa	21.80 Gpa	17.92 Gpa	154.74	120.79	89.98
Magnesite	13.90 Gpa	11.20 Gpa	9.12 Gpa	92.84	83.58	70.67

Figure 5a,b shows the stress–strain curves of forsterite and magnesite under the uniaxial compression test, respectively, under the same conditions as for the uniaxial tensile test. The stress–strain relations under this compressive force represent a trend similar to that under tensile force at different temperatures. Since compressive force usually provides more strain energy to materials, the ultimate stress loading stage occurs earlier than in the case of tensile force. This increasing strain energy results in the material entering the plastic stage. For forsterite (Figure 5a), the ultimate stress point is observed at 17.28 GPa, at 300 K, and at a strain value of 0.143; the calculated Young's modulus is about 120.457 GPa. Increasing the temperature to 500 K and 700 K reduces the maximum stress point to approximately 15.12 GPa and 13.45 GPa, respectively. These results indicate that the effect of temperature on stress–strain properties under compressive forces is similar to those for tensile forces. The only difference is that, due to the contraction under compressive forces applied to the crystal material, the elastic stage ends sooner compared to the elongation for tensile forces. For example, under compressive stress at 300 K, the elastic stage comes to an end at a maximum stress point of 17.28 GPa (Figure 5a), whereas, for tensile stress, it reaches 26 GPa (Figure 4b). In magnesite (Figure 5b), the increasing stress loads (up to a strain value of 0.107) cannot provide any significant change in the elastic stage under compressive force for each temperature studied (300 K, 500 K, and 700 K). However, the maximum values of the stress load of the magnesite are lower for increasing temperature. This behavior of the crystals under tensile and compressive forces has indicated that increasing temperature (300 K) results in deformation for both crystals as high temperatures contribute to initializing local stress for propagation/stretching out of the crystals. However, the results also show that increasing the temperature from 300 K to 500 K and 700 K under compressive force reduces the Young's modulus of forsterite (from 120.457 GPa to 79.98 GPa, respectively) more than magnesite (from 77.04 GPa to 60.88 GPa, respectively).

The effect of increasing temperature is the same for both polycrystals under two different loading velocities. In both cases, increasing temperature leads to smaller peaks of stress, but with higher ductility (for forsterite) and lower ductility (for magnesite) up to the failure point of the polycrystals. Forsterite shows more ductility at elevated temperatures than magnesite. The results obtained from the uniaxial tensile and compressive force tests for the two polycrystals at a fixed temperature (600 K) are shown in Figure 6 Table 5. In addition, Figure 7a,b shows the evolution of strain energy as a function of strain for a constant temperature (at 600 K) to understand the change in energy during the application of force. As seen in Figure 7b, the strain energy of the microstructure (after the relaxation period) changes with strain values. For both polycrystals, each curve has an inflection point indicating the maximum strain energy the minerals can tolerate. This peak represents the polycrystals' maximum stress point (ultimate strength). After crossing the peak, the system starts dissipating energy which results in the deformation of the polycrystals. This dissipation of energy moves the strain energy curve in the downward direction for additional strain values. Similar findings have been noted in the strain energy curve. Forsterite shows higher ductility and elastic properties for both types of applied forces (strain energy ranges from 12.02–15.05 meV/A^3) compared to magnesite.

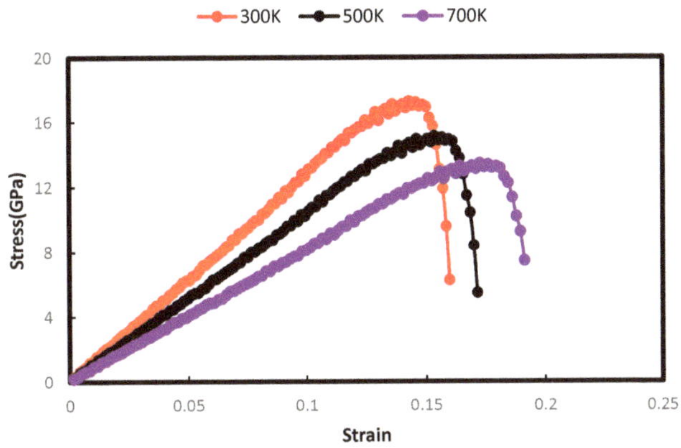

(**a**)

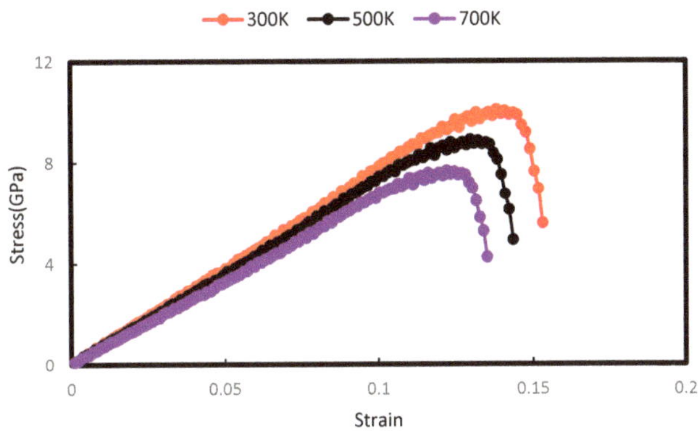

(**b**)

Figure 5. (**a**) The stress–strain curve of forsterite for different temperatures under compression test (at a strain rate of 0.01 ps^{-1}). (**b**) The stress–strain curve of magnesite for different temperatures under compression test (at a strain rate of 0.01 ps^{-1}).

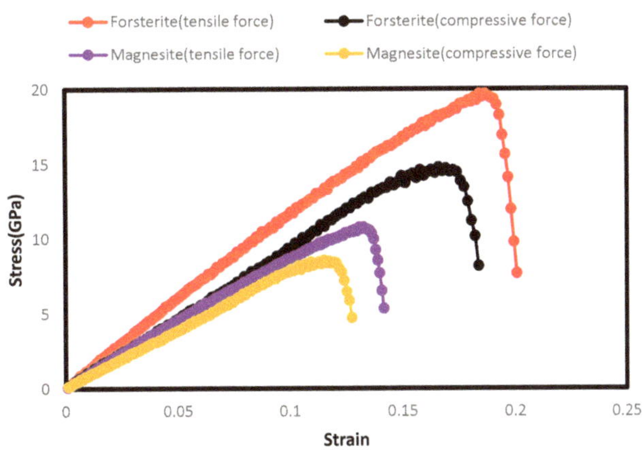

Figure 6. The stress–strain curves for a constant temperature (600 K) under different test conditions.

Table 5. Comparison of the obtained results at 600 K for two different applied forces.

Material Type	Temperature	Maximum Stress Reduction from Tensile Force to Compressive Force (%)	Strain Variation in Achieving Maximum Stress Loading Capacity for Different Applied Forces (%)
Forsterite	600 K	25.01	11.18
Magnesite	600 K	20.12	13.26

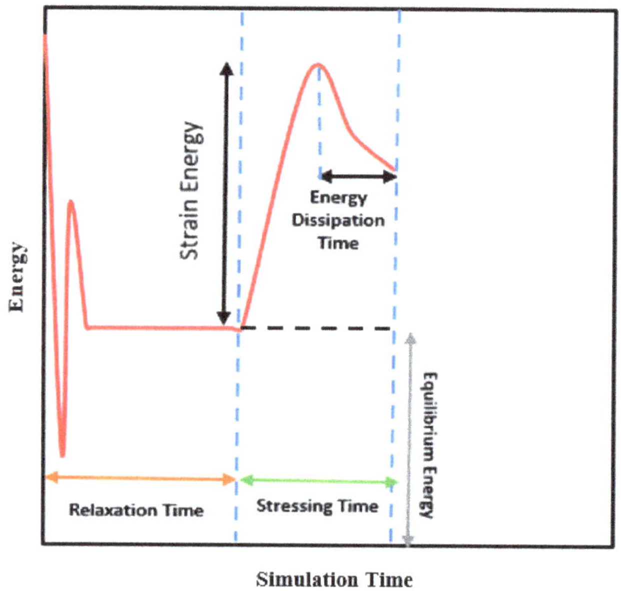

(a)

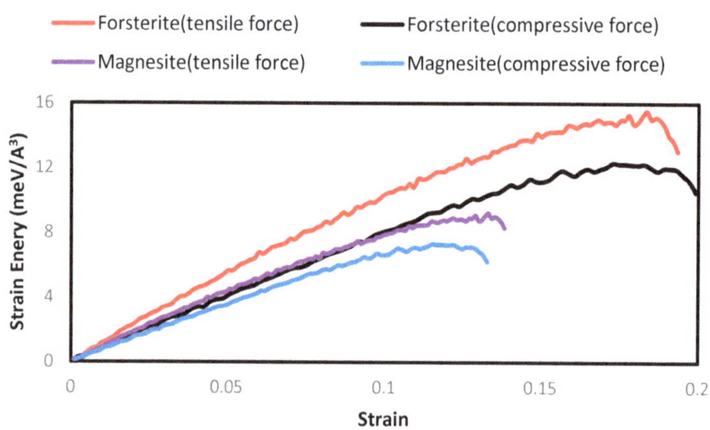

(b)

Figure 7. Evolution of strain energy as a function of strain: (**a**) a generalized curve, (**b**) for both forsterite and magnesite at 600 K.

3.2. Stress–Strain Behavior of the Geomaterials for Different Strain Rates

In this section, different strain rates are considered to study the impact of loading velocity on the mechanical deformation of both polycrystals. For this purpose, three different loading rates (0.001, 0.01, and 0.05 ps^{-1}) are applied under both tensile and compressive forces. The effect of different loading rates has been studied previously on different crystals using molecular dynamics simulation. It should be noted that the polycrystals were in a relaxation state before applying both tensile and compressive forces. Therefore, the stress values as a function of strain values start from the undeformed state of the polycrystal (at a strain value of zero) either for increasing or decreasing loading velocities. Figure 8a,b shows the stress–strain curves for forsterite and magnesite at a constant temperature of 300 K (under tensile force), respectively, for the three abovementioned loading rates. According to the results demonstrated in the figures, increasing strain rate increases the maximum stress points for both polycrystals. Particularly, the elastic stage is increased for increasing strain rates with negligible impact on the plastic stage. The yield points are only changed for different loading velocities. In addition, the strength of non-viscoelastic materials like crystals greatly depends on the loading rate; a higher strain rate indicates a higher strength of the material. For forsterite (Figure 8a), increasing the strain rate from 0.01 ps^{-1} to 0.05 ps^{-1} increases the maximum stress point by 10.71%, and decreasing the strain rate to 0.001 ps^{-1} decreases it by 11.43%. For magnesite (Figure 8b), the maximum stress point moves to 15.21 Gpa for the loading rate of 0.05 ps^{-1}, whereas the peak stress point is decreased by 18.30% to 0.001 ps^{-1} when the strain rate decreases. Under application of compressive force, the changes in the stress–strain relationship follow a similar trend for both materials like the tensile force. Figure 9a shows the results for forsterite, whereas the changes for magnesite are shown in Figure 9b.

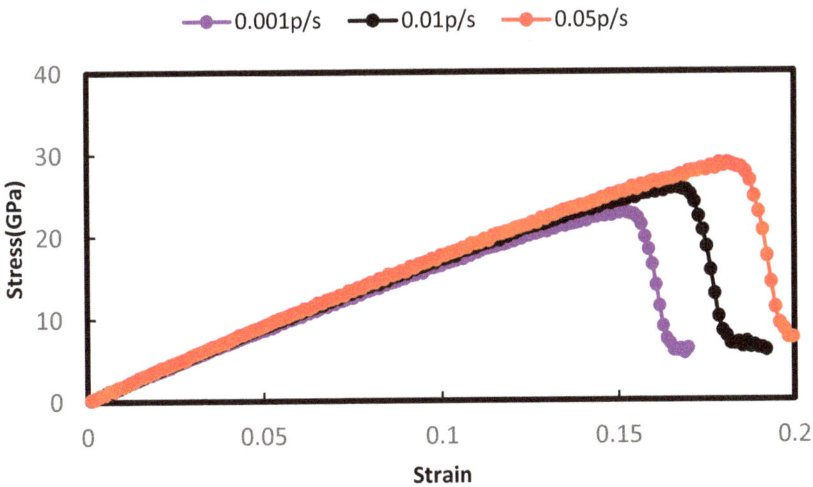

(a)

Figure 8. *Cont.*

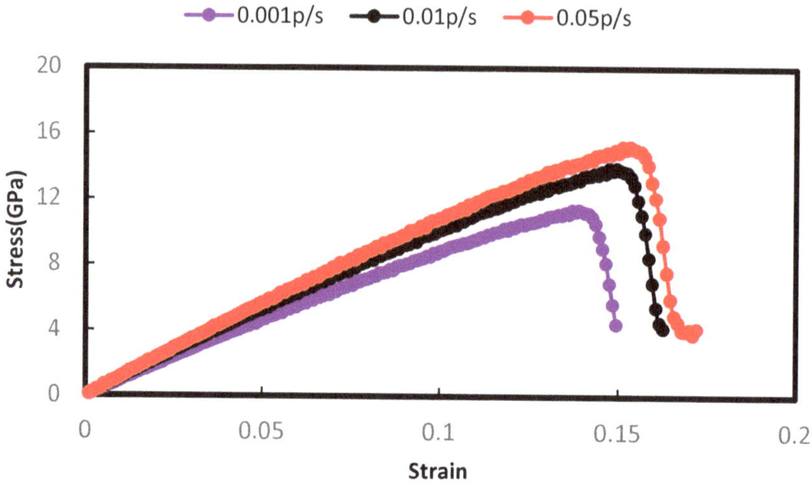

(b)

Figure 8. (**a**) The stress–strain curve of forsterite for different strain rates under tensile test (at 300 K). (**b**) The stress–strain curve of magnesite for different strain rates under tensile test (at 300 K).

Here, for higher strain rates, the values of strain energy increase for both forsterite and magnesite to initiate plastic deformation. This is contradictory to the case of higher temperatures. The higher strain rate provides a smoother trend of the stress–strain relationship (more linear) which leads to shortening the breaking time during deformation. This higher strain rate makes the elastic region more linear, which is also described by Hooke's law [34]. This required energy is higher for tensile forces than for compressive forces. Hence, forsterite shows more strength at a higher strain rate compared to magnesite. The results of the stress–strain relationship between these two polycrystals obtained under both types of forces are compared for a constant strain rate (at 300 K and a strain rate of 0.03 K) and are shown in Figure 10.

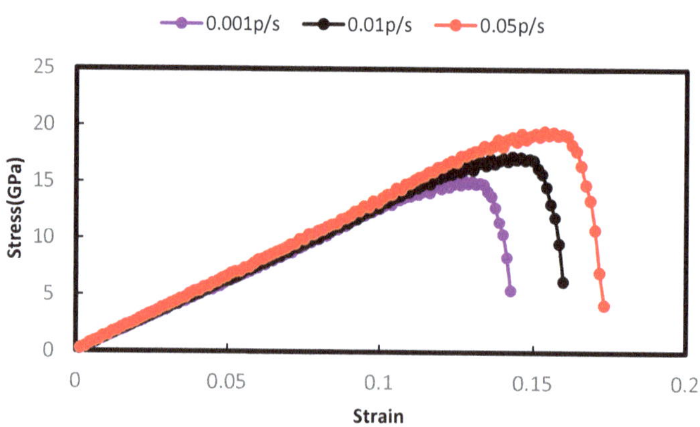

(a)

Figure 9. *Cont.*

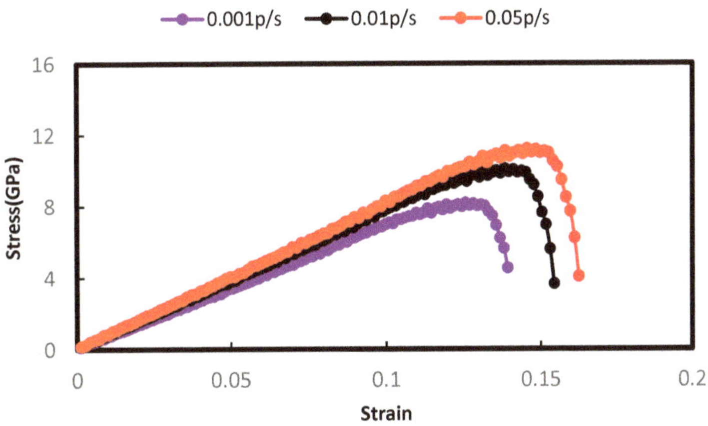

(**b**)

Figure 9. (**a**) The stress–strain curve of forsterite for different strain rates under compression test (at 300 K). (**b**) The stress–strain curve of magnesite for different strain rates under compression test (at 300 K).

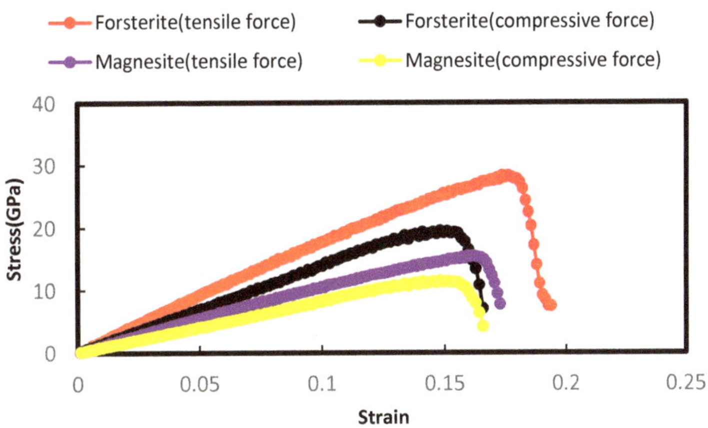

Figure 10. The stress–strain curves for a constant strain rate under different tests.

3.3. *Strain Rate Sensitivity*

The strain rate sensitivity (SRS) measures the change in yield pressure for different strain rates. Sometimes, applied force alters the characteristics of the material which can be understood by SRS calculation. This is a relationship wherein a material's tensile strength depends on different loading rates. Based on the conditions of a steady-state process, the SRS of a material in loaded tension can be demonstrated (as a differential form) by Hart [35] as follows:

$$m = \frac{\partial \log(\sigma)}{\partial \log(\acute{\varepsilon})} \tag{4}$$

where m is the strain rate sensitivity, σ is the tensile strength, and $\acute{\varepsilon}$ is the strain rate.

Hence, Equation (4) states that the SRS values can be obtained from the slopes of the logarithmic plot of tensile stresses and corresponding strain rates. Hart [35] exhibited that stress rates are proportional to strain rates up to the ultimate/maximum strength point, and during plastic deformation, the material is less sensitive to the extra load. This sensitivity impacts the material's stability during deformation due to strain localization. The SRS shows the ability of a material to resist necking and maintain its stability during deformation. However, higher temperatures reduce the SRS of materials, and the higher the values of SRS are, the higher the strength of materials is. For higher SRS, materials are able to distribute force more evenly through their structure, rather than localizing it in one area.

Here, simulations were performed for three different temperatures (300 K, 500 K, and 700 K) and three different strain rates (0.001, 0.01, and 0.05 ps^{-1}) to determine and compare the obtained SRS results for forsterite and magnesite. Here, steady strain rate sensitivity is considered, where the strain rate sensitivity coefficient m is computed using the above-mentioned equation for different strain rates. The negative sign is neglected here in the obtained values of m. The stability of the material is proportional to the strain rate sensitivity. A higher value of m indicates a large change in the flow stress for a change in strain rate. Figure 11a,b shows the variation in m values with strain rates for forsterite and magnesite, respectively, for different temperatures and strain rates. For forsterite and magnesite (Figure 11a,b), at a constant strain rate (0.01 ps^{-1}), increasing the temperature from 300 K $\rightarrow$ 500 K $\rightarrow$ 700 decreases the m value from 0.70764 to 0.66992, followed by 0.62671 and from 0.5716 to 0.52972 and followed by 0.47712. Both forsterite and magnesite become less stable at higher temperatures. On the other hand, for a constant temperature (300 K), when the strain rates increase, the m values increase, for both forsterite and magnesite (known as strain rate hardening). Hence, both materials show high sensitivity to increasing strain and more stability for higher strain rates. In short, higher temperatures decrease the sensitivity of materials to strain rate and decrease the yield strength.

From the above results, it has been found that forsterite shows higher sensitivity than magnesite, which means that magnesite is comparatively less stable. The comparison of the results for both polycrystals is shown in Figure 12. At a constant temperature (500 K), the SRS values of forsterite and magnesite are 0.43128 and 0.32405, 0.66922 and 0.52972, and 1.0473 and 0.8618 for the strain rates of 0.001 ps^{-1}, 0.01 ps^{-1} and 0.05 ps^{-1}, respectively. However, despite having a lower strain rate effect at lower temperatures, the effect of temperature on SRS is significantly stronger for lower strain rates as the sensitivity arises from the inertness of the defective structure evolution of materials for lower strain rates [36]. Since the strength of the materials is proportional to the SRS, both polycrystals are prompted to deform earlier when the SRS values are lower for corresponding temperatures and strain rates. The impact of temperature is found to be higher in tensile strength for magnesite than forsterite which agrees with the stress–strain curves in the previous section. In short, less sensitivity to applied forces makes magnesite weaker than forsterite.

3.4. *Effect of Grain Size*

The impact of grain size on both polycrystals is studied here. Polycrystals are comprised of multiple grains, and these grains are bounded by some interfacial defects in the grain boundaries of those polycrystals. These grains' structure and energy provide microscopic insight into the mechanical deformation of the polycrystals. Particularly, the morphology, size distribution, and nature of these grains and grain boundaries are important features of the polycrystals [37–39]. Also, the grain size and boundaries resist the elongation of the polycrystals and reduce the yield strength and ductility of the materials. However, there is a significant correlation between grain size and the mechanical strength properties of minerals (particularly in sedimentary rocks) which has been proven by previous studies [40]. Studies on grain size microstructure provide insights into materials' toughness, corrosion resistance, thermal conductivity, and magnetic susceptibility.

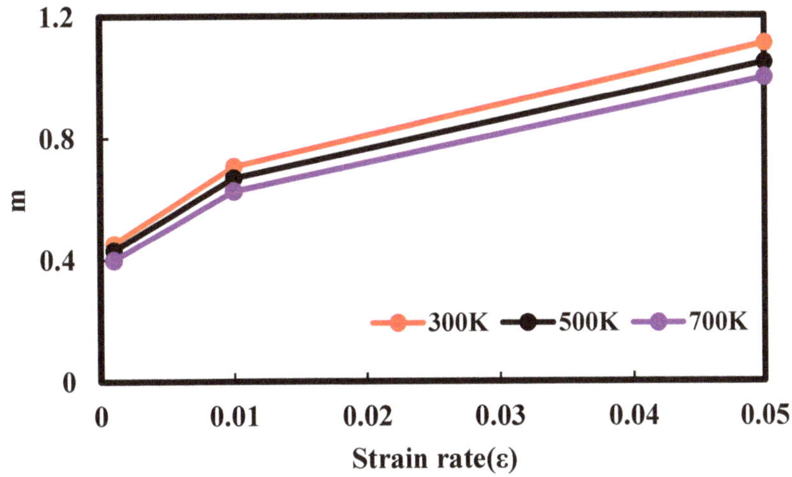

(**a**)

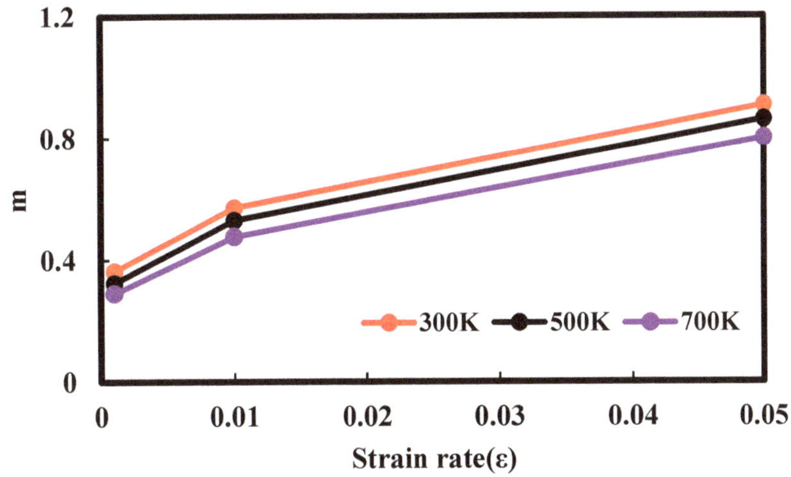

(**b**)

Figure 11. Strain rate sensitivity of (**a**) forsterite and (**b**) magnesite, for different temperatures (300 K, 500 K, and 700 K) and strain rates (0.001 ps^{-1}, 0.01 ps^{-1}, and 0.05 ps^{-1}).

Since there is a relationship (nonproportional) between grain size and density of the materials, in this study, the number of atoms is increased for both minerals to study the impact of grain size. This is because a greater number of atoms are more closely packed and result in higher densities of materials, manifesting as smaller grain sizes [41]. Here, the grain size is calculated by measuring the average distance from the two adjacent peaks in the radial distribution function for both studied polycrystals. Figure 13 shows the yield strengths of both forsterite and magnesite at a constant strain rate (0.01 ps^{-1})

for 5600 atoms and 4850 atoms, respectively. This figure shows that the polycrystals are more deformative for growing numbers of atom. The yield strengths of the forsterite and magnesite dropped by 7.89% and 9.09% compared to the initial systems. These results show an opposite correlation between compressive strength and number of atoms. From the figure, it is observed that the grain size also impacts the Young's modulus values of the two polycrystals (dropped to 15.92 and 9.17 GPa, respectively). In the case of the two studied polycrystals, the only components bearing the applied stress are grains. In addition, the distribution of the grains becomes uniform for higher numbers of atoms, hence the deformation of the polycrystals occurring earlier under pristine conditions with a low number of atoms. Shear velocity, wave velocity, relatively low energy, and partial double-bond character in Si–O and C–O bonds also cause a reduction in strength properties for lower grain sizes; however, these discussions are not part of our current study.

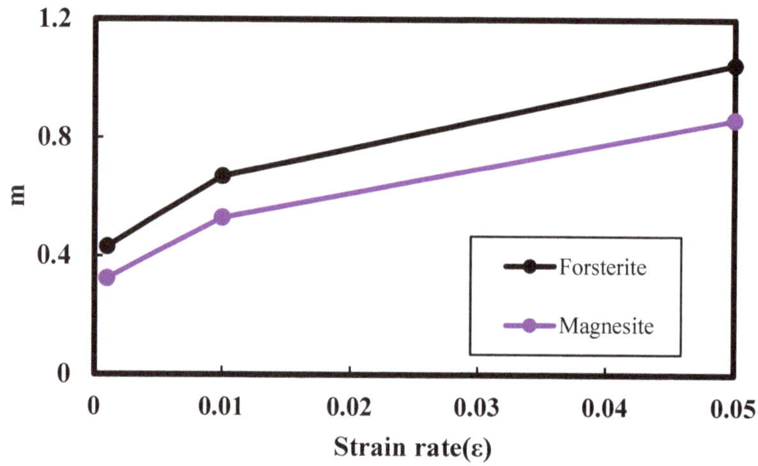

Figure 12. Strain rate sensitivity of both forsterite and magnesite at 500 K.

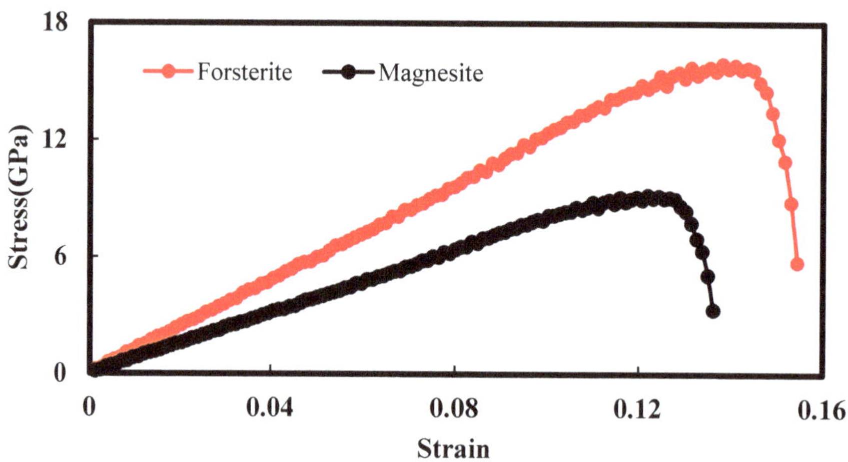

Figure 13. Compressive strength curve of forsterite and magnesite for lower grain sizes.

3.5. Elastic Properties

Elastic properties are crucial parameters of Earth's minerals utilized in geophysics and geochemistry to understand and explain the interior information of the Earth. These properties are indispensable for characterizing the rheology of geotectonics and constructing seismic acoustic waves in solid-state physics. The elastic properties represent the ability of materials to deform for a small change in stress. The crystal structure, composition, and microstructure of any material can be illustrated from these properties. Based on these properties, compositional changes, as well as elongations in materials, can be observed.

The elastic constant is the ratio of the second derivative of energy density concerning the strain and can be expressed simply as in Equation (5) [42].

$$C_{ij} = \frac{1}{V}\frac{\partial^2 E}{\partial \varepsilon_i \partial \varepsilon_j} \tag{5}$$

where C_{ij} is the stiffness coefficient, V is the volume of the unit cell, E is the energy, and ε is the strain.

Here, for calculating the elastic constants, a three-dimensional fourth-order Voigt notion model was used which considers the basis of Hooke's law, finite element analysis, and diffusion MRI [42]. This model is convenient for providing elastic constants of a material as a 6-by-6 matrix based on pressure tensors. Here, only three pressure- and temperature-dependent elastic stiffness constants (C_{11}, C_{33}, and C_{44}) for both forsterite and magnesite polycrystals have been considered to study the mechanical deformation of the minerals. These C_{11}, C_{33}, and C_{44} coefficients are known as the pressure tensors (part of the 6-by-6 matrix) and represent the specific components of the stress and strain relationship. C_{11} represents the elastic stiffness in the direction parallel to x axis, C_{33} in the direction parallel to z-axis, followed by C_{44} in the y–z plane (shear stiffness).

Figures 14 and 15 show the pressure- and temperature-dependent elastic constants for both forsterite and magnesite polycrystals. The obtained results agree well with those of previous works. For both forsterite and magnesite, the elastic constant values of C_{11}, C_{33}, and C_{44} at 1 GPa and 300 K are validated with results from previous works in Table 6 [21,43]. As can be seen in Figures 14 and 15, the pressure and temperature changes noticeably affect the elastic constants of the two polycrystals. For forsterite (Figure 14a), the changes in pressure from 1 GPa to 100 GPa increased the values of C_{11}, C_{33}, and C_{44} to 344.73, 620.59, and 859.89 GPa, respectively. On the other hand, these values were found to be approximately 220.61, 347.25, and 462.89 GPa, respectively, for magnesite (Figure 15a). So, it can be observed that the elasticity is linearly dependent on the applied pressure. At higher pressures, both forsterite and magnesite undergo a phase transition to a denser crystal structure. Also, Young's modulus values for these minerals are higher. Further, the higher pressure changes the bonding between the atoms. These phenomena cause the elastic properties of the minerals to be higher. In short, higher pressure and elasticity result in higher tensile strength for the two minerals.

However, increasing temperature increases the atomic vibration through the crystal structure and continues until it reaches a value where the atomic bonds become weak. This causes disorder and irregularity in the crystal lattice. As a result, the elastic properties of the minerals decrease and the minerals become less resistant to deformation. Hence, the temperature increasing from 300 K to 700 K decreases the values of C_{11}, C_{33}, and C_{44} for forsterite by 35.94%, 45.33%, and 49.10% (Figure 14b) and for magnesite by 41.98%, 30.76%, and 47.09% (Figure 15b), respectively. Consequently, the temperature is reversely correlated to the elasticity of the materials. The reason is at the higher temperature, the higher the thermal vibration the materials have to go through which increases thermal expansion inside the materials but reduces the tensile strength and corresponding lattice constants. The study of these elastic properties for different pressures and temperatures is essential for investigating the phase transition of minerals (including mineralogy and geology). This phase transition is very helpful in interpreting the geophysical data and design of the studied minerals for their desired properties. This study can give insights into developing a model of the studied minerals for predicting behaviors under extreme conditions.

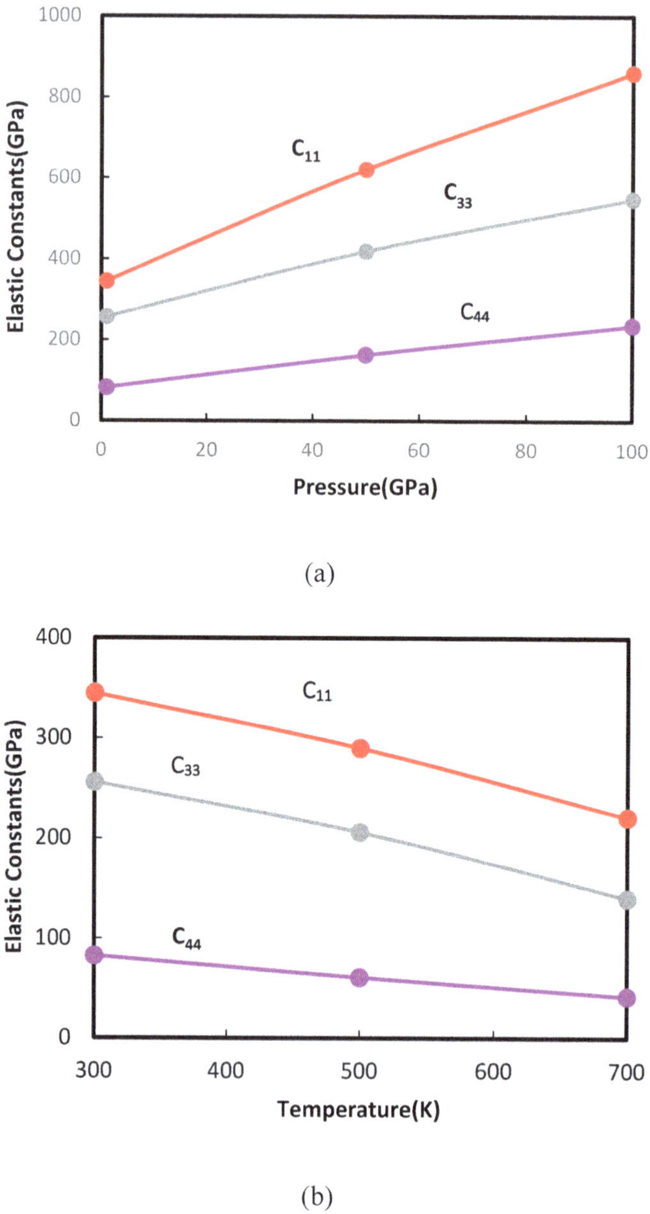

Figure 14. Elastic constants of forsterite: (**a**) effect of pressure and (**b**) effect of temperature.

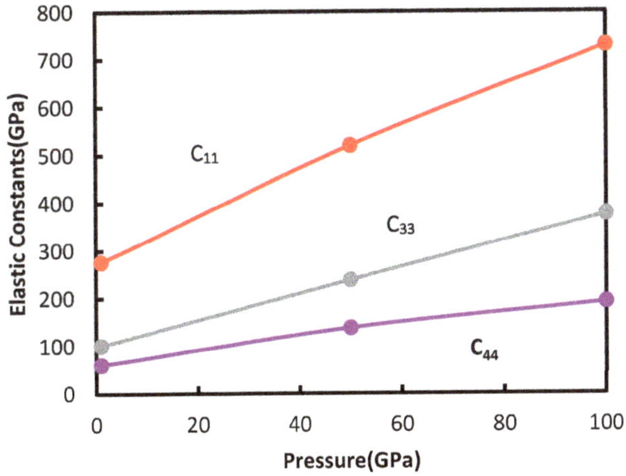

(a)

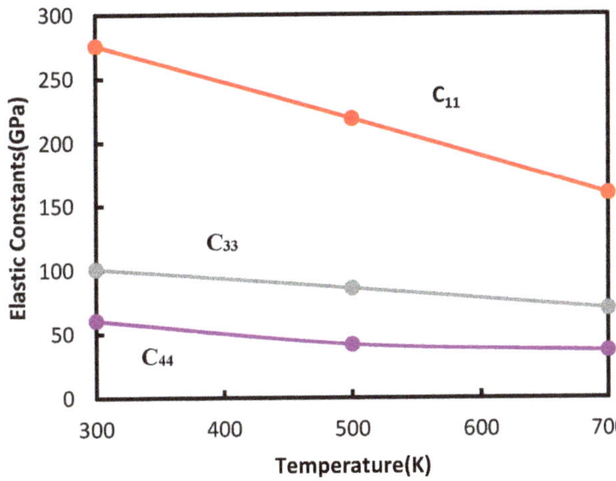

(b)

Figure 15. Elastic constants of magnesite: (**a**) effect of pressure and (**b**) effect of temperature.

Table 6. Elastic stiffness coefficients at 300 K.

	Forsterite			Magnesite		
	C_{11}(GPa)	C_{33}(GPa)	C_{44}(GPa)	C_{11}(GPa)	C_{33}(GPa)	C_{44}(GPa)
MD work (current study)	345.0	256.1	82.52	275.85	101.79	60.81
Experimental work [21,41]	342.0	253.1	79.49	272.38	102.55	58.81

3.6. Radial Distribution Function

The radial distribution function (RDF) of any bonded simulated system is calculated mainly to determine the pairwise interaction and coordination number between groups of nearest-neighbor atoms. The RDF represents the probability of finding a particle at a specific distance (r) from a reference particle in the mineral. Particularly, this function determines the distribution of neighboring particles around a center particle. The RDF for any bonded system can be expressed as follows:

$$g(r) = \frac{\rho(r)}{\rho} \tag{6}$$

where $\rho(r)$ is the average local number density of particles at a distance r, and ρ is the bulk density of the particles.

The histogram form represents the RDF calculation by binning pairwise interactions into a distance of several bins. This distance is specified as the pair cutoff distance r for a particular potential field. The RDF is counted only for the specified cutoff distance, and the coordination of any atoms beyond this distance is out of consideration. The RDF function shows better results (with multiple peaks in the histogram) if the system is uniform and well equilibrated. If the system is neither uniform nor well equilibrated, there is a sharp change in the coordination of the atoms with one single peak due to less interaction.

Here, Figure 16 shows the pair of RDFs for the disordered structures of both forsterite and magnesite polycrystals. Only the pairwise interactions between Si–O and C–O atoms are considered for this RDF calculation since the bonded interactions of these atoms are the most contributing factors for providing strength in the respective polycrystals. Even though several peaks can be seen in the figure, only the sharp peak is considered the strongest interaction between the pairs of atoms. The red curve is for the Si–O atoms whereas the black curve is for C–O atoms. These two RDF curves are obtained from the trajectories of the coordination numbers of their bonded atoms. The highest peak of the first curve is found around 2.54 angstrom which indicates the strongest interaction between the bonded Si–O atoms. On the other hand, the highest peak position is seen at 3.26 angstrom for bonded C–O atoms. In addition, the densities of both bonded Si–O and C–O atoms are higher at those peak points of the curves. These two curves' peak intensity indicates that the oxygen atoms usually attract the Si atoms more strongly than the C atoms. However, from the RDF calculation for these two polycrystals, it is observed that the weaker interaction between the carbon and oxygen atoms compared to the silicon and oxygen atoms makes magnesite less strong than forsterite.

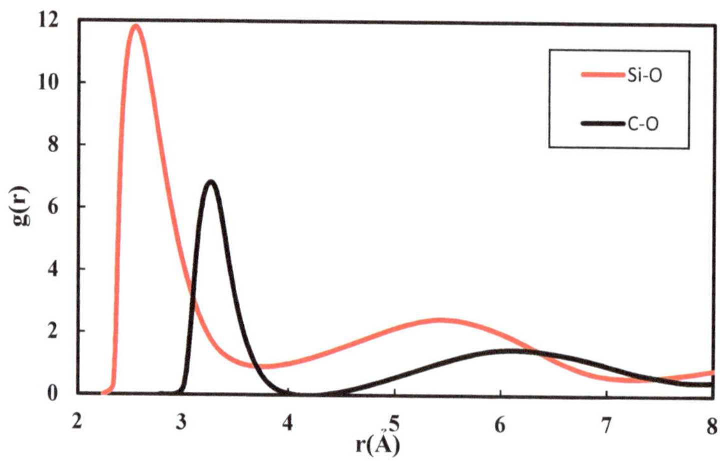

Figure 16. Pairwise Radial Distribution Function of forsterite and magnesite samples.

4. Conclusions

At the microscale, the presence of many mineral phases as well as microstructural defects distinguishes geomaterials, which are known to be heterogeneous and discontinuous. In summary, this study examines the changes in mechanical properties of forsterite and magnesite polycrystals using molecular dynamics simulation under different conditions. A direct comparison can be made between each computation as the MD simulation allows for the splitting of energy data calculations into individual parts such as surface energy, elastic energy, and plastic energy. In this work, the effects of strain rate and temperature on the stress–strain properties of the polycrystal models are studied.

The highlights of this research are listed below:

- The Young's modulus values of forsterite and magnesite are found to be approximately 154.7451 GPa and 92.84 GPa under tensile force, while these values are found to be around 120.457 GPa (forsterite) and 77.04 GPa (magnesite) under compressive force. Increasing temperature reduces the maximum strength of the polycrystals. For higher temperatures, forsterite shows higher ductility than magnesite;
- Higher strain rates require higher strain energy to initiate plastic deformation in the polycrystals. This effect is the opposite for the case of increasing temperature;
- According to the strain rate sensitivity results, magnesite shows less sensitivity to applied force than forsterite. At 300 K and a strain rate of 0.01 ps^{-1}, the SRS values of forsterite and magnesite are found to be approximately 0.70764 and 0.5716, respectively. The results also state that the impact of temperatures on SRS is higher for lower strain rates;
- Decreasing grain size (or increasing numbers of atoms) reduces the mechanical strength properties of the polycrystals. The yield strengths of the forsterite and magnesite dropped by 7.89% and 9.09% compared to the initial systems;
- Increasing pressure induces phase transition and increases the elastic properties of the polycrystals. On the other hand, increasing the temperature increases the atomic vibration through the crystal structure and this causes disorder and irregularity in the crystal lattice;
- In addition, from the RDF results, it is observed that the peak intensity of pairwise interaction between Si–O is higher than that for Mg–O.

Finally, this study has found that magnesite, which is the product of mineral carbonation of forsterite, is a favorable rock type for comminution. Magnesite shows less ductility at higher temperatures compared to forsterite. Our results imply that mineral carbonation impacts the energy requirements of minerals for comminution and serves as an energy-saving approach for mineral processing in addition to its effect on reducing greenhouse gases in the atmosphere.

Author Contributions: A.T.: methodology, investigation, validation, writing—original draft. B.N.: conceptualization, supervision, writing—review and editing. All authors have read and agreed to the published version of the manuscript.

Funding: This research received no external funding.

Institutional Review Board Statement: Not applicable.

Informed Consent Statement: Not applicable.

Data Availability Statement: Data will be made available upon request.

Conflicts of Interest: The authors declare no conflict of interest.

References

1. Wang, F.; Dreisinger, D. Carbon mineralization with concurrent critical metal recovery from olivine. *Proc. Natl. Acad. Sci. USA* **2022**, *119*, e2203937119. [CrossRef] [PubMed]
2. Hills, C.D.; Tripathi, N.; Carey, P.J. Mineralization Technology for Carbon Capture, Utilization, and Storage. *Front. Energy Res.* **2020**, *8*, 142. [CrossRef]

3. Gadikota, G. Carbon mineralization pathways for carbon capture, storage and utilization. *Commun. Chem.* **2021**, *4*, 23. [CrossRef] [PubMed]
4. Veetil, S.P.; Hitch, M. Recent developments and challenges of aqueous mineral carbonation: A review. *Int. J. Environ. Sci. Technol.* **2020**, *17*, 4359–4380. [CrossRef]
5. Power, I.M.; Harrison, A.L.; Dipple, G.M.; Wilson, S.A.; Kelemen, P.B.; Hitch, M.; Southam, G. Carbon Mineralization: From Natural Analogues to Engineered Systems. *Rev. Miner. Geochem.* **2013**, *77*, 305–360. [CrossRef]
6. Renforth, P.; Washbourne, C.-L.; Taylder, J.; Manning, D.A.C. Silicate Production and Availability for Mineral Carbonation. *Environ. Sci. Technol.* **2011**, *45*, 2035–2041. [CrossRef]
7. Matter, J.M.; Kelemen, P.B. Permanent storage of carbon dioxide in geological reservoirs by mineral carbonation. *Nat. Geosci.* **2009**, *2*, 837–841. [CrossRef]
8. Huijgen WJ, J.; Comans, R.N.J. *Carbon Dioxide Sequestration by Mineral Carbonation. Literature Review Update 2003–2004;* IAEA: Vienna, Austria, 2005.
9. Xu, T.; Apps, J.A.; Pruess, K. *Analysis of Mineral Trapping for CO_2 Disposal in Deep Aquifers;* University of California: Oakland, CA, USA, 2001.
10. Mazzotti, M.; Abanades, J.C.; Allam, R.; Lackner, K.S.; Meunier, F.; Rubin, E.; Sanchez, J.C.; Yogo, K.; Zevenhoven, R. Mineral carbonation and industrial uses of carbon dioxide. In *IPCC Special Report on Carbon Dioxide Capture and Storage;* Cambridge University Press: Cambridge, UK, 2005.
11. Snæbjörnsdóttir, S.Ó.; Sigfússon, B.; Marieni, C.; Goldberg, D.; Gislason, S.R.; Oelkers, E.H. Carbon dioxide storage through mineral carbonation. *Nat. Rev. Earth Environ.* **2020**, *1*, 90–102. [CrossRef]
12. Kelemen, P.B.; Matter, J. In situ carbonation of peridotite for CO_2 storage. *Proc. Natl. Acad. Sci. USA* **2008**, *105*, 17295–17300. [CrossRef]
13. Gerdemann, S.J.; O'Connor, W.K.; Dahlin, D.C.; Penner, L.R.; Rush, H. Ex Situ Aqueous Mineral Carbonation. *Environ. Sci. Technol.* **2007**, *41*, 2587–2593. [CrossRef] [PubMed]
14. Romanov, V.; Soong, Y.; Carney, C.; Rush, G.E.; Nielsen, B.; O'Connor, W. Mineralization of Carbon Dioxide: A Literature Review. *ChemBioEng Rev.* **2015**, *2*, 231–256. [CrossRef]
15. Qomi, M.J.A.; Miller, Q.R.S.; Zare, S.; Schaef, H.T.; Kaszuba, J.P.; Rosso, K.M. Molecular-scale mechanisms of CO_2 mineralization in nanoscale interfacial water films. *Nat. Rev. Chem.* **2022**, *6*, 598–613. [CrossRef] [PubMed]
16. Tromans, D. Mineral comminution: Energy efficiency considerations. *Miner. Eng.* **2008**, *21*, 613–620. [CrossRef]
17. Parapari, P.S.; Parian, M.; Rosenkranz, J. Breakage process of mineral processing comminution machines—An approach to liberation. *Adv. Powder Technol.* **2020**, *31*, 3669–3685. [CrossRef]
18. King, R. Comminution and liberation of minerals. *Miner. Eng.* **1994**, *7*, 129–140. [CrossRef]
19. Holyoke, C.W.; Kronenberg, A.K.; Newman, J.; Ulrich, C. Rheology of magnesite. *J. Geophys. Res. Solid Earth* **2014**, *119*, 6534–6557. [CrossRef]
20. Liu, Z.-J.; Sun, X.-W.; Song, T.; Guo, Y.; Zhang, C.-R.; Zhang, Z.-R. Atomistic simulation of the structural and elastic properties of magnesite. *Bull. Mater. Sci.* **2016**, *39*, 1319–1325. [CrossRef]
21. Yao, C.; Wu, Z.; Zou, F.; Sun, W. Thermodynamic and Elastic Properties of Magnesite at Mantle Conditions: First-Principles Calculations. *Geochem. Geophys. Geosyst.* **2018**, *19*, 2719–2731. [CrossRef]
22. Martínez-González, J.; Navarro-Ruiz, J.; Rimola, A. Multiscale Computational Simulation of Amorphous Silicates' Structural, Dielectric, and Vibrational Spectroscopic Properties. *Minerals* **2018**, *8*, 353. [CrossRef]
23. Gouriet, K.; Carrez, P.; Cordier, P. Ultimate Mechanical Properties of Forsterite. *Minerals* **2019**, *9*, 787. [CrossRef]
24. Choudhary, R.; Venkatraman, S.K.; Bulygina, I.; Chatterjee, A.; Abraham, J.; Senatov, F.; Kaloshkin, S.; Ilyasov, A.; Abakumov, M.; Knyazeva, M.; et al. Impact of forsterite addition on mechanical and biological properties of composites. *J. Asian Ceram. Soc.* **2020**, *8*, 1051–1065. [CrossRef]
25. Ajalloeian, R.; Mansouri, H.; Baradaran, E. Some carbonate rock texture effects on mechanical behavior, based on Koohrang tunnel data, Iran. *Bull. Eng. Geol. Environ.* **2017**, *76*, 295–307. [CrossRef]
26. Meng, Z.; Pan, J. Correlation between petrographic characteristics and failure duration in clastic rocks. *Eng. Geol.* **2007**, *89*, 258–265. [CrossRef]
27. Hansson, T.; Oostenbrink, C.; van Gunsteren, W. Molecular dynamics simulations. *Curr. Opin. Struct. Biol.* **2002**, *12*, 190–196. [CrossRef]
28. Atomic, L.S.; Simulator, M.M.P. Lammps. 2013. Available online: http://lammps.sandia.gov (accessed on 17 April 2023).
29. Ogorodnikov, V.V.; Rogovoi, Y.I. Buckingham-potential parameters and relations between them for solids. *Sov. Powder Met. Met. Ceram.* **1995**, *33*, 327–333. [CrossRef]
30. Mahendran, S.; Carrez, P.; Groh, S.; Cordier, P. Dislocation modelling in Mg_2SiO_4 forsterite: An atomic-scale study based on the THB1 potential. *Model. Simul. Mater. Sci. Eng.* **2017**, *25*, 054002. [CrossRef]
31. Momenzadeh, L.; Moghtaderi, B.; Liu, X.F.; Sloan, S.W.; Belova, I.V.; Murch, G.E. The thermal conductivity of magnesite, dolomite and calcite as determined by molecular dynamics simulation. In *Diffusion Foundations;* Trans Tech Publications Ltd.: Stafa-Zurich, Switzerland, 2019; Volume 19, pp. 18–34.
32. Jewett, A.I.; Zhuang, Z.; Shea, J.-E. Moltemplate a Coarse-Grained Model Assembly Tool. *Biophys. J.* **2013**, *104*, 169a. [CrossRef]

33. Subramaniyan, A.K.; Sun, C.T. Continuum interpretation of virial stress in molecular simulations. *Int. J. Solids Struct.* **2008**, *45*, 4340–4346. [CrossRef]
34. Bohlke, T.; Bruggemann, C. Graphical representation of the generalized Hooke's law. *Tech. Mech.* **2001**, *21*, 145–158.
35. Goble, D.L.; Wolff, E.G. Strain-rate sensitivity index of thermoplastics. *J. Mater. Sci.* **1993**, *28*, 5986–5994. [CrossRef]
36. Borodin, E.N.; Gruzdkov, A.A.; Mayer, A.E.; Selyutina, N.S. Physical nature of strain rate sensitivity of metals and alloys at high strain rates. *J. Phys. Conf. Ser.* **2018**, *991*, 012012. [CrossRef]
37. Ma, Z.; Gamage, R.P.; Zhang, C. Effects of temperature and grain size on the mechanical properties of polycrystalline quartz. *Comput. Mater. Sci.* **2020**, *188*, 110138. [CrossRef]
38. Gong, F.; Zhi, D.; Zhou, Y.; Zeng, Q.; Wang, Z. Empirical modeling of pore size distribution for rock materials with its impact on pore water freezing. *Cold Reg. Sci. Technol.* **2022**, *201*, 103619. [CrossRef]
39. Gong, F.; Wang, Z.; Zhou, Y.; Wang, J.; Yang, L.; Ueda, T. Mesoscale simulation of frost damage to rock material based on Rigid Body Spring Method. *Cold Reg. Sci. Technol.* **2022**, *201*, 103621. [CrossRef]
40. Blott, S.J.; Pye, K. GRADISTAT: A grain size distribution and statistics package for the analysis of unconsolidated sediments. *Earth Surf. Process. Landf.* **2001**, *26*, 1237–1248. [CrossRef]
41. Eberhardt, E.; Stimpson, B.; Stead, D. Effects of Grain Size on the Initiation and Propagation Thresholds of Stress-induced Brittle Fractures. *Rock Mech. Rock Eng.* **1999**, *32*, 81–99. [CrossRef]
42. Garber, J.A.; Granato, A.V. Fourth-order elastic constants and the temperature dependence of second-order elastic constants in cubic materials. *Phys. Rev. B* **1975**, *11*, 3998–4007. [CrossRef]
43. Suzuki, I.; Anderson, O.L.; Sumino, Y. Elastic properties of a single-crystal forsterite Mg_2SiO_4, up to 1200 K. *Phys. Chem. Miner.* **1983**, *10*, 38–46. [CrossRef]

sustainability

Article

Study on the Dynamic Mechanism of the Desiccation Crack Initiation and Propagation in Red Clay

Aijun Chen, Chaohua Li, Shanshan Zhao *, Bai Yang * and Chuanyang Ding

School of Architecture and Transportation Engineering, Guilin University of Electronic Technology, Guilin 541004, China; caj3026@163.com (A.C.); 21152303013@mails.guet.edu.cn (C.L.); 21152303004@mails.guet.edu.cn (C.D.)
* Correspondence: shanshanzhao@guet.edu.cn (S.Z.); ayangbai@163.com (B.Y.)

Abstract: Red clay is susceptible to cracking in desiccating environments, with resulting crisscrossing cracks that compromise the soil structure and increase the likelihood of geological hazards. To investigate the dynamic mechanism of the initiation and propagation of soil desiccation cracks under natural hygrothermal conditions, a desiccation test was conducted on a red clay slurry using three-dimensional digital image correlation (3D DIC) technology. The evolution behaviour of desiccation cracks was analysed, and the dynamic relationships between moisture content, displacement field, strain field, and soil desiccation cracking were explored. The test results showed that the Atterberg limits of red clay are correlated with desiccation cracking. Cracks tend to initiate in areas where tensile strain is concentrated or significant displacement differences exist. Following crack initiation, the surrounding strain and displacement fields redistribute, influencing the propagation direction, development rate, and morphology of subsequent cracks nearby. Additionally, the relative displacement and strain at the edges of cracks are related to the crack propagation direction. Earlier crack initiation usually corresponds to a larger relative displacement and strain at the crack edges, while the displacement and strain at the soil clod centre are typically smaller than those at the crack edges. DIC technology can quickly and accurately obtain dynamic information about displacement and strain fields, providing feasible technical support for analysing the dynamic mechanism behind soil desiccation cracking. It has potential value in engineering hazard prevention and sustainable development.

Keywords: red clay; desiccation cracking mechanism; digital image correlation technology; strain; displacement

Citation: Chen, A.; Li, C.; Zhao, S.; Yang, B.; Ding, C. Study on the Dynamic Mechanism of the Desiccation Crack Initiation and Propagation in Red Clay. *Sustainability* **2023**, *15*, 11156. https://doi.org/10.3390/su151411156

Academic Editors: Jian Zhou, Mahdi Hasanipanah and Danial Jahed Armaghani

Received: 6 June 2023
Revised: 21 June 2023
Accepted: 12 July 2023
Published: 18 July 2023

1. Introduction

Red clay is susceptible to cracking in desiccating environments due to its sensitivity to ambient humidity. The resulting desiccation cracks threaten the strength and stability of the soil, negatively impacting its permeability and hydraulic characteristics, leading to a range of engineering challenges. The crack initiation and propagation facilitate rainwater infiltration into the soil, causing engineering hazards such as foundation deformation, slope collapse, landslides, failure of landfill barrier function, and dam instability [1–8]. The existence of soil cracks in agricultural engineering diminishes moisture and fertilizer utilisation efficiency, amplifies soil erosion and pollution, and deepens soil weathering, consequently reducing crop yields. In environmental science, cracks facilitate the weathering and erosion of rocks and soils, resulting in land desertification and ecological environment disruption, compromising the sustainability of human and natural development. Therefore, a systematic study of the dynamic mechanism of desiccation cracking in red clay is crucial for effectively preventing and controlling engineering problems caused by cracking, thereby promoting sustainable development.

Desiccation cracking of clayey soil is the result of soil shrinkage caused by moisture evaporation, which is influenced by boundary restriction and uneven shrinkage [9], resulting in the formation and development of the soil stress–strain field. Once the tensile stress exceeds the maximum tensile strength of the soil, cracks initiate and continue to propagate during evolution [10,11]. The resulting cracks increase the soil surface area, promoting moisture evaporation, exacerbating the desiccation, shrinkage, and further development of cracks [12,13]. Thus, crack development, moisture content, and shrinkage deformation are mutually influential and dynamically develop.

Moisture evaporation induces desiccation cracking. Liu et al. [14] investigated the failure mechanism of red clay slopes. They found that soil shrinkage and cracking result from matric suction differences between the upper and lower soil layers induced by the moisture content gradient. Moisture content changes play a crucial role in the initiation and propagation of cracks. Researchers have identified the critical moisture content (wc) as a constant parameter for soil cracking [15,16]. They have also explored the relationship between crack evolution and moisture content [17,18]. However, wc is related to soil properties and external environmental factors, and further comprehensive study is needed to explore the relationship between moisture evaporation and crack evolution.

Researchers have investigated the mechanisms underlying the initiation of desiccation cracks [19–21]. It is commonly accepted that the uneven tensile stress caused by soil desiccation results in stress concentration in areas where soil deformation is limited, leading to cracking when the tensile stress exceeds the maximum tensile strength. However, conventional methods for determining soil tensile strength have limitations. Conventional tensile tests measure the tensile strength of the weakest point in a compacted sample at a specific moisture content and dry density [22,23]. Actually, matric suction increases as moisture evaporates, causing a redistribution of moisture and pores that may alter the initial conditions for desiccation crack initiation. Furthermore, tensile strength is a macroscopic mechanical indicator, and the uneven shrinkage deformation of clayey soils makes it difficult for conventional tests to accurately reflect the unevenness of the interparticle cohesive force at the micro level.

Previous researchers have attempted to establish relationships between strain, displacement, and crack evolution. For example, they conducted shrinkage tests and used Vernier callipers to measure soil strain, evaluating the total tensile stress when soil cracks initiate [24]. They also monitored the real-time stress and displacement of slopes using pressure gauges and displacement meters to examine the evolution of slope cracks from the perspective of crack–stress–displacement [25]. However, these methods require direct contact between the measuring instruments and soil, leading to the disturbance of the original structure and constraint conditions of the soil, and cannot provide complete information about soil stress, strain, and displacement. Therefore, the development and adoption of high-precision, contactless, fast, and dynamic testing techniques offer a promising direction to address these limitations.

Digital image correlation (DIC) technology is an optical non-contact technique that can measure the 3D displacement and strain of an object surface [26]. It can monitor the evolution of all cracks in a testing area and detect subtle changes in civil engineering materials in real time, allowing for the analysis of displacement and strain evolution. DIC technology is mostly applied to materials such as metals, reinforced concretes, rocks, polymers, and biological tissues. Studies have reported the real-time measurement of the surface strain and displacement of these materials on the micro- and macroscale. Lei et al. [27] proposed an approach for investigating the planar strain distribution in polycrystalline materials at the grain scale by combining DIC technology with microscopy and verified its feasibility for studying the micromechanical behaviour of a Ni-based alloy specimen. Francic et al. [28] compared the accuracy of 2D DIC technology and LVDT sensors in measuring the crack width and vertical displacement of reinforced concrete and found DIC technology had higher precision. Han et al. [29] used a split Hopkinson pressure bar (SHPB) system and DIC technology to monitor the deformation, vertical displacement, and maximum principal strain of sandstone during the impact loading process and conducted dynamic fracture analysis. Wang et al. [30] used DIC technology to obtain the heterogeneous deformation fields of polyurethane foam samples during compression and revealed the evolution of surface deformation textures of open-cell polyurethane foam samples with different relative densities. Zhang et al. [31] used DIC technology to measure the mechanical properties of biological tissues. Despite DIC technology having been applied to various materials, there are limited studies on its application in the research on soil crack evolution, and few reports discuss the relationship between crack evolution and deformation (displacement and strain). Consequently, applying DIC technology to study the relationship between crack evolution and soil deformation presents significant advantages and provides an opportunity for a comprehensive analysis of the dynamic mechanism of crack initiation and propagation.

In this study, 3D DIC technology was utilised to dynamically monitor red clay desiccation cracking in indoor desiccating tests. The relationships between moisture content, strain, displacement, and crack evolution were analysed with soil moisture monitoring and surface crack identification. The dynamic mechanism of soil desiccation crack initiation and propagation was studied. These study results can contribute to a deeper understanding of the dynamic mechanism of soil desiccation cracking.

2. Materials and Methods

2.1. Soil Samples

The investigated soil was red clay at a depth of 2.5–4.0 m, which was recovered from Shaoyang, Hunan province, China. According to the Test Methods of Soils for Highway Engineering in China (JTG3430-2020) [32], the basic physical properties were determined by laboratory tests and are given in Table 1. The soil liquid limit is higher than 50%, the plasticity index is higher than 35%, and the percentage of soil particles larger than 0.075 mm is less than 25%. The particle composition and plasticity index indicate that the soil belonged to the high-liquid-limit clay group.

Table 1. Physical properties of red clay for testing.

Specific Gravity /(g/cm^3)	Liquid Limit /(%)	Plastic Limit /(%)	Shrinkage Limit /(%)	Plasticity Index /(%)	Optimum Moisture Content /(%)	Maximum Dry Density (g/cm^3)	Clay Content /(%)
2.72	67.7	28.3	18.8	39.4	18.5	1.86	62.8

2.2. Sample Preparation

The soil was air-dried, crushed with a rubber mallet, and sieved through a 2 mm sieve. The sieved soil was then placed into a 200 mm diameter circular container, and distilled water was added to prepare a saturated slurry with a moisture content of 120%. Air bubbles were removed from the slurry sample by stirring and vibrating and leaving it to set for at least 48 h for sedimentation. Finally, the pure water above the slurry surface was removed, resulting in a circular slurry sample with an initial thickness of 15 mm and an initial moisture content of 87.3%.

To obtain soil surface information using DIC technology, the soil surface must have recognisable characteristics [33]. According to the characteristics of the tested soil and the requirements for crack identification using DIC technology, white paint and coal particles were used as base materials and speckles, respectively. A layer of white paint was uniformly applied to cover the soil surface, followed by the random placement of coal particles (with diameters ranging from 0.2 to 1 mm) on the white paint. The thickness of the white paint layer was approximately 0.1 mm, which did not increase the soil cohesive strength and affect soil-cracking behaviour. The influence of the coal particles on the experimental characteristics of the sample was considered negligible due to their small diameter.

2.3. Experimental Device and Procedures

A self-designed device (Figure 1) was used to desiccate the soil sample in this test. This device can also obtain the sample mass, surface crack morphology, displacement, and strain at different time intervals.

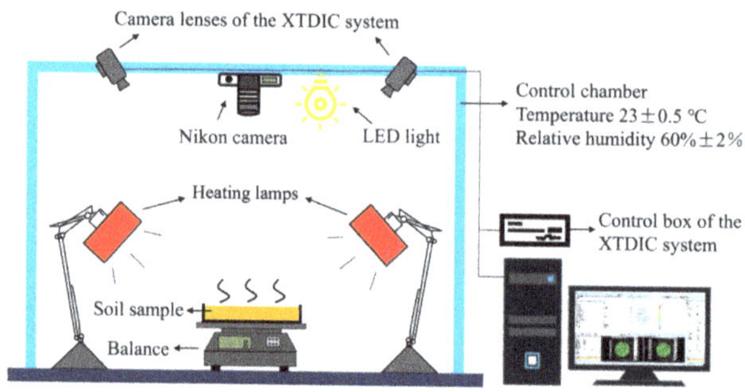

Figure 1. Layout of the desiccation test device.

The 3D optical non-contact deformation measurement system (XTDIC) was developed in China by Xi'an Xintuo Three-Dimension Photometry Technology Corporation (Xi'an, China). The system comprised two adjustable high-precision camera lenses, a control box, and a high-performance computer. The two lenses and LED light of the XTDIC system were mounted on a horizontal beam secured with an adjustable-height steel bracket. It combined DIC technology and binocular stereo vision technology [33]. Two cameras positioned at different angles were used to capture speckle patterns on the sample's surface, and 3D spatial information was obtained through calibration. The captured images were analysed using the DIC algorithm to reconstruct the 3D coordinates of the points of the sample's surface before and after deformation and to obtain the dynamic 3D deformation information of the sample. The sample was illuminated by two heating lamps with a design illumination intensity of 600 W/m^2. The laboratory temperature was maintained at 23 ± 0.5 °C, and the relative humidity was 60 ± 2%. A camera (Nikon D5100, Nikon corporation, Tokyo, Japan) was placed above the soil sample to capture the sample for the extraction of crack parameters.

The sample initial mass was measured using an electronic balance (accuracy 0.1 g), and the sample initial state was simultaneously captured using the XTDIC system and camera. During the test, the heating lamps illuminated the sample surface, and the electronic balance automatically recorded the sample mass. The XTDIC system and camera automatically captured images at intervals of 10 min. The test was completed when the electronic balance showed that the difference between two adjacent readings was less than 0.5 g.

MATLAB was used to perform image digitalisation on crack photos taken at different times [34], including image binarisation, morphological processing, filtration, skeleton extraction, and deburring, as shown in Figure 2. Image binarisation enables the differentiation of crack and non-crack areas by assigning black and white pixels. Morphological processing is used to modify pixel values to enhance the cracks in the image. Filtration effectively eliminates image noise, such as coal particles, to obtain the crack area of the image. Skeleton extraction refines the cracks and extracts the skeleton of single-pixel-width cracks. Deburring yields a smooth crack photo, providing the total crack length. From these processes, the crack ratio [35], average crack width, and total crack length versus time were obtained. In addition, the moisture content of the sample versus time was obtained by analysing the electronic balance readings at different times.

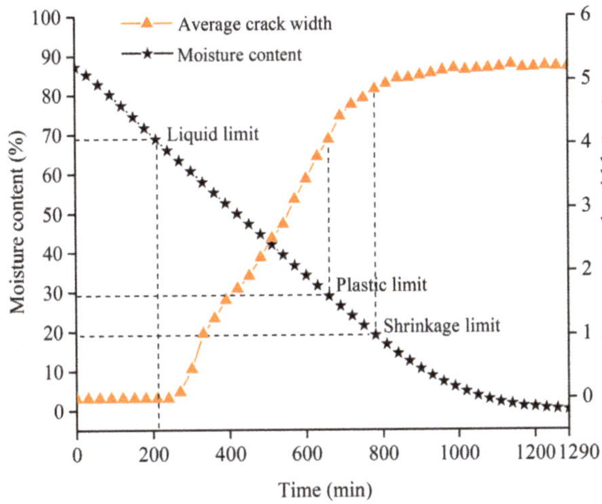

Figure 2. The relationship between moisture content, average crack width, and time.

3. Results and Discussion

3.1. The Influence of Moisture Content on Crack Development

The various relationships of moisture content and crack parameters versus time are shown in Figures 2–4. The relationship between moisture content (w) in percentage and time (t) in hours can be expressed as a function [36]: $w(t) = (120 - 0.5).erfc(t/20) + 0.5$. The moisture content curve presented in Figure 3 can be divided into the constant rate stage (0–800 min), the falling rate stage (800–1200 min), and the stable stage (1200–1290 min). The moisture content decreased linearly versus time before 800 min. The evaporation rate was mainly affected by environmental conditions. The decreasing trend of moisture content slowed from 800 min to 1200 min. After 1200 min, the moisture content tended to stabilise, with a residual moisture content of 0.5%. An earlier study by An et al. [37] on the moisture content curve of clay slurry samples of different thicknesses obtained similar results, albeit with a different residual moisture content due to differences in soil properties and external temperature conditions.

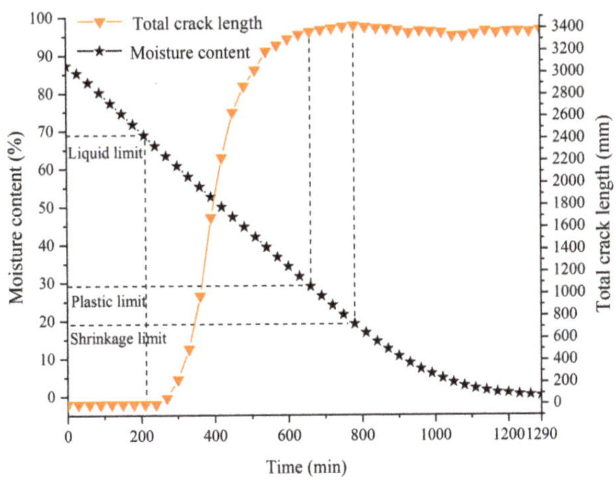

Figure 3. The relationship between total crack length and time.

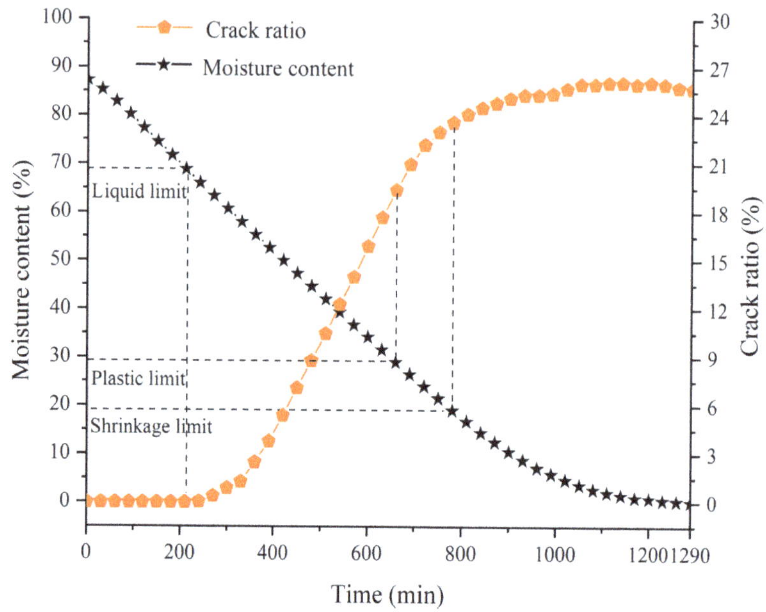

Figure 4. The relationship between crack ratio and time.

As shown in Figures 2–4, there were no significant changes in the curves of total crack length, average crack width, and crack ratio before 270 min. Subsequently, as cracks initiated and propagated, the total length, average crack width, and crack ratio increased rapidly before stabilising, but the increase rate was not constant. The total crack length stabilised earliest at 660 min, while the crack ratio and average crack width continued to increase at a slower rate until stabilising at 800 min, indicating that the increase in crack ratio was primarily attributed to the increase in crack width. The crack parameters are influenced by the sample size and initial moisture content [38], yet the overall development trend is still similar.

There was a certain relationship between the Atterberg limits and crack development. Such that the cracks began to initiate when the moisture content dropped to the liquid limit (67.7%), the crack length gradually stabilised when the moisture content approached the plastic limit (28.3%), and the crack length did not significantly increase when the moisture content fell below the shrinkage limit (18.8%). Combined with Figures 2–4 and the crack evolution (Figure 5(a1–a6)), the relationship between crack evolution and moisture content can be verified. The first crack appeared around 270 min when the moisture content was near the liquid limit. The moisture content approached the plastic limit at 650 min. Compared to Figure 5(a5,a6), crack development slowed, crack length was basically stable, and no new cracks were formed. Instead, with moisture evaporating, crack width and depth continued to increase. The moisture content neared the shrinkage limit at 800 min, the crack length stabilised, and the crack width increase was small, indicating that crack development had gradually stabilised. In fact, the relationship between crack development and moisture content is influenced by the coupling effect of sample shape, thickness, and interface roughness [39]. Therefore, there may be discrepancies between laboratory tests and field tests.

In addition, when the initial moisture content is greater than a certain value, overall shrinkage will occur before the cracks initiate [40]. This study took a circular slurry sample with a 200 mm diameter and 15 mm thickness as the research object. Because the cohesive force between the sample and the bottom surface of the container was greater than the shrinkage stress, there was no overall centripetal shrinkage during the desiccation process.

Hence, it is speculated that the critical moisture content is meaningful only for saturated soil, and the initial moisture content of unsaturated soil is the critical moisture content.

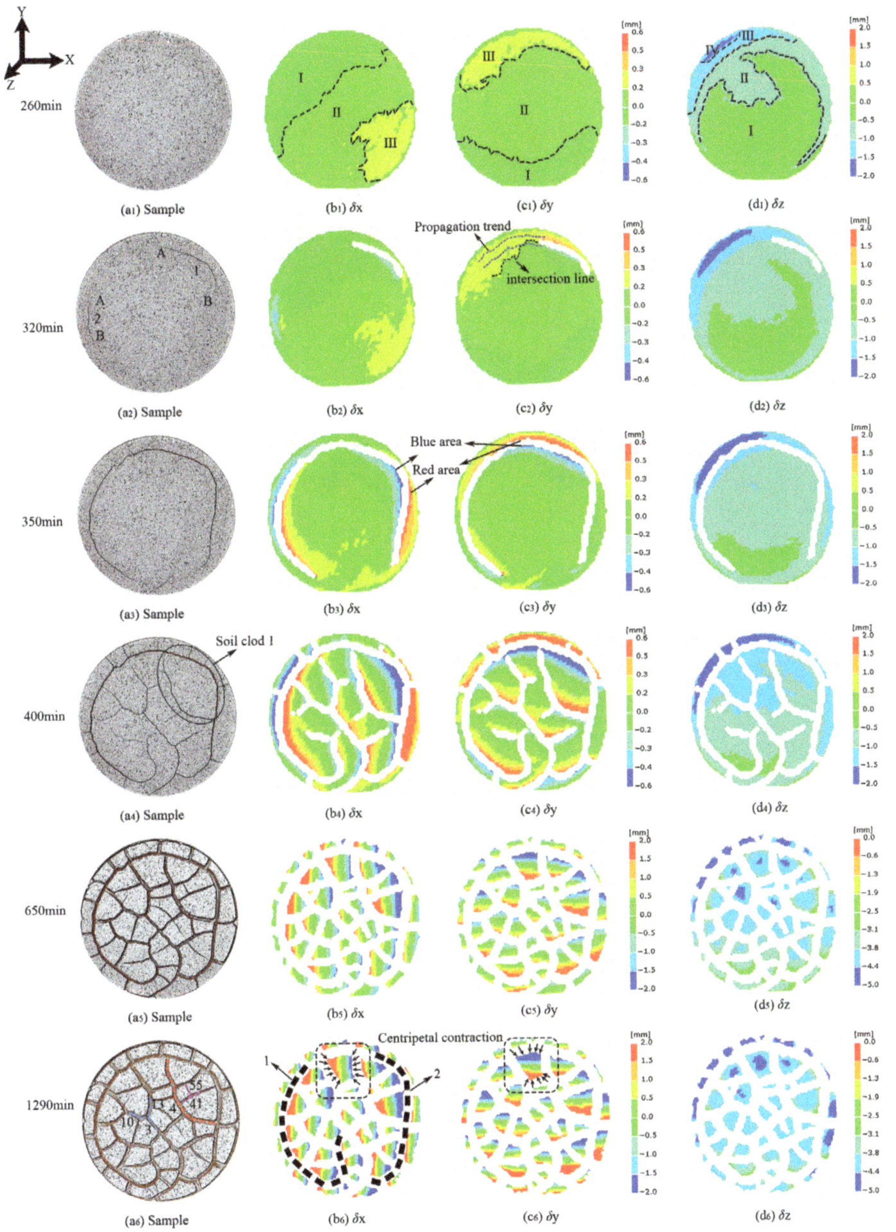

Figure 5. Crack photos and displacement images of the sample at different times. In the colour bar, different colours may represent different displacement ranges in different figures. I, II, III, and IV represent the division of different displacement fields. 1 and 2 represent crack 1 and 2, respectively. A and B represent crack ends.

3.2. Interaction between Displacement Changes and Crack Evolution

Figure 5 shows the crack photos and displacement images of the sample at different times. The horizontal transverse displacement, horizontal longitudinal displacement, and vertical displacement are represented by δx, δy, and δz, respectively. The *X*-axis represents the horizontal transverse direction, the *Y*-axis represents the horizontal longitudinal direction, and the *Z*-axis represents the vertical direction (positive upwards and negative downwards).

As shown in Figure 5(b1–d1), cracks have not yet initiated at 260 min, and the displacement fields could be divided into 3–4 areas which distributed more evenly than in later stages. The vertical displacement (δz) was negative, indicating a single-directional contraction trend, with a greater magnitude compared to the horizontal displacements (δx and δy). From 320 min to 650 min, opposite-directional horizontal displacements appeared on both sides of the cracks as cracks widened, which increased rapidly with time. The red and blue areas on both sides of the cracks in Figure 5(b3,c3) showed a faster increase in displacement compared to the non-cracked areas. At 1290 min, multiple displacement fields with different colours existed on the small soil clod when the displacement field stabilised (Figure 5(b6–d6)). The horizontal displacement decreased from the edges of the soil clod towards the centre, while the vertical displacement showed an opposite trend. Such as the small soil clod in Figure 5(b6,c6), the arrows represent the direction of the soil clod shrinkage. According to the displacement field evolution of the small soil clod, it can be observed that the shrinkage deformation of the small soil clod could present the phenomenon of centripetal shrinkage, and the shrinkage centre of the small soil clod also changed with time [41].

The direct cause of crack initiation is the relative displacement between adjacent displacement fields caused by uneven soil shrinkage, and crack initiation can be explored by analysing the relative displacement. Soil desiccation cracks were typical tensile cracks that initiated when the relative displacement at the intersection of different displacement fields exceeded the soil limiting value. As the tensile stress is greater in the upper layer than the lower layer of the soil, cracks usually propagate from the surface downwards [42]. By analysing the relative displacement on both sides of the cracks and the potential positions of cracks, the minimum relative displacement for crack initiation in the tested soil sample was preliminarily determined to be 0.2–0.25 mm. As shown in Figure 5(c1), the δy on both sides of crack 1 (Figure 5(a2)) was approximately 0.2 mm before it initiated.

The relative displacements in different directions between adjacent displacement fields influence the trend and rate of soil crack development. Comparing Figure 5(b2–d2), the initiation of crack 1 at the intersection of the displacement fields in Figure 5(a2) was mainly due to the influence of δy, but it was also influenced by δx and δz during its development process because crack 1 did not always develop along the intersection line of the yellow and green displacement fields in the Y direction shown in Figure 5(c2). The larger the relative displacement, the faster the crack develops. Near the A-ends of cracks 1 and 2 in Figure 5(a2), the relative displacement in all directions was greater than near the B-ends, leading to the A-ends developing faster than the B-ends. When the soil displacements on both sides of the crack were different, the crack extended towards the larger displacement side, resulting in a Y-shaped or arc-shaped crack, such as arc crack 4 (red) and the Y-shaped crack (blue) formed by cracks 3, 10, and 13 in Figure 5(a6). Conversely, when the soil displacements on both sides of the crack were equal, the crack propagated in a straight line and intersected with the previous cracks to form a T-shaped crack, such as the T-shaped crack (purple) formed by cracks 41 and 55 in Figure 5(a6).

The initiation and propagation of cracks cause changes in the distribution and magnitude of the displacement fields around the cracks. As cracks propagate, moisture evaporates faster from the edges of soil clods and induces more pronounced centripetal shrinkage, leading to greater changes in the displacement fields on both sides of cracks than in the centre of small soil clods (Figure 5(b4–b6,c4–c6)). However, this effect appears to have little influence on vertical displacement fields (Figure 5(d4–d6)). The increase rate of horizontal displacement around the crack is greater when the crack initiates earlier, requiring more time to reach stability and resulting in larger values after stabilisation. For example, the δx and δy of early soil clods around cracks 1 and 2 are generally larger than those of late soil clods in other areas shown in Figure 5(b6). Moreover, the shape, size, and centre position of soil clods change with time, and early formed soil clods may form cracks and divide into smaller soil blocks later. Only the displacements of the central area of the final formed soil clods were smaller than that of their edges. Refer to the displacement field evolution of small soil clod 1 in Figure 5(a4).

3.3. Interaction between Strain Changes and Crack Evolution

Figure 6 shows the crack photos and strain images at different times. The horizontal transverse strain, horizontal longitudinal strain, shear strain, and maximum principal strain are represented by εx, εy, εxy, and εm, respectively. Positive values indicate a tensile strain, while negative values indicate a compressive strain.

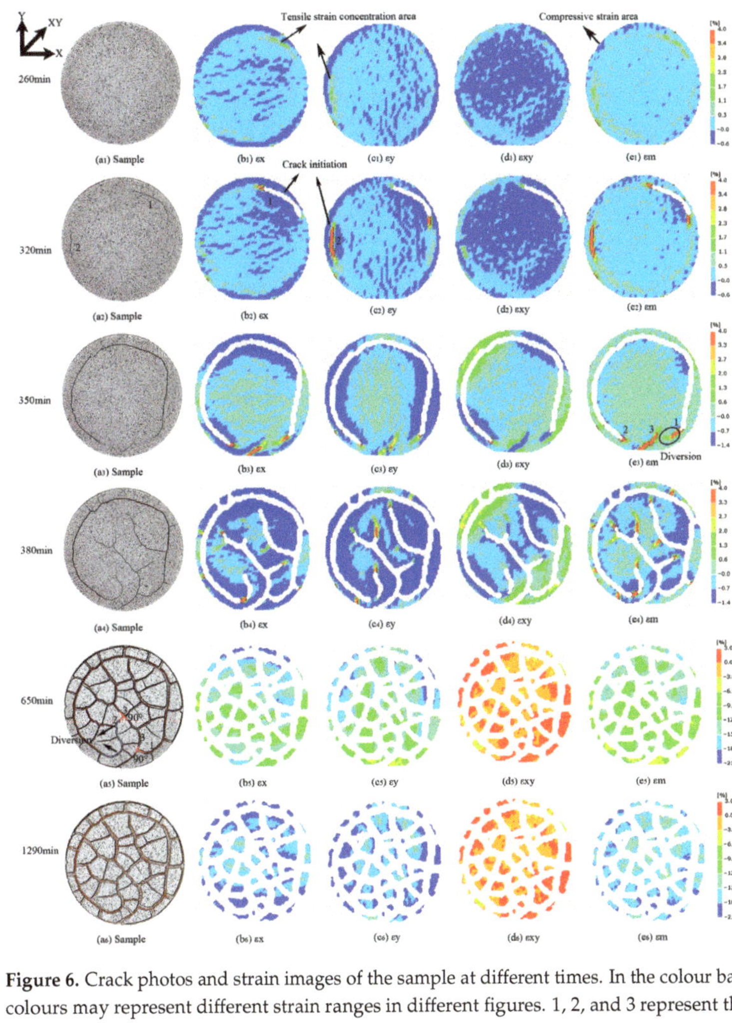

Figure 6. Crack photos and strain images of the sample at different times. In the colour bar, different colours may represent different strain ranges in different figures. 1, 2, and 3 represent the order of the cracks.

At 260 min, tensile strain predominated on the soil surface, while compressive strain was concentrated in a few areas along the container edge (Figure 6(e1)). Starting from 320 min, tensile strain gradually increased and was quickly concentrated at the crack tip. Beginning at 350 min, new cracks began to propagate towards the sample centre, accompanied by high compressive strain around the cracks. The tensile strain gradually changed to compressive strain until it stabilised at 650 min, where the entire soil surface was under compressive strain (Figure 6(e5)). The changes in εx and εy were similar to εm, and εxy changed from a compressive state to a tensile state.

The strain parallel to the crack direction was greater than that perpendicular to the crack direction. Such as crack 2 was parallel to the *Y*-axis, so εy was significantly larger than εx. The strain distribution was usually uneven due to the uneven soil shrinkage, leading to concentration areas of tensile and compressive strain in the strain images. When the strain stabilised, εxy was significantly smaller than εx and εy.

According to the concentration areas of tensile strain, the subsequent crack initiation and propagation can be preliminarily predicted. In Figure 6(b1,c1), two green areas of tensile strain concentration appeared approximately parallel to the container wall due to boundary effects [43]. Subsequently, cracks 1 and 2 initiate in this area, as shown in Figure 6(a2). It can be inferred from Figure 6(e1–e5) that cracks may initiate when the εm in an area exceeds 2.3%. Once the cracks initiated, the accumulated local tensile strain energy was released, with tensile stress concentrating on the crack tip and directing the crack propagation. For example, εm near the tip of cracks 1 and 2 in Figure 6(e1) is close to 2.3%.

The crack initiation and propagation were influenced by the surrounding strain fields, which also reacted on the evolution of the surrounding strain fields. Crack 1 was influenced by the strain field near the newly formed crack 3 (Figure 6(e3)), resulting in a change in its original circular propagation trend and intersecting with crack 3 at 90°, whereas crack 2 intersected with crack 3 near 90° after several diversions (Figure 6(a5)). When the sample thickness exceeds 4 mm, the cracks intersect at 90°, whereas if it is less than 4 mm, cracks typically intersect at 120° [44].

The evolution of the maximum principal strain near cracks 1–3 in Figure 6(e3,e4) shows that the initiation of adjacent crack 3 releases the previously accumulated strain energy, leading to the local rearrangement of the strain field [45], which influences the propagation trends of cracks 1 and 2 by altering the stress fields around them. As moisture evaporates, the cracks widen causing the surrounding soil to be compressed. The tensile strain areas on the sides of the cracks gradually transform into compressive strain areas, but the influence of this transition does not extend to the strain fields around the non-crack areas. Therefore, the tensile strain areas gradually decrease as the number of cracks increases, and the compressive strain areas expand.

3.4. Relationship between the Temporal and Spatial Evolution of Soil Surface Deformation and Cracks

In Figure 7, eight positions where cracks initiated at different times are presented, and sets of relative displacement feature points (Gi) are selected on both sides of each position, including the left feature point (Li) and the right feature point (Ri). The relative displacements in each direction for each set of feature points are calculated (i.e., the displacement of Ri minus the displacement of Li in the same direction). The horizontal transverse relative displacement, horizontal longitudinal relative displacement, and vertical relative displacement are represented by Δx, Δy, and Δz, respectively. The X-, Y-, and Z- axes represent the horizontal transverse direction, horizontal longitudinal direction, and vertical direction. Two areas of soil samples were selected for analysis (feature points G1–G8, differentiated by red and blue, where cracks near G1–G8 initiated at 320 min, 330 min, 360 min, 370 min, 370 min, 380 min, 390 min, and 510 min). The relationship between the relative displacement of each point and time is shown in Figures 8–10.

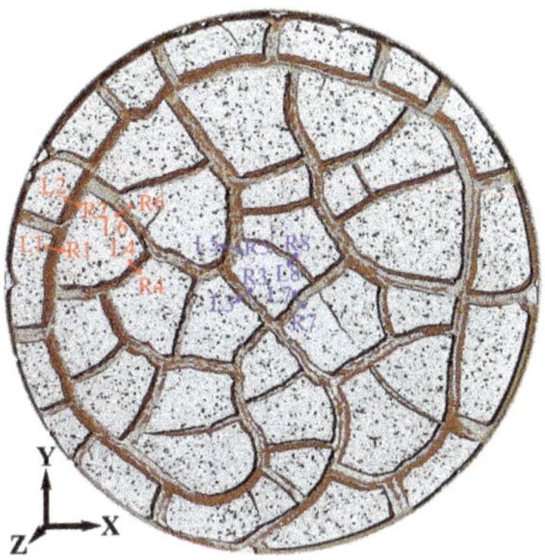

Figure 7. Distribution of relative displacement feature points on the soil surface. L1–L8 represent the left feature point of cracks at different positions. R1–R8 represent the right feature point of cracks at different positions.

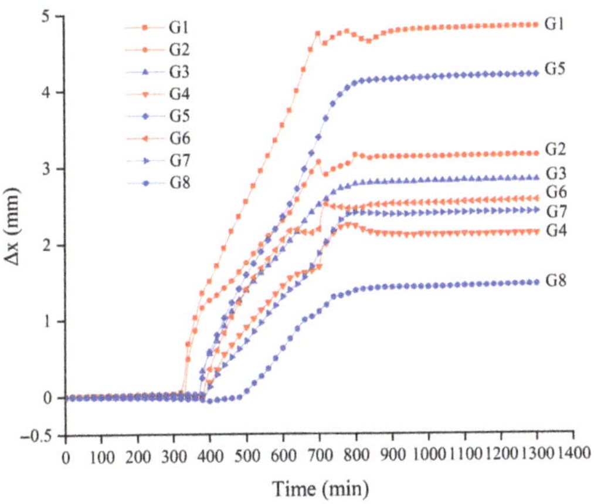

Figure 8. Variations of Δx versus time.

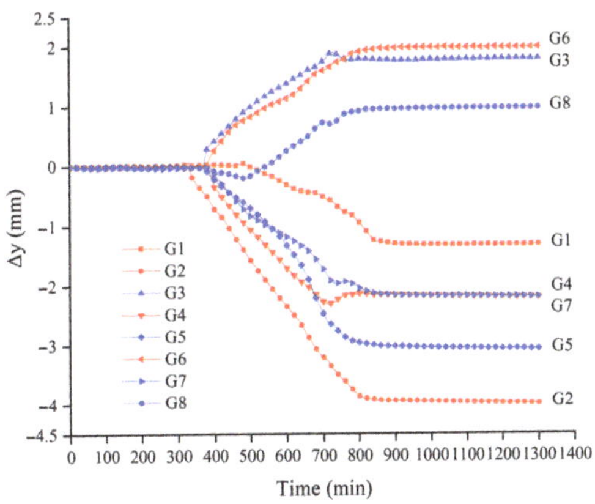

Figure 9. Variations of Δy versus time.

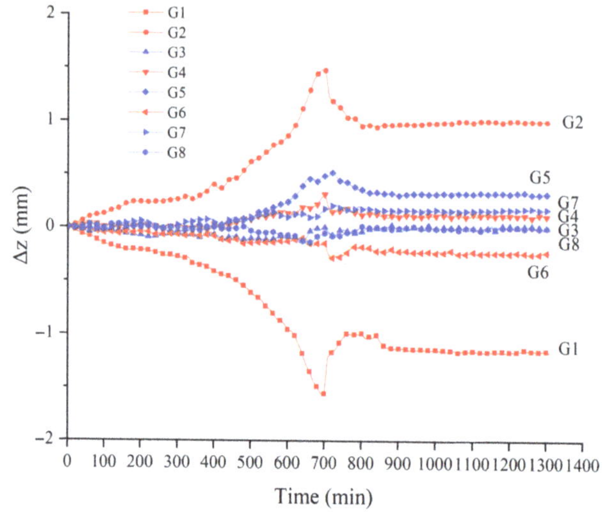

Figure 10. Variations of Δz versus time.

The relative displacement curves can be roughly divided into three stages. In the first stage (before crack initiation), Δx and Δy were not significant, and Δz of G1 and G2 increased with time, but Δz of other feature points was very small. In the second stage (the crack propagation stage after crack initiation), Δx and Δy rapidly increased when the cracks were about to initiate. The increase rate of relative displacement can be calculated by taking the derivative of the relative displacement. The increase rates of Δx, Δy, and Δz were 1.0–1.2 $\times 10^{-2}$ mm/min, 0.4–0.8 $\times 10^{-2}$ mm/min, and 0.1–0.3 $\times 10^{-2}$ mm/min, respectively. This indicates that the change trend of Δz was relatively small, and Δx and Δy were the primary cause of soil surface cracking. This is attributed to the sample thickness being much smaller than its diameter, and soil vertical shrinkage exhibits less magnitude compared to horizontal shrinkage. In the late second stage, crack propagation basically completed, Δx and Δy had no significant changes, but Δz decreased. Combined with Figure 3, it can be inferred that the crack length was generally stable in this stage, indicating that the crack network was formed roughly. When the shrinkage stress exceeded the cohesive force between the soil clod bottom and container surface, the soil clod separated from the container bottom, causing the soil on both sides of the crack to arch upwards [46]. In the third stage (the crack stable stage), Δx, Δy, and Δz were basically stable. The earlier the crack initiates, the longer the crack development duration and the larger the crack width.

There are differences in the relative displacement between different feature points. The cracks at G4 and G5 initiate simultaneously, but the changes in the all-direction displacement curve at G5 is significantly larger than that at G4. because the crack near G4 is a secondary crack resulting from the bifurcation of the primary crack, whereas the crack at G5 is the primary crack. Δx for each set of feature points is positive, while Δy and Δz can be positive or negative. Small soil clods exhibit centripetal shrinkage, resulting in the opposite displacement directions of feature points on the left and right sides of a crack. Different shrinkage degrees lead to different positive and negative relative displacements because there is spatial anisotropy in the soil shrinkage deformation on both sides of the crack [47].

The relative displacement is related to the direction of crack propagation. The propagation direction and width of cracks near G3 and G6, as well as G4 and G7, are roughly similar, and the development trend of the relative displacement curves is also similar. The crack propagation direction impacts the change in the relative displacement of feature points. Such as when the crack at G1 developed along the Y direction, Δx began to increase at 330min, whereas Δy started to increase at 480min. Therefore, the earlier the crack initiates near the feature point, the more likely the relative displacement between the two feature points may not always be larger when the crack is stable. For example, Δx and Δz of G1 are the largest, while Δy is very small, due to the relative displacement parallel to the crack direction on the sample being smaller than that perpendicular to the crack direction.

Figure 11 shows several feature points selected from soil clods formed at different times, with red and blue colours used to distinguish feature points on two soil clods. C1 and C2 represent the centre points of the soil clods, while P1–P6 represent the edge points of the cracks. The crack initiation times near P1–P6 were 330 min, 360 min, 370 min, 380 min, 390 min, and 510 min, respectively. The strain curves over time are shown in Figures 12–15.

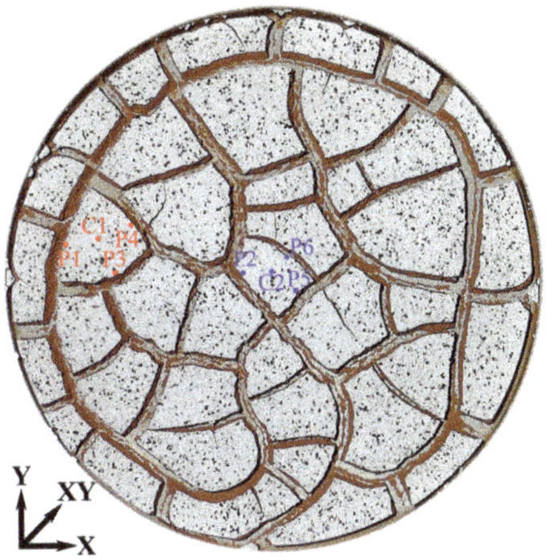

Figure 11. Distribution of strain feature points on the soil surface.

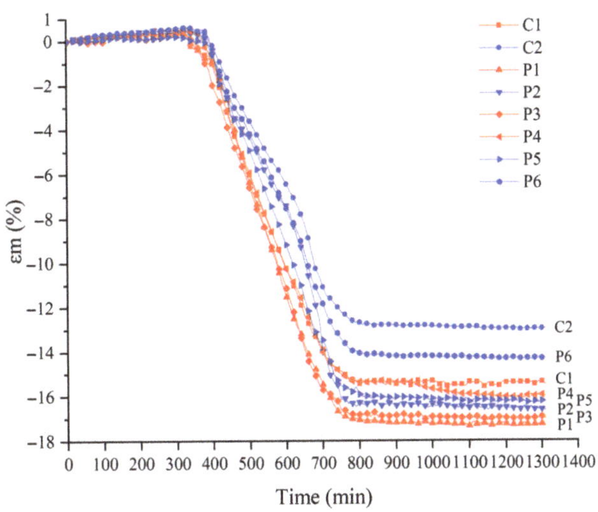

Figure 12. Variations of εm versus time.

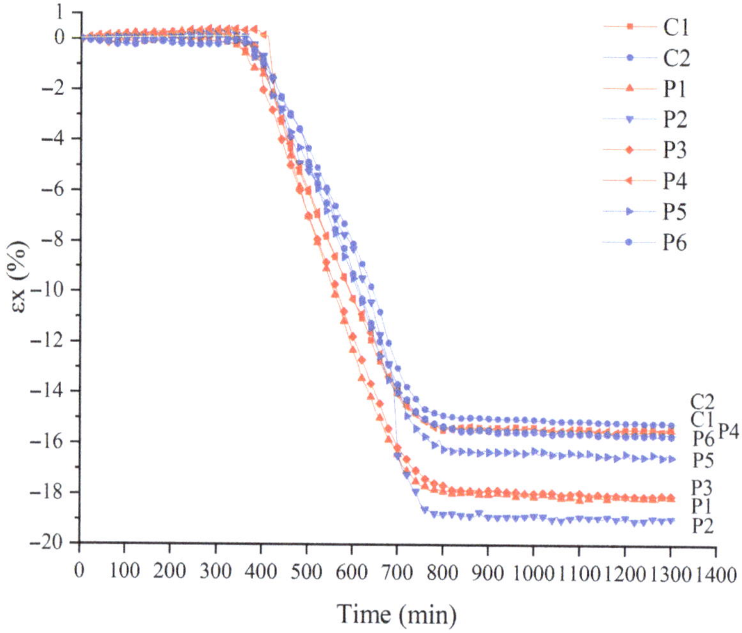

Figure 13. Variations of εx versus time.

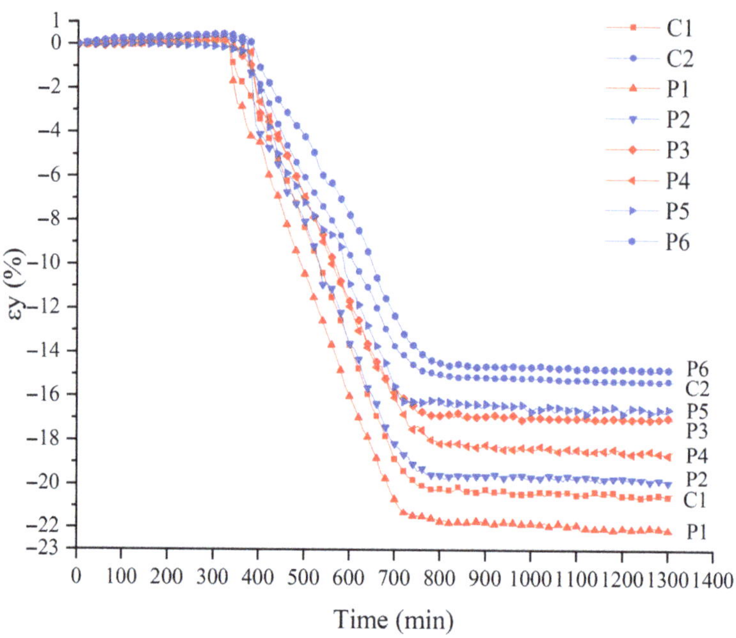

Figure 14. Variations of εy versus time.

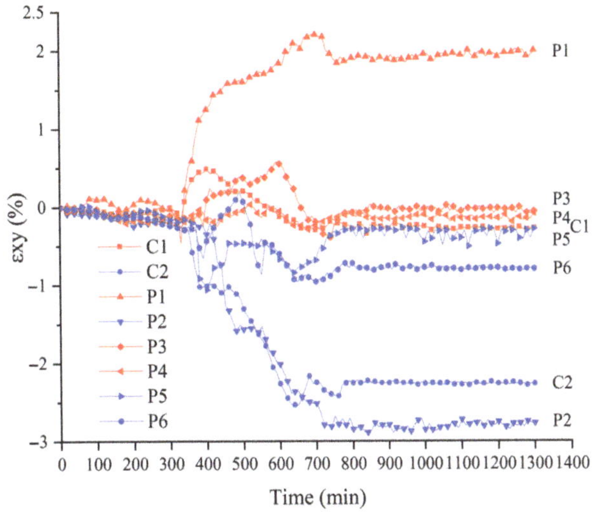

Figure 15. Variations of εxy versus time.

The strain field evolution is related to the crack development, as strain changes cause the initiation and propagation of cracks. In the first stage (before crack initiation), feature point strains remain small, and the strain changes are not significant. In the second stage (crack initiation and propagation stage), εx, εy, and εm increase linearly with time, with increasing rates of 3.2–3.8 $\times$ 10^{-2}%/min, 3.2–5.3 $\times$ 10^{-2}%/min, and 3.6–2.8 $\times$ 10^{-2}%/min, respectively, whereas εxy changes are much smaller than those of εx and εy. The strain increase rate at the crack edge points in the same soil clod is generally greater than that at the centre points of the soil clod, indicating that the shrinkage of the centre of the soil clod is smaller than that of the edges. In the third stage (the crack stable stage), cracks near the feature points no longer develop significantly, and the strain curves tend to stabilise. The maximum principal strain is ranked from smallest to largest: C2 (12.9%) < P6 (14.3%) < C1 (15.5%) < P4 (16.0%) < P5 (16.2%) < P2 (16.6%) < P3 (17.0%) < P1 (17.3%). After stabilisation, εx, εy, and εxy also show that the earlier the crack initiates, the larger the strain of the crack edges usually is. In the same soil clod, the strain of the soil clod centre is usually smaller than that of the crack edges.

With the exception of εxy, εx and εy are positive before crack initiation and become negative after crack initiation. The primary reason is that soil moisture evaporation leads to a tensile stress field inside the soil, and cracks initiate when the tensile stress exceeds the soil limiting value. The soil around the crack changes from being under tension to being under compression, and εx and εy gradually become negative. εxy is negative at blue feature points and positive at red feature points, indicating that feature points are subjected to opposite directions of shear stress.

4. Conclusions

This study conducted a desiccation cracking test on a 200 mm diameter and 15 mm thick red clay sample under simulated natural hygrothermal conditions, and dynamic information on surface displacements and strains was obtained using DIC technology without disturbing the soil. The relationships between moisture content, surface displacement, strain, and cracks were explored to investigate the dynamic mechanism of desiccation crack initiation and propagation in red clay. The following conclusions were drawn:

1. Soil desiccation crack evolution was related to the Atterberg limit of red clay. Cracks initiated when the moisture content approached the liquid limit of 67.7%, developed slowly when the moisture content reached the plastic limit of 28.3%, and tended to stabilise when the moisture content was less than the shrinkage limit of 18.8%.
2. Changes in strain and displacement interacted with crack initiation and propagation. The initiation and evolution of subsequent cracks could be predicted by strain and displacement images, as cracks tended to initiate in areas with concentrated tensile strain or obvious horizontal relative displacement. The shape and propagation trend of cracks were under the influence of the distribution of the surface displacement and strain fields. Crack initiation caused the redistribution of the displacement and strain fields around them; the displacement and strain gradually increased with crack propagation, and the displacement and strain values on both sides of the cracks increased faster than in non-crack areas.
3. The ratio of sample to diameter thickness is 40:3, with the diameter being significantly larger than the thickness. Hence, the horizontal relative displacement (Δx and Δy) on both sides of the cracks was significantly greater than the vertical relative displacement (Δz), and uneven shrinkage-induced horizontal relative displacement was the primary cause of soil cracking. Additionally, the numerical values of the relative displacement were also related to the crack propagation direction.
4. Horizontal transverse strain and horizontal longitudinal strain (εx and εy) were greater than the shear strain (εxy), and cracks may initiate when the maximum principal strain (εm) exceeds 2.3%. The earlier the cracks initiated, the longer the development duration and the larger the strain at the crack edges.
5. DIC technology can obtain dynamic information on the surface displacement and strain of the sample without disturbing the soil. It can analyse the critical strain and displacement when cracks are about to initiate and study the deformation behaviour during soil desiccation cracking. This is helpful in analysing the dynamic mechanism of soil desiccation crack initiation and propagation.

Under the influence of hot and arid weather, cracks initiate when soil loses moisture to the critical moisture content. The study results indicate that the majority of cracks initiate and propagate rapidly during the constant rate stage. Therefore, it is necessary to implement effective measures to maintain a uniform distribution of soil moisture and mitigate crack initiation. Through the utilisation of 3D DIC technology, it becomes feasible to preliminarily predict the initial location and subsequent propagation direction of desiccation cracks. This provides advantages for optimising the arrangement of irrigation systems during crop cultivation and growth, as proper irrigation contributes to maintaining soil moisture and limiting downward crack propagation. Moreover, the stability of soil slopes is heavily influenced by the development of soil cracks. By implementing reinforcement measures that restrict excessive strain and displacement, based on critical thresholds for soil crack initiation, the occurrence of hazards resulting from cracks can be minimised.

Author Contributions: Conceptualization, A.C.; Methodology, S.Z. and B.Y.; Software, C.L. and C.D.; Formal analysis, C.D.; Investigation, S.Z.; Data curation, C.L. and C.D.; Writing—original draft, C.L.; Writing—review & editing, S.Z.; Visualization, C.D.; Supervision, A.C. and B.Y.; Project administration, B.Y.; Funding acquisition, A.C. All authors have read and agreed to the published version of the manuscript.

Funding: This work was supported by the Guangxi Science and Technology Program (2022GXNSFAA035485), and by the Foundation for Science and Technology Base and Talents of Guangxi Provence of China (GUIKEAD 20238063).

Institutional Review Board Statement: Not applicable.

Informed Consent Statement: Not applicable.

Data Availability Statement: Data can be requested from the corresponding author.

Conflicts of Interest: The authors declare no conflict of interest.

References

1. Yang, B.B.; Zhang, Z.P.; Ma, W.L. Effect of tea waste on cracking of foundation soil. *Adv. Mater. Sci. Eng.* **2021**, *2021*, 7525811. [CrossRef]
2. Izzo, M.Z.; Miletic, M. Sustainable improvement of the crack resistance of cohesive soils. *Sustainability* **2019**, *11*, 5806. [CrossRef]
3. Zhang, J.M.; Luo, Y.; Zhou, Z. Effects of preferential flow induced by desiccation cracks on slope stability. *Eng. Geol.* **2021**, *288*, 106164. [CrossRef]
4. Liu, J.X.; Yang, C.H.; Gan, J.J. Stability analysis of road embankment slope subjected to rainfall considering runoff-unsaturated seepage and unsaturated fluid-solid coupling. *Int. J. Civ. Eng.* **2017**, *15*, 865–976. [CrossRef]
5. Tay, Y.Y.; Stewart, D.I.; Cousens, T.W. Shrinkage and desiccation cracking in bentonite-sand landfill liners. *Eng. Geol.* **2001**, *60*, 263–274. [CrossRef]
6. Peron, H.; Laloui, L.; Hueckel, T. Desiccation cracking of soil. *Eur. J. Environ. Civ. Eng.* **2009**, *13*, 869–888. [CrossRef]
7. Xu, G.Y.; Han, X.; Zhang, Y.W. Dam crack image detection model on feature enhancement and attention mechanism. *Water* **2023**, *15*, 64. [CrossRef]
8. Ding, J.X.; Meng, Y.W. Review of applied study for red clay in highway subgrades. In Proceedings of the 4th International Conference on Structures and Building Materials (ICSBM), Guangzhou, China, 15–16 March 2014.
9. Towner, G.D. The mechanics of cracking of drying clay. *J. Agric. Eng. Res.* **1987**, *36*, 115–124. [CrossRef]
10. El Hajjar, A.; Ouahbi, T.; Hamrouni, I. Experimental characterization of desiccation cracking mechanisms in clayey soils. *Eur. J. Environ. Civ. Eng.* **2022**, 1–22. [CrossRef]
11. Hirobe, S.; Oguni, K. Modeling and numerical investigations for hierarchical pattern formation in desiccation cracking. *Phys. D Nonlinear Phenom.* **2017**, *359*, 29–38. [CrossRef]
12. Poulsen, T.G.; Cai, W.L.; Garg, A. Water evaporation from cracked soil under moist conditions as related to crack properties and near-surface wind speed. *Eur. J. Soil Sci.* **2020**, *71*, 627–640. [CrossRef]
13. Chen, K.S. Field test research on red clay slope under atmospheric action. *Adv. Civ. Eng.* **2021**, *2021*, 6668979. [CrossRef]
14. Liu, Y.; Chen, K.S.; Lyu, M.F. Study on failure of red clay slopes with different gradients under dry and wet cycles. *Bull. Eng. Geol. Environ.* **2020**, *79*, 4609–4624. [CrossRef]
15. Nahlawsi, H.; Kodikara, J.K. Laboratory experiments on desiccation cracking of thin soil layers. *Geotech. Geol. Eng.* **2006**, *24*, 1641–1664. [CrossRef]
16. Tang, C.S.; Cui, Y.J.; Tang, A.M. Experiment evidence on the temperature dependence of desiccation cracking behavior of clayey soils. *Eng. Geol.* **2010**, *114*, 261–266. [CrossRef]
17. Rodriguez, R.; Sanchez, M.; Ledesma, A. *Exp.* and numerical analysis of desiccation of a mining waste. *Can. Geotech. J.* **2017**, *44*, 644–658. [CrossRef]
18. Peron, H.; Laloui, L.; Hueckel, T. Experimental study of desiccation of soil. In *UNSAT 2006*; ASCE Geotechnical Special Publication; ASCE: Reston, VA, USA, 2006. [CrossRef]
19. Xie, Y.H.; Zhang, B.H.; Liu, B.C. Shrinkage cracking and strength deterioration of red clay under cyclic drying and wetting. *Alex. Eng. J.* **2022**, *61*, 2574–2588. [CrossRef]
20. Zeng, H.; Tang, C.S.; Zhu, C. Investigating soil desiccation cracking using an infrared thermal imaging technique. *Water Resour. Res.* **2021**, *58*, e2021WR030916. [CrossRef]
21. Costa, S.; Kodikara, J.; Shannon, B. Salient factors controlling desiccation cracking of clay in laboratory experiments. *Geotechnique* **2013**, *63*, 18–29. [CrossRef]
22. Akin, I.D.; Likos, W.J. Brazilian tensile strength testing of compacted clay. *Geotech. Test. J.* **2017**, *40*, 608–617. [CrossRef]
23. Sawada, M.; Sumi, Y.; Mimura, M. Measuring desiccation-induced tensile stress during cracking process. *Soils Found.* **2021**, *61*, 915–928. [CrossRef]
24. Peron, H.; Herchel, T.; Laloui, L. Fundamentals of desiccation cracking of fine-grained soils: Experimental characterization and mechanisms identification. *Can. Geotech. J.* **2009**, *46*, 1177–1201. [CrossRef]
25. Hou, H.J.; Zhu, Z.W.; Wang, B. A practical model study on the mechanism of clay landslide under static loads: From the perspective of major crack-stress-displacement. *Appl. Sci.* **2022**, *12*, 3224. [CrossRef]
26. Tang, C.S.; Shi, B.; Liu, C. Influencing factors of geometrical structure of surface shrinkage cracks in clayey soils. *Eng. Geol.* **2008**, *101*, 204–217. [CrossRef]
27. Lei, D.; Hou, F.; Gong, X. Investigation of deformation at the grain scale in polycrystalline materials by coupling digital image correlation and digital microscopy. *Exp. Tech.* **2012**, *36*, 24–31. [CrossRef]
28. Francic Smrkic, M.; Koscak, J.; Damjanovic, D. Application of 2D digital image correlation for displacement and crack width measurement on RC elements. *Gradevinar* **2018**, *70*, 771–781. [CrossRef]
29. Han, Z.Y.; Xie, S.J.; Li, D.Y. Dynamic fracture analysis of sandstone specimens containing different inclusions. *Theor. Appl. Fract. Mech.* **2022**, *122*, 103623. [CrossRef]
30. Wang, Y.; Cuitino, A.M. Full-field measurements of heterogeneous deformation patterns on polymeric foams using digital image correlation. *Int. J. Solids Struct.* **2002**, *39*, 13–14. [CrossRef]
31. Zhang, D.S.; Arola, D.D. Applications of digital image correlation to biological tissues. *J. Biomed. Opt.* **2004**, *9*, 691–699. [CrossRef]
32. *JTG3430-2020*; Test Methods of Soils for Highway Engineering. Ministry of Transport of the People's Republic of China: Beijing, China, 2020.

33. Ellis, C.L.; Hazell, P. Visual methods to assess strain fields in armour materials subjected to dynamic deformation. *Appl. Sci.* **2020**, *18*, 2644. [CrossRef]
34. Wang, C.; Zhang, Z.Y.; Qi, W. Morphological approach to quantifying soil cracks: Application to dynamic crack patterns during wetting-drying cycles. *Soil Sci. Soc. Am. J.* **2018**, *82*, 757–771. [CrossRef]
35. Miller, C.J.; Mi, H.; Yesiller, N. Experimental analysis of desiccation crack propagation in clay liners. *J. Am. Water Resour. Assoc.* **1998**, *34*, 677–686. [CrossRef]
36. Trabelsi, H.; Chebbi, M.; Guiras, H. Stabilization of clayey soil using fibre reinforcement. In *Unsaturated Soils: UNSAT 2018, Proceedings of the 7th International Conference on Unsaturated Soils, Hong Kong, China, 3–5 August 2018*; The Hong Kong University of Science and Technology (HKUST): Hong Kong, China, 2018; pp. 545–550.
37. An, N.; Tang, C.S.; Xu, S.K. Effects of soil characteristics on moisture evaporation. *Eng. Geol.* **2018**, *239*, 126–135. [CrossRef]
38. Wang, W.W.; Li, W.; Yi, Y. Study on the cracking law of Nanyang expansive soil. *Chin. J. Undergr. Space Eng.* **2015**, *11*, 1437–1443.
39. Zeng, H.; Tang, C.S.; Liu, C.L. Effects of boundary friction and layer thickness on desiccation cracking behaviors of soils. *Chin. J. Geotech. Eng.* **2019**, *41*, 544–553.
40. Lin, L.; Tang, C.S.; Cheng, Q. Desiccation cracking bebavior of soils based on digital image correlation technique. *Chin. J. Geotech. Eng.* **2019**, *41*, 1311–1318. [CrossRef]
41. Wang, L.L.; Tang, C.S.; Shi, B. Nucleation and propagation mechanisms of soil desiccation cracks. *Eng. Geol.* **2018**, *238*, 27–35. [CrossRef]
42. Levatti, H.U.; Prat, P.C.; Ledesma, A. Numerical and experimental study of initiation and propagation of desiccation cracks in clayey soils. *Comput. Geotech.* **2019**, *105*, 155–167. [CrossRef]
43. Lakshmikantha, M.R.; Prat, P.C.; Ledesma, A. Boundary effects in the desiccation of soil layers with controlled environmental conditions. *Geotech. Test. J.* **2018**, *41*, 675–697. [CrossRef]
44. Yin, P.H.; Vanapalli, S.K.; Yu, S. Morphological characteristics of desiccation-induced cracks in cohesive soils: A critical review. *Bull. Eng. Geol. Environ.* **2022**, *81*, 503. [CrossRef]
45. Lecocq, N.; Vandewalle, N. Dynamics of crack opening in a one-dimensional desiccation experiment. *Phys. A Stat. Mech. Its Appl.* **2003**, *321*, 431–441. [CrossRef]
46. Du, C.C.; Zhu, Y.B.; Miao, S.S. The evolution of cracks in the dewatering shrinkage process of hipparion red soil. *Rock Soil Mech.* **2019**, *40*, 3019–3027. [CrossRef]
47. Wang, C.; Zhang, Z.Y.; Liu, Y. Geometric and fractal analysis of dynamic cracking patterns subjected to wetting-drying cycles. *Soil Tillage Res.* **2017**, *170*, 1–13. [CrossRef]

sustainability MDPI

Article

Damping of Dry Sand in Resonant Column-Torsional Simple Shear Device

Majd Ahmad * and Richard Ray

Structural and Geotechnical Engineering Department, Faculty of Architecture, Civil and Transport Sciences Széchenyi István University, 9026 Győr, Hungary; ray@sze.hu
* Correspondence: ahmad.majd@hallgato.sze.hu

Abstract: The damping ratio values of three different Danube sands were measured in the Resonant Column-Torsional Simple Shear device (RC-TOSS). The distinctive configuration of the RC-TOSS device employed in this investigation enabled the performance of both tests using a single sample. This research estimates and compares the damping ratio values measured with three distinct methods (two of which are in the RC test): The Free Vibration Decay (FVD), the Steady-State Vibration (SSV) methods, and the method of calculating the damping ratio from the hysteretic loops generated in the TOSS test. Both dense and loose samples were tested up to a peak-to-peak amplitude shear strain of 1%. The device provides measurements over a wide range of shear strain amplitudes. The results support the employment of the SSV methods at low strains (below 0.005%), while the FVD method gives a better estimate at higher strains (above 0.03%). The two methods and the TOSS results are in agreement with each other between 0.005% and 0.03%. The effect of the number of cycles on the damping ratio was investigated where a significant decrease was observed in the damping ratio with an increasing number of cycles. A parameter is introduced to describe the rate of this decrease, which should be considered during the structural design to reduce maintenance and life-cycle costs and enhance sustainability.

Keywords: resonant column test; torsional simple shear test; damping ratio; steady-state vibration; free vibration decay

Citation: Ahmad, M.; Ray, R. Damping of Dry Sand in Resonant Column-Torsional Simple Shear Device. *Sustainability* **2023**, *15*, 11060. https://doi.org/10.3390/su151411060

Academic Editors: Jian Zhou, Mahdi Hasanipanah and Danial Jahed Armaghani

Received: 31 May 2023
Revised: 3 July 2023
Accepted: 12 July 2023
Published: 14 July 2023

1. Introduction

When waves propagate through soil during cyclic loading, complex mechanisms contribute to the loss of energy in the material. Fluid flow loss and inelastic friction loss are mainly responsible for energy dissipation in soil [1]. The dominant mechanism depends on the strain level and nonlinear behavior of soil. The damping ratio describes energy dissipation in a system. This parameter plays an important role when modeling dynamic geotechnical and structural problems (e.g., soil response analysis and soil–structure interaction problems). Thus, a substantial effort has been made in the last few decades to introduce experimental dynamic laboratory procedures and in situ methods to study the damping behavior of soils for a wide range of shear strain amplitudes. Resonant Column (RC) and Torsional Simple Shear (TOSS) tests have been developed and improved to measure the damping ratio with reasonable accuracy. The effects of various soil parameters on damping have been further investigated over the years. The most impactful factors are the void ratio, confining pressure, particle shape, and number of loading cycles [2–4].

Damping in materials may take several forms. In soils, geotechnical engineers are mostly concerned with hysteretic and viscous damping, which are proportional to displacement and velocity, respectively. While vibration frequency affects viscous damping, it seems to have no impact on the hysteretic damping. These two forms of damping are results of distinct mechanisms.

Soils have a tendency to dissipate energy even when subjected to low levels of strain [5]. Soils loaded cyclically below the linear strain threshold (γ_{tl}) exhibit an elastic behavior.

Hysteresis loops do not generate at strain amplitudes lower than that threshold. According to Kokusho [6], damping is a consequence of the fluids present in the voids, suggesting that the loss of fluid (viscous damping) is the main mechanism behind the soil's damping at small strain levels (below γ_{tl}). Conversely, beyond γ_{tl}, the shear stress–strain curves start to form hysteresis loops when soil is cyclically loaded. This is attributed to the nonlinear behavior of soil at higher strains. A rise in hysteretic damping is observed as the strain amplitude increases, and most of the energy dissipation is a result of inelastic friction. Thus, viscous damping is marginal, and the nature of damping is hysteretic [7,8].

Resonant column testing typically involves two distinct methods for measuring the damping of materials: The Steady-State Vibration (SSV) and the Free Vibration Decay (FVD) [9,10]. Several studies compared these two methods and provided suggestions on their scopes and limitations [11–13]. However, the unique design of the RC-TOSS device used in this study allows for conducting both tests on the same sample. This feature provided an accurate comparison between the three methods of measuring the damping ratio. These three methods are the SSV, FVD, and the method of calculating the damping ratio from the hysteretic loops generated in the TOSS test. This eliminates any potential effect of sample preparation on the results, which facilitates a more precise comparison. Since the damping ratio of soil influences the dynamic response of structures, a better understanding and more accurate estimation of the damping ratio can lead to a reduction in the structural demands and potentially increase the safety and resilience of the design. This can contribute to the long-term sustainability of the structure by preventing or minimizing damage.

2. Testing Device, Materials, and Methods

Hollow, cylindrical soil samples were tested using a combined Resonant Column-Torsional Simple Shear device (RC-TOSS). The samples have an outer diameter of 6 cm, an inner diameter of 4 cm, and a height of 14 cm. A drive head applies torsional loading on the top of the sample (the free end), while the bottom of the specimen is fixed, as shown in Figure 1. The drive system consists of a set of two magnets inserted inside four coils. Two proximitors are fixed on the measurement post to measure the gap between the sensors and targets attached to the specimen. The difference between the two gaps gives the rotation from which the shear strain is found at each loading step. Torque is calculated from the known current passing through the coils during the test. The RC test requires displacement measurements with very high accuracy at very low amplitude ($\gamma = 10^{-4}\%$). An accelerometer mounted on the drive head provides such accurate measurements. A multimeter and an oscilloscope read and record the response curve, resonance frequency, and decaying response of the sample. The device is capable of conducting RC and TOSS tests on the same sample. A detailed description of the device and the methods of calibration are present in [3,14].

Three different types of Danube sands were selected for testing. Samples A and B retained a very low percentage of fines, while the fine content for sample C was 21.11%. The soil comprised fluvial sediments of the Danube River, extracted from a location near the channel of the river. The particles were sub-angular to rounded. The properties of the samples are shown in Table 1.

A vacuum applied a confining pressure of 97 KPa on the tested samples. The preparation of dense specimens was performed via the pluviation method. The soil was poured slowly from a height of 50 cm to achieve the thinnest thread of soil possible. On the other hand, to prepare samples in a loose state, the dry filling with tamping method was applied. Dry soil was poured through a glass funnel with a 14 cm spout into the mold. Pouring started with the spout touching the base ring, and then the spout was constantly and carefully raised and moved around to touch the surface of the soil. The testing plan appears in Table 2.

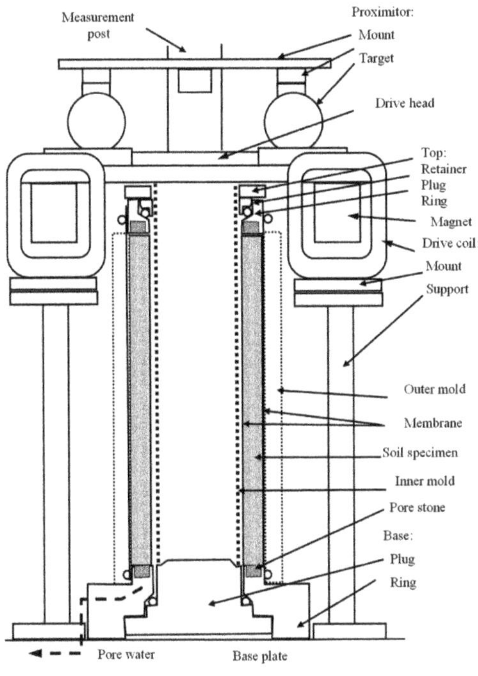

Figure 1. Cross-section of the RC-TOSS testing device.

Table 1. Tested soil properties.

Sample ID	Mean Particle Diameter	Effective Particle Diameter	Uniformity Coefficient	Fines Content	Max Void Ratio	Min Void Ratio	Liquid Limit for Fines	Plastic Limit for Fines	Plastic Index for Fines
	d_{50} [mm]	d_{10} [mm]	C_u [-]	FC [%]	e_{max} [-]	e_{min} [-]	w_l [%]	w_p [%]	I_p [%]
A	0.211	0.109	2.06	7.56	0.81	0.52	-	-	-
B	0.243	0.130	2.18	5.69	0.79	0.516	-	-	-
C	0.107	0.013	9.85	21.11	0.9	0.524	30.4	19.7	10.7

Table 2. Testing program.

Test #	Sample ID	Void Ratio	Relative Density	Maximum Shear Stress
		e [-]	D_r [-]	G_{max} [KPa]
1	A	0.77	0.14	84,500
2	A	0.57	0.83	115,000
3	B	0.73	0.22	79,700
4	B	0.58	0.77	101,500
5	C	0.85	0.13	69,300
6	C	0.63	0.72	88,000

2.1. Damping in the Torsional Simple Shear Device

Around load reversal points throughout cyclic torsional loading, the soil recovers its stiffness due to the reengagement and interlocking of the previously slipped contacts between particles in the opposite direction. The shear stress–strain path forms a hysteresis loop as demonstrated in Figure 2a. The gradient of the straight line that joins the endpoints of the hysteresis loop indicates the typical level of shear stiffness exhibited by the soil, which is also referred to as the secant shear modulus (*Gsec*). Although soil damping is recognized to be hysteretic in nature, it is commonly substituted with equivalent viscous damping in most analyses for the sake of mathematical convenience. "The equivalent viscous damping is determined in such a manner as to yield the same dissipation of energy per cycle as that produced by the actual damping mechanism" [15]. The damping ratio (*D*) represents the ratio of the energy absorbed in one cycle of vibration to the potential energy at maximum displacement in that cycle [16]. The following equation calculates the equivalent damping ratio *D*:

$$D = \frac{A_L}{4\pi * A_T} \tag{1}$$

where A_L is the area of the loop (shadowed in Figure 2a) and A_T is the area of the triangle bounded by a straight line defining the secant modulus at the point of maximum strain (Figure 2a).

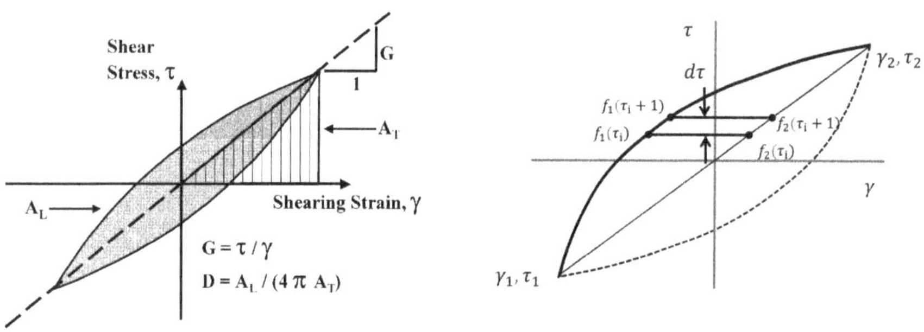

(**a**) Dynamic properties from hysteretic loops (**b**) Integration element for half a loop

Figure 2. Hysteretic loops due to cyclic loading in TOSS test.

Subroutines in Visual Basic for Applications (VBA) in Excel were created to calculate the area between the loop and line that connects its two ends. The subroutine divides this area into very small slices that can be considered trapezoids (Figure 2b). The sum of the areas multiplied by two denotes A_L in Equation (1).

The proximitors incorporated in the TOSS test have low accuracy compared to the accelerometer measurements in the RC test. Therefore, TOSS tests in this study allow for damping ratio measurements at strain amplitudes above 0.02%.

2.2. Damping in the Resonant Column Device

2.2.1. The Free Vibration Decay Method (FVD)

The resonant column test starts by applying a torsional oscillation on the top of the specimen with an increasing amplitude sine wave. For each tested strain amplitude, the frequency progressively increases to obtain the dynamic response of the specimen. The frequency that corresponds to the maximum response amplitude is the resonance frequency of the sample. The dynamic shear modulus is a function of the fundamental frequency based on wave propagation and elasticity theories. Once the operator finds the resonance frequency, they cut the power and track the decaying response on a storage oscilloscope.

A multimeter records the decaying curve on an Excel sheet where damping is determined from the logarithmic decrement equation [5]:

$$\delta = \frac{1}{N} ln \frac{Z_1}{Z_{1+N}} = \frac{2\pi D}{\sqrt{1 - D^2}} \quad (2)$$

where δ is the logarithmic decrement, N is the number of cycles, Z_1 is the first amplitude, Z_{1+N} is the amplitude after N cycles (Figure 3a), and D is the damping ratio.

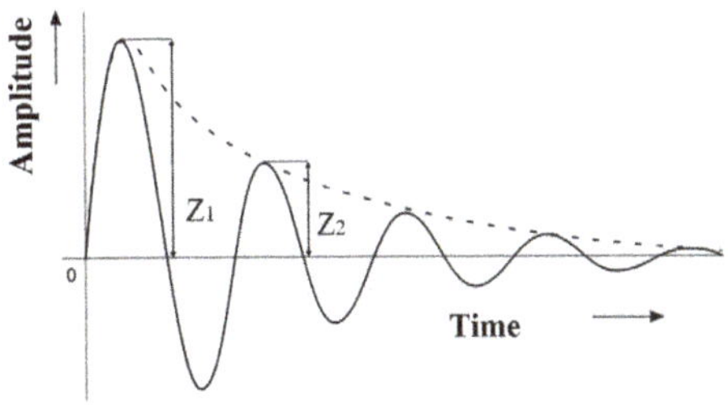

(**a**) Decay-response curve for FVD method

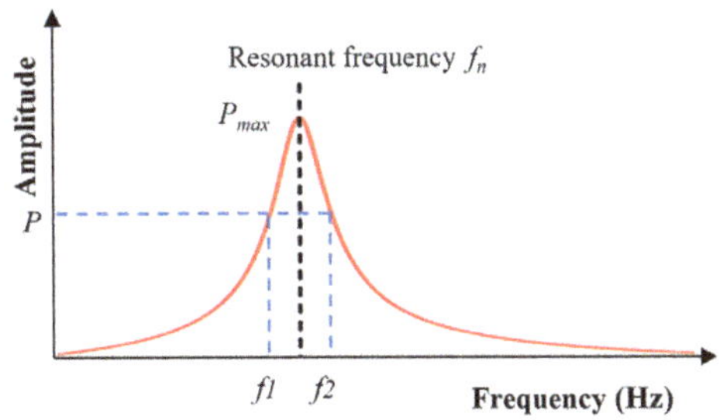

(**b**) Steady-State Vibration method

Figure 3. Damping in Resonant Column test.

For small values of the damping ratio, as found in soil, $\sqrt{1 - D^2} \cong 1$, the damping expression becomes:

$$D = \frac{\delta}{2\pi} = \frac{1}{2N\pi} ln \frac{Z_1}{Z_{1+N}} \quad (3)$$

Ray [14] suggests using a small number of cycles (N) when driving at high amplitude. A larger N would cause a reduction in strain amplitude by a factor of about 3 over the measurement interval. ASTM D4015 [17] recommends the use of less than 10 cycles. Mog and Anbazhagan [13] conducted resonant column tests using the GCTS device to investigate the effect of the number of successive cycles (N) used in measuring the damping ratio. They observed an increase in the damping ratio when increasing the number of cycles up to 10 cycles. Nonetheless, beyond 10 cycles, the damping ratio starts to decrease for a higher number of cycles in the measurements (i.e., for 20, 30, and 50 cycles). Damping ratio measurements determined by this method scatter significantly. Therefore, 2 or 3 successive cycles in the decay response curve should determine the damping ratio in the RC test.

2.2.2. The Steady-State Vibration Method (SSV)

Initially, the SSV method (also termed the half-power bandwidth method) was created to calculate the modal damping ratio (ξ) of a structure by measuring the width of peaks in its frequency response function. However, it could be used to measure the damping ratio of soil during the RC test. In this technique, the width of the frequency-response curve near the resonance determines the logarithmic decrement (δ). The half-power bandwidth ($\Delta\omega$) is the width of the peak where the magnitude of the frequency corresponds to $1/\sqrt{2} * P_{max}$ [18]. Equation (4) determines δ (GCTS-CATS, 2007):

$$\delta = \frac{\pi \left(f_2^2 - f_1^2\right)}{2 f_r^2} * \sqrt{\frac{P^2}{P_{max}^2 - P^2}} \frac{\sqrt{1 - 2D^2}}{1 - D^2} \quad (4)$$

where f_1 and f_2 are frequencies below and above the resonance where the strain amplitude is P, P_{max} is the maximum amplitude (or resonant amplitude), and f_r is the resonant frequency (Figure 3b).

Since the damping in soils is small and the amplitude P is $P_{max}/\sqrt{2}$, Equation (4) becomes:

$$\delta \cong \frac{\pi(f_2 - f_1)}{f_r} \tag{5}$$

Then, the damping ratio can be expressed as:

$$D \cong \frac{(f_2 - f_1)}{2f_r} \tag{6}$$

According to Mog and Anbazhagan [13], at strain levels below 0.005%, there can be a variation of up to 15% in the damping ratio when measured using the SSV method and FVD method with two consecutive cycles. However, the variation can be even greater (up to 50%) when a higher number of successive cycles (3, 7, and 10 cycles) are used. These results agree with Senetakis et al. [11] where a scatter of 15% was also reported. The ambient noise and the number of applied cycles during the RC test justify this scatter. A much higher number of cycles is needed in the SSV method to plot the frequency-response curve [19].

3. Results and Discussions

Every test starts with very small strain RC measurements to obtain the maximum shear modulus (G_{max}) and minimum damping ratio (D_{min}). The RC tests continue with ascending strain amplitudes up to the volumetric-threshold shearing strain (γ_{tv}), which is around 0.01% for sand. Below γ_{tv}, the behavior is nonlinear but still elastic and there is no effect of the cyclic loading on the dynamic behavior of soil [3,20,21]. Next, cyclic TOSS tests load the sample for two cycles at progressively higher stress levels (5–10–15–20–25–30–35–40–45–50 KPa). RC tests are continued after reaching the maximum shear strain that can be measured using the proximitors (just below 1% peak-to-peak strain). The six RC tests provided a total of 115 damping measurements. This allowed for comparison with 54 data points from the TOSS test.

Typical response curves for the SSV method appear in Figure 4 for ascending strain levels. At higher strain amplitudes (above 0.1%), dynamic instability prevented precise response data from being recorded by the multimeter. For the FVD method, however, very small strain levels inhibited accurate damping measurements, and relatively accurate decay-response curves required several attempts. During the disconnection from drive coils, an electric pulse often produces a strong transient. This pulse creates noise that distorts the readings of the most critical first few cycles, since damping is calculated from the first three cycles before strain levels decrease substantially. This noise did not affect the response at strain amplitudes higher than 0.005%, and the peaks could be spotted easily for the first few cycles. Figure 5 compares the damping ratio measurements of the three methods.

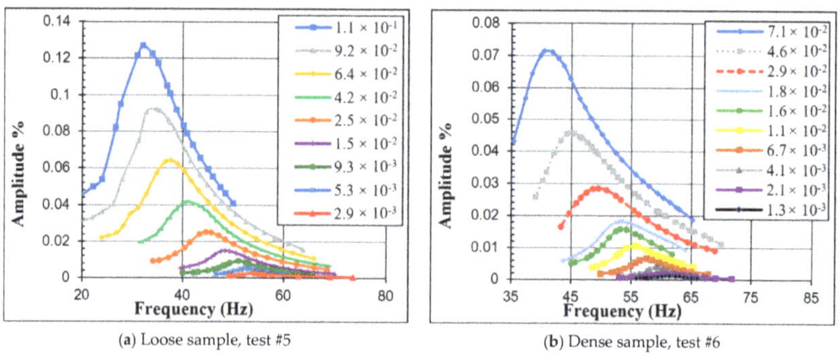

Figure 4. Response curves in the RC test for Sample C.

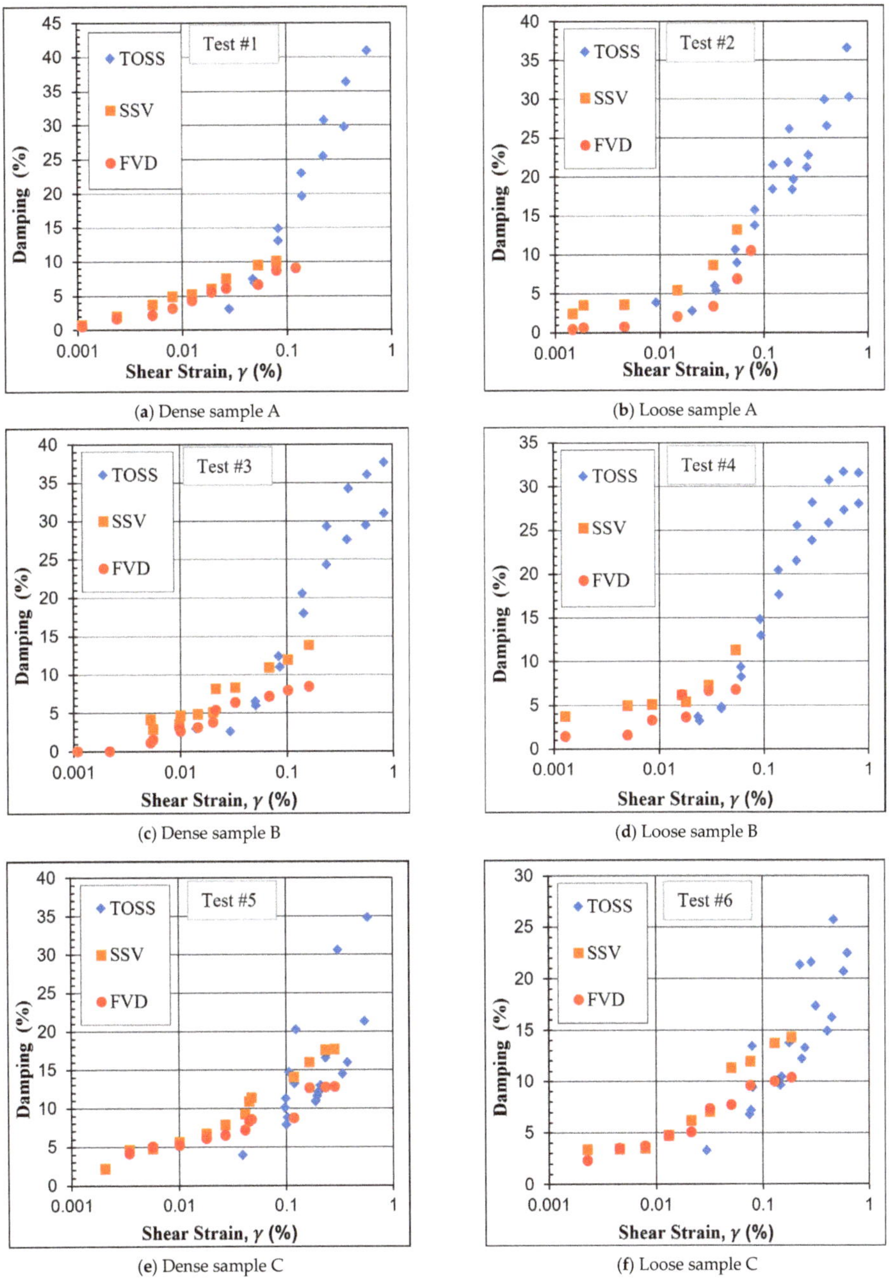

(**a**) Dense sample A (**b**) Loose sample A (**c**) Dense sample B (**d**) Loose sample B (**e**) Dense sample C (**f**) Loose sample C

Figure 5. Damping results of the six tests with the three methods.

The SSV method generally provides higher damping values than the FVD method, especially at higher strain levels. At the same time, there is a closer agreement between the two methods at medium strain amplitudes (0.005–0.03%). At higher strain amplitudes, the frequency-response curve becomes asymmetric, compromising the accuracy of the damping ratio calculations in the SSV method and causing an overestimation of their values. Such behavior also occurs when comparing with the TOSS tests, as damping values obtained from the TOSS tests agree more with the FVD method in the RC test at the strain levels where the SSV method is questionable.

Due to the nature of the TOSS test measurements and insufficient accuracy of the proximitors at very small strain amplitudes, comparisons between the RC and TOSS tests become difficult at strain levels below 0.02%. However, the two tests are in agreement and their combined response can generate damping-ratio curves for a wide range of strains.

Multiple effects and uncertainties cause the scatter seen in the damping measurements. The sources of variability include the decreasing strain amplitude with time (FVD), response curve asymmetry (SSV), and the number of cycles and stiffening behavior in the TOSS test. Such effects cause difficulties when comparing the damping ratio obtained from different methods. As a result, they produce a scatter that may exceed 40% in the data when comparing the SSV and FVD methods, as seen in Figure 6.

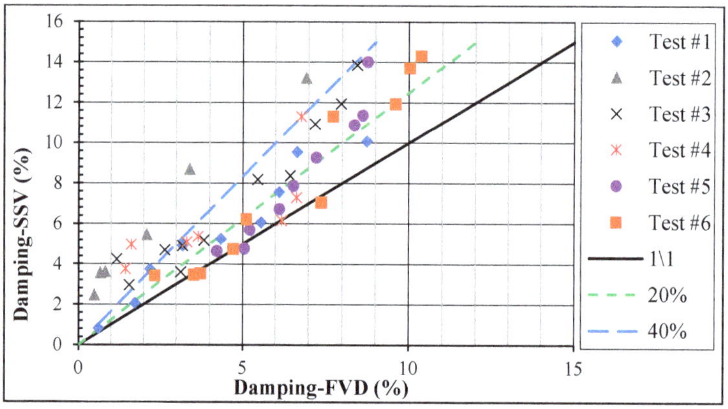

Figure 6. Comparison of damping between SSV and FVD for all samples.

The best fit for all the damping ratio data points from the six tests using the three methods appears in Figure 7, along with a comparison with two earlier studies [22,23] on dry sand.

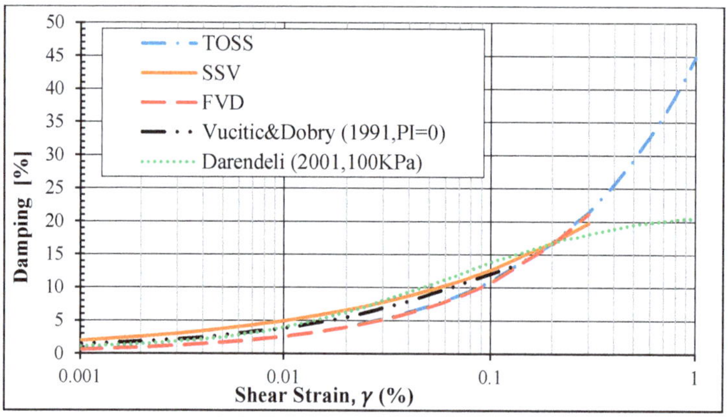

Figure 7. A fit of damping data points for each method.

3.1. Damping Ratio Correlation

To estimate the damping ratio with correlations, it is common practice to relate it with the normalized shear modulus using a second-degree polynomial of (G/G_{max}) [11,24,25].

The modified hyperbolic model based on the Hardin–Drnevich model provides an excellent presentation of the modulus-reduction curve. This model employs two curve-fitting parameters:

$$\frac{G}{G_{max}} = \frac{1}{1 + \left(\frac{\gamma}{\gamma_r}\right)^a} \tag{7}$$

The reference shear strain γ_r in this model is defined as the strain amplitude when the shear modulus reduces to one-half of G_{max} [23]. The fit for all the samples is shown in Figure 8a for γ_r = 0.1 and a = 0.974.

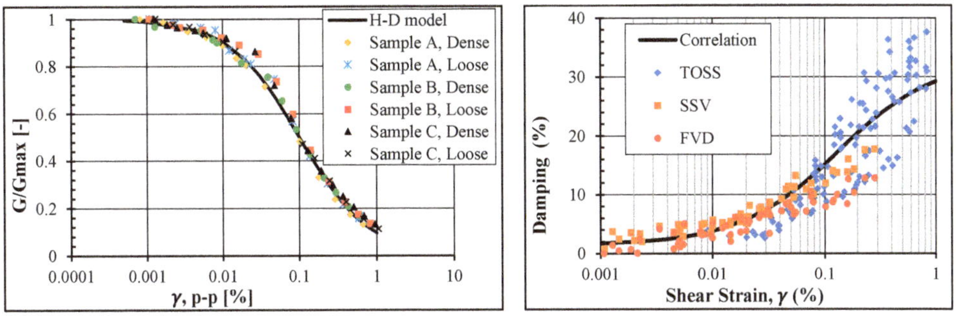

(**a**) Shear modulus (**b**) Damping ratio

Figure 8. Shear modulus and damping ratio correlations fit with test data.

After finding the correlation for the shear-modulus-degradation curve, the following equation models the damping ratio increase with increasing strain amplitude:

$$D = C_1\left(\frac{G}{G_{max}}\right)^2 + C_2\left(\frac{G}{G_{max}}\right) + C_3 \tag{8}$$

where C_1, C_2, and C_3 are curve-fitting constants. The constants are found using the least square method, by minimizing the summation of the squared errors between the equation and the lab measurements. For the damping ratio measurement in this study for all tested samples, the equation becomes:

$$D = 8.5\left(\frac{G}{G_{max}}\right)^2 - 40\left(\frac{G}{G_{max}}\right) + 33 \tag{9}$$

Figure 8b demonstrates how the equation fits the RC-TOSS damping measurements.

3.2. Effect of Torsional Cyclic Loading on Damping

After exceeding the volumetric-shear-strain threshold (γ_{tv}), the cyclic loading causes an increase in soil stiffness [26–28]. Figure 9 of the shear-stress–strain curves for 100 cycles shows how the hysteresis cycles decrease in size. Therefore, the damping ratio decreases with the increasing number of cycles. Several authors have detected this effect [6,29,30]. However, most studies have focused on the effect of low-strain cyclic vibration with a high number of cycles on the dynamic properties of soil. No agreement has been found regarding the effect of high-strain-amplitude cyclic loading. Thus, in this study, cyclic TOSS tests were conducted with increasing stress levels ranging between 10 KPa to 60 KPa with a maximum peak-to-peak strain amplitude of 1%. Each test applies 100 cycles where the damping ratio is calculated for all cycles.

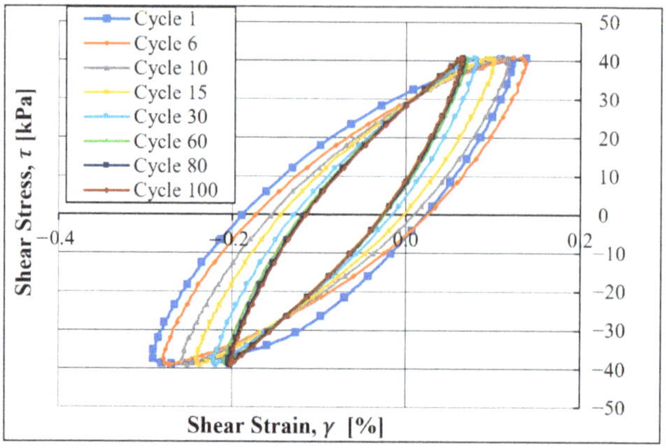

Figure 9. The stiffening behavior of the sample during cyclic torsional loading.

The decrease in the damping ratio with the increasing number of cycles (N) at different stress levels is shown in Figure 10 for sample C at two different states; (a) is a loose sample and (b) is a dense sample. It is difficult to estimate the rate of change for each stress level due to the sequence of the TOSS test and cyclic loading that the sample has experienced at lower stress levels. However, the higher the stress level, the higher the effect of (N) on damping up to a specific limit. Beyond this, the stress level seems to have a minor influence on the rate of decrease in the damping ratio. Specimens with a higher void ratio or lower relative density exhibit a more significant reduction in damping ratio. For instance, at stress levels of 20 KPa and 30 KPa, the damping ratio decreased by 50% and 54% after 100 cycles for a dense sample, while a loose sample decreased by 61% and 64% (Figure 11). D_N is the damping ratio calculated from a cyclic TOSS test at the Nth cycle, and D_1 is the damping ratio of the first cycle (maximum damping ratio at that stress level). Results show that $\log(D_N/D_1)$ is linearly proportional to $\log(N)$, and the slope of $\log(D_N/D_1) - \log(N)$ plot represents a parameter (r) that describes the rate of decrease in damping ratio with the number of cycles.

$$r = \frac{log(D/D_1)}{log(N)} \tag{10}$$

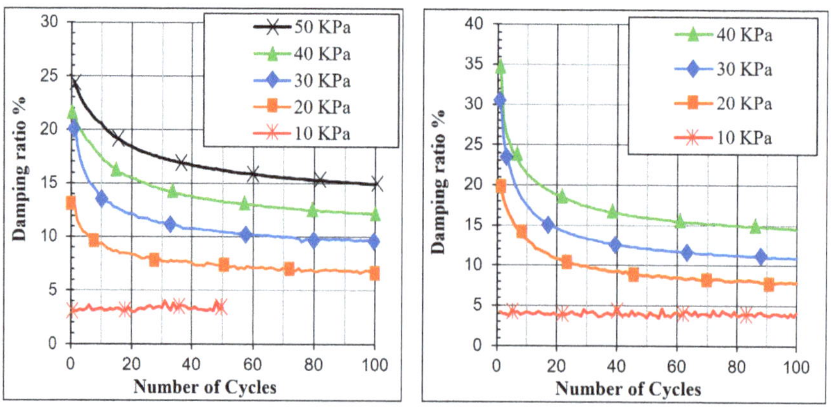

(**a**) Dense sample C (**b**) Loose sample C

Figure 10. The decrease in damping with increasing number of cycles in TOSS test.

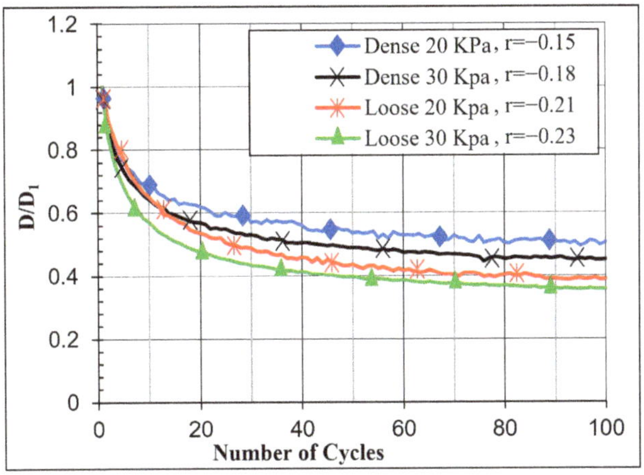

Figure 11. The rate of decrease in damping during cyclic loading for Sample C.

The values of the parameter (*r*) also vary in a manner similar to damping ratio measurements, and it is difficult to find a pattern in the change in (*r*) with the increasing strain amplitude. Generally, (*r*) takes values between (−0.15 and −0.24) for strain amplitudes above 0.1%.

4. Conclusions

Several methods are in use nowadays to measure the damping ratio in practical applications. The scatter in the measurements is typically present in the literature due to uncertainties in the strain amplitudes, frequency-response curve symmetry, and the effect of the number of cycles. Luckily, the impact of this scatter is marginal in dynamic problems, such as site response analysis. Previous studies have not provided the opportunity to compare the measurement methods for damping ratios in both RC and TOSS tests. However, this study benefits from the exceptional design of the device, which enables the execution of both tests on a single sample. This allows for a direct comparison of the results without the interference of external factors, disturbances, or sample preparation. Steady-State Vibration (SSV), Free Vibration Decay (FVD), and damping in the Torsional Simple Shear device have been investigated and compared for three types of Danube sands with different relative densities. Based on the results, it is recommended to use the SSV method at very low strain amplitudes (below 0.005%) where the response curve is symmetrical. For this strain range, the FVD measurements are unreliable due to the noise created by the device when turning off the drive power. At medium strain levels (0.005%–0.03%), the two methods and the TOSS results agree with each other. However, at higher strain amplitudes (above 0.03%), it is advised to employ the FVD method to measure the damping ratio, due to the reduced accuracy of the SSV method. The SSV-response curves are not symmetrical anymore, which compromises the use of the half-bandwidth method, causing an overestimation in the damping ratio values. Achieving a more precise estimation of the damping ratio of soil leads to improving the performance and safety of structures under dynamic loads, which can contribute to the longevity of the design, reducing the overall environmental footprint associated with maintenance.

The effect of the shear stiffening behavior during cyclic loading in the TOSS test was investigated. After exceeding the volumetric shear strain threshold, a considerable decrease in the damping ratio was observed. This trend is more evident in loose samples where damping can decrease by up to 64% after 100 cycles compared to a 54% decrease in dense samples for the same stress level of the applied cycles. Considering this decrease during the design phase can improve the analysis of the dynamic response of structures to reach a more resilient design of the infrastructure when subjected to dynamic loading, this can reduce the need for repairs or reconstruction after seismic events, thereby minimizing resource consumption and waste generation, contributing to long-term sustainability.

Author Contributions: Conceptualization, M.A. and R.R.; methodology, M.A. and R.R.; software, R.R.; investigation, M.A.; resources, R.R.; writing—original draft preparation, M.A.; writing—review and editing, R.R.; visualization, M.A.; supervision, R.R. All authors have read and agreed to the published version of the manuscript.

Funding: This research is funded by the Stipendium Hungaricum Scholarship, grant number 192800.

Data Availability Statement: The data presented in this study can be provided by the author upon request.

Conflicts of Interest: The authors declare no conflict of interest.

References

1. White, J.E. *Underground Sound: Applications of Seismic Waves*; Elsevier: Amsterdam, The Netherlands, 1983; pp. 83–137.
2. Payan, M.; Senetakis, K.; Khoshghalb, A.; Khalili, N. Influence of particle shape on small-strain damping ratio of dry sands. *Geotechnique* **2016**, *66*, 610–616. [CrossRef]
3. Szilvágyi, Z. Dynamic Soil Properties of Danube Sands. Ph.D. Thesis, Széchenyi István University, Győr, Hungary, 2017.
4. Upreti, K.; Leong, E.C. Effect of mean grain size on shear modulus degradation and damping ratio curves of sands. *Géotechnique* **2021**, *71*, 205–215.
5. Kramer, S.L. *Geotechnical Earthquake Engineering*; Prentice Hall: Upper Saddle River, NJ, USA, 1996; p. 643, ISBN 978-0133749434.
6. Kokusho, T. In Situ Dynamic Soil Properties and Their Evaluations. In Proceedings of the 8th Asian Regional Conference on Soil Mechanics and Foundation Engineering, Kyoto, Japan, 20–24 July 1987.
7. Hardin, B. The Nature of Damping in Sands. *J. Soil Mech. Found. Div.* **1965**, *91*, 63–98. [CrossRef]
8. Dobry, R. *Damping in Soils: Its Hysteretic Nature and the Linear Approximation*; M.I.T. Department of Civil Engineering, Inter-American Program: Cambridge, MA, USA, 1970.
9. Facciorusso, J. An archive of data from resonant column and cyclic torsional shear tests performed on Italian clays. *Earthq. Spectra* **2020**, *73*, 545–562. [CrossRef]
10. Xu, Z.; Tao, Y.; Hernandez, L. Novel Methods for the Computation of Small-Strain Damping Ratios of Soils from Cyclic Torsional Shear and Free-Vibration Decay Testing. *Geotechnics* **2021**, *1*, 330–346. [CrossRef]
11. Senetakis, K.; Anastasiadis, A.; Pitilakis, K. A comparison of material damping measurements in resonant column using the steady-state and free-vibration decay methods. *Soil Dyn. Earthq. Eng.* **2015**, *74*, 10–13. [CrossRef]
12. Yu, S.; Shan, Y. Experimental Comparison and Study on Small-Strain. *Geotech Geol. Eng.* **2017**, *35*, 2479–2483. [CrossRef]
13. Mog, K.; Anbazhagan, P. Evaluation of the damping ratio of soils in a resonant column using different methods. *Soils Found.* **2022**, *62*, 101091. [CrossRef]
14. Ray, R.P. Changes in Shear Modulus and Damping in Cohesionless Soils Due to Repeated Loading. Ph.D. Thesis, University of Michigan, Ann Arbor, MI, USA, 1984; p. 417. [CrossRef]
15. Bae, Y.-S. Modeling Soil Behavior in Large Strain Resonant Column and Torsional Shear Tests. Ph.D. Thesis, Utah State University, Logan, UT, USA, 2007.
16. Richart, F.E.; Hall, J.R.; Woods, R.D. Vibrations of soils and foundations. Prentice-Hall: Upper Saddle River, NJ, USA, 1970.
17. *ASTM D4015*; Standard Test Methods for Modulus and Damping of Soils by Resonant-Column Method. Annual Book of ASTM Standards, USA; ASTM International: West Conshohocken, PA, USA, 1992.
18. Chopra, A.K.; Structures, D.O. *Theory and Applications to Earthquake Engineering*, 3rd ed.; Prentice Hall: Upper Saddle River, NJ, USA, 2007.
19. Meng, J. The Influence of Loading Frequency on Dynamic Soil Properties. Ph.D. Thesis, Georgia Institute of Technology, Atlanta, GA, USA, 2003.
20. Georgiannou, V.N.; Rampello, S.; Silvestri, F. Static and dynamic measurements of undrained stiffness of natural overconsolidated clays. In *Deformation of Soils and Displacement of Structures*; Balkema: Florence, Italy, 1991.
21. Vucetic, M. Cyclic threshold shear strains in soils. *J. Geotech. Eng.* **1994**, *120*, 2208–2228. [CrossRef]
22. Vucetic, M.; Dobry, R. Effect of soil plasticity on cyclic response. *J. Geotech. Eng.* **1991**, *117*, 89–107. [CrossRef]
23. Darendeli, B. Development of a New Family of Normalized Modulus Reduction and Material Damping Curves. Ph.D. Thesis, University of Texas at Austin, Austin, TX, USA, 2001; pp. 1–362.
24. Ishibashi, I.; Zhang, X. Unified dynamic shear modulus and damping ratios of sand and clay. *Soils Found. Jpn. Soc. Soil Mech. Found. Eng.* **1993**, *33*, 182–191. [CrossRef]
25. Zhang, J.; Andrus, R.; Juang, C. Normalized Shear Modulus and Material Damping Ratio Relationships. *J. Geotech. Geoenvironmental. Eng.* **2005**, *131*, 453–464. [CrossRef]
26. Hardin, B.; Drnevich, V. Shear modulus and damping in soils: Design equations and curves. *J. Soil Mech. Found. Div.* **1972**, *98*, 667–692. [CrossRef]

27. Iwasaki, T.; Tatsuoka, F.; Takagi, Y. Shear moduli of sands under cyclic torsional shear loading. *Soils Found.* **1978**, *18*, 39–50. [CrossRef]
28. Ray, R.; Woods, R. Modulus and damping due to uniform and variable cyclic loading. *J. Geotech. Eng.* **1988**, *114*, 861–876. [CrossRef]
29. Silver, M.L.; Seed, H.B. Volume changes in sands during cyclic loading. *J. Soil Mech. Found. Div.* **1971**, *97*, 1171–1182. [CrossRef]
30. Edil, T.B.; Luh, G.F. Dynamic modulus and damping relationships for sands. In Proceedings of the ASCE Geotechnical Engineering Division Specialty Conference, Pasadena, CA, USA, 19–21 June 1978.

sustainability

MDPI

Article

Appraisal of Different Artificial Intelligence Techniques for the Prediction of Marble Strength

Muhammad Saqib Jan [1,2,†], Sajjad Hussain [1,2,†], Rida e Zahra [2], Muhammad Zaka Emad [3], Naseer Muhammad Khan [4,*], Zahid Ur Rehman [2], Kewang Cao [1,5,*], Saad S. Alarifi [6], Salim Raza [2], Saira Sherin [2] and Muhammad Salman [7]

1 School of Art, Anhui University of Finance & Economics, Bengbu 233030, China
2 Department of Mining Engineering, University of Engineering and Technology, Peshawar 25000, Pakistan
3 Department of Mining Engineering, University of Engineering and Technology, Lahore 39161, Pakistan
4 Department of Sustainable Advanced Geomechanical Engineering, Military College of Engineering, National University of Sciences and Technology, Risalpur 23200, Pakistan
5 School of Civil, Environmental and Architectural Engineering, Korea University, 145, Anam-ro, Seongbuk-gu, Seoul 02841, Republic of Korea
6 Department of Geology and Geophysics, College of Science, King Saud University, P.O. Box 2455, Riyadh 11451, Saudi Arabia
7 Department of Civil Engineering, University of Engineering and Technology, Peshawar 25000, Pakistan
* Correspondence: nmkhan@mce.nust.edu.pk (N.M.K.); tb18220001b0@cumt.edu.cn (K.C.)
† These authors contributed equally to this work.

Abstract: Rock strength, specifically the uniaxial compressive strength (UCS), is a critical parameter mostly used in the effective and sustainable design of tunnels and other engineering structures. This parameter is determined using direct and indirect methods. The direct methods involve acquiring an NX core sample and using sophisticated laboratory procedures to determine UCS. However, the direct methods are time-consuming, expensive, and can yield uncertain results due to the presence of any flaws or discontinuities in the core sample. Therefore, most researchers prefer indirect methods for predicting rock strength. In this study, UCS was predicted using seven different artificial intelligence techniques: Artificial Neural Networks (ANNs), XG Boost Algorithm, Random Forest (RF), Support Vector Machine (SVM), Elastic Net (EN), Lasso, and Ridge models. The input variables used for rock strength prediction were moisture content (MC), P-waves, and rebound number (R). Four performance indicators were used to assess the efficacy of the models: coefficient of determination (R^2), Root Mean Square Error (RMSE), Mean Square Error (MSE), and Mean Absolute Error (MAE). The results show that the ANN model had the best performance indicators, with values of 0.9995, 0.2634, 0.0694, and 0.1642 for R^2, RMSE, MSE, and MAE, respectively. However, the XG Boost algorithm model performance was also excellent and comparable to the ANN model. Therefore, these two models were proposed for predicting UCS effectively. The outcomes of this research provide a theoretical foundation for field professionals in predicting the strength parameters of rock for the effective and sustainable design of engineering structures

Keywords: marble strength; direct and indirect methods; correlations analysis; artificial intelligence techniques; performance indicators

Citation: Jan, M.S.; Hussain, S.; e Zahra, R.; Emad, M.Z.; Khan, N.M.; Rehman, Z.U.; Cao, K.; Alarifi, S.S.; Raza, S.; Sherin, S.; et al. Appraisal of Different Artificial Intelligence Techniques for the Prediction of Marble Strength. *Sustainability* **2023**, *15*, 8835. https://doi.org/10.3390/su15118835

Academic Editors: Zubair Baig and Gwanggil Jeon

Received: 22 February 2023
Revised: 17 May 2023
Accepted: 26 May 2023
Published: 30 May 2023

1. Introduction

Uniaxial compressive strength (UCS) is an integral parameter significantly used in the effective and sustainable design of tunnels and other engineering structures in civil and mining engineering [1–12]. UCS is determined using direct and indirect methods. The direct method proceeded according to ISRM and ASTM [13–18] standards on rock samples in the laboratory that involve (1) acquisition of a standard high-quality or NX core sample (54 mm dia) of rock (does not contain any cracks), (2) preparation of the core sample for flatness of both ends within 0.02 mm and parallelism of the upper and lower surfaces within 0.05 mm, and (3) applying load on the sample using a Universal Testing Machine (UTM) machine at the rate of 0.5–1.0 MPa/s. However, extracting high-quality core samples from a weak and jointed rock is challenging, and the arduous laboratory testing procedure may provide hurdles during the determination of UCS [19]. It takes time and is costly to safely execute UCS in a lab test. The results obtained may be questionable in the case of the presence of discontinuities in the core sample [20,21]. Therefore, most researchers often prefer indirect methods for estimating UCS. Indirect methods include predictive models developed by different researchers based on mineralogical–petrographic analyses and physical and index properties of rock for estimation of UCS [22–25]. When compared to the direct technique, indirect methods for estimating UCS are quicker, more convenient, and less costly [26]. For easy understanding, indirect methods are divided into conventional predictive methods and soft computing methods for the estimation of UCS [27]. The conventional predictive methods use statistical techniques, i.e., simple and multilinear regression modeling. These have been used successfully for predicting UCS [28] and simultaneously correlate inputs with output (i.e., multilinear regression) [29]. Linear and multilinear regression have the problem that they can only predict the mean values, and when the data are bigger, they cannot accurately predict the values that are needed. As a result, these approaches are less suitable for application in nonlinear and multivariable engineering issues [30–37].

Soft computing methods include Artificial Neural Networks (ANNs), Adaptive Network-Based Fuzzy Inference Systems (ANFIS), Relevance Vector Machines (RVMs), and other techniques are some of its examples. These methods are now widely used in rock mechanics because of how easily and adaptably they can anticipate the required values depending on a variety of inputs. These techniques are well suited for usage when conventional statistical techniques are less accurate in making predictions [38–42]. In rock mechanics, these methods are getting more attention now because they are flexible and make it easy to predict required values based on the input variables. In situations when using traditional statistical approaches for prediction is not as convenient, these methods are particularly appropriate to be applied [43–46].

Various research at national and international levels has been conducted using different artificial intelligence techniques to predict the UCS of rock. In this regard, Shahani and Zheng [47] predicted UCS using dry-density Brazilian tensile strength (BTS) and the point load index as input variables using ANNs and a multilinear regression model (MLRM). They found that the ANN models show better performance, i.e., R^2 0.99 than the MLRM. Manouchehrian et al. [48] used texture as an input variable in ANN and multivariate modeling for the prediction of UCS. They predicted UCS values effectively, and the performance of the ANNs was better compared to the multivariate statistics. Similarly, Torabi-Kaveh, Naseri, Abdi, Garavand [49,50] used various input variables such as porosity (η), P-wave (P_V), and density (ρ) to predict UCS and E_S using ANNs and ANFIS. Dehghan et al. [1] used ANNs and MLR to predict UCS and the Static Young modulus (Es) based on the P_V, point load index, Schmidt hammer rebound number, and η as input variables. Due to the advancement in artificial intelligence, some of the latest and most effective algorithms have been developed to predict UCS. In this regard, Zhang et al. [51] created a Random Forest (RF) model based on the beetle antennae search (BAS) method for estimating the UCS of lightweight self-compacting concrete (LWSSC) with high precision and efficiency. Matin et al. [52] used just a few rock parameters and indices such as porosity (η), water content (Wc), Is (50), P-wave velocity (P_V), and rebound numbers (Rns) that were determined using an RF model. Based on these variables, an effective model for predicting UCS was created. Suthar [53] used the M5 model tree, RF, ANN, Support Vector Machine (SVM), and Gaussian processes to forecast the UCS of stabilized pond ashes with lime and lime sludge (GPs). Wang et al. [54] developed an effective model for predicting UCS based on RF-selected variables. Ren et al. [55] created k-nearest neighbors (KNNs), naive Bayes, RF, ANN, and (SVM) as machine-learning (ML) methods to precisely predict rock UCS using multiple input parameters. Ghasemi et al. [56] built a tree-based method for predicting the UCS and Young Modulus (E) of carbonate rocks. They found that the applied method gives promising results. Saedi et al. [57] predicted UCS and E of migmatite rocks using ANNs and multivariate regression (MVR). They found that ANN and ANFIS show better prediction performance. Shahani et al. [58] suggested an XGBoost model for predicting the UCS and E of sedimentary rock that is still intact. To estimate the UCS and E of various rocks, Armaghani et al. [59] created a hybrid model based on an ANN and an imperialist competitive algorithm (ICA). This literature review provides insight into the use of different artificial intelligence techniques for predicting UCS. While ANNs have shown promising results, their prediction performance is still a matter of debate. Moreover, the literature highlights that there has been no analysis of input variables using statistical methods to identify the most suitable input variables, which could enhance the prediction performance of both artificial intelligence and statistical techniques. In addition, the latest artificial intelligence techniques, such as the XG Boost algorithm, Random Forest (RF), Elastic Net (EN), Lasso, and Ridge, have not been adequately explored for effective prediction of UCS. Therefore, accurate prediction of rock UCS is vital for ensuring the safe and efficient stability analysis of engineered structures in a rock mass environment.

2. Materials and Methods

2.1. Design of Experimental Works

Representative samples of marble were collected from seven distinct areas, namely Afghanistan, Mardan, Chitral, Mohmand Agency (reddish brown), Buner, Chitral, and Mohmand (super white). The samples were grouped into seven categories represented by A, B, C, D, E, F, and G, respectively. Each group consisted of 10 samples that were tested for various geo-mechanical properties of marble. The tests included moisture content (MC (%)), bulk density (gm/mL)), dry density (gm/mL), water absorption (%), P-wave (km/sec), S-wave (km/sec), slake and durability (Id2), rebound number (R), porosity (η), void ratio (e), and uniaxial compressive strength (UCS). To conduct the tests in line with the International Standard of Rock Mechanics (ISRM), cylindrical core samples with a diameter of 54 × 108 mm were prepared. The two ends of each core sample were meticulously ground with a grinder and sandpaper to ensure parallelism in the upper and lower surfaces within 0.05 mm and the flatness of the surface within 0.02 mm. The samples were prepared and tested as presented in Figure 1 according to the ASTM standard [60,61].

A summary of the various tests mentioned in the above paragraph is presented below. These tests were conducted according to the ISRM standard [60,61].

The method for determining the characteristics of marble involved various techniques and apparatuses. To measure the bulk density of the rock, its weight was measured with a digital balance, and the volume it displaced was measured using a graduated/volumetric cylinder. The dry density was obtained by drying the specimens in an oven, and the dry weight was determined using a digital balance. The volume displaced by the rock was measured using a graduated cylinder.

To determine the moisture content and water absorption test of the marble, the wet and dry weights of the specimens were measured using an oven and a digital balance. The slake durability index was determined by subjecting the specimens to four wetting and drying cycles using a testing apparatus. The porosity and void ratio of the marble were determined using a volumetric cylinder and an oven to dry the samples.

An ultrasonic wave transducer apparatus was used to determine the primary and secondary wave velocities (P-wave and S-wave velocities) of the marble, while the Schmidt hammer test was used to determine the rebound number using the Schmidt rebound hammer or a concrete test hammer.

For the direct testing of the uniaxial compressive strength (UCS) of marble, core samples of marble were prepared in a cylindrical form with a length/diameter ratio of 2.5–3 according to the ISRM. The specimens were carefully ground and covered with a polyethylene plastic cover to protect them from moisture. The UCS was determined using an electrohydraulic servo universal testing machine of model C64.106* with a maximum load of 1000 kN. The machine was set to load at an equal displacement of 0.1 mm/min with a collection rate of 10 times/s.

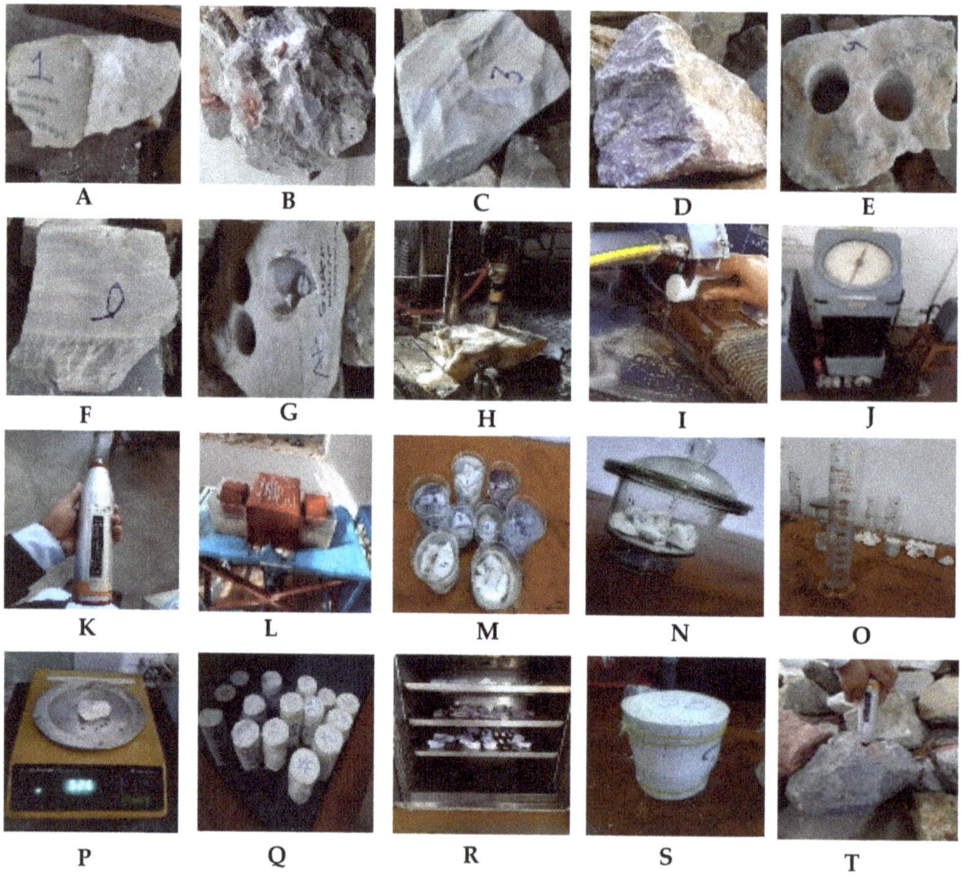

Figure 1. (**A**) Afghanistan marble, (**B**) Mardan (spin kala) marble, (**C**) Chitral marble, (**D**) Mohmand marble, (**E**) Bunir Bumbpoha, (**F**) Chitral marble, and (**G**) Mohmand (super white) marble. (**H**) Core drilling, (**I**) core cutting, (**J**) triaxial testing machine, (**K**) Schmidt hammer apparatus, (**L**) slake durability apparatus, (**M**) sample dipped for 24 h, (**N**) desiccator used to cool sample, (**O**) volumetric cylinder used to find volume, (**P**) sampling weight, (**Q**) core samples, (**R**) oven for drying samples, (**S**) core failure after UCS testing, and (**T**) sampling using a Schmidt hammer.

2.2. *Predictive Models*

XG Boost Algorithm

XGBoost stands for Extreme Gradient Boosting, which was proposed by Tianqi Chen and Guestrin, and is an efficient gradient-boosted decision tree (GBDT) ML library that is portable and scalable [62]. XG boost is an extension of boosting that formalizes the additive creation of weak models, and it uses the gradient descent method over an objective function. XGBoost uses the loss function assessment as a starting point and matches the results to the standard gradient boosting techniques quickly and efficiently [63].

$$Obj\,(\theta) = \frac{1}{n}\sum\nolimits_{i}^{n} Z\,(y_i - Y_i) + \sum\nolimits_{j=1}^{j} \Omega\,(f_j) \quad (1)$$

In Equation (1), Z represents the training loss function, which is used to evaluate how well a model performs when trained on data. Ω is a regularization term intended to limit model complexity by preventing overfitting. f_j denotes a *j*th tree prediction in the formula [63]. Figure 2 shows the XGBoost model structure [64].

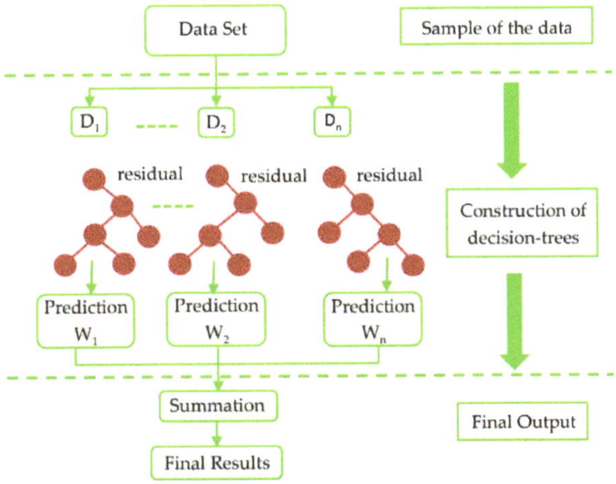

Figure 2. Structure of a XGBoost model.

The boosting methodology improves the framework evaluation accuracy by creating multiple trees as alternatives to produce an addressed tree, then connecting them to determine a methodical predictive algorithm [64]. In addition to parallel tree boosting, it is the top machine-learning library for tackling regression, classification, and ranking problems [65].

Random Forest

The Random Forest regression model, introduced by Breiman in 2001, is one of the machine learning ensemble approaches [64]. It is one of the tree-based techniques used for classification and regression analysis. RF trees are constructed using a subset of the random variables chosen independently and replaced with the original data set. When solving forecasting problems, it incorporates categorical as well as numerical variables [66]. The basic architecture of an RF is given in Figure 3.

Random Forest is an ultra-modern technique for bagging. The built-in cross-validation function for the Random Forest allows ranking explanatory factors from the most effective to the least associated with the outcome variable. As a result, feature extraction is more valuable when examining data from various sources [66]. Among other widely accepted forms of AI computing, RF observes a singular association between model embodiment and the predictive accuracy [64]. The Random Forest algorithm may be expressed as [67]:

$$Y = \frac{1}{N} \sum_{i=1}^{N} F_i(X) \quad (2)$$

In Equation (2), *X* represents the input parameter factor, *Y* represents the prediction result and *N* shows the number of regression trees formed. Figure 3 shows the basic structure of a Random Forest Regression (RFR) model [64].

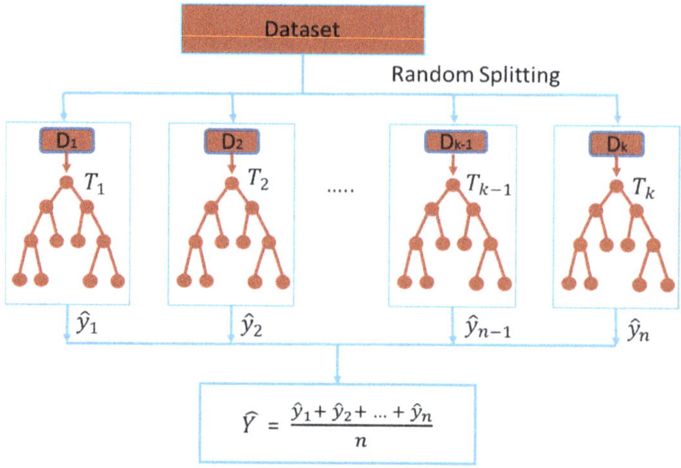

Figure 3. Fundamental Structure of a Random Forest Regression Model.

Support Vector Machine

A type of supervised learning known as Support Vector Machines (SVMs) was first proposed by Vapnik et al. in 1997 [68]. The fundamental concept of SVMs is that neurons are grouped in two layers, just like in ANNs. SVMs with a sigmoid kernel function are equivalent to a two-layer perceptron neural network. SVMs are alternative training methods for polynomial, radial basis function, and multilayer perceptron classifiers in which the network weights are determined by solving a quadratic programming problem with linear constraints [69].

Support Vector Machines are capable of solving classification and complicated non-linear regression issues. The basic goal when applied to regression problems is to create a perfect classification surface that reduces the error in all training samples obtained from that surface [67].

Figure 4 illustrates the architecture of a Support Vector Machine (SVM). The signal vector input is present in the input layer. In the buried layer, an inner-product kernel is created between the input signal vector (x) and the support vector (s_i). The output neuron receives an addition of the buried layer's neuron linear outputs. The output neuron is biased [70].

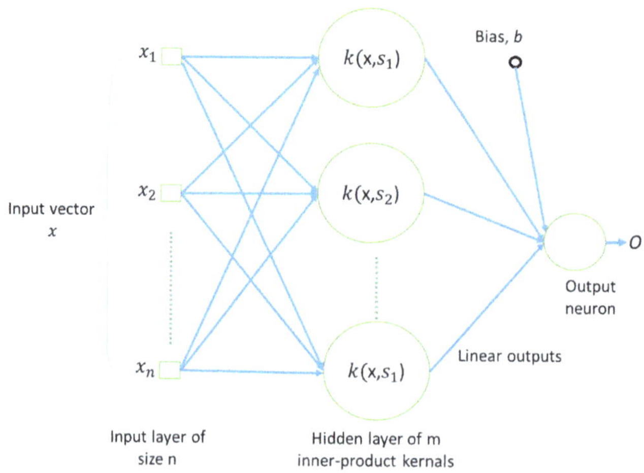

Figure 4. General Architecture of a Support Vector Machine [70].

Lasso Regression

Lasso regression was introduced to geophysics between 1986 and 1996 [71,72]. It carries out feature selection and regularization penalties to raise the accuracy of prediction. However, multicollinearity is avoided by selecting the most important predictor from a group of highly correlated independent variables and neglecting the others. An L1-norm penalty term was used to reduce regression coefficients, some to zero, ensuring that the most significant explanatory variables were selected. Another benefit of Lasso is that it can only choose n parameters when a data set of size n is fitted to a regression model with p parameters and $p > n$ (p represents predictor variables and n represents the number of observations) [72].

Ridge Regression

Ridge regression was developed to improve the predictability of a regression model. In Ridge regression, the L2-norm penalty term is used to reduce the regression coefficients to nonzero values to prevent overfitting, but it does not serve as a feature-selection mechanism [72]. If there are numerous predictors, all of which have non-zero coefficients and are selected from a normal distribution, ridge regression is the best option. It specifically performs well when there are many predictors, each with little influence, and avoids poorly defined and large variance coefficients in linear regression models with numerous correlated variables [71].

Elastic Net

The Elastic Net is a variant of the Lasso that can withstand large levels of inter-predictor correlation. When predictors are highly correlated, the Lasso solution routes may become unstable (like SNPs in high linkage disequilibrium). The Elastic Net (ENET) was suggested for high-dimensional data processing to address this problem [71].

Elastic Net is a member of a group of regression algorithms that use L1-norm and L2-norm regularization penalty terms; the tuning parameter regulates the potency of these penalty terms [72]. Automatic variable selection is performed using the L1 component of the ENET, and clustered selection is encouraged with the L2 component, which also stabilizes the solution routes concerning random sampling to enhance prediction. By creating a grouping effect during the variable selection process, the ENET can choose groups of correlated features even when the groups are unknown. This is because a group of strongly correlated variables tends to have coefficients with comparable magnitude. When $p > n$ (p represents predictor variables and n represents the number of observations), the Elastic Net chooses more variables than n, in contrast to the Lasso. However, the elastic net is free of the oracle characteristic [73].

Artificial Neural Networks (ANNs)

One of the supervised machine learning (ML) techniques that are frequently used is the Artificial Neural Network (ANN). ANN computational models have been used to solve a wide range of issues in various disciplines [72]. A model comprises several little processing units (neurons) that are capable of handling complicated data processing and knowledge representation [74]. An ANN has three main components that include input layers, hidden layers, and an output layer [72]. Since it essentially maps the input and output values, it has good interpolation capabilities, particularly when the input data are noisy. Neural networks can be used in place of auto-correlation, multivariable regression, linear regression, trigonometric analysis, and other statistical analysis approaches [74].

For any regression model in an ANN, a supervised learning method is required during training to provide the highest levels of accuracy and efficiency. In network training, the BP algorithm uses a sequence of instances to establish connections between nodes, as well as to determine the parameterized function [75]. Many networks are trained using the BP method. According to the available literature, the BP algorithm performs the NN operation by evaluating and implying random variables. There is a need to train the model, and research studies have been conducted to complete this in a better way [76].

Equation (3) gives a mathematical expression for an ANN.

$$Basic\ network = f(wx + bias) \tag{3}$$

where w and x indicate weights and input, respectively. The weight and input for n numbers are presented as

a. $w = w_1, w_2, w_3, w_4, \ldots\ldots, w_n;$
b. $x = x_1, x_2, x_3, x_4, \ldots\ldots, x_n.$

The ANNs used Equation (4) to predict the values.

$$net = \sum_{i=1}^{n} f(w_i x_i + b) \tag{4}$$

The tangent sigmoid function described in Equation (5) was used as the transferred function in this investigation.

$$y = tanh\ (net) \tag{5}$$

Using Equation (6), the output of the network represented by "y" may be computed.

$$output\ of\ the\ network = y = tanh(net) = tanh\left(\sum_{i=1}^{n} f(w_i + b))\right) \tag{6}$$

The network error is defined as the "calculated values ($V_{Calculated}$) minus estimated values ($V_{Predicted}$) of the network". By increasing or decreasing the neuron's weight, it is possible to reduce the error in this network to some extent. Equation (7) represents the inaccuracy of networks in their mathematical form.

$$E_m = V_{Calculated} - V_{Predicted} \tag{7}$$

Moreover, the total error in a network can be calculated using Equation (8).

$$E_{Total} = \frac{1}{2} \sum_m E_m^2 \tag{8}$$

Code Development for ANNs using MATLAB

Figure 5 shows an example of the self-generated ANN code used in this study for n networks using a similar training and activation function for a single loop. An internal loop in this program can be used to process data for as many networks as desired. The activation function for the code was static, although the structure of the data was likely to vary. Here, one algorithm run was used to process 100 networks. As a result, network1 contains one neuron, network2 contains two, and so forth. Although there are numerous ANN approaches, Khan et al. utilized the Levenberg–Marquardt algorithm and suggested BP [77,78]. Khan et al. [77,78] discovered that the Levenberg–Marquardt (LM) method is better than other algorithms and is more time-effective. As a result, LM was used in the current model for both the hidden and output layers. The fundamental ANN structure in this study consists of three inputs (moisture content, P-waves, and rebound number as input variables) and one output, i.e., UCS, as shown in Figure 6. The data were classified into three classes: training (70%), testing (15%), and validation (15%).

The Artificial Neural Network designed to estimate the UCS in the present work is presented in Figure 6.

In order to determine the relationship between the model input variables and the corresponding outputs, ANNs learn from the samples of data that are presented to them and utilize those samples to alter their weights. As a result, ANNs do not require prior knowledge regarding the nature of the relationship between the input and output variables, which is one advantage they have over most empirical and statistical methods. If the relationship between x and y is non-linear, regression analysis can only be used successfully if the nature of the non-linearity is known beforehand. On the other hand, ANN models do not require this prior understanding of the type of non-linearity [21]. Generally, several machine learning models work based on the above statement.

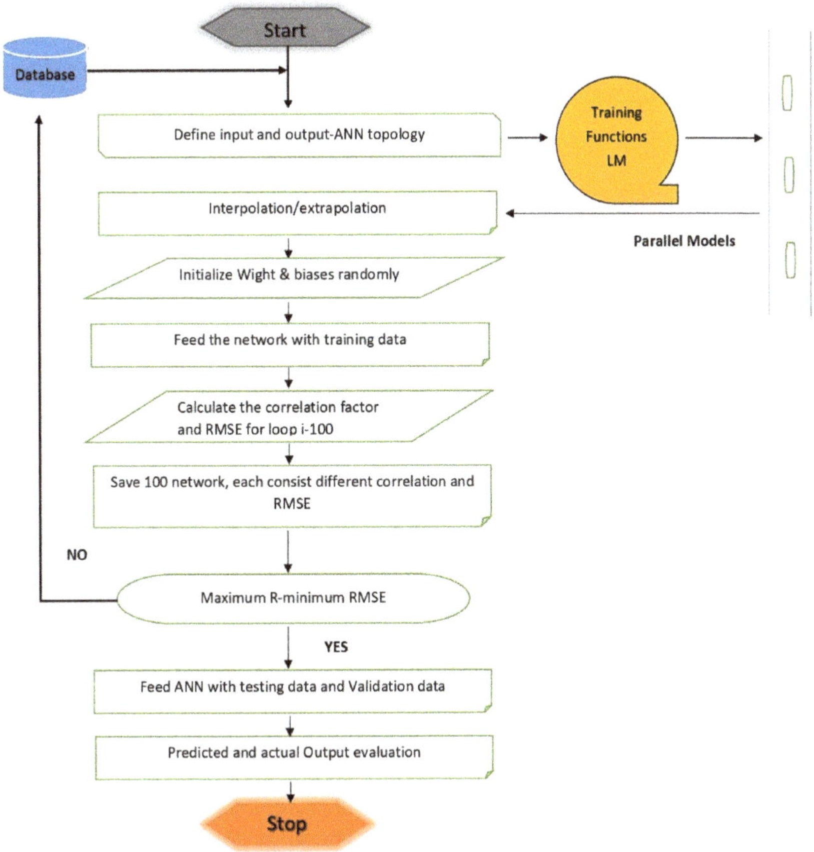

Figure 5. Flow chart representing the procedure for developing ANN code using MATLAB.

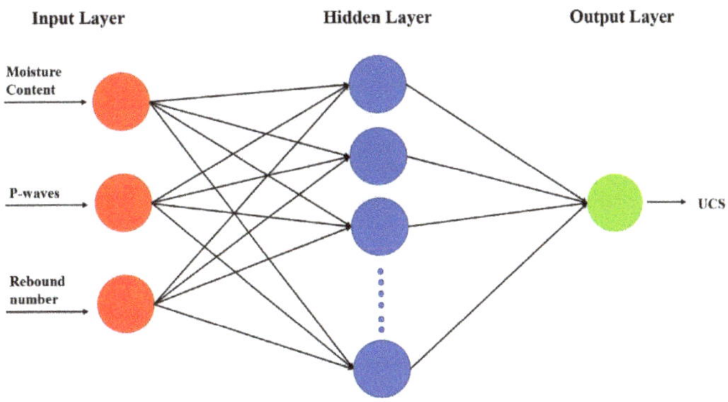

Figure 6. Artificial Neural Network framework for estimating UCS.

2.3. Data Analysis for Selecting the Most Appropriate Input Variables

In this study, various parameters were determined in the laboratory using direct methods, i.e., moisture content, bulk density, dry density, water absorption, slake and durability, rebound number, P-wave, S-wave, porosity, void ratio, and UCS. The descriptive statistical analysis of these variables was carried out for a better understanding of their statistical behavior, and the results are presented in Table 1. Furthermore, these variables were analyzed using pairwise correlation with output, and correlation matrix analysis to choose the most appropriate input variables for predicting uniaxial compressive strength (UCS) using different artificial intelligence techniques.

Table 1. Descriptive statistics for the input and output variables.

S.No	Input and Output	N total	Mean	Standard Deviation	Sum	Min	Median	Max
1	bulk density (g/mL)	70.00	2.73	0.27	191.34	2.12	2.69	3.53
2	dry density (g/mL)	70.00	2.67	0.24	187.16	2.12	2.65	3.61
3	moisture content (MC (%))	70.00	0.36	0.19	25.46	0.00	0.35	0.99
4	water absorption (%)	70.00	0.36	0.24	25.28	0.00	0.34	1.20
5	slake durability index (Id_2)	70.00	97.08	3.21	6795.85	83.24	98.25	99.11
6	rebound number (R)	70.00	45.88	6.31	3211.57	34.70	44.82	64.14
7	porosity (η)	70.00	0.36	0.24	25.28	0.00	0.34	1.20
8	void ratio (e)	70.00	1.15	3.03	80.25	0.00	0.0034	0.012
9	P-wave (km/s)	70.00	4.74	0.20	331.52	4.43	4.70	5.49
10	S-wave (km/s)	70.00	3.02	0.01	211.14	2.98	3.02	3.03
11	UCS (Mpa)	70.00	52.17	12.10	3651.59	34.89	49.51	93.76

The correlation matrix analysis of the input and output was also carried out to select the most effective input variables to eliminate multicollinearity in the prediction. To better understand the variance and covariance in the regressions used in the prediction model, a correlation matrix can be used as a descriptive statistical tool. It is often used in conjunction with other matrices in statistical analysis. Conversely, correlation explains the interaction between the regression variables used in predictive analyses. Figures 7 and 8 show how the correlation matrix often describes the variance in each parameter. There are both positive and negative correlations among some of them. This will allow the researcher to see how different factors affect the predicted model's final outcomes. As the strength of the negative or positive correlation increases, so too will the significance of model efficiency. The criteria for selecting parameters are (a) check the high relationship (negative or positive) parameters with the output (UCS), (b) check the input parameter relationships with each other, (c) check the high correlation parameters with the output and with each other. If there are two parameters that have a high correlation with the output and also a high correlation with each other, then select the one input parameter. For example, Figures 7 and 8 show that there are four input parameters that show a high correlation with the output, i.e., moisture content, dry density, rebound number, and P-wave. In addition, rebound number and P-wave have a high correlation with each other, so in both, one parameter will be considered as the input for better model learning. Therefore, moisture content, P-wave, and rebound number were selected for the input and the others were discarded during the AI model evolution. These three parameters were selected for the prediction of UCS using seven different AI techniques.

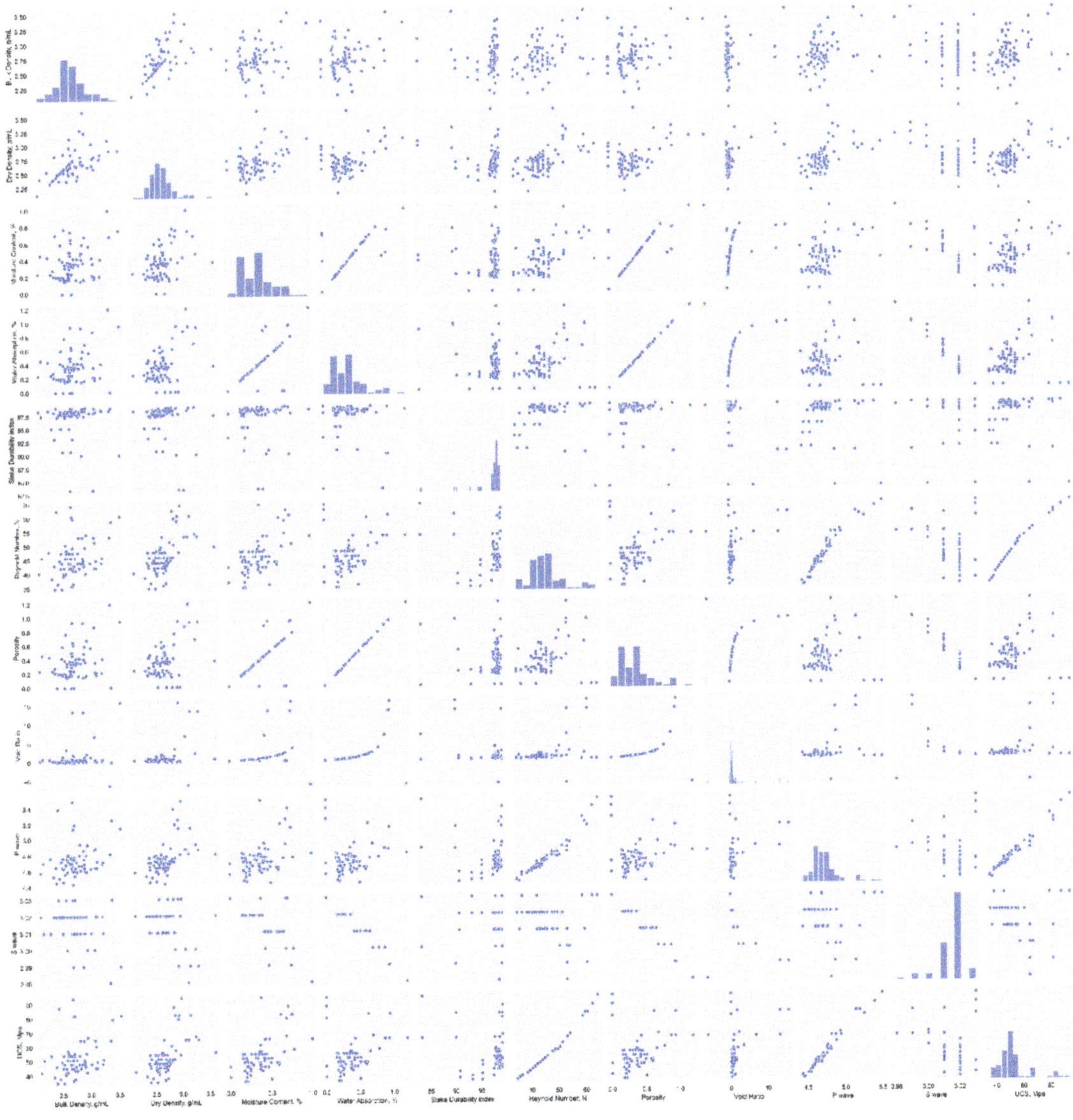

Figure 7. Scatter plot showing the input parameters with uniaxial compressive strength (UCS).

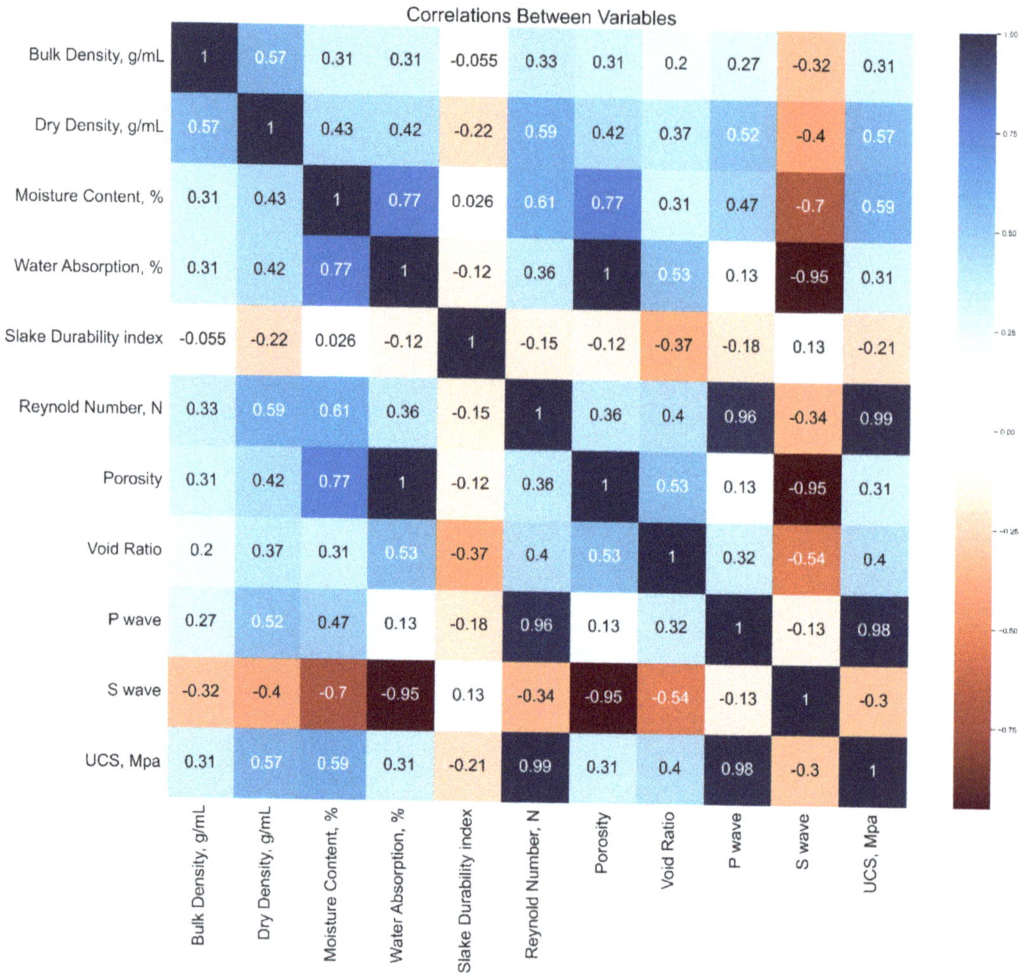

Figure 8. Correlation between input variables and UCS.

2.4. Performance Indicator

Various performance indicators such as the coefficient of determination R^2, Root Mean Square Error (RMSE), Mean Square Error (MSE), and Mean Absolute Error (MAE) were used to evaluate the prediction performance of the Artificial Neural Network (ANN), XG Boost algorithm, Random Forest Regression (RFR), Elastic Net (EN), Lasso, Support Vector Machine (SVM), and Ridge models. The following formulas as mentioned in Equations (9)–(12) were used [43,64]:

$$R^2 = 1 - (RSS/TSS) \tag{9}$$

where:

RSS = sum of the square of the residual;
TSS = total sum of the square.

$$RMSE = \sqrt{MSE} = \sqrt{\frac{1}{T}\sum_{n=1}^{T}(s_n - \hat{s})^2} \tag{10}$$

where:

T = total no of observations;
s_n = actual value of nth observation;
$\hat{s}$ = predicted value of s.

$$MSE = \frac{1}{T}\sum_{n=1}^{T}(s_n - \hat{s})^2 \tag{11}$$

where:
T = total no of observations;
s_n = actual value of nth observation;
$\hat{s}$ = predicted value of s.

$$MAE = \frac{1}{T}\sum_{n=1}^{T}|s_n - \hat{s}| \tag{12}$$

where:
T = total no of observations;
s_n = actual value of nth observation;
$\hat{s}$ = predicted value of s.

3. Analysis of Results

3.1. Model Hyperparameter Optimization

It is important to determine the optimal combination of hyperparameters in machine-learning models when attempting to improve the predictability of the model. Hyperparameters determine how well the model can learn. A tuning technique known as "grid search" was used in the current study, which searched exhaustively for all optimum values for the user-specified hyperparameter combinations using this technique. Additionally, to overcome the problem of overfitting, standard k-fold cross-validation was used as part of the process. In order to carry out the k-fold cross-validation, the procedure outlined below must be followed [79]:

a. In order to train the data set, the training data set needs to be divided into k folds.
b. The (k-1) fold is used for training out of all k folds.
c. The remaining last k-fold is used for validation.
d. In order to train the model with specific hyperparameters, training data (k-1 folds) are used, and validation data are used as 1-fold. For each fold, the model's performance is recorded.
e. K-fold cross-validation refers to the process of repeating the steps above until each k-fold is used for validation purposes. That is why this process is known as "K-fold cross-validation".
f. After calculating each model score for each model in step d, the mean and standard deviation of the model performance are computed.
g. It is necessary to repeat steps b to f for different values of the hyperparameters.
h. The hyperparameters associated with the best mean and standard deviation of the model scores are then selected.
i. Using the entire training data set, the model is trained, and its performance is evaluated on the basis of the test data set.

In this study, grid search is combined with a 10-fold cross-validation (k = 10) in order to optimize the hyperparameters in the classification algorithms as a result of the grid search, as shown in Figure 9. The optimum hyperparameter for all seven AI techniques is described in Table 2.

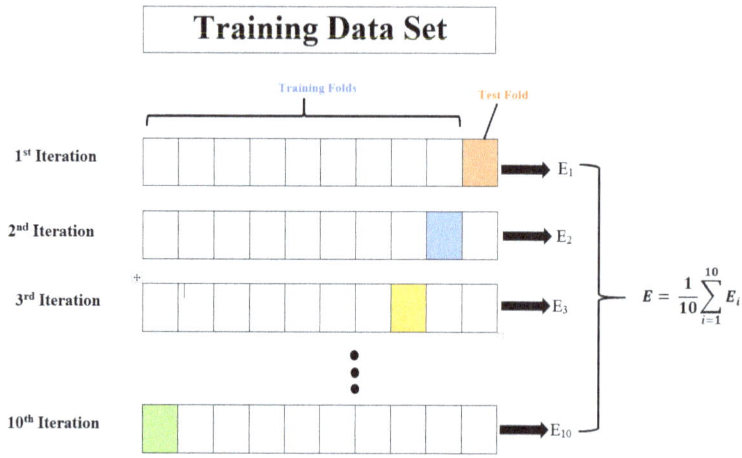

Figure 9. Hyperparameter tuning using grid search and 10-fold cross-validation.

Table 2. Optimized hyperparameters for all models.

Output	Model	Parameters
UCS (Mpa)	Artificial Neural Network	Neuron = 48
	XG Boost Regressor	learning_rate = 0.01, max_depth = 3, n_estimators = 100
	Support Vector Machine	n_split = 10, n_repeats = 5, random state = 42, C = 1 function = SVR (kernal 'rbf')
	Random Forest Regression	n_split = 10, n_repeats = 5, random state = 42, max_depth = 3
	Lasso	Alpha = 0.01, n_split = 10, n_repeats = 5, random state = 42
	Elastic Net	Alpha = 0.01, l1_ratio = 0.95, n_split = 10, n_repeats = 5, random state = 42
	Ridge	Alpha = 0.1, n_split = 10, n_repeats = 5, random state = 42

3.2. Prediction of UCS using Artificial Neural Networks

Figure 10 shows the regression data and the ANN training, validation, and testing phases for the UCS model. A good regression is obtained between the predicted and measured UCS values during training, validation, and testing. Figure 10 shows the results of the ANN model draws from the plot graph process. Figure 11 shows a very good R^2 value of 0.9995 between the predicted and measured UCS.

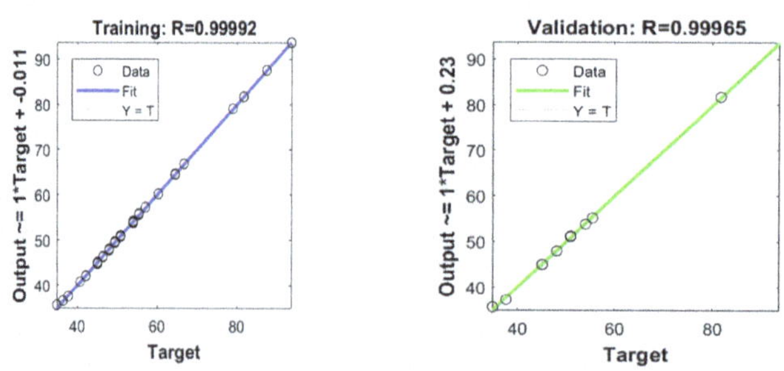

Figure 10. *Cont.*

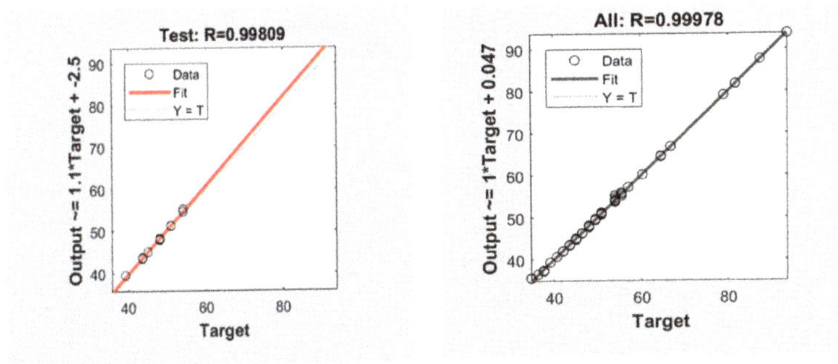

Figure 10. ANN phases of training, validation, and testing and the regression coefficient of determination for UCS.

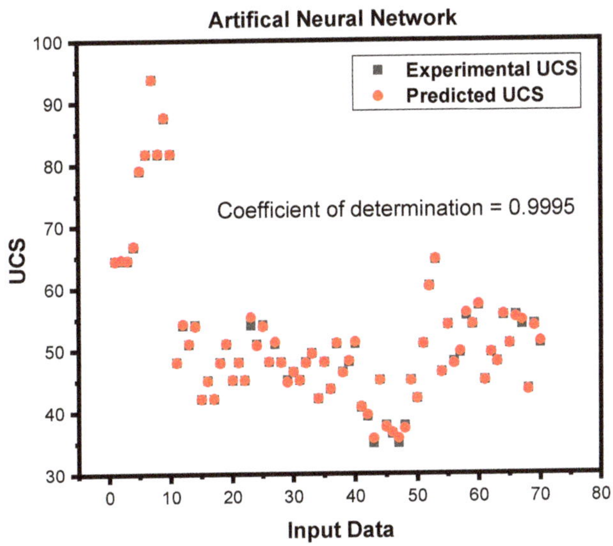

Figure 11. Coefficient of determination between the measured and predicted UCS.

Ridge Regression

The used Python library is an open-source software package with utility functions for engineering, especially machine learning. These libraries are used by the user to predict their metric of interest. Python's Scikit-learn includes a free and open-source machine-learning library, which includes Ridge regression, Elastic Net, Lasso regression SVR, RFR, and XG Boost. For the Ridge regression, the model was executed on the training set (80%) and the testing set (20%). Figure 12 shows a graph between the predicted values and actual values with the correlation coefficient (R^2 = 0.9790).

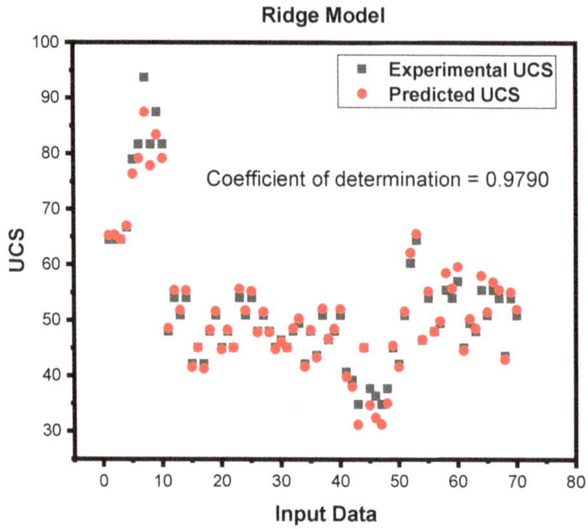

Figure 12. Ridge regression model with the coefficient of determination between the measured and predicted UCS.

Elastic Net

For the Elastic Net, the model was executed on the training set (80%) and the testing set (20%). Figure 13 shows a graph between the predicted values and actual values with the correlation coefficient (R^2 = 0.9755).

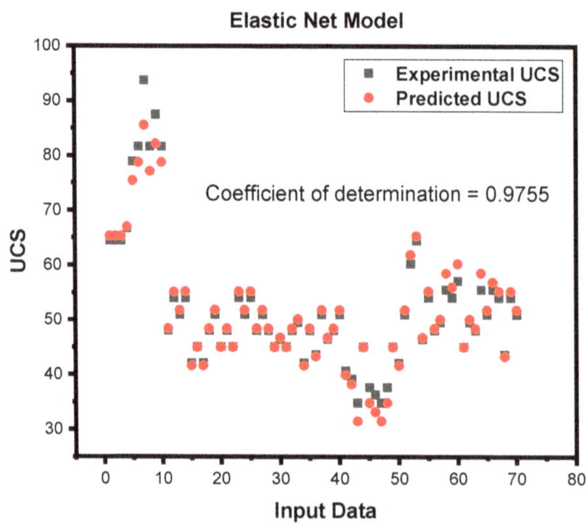

Figure 13. Elastic Net model with the coefficient of determination between the measured and predicted UCS.

Lasso Regression

For the Lasso regression, the model was executed on the training set (80%) and the testing set (20%). Figure 14 shows us a graph between the predicted values and actual values with the correlation coefficient (R^2 = 0.9755).

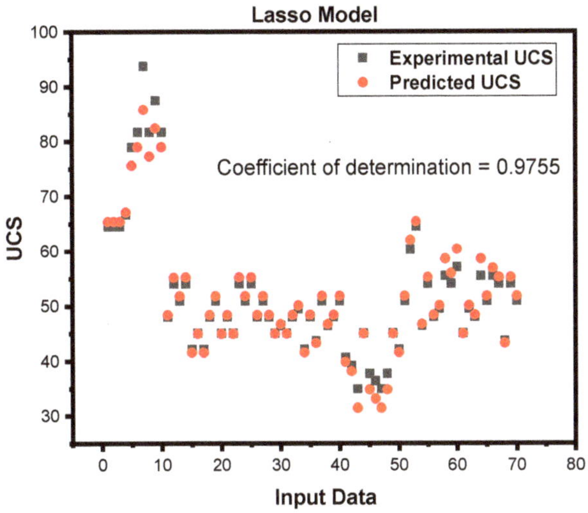

Figure 14. Lasso regression with the coefficient of determination between the measured and predicted UCS.

Support Vector Machine

For the Support Vector Machine, the model was executed on the training set (80%) and the testing set (20%). The SVR is similar to the ANN, which included an input layer, hidden layer, and output layer. The SVR model estimates the average of the prediction values. Figure 15 shows us a graph between the predicted values and actual values having the correlation coefficient (R^2 = 0.9573).

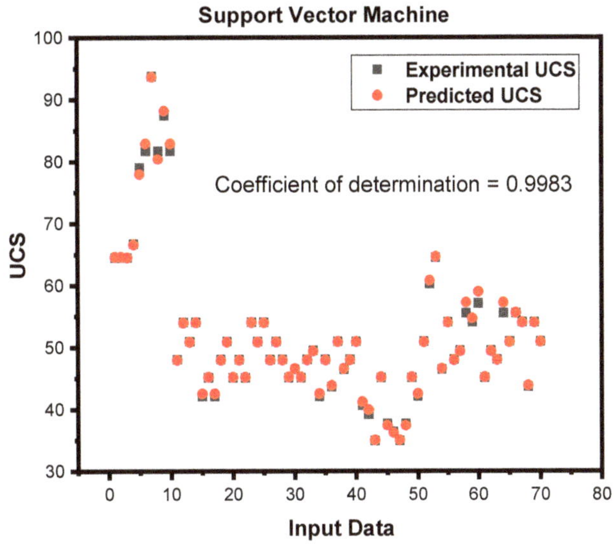

Figure 15. Support Vector Machine model with the coefficient of determination between the measured and predicted UCS.

Random Forest

The model was applied to the training set (80%) and testing set (20%) for the Random Forest Regression, and the n estimated and max depth parameters of this RFR model were determined. To determine the higher average of the forecast, the number of estimators is the same as the number of decision trees (DTs) produced using the Random Forest Regression (RFR) model. Figure 16 shows us a graph between the predicted values and actual values with the correlation coefficient (R^2 = 0.9949).

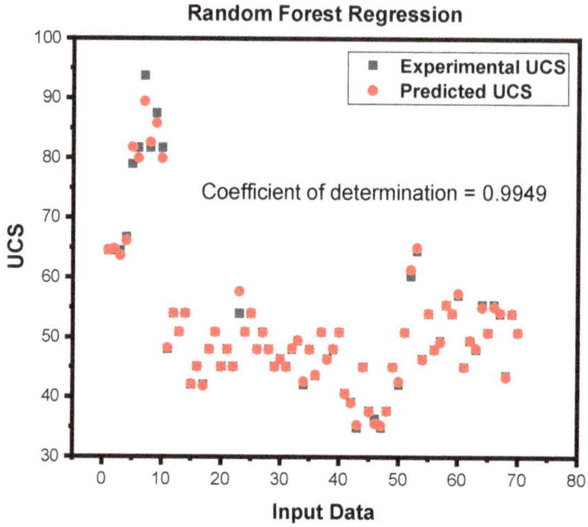

Figure 16. Random Forest Regression with the coefficient of determination between the measured and predicted UCS.

XG Boost Algorithm

A Python module called XG Boost Algorithm is used to build machine learning models. For the XG Boost, the model was applied to the training set (80%) and the testing set (20%). The XG boost algorithm is a highly interpretable model. After creating a tree model, the predicted values were directly obtained. Figure 17 shows us a graph between the predicted and actual values with the correlation coefficient (R^2 = 0.9990).

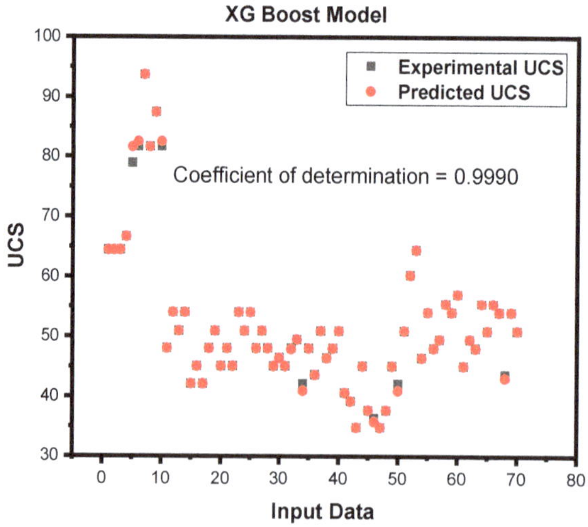

Figure 17. XG Boost Model with the coefficient of determination between the measured and predicted UCS.

The results obtained using the above performance indicator for evaluating the efficacy of each predictive model are shown in Table 3.

Table 3. Comparative analysis of the performance of different AI techniques.

S.no	Models	Training Accuracy				Testing Accuracy			
		R^2	MAE	MSE	RMSE	R^2	MAE	MSE	RMSE
1	Artificial Neural Network	0.9990	0.1428	0.0782	0.2796	0.9995	0.1642	0.0694	0.2634
2	XG Boost Regressor	0.9989	0.5694	0.8664	0.9308	0.9990	0.1145	0.1732	0.4162
3	Support Vector Machine	0.9987	0.3649	0.3022	0.5498	0.9983	0.2891	0.2595	0.5094
4	Random Forest Regression	0.9943	0.7176	1.3294	1.1530	0.9949	0.3555	0.6584	0.8114
5	Lasso	0.9887	1.3670	3.0666	1.7512	0.9755	1.8918	3.5788	1.2555
6	Elastic Net	0.9887	1.3751	3.2071	1.7908	0.9755	1.2410	3.6308	1.9055
7	Ridge	0.9876	1.3906	3.0492	1.7462	0.9790	1.2149	3.0010	1.7347

To compare their performance, the training and testing accuracies of the seven different models are listed in Table 2. Among the various models, the Artificial Neural Network gave the most accurate prediction on the training and testing data sets (99%), while the Support Vector Machine model showed the lowest predicted performance on the testing and training data sets. The R^2, MAE, MSE, and RMSE for the ANN model were 0.999, 0.1428, 0.0782, and 0.2796, respectively, on the training data set, while they were 0.9995, 0.6420,0.0694, and 0.2634 on testing data, respectively, which shows that the performance of the ANN model is greater than all the other predictive models. For the XG Boost Regressor, the value of the performance indicator R^2 was 0.9989, MAE was 0.5694, MSE was 0.0782, and RMSE was 0.2796 for the training data set, while for the testing data set, the R^2 was 0.9990, MAE was 0.1145, MSE was 0.0694, and RMSE was 0.4162. For the Random Forest Regression, the performance indicator R^2 was 0.9943, MAE was 0.7176, MSE was 1.3294, and RMSE was 1.1530 for the training data set, while for the testing data set, the R^2 was 0.9949, MAE was 0.3555, MSE was 0.6584, and RMSE was 0.8114. For the Lasso, the performance indicators R^2, MAE, MSE, and RMSE were 0.9887, 1.367, 3.0666, and 1.7512 for the training data set, respectively, while for the testing data set, the R^2 was 0.9755, MAE was 1.8918, MSE was 3.5788, and RMSE was 1.2555. For the Ridge model, the R^2, MAE, MSE, and RMSE were 0.9876, 1.3906, 3.0492, and 1.7462, respectively, for the training data set, while for the testing data, the performance indicators R^2, MAE, MSE, and RMSE were 0.979, 1.2149, 3.001, and 1.7347, respectively. For the Elastic net model, the performance indicator R^2 was 0.9887, MAE was 1.3751, MSE was 3.2071, and RMSE was 1.7908 for the training data set, while for the testing data set, the R^2 was 0.9755, MAE was 1.241, MSE was 3.6308, and RMSE was 1.9055. Similarly, for the Support Vector Machine, the R^2, MAE, MSE, and RMSE were 0.9826, 9.4444, 187.2607, and 13.68, respectively, for the training data set, and for the testing data set, the value of R^2, MAE, MSE, and RMSE were 0.9573, 6.5449, 111.4614, and 10.5575, respectively. According to Table 2, the ANN model had values of 0.9995, 0.2634, 0.0694, and 0.1642 for R^2, RMSE, MSE, and MAE, respectively. This highlights that the ANNs model's performance was better than that of any other prediction model. However, the hierarchy of the mentioned predictive models in terms of their efficacy based on the performance indicators in predicting the UCS can be ANN > XG Boost Regressor > SVR > Random Forest Regressor > Lasso > Elastic Net > Ridge.

4. Discussion

This study conducted a data analysis to select the most appropriate input variables for predicting the uniaxial compressive strength (UCS) using different artificial intelligence techniques. Various laboratory parameters were determined using direct methods, including moisture content, bulk density, dry density, water absorption, slake and durability, rebound number, P-wave, S-wave, porosity, void ratio, and UCS. A descriptive statistical analysis of these variables was carried out, including a p-value significance analysis, pairwise correlations with the output, and correlation matrix analysis to choose the most appropriate input variables. The statistical analysis showed that the p-value for the rebound number, P-wave, and moisture content had a positive coefficient of less than 0.05, which indicated a strong correlation with the UCS. The other input variables, such as dry density, bulk density, water absorption, and the slake durability index, showed a negative correlation with the UCS, and, therefore, were not selected as input variables. The porosity and void ratio showed an invalid p-value and were also not selected. Additionally, the correlation matrix analysis was carried out to select the most effective input variables and eliminate multicollinearity in the prediction. The results of the correlation matrix analysis indicated that moisture content, rebound number, and P-wave had a strong correlation with the UCS, as presented in Figures 7 and 8. Therefore, these variables were selected as appropriate input variables for the prediction of UCS.

The above analysis shows the performance of various machine-learning models when predicting the UCS of rock samples, as presented in Section 3. The ANN model achieved an impressive coefficient of determination (R^2) of 0.9995, indicating a strong correlation between predicted and measured UCS values. This suggests that the ANN model can be used to predict UCS values accurately. Among the other models, the Random Forest Regression (RFR) performed well with an R^2 value of 0.9949. This suggests that RFR can also be used as an alternative method for predicting UCS values. The XG Boost algorithm also performed well, with an R^2 value of 0.9990, which is similar to the ANN model. The Ridge regression, Elastic Net, and Lasso regression models also showed good performance with R^2 values ranging from 0.9887 to 0.9886. However, their performance was slightly lower than that of the ANN, XG Boost, SVM, and RFR models. Overall, the analysis suggests that the ANN model, followed by XG Boost, SVM, and RFR are the best models for predicting UCS values, while Ridge regression, Elastic Net, and Lasso regression are also good alternatives. The SVM model may not be the best option for predicting UCS values. This study considers a small data set due to limited resources. In future studies, the authors will use the application of infrared radiation (IR) technology and AI together to avoid such a large parameter determination in the field as used in this study training. The IR and AI together will make the prediction more reliable and applicable. Moreover, in the future, the given 70 sample data set can be increased using the harmony search optimization algorithm [80].

5. Conclusions

The strength property (uniaxial compressive strength) of rock is a fundamental parameter significantly used in the effective and sustainable design of the tunnel and other engineering structures. In this research, the UCS was predicted using seven different artificial intelligence (AI) metaheuristics techniques, i.e., Artificial Neural Networks (ANNs), the XG Boost algorithm, Random Forest Regression (RFR), Support Vector Machine (SVM), Elastic Net (EN), Lasso, and Ridge models using moisture content, P-waves, and rebound number as input parameters in order to choose the best prediction model. The efficacy of the models was evaluated using four performance indicators, i.e., the coefficient of determination (R^2), Root Mean Square Error (RMSE), Mean Square Error (MSE), and Mean Absolute Error (MAE). The results show that the performance indicators for the ANN were 0.9995, 0.2634, 0.0694, and 0.1642, respectively. The comparative analysis based on the performance indicators revealed that the ANN model has greater prediction efficacy compared to the other AI models; however, the ANN model gives approximately a similar performance as the XG Boost Regressor model. Furthermore, it was noticed that SVM, RFR, Ridge, Lasso, and Elastic Net models give acceptable prediction performance; however, they are less effective in performance than the ANN and XG Boost Regressor models when predicting UCS. Therefore, the ANN and XG Boost Regressor are recommended to be used as the most effective predictive models for the prediction of UCS. Since this research work was conducted using a limited number of rock samples, it would be beneficial to extend the data set in order to refine the findings. Additionally, since this study was focused on marble only, it would be necessary to carry out further fine-tuning of the models before applying them to any other type of rock mass environment to ensure the best possible results. Further research needs to be carried out to explore the applications of the various AI techniques for the effective prediction of the UCS. The outcomes of this research will provide a theoretical foundation for field professionals in the prediction of the strength parameters of rock for an effective and sustainable design of engineering structures.

Author Contributions: Contributed to this research, designed experiments, and wrote this paper: M.S.J., R.e.Z., S.H., N.M.K. and Z.U.R.; conceived this research and were responsible for this research: S.H., K.C., M.Z.E. and S.R.; analyzed data, S.S.A., S.S. and M.S. All authors have read and agreed to the published version of the manuscript.

Funding: This research was supported by the Researchers Supporting Project number (RSP2023R496), King Saudi University, Riyadh, Saudi Arabia. And also supported by Anhui Provincial Scientific Research Preparation Plan Project (2022AH050596).

Institutional Review Board Statement: Not applicable.

Informed Consent Statement: Informed consent was obtained from all subjects involved in this study.

Data Availability Statement: The data that support the findings of this study are available from the corresponding author upon reasonable request.

Conflicts of Interest: The authors declare no conflict of interest.

References

1. Dehghan, S.; Sattari, G.; Chelgani, S.C.; Aliabadi, M.J.M.S. Technology. Prediction of uniaxial compressive strength and modulus of elasticity for Travertine samples using regression and artificial neural networks. *Min. Sci. Technol.* **2010**, *20*, 41–46.
2. Bieniawski, Z.T. Estimating the strength of rock materials. *J. S. Afr. Inst. Min. Metall.* **1974**, *74*, 312–320. [CrossRef]
3. Mahdiabadi, N.; Khanlari, G.J. Prediction of uniaxial compressive strength and modulus of elasticity in calcareous mudstones using neural networks, fuzzy systems, and regression analysis. *Period. Polytech. Civ. Eng.* **2019**, *63*, 104–114. [CrossRef]
4. Khan, N.M.; Cao, K.; Emad, M.Z.; Hussain, S.; Rehman, H.; Shah, K.S.; Rehman, F.U.; Muhammad, A.J. Development of Predictive Models for Determination of the Extent of Damage in Granite Caused by Thermal Treatment and Cooling Conditions Using Artificial Intelligence. *Mathematics* **2022**, *10*, 2883. [CrossRef]
5. Wu, H.; Ju, Y.; Han, X.; Ren, Z.; Sun, Y.; Zhang, Y.; Han, T. Size effects in the uniaxial compressive properties of 3D printed models of rocks: An experimental investigation. *Int. J. Coal Sci. Technol.* **2022**, *9*, 83. [CrossRef]
6. Gao, H.; Wang, Q.; Jiang, B.; Zhang, P.; Jiang, Z.; Wang, Y. Relationship between rock uniaxial compressive strength and digital core drilling parameters and its forecast method. *Int. J. Coal Sci. Technol.* **2021**, *8*, 605–613. [CrossRef]
7. Kim, B.-H.; Walton, G.; Larson, M.K.; Berry, S. Investigation of the anisotropic confinement-dependent brittleness of a Utah coal. *Int. J. Coal Sci. Technol.* **2021**, *8*, 274–290. [CrossRef]
8. Li, Y.; Mitri, H.S. Determination of mining-induced stresses using diametral rock core deformations. *Int. J. Coal Sci. Technol.* **2022**, *9*, 80. [CrossRef]
9. Li, Y.; Yang, R.; Fang, S.; Lin, H.; Lu, S.; Zhu, Y.; Wang, M. Failure analysis and control measures of deep roadway with composite roof: A case study. *Int. J. Coal Sci. Technol.* **2022**, *9*, 2. [CrossRef]
10. Liu, B.; Zhao, Y.; Zhang, C.; Zhou, J.; Li, Y.; Sun, Z. Characteristic strength and acoustic emission properties of weakly cemented sandstone at different depths under uniaxial compression. *Int. J. Coal Sci. Technol.* **2021**, *8*, 1288–1301. [CrossRef]
11. Liu, T.; Lin, B.; Fu, X.; Liu, A. Mechanical criterion for coal and gas outburst: A perspective from multiphysics coupling. *Int. J. Coal Sci. Technol.* **2021**, *8*, 1423–1435. [CrossRef]
12. Ma, D.; Duan, H.; Zhang, J.; Bai, H. A state-of-the-art review on rock seepage mechanism of water inrush disaster in coal mines. *Int. J. Coal Sci. Technol.* **2022**, *9*, 50. [CrossRef]
13. Ulusay, R.; Hudson, J.A. *The Complete ISRM Suggested Methods for Rock Characterization, Testing and Monitoring, 1974–2006*; International Society for Rock Mechanics (ISRM): Ankara, Turkey; Pergamon, Turkey; Oxford, UK, 2007.
14. *Standard Test Method for Unconfined Compressive Strength of Intact Rock Core Specimens*; ASTM 2938. ASTM International: West Conshohocken, PA, USA, 1995.
15. Ali, Z.; Karakus, M.; Nguyen, G.D.; Amrouch, K. Effect of loading rate and time delay on the tangent modulus method (TMM) in coal and coal measured rocks. *Int. J. Coal Sci. Technol.* **2022**, *9*, 81. [CrossRef]
16. Bai, Q.; Zhang, C.; Paul Young, R. Using true-triaxial stress path to simulate excavation-induced rock damage: A case study. *Int. J. Coal Sci. Technol.* **2022**, *9*, 49. [CrossRef]

17. Chen, Y.; Zuo, J.; Liu, D.; Li, Y.; Wang, Z. Experimental and numerical study of coal-rock bimaterial composite bodies under triaxial compression. *Int. J. Coal Sci. Technol.* **2021**, *8*, 908–924. [CrossRef]
18. Chi, X.; Yang, K.; Wei, Z. Breaking and mining-induced stress evolution of overlying strata in the working face of a steeply dipping coal seam. *Int. J. Coal Sci. Technol.* **2021**, *8*, 614–625. [CrossRef]
19. Cavaleri, L.; Barkhordari, M.S.; Repapis, C.C.; Armaghani, D.J.; Ulrikh, D.V.; Asteris, P.G. Convolution-based ensemble learning algorithms to estimate the bond strength of the corroded reinforced concrete. *Constr. Build. Mater.* **2022**, *359*, 129504. [CrossRef]
20. Ceryan, N. Application of support vector machines and relevance vector machines in predicting uniaxial compressive strength of volcanic rocks. *J. Afr. Earth Sci.* **2014**, *100*, 634–644. [CrossRef]
21. Asadi, A. Application of artificial neural networks in prediction of uniaxial compressive strength of rocks using well logs and drilling data. *Procedia Eng.* **2017**, *191*, 279–286. [CrossRef]
22. Skentou, A.D.; Bardhan, A.; Mamou, A.; Lemonis, M.E.; Kumar, G.; Samui, P.; Armaghani, D.J.; Asteris, P.G. Closed-Form Equation for Estimating Unconfined Compressive Strength of Granite from Three Non-destructive Tests Using Soft Computing Models. *Rock Mech. Rock Eng.* **2022**, *56*, 487–514. [CrossRef]
23. Zhang, L.; Ding, X.; Budhu, M. A rock expert system for the evaluation of rock properties. *Int. J. Rock Mech. Min. Sci.* **2012**, *50*, 124–132. [CrossRef]
24. Singh, T.N.; Verma, A.K. Comparative analysis of intelligent algorithms to correlate strength and petrographic properties of some schistose rocks. *Eng. Comput.* **2012**, *28*, 1–12. [CrossRef]
25. Gokceoglu, C.; Sonmez, H.; Zorlu, K. Estimating the uniaxial compressive strength of some clay-bearing rocks selected from Turkey by nonlinear multivariable regression and rule-based fuzzy models. *Expert Syst.* **2009**, *26*, 176–190. [CrossRef]
26. Sarkar, K.; Vishal, V.; Singh, T. An empirical correlation of index geomechanical parameters with the compressional wave velocity. *Geotech. Geol. Eng.* **2012**, *30*, 469–479. [CrossRef]
27. Shan, F.; He, X.; Armaghani, D.J.; Zhang, P.; Sheng, D. Success and challenges in predicting TBM penetration rate using recurrent neural networks. *Tunn. Undergr. Space Technol.* **2022**, *130*, 104728. [CrossRef]
28. Verwaal, W.; Mulder, A. Estimating rock strength with the Equotip hardness tester. *Int. J. Rock Mech. Min. Sci. Geomech. Abstr.* **1993**, *30*, 659–662. [CrossRef]
29. Yagiz, S.; Sezer, E.; Gokceoglu, C. Artificial neural networks and nonlinear regression techniques to assess the influence of slake durability cycles on the prediction of uniaxial compressive strength and modulus of elasticity for carbonate rocks. *Int. J. Numer. Anal. Methods Géoméch.* **2012**, *36*, 1636–1650. [CrossRef]
30. Grima, M.A.; Babuška, R. Fuzzy model for the prediction of unconfined compressive strength of rock samples. *Int. J. Rock Mech. Min. Sci.* **1999**, *36*, 339–349. [CrossRef]
31. Indraratna, B.; Armaghani, D.J.; Correia, A.G.; Hunt, H.; Ngo, T. Prediction of resilient modulus of ballast under cyclic loading using machine learning techniques. *Transp. Geotech.* **2023**, *38*, 100895. [CrossRef]
32. Feng, F.; Chen, S.; Zhao, X.; Li, D.; Wang, X.; Cui, J. Effects of external dynamic disturbances and structural plane on rock fracturing around deep underground cavern. *Int. J. Coal Sci. Technol.* **2022**, *9*, 15. [CrossRef]
33. Gao, R.; Kuang, T.; Zhang, Y.; Zhang, W.; Quan, C. Controlling mine pressure by subjecting high-level hard rock strata to ground fracturing. *Int. J. Coal Sci. Technol.* **2021**, *8*, 1336–1350. [CrossRef]
34. Gorai, A.K.; Raval, S.; Patel, A.K.; Chatterjee, S.; Gautam, T. Design and development of a machine vision system using artificial neural network-based algorithm for automated coal characterization. *Int. J. Coal Sci. Technol.* **2021**, *8*, 737–755. [CrossRef]
35. He, S.; Qin, M.; Qiu, L.; Song, D.; Zhang, X. Early warning of coal dynamic disaster by precursor of AE and EMR "quiet period". *Int. J. Coal Sci. Technol.* **2022**, *9*, 46. [CrossRef]
36. Jangara, H.; Ozturk, C.A. Longwall top coal caving design for thick coal seam in very poor strength surrounding strata. *Int. J. Coal Sci. Technol.* **2021**, *8*, 641–658. [CrossRef]
37. Nikolenko, P.V.; Epshtein, S.A.; Shkuratnik, V.L.; Anufrenkova, P.S. Experimental study of coal fracture dynamics under the influence of cyclic freezing–thawing using shear elastic waves. *Int. J. Coal Sci. Technol.* **2021**, *8*, 562–574. [CrossRef]
38. Demir Sahin, D.; Isik, E.; Isik, I.; Cullu, M. Artificial neural network modeling for the effect of fly ash fineness on compressive strength. *Arab. J. Geosci.* **2021**, *14*, 2705. [CrossRef]
39. Chen, S.; Xiang, Z.; Eker, H. Curing Stress Influences the Mechanical Characteristics of Cemented Paste Backfill and Its Damage Constitutive Model. *Buildings* **2022**, *12*, 1607. [CrossRef]
40. Köken, E. Assessment of Los Angeles Abrasion Value (LAAV) and Magnesium Sulphate Soundness (Mwl) of Rock Aggregates Using Gene Expression Programming and Artificial Neural Networks. *Arch. Min. Sci.* **2022**, *67*, 401–422.
41. Şahin, D.D.; Kumaş, C.; Eker, H. Research of the Use of Mine Tailings in Agriculture. *JoCREST* **2022**, *8*, 71–84.
42. Strzałkowski, P.; Köken, E. Assessment of Böhme Abrasion Value of Natural Stones through Artificial Neural Networks (ANN). *Materials* **2022**, *15*, 2533. [CrossRef]
43. Hussain, S.; Muhammad Khan, N.; Emad, M.Z.; Naji, A.M.; Cao, K.; Gao, Q.; Ur Rehman, Z.; Raza, S.; Cui, R.; Salman, M. An Appropriate Model for the Prediction of Rock Mass Deformation Modulus among Various Artificial Intelligence Models. *Sustainability* **2022**, *14*, 15225. [CrossRef]
44. Chen, L.; Asteris, P.G.; Tsoukalas, M.Z.; Armaghani, D.J.; Ulrikh, D.V.; Yari, M. Forecast of Airblast Vibrations Induced by Blasting Using Support Vector Regression Optimized by the Grasshopper Optimization (SVR-GO) Technique. *Appl. Sci.* **2022**, *12*, 9805. [CrossRef]

45. Zhou, J.; Lin, H.; Jin, H.; Li, S.; Yan, Z.; Huang, S. Cooperative prediction method of gas emission from mining face based on feature selection and machine learning. *Int. J. Coal Sci. Technol.* **2022**, *9*, 51. [CrossRef]
46. Huang, F.; Xiong, H.; Chen, S.; Lv, Z.; Huang, J.; Chang, Z.; Catani, F. Slope stability prediction based on a long short-term memory neural network: Comparisons with convolutional neural networks, support vector machines and random forest models. *Int. J. Coal Sci. Technol.* **2023**, *10*, 18. [CrossRef]
47. Vagnon, F.; Colombero, C.; Colombo, F.; Comina, C.; Ferrero, A.M.; Mandrone, G.; Vinciguerra, S.C. Effects of thermal treatment on physical and mechanical properties of Valdieri Marble-NW Italy. *Int. J. Rock Mech. Min. Sci.* **2019**, *116*, 75–86. [CrossRef]
48. Manouchehrian, A.; Sharifzadeh, M.; Moghadam, R.H. Application of artificial neural networks and multivariate statistics to estimate UCS using textural characteristics. *Int. J. Min. Sci. Technol.* **2012**, *22*, 229–236. [CrossRef]
49. Torabi-Kaveh, M.; Naseri, F.; Saneie, S.; Sarshari, B. Application of artificial neural networks and multivariate statistics to predict UCS and E using physical properties of Asmari limestones. *Arab. J. Geosci.* **2015**, *8*, 2889–2897. [CrossRef]
50. Abdi, Y.; Garavand, A.T.; Sahamieh, R.Z. Prediction of strength parameters of sedimentary rocks using artificial neural networks and regression analysis. *Arab. J. Geosci.* **2018**, *11*, 587. [CrossRef]
51. Prabakar, J.; Dendorkar, N.; Morchhale, R. Influence of fly ash on strength behavior of typical soils. *Constr. Build. Mater.* **2004**, *18*, 263–267. [CrossRef]
52. Matin, S.; Farahzadi, L.; Makaremi, S.; Chelgani, S.C.; Sattari, G. Variable selection and prediction of uniaxial compressive strength and modulus of elasticity by random forest. *Appl. Soft Comput.* **2018**, *70*, 980–987. [CrossRef]
53. Suthar, M. Applying several machine learning approaches for prediction of unconfined compressive strength of stabilized pond ashes. *Neural Comput. Appl.* **2020**, *32*, 9019–9028. [CrossRef]
54. Wang, M.; Wan, W.; Zhao, Y. Prediction of the uniaxial compressive strength of rocks from simple index tests using a random forest predictive model. *Comptes Rendus Mécanique* **2020**, *348*, 3–32. [CrossRef]
55. Ren, Q.; Wang, G.; Li, M.; Han, S. Prediction of rock compressive strength using machine learning algorithms based on spectrum analysis of geological hammer. *Geotech. Geol. Eng.* **2019**, *37*, 475–489. [CrossRef]
56. Ghasemi, E.; Kalhori, H.; Bagherpour, R.; Yagiz, S. Model tree approach for predicting uniaxial compressive strength and Young's modulus of carbonate rocks. *Bull. Eng. Geol. Environ.* **2018**, *77*, 331–343. [CrossRef]
57. Saedi, B.; Mohammadi, S.D. Prediction of uniaxial compressive strength and elastic modulus of migmatites by microstructural characteristics using artificial neural networks. *Rock Mech. Rock Eng.* **2021**, *54*, 5617–5637. [CrossRef]
58. Shahani, N.M.; Zheng, X.; Liu, C.; Hassan, F.U.; Li, P. Developing an XGBoost regression model for predicting young's modulus of intact sedimentary rocks for the stability of surface and subsurface structures. *Front. Earth Sci.* **2021**, *9*, 761990. [CrossRef]
59. Armaghani, D.J.; Tonnizam Mohamad, E.; Momeni, E.; Monjezi, M.; Sundaram Narayanasamy, M.S. Prediction of the strength and elasticity modulus of granite through an expert artificial neural network. *Arab. J. Geosci.* **2016**, *9*, 48. [CrossRef]
60. Fairhurst, C.; Hudson, J.A. Draft ISRM suggested method for the complete stress-strain curve for intact rock in uniaxial compression. *Int. J. Rock Mech. Min. Sci. Geomech. Abstr.* **1999**, *36*, 279–289.
61. Małkowski, P.; Niedbalski, Z.; Balarabe, T. A statistical analysis of geomechanical data and its effect on rock mass numerical modeling: A case study. *Int. J. Coal Sci. Technol.* **2021**, *8*, 312–323. [CrossRef]
62. Ramraj, S.; Uzir, N.; Sunil, R.; Banerjee, S. Experimenting XGBoost algorithm for prediction and classification of different datasets. *Int. J. Control. Theory Appl.* **2016**, *9*, 651–662.
63. Chandrahas, N.S.; Choudhary, B.S.; Teja, M.V.; Venkataramayya, M.; Prasad, N.K. XG Boost Algorithm to Simultaneous Prediction of Rock Fragmentation and Induced Ground Vibration Using Unique Blast Data. *Appl. Sci.* **2022**, *12*, 5269. [CrossRef]
64. Shahani, N.M.; Zheng, X.; Guo, X.; Wei, X. Machine learning-based intelligent prediction of elastic modulus of rocks at thar coalfield. *Sustainability* **2022**, *14*, 3689. [CrossRef]
65. Choi, H.-y.; Cho, K.-H.; Jin, C.; Lee, J.; Kim, T.-H.; Jung, W.-S.; Moon, S.-K.; Ko, C.-N.; Cho, S.-Y.; Jeon, C.-Y. Exercise therapies for Parkinson's disease: A systematic review and meta-analysis. *Park. Dis.* **2020**, *2020*, 2565320. [CrossRef] [PubMed]
66. Ogunkunle, T.F.; Okoro, E.E.; Rotimi, O.J.; Igbinedion, P.; Olatunji, D.I. Artificial intelligence model for predicting geomechanical characteristics using easy-to-acquire offset logs without deploying logging tools. *Petroleum* **2022**, *8*, 192–203. [CrossRef]
67. Yang, Z.; Wu, Y.; Zhou, Y.; Tang, H.; Fu, S. Assessment of machine learning models for the prediction of rate-dependent compressive strength of rocks. *Minerals* **2022**, *12*, 731. [CrossRef]
68. Gu, J.-C.; Lee, S.-C.; Suh, Y.-H. Determinants of behavioral intention to mobile banking. *J. Agric. Food Res.* **2009**, *36*, 11605–11616. [CrossRef]
69. Qin, P.; Wang, T.; Luo, Y. A review on plant-based proteins from soybean: Health benefits and soy product development. *J. Agric. Food Res.* **2022**, *7*, 100265. [CrossRef]
70. Frimpong, E.A.; Okyere, P.Y.; Asumadu, J. Prediction of transient stability status using Walsh-Hadamard transform and support vector machine. In Proceedings of the 2017 IEEE PES PowerAfrica, Accra, Ghana, 27–30 June 2017; pp. 301–306.
71. Hassan, M.Y.; Arman, H. Comparison of six machine-learning methods for predicting the tensile strength (Brazilian) of evaporitic rocks. *Appl. Sci.* **2021**, *11*, 5207. [CrossRef]
72. Ogutu, J.O.; Schulz-Streeck, T.; Piepho, H.-P. Genomic selection using regularized linear regression models: Ridge regression, lasso, elastic net and their extensions. *BMC Proc.* **2012**, *6*, S10. [CrossRef]

73. Ozanne, M.; Dyar, M.; Carmosino, M.; Breves, E.; Clegg, S.; Wiens, R. Comparison of lasso and elastic net regression for major element analysis of rocks using laser-induced breakdown spectroscopy (LIBS). In Proceedings of the 43rd Annual Lunar and Planetary Science Conference, The Woodlands, TX, USA, 19–23 March 2012; p. 2391.
74. Sarkar, K.; Tiwary, A.; Singh, T. Estimation of strength parameters of rock using artificial neural networks. *Bull. Eng. Geol. Environ.* **2010**, *69*, 599–606. [CrossRef]
75. Tayarani, N.S.; Jamali, S.; Zadeh, M.M. Combination of artificial neural networks and numerical modeling for predicting deformation modulus of rock masses. *Arch. Min. Sci.* **2020**, *65*, 337–346.
76. Lawal, A.I.; Kwon, S.J. Application of artificial intelligence to rock mechanics: An overview. *J. Rock Mech. Geotech. Eng.* **2021**, *13*, 248–266. [CrossRef]
77. Ma, L.; Khan, N.M.; Cao, K.; Rehman, H.; Salman, S.; Rehman, F.U. Prediction of Sandstone Dilatancy Point in Different Water Contents Using Infrared Radiation Characteristic: Experimental and Machine Learning Approaches. *Lithosphere* **2022**, *2021*, 3243070. [CrossRef]
78. Khan, N.M.; Ma, L.; Cao, K.; Hussain, S.; Liu, W.; Xu, Y. Infrared radiation characteristics based rock failure indicator index for acidic mudstone under uniaxial loading. *Arab. J. Geosci.* **2022**, *15*, 343. [CrossRef]
79. Kutty, A.A.; Wakjira, T.G.; Kucukvar, M.; Abdella, G.M.; Onat, N.C. Urban resilience and livability performance of European smart cities: A novel machine learning approach. *J. Clean. Prod.* **2022**, *378*, 134203. [CrossRef]
80. Bekdaş, G.; Cakiroglu, C.; Kim, S.; Geem, Z.W. Optimal dimensioning of retaining walls using explainable ensemble learning algorithms. *Materials* **2022**, *15*, 4993. [CrossRef]

sustainability

Article

Salt Cavern Thermal Damage Evolution Investigation Based on a Hybrid Continuum-Discrete Coupled Modeling

Kai Feng [1,2,3], Wenjing Li [3,*], Xing Nan [4] and Guangzhi Yang [5]

1 State Key Laboratory of Shale Oil and Gas Enrichment Mechanisms and Effective Development, Beijing 100083, China
2 State Energy Center for Shale Oil Research and Development, Beijing 100083, China
3 Key Laboratory of Petroleum Engineering Beijing, Ministry of Education, Beijing Key Laboratory of Urban Oil & Gas Distribution Technology, China University of Petroleum, Beijing 102249, China
4 Sinochem Energy Logistics Corporation, Ltd., Beijing 100031, China; nanxing@sinochem.com
5 China Petroleum Engineering & Construction Corp. Beijing Engineering Company, Beijing 100085, China; yangguangzhi@cnpc.com.cn
* Correspondence: wenjing.li@cup.edu.cn

Abstract: The integrity and stability of salt caverns for natural gas storage are subjected to a gas cycling loading operation. The coupled effect of confining pressure and temperature on the response of the salt cavity surrounding the wall is essential to stability analysis. In this study, a hybrid continuum-discrete model accounting for the thermal-mechanical process is proposed to investigate the thermal-damage evolution mechanism towards a field case with blocks falling off the salt cavity. The salt cavity is modeled by continuum zones, and the potential damage zones are simulated by discrete particles. Three specimens at different locations around the surrounding wall are compared in the context of severe depressurization. The dynamic responses of rock salt, including temperature spatiotemporal variation, microscopic cracking patterns, and energy evolution exhibit spatial and confinement dependence. A series of numerical simulations were conducted to study the influence of microproperties and thermal properties. It is shown that the evolution of cracks is controlled by (1) the thermal-mechanical process (i.e., depressurization and retention at low pressure) and (2) the anomalous zone close to the brim of the salt cavity surrounding the wall. The zone far away from the marginal surrounding wall is less affected by temperature, and only the mechanical conditions control the development of cracks. This continuum/discontinuum approach provides an alternative method to investigate the progressive thermal damage and its microscopic mechanism.

Keywords: salt cavern; underground gas storage; continuum-discrete coupled method; thermal-mechanical coupling; thermal damage

Citation: Feng, K.; Li, W.; Nan, X.; Yang, G. Salt Cavern Thermal Damage Evolution Investigation Based on a Hybrid Continuum-Discrete Coupled Modeling. *Sustainability* **2023**, *15*, 8718. https://doi.org/10.3390/su15118718

Academic Editors: Mahdi Hasanipanah, Danial Jahed Armaghani and Jian Zhou

Received: 19 April 2023
Revised: 24 May 2023
Accepted: 24 May 2023
Published: 28 May 2023

1. Introduction

Salt rock is an attractive candidate for hosting energy storage, due to its favorable low permeability [1–4]. The salt caverns, which are constructed by the solution mining process, have been used as gas storage for several decades. However, when the salt cavern is subjected to gas-cycling loading, there are potential risks of fractures generation, block fall, and even collapse on the cavern roof [5–7]. The thermal damage evolution induced by the cycling loading process is still unknown and is worth investigating by innovative methods.

1.1. Problem Statement

A field case with blocks falling off from the roof of a salt cavern in Jintan, Jiangsu province of China has been demonstrated by Li et al. [8]. Compared with the sonar results between the four years, there is a 4-m displacement at the shoulder of the salt cavity, indicating the cavern geometry has been altered (see Figure 1). The field engineers assumed that it was the consequence of the thermal effects induced by the gas-cycling loadings. They suspected that the thermal stress led to spalling and resulted in the collapse of the roof eventually, after a 4-year operation. A proposed thermal-mechanical modeling in FLAC3D has been established, and the numerical work confirmed the thermal effect had impacted the stability of the salt cavern to some extent. Additionally, the operation conditions triggering the roof collapse are concluded. Despite the previous research achievements, some fundamental mechanisms of the thermal-dynamic response of rock salt are yet to be understood, particularly the onset and propagation of thermal micro-cracks, i.e., the thermal damage process induced by the thermal effect during gas-cycling loading in a salt cavern.

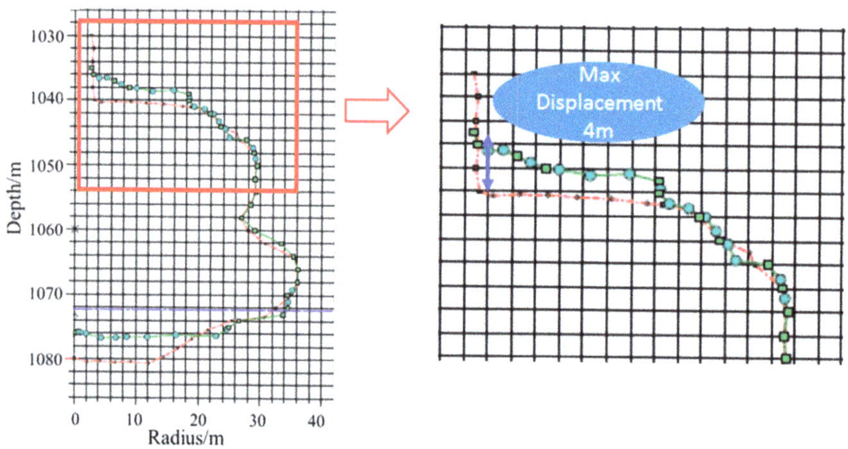

(**a**) (**b**)

Figure 1. A field case in Jintan, sonar results of Cavern L in the years 2009 and 2015, respectively [8]. (**a**) Sonar monitoring results of the cavity; (**b**) Displacement difference before and after cavity collapse.

1.2. Micromechanism of Thermal Damage

Thermo-mechanical responses of containment rocks are critical to the design and safe operation of underground energy storage [9]. The thermal effect induced by a gas injection-and-withdrawal process in a salt cavern was discussed comprehensively [10] (2019), who indicated that a tensile crack is possibly created at the surrounding rock of the salt cavity. Following Bérest's work, both experimental and numerical investigations were conducted to investigate the thermo-mechanical response of salt caverns during rapid cooling [11–17]. To ensure the integrity and stability of salt caverns, fractures, and rock damage should be avoided [18–21]. Rock damage is defined as the degradation of the macroscopic properties, such as strength, stiffness, etc [22–24]. The damage is the consequence of microcracks propagation, coalescence [25–29], and stiffness degradation [30]. The damage mechanics in rock engineering studies the evolution of damage that starts from microcracks and results in rupture failure in the macroscale of the structure [31]. Creep, one of the features of rock salt, is accompanied by microfractures [2]. Under cyclic loading, the fatigue-induced damage of salt rock at first is relatively small and then increases rapidly when it is close to failure [32]. Quick cyclic loading is prone to damage [33]. Ding et al. [34] investigated the grain-scale micromechanisms of the deformation of salt rock and concluded that viscoelastic and hysteretic behaviors are associated with the microprocesses at grain boundaries. Li et al. [35] investigated the damage pattern of rock salt subjected to cycling loadings for CAES and concluded that different responses of the internal structure to fatigue and creep lead to the interaction between creep and fatigue. The micromechanisms of thermal damage in rock salt are even more complicated and challenging.

1.3. Development of Hybrid Modeling

Numerical simulation is considered an important method to study the stability performance during the operation's full life cycle and the associated mechanisms of salt caverns for energy storage [36]. The finite element method (FEM) has been utilized extensively for the salt-cavity integrity analysis. Many constitutive damage-mechanics models have been developed for salt rock. However, those constitutive models are not able to predict damage and other dynamic behaviors associated with microfractures.

The discrete element method (DEM) [37] was proposed later than FEM, as an alternative method, and has its own advantages over other methods. DEM treats rock materials as an assembly of rigid particles bonded with certain segregated contact modes. Discrete elements are independent and allow departing from the rock mass. When forces acting on the particles exceed their bond strength, the contact bond breaks. In DEM, the fracture is deciphered explicitly, the damage progress is reproduced as microcracks coalesce into macrofractures, and the dynamic process can be simulated simultaneously. When compared with FEM, the damage mechanism in DEM is not based on complex damage constitutive correlations; instead, the breakage between particles is simple, while the macroscopic damage is the consequence of individual "breakage", the assembling of discrete dynamic behavior, and the properties of the individual particle. Zhao et al. [38] proposed a grain texture model (GTM) with DEM, and, for the first time, this model can capture the major macromechanical characteristics of textured rock, including the failure process. Despite the unique features of DEM in dealing with the dynamic process, the computation efficiency is limited owing to the huge number of particles. Usually, computation efficiency depends on the number of particles and the size of the domain.

In order to overcome the shortcoming of computational efficiency in DEM, and meanwhile to investigate the dynamic behavior of rock engineering problems, hybrid models based on the continuum-discrete method are adopted. Hu et al. [39] (2021) employed a 3D continuum-discrete coupled method to establish the triaxial Hopkinson bar system, in which the steel bars and a cubic specimen were modeled by continuum zones and bonded-particle sections, respectively. This model was able to simulate the dynamic responses of the rock under different load conditions. Zhang et al. [40,41] (2017, 2019) demonstrated the capability of hybrid discrete-continuum modeling to simulate hydraulic fracturing propagation and interactions with natural fractures. The continuum-discrete coupled model has advantages in reproducing confinement loading via continuum and, meanwhile, accounts for the microstructure and the dynamic microbehavior of the target specimen via a discrete method. In general, the hybrid method is promising in addressing dynamic deformation, the fracturing behaviors of rock, and the related dynamic problems. Particularly, the continuum-discrete hybrid model established a correlation between macroscopic performance and the microscopic mechanism in the damage of rock.

The field case with blocks falling off from the roof of a salt cavern in Jintan, China was investigated with a thermal-mechanical model in FLAC3D (Li et al., 2021) [8]. However, the thermal effects on the micromechanism and microcracking evolution are lacking, and the progressive damage mechanism during the operation is still unknown. In this study, a 3D continuum-discrete coupled hybrid model is established to investigate the thermal-mechanical dynamic behavior of the surrounding rock of the salt cavity subjected to gas cycling loadings. The salt cavern is represented by continuum zones, while the rock specimens on the roof with potential damage risks are simulated by discrete-element modeling. The hybrid model accounts for the influence of temperature variation. The thermal-mechanical coupling mechanism in the hybrid model is described in Section 2 of this paper. In Section 3, three specimens at different locations around the surrounding wall of the salt cavern are selected to understand the thermal damage process in the context of severe depressurization during operation. A parametric study is performed in Section 4 to discuss the influence of confining loading and the effects of rock properties parameters on cracking development. Based on this, the conclusions are presented in Section 5.

2. Numerical Method

A 3D continuum-discrete coupled hybrid model is presented in this section. This model is an improved one after Li et al. (2021) [8]. The salt cavern is represented by continuum zones using fast Lagranian analysis of continua (FLAC3D), while the rock specimens on the roof where the collapse occurs are represented by discrete element modeling, implemented by particle flow code (PFC3D), shown in Figure 2. The coupling methodology is described as follows in detail.

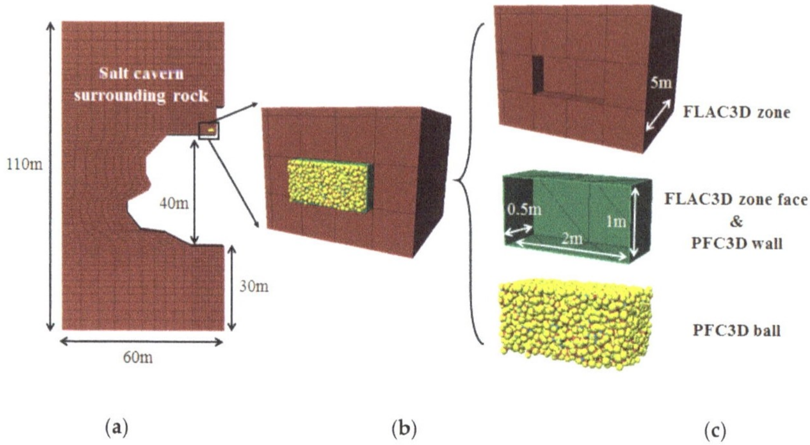

Figure 2. Hybrid continuum-discrete model for a salt cavern subjected to gas-cycling loading. (**a**) Salt cavern surrounding rock continuum part; (**b**) Selected continuous-discrete coupling part; (**c**) The size of the discrete part of the ball ranges from 0.04 to 0.06 m.

2.1. Model Description

To investigate the thermal damage induced by the operation condition, a 3D continuum-discrete coupled model is adopted using FLAC3D-PFC3D. As shown in Figure 3, the salt cavern surrounding rock is a continuous medium and FLAC3D is employed. To improve the computational efficiency, the simulation objective is one quarter of the entire salt cavity. The geometry of the FLAC3D domain is a length of 60 m, a width of 30 m, and a height of 110 m. The FLAC3D zone face and PFC3D wall are the interface, the geometry is 0.5 m × 1 m × 2 m. The rock properties used in the FLAC3D continuum-based method are listed in Table 1.

The PFC3D domain is embedded in a FLAC3D domain, consisting of 10,607 particles, and is located at the cavity shoulder. The location is selected due to its potential damage risk, where the block falls apart from the surrounding rock. For those selected sections, PFC is employed to mimic the dynamic and irreversible damage behavior of the rock salt, aiming at reproducing the evolution of microfractures. From the view of the discrete element method, the rock specimen is regarded as assemblies of discrete rigid particles connected with certain contacts. The movements of particles are governed by Newton's second law. The contact bond breaks when the contact force exceeds the tensile or shear strength of the contact bonds, caused by motion between adjacent particles. The input parameters of the FLAC3D and PFC3D models are listed in Tables 1 and 2, respectively.

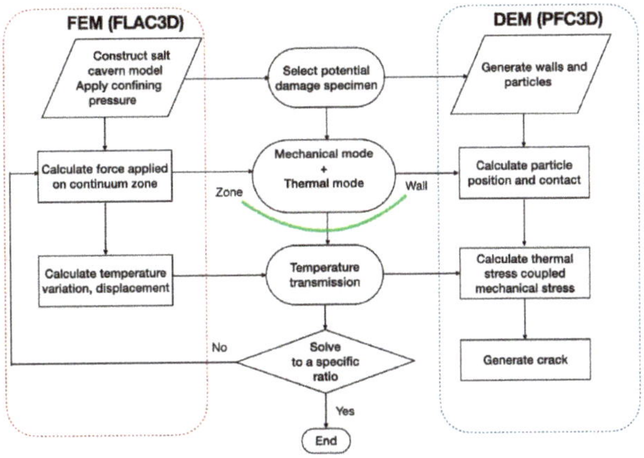

Figure 3. The workflow for simulation of the hybrid continuum-discrete model.

Table 1. Rock mass properties for rock salt in FLAC3D.

Parameters	Units	Values
Young's modulus	GPa	30
Poisson's ratio	/	0.3
Density	kg/m^3	2160
Friction angle, φ	degrees	45
Tensile strength	MPa	4
Cohesion strength	MPa	4
Thermal conductivity	W/m·°C	6.5
Specific heat	J/kg·°C	880
Linear thermal expansion coefficient	°C^{-1}	5×10^{-5}

Table 2. Model parameters in PFC3D.

Parameters	Units	Values
Particle density	kg/m^3	2160
Coefficient of interparticle friction	/	0.3/0.4/0.5/0.6
Normal-to-shear stiffness ratio	/	1.0/1.2/1.4/1.6
Thermal conductivity	W/m·*K*	6.5/7.5/8.5/9.5
specific heat	J/kg·°C	1000/2000/3000/4000
Thermal expansion coefficient	1/°C	$1 \times 10^{-5}/0.7 \times 10^{-5}/0.3 \times 10^{-5}/1 \times 10^{-6}$

2.2. Coupling Mechanism of FLAC-PFC

Figure 3 illustrates the workflow of the DEM-PFC coupled simulation for the thermal-mechanical process of a salt cavern subjected to gas cycling loading. The continuum behavior of the salt cavity is simulated with FLAC, and the DEM model of the rock salt specimen enclosed by a surrounding wall is established by using the commercial code PFC3D. The thermal mode is coupled with the mechanical mode in each computation step. The thermal-mechanical interactive interface between FLAC and PFC is developed to account for the temperature-boundary settings in the FLAC zone, the wall–zone interface, and the particle wall in PFC. The thermal-mechanical coupling process occurs only in one direction (Figure 4): the changes in temperature induce the thermal strains which lead to the change of the mechanical stress, while the influence of mechanical changes on the heat conduction calculation is not considered. The contact forces between particles, displacement, and the distribution of cracks are updated. The model is solved to a predetermined ratio.

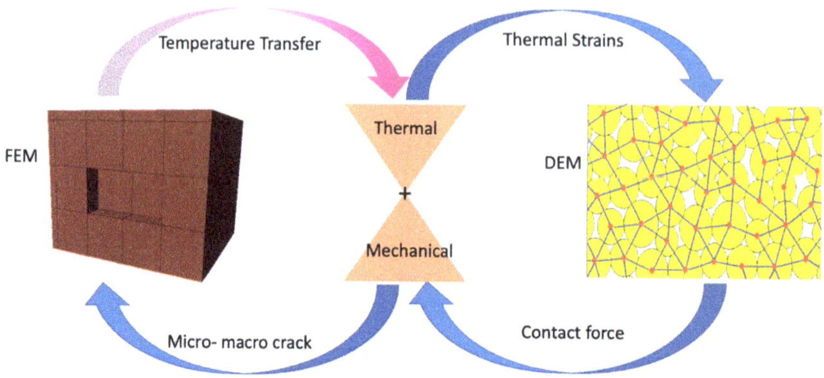

Figure 4. Coupled continuum-discrete model for thermal-mechanical coupling process.

The DEM model accounts for the thermal expansion of particles with linear parallel-bond contacts (Itasca, 2017; Li et al., 2016; Li et al., 2017) [26,42,43]. Figure 4 illustrates the heat conduction in a network composed of thermal reservoirs and pipes: yellow disks represent particles; red dots indicate the heat source; blue lines passing through the contact points of two circular particles are the active thermal pipes. For a bonded linear parallel-bond contact, its mechanical contact is associated with thermal contact. Two consequences can be induced by the thermal-mechanical coupling: first, the particle size is modified due to thermal strain; second, the normal component of the contact force is affected by the temperature changes. The corresponding increment of particle radius ΔR induced by a temperature increment ΔT is:

$$\Delta R = \alpha R \Delta T \tag{1}$$

where α is the coefficient of linear thermal expansion, in the unit of 1/°C. It is a micro property associated with the particle material.

The normal component of the force vector carried by the bond is assumed to be affected by the change in temperature. The relationship between the present parallel bond and active thermal pipe is expressed as:

$$\Delta \overline{F}^{n} = -\overline{k}^{n} A \Delta U^{n} = -\overline{k}^{n} A \left(\overline{a} \overline{L} \Delta T \right) \tag{2}$$

where $\overline{k}^{n}$ is the bond's normal stiffness, A is the area of the bond's cross-section, $\overline{a}$ is the expansion coefficient of bond material, $\overline{L}$ is the bond length, and ΔT is the temperature increment, which equals the average temperature change of the two particles at two ends of the pipe associated with the contact bond.

Figure 5 demonstrates the interaction between the continuum FLAC3D zone and discrete particles. The interface (wall zone) consists of FLAC3D zone surfaces and PFC3D walls, which are created coinciding with the zone faces. The PFC walls are composed of edge-connected triangular faces, and the balls are in contact with the wall facet, wrapping the zone face. The coupling mechanism for FLAC-PFC works by updating the force system at facet vertices in PFC, which is determined by contact forces and moments at each ball–facet contact. The forces along with stiffness are communicated at grid points/nodes. The acting contact force and movement are distributed at grid points/nodes by equivalent forces. The grid points/nodes at the coupling wall zone and grids of zones move synchronously, and the updating force involves continuum zone computation in FLAC. Similarly, the deformation of the continuum zone leads to the movement of the coupling wall zone. In response to the forces and velocities acting at the coupling wall zone, in DEM the particles displace and generate cracks when the stress at contact bonds exceeds the prescribed tensile or shear stress.

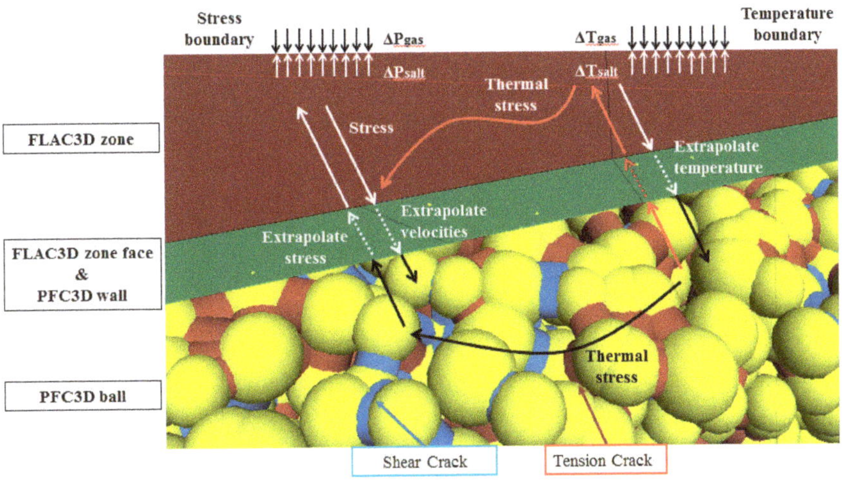

Figure 5. Coupled thermal-mechanical interaction at the FLAC-PFC interface.

3. Thermal Progressive Damage Evolution

Figure 6 illustrates the cycling gas-loading process over the 5-year period. Five years is the time period for sonar monitoring for salt-cavity volume convergence. Particularly, at the time of 3.14 years, the temperature and pressure drop abruptly (from 16 MPa to 8 MPa) when there is gas withdrawal. However, not all the stages of the process of gas injection and withdrawal are xposure to the risk of thermal damage. Li et. al. (2021) [8] investigated cavern L and found that when the gas withdrawal is fast and followed by retaining low pressure, thermal cracking or even fractures occur. Therefore, we focus on the cracking development after 3.14-years of operation by comparing three different locations of surrounding wall of the salt cavity.

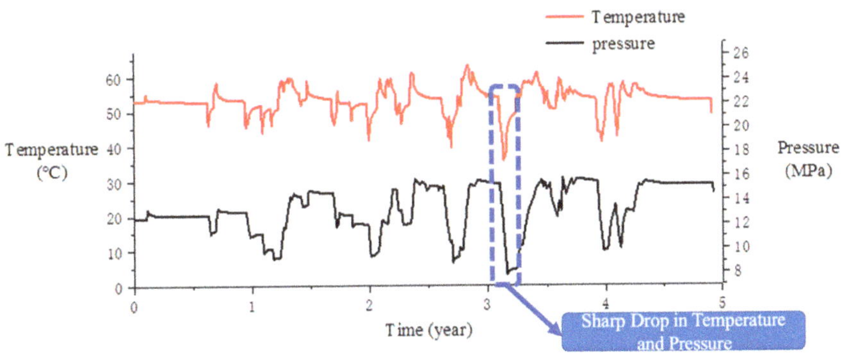

Figure 6. Gas temperature and pressure variations during five years of gas-cycling loading operation (Li et al., 2021) [8].

3.1. Thermal Effect at Three Observed Locations

For the field case of cavern L (Li et al., 2021) [8], a sharp pressure drop occurs due to gas withdrawal at 3.14-years of operation, before and after, the pressure drop reaches

7 MPa (Figure 7a). The three selected monitoring areas and corresponding locations in the hybrid model are (1) at the knee point of the cavity shoulder, i.e., the convexity; (2) the right part of the knee point; and (3) 11.8 m away from the cavity surrounding wall brim.

The initial temperature of rock-salt formation changes linearly with the depth at a temperature gradient of 2.55 °C/100 m. The salt cavern is located from −1030 m to −1080 m, assuming the ground temperature is 20 °C and the temperature of the cavity is approximately 45 °C. In response to the gas withdrawal and the consequent sharp pressure drop at 3.14 years of operation, the temperature decreases correspondingly. The temperature decreases gradually from the surface to the neighboring deeper domain in the formation. Overall, the temperature distribution of the monitoring areas after sharp gas depressurization exhibits obvious confinement dependence on the locations. Location 1 is the closest spot to the brim of the surrounding wall with the lowest temperature. Location 2 has the highest temperature; however, the temperature variation range is small. Locations 2 and 3 are less affected by the thermal effect. The closer to the salt cavity surrounding the wall brim, the more distinct the thermal influence it receives.

To assess the thermal effect on the response of the salt cavern surrounding wall, we proposed a thermal-mechanical factor TMF, which is defined as:

$$TMF = \frac{Crack(Thermal - Mechanical) - Crack(Mechanical)}{Total\ Crack\ (Thermal - Mechanical)} \tag{3}$$

TMF is the ratio of the difference between the crack number induced by the coupled thermal-mechanical effect and the crack number only induced by the mechanical effect to the total crack number. The higher TMF indicates that the thermal effect is dominant. TMF is close to 0.5, indicating both the thermal and mechanical working together. If TMF is approximately 0, the cracking is simply induced by the mechanical effect and is irrelevant to the thermal effect.

Figure 8 illustrates the crack development comparison between the thermal-mechanical coupled effect and the mechanical effect only for three different locations. The different particle colors represent the different detached broken particles formed as a result of cracking. At locations 1, 2, and 3, the $TMF = 0.65$, 0.18, and 0.58, respectively. Location 1, nearest to the cavity, is affected by the thermal effect most. Location 1 is sensitive to the thermal effect and the microcracking is controlled by the ball heat capacity, coefficient of thermal expansion, zone conductivity, conductivity coefficient of thermal contact mode, etc.

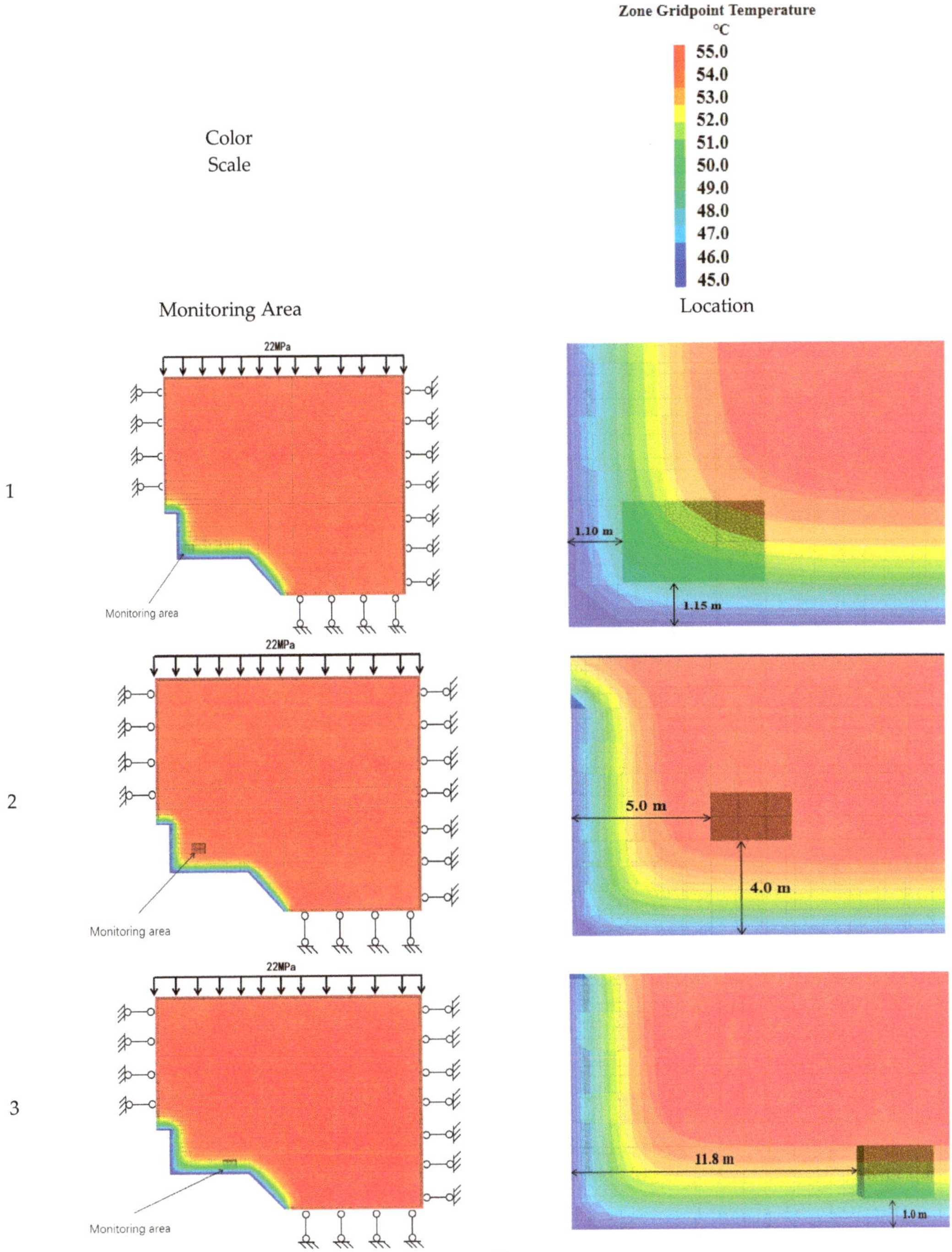

(**a**)

Figure 7. *Cont.*

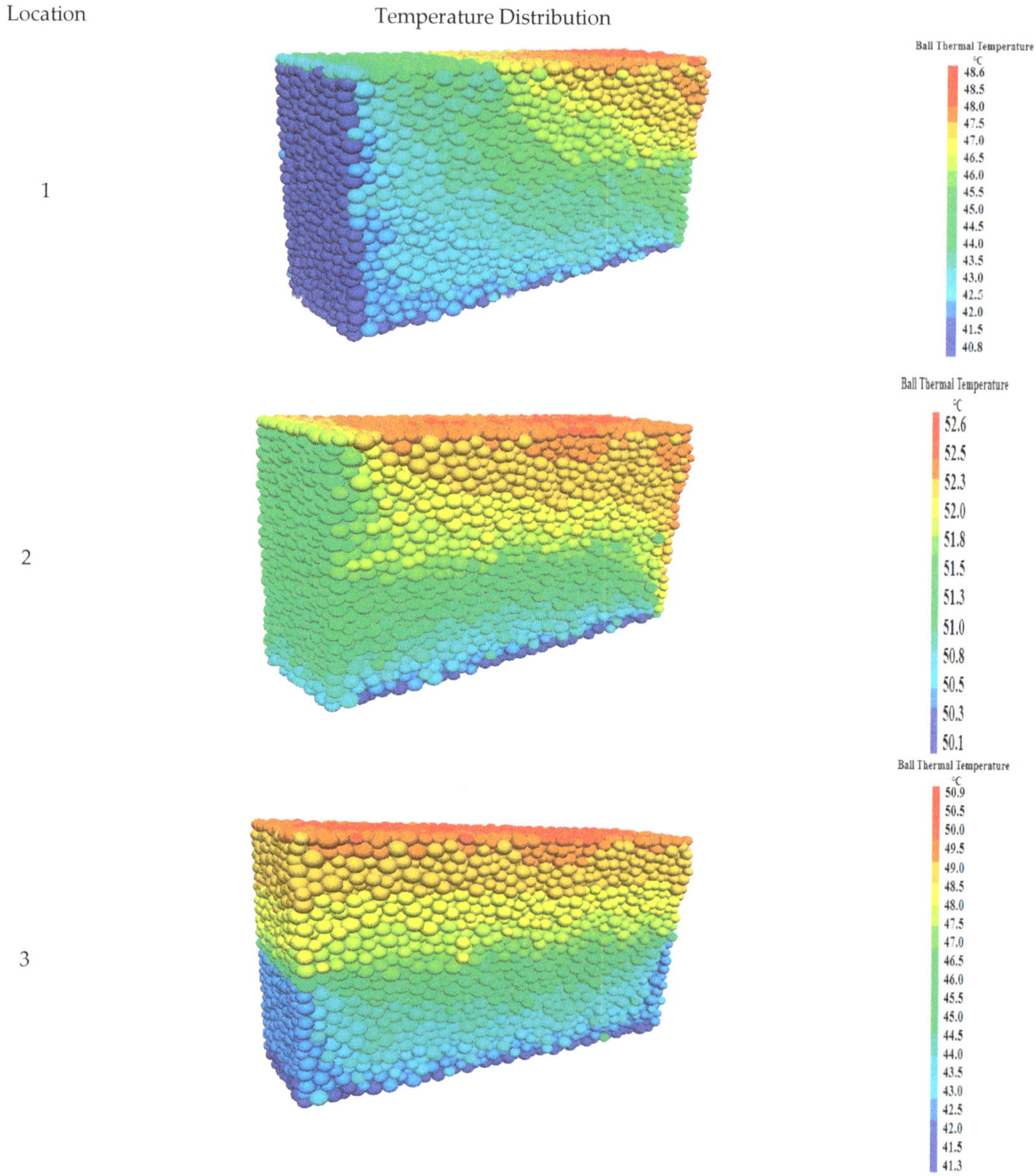

(**b**)

Figure 7. Three selected locations for observation of microcrack development in PFC (**a**); and the corresponding temperature (**b**) after sharp pressure drop due to gas withdrawal at 3.14 years of operation.

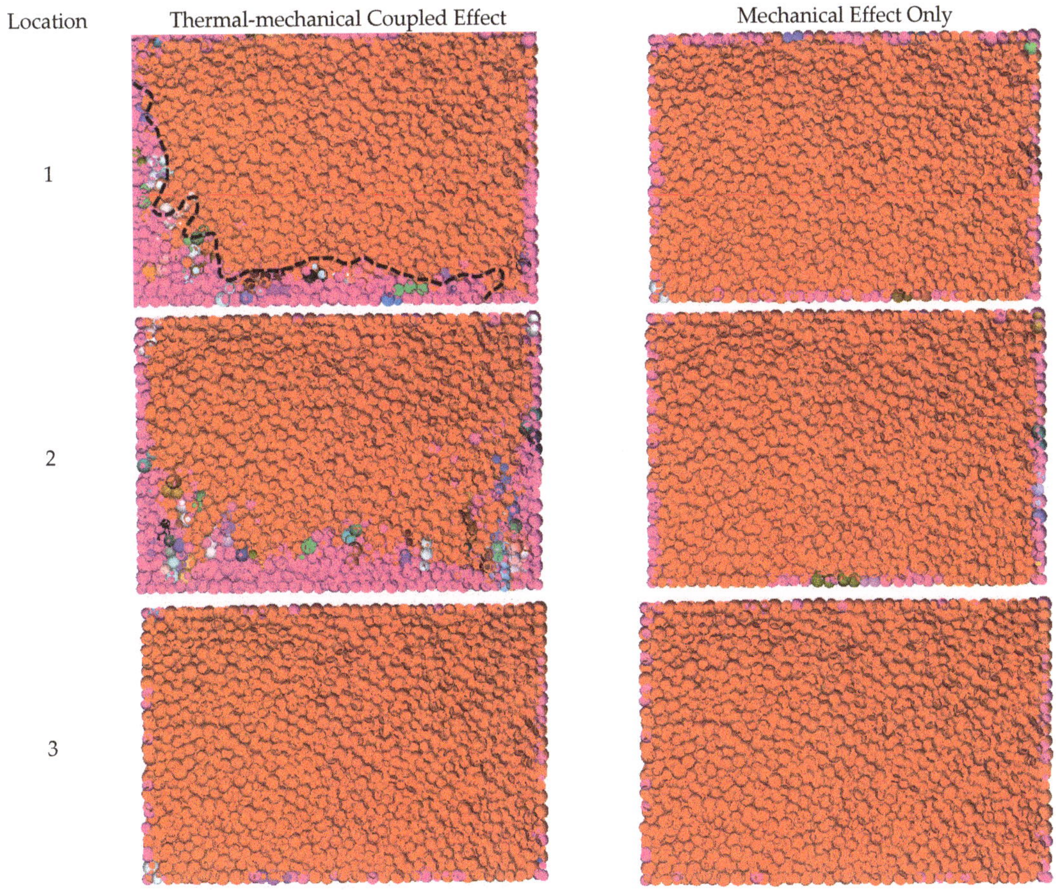

Figure 8. Comparison of crack development at three locations for coupled thermal-mechanical effect and mechanical effect only.

3.2. *Dynamic Response to Depressurization and Progressive Damage Characteristics*

Crack increments at three different locations are compared (Figure 9). The crack numbers at locations 1 and 3 are much higher than the number at location 2. The closer to the center of the cavern, the more affected by the thermal effect, and the more cracks formed. Particularly for location 1, close to the section of the most irregular geometrical of the caverns, which is the concentration of high stress, and has the potential local failure zones resulting from microcracking and might lead to damage. The tension crack is the dominant crack mode, as the tension cracking is the failure consequence of tension force (Equation (2)) induced by the thermal effect. Only the normal component of contact force between particles is affected by thermal expansion, resulting in tension failure. However, the influence of temperature is limited to some extent. The thermal effects are negligible beyond 10 m away from the brim of the surrounding wall of the salt cavity. At location 2, it is noticed that the increasing rate of tension crack decreases with time. Figure 10 illustrates the comparison of cracks development due to severe pressure drop, before and after. It can be seen clearly from the side view of the DEM specimen. After the depressurization, it is observed at location 1 that the cracks are generated and propagated starting from the lower left corner close to the cavity brim, the shear cracks are dominant. The sideview shows that the microcracking induced the discontinuities and was prone to particles falling off.

Location 3 is much closer to the brim of the cavity along the edge and cracks are formed and propagate on both sides of the DEM specimen. Location 2 is farther away from the cavity wall, hence there are fewer cracks, and the cracks propagate from the left lower corner dominated by tension cracks.

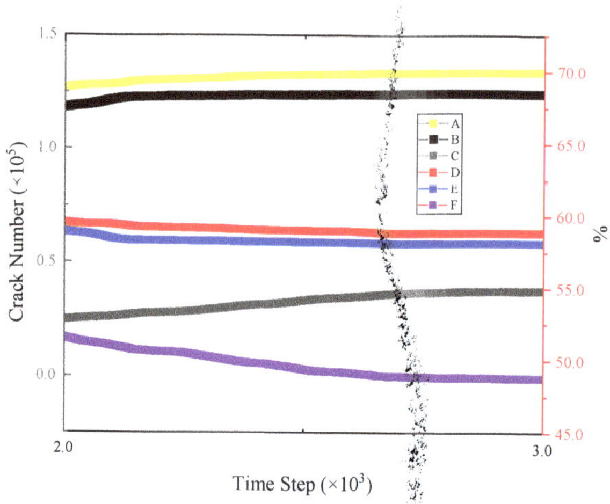

Figure 9. Temporal and spatial evolution of crack increments: A—Location 1; B—Location 3; C—Location 2; D—The ratio of tensile cracks to total cracks at Location 1; E—The ratio of tensile cracks to total cracks at Location 3; F—The ratio of tensile cracks to total cracks at Location 2.

In the PFC3D model, the fracture mass density is defined as the total fracture surface area per unit volume. The definition of fracture area is as followings:

If the domain is cubic, L is the length of the side of the cube. The number of fractures with sizes between l_1 and l_2 is given by:

$$n(l_1 \le l \le l_2) = \int_{l_1}^{l_2} n(l) \cdot L^3 dl = a\left(\frac{l_2^{1-a} - l_1^{1-a}}{1-a} \cdot L^3\right) \tag{4}$$

The term a fixes the total fracture density by a range of fracture sizes.

$$d_m(l_c) \cong \int_{l_c}^{\infty} n(l) \cdot l^2 \cdot L^{-3} \cdot dl \tag{5}$$

The damage process is associated with cracking induced by cycles of gas pressurization and depressurization in the salt cavern.

To better evaluate the distribution of microcracks in discrete modeling, a normalized relative index c_d is employed. The base point is 1.25×10^{-3} m^2/m^3, as it is the value of the fracture surface area per unit volume. The absolute index c_d is the fracture mass density, which is defined as the fracture surface area per unit volume. The normalized concentration degree of crack index c_d ranging from 0 to 1 is defined as the absolute index normalized by the maximum to minimum value among all the cracks. Figure 11 shows the comparison of three different locations of the surrounding wall of the salt cavity, which are: (a) location 1—at the knee point of the cavity shoulder, i.e., the convexity closest to the rim; (b) location 2—right part off the knee point, deep in the rock salt formation; (c) location 3—11.8 m away from the cavity surrounding the wall edge, the most farthest from the centerline of the cavity. The microcrack forms and distributes into the entire specimen in location 1. The crack distribution is more concentrated compared with the other two locations far away from the convexity. The microcracks become less intensive in the region that is not close to the inner cavity with no convexity–concavity. The damage failure exhibits operation confinement and shape dependence. Figure 12 indicates the trend of the crack number using the normalized relative index c_d. It shows that most of the cracks are with $c_d = 0.2$. To investigate the heterogeneities of thermal cracking in rock salt, Figure 13 illustrates the rose diagrams of tensile and shear cracks at location 1, location 2, and location 3, respectively. The radial length of each bin indicates the number of shear or tensile cracks oriented within the corresponding angles. It shows that tensile cracks tend to initiate in the horizontal orientation at location 1 which is close to the cavity surrounding the wall. While the orientation of shear cracks in all three selected locations is uniformly distributed.

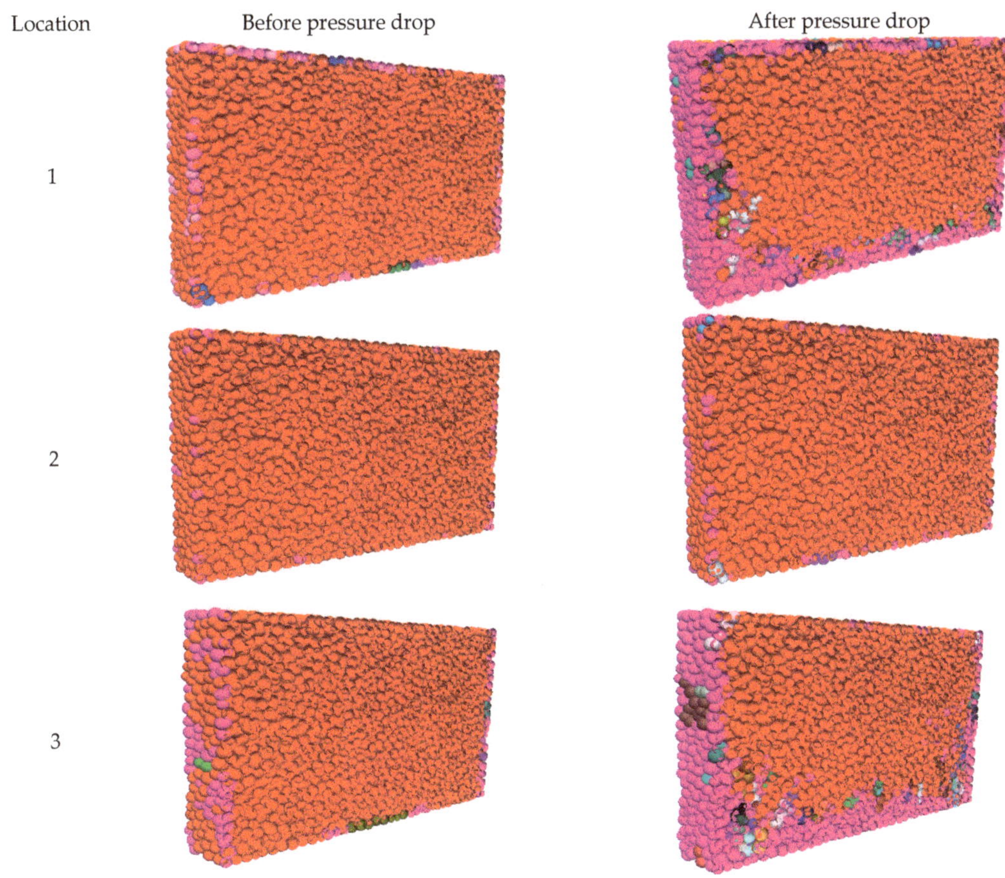

Figure 10. Comparison of crack development at three locations before and after the sharp pressure drop.

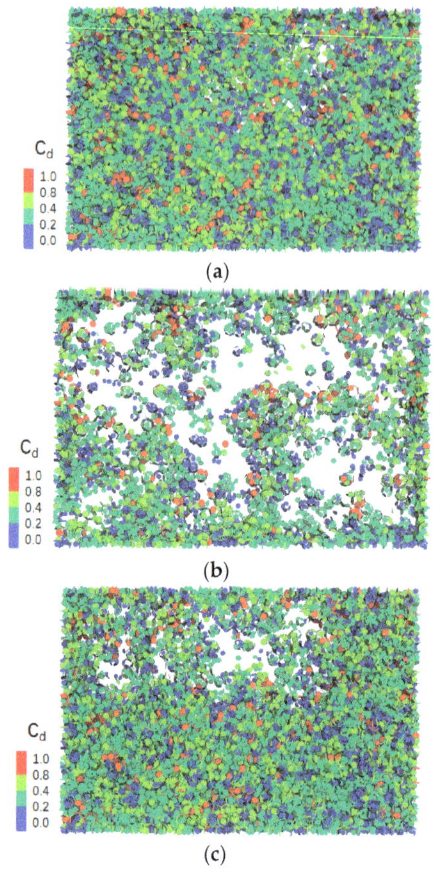

Figure 11. Concentration degree of distribution of microcracks at (**a**) Location 1; (**b**) Location 2; (**c**) Location 3.

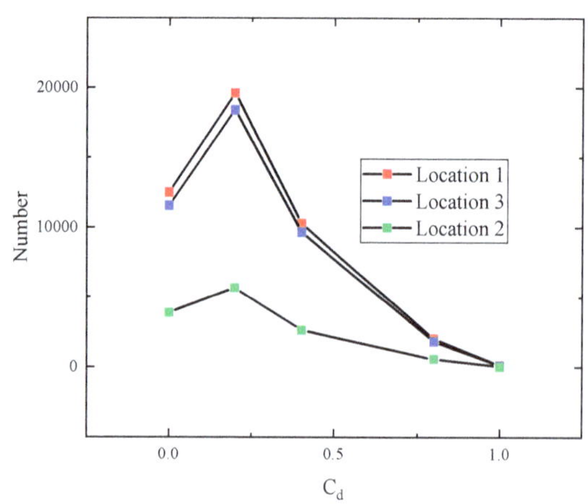

Figure 12. Crack number variation with concentration degree normalized relative index c_d.

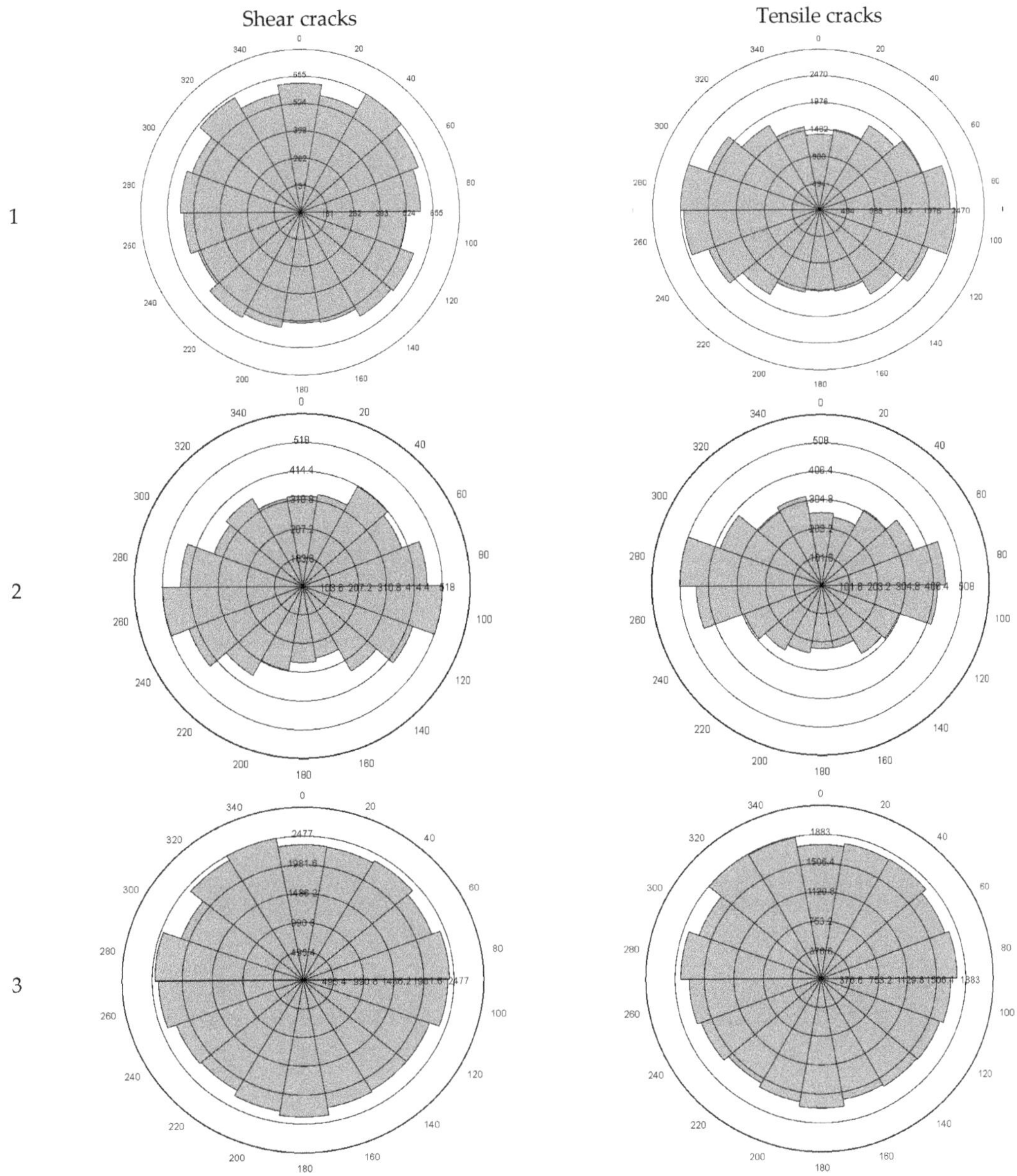

Figure 13. Orientation distribution of shear/tension cracks at different locations.

3.3. Energy Tracking

When rock is subjected to loading, the energy dissipates with the elastic process. Numerical simulation is used to track the development of the strain energy, the slip energy, and the kinetic energy. Those energy partitions reflect the energy evolution during fractures propagation subjected to the thermal-mechanical coupling process. When the thermal effects are taken into account, the total stress is the summation of the mechanical stress and the thermal stress.

Strain energy, E_K, is defined as the energy stored in the linear springs (Itasca Inc., 2019).

$$E_K = \frac{1}{2}\left(\frac{\left(F_n{}^l\right)^2}{K_n} + \frac{\| \mathrm{F}^{\mathrm{l}}{}_{\mathrm{s}} \|^2}{K_S}\right) \tag{6}$$

$F_n{}^l$: linear normal force;
$\mathrm{F}^{\mathrm{l}}{}_{\mathrm{s}}$: linear shear force;
K_n: Normal stiffness;
K_S: Shear stiffness;
Slip energy E_μ, is defined as the total energy dissipated by frictional slip.

$$E_\mu := E_\mu - \frac{1}{2}\left(\left(\mathrm{F_s}^1\right)_o + \mathrm{F_s}^1\right) \cdot \Delta\delta^{\mu} \tag{7}$$

$\left(\mathrm{F_s}^1\right)_o$: linear shear force at the beginning of the timestep;
$\mathrm{F}^{\mathrm{l}}{}_{\mathrm{s}}$: linear shear force;
$\Delta\delta^{\mu}$: shear displacement component decomposed into slip component;
E_{kinetic}: kinetic energy of the particle.

$$E_{\text{kinetic}} = \frac{1}{2}mv^2 \tag{8}$$

m: particle mass;
V: particle velocity.

The evolution of these energy partitions is significantly impacted by the sharp pressure drop, as shown in Figure 14. At location 1, the strain energy decreases slightly followed by a dramatic rise and an increase of crack numbers. The initiation of cracks is positively correlated to the dissipation of strain energy, as the stain energy is stored in the contacts of neighboring particles. The higher rate of energy dissipation means that more cracks initiate. Slip energy rises gradually, indicating the increase of the relative deformation between particles. There is fluctuation in kinetic energy at all the three locations. The particle motion leads to the accumulation of energy to create new cracks. At first, the kinetic energy along the slip energy starts to rise rapidly, probably due to the thermal effect. Meanwhile, the strain energy, E_{strain}, assumed stored majority input work gradually increases. The breakage of contact bonds indicates more microcracks are initiated. With the reduction of kinetic energy, the kinetic energy is converted to other forms of energy, which leads to the consistent growth of the strain energy. Then, fewer new micro-cracks are developed, and the relative particle motion reduces. As a result, kinetic energy is decreasing gradually. The energy evolution can be attributed to the confinement condition (i.e., the pressure drop in this case), and is clearly associated with fracture generation and propagation.

To evaluate the thermal effect on the energy of the microscopic particle system, we proposed a thermal-mechanical energy factor. TMK, TML, TMS, and TMK are defined as:

$$TMK = \frac{A_K - B_K}{A_K} \tag{9}$$

where A_K is the kinetic energy of particles under thermal-mechanical force and B_K is the kinetic energy of particles under mechanical force only.

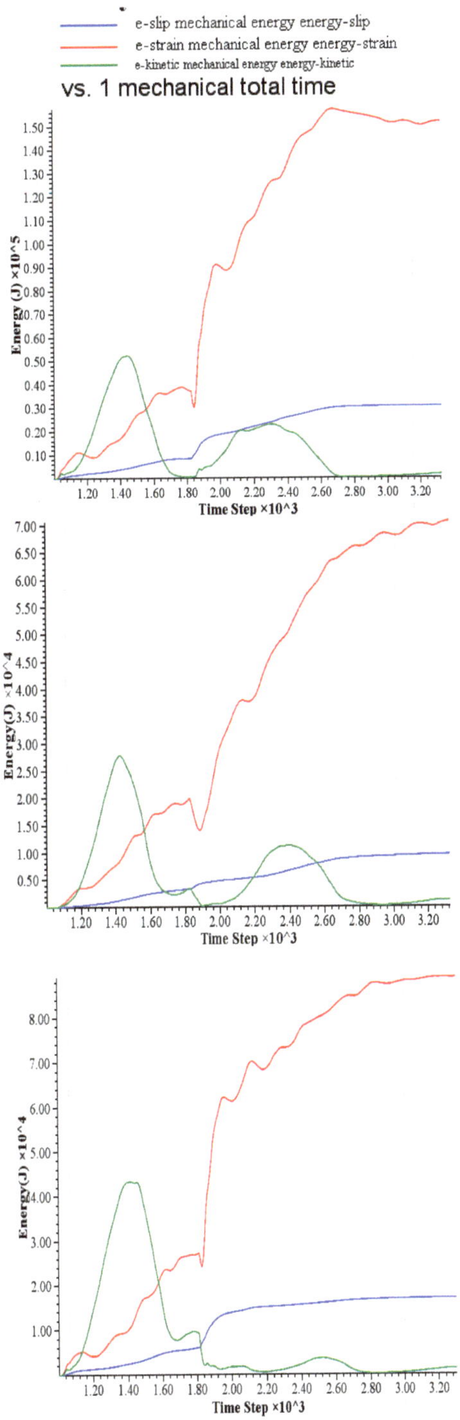

Location 1

Location 2

Location 3

Figure 14. Energy evolution at three different locations.

The larger the *TMK* is, the greater the influence of the thermal effect on the kinetic energy of particles, and the closer *TMK* is to one, indicating that the kinetic energy of particles is dominated by the thermal effect. *TMK* is close to zero, indicating that the kinetic energy of particles is affected by dynamic mechanical force only.

$$TML = \frac{A_L - B_L}{A_L} \tag{10}$$

where A_L is the sliding energy of particles under thermal-mechanical force and B_L is the sliding energy of particles under mechanical force only.

$$TMS = \frac{A_S - B_S}{A_S} \tag{11}$$

where A_S is the strain energy of particles under thermal-mechanical force and B_S is the strain energy of particles under mechanical force only.

From Figure 15, it is not difficult to see that the *TMLs* of position 1 and position 2 are near the red line, indicating that the thermal effect has an influence on the sliding energy. According to the slope of each line, it can be seen that from location 1 to location 2 and then to location 3, which is, the farther away from the sensitive brim of the salt cavity, the influence of the thermal effect gradually attenuates.

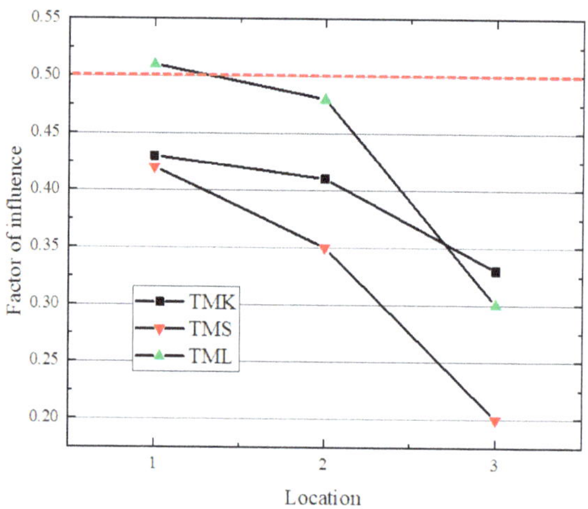

Figure 15. Thermal effect on energy of microscopic particle system at three different locations.

4. Results and Discussion

4.1. Influence of Confining Pressure

The thermodynamic response of the surrounding wall in a salt cavern subjected to a gas-cycling loading process is further investigated in the context of different confinement conditions. The operation pressure range of cavern L (Li et al., 2021) [8] is 16 MPa to 8 MPa. Three different confining pressures are schemed, which are 16 MPa, 12 MPa, and 8 MPa, respectively.

Figure 16 shows the influence of the confinement pressure at location 1 (see Figure 7), which is at the knee point of the cavity shoulder (i.e., the closest location to the surrounding wall brim). The reduction of confining pressure (from 16 MPa to 8 MPa) augments the cracks generation during the depressurization process. The propagation and accumulation of microcracks depend on the variation of the confining pressure. Lower confinement enhances the thermal damage of the salt cavity.

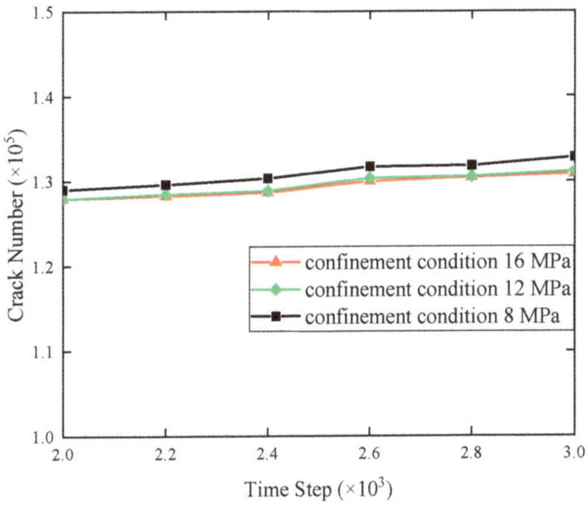

(**a**)

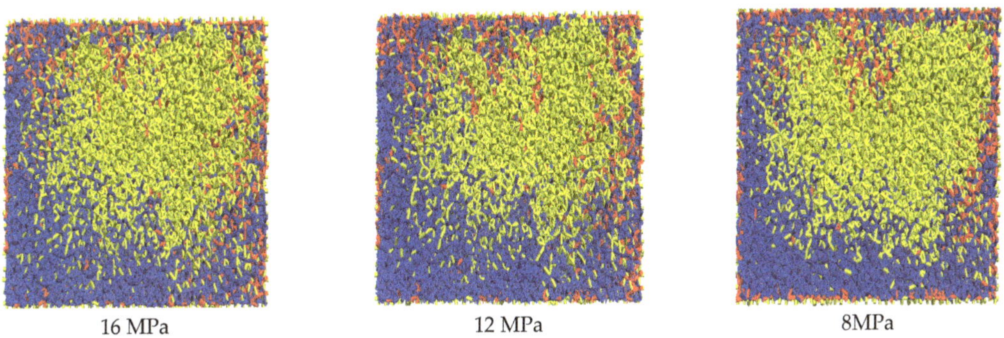

(**b**)

Figure 16. Influence of confining pressure on (**a**) crack increments; (**b**) failure mode; Yellow represents the intact particles; Blue represents the detached particles induced by tension force; Red represents the detached particles induced by shear force.

The pressure is first increased to 8 MPa and then is retained at 8 MPa for a period. As shown in Figure 17, more microcracks initiate when the pressure is retained at 8 MPa. Tensile cracking is the dominant crack mode during the lower-pressure operation. However, the proportion of tensile cracks reduces with time due to the limitation of temperature extension. The more damage accumulates at that lower pressure and increases with the retaining time. Low pressure coupled with retaining time enhances the development of microcracks, ultimately resulting in a damage zone.

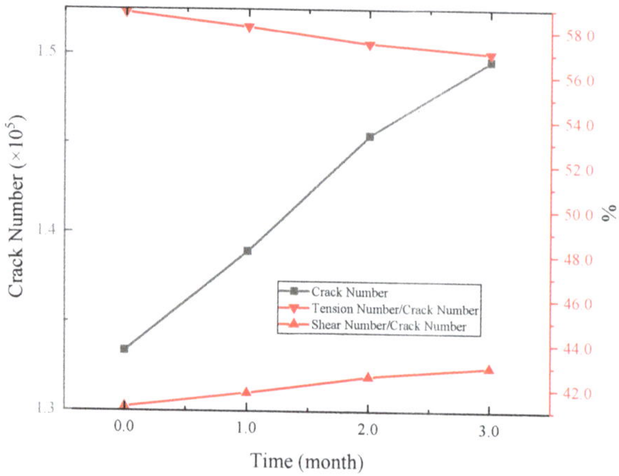

(a)

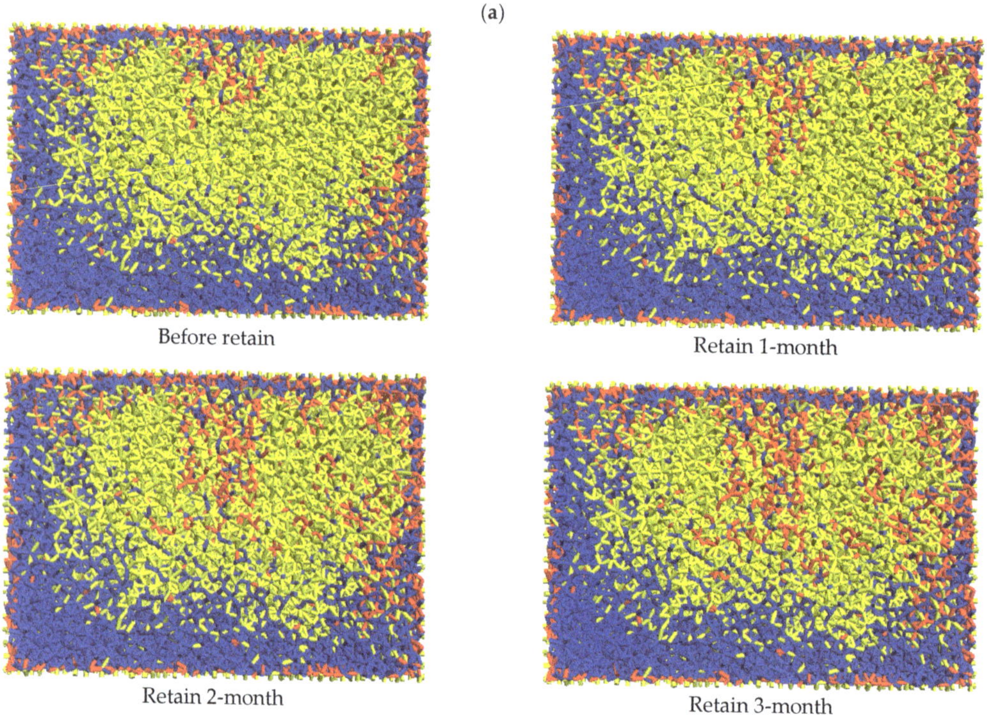

(b)

Figure 17. Temporal and spatial evolution of microcracking at operation pressure of 8 MPa: (**a**) crack increment; (**b**) damage propagation process with retain time.

4.2. Effect of Particle Microproperties

The development of cracks is affected by the microproperties of the particle, including the normal-to-shear stiffness ratio and the coefficient of interparticle friction. As shown in Figure 18, with the increase of the normal-to-shear stiffness ratio, more cracks initiate, leading to the lower yield strength.

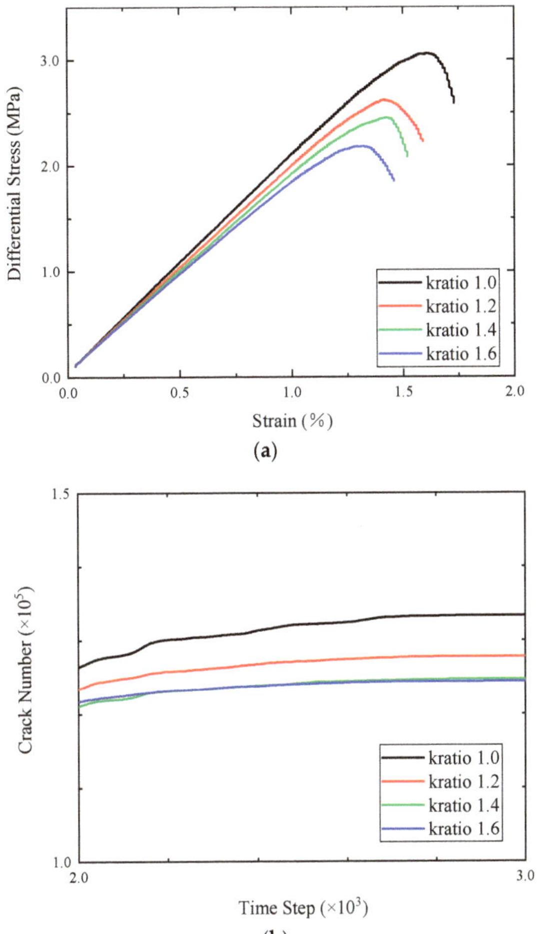

Figure 18. Effect of normal-to-shear ratio on (**a**) stress–strain curves; (**b**) microcrack increments.

Figure 19 illustrates the effect of the interfacial friction coefficient. In general, the friction coefficient has a slight effect on stress–strain curve. With the increase of the friction coefficient, the deformation resistance of particles is enhanced. It is shown that when the friction coefficient increases from 0.5 to 0.8, the larger friction between the particles makes it difficult to exceed the tensile or shear strength of the bonds. Therefore, fewer cracks are generated.

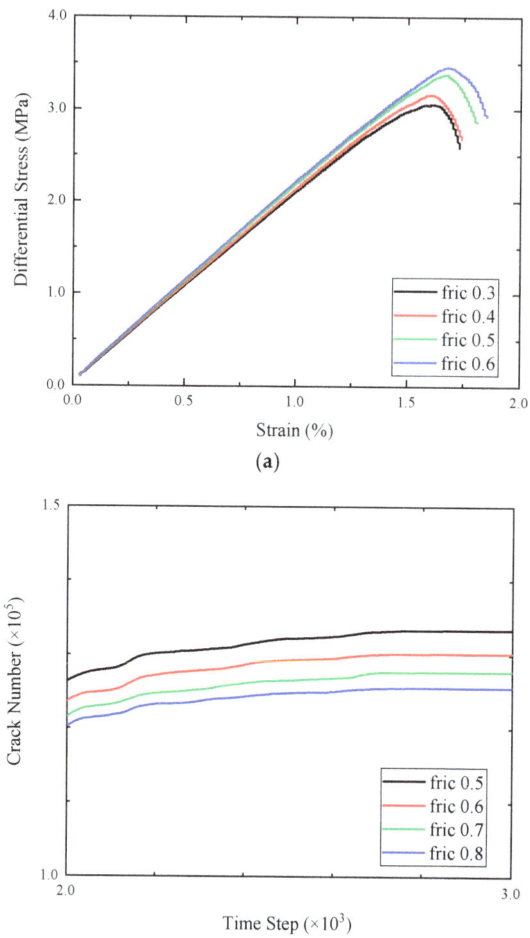

Figure 19. Effect of interfacial friction coefficient on (**a**) stress–strain curves; (**b**) microcrack increments.

4.3. *Effect of Thermal Properties*

Simulations with various thermal conductivities, particle-specific heats, and thermal expansion coefficients are carried out to understand the effect of the thermal properties. All the other parameters are set to be the same as the case with pressure equal to 16 MPa at operation. As shown in Figure 20, the increase in thermal conductivity leads to an increment in crack number. On the other hand, a larger particle-specific heat enhances the energy stored between particles and slows down the heat conduction and the thermal expansion of the particles. Thus, fewer cracks are generated. The thermal-expansion coefficient is positively correlated with the number of cracks by augmenting the force carried by the bond. Hence, the thermal properties of the particles have an important impact on the crack generation and propagation, and the macro-behavior of the surrounding rock wall of a salt cavern as well.

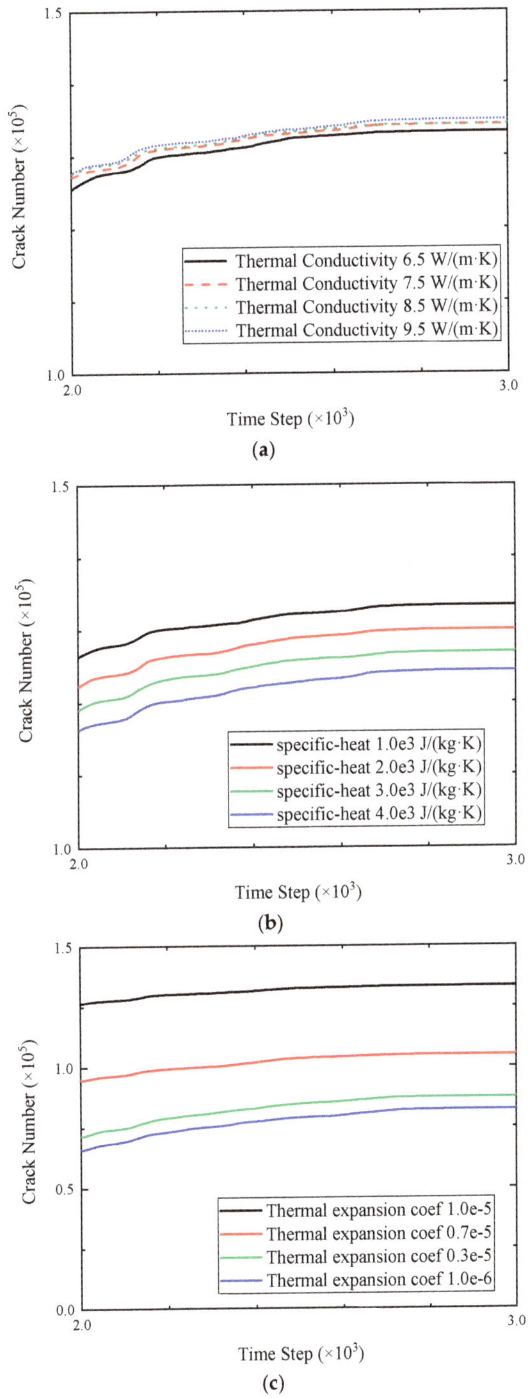

Figure 20. Microcrack increments with variations of (**a**) thermal conductivity; (**b**) particle-specific heat; and (**c**) thermal expansion coefficient.

5. Conclusions

A thermal-mechanical model was proposed to study a field case with blocks falling off from the salt cavern roof using a 3D hybrid continuum-discrete method. This hybrid model has advantages in allowing for implementing the gas-loading operation conditions via continuum modeling; meanwhile, the thermal-mechanical effects and processive damage were systematically investigated with the proposed micromechanical numerical framework of the hybrid model. The main conclusions are as the followings:

Three specimens at different locations around the surrounding wall of a salt cavern were selected to understand the thermal-damage evolution process under the severe depressurization condition. The crack microscopic patterns were subjected to the cavern's anomalous geometry. The energy evolution associated with fracture creation exhibited spatiotemporal confinement dependence.

Cycling loading at a lower confining pressure and the longer retaining time of low-pressure led to progressive microcracking, which further resulted in the development of macrofractures and the formation of the damage zone. The tensile crack was the dominant crack mode. The anisotropic distribution of crack orientation was observed in the zone close to the edge of the surrounding wall, which was the consequence of tension failure induced by the thermal effect. The thermal effect induced by gas injection and withdrawal was limited. The zone far away from the marginal surrounding wall was less affected by temperature and only the mechanical conditions controlled the development of cracks.

The proposed 3D continuum-discrete coupled thermal-mechanical hybrid model overcomes the limitation of the continuum method and was capable of analysis of rock heterogeneity by accounting for the thermal-mechanical effect and naturally capturing the microcrack initiation and propagation by discrete modeling. The hybrid model was successfully applied in the assessment of thermal-damage evolution. This study can shed light on the analysis of thermal-mechanical coupling and the understanding of the influence of microcrack development on the macroscopic behavior of salt rock. In the future, the validity of the continuum-discrete coupled thermal-mechanical hybrid model should be verified by more field data and the thermal effect of the gas injection and withdrawal process needs to be investigated for energy storage in the deep stratum.

Author Contributions: Conceptualization, W.L.; Investigation, K.F. and W.L.; Methodology, X.N.; Software, K.F.; Supervision, G.Y.; Writing—original draft, W.L.; Writing—review and editing, K.F. All authors have read and agreed to the published version of the manuscript.

Funding: This study is funded by the National Natural Science Foundation of China project (Grant NO. 42102304). Open Research Fund of State Key Laboratory of Shale Oil and Gas Enrichment Mechanisms and Effective Development—State Energy Center for Shale Oil Research and Development.

Institutional Review Board Statement: Not applicable.

Informed Consent Statement: Not applicable.

Data Availability Statement: Not applicable.

Acknowledgments: The authors also thank Jintan Underground Gas Storage, China Oil and Gas Piping Network Corporation for their advice on this study.

Conflicts of Interest: The authors declare that they have no known competing financial interests or personal relationships that could have appeared to influence the work reported in this paper.

References

1. Li, S.-Y.; Urai, J.L. Rheology of rock salt for salt tectonics modeling. *Pet. Sci.* **2016**, *13*, 712–724. [CrossRef]
2. Jackson, M.P.; Hudec, M.R. *Salt Tectonics: Principles and Practice*; Cambridge University Press: Cambridge, UK, 2017.
3. Arson, C. Micro-macro mechanics of damage and healing in rocks. *Open Géoméch.* **2020**, *2*, 1–41. [CrossRef]
4. AbuAisha, M.; Rouabhi, A.; Billiotte, J.; HadjeHassen, F. Noneisothermal twoephase hydrogen transport in rock salt during cycling in underground caverns. *Int. J. Hydrogen Energy* **2021**, *46*, 6632–6647. [CrossRef]
5. Bérest, P.; Brouard, B.; Djizanne, H.; Hévin, G. Thermomechanical effects of a rapid depressurization in a gas cavern. *Acta Geotech.* **2014**, *9*, 181–186. [CrossRef]

6. Sicsic, P.; Bérest, P. Thermal cracking following a blow out in a gas-storage cavern. *Int. J. Rock Mech. Min. Sci.* **2014**, *7*, 320–329. [CrossRef]
7. Böttcher, N.; Görke, U.-J.; Kolditz, O.; Nagel, T. Thermo-mechanical investigation of salt caverns for short-term hydrogen storage. *Environ. Earth Sci.* **2017**, *76*, 98. [CrossRef]
8. Li, W.; Nan, X.; Chen, J.; Yang, C. Investigation of thermal-mechanical effects on salt cavern during cyclying loading. *Energy* **2021**, *232*, 120969. [CrossRef]
9. Wu, W.; Lu, D.; Romagnoli, A. A temperature gradient test system for investigating thermo-mechanical responses of containment materials of underground storage facilities. *Rock Mech. Bull.* **2023**, *2*, 100043. [CrossRef]
10. Bérest, P. Heat transfer in salt caverns. *Int. J. Rock Mech. Min. Sci.* **2019**, *120*, 82–95. [CrossRef]
11. Serbin, K.; Ślizowski, J.; Urbańczyk, K.; Nagy, S. The influence of thermodynamic effects on gas storage cavern convergence. *Int. J. Rock Mech. Min. Sci.* **2015**, *79*, 166–171. [CrossRef]
12. Khaledi, K.; Mahmoudi, E.; Datcheva, M.; Schanz, T. Stability and serviceability of underground energy storage caverns in rock salt subjected to mechanical cyclic loading. *Int. J. Rock Mech. Min. Sci.* **2016**, *86*, 115–131. [CrossRef]
13. Khaledi, K.; Mahmoudi, E.; Datcheva, M.; Schanz, T. Analysis of compressed air storage caverns in rock salt considering thermo-mechanical cyclic loading. *Environ. Earth Sci.* **2016**, *75*, 1149. [CrossRef]
14. Blanco-Martín, L.; Rouabhi, A.; Billiotte, J.; Hadj-Hassen, F.; Tessier, B.; Hévin, G.; Hertz, E. Experimental and numerical investigation into rapid cooling of rock salt related to high frency cycling of storage caverns. *Int. J. Rock Mech. Min. Sci.* **2018**, *102*, 120–130. [CrossRef]
15. Soubeyran, A.; Rouabhi, A.; Coquelet, C. Thermodynamic analysis of carbon dioxide storage in salt caverns to improve the Power-to-Gas process. *Appl. Energy* **2019**, *242*, 1090–1107. [CrossRef]
16. Li, W.; Zhu, C.; Han, J.; Yang, C. Thermodynamic response of gas injection-and-withdrawal process in salt cavern for underground gas storage. *Appl. Therm. Eng.* **2019**, *163*, 114380. [CrossRef]
17. Li, W.; Miao, X.; Yang, C. Failure analysis for gas storage salt cavern by thermo-mechanical modelling considering rock salt creep. *J. Energy Storage* **2020**, *32*, 102004. [CrossRef]
18. Warren, J.K. Salt usually seals, but sometimes leaks: Implications for mine and cavern stabilities in the short and long term. *Earth-Sic. Rev.* **2017**, *165*, 302–341. [CrossRef]
19. Wang, T.; Li, J.; Jing, G.; Zhang, Q.; Yang, C.; Daemen, J. Determination of the maximum allowable gas pressure for an underground gas storage salt cavern–A case study of Jintan, China. *J. Rock Mech. Geotech. Eng.* **2019**, *11*, 251–262. [CrossRef]
20. Habibi, R.; Moomivand, H.; Ahmadi, M.; Asgari, A. Stability analysis of complex behavior of salt cavern subjected to cyclic loading by laboratory measurement and numerical modeling using LOCAS (case study: Nasrabad gas storage salt cavern). *Environ. Earth Sci.* **2021**, *80*, 317. [CrossRef]
21. Yin, H.; Yang, C.; Ma, H.; Shi, X.; Zhang, N.; Ge, X.; Li, H.; Han, Y. Stability evaluation of underground gas storage salt caverns with micro-leakage interlayer in bedded rock salt of Jintan, China. *Acta Geotech.* **2020**, *15*, 549–563. [CrossRef]
22. Fairhurst, C. *The Waste Isolation Pilot Plant: A Potential Solution for the Disposal of Transuranic Waste*; National Academy of Sciences: Washington, DC, USA, 1996.
23. Hunsche, U.; Hampel, A. Rock salt—The mechanical properties of the host rock material for a radioactive waste repository. *Eng. Geol.* **1999**, *52*, 271–291. [CrossRef]
24. Ngo, D.; Pellet, F. Numerical modeling of thermally-induced fractures in a large rock salt mass. *J. Rock Mech. Geotech. Eng.* **2018**, *10*, 844–855. [CrossRef]
25. Zhu, C.; Arson, C. A thermo-mechanical damage model for rock stiffness during anisotropic crack opening and clusure. *Acta Geotech.* **2014**, *9*, 847–867. [CrossRef]
26. Li, W.; Soliman, M.; Han, Y. Microscopic numerical modeling of Thermo-Hydro-Mechanical mechanisms in fluid injection process in unconsolidated formation. *J. Pet. Sci. Eng.* **2016**, *146*, 959–970. [CrossRef]
27. Ding, J.; Chester, F.; Chester, J.; Shen, X.; Arson, C. Microcrack Network Development in Salt-Rock During Cyclic Loading at Low Confining Pressure. In Proceedings of the 51st US Rock Mechanics/Geomechanics Symposium, San Francisco, CA, USA, 25–28 June 2017; ARMA, American Rock Mechanics Association.
28. Shen, X.; Arson, C.; Ding, J.; Chester, F.M.; Chester, J.S. Mechanisms of Anisotropy in Salt Rock Upon Microcrack Propagation. *Rock Mech. Rock Eng.* **2020**, *53*, 3185–3205. [CrossRef]
29. Wu, X.; Li, Y.; Tang, C. Comparative study on heat extraction performance of three enhanced geothermal systems. *Rock Mech. Bull.* **2023**, *2*, 100041. [CrossRef]
30. Shen, X.; Ding, J.; Arson, C.; Chester, J.S.; Chester, F.M. Micromechanical modeling for rate-dependent behavior of salt rock under cyclic loading. *Int. J. Numer. Anal. Methods Géoméch.* **2020**, *45*, 28–44. [CrossRef]
31. Shojaei, A.; Shao, J. *Porous Rock Fracture Mechanics with Application to Hydraulic Fracturing, Drilling and Structural Engineering*; Woodhead Publishing: Duxford, UK, 2017.
32. Wang, J.; Zhang, Q.; Song, Z.; Liu, X.; Wang, X.; Zhang, Y. Microstructural variations and damage evolvement of salt rock under cyclic loading. *Int. J. Rock Mech. Min. Sci.* **2022**, *152*, 105078. [CrossRef]
33. Han, Y.; Ma, H.; Cui, H.; Liu, N. The Effects of Cycle Frequency on Mechanical Behavior of Rock Salt for Energy Storage. *Rock Mech. Rock Eng.* **2022**, *55*, 7535–7545. [CrossRef]

34. Ding, J.; Chester, F.M.; Chester, J.S.; Shen, X.; Arson, C. Coupled Brittle and Viscous Micromechanisms Produce Semibrittle Flow, Grain-Boundary Sliding, and Anelasticity in Salt-Rock. *J. Geophys. Res. Solid Earth* **2021**, *126*, e2020JB021261. [CrossRef]
35. Li, Z.; Suo, J.; Fan, J.; Fourmeau, M.; Jiang, D.; Nelias, D. Damage evolution of rock salt under multilevel amplitude creep-fatigue loading with acoustic emission monitoring. *Int. J. Rock Mech. Min. Sci.* **2023**, *164*, 105346. [CrossRef]
36. Zhu, C.; Shen, X.; Arson, C.; Pouya, A. Numerical study of thermo-mechanical effects on the viscous damage behavior of rock salt caverns. In Proceedings of the 51st US Rock Mechanics/Geomechanics Symposium, San Francisco, CA, USA, 25–28 June 2017.
37. Jing, L.; Stephansson, O. *Fundamental of Discrete Element Methods for Rock Engineering Theory and Applications*; Elsevier Science: Amsterdam, The Netherlands, 2007.
38. Zhao, X.; Elsworth, D.; He, Y.; Hu, W.; Wang, T. A grain texture model to investigate effects of grain shape and orientation on macro-mechanical behavior of crystalline rock. *Int. J. Rock Mech. Min. Sci.* **2021**, *148*, 104971. [CrossRef]
39. Hu, W.; Liu, K.; Potyondy, D.; Zhang, Q. 3D continuum-discrete coupled modelling of triaxial Hopkinson bar tests on rock under multiaxial static-dynamic loads. *Int. J. Rock Mech. Min. Sci.* **2020**, *134*, 104448. [CrossRef]
40. Zhang, F.; Dontsov, E.; Mack, M. Fully coupled simulation of a hydraulic fracture interacting with natural fractures with a hybrid discrete-continuum method. *Int. J. Numer. Anal. Methods Géoméch.* **2017**, *41*, 1430–1452. [CrossRef]
41. Zhang, F.; Damjanac, B.; Maxwell, S. Investigating Hydraulic Fracturing Complexity in Naturally Fractured Rock Masses Using Fully Coupled Multiscale Numerical Modeling. *Rock Mech. Rock Eng.* **2019**, *52*, 5137–5160. [CrossRef]
42. Itasca Consulting Group, Inc. *FLAC3D Users Manual*; Itasca Consulting Group, Inc.: Minneapolis, MN, USA, 2017.
43. Li, W.; Han, Y.; Wang, T.; Ma, J. DEM micromechanical modeling and laboratory experiment on creep behavior of salt rock. *J. Nat. Gas Sci. Eng.* **2017**, *46*, 38–46. [CrossRef]

sustainability

Article

Infrared Precursor Experiment to Predict Water Inrushes in Underground Spaces Using a Multiparameter Normalization

Kewang Cao [1,2], Furong Dong [1,2], Liqiang Ma [2,*], Naseer Muhammad Khan [3,*], Tariq Feroze [3], Saad S. Alarifi [4], Sajjad Hussain [5] and Muhammad Ali [6]

1 School of Art, Anhui University of Finance and Economics, Bengbu 233030, China
2 School of Mines, China University of Mining and Technology, Xuzhou 221116, China
3 Department of Sustainable Advanced Geomechanical Engineering, Military College of Engineering, National University of Sciences and Technology, Risalpur 23200, Pakistan
4 Department of Geology and Geophysics, College of Science, King Saud University, P.O. Box 2455, Riyadh 11451, Saudi Arabia
5 Department of Mining Engineering, University of Engineering & Technology, Peshawar 25000, Pakistan
6 Australia Education Management Group, 635 Canterbury Road, Surrey Hills, VIC 3127, Australia
* Correspondence: ckma@cumt.edu.cn (L.M.); nmkhan@mce.edu.pk (N.M.K.)

Abstract: Rock failure is the root cause of geological disasters such as slope failure, civil tunnel collapse, and water inrush in roadways and mines. Accurate and effective monitoring of the loaded rock failure process can provide reliable precursor information for water inrushes in underground engineering structures such as in mines, civil tunnels, and subways. The water inrush may affect the safe and efficient execution of these engineering structures. Therefore, it is essential to predict the water inrush effectively. In this paper, the water inrush process of the roadway was simulated by laboratory experiments. The multiparameters such as strain energy field and infrared radiation temperature field were normalized based on the normalization algorithm of linear function transformation. On the basis of analyzing the variation characteristics of the original parameters, the evolution characteristics after the parameters normalization algorithm were studied, and the precursor of roadway water inrush was predicted comprehensively. The results show that the dissipation energy ratio, the infrared radiation variation coefficient (IRVC), the average infrared radiation temperature (AIRT), and the variance of successful minor infrared image temperature (VSMIT) are all suitable for the prediction of roadway water inrushes in the developing face of an excavation. The intermediate mutation of the IRVC can be used as an early precursor of roadway water inrush in the face of an excavation that is being developed. The inflection of the dissipation energy ratio from a declining amount to a level value and the mutation of VSMIT during rock failure can be used as the middle precursor of roadway water inrush. The mutation of AIRT and VSMIT after rock failure can be used as the precursor of roadway imminent water inrush. Combining with the early precursor and middle precursor of roadway water inrush, the graded warning of "early precursor–middle precursor–final precursor" of roadway water inrush can be obtained. The research results provide a theoretical basis for water inrush monitoring and early warning in the sustainable development of mine, tunnel, shaft, and foundation pit excavations.

Keywords: water inrush; precursor; strain energy; infrared radiation; normalized; sustainable mine

Citation: Cao, K.; Dong, F.; Ma, L.; Khan, N.M.; Feroze, T.; S. Alarifi, S.; Hussain, S.; Ali, M. Infrared Precursor Experiment to Predict Water Inrushes in Underground Spaces Using a Multiparameter Normalization. *Sustainability* **2023**, *15*, 7570. https://doi.org/10.3390/su15097570

Academic Editor: Jianjun Ma

Received: 16 February 2023
Revised: 26 April 2023
Accepted: 1 May 2023
Published: 5 May 2023

1. Introduction

Due to the rapid development of underground spaces such as subways, tunnels, and caverns, large-scale geological disasters such as water inrushes, mud rushes, and rock collapses often occur in the construction stage [1–7]. In coal mining, water inrush accidents often occur, resulting in serious and often irreparable property losses and casualties, seriously affecting the normal production of coal mines [8,9]. It is therefore important to carry out relevant underground development face water inrush laboratory experiments to determine the variation characteristics of water inrush precursor information; this can not only improve the reliability of water inrush prediction in underground space development but also meet the needs of sustainable protection of mining water resources. Water resource protection is the theme of sustainable development of mining ecological environments and a key way to achieve green mining.

Mining and underground civil activities inevitably lead to the redistribution of underground rock stress and fracture damage of rock, which greatly change the permeability of surrounding rock. This can lead to water inrush mainly in the roof, through stress-induced natural fractures or geology features, causing safety accidents.

In subway construction, damage to the lining of shield tunnels may cause water inrush in the subway. This is caused by nonwatertight drilling of through-holes near the intersection of two subway tunnels. Under high hydraulic gradients, weak permeable areas form and extend, leading to soil water slurry explosions [10–13]. The infrared radiation information of rock changes during the process of stress redistribution and consequent fracture damage process [14–20]. Monitoring the infrared radiation released to the outside during the process of rock loading can predict the characteristics and process of rock deformation and failure. This provides reliable information for the establishment of rock failure precursors [21–29]. In recent years, many scholars have carried out considerable research using the infrared radiation characteristics of rock fracture and water seepage. Asakura et al. [30] studied the infrared radiation monitoring of water leakage in tunnel lining and proved the feasibility of monitoring water leakage by infrared technology. Liu et al. [31] studied the characteristics of infrared radiation in the process of concrete fracture and water seepage and found that the "initial increase followed by a decrease" in the curve of infrared radiation temperature was an abnormal precursor of the infrared radiation of concrete fracture and water seepage. Dou et al. [32] carried out the infrared radiation observation experiments of tunnel leakage, studied the infrared radiation variation characteristics, and wrote the MATLAB image processing program to extract the infrared image characteristics during the process of concrete leakage. Zhang [33] studied the characteristics of infrared radiation in the process of sandstone fracture and water inrush, in which the sudden decrease of infrared radiation temperature predicted the occurrence of roadway water inrush, and the accelerated rate of infrared radiation temperature could be used as a precursor for water inrushes.

To quantitatively analyze the characteristics of infrared radiation in the process of rock fracture and water seepage, Wu et al. [34] first proposed the average infrared radiation temperature (AIRT) index of rock surface. Liu [35] used the AIRT index to analyze the infrared radiation characteristics of dry and water saturated rocks during uniaxial loading, and found that water can promote the AIRT of a rock surface. However, different areas of the rock surface may simultaneously heat up and cool down during the unstable development crack development stage in the rock, resulting in no change in overall AITR index [36]. Therefore, Liu [37], Ma et al. [38], and Yang [39] proposed the infrared radiation variance (*IRV*), variance of successful minor infrared image temperature (*VSMIT*), and infrared radiation variation coefficient (*IRCV*) of rock surface. The results show that the above indexes can well reflect the differentiation characteristics of infrared radiation on the rock surface. For example, Ma and Zhang [38] studied the internal relationship between stress adjustment (due to excavating) and *VSMIT* index during the loading process of dry and water-saturated rocks, and found that *VSMIT* index can correlate well with the rock failure and water has an amplification effect on the mutation characteristics of *VSMIT* index.

Although scholars have carried out a lot of research work on the monitoring and warning of water inrush during underground development, the results to date have not proved reliable for detecting an imminent water inrush. This is mainly due to the localized complex geology and hydrology. Based on the analysis of the physical parameters such as stress, infrared radiation, and strain energy in the process of predicting roadway water inrushes, this paper normalizes each physical parameter, then comprehensively analyzes the multielement information evolution characteristics and their correlation with roadway water inrushes, and studies the comprehensive precursor characteristics of them. This study proposes an early warning precursor and monitoring the occurrence of roadway water inrush. The research findings will provide a theoretical basis for monitoring and early warning of water inrush in underground spaces for their safe and efficient development.

2. Experimental Principle

Stefan–Boltzmann's law states that any object above absolute zero will radiate electromagnetic waves to the outside world. Due to this, the radiation intensity and the temperature of the object satisfy the following formula [39]:

$$J^* = \varepsilon\sigma T^4 \quad (1)$$

where J^* represents the total energy radiated by the object per unit area, $W{\cdot}m^{-2}$; ε represents the surface emissivity of the object, $0 < \varepsilon < 1$; σ is the Stefan–Boltzmann constant, $5.670373 \times 10^{-8}\ W{\cdot}m^{-2}{\cdot}K^{-4}$; T is the absolute temperature of the surface, K.

This law explains how the infrared radiation on the surface of the object at room temperature is affected. Further, the relationship between the radiation intensity and the temperature of the object satisfies the fourth power. The force exerted on a solid causes changes in the distance between internal particles, resulting in thermodynamic changes and, thus, temperature changes. This phenomenon of temperature change caused by heat generated due to force can be referred to as the thermal–mechanical coupling effect [36]. Materials with different mechanical properties (such as elastic materials, elastoplastic materials, viscoelastic materials, etc.) and the same material have different thermal and mechanical coupling effects at different stress stages. These different thermodynamic coupling effects have different characteristics and laws due to their different microscopic mechanisms. In the elastic range, the object undergoes the process of tension or compression accompanied by the reversible conversion of heat. In the adiabatic environment, the sum of temperature and principal stress satisfies the linear relationship [40]:

$$\Delta T = \frac{\alpha}{\rho C_\sigma} TS \quad (2)$$

where ΔT and T are the change of object temperature and object temperature, respectively; α is the linear coefficient of thermal expansion; C_σ is the specific heat coefficient under constant stress; S is the sum of the principal stresses.

The change of principal stress during uniaxial rock loading is only related to σ_1. The fourth term represents the internal dissipated energy of the material, which is manifested in the thermoelastoplastic comprehensive effect at this stage. The work carried out by the external force is not all transformed into the internal thermal energy of the material but is mostly consumed in the process of internal microstructure change. Plastic deformation in the process of energy dissipation and thermal energy conversion is not reversible, and ΔE mainly includes the following three parts in the process of energy consumption [36]:

$$\Delta E = \Delta E_1 + \Delta E_2 + \Delta E_3 \quad (3)$$

ΔE_1 is the energy carried by the escape process of pore gas, in general, $\Delta E_1 < 0$. ΔE_2 is the energy consumed by the expansion of pores, fractures, and joints and the generation of new fractures in the rock. The internal pores, fractures, and joint weak surfaces will first contract and close with the increase of stress. The pores will then collapse, and the primary fractures and joints will further expand, penetrate, and merge. Moreover, the new fractures will be generated by the increase in stress. The ΔE_2 is less than 0 due to the consumption of energy in this process. ΔE_3 is the energy generated by friction because there are friction behaviors among the pores, fissures, joints, and rock particles along all directions in the interior of the rock. Two factors influence the process of frictional heat generation: one is the positive pressure on the contact surface inside the rock, and the other is the friction coefficient. When the friction coefficient is fixed, the friction force is positively related to the normal stress on the contact surface. The larger the friction force, the more work will be carried out to overcome the friction force in the process of crack and particle sliding, resulting in the higher energy consumption. It is important to note that in this process the temperature of the contact surface due to friction heat production will increase, therefore ΔE_3 is greater than zero.

3. Experimental Design

3.1. Experimental Equipment

The experimental loading equipment used an SANS electronic universal testing machine system with a maximum vertical load of 1000 kN. The water pressure loading equipment adopts a Shanghai SB water pressure pump, the maximum working pressure is 10 MPa, and the water pressure is set to 0.5 MPa. The infrared radiation detection device adopts the American FLIRA615 infrared thermal imager, whose thermal sensitivity is 0.025 °C, and the wavelength range is 7.5–14.0 m. The image acquisition rate was set at 25 frames/s.

3.2. Rock Samples

The representative samples of sandstone collected from a coal mine in Shandong province were used in the laboratory experimental process. All samples were obtained from the same rock sample. The specimen design specification is a cuboid of 100 × 100 × 150 mm. The diameter and depth of the observation hole in the test block are 50 mm, and the diameter and depth of the water injection hole are 50 mm and 50 mm. A total of five specimens were prepared, represented by A1, A2, A3, A4, and A5. The actual measurement specifications of these specimens are shown in Table 1. The rock sample model is shown in Figure 1a, and the processed specimen is shown in Figure 1b. The water gushing from the tunnel mainly comes from the rich water in front of the tunnel face, so this special test piece shape is designed. The first section of the water injection hole of the rock sample is bonded together with the iron block of the fixed abrasive tool using strong adhesive, and then the fixed abrasive tool is reinforced by electric welding to resist the water pressure in the water injection hole after the water pump is running. The water injection pipe and the abrasive tool are tightened and fixed by screws to ensure that there is no water leakage on the water injection side during the experiment. The following figure shows the fixed mold of the rock sample.

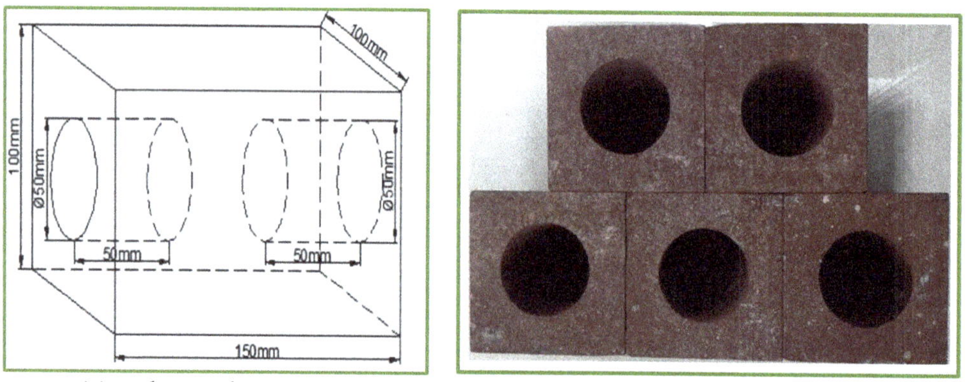

(**a**) rock sample structure (**b**) rock sample picture

Figure 1. Schematic diagram of rock sample.

Table 1. Actual measurement specifications of the specimen.

Number	Length × Width × Height (mm)	Observation Hole Diameter × Depth (mm)	Water Injection Hole Diameter × Depth (mm)
A_1	150.32 × 100.42 × 99.85	φ50.24 × 50.53	φ50.23 × 50.66
A_2	150.31 × 100.48 × 100.38	φ50.22 × 50.68	φ50.23 × 50.97
A_3	150.10 × 100.12 × 99.78	φ50.22 × 50.34	φ50.25 × 50.35
A_4	149.97 × 100.10 × 99.25	φ50.25 × 50.58	φ50.24 × 50.59
A_5	150.35 × 100.31 × 99.77	φ50.23 × 50.73	φ50.22 × 50.67

3.3. *Experiment Process*

The rock specimen was loaded uniaxially with a closure rate of 0.1 mm/min. The data acquisition frequency of the testing machine was set as 10 times/s. The layout of the experimental equipment is shown in Figure 2. To facilitate the sorting and analysis of test data, the water pressure of the water pump was set as 0.5 MPa before the test, and the thermal imager was installed about 1 m away from the sample to observe the infrared temperature field changes on the sample surface. The experiment was started after the infrared radiation temperature on the rock surface remained stable. We synchronously calibrated the time of all test equipment. In addition, the start and end times of each equipment remained the same. Then, uniaxial loading was applied to the rock sample until inrush water appeared.

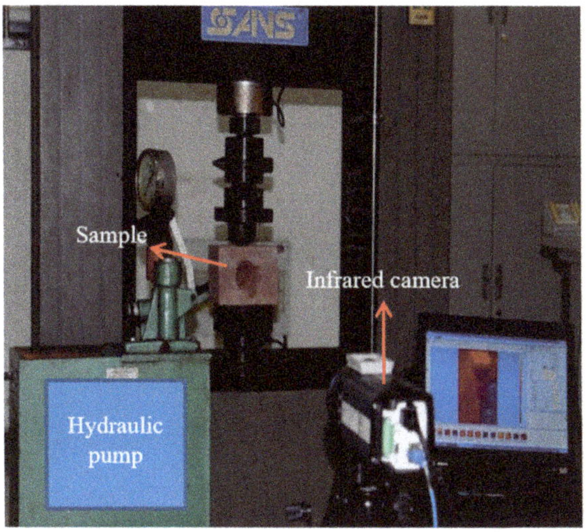

Figure 2. Experimental layout.

4. Indicators

4.1. *Strain Energy*

Assuming that a rock unit deforms under the action of external forces and the process occurs in a closed system, according to the first law of thermodynamics, the following can be obtained:

$$U = U^d + U^e \tag{4}$$

where U is the total strain energy, which is determined by the stress–strain curve and the area around the horizontal axis, U^d is the dissipated strain energy of the unit, and U^e is the elastic strain energy released by the unit.

Figure 3 shows the relationship between U^d and U^e during uniaxial rock loading. The releasable elastic strain energy U^e in the uniaxial loading process of rock can be rewritten as follows [41,42]:

$$U^e = \frac{1}{2E_u}\sigma^2 \tag{5}$$

where E_u is the unloading modulus of elasticity.

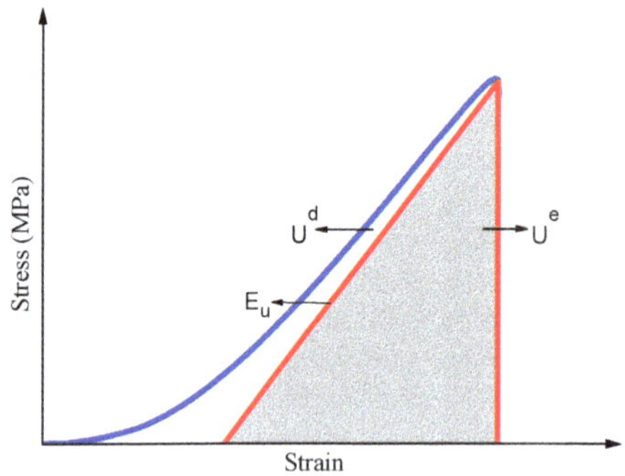

Figure 3. Relationship between dissipated strain energy and elastic strain energy in the stress–strain curve.

In order to facilitate calculation, the elastic modulus E_0 is generally used instead of E_u. In this paper, the average modulus is used to calculate the elastic modulus, and the stress–strain formula is assumed to be $\sigma = f(\varepsilon)$. The following formula obtains the elastic modulus

$$E = \frac{f(\varepsilon_2) - f(\varepsilon_1)}{\varepsilon_2 - \varepsilon_1} \tag{6}$$

where E is the elastic modulus, and $f(\varepsilon_1)$ and $f(\varepsilon_2)$ are the stress values corresponding to the starting and ending points of the elastic phase, respectively.

4.2. Infrared Thermal Image

Infrared thermal image is a series of object surface temperature distribution images output by an infrared thermal imager. The two-dimensional temperature matrix of frame p in the original infrared thermal image is [43]:

$$f_p(x, y) \tag{7}$$

where p is the frame number index of the infrared thermal image sequence; x and y represent the row and column numbers of the thermal imager temperature matrix, respectively.

4.3. VSMIT

VSMIT can reflect the dispersion degree of infrared radiation temperature value of the entire rock sample surface, which is defined as follows [43]:

$$VSMIT = \frac{1}{M}\frac{1}{N}\sum_{y=1}^{N}\sum_{x=1}^{M}[f_p(x, y) - AIRT_p]^2 \tag{8}$$

Among them, $AIRT_p = \frac{1}{M}\frac{1}{N}\sum_{y=1}^{N}\sum_{x=1}^{M} f_p(x, y)$; M and N are the maximum numbers of rows and columns for x and y, respectively.

4.4. *IRVC*

The *IRVC* can measure the dispersion degree of the infrared radiation temperature field on the rock surface, which is defined as follows [39]:

$$IRVC = \sigma / AIRT \tag{9}$$

where σ is the standard deviation of infrared radiation temperature on the rock surface.

5. Experimental Results

5.1. *Strain Energy*

The deformation and failure of rock is a process of energy input, elastic energy accumulation, energy dissipation, and energy release from the point of view of strain energy. Energy dissipation is mainly used for crack initiation and propagation, and energy release is the internal cause of the sudden failure of the rock mass. The elastic strain energy accumulated by the rock mass before excavation is the main source of the energy released by the ultimate failure of the surrounding rock mass, especially the deep hard brittle rock mass with good energy storage under the condition of high in situ stress. A large amount of elastic strain energy accumulated in the excavation is released instantly due to the excavation unloading effect, promoting the occurrence of rock mass failure and then connecting the water diversion fissure channel, finally resulting in a water inrush accident.

Figure 4 shows the strain energy evolution curve during the loading process of the rock sample. Due to space limitations, samples A_1, A_2, and A_3 were selected for analysis in this paper. As shown in Figure 4, the energy absorbed in the rock at the beginning of loading is mainly dissipated strain energy, because most of the strain energy is consumed by pore and microfracture compaction. The curves of elastic strain energy and dissipated strain energy diverge with the increase of stress, and the growth of elastic strain energy increases continuously, while the dissipated strain energy increases in a nearly straight line, which is used for the formation and expansion of plastic deformation and microcracks in rocks. The elastic strain energy drops sharply, while the dissipated strain energy increases sharply at the peak stress, which indicates that the internal microcrack propagation and penetration rate accelerated and the damage was aggravated. After the peak stress, the bearing capacity of the rock decreases rapidly and maintains a certain residual strength. The strain energy absorbed by the rock is transformed into dissipative strain energy in this period, which is used for the further development of rock fracture and shear deformation along the slip surface. Then, a macroscopic water diversion channel is formed in the rock, and water inrush eventually occurs in the roadway.

The dissipated strain energy ratio of rock refers to the proportion of dissipated strain energy to the total strain energy. The dissipated strain energy ratio curve shows a trend of decline before the peak stress, changes from a decline to a level near the peak stress when approaching the peak stress, and then begins to increase. The curve changes abruptly with the rock failure. After that, the dissipated strain energy ratio curve continues to increase until a water inrush occurs in the rock face. The analysis found that the strain energy dissipation ratio curve of samples experienced a "decline-level" process when the rock approached failure. This is due to the dissipated strain energy used for plastic strain and crack growth increasing with the rapid development of microcracks in the rock; although the elastic strain energy is still accumulating, the rock will reach the maximum energy storage limit. Therefore, the dissipated strain energy ratio of rock declines to the level, and then the elastic strain energy reaches the energy storage limit. The accumulated elastic strain energy is quickly released and causes rock failure. In summary, the turning point of dissipated energy ratio from decline to level can be used as the medium warning information of roadway water inrush.

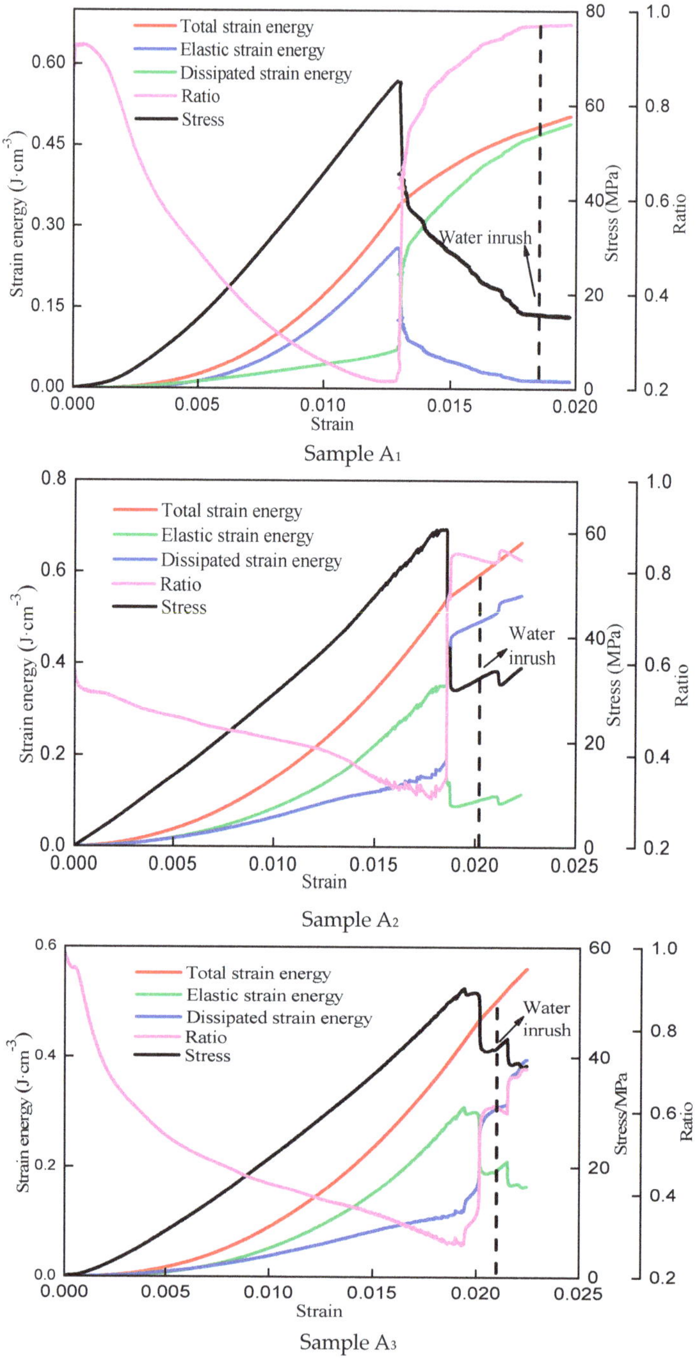

Figure 4. Evolution curves of strain energy during rock sample loading.

5.2. *Infrared Radiation*

Figure 5 shows the photo of the water inrush instant of the roadway in the laboratory. The water diversion channel is formed after the rock in the roadway reaches failure, and then water inrush occurs in the mine. Hence, the hole is selected as the analysis area of infrared radiation data, as shown in Figure 6. The evolution characteristics of *AIRT*, *IRCV*, *VSMIT*, and infrared thermal image in rock holes during the rock loading failure and water inrush were analyzed, and the precursors of roadway water inrush were identified.

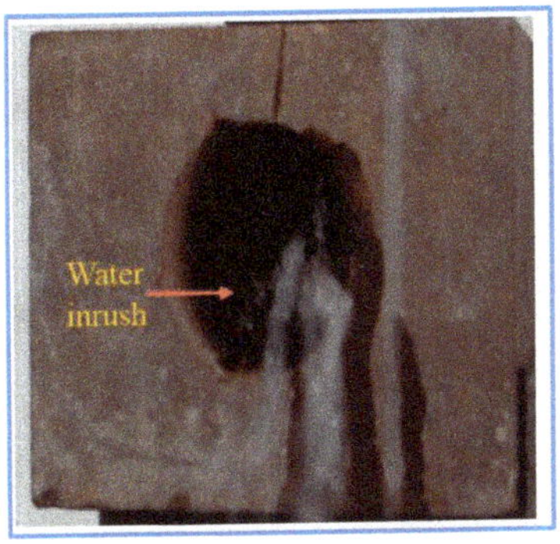

Figure 5. Water inrush in roadway.

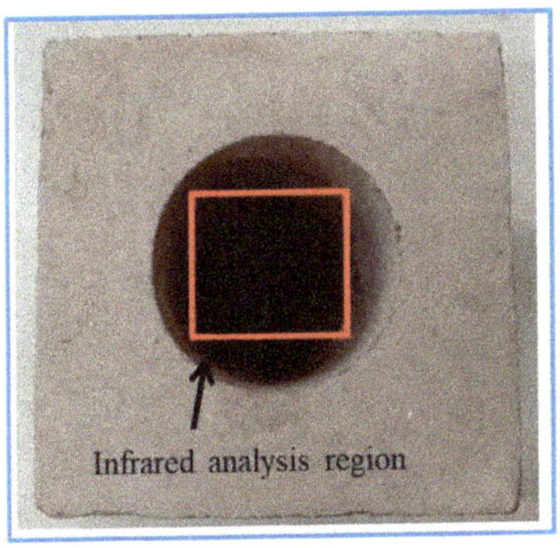

Figure 6. Infrared radiation analysis region.

5.2.1. AIRT

Figure 7 shows the time-varying curves of *AIRT*, *IRCV*, and *VSMIT* in the hole during the uniaxial rock sample loading. As shown in Figure 7, the *AIRT* in the hole of rock under loading showed a trend of stable fluctuation with the increase of stress during the process of rock uniaxial loading. The *AIRT* curves of rock samples A_1 and A_2 showed no obvious abnormal phenomena before rock failure, while the *AIRT* of the rock sample A_3 gradually increased, with a temperature rise of about 0.1 °C. However, the *AIRT* of rock samples A_1, A_2, and A_3 all dropped abruptly before water inrush, with a decrease temperature range of between 0.3~0.6 °C. This is due to the water seepage into the observation surface of the roadway absorbing part of the heat from the rock, resulting in a downward trend of *AIRT*. Therefore, the sudden drop of *AIRT* can be regarded as the precursor of water inrush. Two conditions must be satisfied to take a characteristic of infrared radiation index as a precursor: one is that all infrared radiation indexes of rock samples have this characteristic; the other is that this characteristic is easy to distinguish. The lead time of water inrush precursor of rock *AIRT* is 15~30 s before water inrush, hence *AIRT* decreased slightly and then increased.

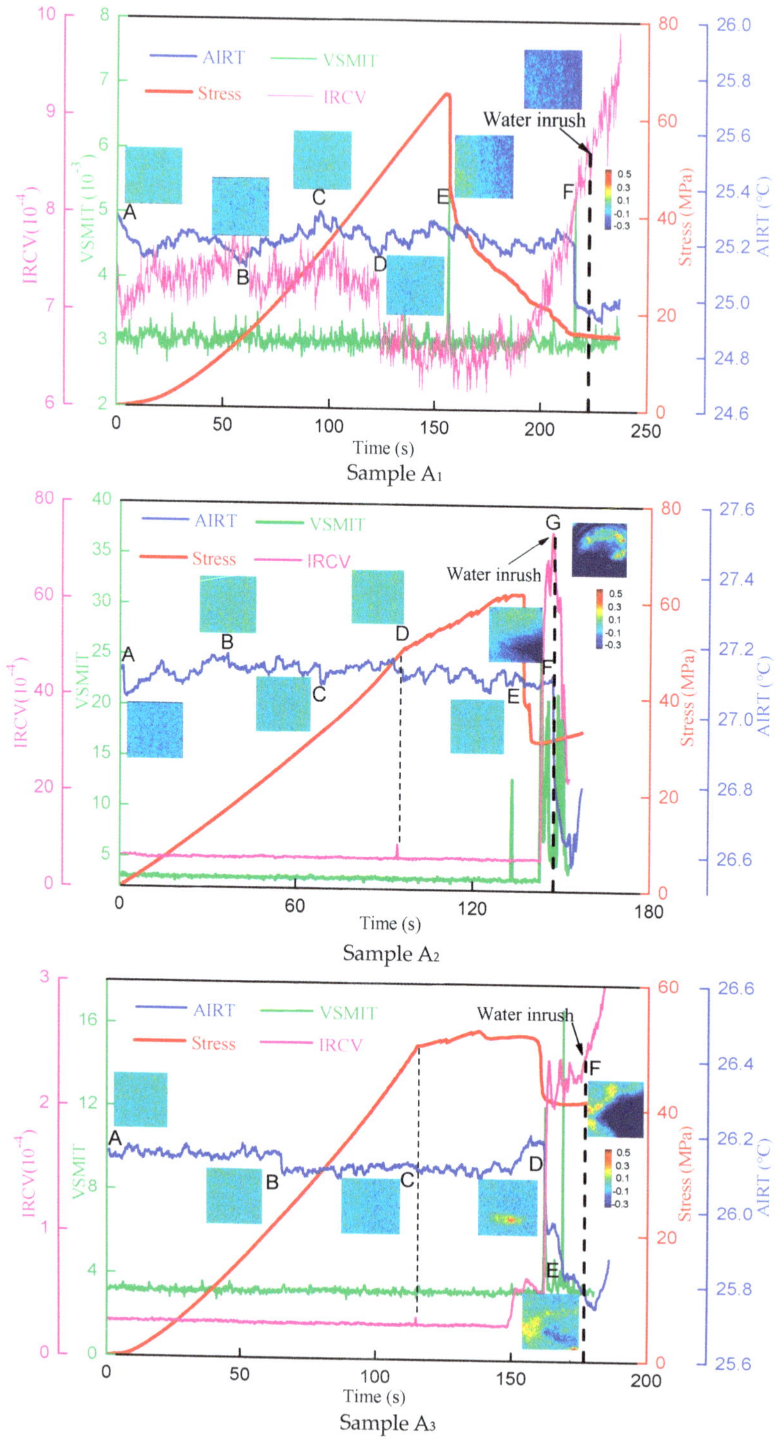

Figure 7. Evolution curves of *AIRT*, *IRCV*, and *VSMIT* during rock sample loading. A–G: different stages.

5.2.2. IRCV

The *IRCV* curve of rock sample A_1 shows a trend of steady fluctuation–decline–rise in the process of water inrush experiment under uniaxial loading, and the curve of *IRCV* mutates when loaded to 123 s, while the *IRCV* curves of rock samples A_2 and A_3 show a nearly horizontal trend, and both of them have a mutation in the middle of loading, with the occurrence moments of 94 s and 115 s, respectively. As shown in Figure 7, the *AIRT* value corresponding to the mutation of *IRCV* in the middle and late stages of loading decreases gradually, so the mutation of *IRCV* is caused by the mutation of infrared radiation standard deviation (σ). The authors propose that this is due to the rock having just entered the stage of unstable crack development, and microfracture events increase. The rock failure is dominated by microcracks induced by tensile failure. The tensile failure area corresponds to a drop in the rock surface temperature, while the shear failure area corresponds to a rise in the rock surface temperature, resulting in a gradual drop in *AIRT*. The increase of microfracture events leads to the occurrence of heating and cooling zones in different regions of the rock surface, and thus the standard deviation of corresponding infrared radiation temperature is suddenly changed. In conclusion, the curve of *IRCV* has suddenly changed in the middle and late stages of rock loading. Therefore, the *IRCV* mutation in the middle and late stages of rock loading can be regarded as the early precursor of water inrush.

As shown in Figure 7, when the *AIRT* of the rock drops suddenly before the water inrush (the precursor of water inrush), the *IRCV* curve of the rock increases or mutates gradually. Specifically, the *IRCV* curve of the rock sample A_1 increases gradually, and the rock samples A_2 and A_3 have a mutation. This feature, therefore, is not suitable as a precursor of roadway water inrush. If the water flows homogeneously into the roadway before water inrush, and the water has an amplification effect on the infrared radiation of rock [34], *AIRT* will drop abruptly, and the dispersion degree of infrared radiation temperature (σ) may also correspond to a sudden drop, which may cause no mutation in *IRCV*. If the water flows into the roadway inhomogeneously before water inrush, the *AIRT* will drop sharply and the differentiation of infrared radiation temperature (σ) will increase sharply, which will cause *IRCV* mutation. To sum up, *IRCV* curve mutation before water inrush is not universal, so it is not suitable as a precursor of roadway water inrush.

5.2.3. VSMIT

The *VSMIT* curve of all the rock samples shows a general horizontal trend during the roadway water inrush test, and the *VSMIT* increases abruptly when the rock failure, with the mutation range, is 0.01~0.03. Due to the universality, synchronism, and significance of *VSMIT* mutation characteristics at the rock failure [26], the first mutation of *VSMIT* can be regarded as a midterm precursor of water inrush. With the process of rock loading, the stress gradually drops to the residual stress. When the *AIRT* of the rock declines abruptly (the precursor of water inrush), the *VSMIT* of all rock samples mutate for the second time, with an increased range of 0.02~0.05. This is due to the water seepage increasing the dispersion degree of infrared radiation temperature of the two adjacent frames, which also indicates that the sudden drop in *AIRT* will be accompanied by the sudden increase of the *VSMIT* index. Therefore, the second mutation of *VSMIT* can be used as a precursor of water inrush.

5.2.4. Infrared Thermal Image

The above infrared radiation characteristics during the process of rock failure and water inrush were all obtained from the quantitative index analysis of infrared radiation, so only the time information of infrared radiation characteristics can be obtained. If the spatial information of infrared radiation during the process of rock failure and water inrush is needed, the infrared thermal image can be analyzed. To highlight the change of infrared radiation caused by the loading of the sample, and reduce the impact of the local radiation rate difference and environmental interference of the sample when processing the infrared radiation experimental data, the thermal image obtained during the loading process is processed as a difference [31], that is, the first thermal image at the beginning of the loading is subtracted from each thermal image, and the change of radiation temperature field is analyzed by using the image after the difference.

As shown in Figure 7, the infrared thermal image of rock sample A_1 before failure shows the changing trend of bright (A)–dark (B)–bright (C)–dark (D), corresponding to the decline (AB)–rise (BC)–decline (CD) of the AIRT curve, and the temperature difference between the left and right sides of the sample is obvious at the rock failure (the temperature on the left side is high while that on the right side is low), and the infrared thermal image at the precursory point (F) of the roadway water inrush becomes dark as a whole. At the beginning of the loading stage (AB) of sample A_2, the infrared thermal image of the rock sample changes from dark to bright, and there is no obvious abnormal change from point B to rock failure (E). The lower part of the rock shows the abnormal low-temperature area near the roadway water inrush precursory point (F), while the upper part of the infrared thermogram shows the abnormal high-temperature area when the roadway water inrush occurs, and the rest is the low-temperature area. At the beginning of loading stage (AB) of rock sample A_3, there is no obvious change in the infrared thermogram, and the infrared thermogram becomes dark at the BC stage, while the abnormal high-temperature area appears at the lower part of the rock sample (D). With the increase of stress, the abnormal high-temperature area becomes an abnormal low-temperature area in the roadway water inrush precursor (E), and the high-temperature area appears on the left side. The high-temperature area extends upward with the loading process, and part of the original high-temperature area is eroded by water, so the right side of the rock sample presents a large area of low temperature.

6. Multiparameter Normalization

6.1. *Define*

If the value range of the sample data is [Min, Max], then the normalized expression of linear function is

$$y = (x - \text{Min})/(\text{Max} - \text{Min}) \tag{10}$$

where x and y are the values before and after conversion, respectively; Max and Min are the maximum and minimum values of samples, respectively.

The linear function normalization has the following properties: (1) the sample size relation remains unchanged; (2) the relative distance of samples remains unchanged. The variation trend of each physical quantity obtained by linear function normalization is consistent with the original data curve, which can well reflect the key information such as fluctuation and mutation points in the original data curve [44].

6.2. *Analysis of Normalized Results*

Based on the analysis of strain energy and infrared radiation characteristics of roadway water inrush, it is found that dissipative energy ratio, *AIRT*, *VSMIT*, and *IRCV* are suitable as the main parameters to predict roadway water inrush. In the practical application of multiparameter joint monitoring of a roadway water inrush disaster, due to the differences in the range and dimension of each parameter, and the different physical parameters in different coordinate systems, it is not conducive to the rapid and intuitive identification of the sequence of the occurrence of the abrupt point of each physical parameter, and this also affects the analysis of the correlation between each physical parameter. Therefore, it is necessary to normalize each parameter in order to comprehensively compare and analyze the variation rule of each parameter in the same scale, which can provide a comprehensive basis for the early warning of roadway water inrush disasters.

Figure 8 shows the normalized curves of stress, dissipated energy ratio, *AIRT*, *VSMIT*, and *IRCV* with time collected in a laboratory experiment of sandstone roadway water inrush. Based on the comprehensive analysis of all rock samples, it can be found that *IRCV* mutation occurs at the stage of 0.55~0.65 TWI (TWI is the time of water inrush) in the middle loading stage, which is the early precursor of water inrush in a sandstone roadway (as shown the EPWI in Figure 8). The dissipated energy ratio curve drops to the lowest point with the increase of stress, which is the first middle precursor of roadway water inrush (as shown by the FMPWI in Figure 8), and the early precursor is 20~33 s earlier than the middle precursor one. As the loading continues, rock sample failure occurs and accompanies the mutation of *VSMIT*, which is the second middle precursor of roadway water inrush (as shown by SMPWI in Figure 8). Thereafter, the stress drops rapidly and *AIRT* decreases abruptly at about 0.95~0.98 TWI after it decreases to residual strength. This is the first precursor of roadway imminent water inrush (as shown by FPIWI in Figure 8), corresponding to the sudden increase of *VSMIT*, which is the second precursor of imminent water inrush (as shown by SPIWI in Figure 8), indicating that roadway water inrush may occur at any time.

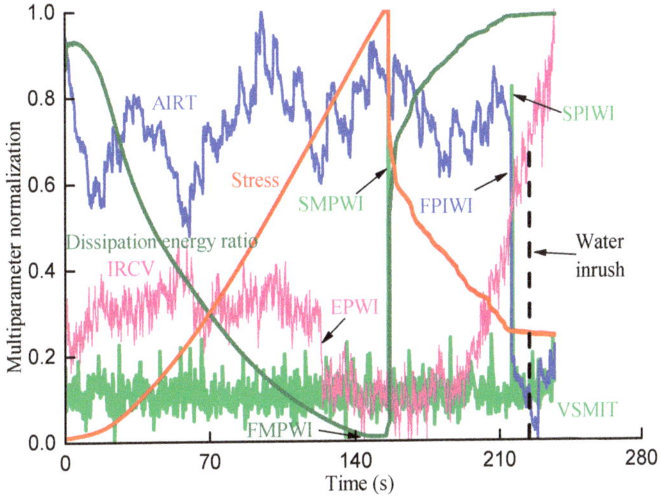

Sample A_1

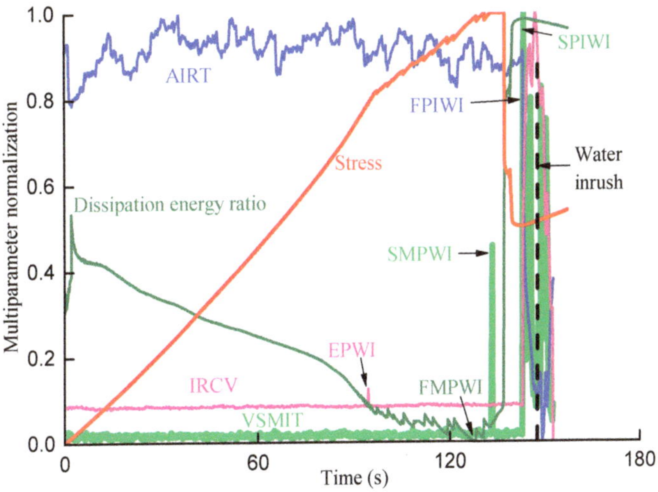

Sample A_2

Figure 8. *Cont.*

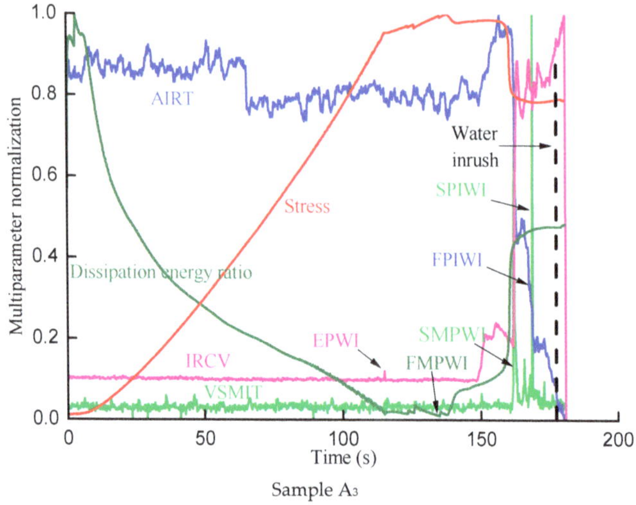

Figure 8. Multiparameter normalization curve in the process of roadway water inrush.

7. Discussion

(1) Due to the nonlinear process of roadway water inrush, the complexity and diversity of influencing factors, and the control of the accuracy of monitoring technology, the prediction of roadway water inrush with a single parameter has great limitations. The precursor of roadway water inrush cannot be accurately and effectively identified in practical application, which may lead to the problem of false alarms or missed alarms. The precursors of roadway water inrush can be identified more quickly by the normalized treatment of each physical parameter in the process of roadway water inrush. At the same time, the correlation between precursor information of roadway water inrush was obtained. Comprehensively considering and analyzing the precursors' information for roadway water inrush and its correlation, the hierarchical warning of a roadway water inrush disaster can be obtained and the accuracy and reliability of roadway water inrush warning can be improved.

(2) *AIRT* reflects the whole infrared radiation intensity of the rock surface, but there may be different heating and cooling zones in the process of rock loading and fracture due to which the *AIRT* remains unchanged. The *IRCV* of rocks reflects the dispersion degree of the original infrared radiation temperature, which has the advantage of avoiding the unit of measurement of data and neglecting the influence of numerical magnitude. Compared with *AIRT* index, *IRCV* can reflect the dispersion characteristics of infrared radiation caused by temperature rise and temperature drop areas. *VSMIT* reflects the dispersion degree of the difference in infrared radiation temperature between two adjacent frames. Compared with *IRCV* index, it eliminates the cumulative heating effect of loaded rock, and is easier to monitor the process of rock failure, instability, and seepage. From the sensitivity of *IRCV* index to the unstable crack development stage of rock, *IRCV* mutation was proposed as the early precursor of roadway water inrush. Based on the feature that *VSIMT* can monitor rock failure, the first mutation of *VSIMT* was proposed as the medium-term precursor of roadway water inrush. Established from the characteristic that *AIRT* and *VSIMT* are sensitive to water, the sudden drop of *AIRT* and the second mutation of *VSIMT* were proposed as the precursor of roadway imminent water inrush. In this paper, combined with the advantages of *AIRT*, *IRCV*, and *VSMIT*, the multiparameters precursory characteristics of roadway water inrush were determined.

(3) Infrared observation technology is a promising new method for monitoring rock samples, which has the advantages of noncontact and strong anti-interference, and can

be used for monitoring and warning the stability of bearing rock and surrounding rock of tunnels in underground engineering. Acoustic emission monitoring technology can detect the time and location of a microfracture in a rock mass. Therefore, acoustic emission monitoring (internal) and infrared radiation monitoring (external) should be combined in subsequent roadway water inrush experiments and underground engineering construction sites. Acoustic emission (AE) will be used to locate the water-conducting fissure passage in the rock, and the change rules and coupling effects of the stress field, infrared radiation temperature field, and seepage field before water inrush in underground engineering will be studied in order to build a multifield coupling model of "stress–temperature–seepage" of roadways based on infrared radiation and reveal the mechanism of water inrush in underground engineering.

8. Conclusions

The forecast of highway water inrush with a single parameter has significant limits because of the nonlinear nature of the roadway water inrush process, the complexity and variety of contributing elements, and the control of the accuracy of monitoring equipment. The issue of false alerts or missed alarms may arise from the inability to precisely and efficiently identify the antecedent of highway water inrush in practical application. The normalized treatment of each physical parameter in the process of highway water inrush may help identify the antecedents of roadway water inrush more promptly. The association between the precursor data of the highway water inrush was discovered concurrently. The following conclusions were drawn:

(1) Dissipative energy ratio, AIRT, VSMIT, and IRCV are suitable as precursor indexes for roadway water inrush prediction, and can be used to monitor and predict the occurrence of roadway water inrush.

(2) The midterm mutation of IRCV can be used as the early precursor information of roadway water inrush. The turning point of dissipation energy ratio from decreasing to level and the sudden change of VSMIT during rock failure can be used as the medium term precursor information of roadway water inrush. AIRT and VSMIT mutation after rock failure can be used as precursor information of roadway imminent water inrush.

(3) By using the normalization of linear function transformation to normalize the multi-physical parameters in the process of roadway water inrush monitoring, this realizes early warning for roadway water inrush as "early precursor–medium precursor–final precursor". In future research, we will select representative mining sections in coal and rock mining damage areas, conduct infrared radiation observations at different mining stages (damage states), compare and analyze onsite monitoring and laboratory test results, and establish a graded precursor warning based on onsite infrared radiation data.

Author Contributions: K.C.: conceptualization, writing—original draft, and experiments; F.D.: conceptualization, supervision, and project administration; L.M.: supervision and project administration; N.M.K.: software and data curation, writing—original draft; methodology; T.F.: visualization; S.S.A.: investigation and data curation; S.H.: Formal analysis; M.A.: experiments. All authors have read and agreed to the published version of the manuscript.

Funding: This research was supported by Anhui Provincial Scientific Research Preparation Plan Project (2022AH050596). The authors also acknowledge the Researchers Supporting Project number (RSP2023R496), King Saud University, Riyadh, Saudi Arabia.

Institutional Review Board Statement: Not applicable.

Data Availability Statement: The data that support the findings of this study are available from the corresponding author upon reasonable request.

Conflicts of Interest: The authors declare no conflict of interest.

References

1. Ma, D.; Duan, H.; Zhang, J.; Bai, H. A state-of-the-art review on rock seepage mechanism of water inrush disaster in coal mines. *Int. J. Coal Sci. Technol.* **2022**, *9*, 50. [CrossRef]
2. Chi, X.; Yang, K.; Wei, Z. Breaking and mining-induced stress evolution of overlying strata in the working face of a steeply dipping coal seam. *Int. J. Coal Sci. Technol.* **2021**, *8*, 614–625. [CrossRef]
3. Guo, P.; Gu, J.; Su, Y.; Wang, J.; Ding, Z. Effect of cyclic wetting–drying on tensile mechanical behavior and microstructure of clay-bearing sandstone. *Int. J. Coal Sci. Technol.* **2021**, *8*, 956–968. [CrossRef]
4. Liu, Z.; Wang, G.; Li, J.; Li, H.; Zhao, H.; Shi, H.; Lan, J. Water-immersion softening mechanism of coal rock mass based on split Hopkinson pressure bar experiment. *Int. J. Coal Sci. Technol.* **2022**, *9*, 61. [CrossRef]
5. Liu, A.; Liu, S.; Liu, P.; Wang, K. Water sorption on coal: Effects of oxygen-containing function groups and pore structure. *Int. J. Coal Sci. Technol.* **2021**, *8*, 983–1002. [CrossRef]
6. Lou, J.; Gao, F.; Yang, J.; Ren, Y.; Li, J.; Wang, X.; Yang, L. Characteristics of evolution of mining-induced stress field in the longwall panel: Insights from physical modeling. *Int. J. Coal Sci. Technol.* **2021**, *8*, 938–955. [CrossRef]
7. Hou, L.; Cao, K.; Muhammad Khan, N.; Jahed Armaghani, D.; Alarifi, S.S.; Hussain, S.; Ali, M. Precursory Analysis of Water-Bearing Rock Fracture Based on The Proportion of Dissipated Energy. *Sustainabilty* **2023**, *15*, 1769. [CrossRef]
8. Vervoort, A. Various phases in surface movements linked to deep coal longwall mining: From start-up till the period after closure. *Int. J. Coal Sci. Technol.* **2021**, *8*, 412–426. [CrossRef]
9. Wang, J.; Yang, S.; Wei, W.; Zhang, J.; Song, Z. Drawing mechanisms for top coal in longwall top coal caving (LTCC): A review of two decades of literature. *Int. J. Coal Sci. Technol.* **2021**, *8*, 1171–1196. [CrossRef]
10. Ma, J.J.; Guan, J.W.; Duan, J.F.; Huang, L.C.; Liang, Y. Stability analysis on tunnels with karst caves using the distinct lattice spring model. *Undergr. Space.* **2021**, *6*, 469–481. [CrossRef]
11. Huang, L.C.; Ma, J.J.; Lei, M.F.; Liu, L.H.; Lin, Y.X.; Zhang, Z.Y. Soil-water inrush induced shield tunnel lining damage and its stabilization: A case study. *Tunn. Undergr. Space Technol.* **2020**, *97*, 103290. [CrossRef]
12. Ma, J.J.; Chen, J.J.; Chen, W.X.; Huang, L.C. A coupled thermal-elastic-plastic-damage model for concrete subjected to dynamic loading. *Int. J. Plasticity.* **2022**, *153*, 103279. [CrossRef]
13. Ma, J.J.; Chen, J.J.; Guan, J.W.; Lin, Y.X.; Chen, W.X.; Huang, L.C. Implementation of Johnson-Holmquist-Beissel model in four-dimensional lattice spring model and its application in projectile penetration. *Int. J. Impact Eng.* **2022**, *170*, 104340. [CrossRef]
14. Cao, K.; Ma, L.; Wu, Y.; Khan, N.; Yang, J. Using the characteristics of infrared radiation during the process of strain energy evolution in saturated rock as a precursor for violent failure. *Infrared Phys. Technol.* **2020**, *109*, 103406. [CrossRef]
15. Yang, D.; Ning, Z.; Li, Y.; Lv, Z.; Qiao, Y. In situ stress measurement and analysis of the stress accumulation levels in coal mines in the northern Ordos Basin, China. *Int. J. Coal Sci. Technol.* **2021**, *8*, 1316–1335. [CrossRef]
16. Feng, F.; Chen, S.; Zhao, X.; Li, D.; Wang, X.; Cui, J. Effects of external dynamic disturbances and structural plane on rock fracturing around deep underground cavern. *Int. J. Coal Sci. Technol.* **2022**, *9*, 15. [CrossRef]
17. He, S.; Qin, M.; Qiu, L.; Song, D.; Zhang, X. Early warning of coal dynamic disaster by precursor of AE and EMR "quiet period". *Int. J. Coal Sci. Technol.* **2022**, *9*, 46. [CrossRef]
18. Zhang, L.; Kan, Z.; Zhang, C.; Tang, J. Experimental study of coal flow characteristics under mining disturbance in China. *Int. J. Coal Sci. Technol.* **2022**, *9*, 66. [CrossRef]
19. Ali, M.; Wang, E.; Li, Z.; Wang, X.; Khan, N.M.; Zang, Z.; Alarifi, S.S.; Fissha, Y. Analytical Damage Model for Predicting Coal Failure Stresses by Utilizing Acoustic Emission. *Sustainabilty* **2023**, *15*, 1236. [CrossRef]
20. Cui, R.; Cao, K.; Li, X.; Khan, R.M.A.; Khan, N.M.; Liu, W.; Gao, Q.; Wang, F. The Infrared Radiation Characteristics of Sandstone Fracture Seepage under Coupled Stress-Hydro Effect. *Sustainabilty* **2022**, *14*, 16454. [CrossRef]
21. Liu, S.; Wu, L.; Wu, Y. Infrared radiation of rock at failure. *Int. J. Rock Mech. Min.* **2006**, *43*, 972–979. [CrossRef]
22. Zhao, Y.; Jiang, Y.J. Acoustic emission and thermal infrared precursors associated with bump-prone coal failure. *Int. J. Coal Geol.* **2010**, *83*, 11–20. [CrossRef]
23. Cao, K.W.; Dong, F.R.; Liu, W.; Naseer, M.K.; Cui, R.Y.; Li, X.C.; Hussain, S.; Saad, S.A.; Niu, D.D. Infrared radiation denoising model of "sub-region-Gaussian kernel function" in the process of sandstone loading and fracture. *Infrared Phys. Technol.* **2023**, *129*, 104583. [CrossRef]
24. Wang, C.; Lu, Z.; Liu, L.; Chuai, X. Predicting points of the infrared precursor for limestone failure under uniaxial compression. *Int. J. Rock Mech. Min. Sci.* **2016**, *88*, 34–43. [CrossRef]
25. Mineo, S.; Pappalardo, G. The use of infrared thermography for porosity assessment of intact rock. *Rock Mech. Rock Eng.* **2016**, *49*, 3027–3039. [CrossRef]
26. Pappalardo, G.; Mineo, S.; Zampelli, S.P.; Cubito, A. Infrared Thermography proposed for the estimation of the Cooling Rate Index in the remote survey of rock masses. *Int. J. Rock Mech. Min. Sci.* **2016**, *83*, 182–196. [CrossRef]
27. Ma, L.; Sun, H.; Zhang, Y.; Zhou, T.; Li, K.; Guo, J. Characteristics of infrared radiation of coal specimens under uniaxial loading. *Infrared Phys. Technol.* **2016**, *49*, 1567–1572. [CrossRef]
28. Sun, X.; Xu, H.; He, M.; Zhang, F. Experimental investigation of the occurrence of rockburst in a rock specimen through infrared thermography and acoustic emission. *Int. J. Rock Mech. Min. Sci.* **2017**, *93*, 250–259. [CrossRef]
29. Sun, H.; Ma, L.; Adeleke, N.; Zhang, Y. Background thermal noise correction methodology for average infrared radiation temperature of coal under uniaxial loading. *Infrared Phys. Technol.* **2017**, *81*, 157–165. [CrossRef]

30. Asakura, T.; Kojima, Y.J.T.; Technology, U.S. Tunnel maintenance in Japan. *Tunn. Undergr. Space Technol.* **2003**, *18*, 161–169. [CrossRef]
31. Liu, S.; Zhang, Y.; Wu, L. Infrared radiation characteristics of concrete rupture and water seepage process. *Chin. J. Rock Mech. Eng.* **2009**, *28*, 53–58.
32. Dou, H.T.; Xue, Y.D. Experimental study on influencing factors of infrared radiation characteristics of tunnel lining leakage. *Chin. J. Rock Mech. Eng.* **2011**, *30*, 9.
33. Zhang, Y.B.; Liang, P.; Liu, X.X.; Tian, B.Z. Study on infrared spatiotemporal evolution characteristics of water inrush in soft roadway. *Coal Sci. Technol.* **2015**, *44*, 6.
34. Wu, L.; Wang, J. Infrared radiation features of coal and rocks under loading. *Int. J. Rock Mech. Min. Sci. Geomech. Abstr.* **1998**, *35*, 332419. [CrossRef]
35. Liu, S.J.; Zhang, Y.B.; Chen, Q.L. The variation characteristics of infrared radiation in the process of wet-rock stress. *J. Northeast Univ.* **2010**, *1*, 4.
36. Ma, L.; Sun, H.; Zhang, Y.; Hu, H.; Zhang, C.J.S. The role of stress in controlling infrared radiation during coal and rock failures. *Strain* **2018**, *54*, e12295. [CrossRef]
37. Liu, S.J.; Huang, J.W.; Wu, L.X.; Zhang, Y.B.; Tian, B.Z. Quantitative analysis methods of infrared radiation temperature field variation in rock loading process. *Chin. J. Rock Mech. Eng.* **2015**, *34*, 9.
38. Ma, L.; Zhang, Y. An Experimental Study on Infrared Radiation Characteristics of Sandstone Samples Under Uniaxial Loading. *Rock Mech. Rock Eng.* **2019**, *52*, 3493–3500. [CrossRef]
39. Yang, S.Q.; Dong, Y.; Wang, G.Y.; Chen, D.C.; Zhang, J.Y. Study on the evolution law of surface infrared radiation during shale deformation. *Chin. J. Undergr. Space Eng.* **2019**, *15*, 7.
40. Liu, S.J.; Wu, L.X. Comparison of infrared radiation characteristics between brittle rock and plexiglass under stress. *Chin. J Rock Mech. Eng.* **2007**, *26*, 4183–4188.
41. Liu, W.; Yang, K.; Zhang, S.; Zhang, Z.N.; Xu, R.J. Energy evolution and water immersion-induced weakening in sandstone roof of coal mines. *Int. J. Coal Sci. Technol.* **2022**, *9*, 53. [CrossRef]
42. Du, F.; Ma, J.; Guo, X.; Wang, T.F.; Dong, X.H.; Li, J.S.; He, S.L.; Nuerjuma, D. Rockburst mechanism and the law of energy accumulation and release in mining roadway: A case study. *Int. J. Coal Sci. Technol.* **2022**, *9*, 67. [CrossRef]
43. Ma, L.; Sun, H.J.I.P. Spatial-temporal infrared radiation precursors of coal failure under uniaxial compressive loading. *Infrared Phys. Technol.* **2018**, *93*, 144–153. [CrossRef]
44. Cao, K.W.; Xu, Y.J.; Naseer, M.K.; Li, X.C.; Cui, R.Y.; Hussain, S.; Danial, J.A.; Alarifi, S.S. A comprehensive model for evaluating infrared radiation and acoustic emission characteristics of sandstone fracture. *Eng. Fract. Mech.* **2023**, *283*, 109217. [CrossRef]

sustainability

MDPI

Article

Scale Effects on Shear Strength of Rough Rock Joints Caused by Normal Stress Conditions

Jiayi Shen [1,2], Chenhao Sun [1], Huajie Huang [1], Jiawang Chen [3] and Chuangzhou Wu [1,2,*]

[1] Institute of Port, Coastal and Offshore Engineering, Zhejiang University, Hangzhou 310015, China
[2] Key Laboratory of Offshore Geotechnics and Material of Zhejiang Province, Zhejiang University, Hangzhou 310058, China
[3] Hainan Institute, Zhejiang University, Sanya 572025, China
* Correspondence: ark_wu@zju.edu.cn

Abstract: Scale effects on the mechanical behavior of rock joints have been extensively studied in rocks and rock-like materials. However, limited attention has been paid to understanding scale effects on the shear strength of rock joints in relation to normal stress σ_n applied to rock samples under direct shear tests. In this research, a two-dimensional particle flow code (PFC2D) is adopted to build a synthetic sandstone rock model with a standard joint roughness coefficient (JRC) profile. The manufactured rock model, which is adjusted by the experiment data and tested by the empirical Barton's shear strength criterion, is then used to research scale effects on the shear strength of rock joints caused by normal stresses. It is found that the failure type can be affected by JRC and σ_n. Therefore, a scale effect index (SEI) that is equal to JRC plus two times σ_n (MPa) is proposed to identify the types of shear failure. Overall, shearing off asperities is the main failure mechanism for rock samples with SEI > 14, which leads to negative scale effects. It is also found that the degree of scale effects on the shear strength of rock joints is more obvious at low normal stress conditions, where σ_n < 2 MPa.

Keywords: rock joints; shear strength; scale effects; normal stress; JRC; PFC simulation

Citation: Shen, J.; Sun, C.; Huang, H.; Chen, J.; Wu, C. Scale Effects on Shear Strength of Rough Rock Joints Caused by Normal Stress Conditions. *Sustainability* **2023**, *15*, 7520. https://doi.org/10.3390/su15097520

Academic Editors: Jian Zhou, Mahdi Hasanipanah and Danial Jahed Armaghani

Received: 3 March 2023
Revised: 12 April 2023
Accepted: 26 April 2023
Published: 4 May 2023

1. Introduction

Rock joints play an important role in the estimation of the shear strength of rock masses [1–3]. Effective design of rock engineering projects, such as underground excavations and open pit slopes, requires precise estimation of the shear strength of rock joints [4–6]. However, it is well known that there is a scale effect on the shear strength of rock joints [7–9]. The main difficulty in determining how the shear strength of rock joints varies with scale is conducting expensive and time-consuming engineering scale in situ testing [10]. Although laboratory scale tests on a small jointed sample cannot generate the precise shear strength of rock joints, they can still reveal the mechanical behavior of jointed rock masses [11]. Therefore, laboratory tests are widely used by researchers to investigate how the shear strength of rock joints is affected by sample sizes. Table 1 presents a review of scale effects on the shear strength of rock joints, which shows conflicting results. The majority of results show that there is a negative scale effect on the shear strength, which means the shear strength decreases with the increase of joint sizes. Some results [12,13] show positive scale effects, which represent the shear strength increases when the joint size increases. While other results [13–15] show no scale effects.

Scale effects on the shear strength of rock joints could be explained in different ways. One explanation is that scale effects occur due to the contact area of joints changing with the increase in joint size [16]. Pratt et al. [14] and Yoshinaka et al. [17] attributed the decrease in shear strength to the smaller contact area of the sample where higher stress was concentrated on these contact surfaces. The other explanation is that the scale effect is associated with the change of undulations and asperities on a joint surface as joint size increases. A longer sample will result in higher undulation amplitude compared to a smaller sample [18]. Barton and Choubey [19] concluded that the shear behavior of larger rock samples is governed by larger and gentler asperities, while the steep and small asperities are the controlling mechanism in smaller rock samples. Giani et al. [20] stated that when the rock joint shear strength depends on the random distribution of asperity, it will produce a positive scale effect. If the rock joint shear strength depends on wavy and rough surfaces, then there is a negative scale effect. Therefore, more research is required to determine the exact nature of the scale effect on the shear strength of joints.

Table 1. Review of scale effects on the shear strength of rock joints [21].

Authors	Rock Types	Sample Size	Normal Stress (MPa)	Scale Effect
Azinfar et al. [13]	Silicon rubber	25–2500 cm^2	0.3, 0.8, 1.4	O, N, P
Barton and Choubey [19]	Granite	9.8 × 4.5, 45 × 50 cm	0.1–2	N
Bandis et al. [22]	plaster	6–36 cm	1	N
Bahaaddini et al. [21]	Sandstone	5–40 cm	0.5	N
Castelli et al. [23]	Cement	100–400 cm^2	0.75, 1.5, 3	N
Fardin [24]	Concrete	5 × 5–20 × 20 cm^2	1, 2.5, 5, 10	N
Hencher et al. [25]	Limestone	44–531 cm^2	0.0245	O
Johansson [26]	Granite	36, 400 cm^2	1	O
Ohnishi et al. [12]	Concrete	100–1000 cm^2	0.26–2.04	P
Pratt et al. [14]	Quartz diorite	60, 142–5130 cm^2	3	N
Ueng et al. [15]	Cement	7.5–30 cm^2	0.3, 0.6, 0.9	O, N
Vallier et al. [27]	-	10–200 cm	2	N
Yoshinaka et al. [17]	Granite	20–9600 cm^2	0.26–2.04	N

"N" means negative scale effect; "P" means positive scale effect; "O" means no scale effect.

Based on the literature review, we noticed that the existing laboratory tests were carried out under various normal stress conditions ranging from 0.0245 to 10 MPa, as shown in Table 1. As we know, the failure mode of rock joints during the direct shear test will be affected by the application of normal stresses. When rock samples are under high normal stress conditions, the tips of asperities could be sheared off; therefore, the shear strength would be relatively higher compared to rock samples that are under low normal stress conditions where sliding is the controlling mechanism of rock failure.

Numerical simulations using the PFC are capable of simulating the asperity damage and degradation process during the shearing tests [28]. It has been proven that the shear strength results acquired from PFC modeling are typically comparable with experimental test results [29]. Therefore, PFC simulations as an alternative to physical testing can be used to reveal the fundamental mechanism of shear behavior of rock joints at various scales.

In this research, a synthetic rock model based on the two-dimensional particle flow code (PFC2D)-based synthetic sandstone rock model is used to study the influence of normal stress on scale effects on the shear strength of rock samples with standard joint roughness coefficient (JRC) profiles, and attempts to answer two questions: (1) Are scale effects on shear behavior affected by normal stresses? (2) What is the degree of scale effects affected by normal stresses?

In this paper, the synthetic rock model for numerical tests is introduced in Section 2. The verification of the synthetic rock model is shown in Section 3. Scale effect investigations are presented and discussed in Sections 4 and 5.

2. Synthetic Rock Model for Numerical Tests

2.1. Synthetic Rock Model Based on PFC2D

PFC2D is a discrete element program. The bonded particle model (BPM), a composite of rounded particles, simulates complete rock and does not require a continuum-scale constitutive model to depict the mechanical behavior of intact rock [30]. The parallel bond model, which can replicate the physical behavior of a substance similar to cement linking the two nearby particles, is one of the most fundamental and often used BPMs in the PFC2D, as illustrated in Figure 1. Bond breaking reduces stiffness because contact and bond stiffness both contribute to stiffness in a parallel bond model. While contact stiffness is active as long as particles are in contact, bond stiffness is instantly gone when a bond breaks [31]. Therefore, the parallel bond model is a more accurate bond model for materials that resemble rocks, since it allows for the possibility of bonds breaking in tension or shearing with a corresponding loss in stiffness.

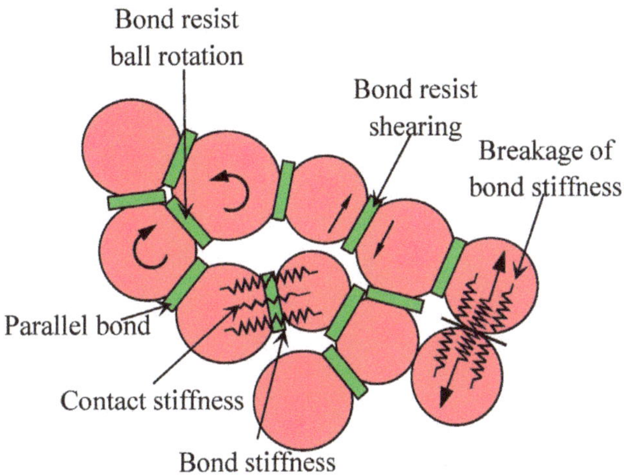

Figure 1. Illustration of the parallel bond model.

On the other hand, by adding joints to a BPM assembly using the smooth joint model (SJM), jointed rock masses can be created. The BPM's original contact microscopic characteristics will be replaced with SJM properties with the names friction coefficient μ_j, shear stiffness k_{sj}, and normal stiffness k_{nj} when the SJM is put into the BPM [32]. The synthetic rock model constructed by the BPM and SJM has the ability to simulate various mechanical responses of jointed rock masses including peak strength [31], scale effect [33], anisotropy [34], and cracking processes [30,33] in rocks and rock-like materials.

The numerical direct shear test used in this research is presented in Figure 2. The specimen (40 mm × 100 mm) is generated using the BPM. The rock joint is created using the SJM. During the direct shear test, the upper block receives the normal force in a vertical direction. The upper block is given a horizontal velocity of 0.03 m/s, while the lower block is held in place.

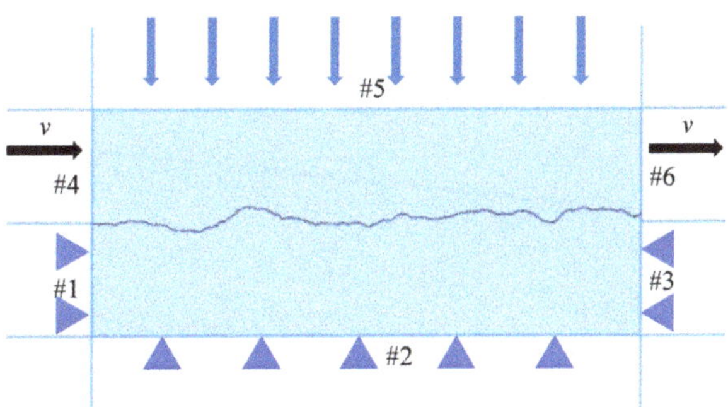

Figure 2. Numerical direct shear test set up.

2.2. Calibration of Numerical Models

In this study, the synthetic rock model was calibrated using laboratory data from Australia's Hawkesbury Sandstone [28]. Firstly, tests for uniaxial compression on rock samples (42 mm × 84 mm) with a loading rate of 0.02 m/s were performed to determine the BPM's parameters after a calibration process [31] to ensure that the mechanical properties of the synthetic rock model are close to laboratory data.

It should be mentioned that one of the key factors influencing the resilience of restricted materials to deformation and strength, such as rocks and cemented soil, is the loading rate [35–38]. In this research, we did not consider such loading rate effects on the mechanical properties of jointed rocks.

Using the calibrated BPM parameters indicated in Table 2 to perform the uniaxial compression test (Figure 3), the values of elastic modulus E, Poisson's ratio v, and UCS produced are comparable to experimental tests, as shown in Table 3.

Table 2. Micro-parameters of the BPM model.

Parameters	Values
Minimum particle radius: R_{min} (mm)	0.28
Maximum particle radius: R_{max} (mm)	0.42
Stiffness ratio: k_n/k_s	2.1
Effective modulus: E_c (GPa)	4.1
Bond tensile strength: Tb (MPa)	11.2
Bond friction angle: Φb (°)	35
Cohesion: cb (MPa)	11.2
Friction coefficient: u	0.2
Porosity ratio: e	0.16

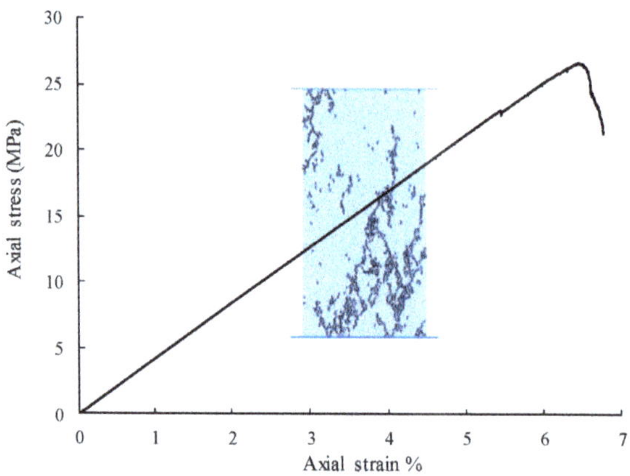

Figure 3. Numerical uniaxial compressive test.

Table 3. Comparison of mechanical properties calculated from the numerical model and tested from laboratory.

Properties	Parameters	Laboratory Test	PFC Model
Intact rock properties	UCS (MPa)	27.40	27.40
	E (GPa)	4.20	4.20
	v	0.20	0.21
Joint properties	K_n (GPa/m)	28.6	28.6
	K_s (GPa/m)	6.40	6.40
	φ_b (°)	37.60	36.10

Then, synthetic rock models (40 mm × 100 mm) with planar joints were constructed. The SJM has the following micro-parameters: friction coefficient μ_j, shear stiffness k_{sj}, and normal stiffness k_{nj}. In this research, the values of k_{nj} = 25 GPa, k_{sj} = 13 GPa, and μ_j = 0.75 were selected using the inverse-modeling calibration approach to ensure that the numerical rock model can give a similar response as that from laboratory tests with joint shear stiffness K_s = 6.4 GPa/m, normal stiffness K_n = 28.6 GPa/m, and joint friction angle φ_b = 37.6°. The calibration procedure was as follows: (1) The normal deformability compression test was carried out to calibrate normal stiffness k_{nj}. (2) The shear test was carried out to calibrate shear stiffness k_{sj} under normal stress of 1 MPa condition. (3) Direct shear tests were undertaken and friction coefficient μ_j was calibrated. Figures 2–4 present the final mechanical responses of the synthetic rock models after the final calibration.

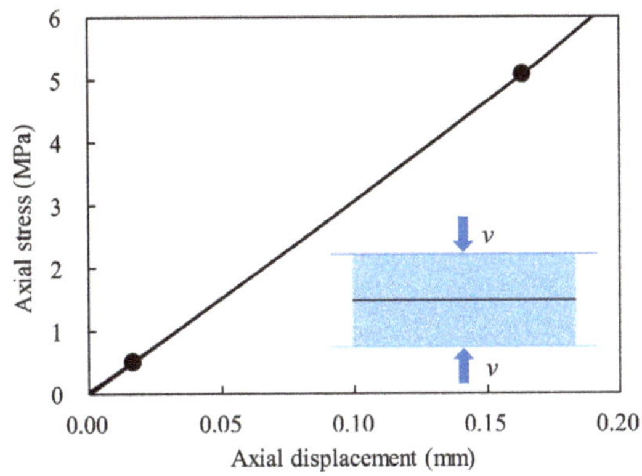

Figure 4. The normal deformability test on the synthetic rock specimen with a planar joint.

Figure 4 shows the axial stress-displacement curves of the synthetic rock specimen (40 mm × 100 mm) under the normal deformability test with the loading rate of 0.02 m/s. The value of joint normal stiffness K_n generated by the synthetic rock specimen is 28.6 GPa/m, which is close to the laboratory test results with K_n = 28.8 GPa/m.

Figure 5 shows the shear stress-displacement curve of the synthetic rock model (40 mm × 100 mm) with a planar joint under a direct shear test (loading rate of 0.03 m/s) with the normal stress of 1 MPa. The value of joint shear stiffness K_s generated by the synthetic rock specimen is 6.4 GPa/m, which is the same as laboratory test results with K_s = 6.4 GPa/m.

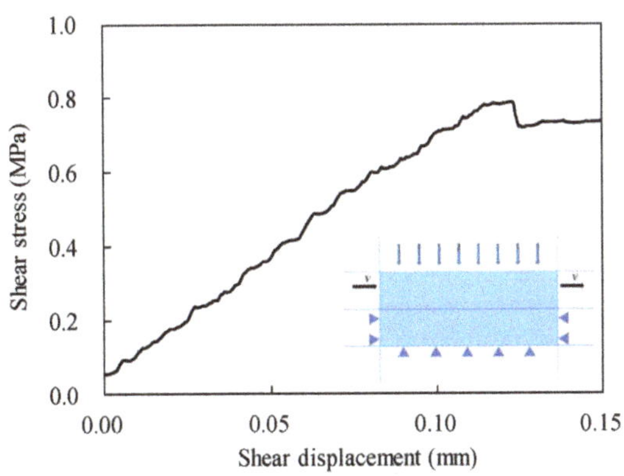

Figure 5. The direct shear test on the synthetic rock specimen with a planar joint under normal stress of 1 MPa.

Figure 6 shows the failure envelope of the synthetic rock model (40 mm × 100 mm) with a planar joint under direct shear tests. The value of joint friction angle φ_b generated by the synthetic rock specimen is 36.1°, which is close to laboratory test results with φ_b = 37.6°.

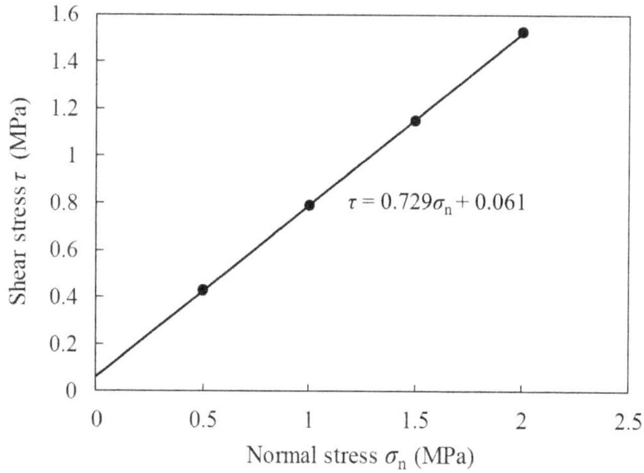

Figure 6. Shear strength of the synthetic rock samples under various normal stresses.

3. Validation of Synthetic Rock Models

To confirm the reliability of the synthetic rock model shown in Section 2, direct shear tests on the synthetic rock models with 10 standard JRC profiles were performed and the shear strength values produced from numerical simulations were compared to those derived from Barton's empirical shear strength model.

3.1. Barton's Shear Strength Model

One of the most widely adopted empirical strength criteria for estimating rock joint shear strength in rock engineering is the Barton's shear strength criterion. Based on the results of a large number of shearing tests on various rock joint profiles, Barton and his co-workers [19,39] proposed an empirical equation to estimate the shear strength of rock joints, as shown in Equation (1).

$$\tau = \sigma_n \tan\left[\varphi_b + \mathrm{JRC}\,\lg\left(\frac{\mathrm{JCS}}{\sigma_n}\right)\right] \quad (1)$$

where φ_b is the joint friction angle. JCS is the joint compression strength, which is equal to UCS of intact rock in this research. JRC stands for joint roughness coefficient and can be calculated using standard joint profiles.

3.2. Numerical Simulation Results

We performed extensive numerical direct shear experiments on synthetic rock models with varied JRC profiles in normal stress levels between 0.5 and 5 MPa. The failure envelopes generated by direct shear tests on synthetic rock models were compared to the empirical Barton's shear strength criterion (Equation (1)) with φ_b = 36.1° and JCS = 27.4 MPa. Figure 7 compares the shear strength obtained from numerical simulations to Barton's model, indicating that the usage of synthetic rock models is capable of generating adequate shear strength of rock joints.

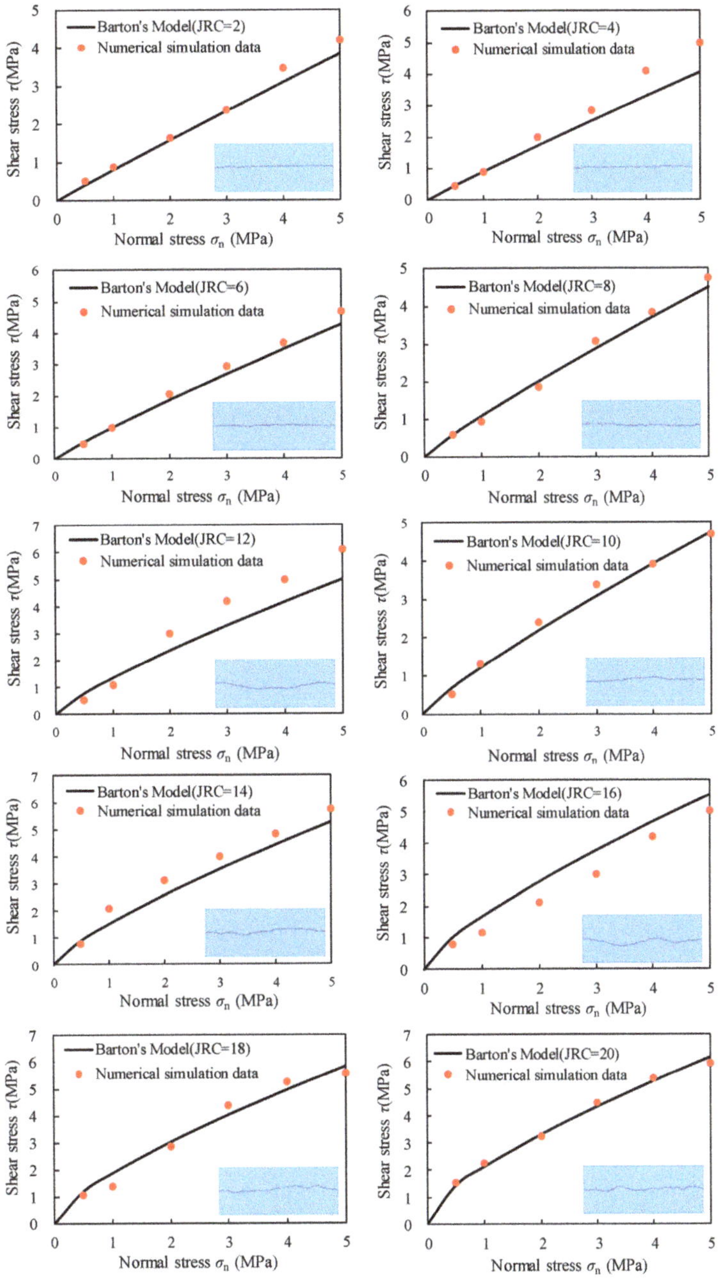

Figure 7. Comparison failure envelopes obtained from numerical simulations and the Barton's empirical model.

4. Configuration of Rock Samples for Scale Effect Investigations

Two methods are widely used for investigating scale effects on the shear strength of rock joints [15]. The first one is to divide a large rock joint into several smaller sizes of rock joints, as shown in Figure 8a, which presents an example of the division of the Barton's typical profile. The geometry characteristics or the values of JRC of smaller sizes of rock joints can be different from that of the original larger rock joint. The other method is the assembly of several repeated 100 mm profiles into larger rock joints many times the original profile length, as shown in Figure 8b. The joint roughness or the value of JRC of assembly samples is the same as that of the original joint surface.

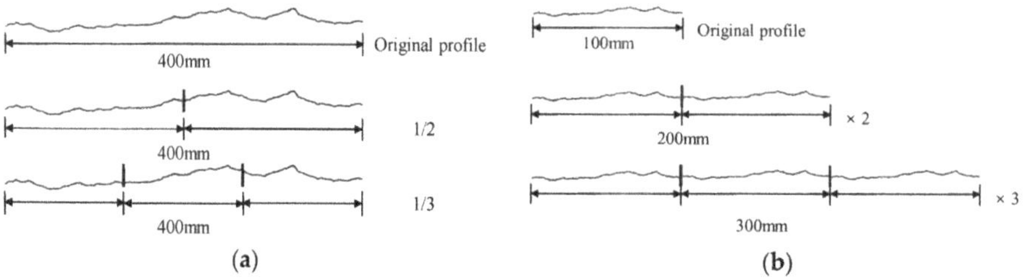

Figure 8. Two types of configuration of samples for scale effect investigations (**a**) division of the Barton's JRC profile (**b**) assembly of repeated Barton's JRC profile.

In fact, the scale effect on the shear strength of rock joints includes two factors, which are the sample size itself and the geometrical characteristics of the joint surface. Rock samples generated by the division method have various sample sizes and geometry characteristics. However, the rock joints generated by the assembly method have various sample sizes but have the same geometry characteristics. It is well known that the geometry characteristics will affect the shear strength of rock joints [8]. Therefore, in this research, we adopt the assembly of a repeated model that has the same geometry characteristics and JRC values to research the influence of pure sample size on the shear strength of rock joints.

5. Results and Discussion

Once synthetic rock models were validated, a series of rock specimens with various JRC profiles and different sizes (40 mm × 100 mm, 80 mm × 200 mm, and 120 mm × 300 mm) were generated to study the effect of sample sizes on rock joint shear strength. The shear strength values of different sizes of rock samples under given normal stresses were calculated and are summarized in Figure 9.

In this research, the index k (see Equation (2)), which is the average slope of three points, was used to identify the types of scale effects.

$$k = \frac{N\sum x_i y_i - \sum x_i \sum y_i}{N\left(\sum x_i^2\right) - \left(\sum x_i\right)^2} \tag{2}$$

where x_i is the joint length of the rock sample, y_i the shear stress of the rock sample, and N is the number of the testing sample. $k > 0$ means the rock joint has a positive scale effect and $k < 0$ means the rock joint has a negative scale effect. The value of k can be calculated using three groups of data. For example, for rock samples with JRC = 2 under the normal stress σ_n = 5 MPa, the shear strength of rock samples with joint lengths l = 100, 200, and 300 mm are 4.2, 4.4, and 4.6 MPa, respectively. Therefore, data (100, 4.2), (200, 4.4), and (300, 4.6) were put into Equation (2) to calculate the value of k. The result shows k = 0.4, which means the scale effect is positive. Table 4 shows comprehensive scale effect results of rock samples with various JRC profiles under different normal stress conditions. In Table 4, P means positive scale effect and N means negative scale effect.

The results presented in Table 4 are also plotted in Figure 10. It is found that the failure mode of rock joints during the direct shear test will be affected by the applying normal stresses and the joint roughness. When rock samples under high normal stress conditions, the tips of asperities with large joint roughness coefficient could be sheared off, therefore, the number of shear crack is relatively higher compared with rock samples with small joint roughness coefficients under low normal stress conditions where sliding is the controlling mechanism of rock failure. For example, when a rock sample with JRC = 2 under the normal stress σ_n = 0.5 MPa, the number of shear cracks is 10 and the scale effect is positive. However, when a rock sample with JRC = 20 under the normal stress σ_n = 5 MPa, the number of shear cracks is 280 and the scale effect is negative.

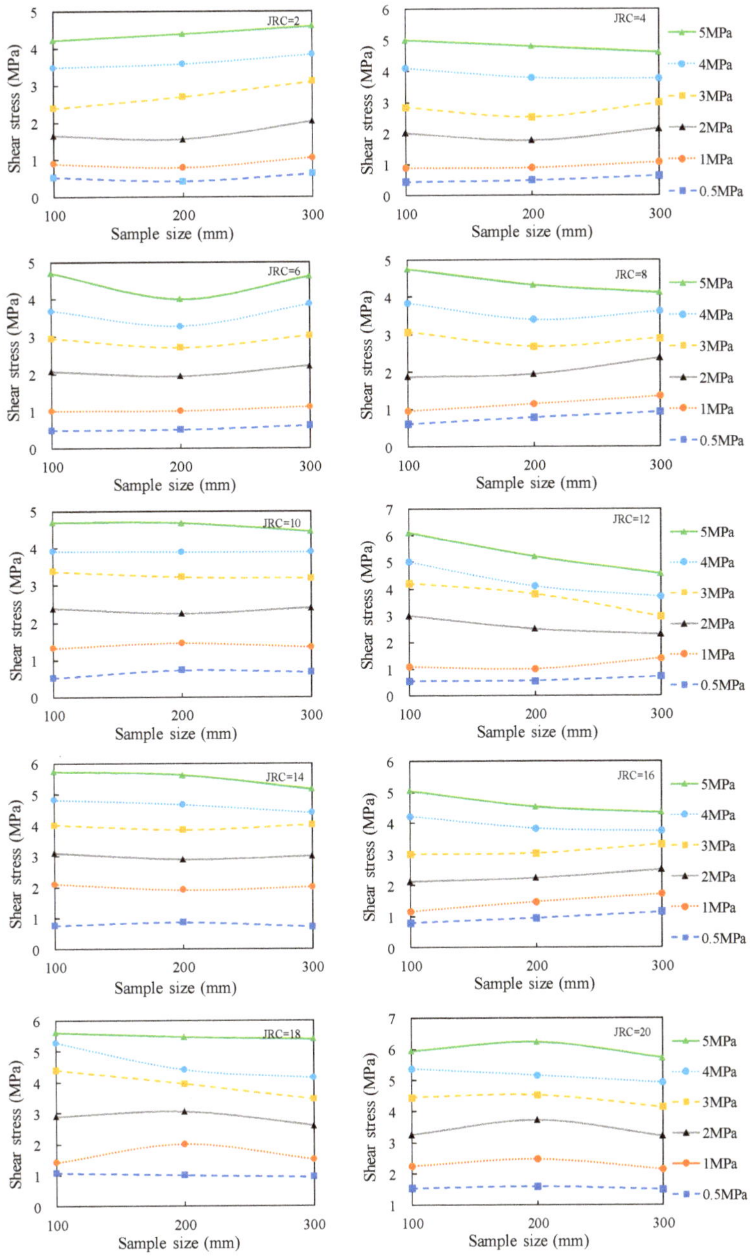

Figure 9. Results of scale effects on the shear strength of rock samples under various normal stress conditions.

Table 4. Results of scale effects of shear strength of rock joints.

JRC	Normal Stress σ_n (MPa)					
	0.5	**1**	**2**	**3**	**4**	**5**
2	P3	P4	P6	P8	P10	P12
4	P5	P6	P8	P10	N12	N14
6	P7	P8	P10	P12	P14	N16
8	P9	P10	P12	N14	N16	N18
10	P11	P12	P14	N16	N18	N20
12	P13	P14	N16	N18	N20	N22
14	N15	N16	N18	N20	N22	N24
16	P17	P18	P20	P22	N24	N26
18	N19	P20	N22	N24	N26	N28
20	N21	N22	N24	N26	N28	N30

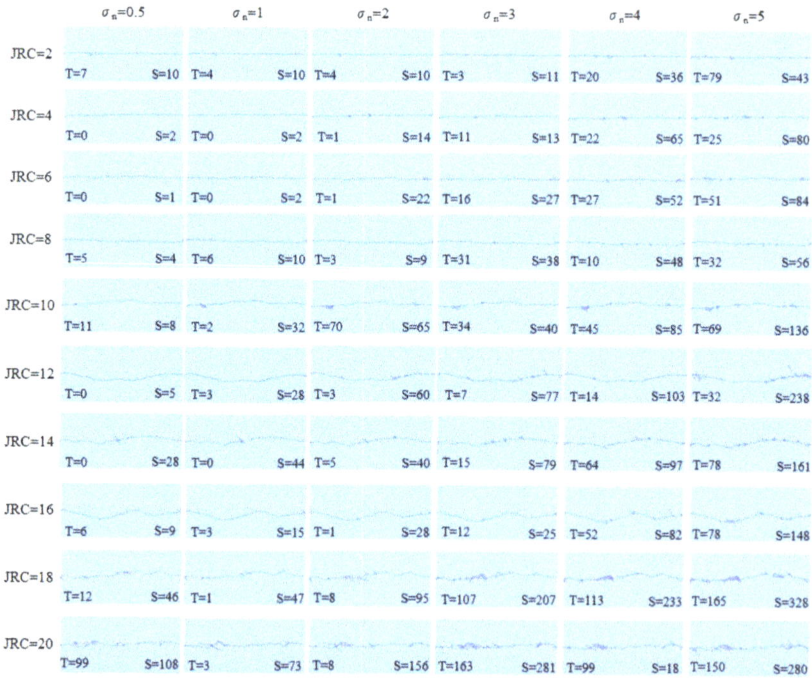

Figure 10. Failure pattern and crack number of rock samples (40 mm × 100 mm) at peak shear strength.

Based on the results of Table 4, a scale effect index (*SEI*) which is equal to *JRC* plus two times of normal stress (MPa), as shown in Equation (3), was proposed to identify the types of scale effects.

$$SEI = JRC + 2\sigma_n \tag{3}$$

The values of *SEI* for rock samples under different normal stress conditions are given in Table 4. For example, for rock samples with *JRC* = 2 under the normal stress σ_n = 0.5 MPa, the value of *SEI* = 2 + 2 × 0.5 = 3. The number P3 in Table 4 means the rock sample with *SEI* = 3 has a positive scale effect. It was found that 20 out of 21 rock samples have negative scale effects when *SEI* > 14, and 29 out of 33 rock samples have positive scale effects when *SEI* < 14.

To find the possible reason why the use of SEI can identify the types of scale effects on shear strength, we monitored the crack number generated in the synthetic rock sample when the stress reaches the peak strength during the direct shear tests. When the parallel link between nearby particles in the PFC rock model is broken, a micro-tensile crack or micro-shear crack can occur. Figure 10 shows the failure pattern corresponding to the shear crack number of each rock sample when the shear stress reaches the peak strength. For example, the S = 10 represents a shear crack number of 10 and T = 7 represents a tension crack number of 7 for a sample with SEI = 3 (JRC = 2 and σ_n = 0.5 MPa). The relations between SEI values and shear crack numbers of rock samples are also plotted in Figure 11. We can find that the number of shear cracks is low when SEI < 14. However, the number of shear cracks dramatically increases when the value of SEI is over 14, where the controlling failure mechanism transforms sliding to shearing off asperities.

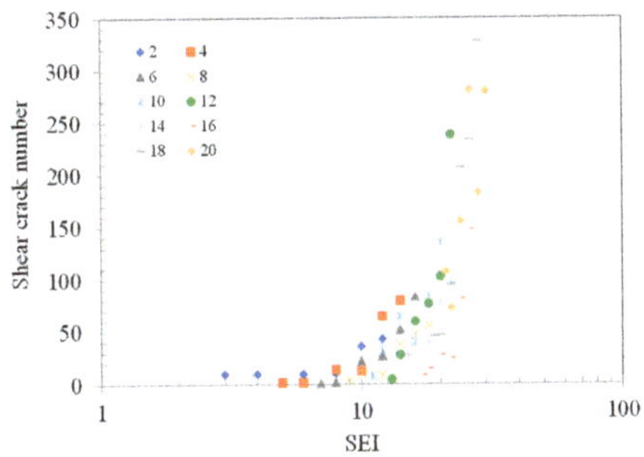

Figure 11. Relations between SEI and shear crack number of rock samples.

Therefore, we can conclude that the results presented in Table 4 and Figure 11 show that the proposed SEI is capable of identifying types of scale effects. When SEI < 14, sliding over joints is the controlling mechanism of rock failure, which leads to positive scale effects; however, shearing off asperities could be the controlling mechanism of rock failure for rock samples with SEI > 14, which leads to negative scale effects.

On the other hand, to further identify the degree of scale effects, the coefficient of variance (CV), which can calculate the value of Standard Deviation/Mean to quantify the random influence of a bunch of data, was further used as a scale effect magnitude index to quantify scale effects on shear strength of rock joints caused by normal stress conditions. The calculation results are presented in Figure 12.

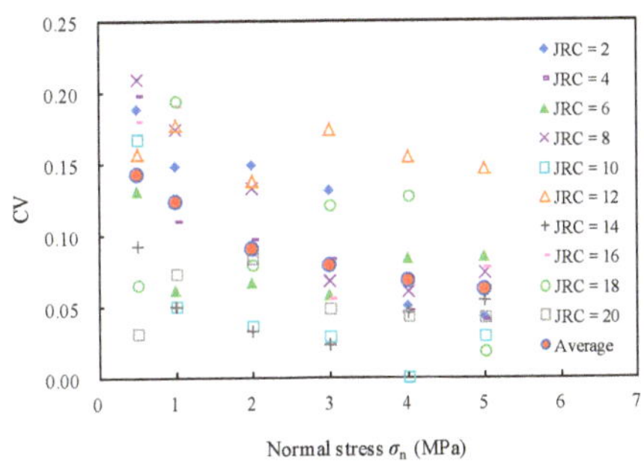

Figure 12. The degree of scale effect caused by normal stresses.

It is found that, in general, the values of CV decrease with the increase of normal stress. We also calculated the average value of CV for a given group of data. Figure 12 shows that the values of average CV decrease when the normal stress increases from 0.5 to 2.0 MPa, then, it tends to be stable with further increase of normal stresses, which means the degree of scale effects on shear strength of rock joints is more obvious at low normal stress conditions where σ_n < 2 MPa.

Such a phenomenon can also be validated by the laboratory data published by Fardin [24], who carried out a laboratory study of the scale effect on the shear strength of concrete replicas with roughness joints. Laboratory test results are shown in Figure 13. The CV values of samples under a specific normal stress condition were calculated and are shown in Figure 14. The value of CV is up to 0.4 when σ_n = 1 MPa, then, it decreases sharply to 0.14 when σ_n increases to 2 MPa. After that, there is a slight change in CV values with a further increase of σ_n from 2 MPa to 10 MPa. Such change in CV values with normal stresses is similar to that of the numerical results in Figure 12.

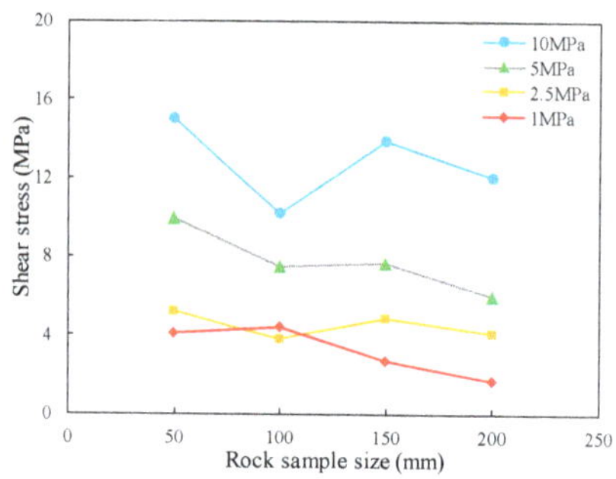

Figure 13. Laboratory data of scale effect on shear stress of rock joints under various normal stress conditions (data from Fardin [24]).

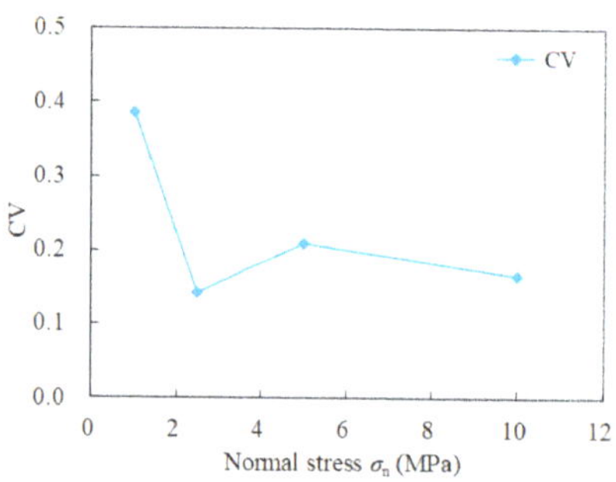

Figure 14. Laboratory investigation on the degree of scale effect caused by normal stresses.

6. Conclusions

Synthetic rock models with standard JRC profiles were constructed in PFC2D to investigate scale effects on the shear strength of rock joints under various normal stress conditions. The capability of the synthetic rock model to simulate the shear behavior of rock joints was tested by comparing numerical simulations with the Barton's shear strength criterion.

Once synthetic rock models were validated, a series of rock specimens of different sizes (40 mm × 100 mm, 80 mm × 200 mm, and 120 mm × 300 mm) were generated to investigate the influence of sample sizes on rock joint shear strength under normal stress ranges from 0.5 to 5 MPa.

Numerical simulation results show that the types of scale effects could be affected by the JRC profiles and normal stresses. Therefore, a scale effect index (SEI) that is equal to JRC plus two times normal stress (MPa), as shown in Equation (3), was proposed to identify the types of scale effects. It is found that for the rock sample with SEI < 14, sliding over joints is the controlling mechanism of rock failure, which leads to positive scale effects. However, shearing off asperities could be the controlling mechanism of rock failure for rock samples with SEI > 14, which leads to negative scale effects.

We also further investigated the influence of normal stress on the degree of scale effects on the shear strength of rock joints. It is discovered that the degree of scale effect is more obvious at low normal stresses conditions where $\sigma_n < 2$ MPa.

Finally, it should be noted that the finding of this research is based on the analysis of test data of Australia Hawkesbury Sandstone. Therefore, the finding of this research is open for further improvements as more shear strength data of various rock types become available.

Author Contributions: Conceptualization, J.S.; methodology, C.S. and H.H.; software, J.S. and H.H.; validation, J.S. and H.H.; formal analysis, J.S. and H.H.; investigation, J.S. and C.S.; resources, J.S. and C.W.; data curation, C.S. and H.H.; writing—original draft preparation, J.S. and H.H.; writing—review and editing, J.S., C.S., J.C. and C.W.; visualization, C.S. and H.H.; supervision, J.S. and C.W. All authors have read and agreed to the published version of the manuscript.

Funding: This research was funded by the Zhoushan Science and Technology Bureau (2022C81001), the Key R&D of Zhejiang Province (2021C03183), and the Scientific Research Fund of Zhejiang University (XY2021011), the Sanya Yazhou Bay Science and Technology City (KYC-2020-01-001), the Finance Science and Technology Project of Hainan Province (ZDKJ202019).

Institutional Review Board Statement: Not applicable.

Informed Consent Statement: Not applicable.

Data Availability Statement: The data presented in this study are available on request from the corresponding author.

Acknowledgments: We express our sincere gratitude to the reviewers for their valuable comments and suggestions.

Conflicts of Interest: The authors declare no conflict of interest.

References

1. Wang, F.; Wang, S.; Yao, W.; Li, X.; Meng, F.; Xia, K. Effect of roughness on the shear behavior of rock joints subjected to impact loading. *J. Rock Mech. Geotech. Eng.* **2023**, *15*, 339–349. [CrossRef]
2. Seungbeom, C.; Byungkyu, J.; Sudeuk, L.; Seokwon, J. Experimental Study on Hydromechanical Behavior of an Artificial Rock Joint with Controlled Roughness. *Sustainability* **2019**, *11*, 1014. [CrossRef]
3. Seungbeom, C.; Sudeuk, L.; Hoyoung, J.; Seokwon, J. Development of a New Method for the Quantitative Generation of an Artificial Joint Specimen with Specific Geometric Properties. *Sustainability* **2019**, *11*, 373. [CrossRef]
4. Yong, R.; Ye, J.; Li, B.; Du, S. Determining the maximum sampling interval in rock join roughness measurements using Fourier series. *Int. J. Rock Mech. Min. Sci.* **2018**, *101*, 77–78. [CrossRef]
5. Hu, J.; Pan, H.; Liu, H.; Li, L.; Fan, H.; Liu, Q. Response of acoustic emission and vibration monitoring data during rock block collapse in the tunnel: Small-and large-scale experiments study. *Tunn. Undergr. Space Technol.* **2023**, *137*, 105121. [CrossRef]
6. Li, S.; Huang, Q.; Hu, B.; Pan, J.; Chen, J.; Yang, J.; Yu, H. Mining Method Optimization of Difficult-to-Mine Complicated Orebody Using Pythagorean Fuzzy Sets and TOPSIS Method. *Sustainability* **2023**, *15*, 3692. [CrossRef]
7. Hencher, S.; Richards, L. Assessing the shear strength of rock discontinuities at laboratory and field scales. *Rock Mech. Rock Eng.* **2015**, *48*, 883–905. [CrossRef]
8. Buzzi, O.; Casagrande, D. A step towards the end of the scale effect conundrum when predicting the shear strength of large in situ discontinuities. *Int. J. Rock Mech. Min. Sci.* **2018**, *105*, 210–219. [CrossRef]
9. Luo, Y.; Wang, Y.; Guo, H.; Liu, X.; Luo, Y.; Liu, Y. Relationship between Joint Roughness Coefficient and Statistical Roughness Parameters and Its Sensitivity to Sampling Interval. *Sustainability* **2022**, *14*, 13597. [CrossRef]
10. Shen, J.; Shu, Z.; Cai, M.; Du, S. A shear strength model for anisotropic blocky rock masses with persistent joints. *Int. J. Rock Mech. Min. Sci.* **2020**, *134*, 104430. [CrossRef]
11. Alejano, L.; Arzúa, J.; Bozorgzadeh, N.; Harrison, J. Triaxial strength and deformability of intact and increasingly jointed granite samples. *Int. J. Rock Mech. Min. Sci.* **2017**, *95*, 87–103. [CrossRef]
12. Ohnishi, Y.; Herda, H.; Yoshinaka, R. Shear strength scale effect and the geometry of single and repeated rock joints. In *Scale Effects in Rock Masses*; CRC Press: Boca Raton, FL, USA, 1993; pp. 167–173.
13. Azinfar, M.; Ghazvinian, A.; Nejati, H. Assessment of scale effect on 3D roughness parameters of fracture surfaces. *Eur. J. Environ. Civ. Eng.* **2019**, *23*, 1–28. [CrossRef]
14. Pratt, H.; Black, A.; Brace, W. Friction and deformation of jointed quartz diorite. In Proceedings of the Third ISRM Congress, Denver, Colorado, 1–7 September 1974; pp. 306–310.
15. Ueng, T.; Jou, Y.; Peng, I. Scale effect on shear strength of computer aided manufactured joints. *J. Geoengin.* **2010**, *5*, 29–37.

16. Pirzada, M.; Roshan, H.; Sun, H.; Oh, J.; Bahaaddini, M. Effect of contact surface area on frictional behaviour of dry and saturated rock joints. *J. Struct. Geol.* **2020**, *135*, 104044. [CrossRef]
17. Yoshinaka, R.; Yoshida, J.; Shimizu, T.; Arai, H.; Arisaka, S. Scale effect in shear strength and deformability of rock joints. In Proceedings of the 7th ISRM International Congress on Rock Mechanics, Aachen, Germany, 16–20 September 1991. [CrossRef]
18. Leal, G. Some new essential questions about scale effects on the mechanics of rock mass joints. In Proceedings of the 10th ISRM Congress: Technology Roadmap for Rock Mechanics, Sandton, South Africa, 8–12 September 2003.
19. Barton, N.; Choubey, V. The shear strength of rock joints in theory and practice. *Rock Mech. Min. Sci.* **1977**, *10*, 1–54. [CrossRef]
20. Giani, G.; Ferrero, A.; Passarello, G.; Reinaudo, L. Scale effect evaluation on natural discontinuity shear strength. *Fract. Jt. Rock Masses* **1992**, *30*, 447–452. [CrossRef]
21. Bahaaddini, M.; Hagan, P.; Mitra, R.; Hebblewhite, B. Scale effect on the shear behaviour of rock joints based on a numerical study. *Eng. Geol.* **2014**, *181*, 212–223. [CrossRef]
22. Bandis, S.; Lumsden, A.; Barton, N. Experimental studies of scale effects on the shear behaviour of rock joints. *Int. J. Rock Mech. Min. Sci. Geomech. Abstr.* **1981**, *18*, 1–21. [CrossRef]
23. Castelli, M.; Re, F.; Scavia, C.; Zaninetti, Z. Experimental evaluation of scale effects on the mechanical behavior of rock joints. In Proceedings of the ISRM Regional Symposium EUROCK 2001, Espoo, Finland, 4–7 June 2001; pp. 205–210.
24. Fardin, N. Influence of structural non-stationarity of surface roughness on morphological characterization and mechanical deformation of rock joints. *Rock Mech. Rock Eng.* **2008**, *41*, 267–297. [CrossRef]
25. Hencher, S.; Toy, J.; Lumsden, A. Scale dependent shear strength of rock joints. In Proceedings of the International Workshop in Scale Effects in Rock Masses, Lisbon, Portugl, 25 June 1993; Volume 93, pp. 233–240.
26. Johansson, F. Influence of scale and matedness on the peak shear strength of fresh, unweathered rock joints. *Int. J. Rock Mech. Min. Sci.* **2016**, *82*, 36–47. [CrossRef]
27. Vallier, F.; Mitani, Y.; Boulon, M.; Esaki, F. A shear model accounting scale effect in rock joints behavior. *Rock Mech. Rock Eng.* **2010**, *43*, 581–595. [CrossRef]
28. Bahaaddini, M.; Hagan, P.; Mitra, R.; Hebblewhite, B. Parametric study of smooth joint parameters on the shear behaviour of rock joints. *Rock Mech. Rock Eng.* **2015**, *48*, 923–940. [CrossRef]
29. Huang, H.; Shen, J.; Chen, Q.; Karakus, M. Estimation of REV for fractured rock masses based on Geological Strength Index. *Int. J. Rock Mech. Min. Sci.* **2020**, *126*, 104–179. [CrossRef]
30. Shen, J.; Zhan, S.; Karakus, M.; Zuo, J. Effects of flaw width on cracking behavior of single-flawed rock specimens. *Bull. Eng. Geol. Environ.* **2021**, *80*, 1701–1711. [CrossRef]
31. Zhou, C.; Xu, C.; Karakus, M.; Shen, J. A systematic approach to the calibration of micro-parameters for the flat-jointed bonded particle model. *Geomech. Eng.* **2018**, *16*, 471–482. [CrossRef]
32. Cundall, P.; Potyondy, D. A bonded-particle model for rock. *Int. J. Rock Mech. Min. Sci.* **2004**, *41*, 1329–1364. [CrossRef]
33. Chen, Y.; Sheng, B.; Xie, S.; Cao, R.; Wang, Y.; Zhao, Y.; Lin, H. Crack propagation and scale effect of random fractured rock under compression-shear loading. *J. Mater. Res. Technol.* **2023**, *23*, 5164–5180. [CrossRef]
34. Wang, P.; Ren, F.; Miao, S.; Cai, M.; Yang, T. Evaluation of the anisotropy and directionality of a jointed rock mass under numerical direct shear tests. *Eng. Geol.* **2017**, *225*, 29–41. [CrossRef]
35. Maqsood, Z.; Koseki, J.; Miyashita, Y.; Xie, J.; Kyokawa, H. Experimental study on the mechanical behaviour of bounded geomaterials under creep and cyclic loading considering effects of instantaneous strain rates. *Eng. Geol.* **2020**, *276*, 105774. [CrossRef]
36. Swan, G.; Cook, J.; Bruce, S.; Meehan, R. Strain rate effects in Kimmeridge Bay shale. *Int. J. Rock Mech. Min. Sci.* **1989**, *26*, 135–149. [CrossRef]
37. Maqsood, Z.; Koseki, J.; Ahsan, M.; Shaikh, M.; Kyokawa, H. Experimental study on hardening characteristics and loading rate dependent mechanical behaviour of gypsum mixed sand. *Constr. Build. Mater.* **2020**, *262*, 119992. [CrossRef]
38. Liu, S.; Wang, W.; Cao, Y.; Li, H.; Shao, J. Numerical study of the influence of loading rate on fracture mechanism in elastoplastic rock-like materials with a modified phase-field model. *Int. J. Rock Mech. Min. Sci.* **2023**, *162*, 105309. [CrossRef]
39. Barton, N.; Bandis, S. Some effects of scale on the shear strength of joints. *Int. J. Rock Mech. Min. Sci.* **1980**, *17*, 69–73. [CrossRef]

sustainability

Article

Slope Failure Risk Assessment Considering Both the Randomness of Groundwater Level and Soil Shear Strength Parameters

Pu Peng [1], Ze Li [1,*], Xiaoyan Zhang [1], Wenlian Liu [2], Sugang Sui [2] and Hanhua Xu [3]

1 Faculty of Civil Engineering and Mechanics, Kunming University of Science and Technology, Kunming 650500, China; 20182110047@stu.kust.edu.cn (P.P.); zhangxiaoyan@kust.edu.cn (X.Z.)
2 Kunming Prospecting Design Institute, China Nonferrous Metals Industry Co., Ltd., Kunming 650051, China; lwenl702@sina.com (W.L.); suisugang2@zskk1953.com (S.S.)
3 Yunnan Key Laboratory of Geotechnical Engineering and Geohazards, Kunming 650051, China; xuhanhua@zskk1953.com
* Correspondence: lize@kust.edu.cn; Tel.: +86-15974807239

Abstract: Conducting research on slope failure risk assessment is beneficial for the sustainable development of slopes. There will be various failure modes considering both the randomness of the groundwater level and soil shear strength parameters. Based on the integrated failure probability (*IFP*), the traditional failure risk analysis needs to count all failure modes, including the failure probability (P_f) and failure risk coefficient (C), one-by-one. A new slope failure risk assessment method that uses the sum of the element failure risk to calculate the overall failure risk is proposed in this paper and considers both the randomness of the groundwater level and soil shear strength parameters. The element failure probability is determined by their location information and failure situation; the element failure risk coefficient is determined by their area. It transforms the complex overall failure risk problem into a simple element failure risk problem, which simplifies the calculation process and improves the calculation efficiency greatly. The correctness is verified with the systematic analysis of a classical case. The results show that the slope failure probability and failure risk are greatly increased from 1.40% to 3.30% and 0.829 m^2 to 2.094 m^2 with rising groundwater level, respectively.

Keywords: failure risk; element failure probability; spatial variability; stochastic groundwater level; upper bound method

Citation: Peng, P.; Li, Z.; Zhang, X.; Liu, W.; Sui, S.; Xu, H. Slope Failure Risk Assessment Considering Both the Randomness of Groundwater Level and Soil Shear Strength Parameters. *Sustainability* **2023**, *15*, 7464. https://doi.org/10.3390/su15097464

Academic Editor: Jianjun Ma

Received: 9 April 2023
Revised: 26 April 2023
Accepted: 28 April 2023
Published: 1 May 2023

1. Introduction

There are many factors that induce slope instability, among which the influences of the mechanical parameters of soil mass and groundwater are particularly important [1]. Due to long-term geological actions, such as sedimentation, post-sedimentation, chemical weathering, and physical denudation, the mechanical parameters of natural soil show certain spatial variability, which affects both the safety performance and failure mode of a slope. However, these effects are often ignored in practical engineering [2,3]. Groundwater mainly affects the safety performance from two aspects: one is making the seepage field change, affecting the stability performance; and the other is reducing shear strength parameters, affecting the safety performance [4]. Generally, the groundwater level inside the slope is not a certain value and is influenced by various uncertain factors. Therefore, it is essential to establish a failure risk analysis model of soil slopes that considers both the spatial variability and the stochasticity of the groundwater level so as to simulate the slope more in line with the real situation and achieve sustainable development of the slope.

In recent years, the slope reliability problem under consideration of spatial variability has received widespread attention by researchers. Using the limit equilibrium method

(LEM), the probability density function (PDF) curve and cumulative probability density function (CDF) curve of the safety factor can be acquired directly by first assuming the sliding surface. However, the LEM only takes the one-dimensional spatial variability of the soil material into account and ignores the impact outside the sliding surface on the slope reliability. In addition, the LEM does not consider the constitutive relation of the soil, so the stress–strain relationship cannot be acquired [5–11]. Using the finite element method (FEM) without assuming the sliding surface, the constitutive relation and spatial variability of materials can be fully considered so as to obtain more reasonable calculation results compared with the LEM [12]. In addition, the FEM can analyze the stress–strain development and progressive slope failure process, but the calculation cost is high [12–16]. Using the limit analysis method (LAM), the range of true solutions and corresponding failure modes can be acquired efficiently and accurately with the upper bound method (UBM) and lower bound method (LBM) [17–24]. Chen Zhaohui et al. combined the stochastic FEM and LAM, taking the spatial variability into account for the slope reliability analysis, and gave the strict frequency distribution range of the safety factor [25]. Peng Pu et al. carried out a study on element failure probability and used affinity propagation (AP) cluster analysis to obtain the failure modes of soil slope [26]. The above research greatly promotes the application prospect of the LAM in slope reliability analysis; however, it is not always reasonable to use only the failure probability to evaluate the slope. For example, there are two failure modes of a slope, corresponding to shallow landslide (P_{f1}, C_1) and deep landslide (P_{f2}, C_2). P_{f1} is larger than P_{f2}, but C_1 is smaller than C_2, so which failure mode has greater impact? To fully consider the impact of P_f and C, researchers proposed the concept of slope failure risk (R) [27,28]. However, whether we use the LEM or FEM to calculate the slope failure risk, the failure modes of the slope need to be identified and counted first, and there is lower computational efficiency with large sample calculations. Zhang Xiaoyan et al., using the slope safety factor and velocity field information to calculate the element failure probability (EFP), provide a new idea for the calculation of slope failure risk [29].

Currently, the groundwater level is usually defaulted at a certain value, and the limit state function does not contain the groundwater level as a stochastic variable. However, due to the stochastic distribution of soil particles and pores inside the slope, the seepage field is uncertain. In addition, because of the uncertainty of the supply by precipitation, surface runoff, underground confluence, and the discharge by evaporation and pumping, the groundwater level is uncertain. It is well known that the groundwater level is directly related to the seepage field, so the stochastic groundwater level will lead to the uncertainty of the seepage field. In recent years, the research about the seepage field mainly focused on the change in the permeability coefficient and paid less attention to the stochastic groundwater level, which leads to the change in the seepage field [30,31]. In addition, autocorrelation functions, such as exponential, Gaussian, logarithmic, and triangular, are frequently applied to characterize the spatial variability of soil mass [32]. In view of this, a slope failure risk assessment method that considers both the randomness of the groundwater level and soil shear strength parameters is proposed in this paper to achieve sustainable development.

2. Methodologies

2.1. Stable Seepage Field of Slope with Stochastic Groundwater Level

In this paper, a slope failure risk assessment that considers both the randomness of the groundwater level and soil shear strength parameters is carried out. Figure 1 shows the stochastic groundwater level model for soil slopes, which includes the velocity of the triangular element, the model of the groundwater level distribution, the shear strength parameters' calculation, and the pore water pressure calculation. The soil mass is discretized by the triangular element, so the three nodes of the element e have horizontal velocity u_i^e vertical velocity v_i^e, cohesive force c_i^e, friction angle φ_i^e, and pore water pressure p_i^e, of which

$i = (1,2,3)$ and $e = (1,\ldots,N_e)$. In addition, there is discontinuity between element a and element b. Nodes ① and ② belong to element a; Nodes ② and ④ belong to element b.

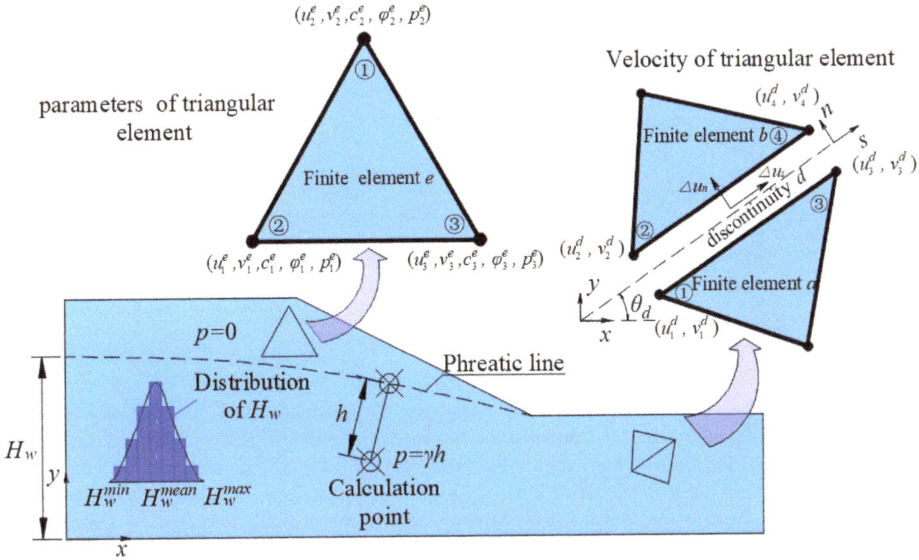

Figure 1. Stochastic groundwater level model for soil slope.

Similar to the literature [33], it is assumed that the stochastic groundwater level H_w follows a truncated normal distribution, where H_w^{min} is the left-truncated tail of the stochastic groundwater level, H_w^{max} is the right-truncated tail of the stochastic groundwater level, and H_w^{mean} is the mean value of the stochastic groundwater level. In addition, this article makes the following assumptions: (1) the seepage field is a saturated and stable seepage field; (2) excess pore water pressure along with the effect on the soil shear strength parameters caused by a sudden rise or drop of groundwater level are not considered; (3) above the phreatic line, the pore water pressure is $p = 0$; and (4) below the phreatic line, the pore water pressure is $p = \gamma h$, of which γ is the volume weight of water, and h is the vertical distance from the point on the phreatic line that is on the same equipotential line as the calculation point to the calculation point. The two-dimensional stable seepage equation is used for the seepage analysis [31]. The specific equation is as follows:

$$k_x \frac{\partial^2 H_w}{\partial x^2} + k_y \frac{\partial^2 H_w}{\partial y^2} = 0 \tag{1}$$

where k_x and k_y are the permeability coefficient in the x and y direction, respectively; and H_w is the stochastic groundwater head function at each point within the soil.

2.2. Stochastic Field of Parameter Spatial Variability

Considering that slope reliability is sensitive to autocorrelation length and insensitive to the form of the autocorrelation function [34,35], this paper selects the exponential autocorrelation function with a simpler mathematical expression for research, which can be expressed with the following equation [35]:

$$\rho_{[(x_a,y_a);(x_b,y_b)]} = \exp\left[-2\left(\frac{x_a - x_b}{L_h}\right) - 2\left(\frac{y_a - y_b}{L_v}\right)\right] \tag{2}$$

where x_a, y_a, x_b, and y_b are the coordinate components of the spatial coordinates; and L_h and L_v are the fluctuation range in the x and y direction, respectively.

Due to the existence of certain mutual correlations between soil parameters and certain autocorrelations of the soil parameters themselves thus involving the discrete process of the associated non-Gaussian fields, this paper adopts a method similar to that of the literature [35] using the midpoint method of Cholesky decomposition for the stochastic field generation, and for the exponential autocorrelation functions, the associated logarithmic stochastic field of the soil materials is as follows:

$$c_i^e = \exp(u_{\ln c} + \sigma_{\ln c} \cdot c_i^{eD}) \tag{3}$$

$$\varphi_i^e = \exp(u_{\ln \varphi} + \sigma_{\ln \varphi} \cdot \varphi_i^{eD}) \tag{4}$$

where c_i^e and φ_i^e are the soil cohesion and internal friction angle of nodes i upon element e in the non-Gaussian stochastic field, respectively; c_i^{eD} and φ_i^{eD} are the soil cohesion and internal friction angle of nodes i upon element e in the Gaussian stochastic field, respectively; $\sigma_{\ln c} = \sqrt{\ln(1 + (\sigma_c{}^2/u_c{}^2))}$; $\sigma_{\ln \varphi} = \sqrt{\ln(1 + (\sigma_\varphi{}^2/u_\varphi{}^2))}$; $u_{\ln c} = \ln u_c - \sigma_{\ln c}{}^2/2$; $u_{\ln \varphi} = \ln u_\varphi - \sigma_{\ln \varphi}{}^2/2$; u_c and σ_c are the mean and standard deviation of the log-normal distribution of c, respectively; u_φ and σ_φ are the mean and standard deviation of the log-normal distribution of φ, respectively; $u_{\ln c}$ and $\sigma_{\ln c}$ are the mean and standard deviation of the corresponding normal variable of $\ln c$, respectively; and $u_{\ln \varphi}$ and $\sigma_{\ln \varphi}$ are the mean and standard deviation of the corresponding normal variable of $\ln \varphi$, respectively.

2.3. Stochastic Programming Model

Sloan et al. [20] discretized the soil mass with triangular elements (as shown in Figure 1) and constructed the kinematically admissible velocity fields (KAVF). It is easy to discern from the UBM that the KAVF is one-to-one, corresponding to the external load, where the minimum is infinitely close to the limit load; therefore, the UBM can be understood for solving the minimization problem. On the basis of previous studies [20,26,29,36], the stochastic programming model that considers both the randomness of the groundwater level and soil shear strength parameters established in this paper is as follows:

$$\begin{cases} Z = k_m - 1 \\ Minimize: \quad k_\gamma = \mathbf{W}_e + \mathbf{W}_d - \mathbf{W}_e^p - \mathbf{W}_d^p \\ Subject\ to: \quad \boldsymbol{a}_e^1 \mathbf{u}_e - \boldsymbol{a}_e^2 \chi_e = 0 \\ \qquad \boldsymbol{a}_d^1 \mathbf{u}_d - \boldsymbol{a}_d^2 \chi_d = 0 \\ \qquad \boldsymbol{a}_b \mathbf{u}_b = 0 \\ \qquad \mathbf{W}_G = 1; \chi_e \geq 0; \chi_d \geq 0 \end{cases} \tag{5}$$

where Z is the limit state function; k_m and k_γ are the safety factor and volume weight overload factor, respectively; χ_e and χ_d are the plastic multiplier of finite element e and node d, respectively; $\boldsymbol{a}_e^1$, $\boldsymbol{a}_e^2$ and $\boldsymbol{a}_d^1$, $\boldsymbol{a}_d^2$ are the plastic flow constraint matrix of the finite element e and velocity discontinuity d, respectively; $\mathbf{u}_e = \begin{bmatrix} u_1^e & v_1^e & u_2^e & v_2^e & u_3^e & v_3^e \end{bmatrix}^T$, of which u_i^e and v_i^e are the horizontal and vertical velocity components of nodes i upon element e, respectively; $\mathbf{u}_d = \begin{bmatrix} u_1^d & v_1^d & u_2^d & v_2^d & u_3^d & v_3^d & u_4^d & v_4^d \end{bmatrix}^T$, of which u_i^d and v_i^d are the horizontal and vertical velocity components of nodes i upon velocity discontinuity d, respectively; $\boldsymbol{a}_b = \begin{bmatrix} \cos\theta & \sin\theta \\ -\sin\theta & \cos\theta \end{bmatrix}$, of which θ is the rotation angle; $\mathbf{u}_b = \begin{bmatrix} u_1^b & v_1^b & \mathrm{K} & \mathrm{K} & u_j^b & v_j^b \end{bmatrix}$, of which u_j^b and v_j^b are the horizontal and vertical velocity components on the boundary, respectively; $\mathbf{W}_e$ and $\mathbf{W}_d$ are the internal power of finite element e and velocity discontinuity d, respectively; $\mathbf{W}_e^p$ and $\mathbf{W}_d^p$ are the external power of finite element e and velocity discontinuity d, respectively; and $\mathbf{W}_G$ is the external power of self-weight [20].

3. Solution Strategy

The $\mathbf{a}_e^2$, $\mathbf{a}_d^2$, $\mathbf{W}_e$, $\mathbf{W}_d$, $\mathbf{W}_e^p$, and $\mathbf{W}_d^p$ matrices in Equation (5) are all associated with the groundwater level as well as the shear strength parameters. Currently, there is no better method for solving this kind of problem; in view of this, on the basis of the Monte Carlo simulation, an iterative method is proposed for solving this issue. The detailed implementation process is as follows:

(1) Assuming that the stochastic groundwater level follows a truncated normal distribution, which is generated with the Monte Carlo simulation method:

$$\begin{cases} H_w(T_w) = Random(Normal, \mu_w, \sigma_w, 1, N_w) \\ H_w^{min} \le H_w(T_w) \le H_w^{max} \end{cases} \quad (6)$$

where $H_w(T_w)$ is the T_wth groundwater level; $T_w = (1, \ldots, N_w)$, N_w is the number of groundwater level; and μ_w and σ_w are the mean and standard deviation of the groundwater level, respectively.

(2) Assuming that the autocorrelation functions of the soil materials are exponential type, using Equations (3) and (4), the midpoint method of Cholesky decomposition for the stochastic field generation $c^e(T_m)$ and $\varphi^e(T_m)$, of which $T_m = (1, \ldots, N_m)$; N_m is the number of random fields for the shear strength parameters.

In this paper, the slope is allowed to reach the limit state by means of capacitive overload, and the capacitive overload factor of the slope when taking the parameters' spatial variability into account under the effect of the stochastic groundwater level is as follows:

$$k_\gamma(T_m, T_w) = \frac{\gamma_c(c^e(T_m), \varphi^e(T_m), H_w(T_w))}{\gamma_a} \quad (7)$$

where $k_\gamma(T_m, T_w)$ is the capacitive overload factor; $\gamma_c(c^e(T_m), \varphi^e(T_m), H_w(T_w))$ is the capacity of the soil in the ultimate state, which is related to $c^e(T_m)$, $\varphi^e(T_m)$, and $H_w(T_w)$; and γ_a is the actual capacity of the soil.

The slope safety factor when taking the parameters' spatial variability into account under the effects of the stochastic groundwater level is as follows:

$$k_m(T_m, T_w) = \frac{c^e(T_m)}{c'^e(T_m)} = \frac{\varphi^e(T_m)}{\varphi'^e(T_m)} \quad (8)$$

where $k_m(T_m, T_w)$ is the safety factor; and $c'^e(T_m)$ and $\varphi'^e(T_m)$ are the cohesion and internal friction angle of finite element e in the T_mth non-Gaussian stochastic field after the strength reduction, respectively.

(3) The stochastic number of T_w groundwater levels generated in step (1) is substituted into the stable seepage field calculation equation to obtain the pore water pressure $p_1^e(T_w)$, $p_2^e(T_w)$, $p_3^e(T_w)$; $e = (1, \ldots, N_e)$; $T_w = (1, \ldots, N_w)$.

(4) From $T_w = 1$ to $T_w = N_w$ cycles, repeat $p_1^e(T_w)$, $p_2^e(T_w)$, $p_3^e(T_w)$, all the finite element nodes' pore water pressure values are successively replaced with the stochastic programming model for the slope reliability analysis; in each cycle from $T_w = 1$ to $T_w = N_w$; $c^e(T_m)$, $\varphi^e(T_m)$ from $T_m = 1$ to $T_m = N_m$ cycles, the number of N_m stochastic fields are brought into Equation (5) and use the dual simplex method to obtain $N_w \times N_m$ numbers of capacity overload factors $[k_\gamma(T_m, T_w)]$ while, at the same time, use the bisection method to obtain $N_w \times N_m$ numbers of the slope safety factor $[k_m(T_m, T_w)]$. Figure 2 shows the specific numerical solution flow.

(5) Calculation of the slope safety factor and plot the related curve.

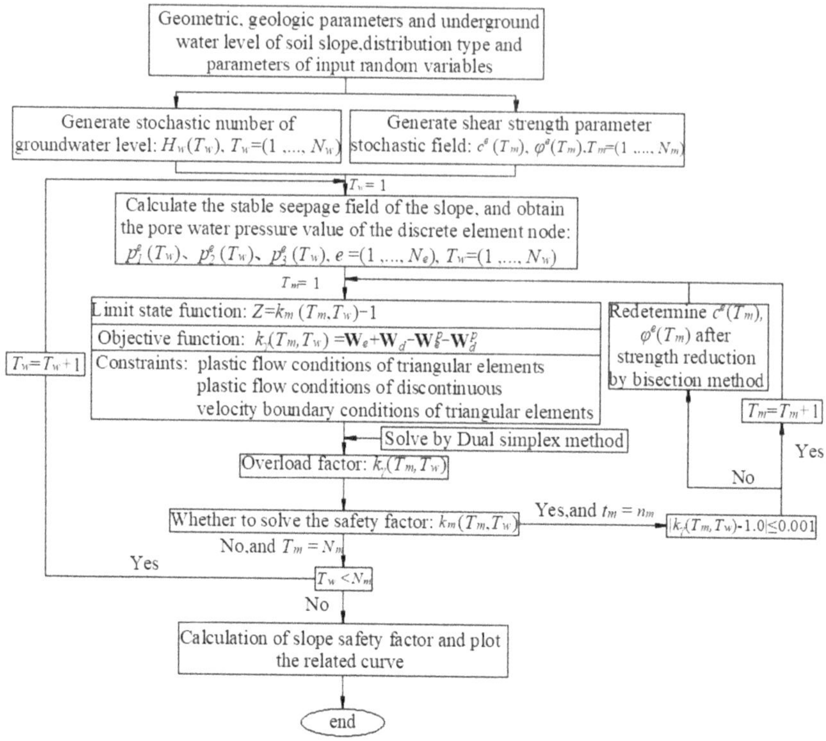

Figure 2. Specific numerical solution flow for slope.

4. Reliability Index for Slope

The traditional *IFP* uses the threshold value of the safety factor to determine whether the slope will fail or not and then conducts the slope reliability analysis accordingly [28]. Based on the *IFP*, the slope failure function is as follows:

$$I(T_m, T_w) = \begin{cases} 0, \text{ if } k_m(T_m, T_w) \geq 1 \\ 1, \text{ if } k_m(T_m, T_w) < 1 \end{cases} \tag{9}$$

where $I(T_m, T_w)$ is the failure function of the slope corresponding to the T_mth stochastic field under T_wth groundwater level acts.

According to the Equation (8) slope failure function, the IFP under T_wth groundwater level acts can be acquired as follows:

$$P_f^{IFP}(T_w) = \frac{\sum\limits_{T_m=1}^{N_m} I(T_m, T_w)}{N_m} \times 100\% \tag{10}$$

where $P_f^{IFP}(T_w)$ is the *IFP* of the slope under T_wth groundwater level acts.

Further, the *IFP* of the slope is as follows:

$$P_f^{IFP} = \sum_{T_w=1}^{N_w} P_f^{IFP}(T_w) \tag{11}$$

where P_f^{IFP} is the *IFP* of the slope under all potential groundwater level acts.

The integrated failure risk (*IFR*) of the slope under T_wth groundwater level acts is as follows:

$$R^{IFP}(T_w) = P_f^{IFP}(T_w)C \tag{12}$$

where $R^{IFP}(T_w)$ is the *IFR* of the slope under T_wth groundwater level acts, and C is the area of the slope failure mode.

Based on the *IFP*, the *IFR* of the slope is calculated as follows [27,28]:

$$R^{IFP} = \sum_{T_w=1}^{N_w} R^{IFP}(T_w) \tag{13}$$

where R^{IFP} is the *IFR* of the slope.

The Equation (13) default is that only one failure mode exists when conducting the slope failure risk analysis, which is contrary to the fact that slopes have multiple failure modes, considering both the randomness of the groundwater level and soil shear strength parameters. For that reason, Huang et al. [37] gave a calculation equation, which considered all failure modes of the slope simultaneously as follows:

$$R_k^{IFP}(T_w) = P_{fk}^{IFP}(T_w)C_k \tag{14}$$

where $R_k^{IFP}(T_w)$ is the *IFR* of the *k*th failure mode of the slope under T_wth groundwater level acts; $P_{fk}^{IFP}(t_w)$ is the *IFP* of the *k*th failure mode of the slope under T_wth groundwater level acts; and C_k is the area of the *k*th slope failure mode.

The IFR of the slope is as follows:

$$R^{IFP} = \sum_{T_w=1}^{N_w} \sum_{k=1}^{N} R_k^{IFP}(T_w) \tag{15}$$

Compared with Equations (13) and (15), it can perform both single and multiple failure mode slope risk assessments; however, both the LEM and FEM require prior work on complex failure mode classification. Zhang Xiaoyan et al. [29] proposed a new concept of *EFP* on the basis of the UBM. They used the slope safety factor and the KVAF to judge the failure elements. The failure function, considering the stochasticity of the groundwater level and spatial variability of soil material, is as follows:

$$I_e(T_m, T_w) = \begin{cases} 0 \text{ if } k_{\mathrm{m}}(T_m, T_w) \geq 1 \\ 0 \text{ if } u_{\mathrm{c}}^e(T_m, T_w) = 0 \text{ and } k_{\mathrm{m}}(T_m, T_w) < 1 \\ 1 \text{ if } u_{\mathrm{c}}^e(T_m, T_w) > 0 \text{ and } k_{\mathrm{m}}(T_m, T_w) < 1 \end{cases} \tag{16}$$

where $I_e(T_m, T_w)$ is the failure function of the element *e* corresponding to the T_mth stochastic field under T_wth groundwater level acts, and $u_{\mathrm{c}}^e(T_m, T_w)$ is the velocity of the element *e* corresponding to the T_mth stochastic field under T_wth groundwater level acts.

$$u_{\mathrm{c}}^e(T_m, T_w) = \sqrt{\left(\sum_{i=1}^{3} \frac{u_i^e(T_m, T_w)}{3}\right)^2 + \left(\sum_{i=1}^{3} \frac{v_i^e(T_m, T_w)}{3}\right)^2} \tag{17}$$

where $u_i^e(T_m, T_w)$ is the horizontal velocity component of nodes *i* upon element *e* corresponding to the T_mth stochastic field under T_wth groundwater level acts, and $v_i^e(T_m, T_w)$ is the vertical velocity component of nodes *i* upon element *e* corresponding to the T_mth stochastic field under T_wth groundwater level acts.

According to the slope element failure function in Equation (16), the slope element failure probability (*EFP*) under T_wth groundwater level acts is as follows:

$$P_{fe}^{EFP}(T_w) = \frac{\sum_{T_m=1}^{N_m} I^e(T_m, T_w)}{N_m} \times 100\% \tag{18}$$

where $P_{fe}^{EFP}(T_w)$ is the *EFP* under T_wth groundwater level acts.

The slope *EFP* under all potential groundwater level acts is as follows:

$$P_{fe}^{EFP} = \frac{\sum_{T_w=1}^{N_w} \sum_{T_m=1}^{N_m} I^e(T_m, T_w)}{N_w \times N_m} \times 100\% \tag{19}$$

The element failure risk (*EFR*) of the slope under T_wth groundwater level acts is as follows:

$$R_e^{EFP}(T_w) = P_{fe}^{EFP}(T_w) C_e \tag{20}$$

where $R_e^{EFP}(T_w)$ is the failure risk of element *e* under T_wth groundwater level acts, and C_e is the area of element *e*.

The element failure risk (*EFR*) of the slope under all potential groundwater level acts is as follows:

$$R_e^{EFP} = \sum_{T_w=1}^{N_w} R_e^{EFP}(Tw) \tag{21}$$

where R_e^{EFP} is the failure risk of element *e* under all potential groundwater level acts.

The *IFR* based on the *EFP* of the slope under T_wth groundwater level acts is calculated in the following equation [20]:

$$R^{EFP}(T_w) = \sum_{e=1}^{N_e} R_e^{EFP}(T_w) \tag{22}$$

where $R^{EFP}(T_w)$ is the *IFR* under T_wth groundwater level acts.

The *IFR* based on the *EFP* of the slope under all potential groundwater level acts is calculated in the following equation:

$$R^{EFP} = \sum_{T_w=1}^{N_w} R^{EFP}(T_w) \tag{23}$$

where R^{EFP} is the *IFR* under all potential groundwater level acts.

A new slope failure risk assessment under consideration of the stochastic groundwater level involves *EFP* and C_e. The *EFP* can be easily obtained by solving the stochastic programming model, and C_e is the area of element *e*, which is constant compared to C_k.

According to the solution strategy of the stochastic programming model, the $N_w \times N_m$ slope safety factors $k_m(T_m, T_w)$ can be easily acquired. Using statistical knowledge, the mean and standard deviation of the slope safety factor can be calculated as follows:

$$\mu_k(T_w) = \frac{\sum_{T_m=1}^{N_m} k_m(T_m, T_w)}{N_m} \tag{24}$$

$$\mu_k = \frac{\sum_{T_w=1}^{N_w} \sum_{T_m=1}^{N_m} k_m(T_m, T_w)}{(N_w \times N_m)} \tag{25}$$

$$\sigma_k(T_w) = \sqrt{\frac{\sum\limits_{T_m=1}^{N_m} (k_m(T_m, T_w) - \mu_k(T_w))}{N_m - 1}} \tag{26}$$

$$\sigma_k = \sqrt{\frac{\sum\limits_{T_w=1}^{N_w} \sum\limits_{T_m=1}^{N_m} (k_m(T_m, T_w) - \mu_k)}{N_w \times N_m - 1}} \tag{27}$$

where $T_w = (1, \ldots, N_w)$; $\mu_k(T_w)$ and $\sigma_k(T_w)$ are the mean and standard deviation of the slope safety factor under T_wth groundwater level acts, respectively; and μ_k and σ_k are the mean and standard deviation of the slope safety factor under all potential groundwater level acts, respectively.

5. Calibration and Application

The UBM program is compiled, and a classic slope calculation example is calculated and analyzed. Comparing the result with the calculation result of the LEM, we verified the correctness of the calculation method.

5.1. Numerical Simulations

Figure 3 shows the homogeneous slope calculation model. The height is 5 m, the width of the top is 10 m, and the ratio is 1:2. Li Dian qing et al. have calculated the failure probability of the slope [35] but have not conducted systematic research on the slope failure risk. In view of this, on the basis of the method proposed in this paper, the slope failure risk is studied. Using the triangular element to discretize the slope, 989 finite elements, 2967 nodes, and 1536 discontinuities are acquired. In addition, there are three pore water pressure monitoring points: P1 (5,5), P2 (10,5), and P3 (15,5). The set soil volume weight γ is 20.0 kN/m^3, and the permeability coefficient is $K = 5 \times 10^{-7}$m/s, both of which are determined values. See Table 1 for the other calculation parameters [26].

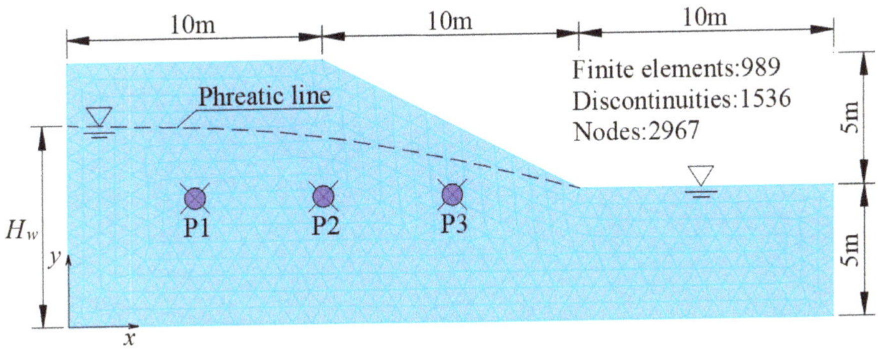

Figure 3. Homogeneous slope calculation model.

Table 1. Statistics of the soil parameters.

Shear Parameter	Mean	Correlation of Variation	Distribution Type	Fluctuation Range	Correlation Coefficient
c(kPa)	10	0.3	Lognormal	L_h = 40 m	$\rho_{c,\varphi} = -0.5$
φ(°)	30	0.2	Lognormal	L_v = 4 m	

The mean and standard deviation of the stochastic groundwater level is 7.5 m and 2.25, respectively, on the basis of the measured groundwater in many projects. The upper boundary of the groundwater level is 8.5 m. The lower boundary of the groundwater level is 6.5 m. The quantity of the stochastic groundwater levels is 50. The groundwater level at the right slope toe is 5 m. According to Equation (6), 50 stochastic numbers of groundwater levels are generated, and their distribution is shown in Figure 4. The stochastic number distribution of the groundwater is relatively close to the mean value of the groundwater level, and relatively small near the relatively high and low groundwater levels.

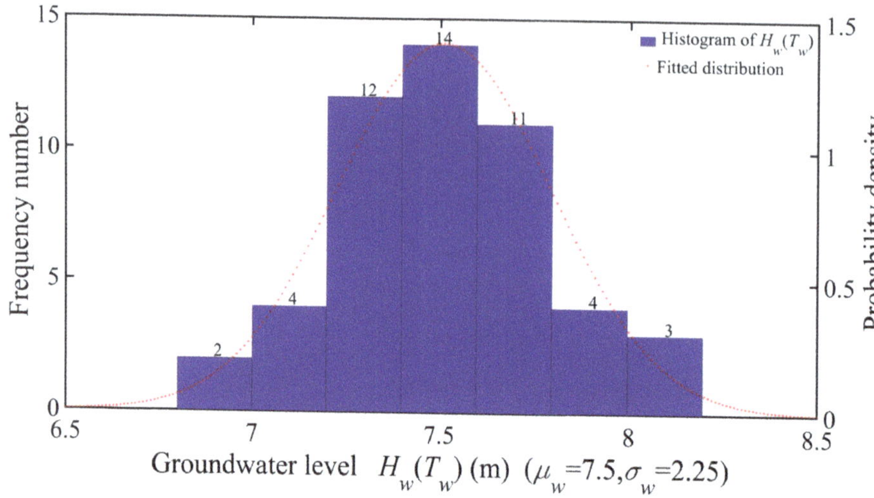

Figure 4. Stochastic groundwater levels distribution histogram.

Figure 5a–c shows the slope stable seepage fields when $T_w = 1$ (H_w = 6.8006 m), $T_w = 25$ (H_w = 7.4848 m), and $T_w = 50$ (H_w = 8.1951 m), respectively. It can be observed that the contours of the pore water pressure become steeper, and the saturated area inside the slope increases as the groundwater level rises. Figure 6 shows the key points pore water pressure. Under the same groundwater level act, the pore water pressure decreases gradually as the coordinates of the key points move to the right. At P1, the mean and standard deviation of the pore water pressure are −19.77 kPa and 2.31 kPa, respectively. At P2, the mean and standard deviation of the pore water pressure are −14.59 kPa and 1.78 kPa, respectively. At P3, the mean and standard deviation of the pore water pressure are −8.94 kPa and 1.13 kPa, respectively.

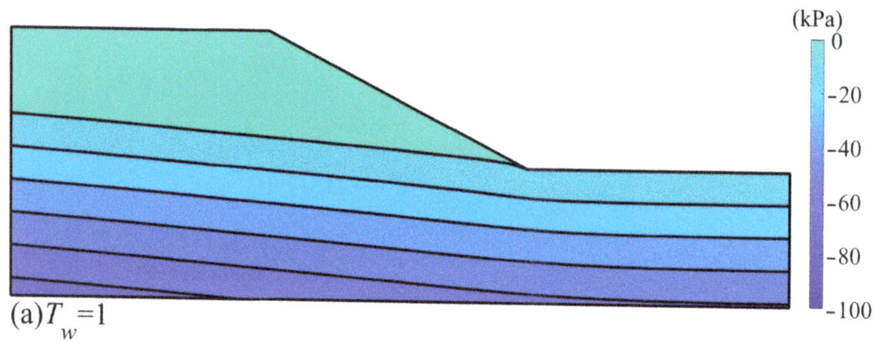

Figure 5. *Cont.*

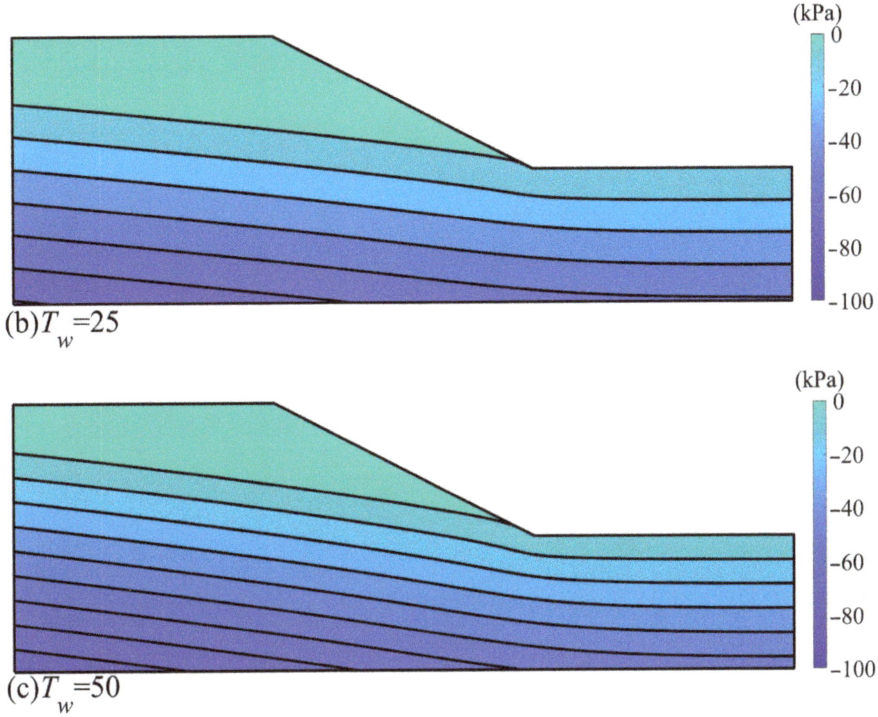

Figure 5. Steady seepage fields of slope: (**a**) $T_w = 1$; (**b**) $T_w = 25$; (**c**) $T_w = 50$.

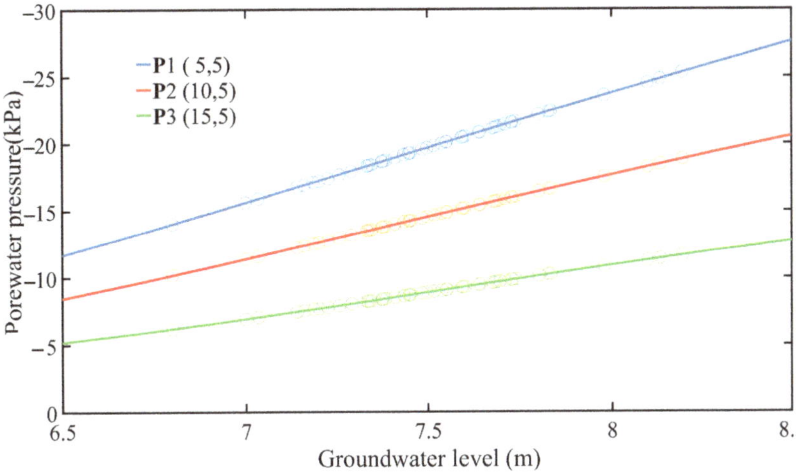

Figure 6. The key points pore water pressure.

Assuming that the autocorrelation functions of the soil materials are exponential type, we used the midpoint method of Cholesky decomposition for the stochastic field generation. The quantity of the stochastic fields of the soil parameters is 2000; according to Equations (3) and (4), 2000 stochastic fields of the slope shear strength parameters are generated. Figure 7a,b shows the stochastic fields of $c_e(500)$ and $\varphi_e(500)$, respectively. The maximum value of soil cohesion and internal friction angle are 18.9084 kPa and 23.3219°, respectively; The minimum value of soil cohesion and internal friction angle are 3.1696 kPa and 9.5851°, respectively. In addition, soil cohesion has a certain negative correlation with internal friction angle in space.

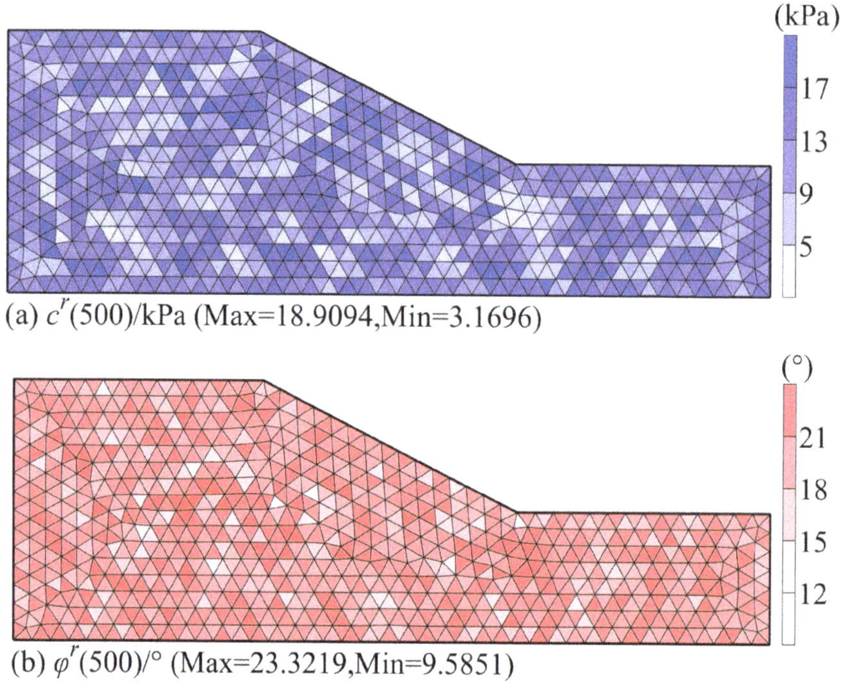

Figure 7. Stochastic fields of the slope shear strength parameters: (**a**) c^r (500); (**b**) φ^r (500).

5.2. Main Results of the Research

This paper selects low, medium, and high groundwater levels ($T_w = 1$, $T_w = 25$, and $T_w = 50$) for comparative analysis to verify the effectiveness of the calculation method. Table 2 shows the calculation results of the reliability index comparison between the UBM and LEM under three groundwater level acts: $T_w = 1$, $T_w = 25$, and $T_w = 50$. Figure 8 shows the distribution characteristics of the slope safety factor under three groundwater level acts. Figure 9 shows the distribution histograms of the slope safety factor under three groundwater level acts.

Table 2. Calculation results of the reliability index.

Groundwater Level	Method	Mean	Standard Deviation	Failure Probability (%)
$T_w = 1$	UBM	1.2956	0.1420	1.40
	LEM	1.2923	0.1488	1.90
$T_w = 25$	UBM	1.2678	0.1394	2.10
	LEM	1.2622	0.1461	2.70
$T_w = 50$	UBM	1.2357	0.1365	3.30
	LEM	1.2310	0.1432	4.65

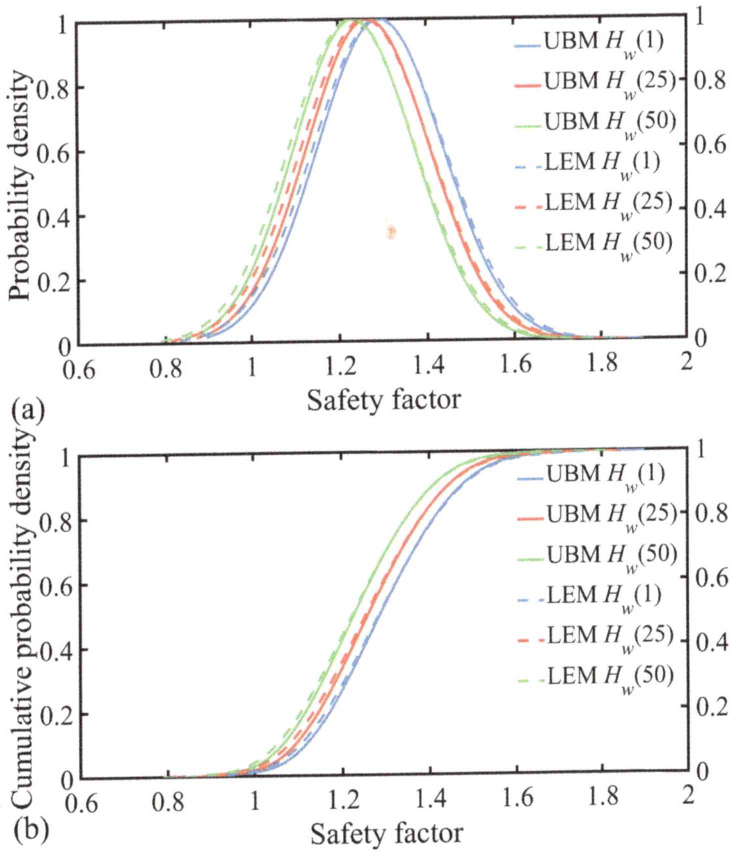

Figure 8. Distribution characteristics of the slope safety factors: (**a**) PDF curve of the safety factor; (**b**) CDF curve of the safety factor.

The calculation results show the following:

(1) When $T_w = 1$, $T_w = 25$, and $T_w = 50$, the mean of the slope safety factors with the UBM is larger than that of the LEM, but the error is small, which conforms to the features of the upper bound solution. In addition, the slope safety factors acquired with the UBM and LEM methods decrease as the groundwater level rises. The slope failure probability increases acquired with the UBM and LEM decrease as the groundwater level rises. On the basis of the upper bound theorem, the slope safety factor acquired with the UBM must be greater than the real solution. Therefore, the UBM will slightly underestimate the failure slope probability.

(2) Figure 8a,b shows the PDF and CDF curves of the slope safety factors acquired with the UBM and LEM under three groundwater level acts, $T_w = 1$, $T_w = 25$, and $T_w = 50$, respectively. It is not difficult to see that the PDF and CDF curves of the slope safety factors acquired with the UBM and LEM are very close with small errors. In addition, the PDF and CDF curves gradually move to the left as the groundwater level rises.

(3) Figure 9a–c are the distribution histograms of the slope safety factors acquired with the UBM under the three groundwater level acts, $T_w = 1$, $T_w = 25$, and $T_w = 50$, respectively. It is not difficult to see that the distribution of the slope safety factors is similar to the stochastic groundwater levels.

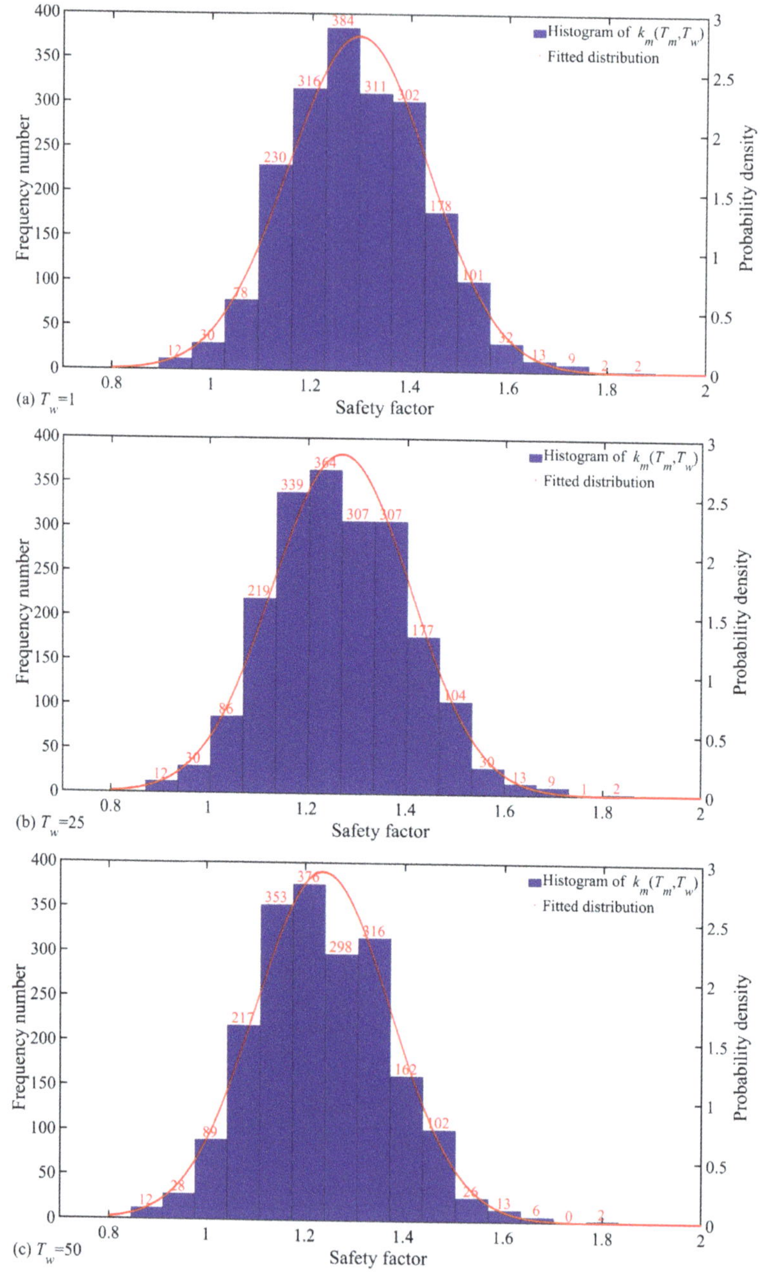

Figure 9. Distribution histograms of the slope safety factors: (**a**) $T_w = 1$; (**b**) $T_w = 25$; (**c**) $T_w = 50$.

On the basis of the stochastic programming model and solution strategy, the quantity of 100,000 slope safety factors was acquired. Figures 10 and 11 are the number of 50 PDF and CDF curves of the slope safety factors, respectively. Figure 12 reflects the mean and standard deviation of the slope safety factors versus the groundwater level. Figure 13 shows the PDF and CDF curves of the slope safety factors under all potential groundwater level acts acquired. The following rules can be acquired through analysis:

(1) The distribution of the slope safety factors is consistent with the normal distribution. The mean of the slope safety factors tends to decrease as the groundwater level rises. The PDF and CDF curves of the slope safety factors gradually move to the left as the groundwater level rises. In addition, the standard deviation of the slope safety factor tends to decrease as the groundwater level rises. The range of the PDF curve and the trend of the CDF curve of the slope safety factors gradually narrow and steepen, respectively.

(2) A polynomial fit is used to acquire the quantitative equation of the mean and standard deviation of the slope safety factors and groundwater level as follows:

$$u_k(T_w) = -0.0439H_w + 1.5963 \tag{28}$$

$$\sigma_k(T_w) = -0.0039H_w + 0.1685 \tag{29}$$

(3) The quantity of 100,000 slope safety factors was acquired from 2000 stochastic fields under 50 groundwater level acts to perform the statistical analysis of all the acquired data; under 50 groundwater level acts, the mean and standard deviation of the slope safety factors are 1.2664 and 0.11, respectively.

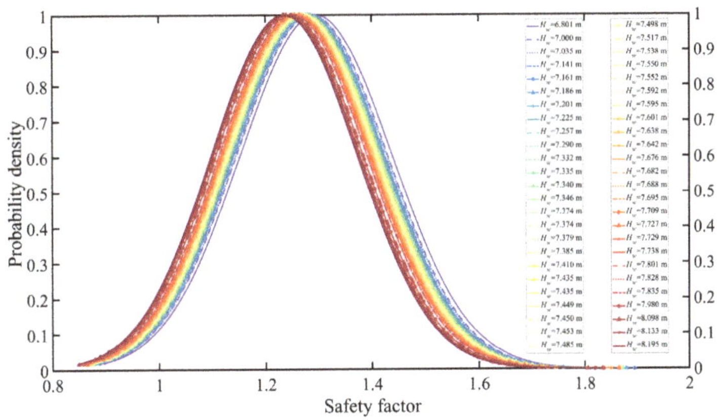

Figure 10. PDF curve of the slope safety factors.

Based on 100,000 Monte Carlo simulations of the slope, the LEM default is that only one failure mode exists (as shown in the LEM sliding surface in Figure 14). The biggest advantage of the UBM compared with the LEM in this paper is that all slope failure modes can be captured according to the failure information of the elements (as shown in the pink failure area in Figure 14) and then the reliability index for the slope can be calculated. Similar to the method in reference 37, the failure areas were used to classify the failure modes; there are six failure modes of the slope under 50 groundwater level acts (as shown in Figure 14). Table 3 lists the failure risk of the above six slope failure modes. Table 4 lists the failure risk corresponding to the failure modes when $T_w = 1$, $T_w = 25$, and $T_w = 50$. The results show that failure modes 1 and 2 have a small failure area that belongs to a shallow landslide, and the failure area is between 28.68 and 58.91 m^2 (failure times is 997). Failure modes 5 and 6 have a large failure area that belongs to a deep landslide, and the failure area is between 89.17 and 119.38 m^2 (failure times is 160). Failure modes 3 and 4 are between a shallow landslide and a deep landslide, and the failure area is between 58.91 and 89.04 m^2 (failure times is 989). When the groundwater level is $T_w = 1$, the slope has only five failure modes; when the groundwater level is $T_w = 25$ and $T_w = 50$, the fifth failure modes occur. The phreatic line moves up, correspondingly, and the saturated area inside the slope increases as the groundwater level rises, thus increasing the probability of the failure mode.

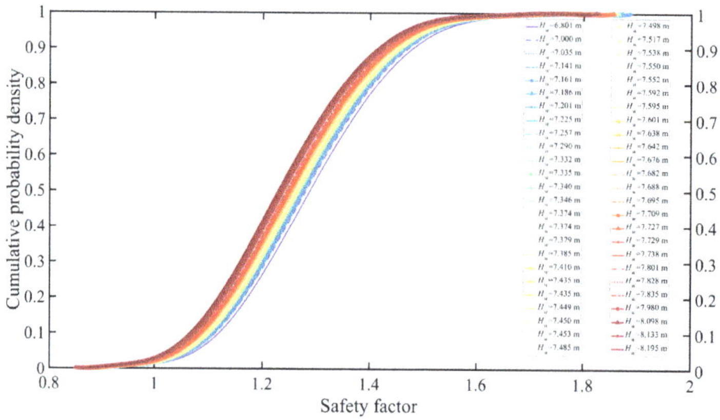

Figure 11. CDF curve of the slope safety factors.

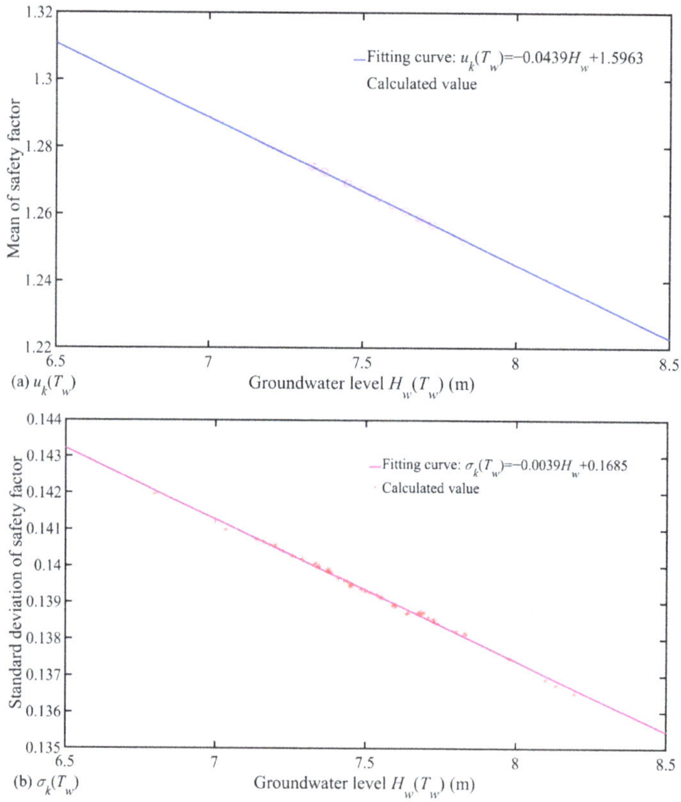

Figure 12. Slope safety factors versus groundwater level: (**a**) u_k (T_w); (**b**) σ_k (T_w).

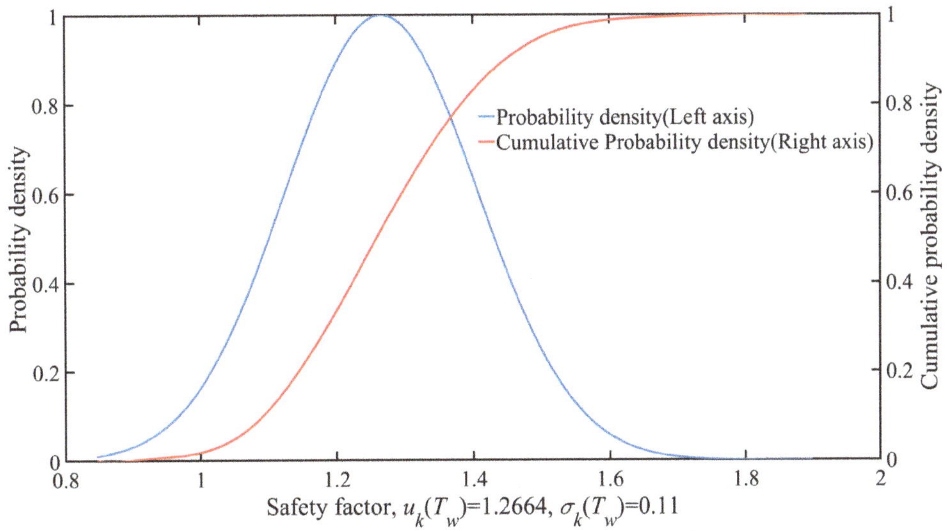

Figure 13. PDF and CDF curves of the slope safety factors under all potential groundwater level acts.

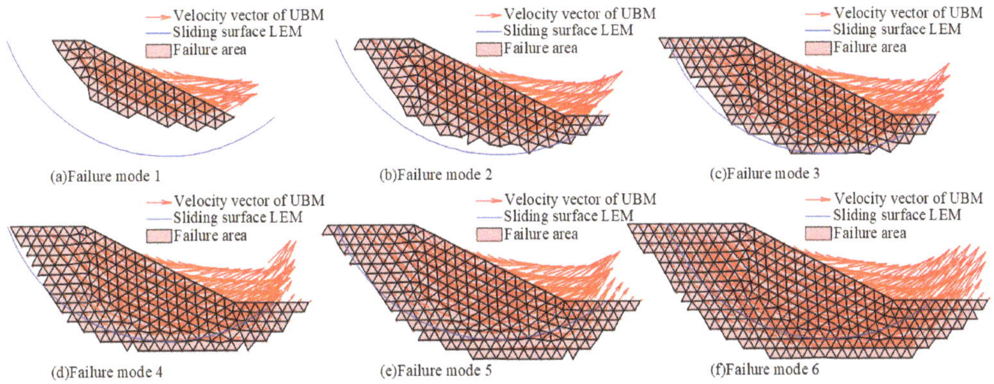

Figure 14. Schematic diagram results of the slope failure modes: (**a**) Failure mode 1; (**b**) Failure mode 2; (**c**) Failure mode 3; (**d**) Failure mode 4; (**e**) Failure mode 5; (**f**) Failure mode 6.

Table 3. Failure risk of the slope.

Failure Mode	Failure Area (m^2)	Failure Times	Failure Probability (%)	Failure Risk (m^2)
Mode 1	28.68–43.70	246	0.246	0.097
Mode 2	43.81–58.91	751	0.751	0.384
Mode 3	58.91–73.96	768	0.768	0.508
Mode 4	74.21–89.04	221	0.221	0.180
Mode 5	89.17–103.70	88	0.088	0.081
Mode 6	106.71–119.38	72	0.072	0.081
Sum	/	2146	2.146	1.332

The failure mode acquired with the LEM is only consistent with failure mode 3 in this paper. The main reason is that, when the LEM is adopted to calculate the slope safety factor, the initial slip surface is assumed in advance, then a constantly repeated search based on the mean of the shear parameters to acquire the critical slip crack surface is performed, and then the slope safety factor is calculated based on the unique critical slip surface. Thus, a difference exists between the critical slip surface obtained with the LEM and the actual slip surface. The UBM constructs a stochastic programming model through finite elements, and then uses the method of stochastic mathematical programming to search the instability area of the slope, so the result is more in line with the real situation. The calculation indicates that the traditional LEM will ignore some failure modes and may miscalculate the slope failure risk, considering both the randomness of the groundwater level and soil shear strength parameters. The UBM can ignore the constitutive relation of the materials and acquire the slope safety factor and failure modes. Its applicability and calculation efficiency are both high.

Table 4. Failure risk of the slope when $T_w = 1$, $T_w = 25$, and $T_w = 50$.

Groundwater Level	Failure Mode	Failure Times	Failure Probability (%)	Failure Risk (m^2)
$T_w = 1$	Mode 1	3	0.15	0.057
	Mode 2	13	0.65	0.334
	Mode 3	10	0.50	0.342
	Mode 4	1	0.05	0.038
	Mode 6	1	0.05	0.057
$T_w = 25$	Mode 1	5	0.25	0.096
	Mode 2	14	0.70	0.366
	Mode 3	16	0.80	0.528
	Mode 4	5	0.25	0.210
	Mode 5	1	0.05	0.045
	Mode 6	1	0.05	0.056
$T_w = 50$	Mode 1	8	0.40	0.160
	Mode 2	19	0.95	0.490
	Mode 3	24	1.20	0.793
	Mode 4	11	0.55	0.445
	Mode 5	2	0.10	0.091
	Mode 6	2	0.10	0.115

Table 5 is the statistical table of the slope failure risk acquired with the three methods. The LEM default is that only one failure mode exists when $T_w = 1$, $T_w = 25$, and $T_w = 50$; the slope failure risk according to the LEM with Equation (12) is 1.121 m^2, 1.593 m^2, and 2.325 m^2, respectively. All failure modes can be acquired with the UBM. When $T_w = 1$, $T_w = 25$, and $T_w = 50$, the slope failure risk according to the UBM with Equation (14) and the UBM Equation (22) are 0.829 m^2, 1.302 m^2, and 2.094 m^2, respectively. It should be noted that Equation (14) has a difference calculation principle from Equation (22). All slope failure modes are required to be counted when Equation (14) is used to calculate the slope failure risk; the EFP for all elements is easy to acquire by solving Equation (5), and the element area is fixed when using Equation (22) to calculate the slope failure risk. From the calculation principle, the proposed method will simplify the calculation process and make the calculation more efficient.

Table 5. Failure risk statistical table of the slope (Unit: m^2).

Groundwater \ Method	LEM with Equation (12)	UBM with Equation (14)	UBM with Equation (22)
$T_w = 1$	1.121	0.829	0.829
$T_w = 25$	1.593	1.302	1.302
$T_w = 50$	2.325	2.094	2.094

The EFP of the slope under T_wth groundwater level acts when $T_w = 1$, $T_w = 25$, and $T_w = 50$ obtained with Equation (18) is shown in Figure 15. The EFP of the slope under all potential groundwater level acts obtained with Equation (19) is shown in Figure 16. It can be observed that the groundwater level influences the EFP of the slope. The white area in the figure is the non-failure element, and the blue area is the failure element. In addition, according to the theory of EFP, the darker the color, the greater the EFP.

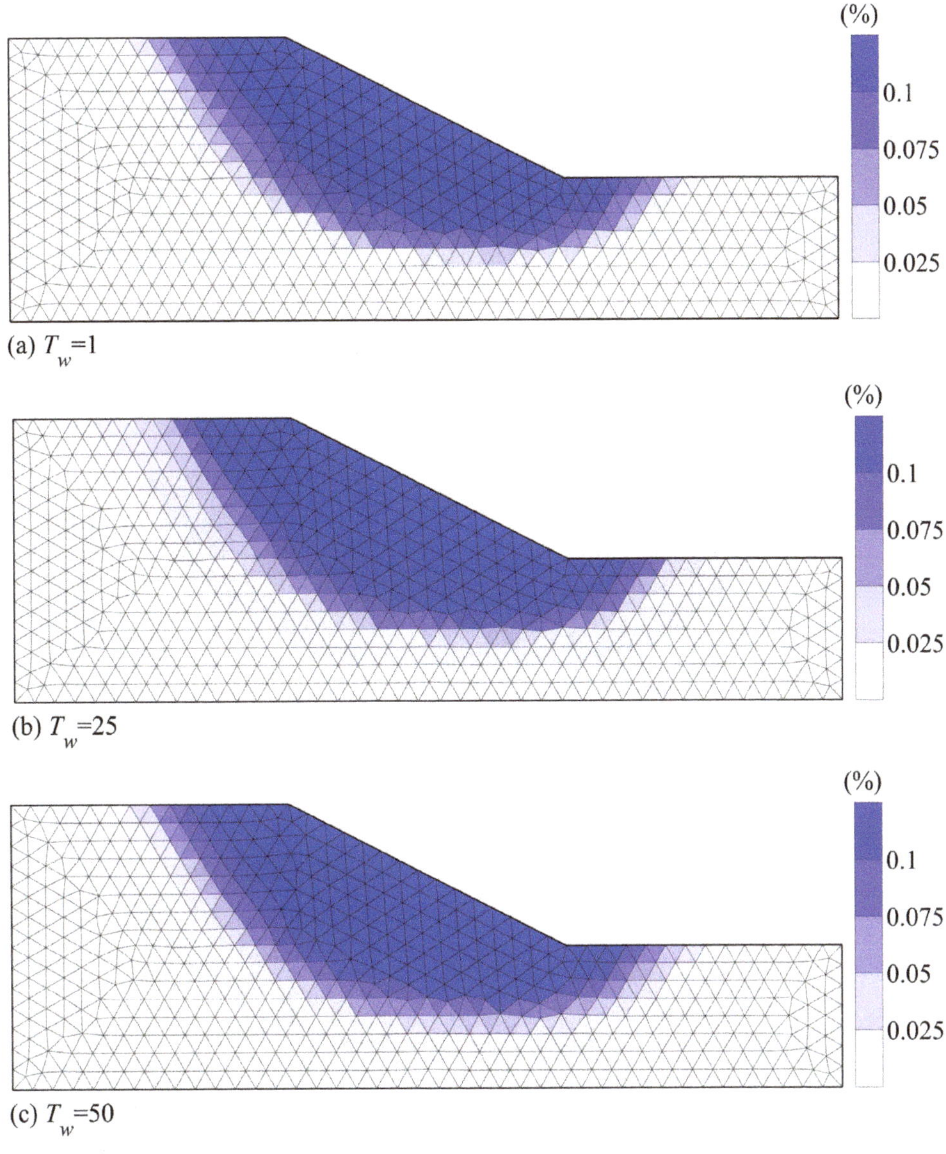

Figure 15. EFP of the slope under T_wth groundwater level acts: (**a**) T_w = 1; (**b**) T_w = 25; (**c**) T_w = 50.

The EFR of the slope under T_wth groundwater level acts when $T_w = 1$, $T_w = 25$, and $T_w = 50$ obtained with Equation (20) is shown in Figure 17. The EFR of the slope under all potential groundwater level acts obtained with Equation (21) is shown in Figure 18. It can be observed that the groundwater level influences the EFR of the slope. When $T_w = 1$, the frequency of EFR between 0 and 0.00062 m^2 is 675, the frequency of EFR between 0.00062 and 0.00124 m^2 is 36, the frequency of EFR between 0.00124 and 0.00186 m^2 is 58, the frequency of EFR between 0.00186 and 0.00248 m^2 is 69, the frequency of EFR between 0.00248 and 0.00310 m^2 is 140, the frequency of EFR between 0.00310 and 0.00372 m^2 is 11, and the slope failure risk is 0.829 m^2. When $T_w = 25$, the frequency of EFR between 0 and 0.00082 m^2 is 670, the frequency of EFR between 0.00082 and 0.00164 m^2 is 32, the frequency of EFR between 0.00164 and 0.00246 m^2 is 54, the frequency of EFR between 0.00246 and 0.00328 m^2 is 56, the frequency of EFR between 0.00328 and 0.00410 m^2 is 165, the frequency of EFR between 0.00410 and 0.00492 m^2 is 12, and the slope failure risk is 1.302 m^2. When $T_w = 50$, the frequency of EFR between 0 and 0.00150 m^2 is 666, the frequency of EFR between 0.00150 and 0.00300 m^2 is 41, the frequency of EFR between 0.00300 and 0.00450 m^2 is 48, the frequency of EFR between 0.00450 and 0.00600 m^2 is 62, the frequency of EFR between 0.00600 and 0.00750 m^2 is 158, and the frequency of EFR between 0.00750 and 0.00900 m^2 is 14. Under all potential groundwater level acts, the frequency of EFR between 0 and 0.04820 m^2 is 674, the frequency of EFR between 0.04820 and 0.09640 m^2 is 30, and the frequency of EFR between 0.09640 and 0.14460 m^2 is 57. The frequency of EFR between 0.014460 and 0.19280 m^2 is 59, the frequency of EFR between 0.19280 and 0.24100 m^2 is 157, the frequency of EFR between 0.24100 and 0.28892 m^2 is 12, and the slope failure risk is 2.094 m^2. The element failure risk comprehensively reflects the contribution of the EFP and the element failure risk coefficient, which can make a quantitative judgment on the slope failure risk of each part. The slope failure risk is 1.332 m^2, obtained by the sum of all element failure risks. It is observed that the slope failure risk assessment method proposed in this paper can avoid the screening and statistical work of failure modes compared with the failure risk calculation results in Table 3.

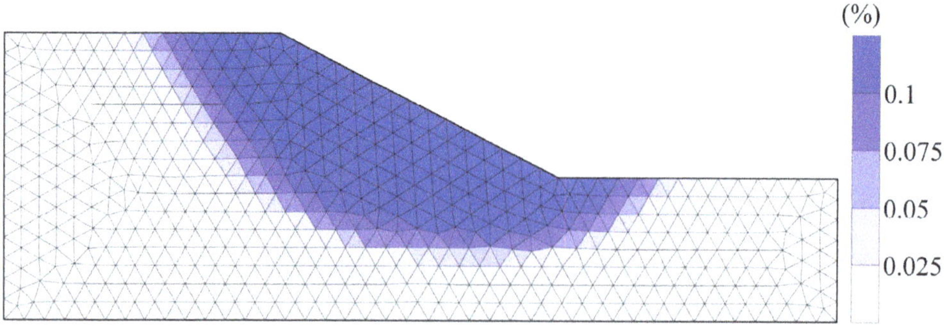

Figure 16. EFP of the slope under all potential groundwater level acts.

The safety performance decreases, and the IFP of the slope increases from 1.40% to 3.30% with the gradual increase in the groundwater level. Through cubic polynomial fitting, the IFP of the slope versus the groundwater level acquired (as shown in Figure 19) is as follows:

$$P_f^{IFP}(T_w) = 0.5271H_w^3 - 11.2957H_w^2 + 81.5794H_w - 196.7446 \quad (30)$$

The EFR of the slope increases from 0.829 m^2 to 2.094 m^2 with the gradual increase in the groundwater level. Through cubic polynomial fitting, the EFR of the slope versus the groundwater level acquired (as shown in Figure 20) is as follows:

$$R^{EFP}(T_w) = 0.1251H_w^3 - 2.4622H_w^2 + 16.6665H_w - 37.9602 \quad (31)$$

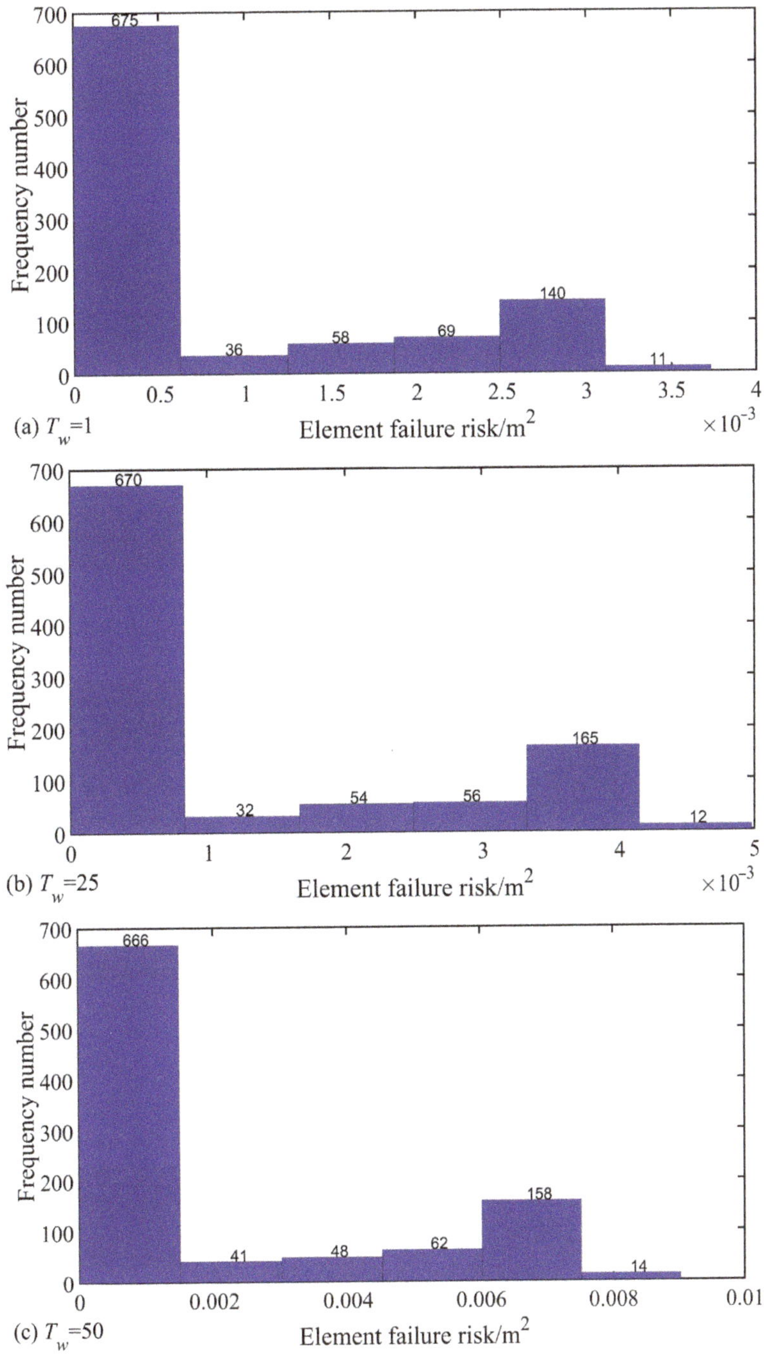

Figure 17. EFR of the slope under T_wth groundwater level acts: (**a**) T_w = 1; (**b**) T_w = 25; (**c**) T_w = 50.

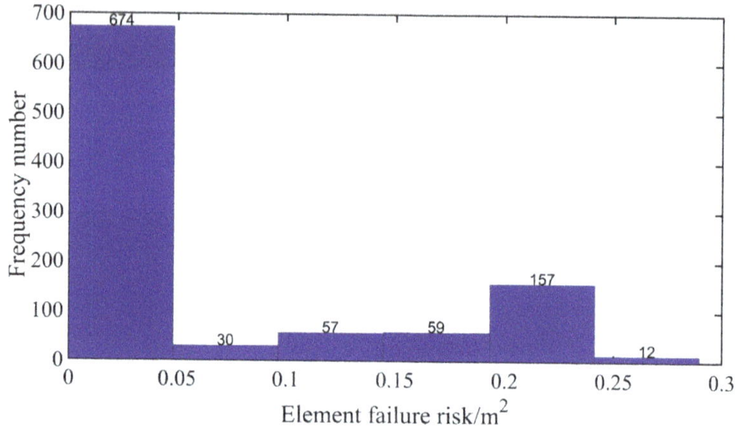

Figure 18. EFR of the slope under all potential groundwater level acts.

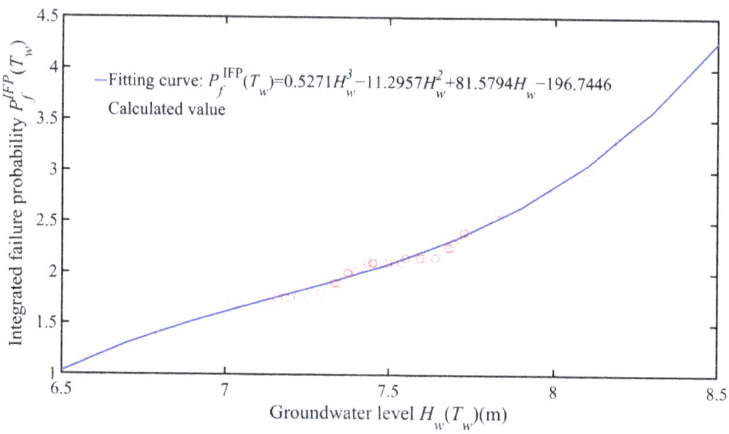

Figure 19. IFP of the slope versus the groundwater level (Unit: %).

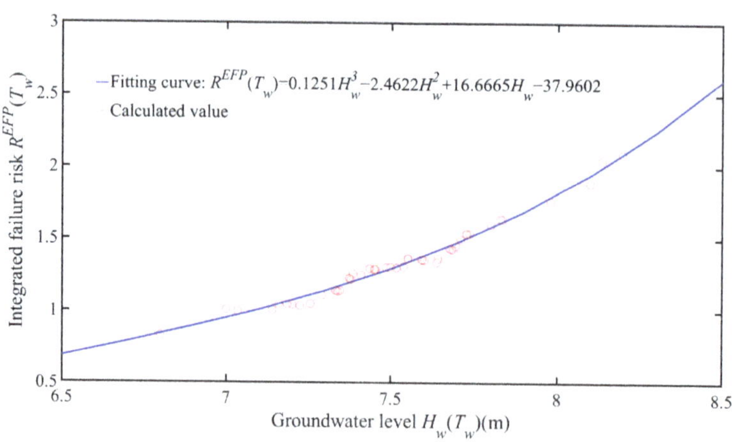

Figure 20. EFR of the slope versus the groundwater level (Unit: m^2).

6. Conclusions

A slope failure risk assessment method that considers both the randomness of the groundwater level and soil shear strength parameters is proposed in this paper. The corresponding calculation program is compiled. The major conclusions are the following:

(1) When the randomness of the groundwater level and soil shear strength parameters are considered comprehensively, the traditional LEM will ignore multiple failure modes and may miscalculate the slope failure risk. However, all failure modes can be acquired with the UBM for seeking the minimum value of the KAVF. Thus, the result is more consistent with the real situation. In addition, the traditional LEM only judges the slope stability by the safety factor, which only reflects the degree of the IPF. The EFP is used to calculate the EFR of the slope, which cannot only reflect the degree of the IFP but, also, the slope failure risk can be accurately acquired. It should be noted that this calculation method can greatly reduce the calculation cost.

(2) The IFP and EFR of the slope are increasing from 1.40% to 3.30% and 0.829 m^2 to 2.094 m^2 with the rise of the groundwater level, respectively. Based on the EFP, the proposed method can accurately obtain the EFR of the slope under each groundwater level act by using the element's location information and failure situation. This will provide engineers with realistic reference values for the slope reinforcement design to achieve sustainable development.

(3) Groundwater level and earthquakes are two important causes of slope instability and failure. However, this study does not consider the impact of earthquakes on slope reliability. Therefore, relevant studies on seismic slope stability will be carried out in the future. In addition, according to the upper bound theory, the upper bound solution is inevitably greater than the true solution. Therefore, the failure probability will be underestimated when using the UBM for slope reliability analysis. To solve this problem, there is a necessity to study the slope reliability calculation method on the basis of the lower bound theory in future research work. The solution of slope failure probability with the UBM and LBM can be obtained at the same time, so the interval range of the real failure probability can be accurately judged, and the reliability index of the slope can be quantified more accurately.

Author Contributions: Conceptualization, P.P., Z.L., X.Z., W.L., S.S. and H.X.; methodology, P.P.; software, Z.L.; validation, W.L.; formal analysis, Z.L.; investigation, Z.L.; resources, W.L.; data curation, W.L.; writing—original draft preparation, Z.L.; writing—review and editing, P.P.; visualization, H.X.; supervision, W.L. and Z.L.; project administration, S.S. and Z.L.; funding acquisition, X.Z. and Z.L. All authors have read and agreed to the published version of the manuscript.

Funding: This research was funded by the National Natural Science Foundation of China grant number 12162018 and 12262016; the Research Foundation of Science and technology plan project of Yunnan Provincial Department of Science and Technology grant number 202201AT070283 and 202305AD160064.

Institutional Review Board Statement: Not applicable for studies not involving humans or animals.

Informed Consent Statement: Not applicable.

Data Availability Statement: Not applicable.

Conflicts of Interest: The authors declare no conflict of interest.

References

1. Dai, F.; Lee, C.; Ngai, Y. Landslide risk assessment and management: An overview. *Eng. Geol.* **2002**, *64*, 65–87. [CrossRef]
2. Li, D.-Q.; Yang, Z.-Y.; Cao, Z.-J.; Zhang, L.-M. Area failure probability method for slope system failure risk assessment. *Comput. Geotech.* **2018**, *107*, 36–44. [CrossRef]
3. Liu, X.; Li, D.-Q.; Cao, Z.-J.; Wang, Y. Adaptive Monte Carlo simulation method for system reliability analysis of slope stability based on limit equilibrium methods. *Eng. Geol.* **2020**, *264*, 105384. [CrossRef]
4. Wang, L.; Wu, C.; Li, Y.; Liu, H.; Zhang, W.; Chen, X. Probabilistic Risk Assessment of unsaturated Slope Failure Considering Spatial Variability of Hydraulic Parameters. *KSCE J. Civ. Eng.* **2019**, *23*, 5032–5040. [CrossRef]

5. Malkawi, A.I.H.; Hassan, W.F.; Abdulla, F.A. Uncertainty and reliability analysis applied to slope stability. *Struct. Saf.* **2000**, *22*, 161–187. [CrossRef]
6. Zhao, H.-B. Slope reliability analysis using a support vector machine. *Comput. Geotech.* **2008**, *35*, 459–467. [CrossRef]
7. Ching, J.; Phoon, K.-K.; Hu, Y.-G. Efficient Evaluation of Reliability for Slopes with Circular Slip Surfaces Using Importance Sampling. *J. Geotech. Geoenviro. Eng.* **2009**, *135*, 768–777. [CrossRef]
8. Tang, X.-S.; Li, D.-Q.; Chen, Y.-F.; Zhou, C.-B.; Zhang, L.-M. Improved knowledge-based clustered partitioning approach and its application to slope reliability analysis. *Comput. Geotech.* **2012**, *45*, 34–43. [CrossRef]
9. Cho, S.E. First-order reliability analysis of slope considering multiple failure modes. *Eng. Geol.* **2013**, *154*, 98–105. [CrossRef]
10. Ji, J.; Zhang, W.; Zhang, F.; Gao, Y.; Lü, Q. Reliability analysis on permanent displacement of earth slopes using the simplified Bishop method. *Comput. Geotech.* **2020**, *117*, 103286. [CrossRef]
11. Wang, W.; Li, D.-Q.; Liu, Y.; Du, W. Influence of ground motion duration on the seismic performance of earth slopes based on numerical analysis. *Soil Dyn. Earthq. Eng.* **2021**, *143*, 106595. [CrossRef]
12. Griffiths, D.V.; Huang, J.; Fenton, G.A. Influence of Spatial Variability on Slope Reliability Using 2-D Random Fields. *J. Geotech. Geoenviron. Eng.* **2009**, *135*, 1367–1378. [CrossRef]
13. Shen, H.; Abbas, S.M. Rock slope reliability analysis based on distinct element method and random set theory. *Int. J. Rock Mech. Min. Sci.* **2013**, *61*, 15–22. [CrossRef]
14. Li, D.-Q.; Jiang, S.-H.; Cao, Z.-J.; Zhou, W.; Zhou, C.-B.; Zhang, L.-M. A multiple response-surface method for slope reliability analysis considering spatial variability of soil properties. *Eng. Geol.* **2015**, *187*, 60–72. [CrossRef]
15. Wang, M.-Y.; Liu, Y.; Ding, Y.-N.; Yi, B.-L. Probabilistic stability analyses of multi-stage soil slopes by bivariate random fields and finite element methods. *Comput. Geotech.* **2020**, *122*, 103529. [CrossRef]
16. Dyson, A.P.; Tolooiyan, A. Comparative Approaches to Probabilistic Finite Element Methods for Slope Stability Analysis. *Simul. Model. Pract. Theory* **2020**, *100*, 102061. [CrossRef]
17. Lysmer, J. Limit Analysis of Plane Problems in Soil Mechanics. *J. Soil Mech. Found. Div.* **1970**, *96*, 1311–1334. [CrossRef]
18. Li, Z.; Chen, Y.; Guo, Y.; Zhang, X.; Du, S. Element Failure Probability of Soil Slope under Consideration of Random Groundwater Level. *Int. J. Géoméch.* **2021**, *21*, 04021108. [CrossRef]
19. Sloan, S.W. Lower bound limit analysis using finite elements and linear programming. *Int. J. Numer. Anal. Methods Géoméch.* **1988**, *12*, 61–77. [CrossRef]
20. Sloan, S.; Kleeman, P. Upper bound limit analysis using discontinuous velocity fields. *Comput. Methods Appl. Mech. Eng.* **1995**, *127*, 293–314. [CrossRef]
21. Li, Z.; Hu, Z.; Zhang, X.; Du, S.; Guo, Y.; Wang, J. Reliability analysis of a rock slope based on plastic limit analysis theory with multiple failure modes. *Comput. Geotech.* **2019**, *110*, 132–147. [CrossRef]
22. Kim, J.; Salgado, R.; Yu, H.S. Limit Analysis of Soil Slopes Subjected to Pore-Water Pressures. *J. Geotech. Geoenviron. Eng.* **1999**, *125*, 49–58. [CrossRef]
23. Li, Z.; Zhou, Y.; Guo, Y. Upper-Bound Analysis for Stone Retaining Wall Slope Based on Mixed Numerical Discretization. *Int. J. Géoméch.* **2018**, *18*, 04018122. [CrossRef]
24. Liu, W.; Xu, H.; Sui, S.; Li, Z.; Zhang, X.; Peng, P. Lower Bound Limit Analysis of Non-Persistent Jointed Rock Masses Using Mixed Numerical Discretization. *Appl. Sci.* **2022**, *12*, 12945. [CrossRef]
25. Chen, Z.H.; Lei, J.; Huang, J.H.; Cheng, X.H.; Zhang, Z.C. Finite element limit analysis of slope stability considering spatial variability of soil strengths. *Chin. J. Geotech. Eng.* **2018**, *40*, 985–993. [CrossRef]
26. Peng, P.; Li, Z.; Zhang, X.Y.; Shen, L.F.; Xu, Y. Research on element failure probability and failure mode of soil slope. *Eng. Mech.* **2022**, *39*, 1–15. [CrossRef]
27. Silva, F.; Lambe, T.W.; Marr, W.A. Probability and Risk of Slope Failure. *J. Geotech. Geoenviron. Eng.* **2008**, *134*, 1691–1699. [CrossRef]
28. Cassidy, M.J.; Uzielli, M.; Lacasse, S. Probability risk assessment of landslides: A case study at Finneidfjord. *Can. Geotech. J.* **2008**, *45*, 1250–1267. [CrossRef]
29. Zhang, X.Y.; Zhang, L.X.; Li, Z. Reliability analysis of soil slope based on upper bound method of limit analysis. *Rock Soil Mech.* **2018**, *39*, 1840–1850. [CrossRef]
30. Zuo, Z.; Zhang, L.; Cheng, Y.; Wang, J.; He, Y. Probabilistic back analysis of unsaturated soil seepage parameters based on Markov chain Monte Carlo method. *Rock Soil Mech.* **2013**, *34*, 2393–2400.
31. Wang, X.; Xia, X.; Zhang, X.; Gu, X.; Zhang, Q. Probabilistic Risk Assessment of Soil Slope Stability Subjected to Water Drawdown by Finite Element Limit Analysis. *Appl. Sci.* **2022**, *12*, 10282. [CrossRef]
32. Cho, S.E.; Park, H.C. Effect of spatial variability of cross-correlated soil properties on bearing capacity of strip footing. *Int. J. Numer. Anal. Methods Géoméch.* **2010**, *34*, 1–26. [CrossRef]
33. Shadabfar, M.; Huang, H.; Kordestani, H.; Muho, E.V. Reliability Analysis of Slope Stability Considering Uncertainty in Water Table Level. *ASCE-ASME J. Risk Uncertain. Eng. Syst. Part A Civ. Eng.* **2020**, *6*, 04020025. [CrossRef]
34. Li, K.S.; Lumb, P. Probabilistic design of slopes. *Can. Geotech. J.* **1987**, *24*, 520–535. [CrossRef]
35. Li, D.Q.; Jiang, S.H.; Zhou, C.B.; Phoon, K.K. Reliability analysis of slopes considering spatial variability of soil parameters using non-intrusive stochastic finite element method. *Chin. J. Geotech. Eng.* **2013**, *35*, 1413–1421.

36. Zhou, J.; Qin, C. Stability analysis of unsaturated soil slopes under reservoir drawdown and rainfall conditions: Steady and transient state analysis. *Comput. Geotech.* **2022**, *142*, 104541. [CrossRef]
37. Huang, J.; Lyamin, A.; Griffiths, D.V.; Krabbenhoft, K.; Sloan, S. Quantitative risk assessment of landslide by limit analysis and random fields. *Comput. Geotech.* **2013**, *53*, 60–67. [CrossRef]

sustainability MDPI

Article

Estimation of Modulus of Deformation Using Rock Mass Rating— A Review and Validation Using 3D Numerical Modelling

Hema Vijay Sekar Bellapu [1,*], Rabindra Kumar Sinha [1] and Sripad Ramchandra Naik [2]

1 Department of Mining Engineering, Indian Institute of Technology, Indian School of Mines, Dhanbad 826 004, India
2 National Institute of Rock Mechanics, Bengaluru 560 070, India
* Correspondence: hema.17dp000199@me.iitism.ac.in

Abstract: The Himalayan region has enormous potential for hydropower development. However, variations in geological and geotechnical conditions pose challenging tasks for the designers. If these variations are not tackled in a timely manner during underground excavations, especially for caverns, instabilities may occur, resulting in time and cost over-runs. For sustainable hydropower development, minimizing these over-runs is necessary. The modulus of deformation (E_d) of a rock mass is an essential input parameter required in the design of underground excavations. This study involves collecting the results of extensive in situ tested values for various hydroelectric projects in the Himalayan regions, along with the rock mass rating (RMR) values at 35 test sites. E_d is estimated empirically based on statistical analysis. Comparisons were made with the empirical equations already available in the literature, using RMR and the proposed equation for estimating E_d. Although different researchers have proposed many equations for estimating the value of E_d using RMR, a gap exists in validating such equations. In this regard, the proposed equation for E_d was verified by carrying out 3D numerical-modelling studies using FLAC3D, an explicit finite-difference software for an underground powerhouse cavern and comparing the displacement values with the field instrumentation data.

Keywords: modulus of deformation; RMR; in situ testing; modelling; instrumentation

Citation: Bellapu, H.V.S.; Sinha, R.K.; Naik, S.R. Estimation of Modulus of Deformation Using Rock Mass Rating—A Review and Validation Using 3D Numerical Modelling. *Sustainability* **2023**, *15*, 5721. https://doi.org/10.3390/su15075721

Academic Editors: Mahdi Hasanipanah, Danial Jahed Armaghani and Jian Zhou

Received: 23 January 2023
Revised: 1 March 2023
Accepted: 9 March 2023
Published: 24 March 2023

1. Introduction

There is a massive shift globally from nonrenewable to renewable energy, i.e., solar, hydropower and wind. Hydropower was ranked as the highest renewable energy in 2019 [1]. However, many hydropower electric projects under construction are delayed due to geological and geotechnical variations. This, in turn, will result in cost and time over-runs. An average of 182% over-run of time was observed in 29 hydroelectric projects located in the Himalayan states of India with an installed capacity of 9840 MW. Of these 29 projects, an average of 114% over-run of cost was observed in 23 projects with an installed capacity of 8138 MW [2]. The over-run of time and cost varied from 49% to 364% and from 14% to 254%, respectively [3]. Hence, the cost and time over-runs must be minimized to complete projects successfully. The completion of the Punatsangchhu II hydroelectric project in Bhutan was delayed due to the collapse of rock mass in the crown of one of the underground caverns [4].

The mountain chain of the Himalayas comprises a complicated fold-and-thrust belt. It can be divided into three units: Sub-Himalaya, Lesser Himalaya, and Higher Himalaya, from south to north [5]. The Sub-Himalayan range is the youngest of the three and has an elevation of about 1200 m. Intracrustal thrusts demarcate the Lesser Himalayan domain, i.e., main boundary thrust (MBT) in the north and main central thrust (MCT) in the south. The Lesser Himalayan range runs parallel to the Sub-Himalayan range and has an elevation of about 2000 m to 5000 m. The Higher Himalayas are the oldest formations out of the three and have an elevation of about 6000 m. The Lesser Himalayas comprises chert, argillaceous, arenaceous, and calcareous units. These complex formations and high tectonic activity in the Himalayas may yield uncertainty in estimating the rock mass parameters required for designing underground excavations.

Hence, underground excavations must be designed considering reliable geotechnical input parameters, such as rock mass strength (compressive, tensile, shear, cohesion, and friction angle), deformation properties, stress regimes, hydrological conditions, and joint characteristics. Out of all these input parameters, the deformation modulus was found to play an essential role in assessing the stability of large caverns in the Himalayas [6].

In designing underground excavations for tunnels and caverns, estimating the expected rock mass deformations around the openings is essential. The modulus of deformation (E_d) of a rock mass typically provides information about the deformation characteristics, i.e., elastic and plastic behavior when the rock mass is subjected to loading and unloading conditions. Joint friction parameters and rock strength play an essential role in the deformation mechanics of rock mass in addition to the E_d value [7]. In recent years, there has been an advancement in numerical tools for analyzing the support system for underground excavations. The output from these numerical tools, however, depends on the reliability of the input data. The E_d value is one of the critical design parameters required for numerical modelling [8] in the design of dam structures and underground excavations.

As per [9,10], E_d is defined as the ratio of stress to strain (elastic and plastic) during the loading of a rock mass, whereas the modulus of elasticity (E_e) is defined as the ratio of stress to strain (elastic) during the unloading of a rock mass. Hence, while carrying out any in situ testing, estimating the E_e value along with E_d is a general practice.

The most-preferred in situ tests for the estimation of E_d are the plate-jacking test (PJT) or uniaxial-jacking test, plate-loading test (PLT), and flat jack test (FJT) carried out in drifts or small tunnels. In contrast, the dilatometer test (DT) and goodman jack test (GJT) are conducted in boreholes of NX size [10]. The size of the drift or gallery required for carrying out the in situ testing needs to be as small as required for carrying out the test. During loading and unloading of the rock mass area, the deformations are measured using a multipoint borehole extensometer (MPBX) in boreholes and a linear variable differential transformer (LVDT) case for measuring surface displacements, which are used in determining the in situ value of E_d. However, in situ tests are complicated, expensive, and time-consuming [6]. In addition, each type of in situ test will result in different values due to differences in test procedures and rock mass damage due to blasting [11,12].

Due to these reasons, several empirical relations were proposed by different researchers for determining the value of E_d based on rock mass classification systems, such as rock mass rating (RMR), tunneling-quality index (Q), geological strength index (GSI); and intact rock properties, such as uniaxial compressive strength (UCS), Young's modulus (Ei), disturbance factor (D), and weathering degree (WD). These empirical equations are developed based on the data collected for a particular location and rock type. Using these equations to estimate the deformation modulus value at other sites may not yield correct values.

Based on the studies carried out by [13], it is noticed that empirical relations proposed for E_d based on intact properties (UCS and Ei) gave less reliable results when compared with those of rock mass classification systems. Although many empirical relations are available in the literature for estimating E_d, only those equations with an RMR value as the input parameter are considered in this study [12,14–31], since this is the most widely accepted method of rock mass characterization. The RMR classification system was developed by Bieniawski (1974, 1989) [32], updating the charts and tables for the six parameters, i.e., intact rock UCS, rock quality designation (RQD), spacing of joint or discontinuity, joint condition, condition of groundwater, and orientation of joint set. The rock masses at the in situ test locations were classified based on the RMR. Drillability studies conducted on rocks also provided insight into the petrophysicomechanical properties that indicated the influence of various petrographic, physical, and mechanical properties of rock [33–35]. From the numerical modelling studies [36], it was observed that when the in situ tested value of E_d is in the range of 1 to 3 GPa, the predicted displacements were almost thrice the measured values. However, suppose the rock mass is too competent, as studied in [37], it can be noted that the in situ value of E_d is higher when compared with that of the back-calculated value from numerical modelling. Depth also was found to influence the E_d values in discontinuum models compared to that of continuum models. At shallow depths, the discontinuities deformed significantly in comparison with that of deeper depths [38]. In addition, studies were carried out for understanding the variations in joint set sizes and orientations on the directional deformation modulus for rock mass [39].

It is understood that E_d is the critical design parameter for the design of large underground excavations, which needs to be determined correctly, and which otherwise has the potential to result in time and cost over-runs. Determining E_d values by in situ testing will have huge financial implications for the project. Thus, this study aims to develop a predictor model for estimating the E_d value using the RMR, which can be useful to the designers or project authorities for design of underground excavations if there is a lack of in situ tested data for projects in the Himalayan region.

2. Methodology

The present study reviews the prediction of E_d values based on the existing empirical relations using the values of the rock mass rating. A new empirical equation is proposed to be developed considering the available in situ tested data from the projects constructed in the Himalayan region. A comparison is made for the value of E_d concerning the existing equations and the newly proposed equation. Finally, 3D numerical modelling studies are carried out considering the value of E_d determined in situ and the value obtained from the proposed equation and comparing the model displacement values with that of the measured values. The empirical equations considered in the study, along with their limitations, coefficient of regression (R^2), number of data sets considered by worldwide researchers, range of RMR values, country of origin, and the lithology considered while developing the relations, are given in Table 1.

The datasets considered in this study involve collecting the in situ tested values of E_d and E_e for 35 test locations in the Himalayan region spanning over India, Bhutan, and Nepal from the published literature [40–45] and the National Institute of Rock Mechanics (NIRM) reports [46–50]. In situ tests conducted at the study locations are PLT (deformations measured at the surface), PJT (deformation measured inside the boreholes), carried out in drifts, and the Goodman jack test, carried out in boreholes. The in situ test locations from where data are collected are shown in Figure 1. The in situ tested values of E_d, E_e, and the corresponding RMR values for the identified 35 site locations are shown in Figure 2 (a) and (b), respectively.

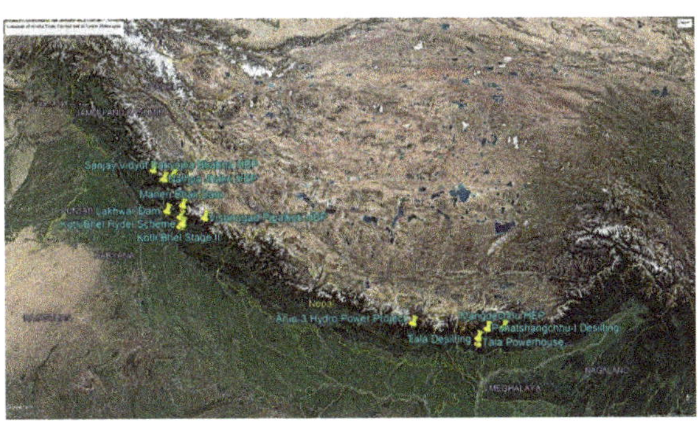

Figure 1. Google Earth image showing the location of the in situ tests carried out in different hydroelectric projects situated in the Himalayan region.

Table 1. Empirical equations for estimating E_d using RMR.

Equation No.	Ref.	Year	Equation	Type of Equation	R^2	Limitations	Data Sets Used	RMR Range	Country of Origin	Lithology
(1)	[12]	1978	$E_d = 2RMR - 100$	Linear	-	RMR > 50	3 Sites	51–85	South Africa	Shale, siltstone, dolerite, mudstone, and sandstone (hard rocks).
(2)	[14]	1983	$E_d = 10^{(RMR-10)/40}$	Power	-	RMR ≤ 50	15	26–83	-	Dolerite, sandstone, mudstone, shale, siltstone, gneiss, and granite (soft rocks).
(3)	[24]	1992	$E_d = 10^{(RMR-20)/38}$	Power	0.91	-	120	-	India	
(4)	[25]	1993	$E_d = 0.03e^{0.07RMR}$	Exponential	-	-	-	-	-	
(5)	[26]	1996	$E_d = e^{(4.407+0.081RMR)}$	Exponential	-	-	-	-	Croatia	Limestone
(6)	[27]	1997	$E_d = 0.0000097RMR^{3.54}$	Power	-	-	-	-	-	Gneiss, granite, and sandstone.
(7)	[28]	1999	$E_d = 0.1\left(\frac{RMR}{10}\right)^3$	Power	-	-	15	26–83	New Zealand	Graywacke, sandstones, and mudstones.
(8)	[29]	1999	$E_d = (7 \pm 3)\left(10^{(RMR-44)/21}\right)^{0.5}$	Non-linear	-	-	-	-	Various	
(9)	[30]	2003	$E_d = 0.0736e^{(0.0755RMR)}$	Exponential	0.62	-	115	20–85	Various	Quartzdiorite, limestone, and shale.
(10)	[31]	2003	$E_d = 19.43\ln RMR - 69.03$	Logarithm	-	-	57	38–84	Turkey	Grey and pinky quartzdiorite.
(11)	[15]	2006	$E_d = 0.3228e^{(0.0485RMR)}$	Exponential	0.36	-	8 Sites	-	Korea	
(12)	[16]	2008	$E_{rm} = 6.7RMR - 103.06$	Linear	0.94	RMR ≥ 27	9	27–61	Turkey	Graywacke
(13)	[17]	2010	$E_d = 0.0003RMR^3 - 0.0193RMR^2 + 0.315RMR + 3.4065$	Polynomial	0.8446	-	42	10–85	Iran	Limestone and marble
(14)	[18]	2012	$E_d = 110e^{-\left(\frac{RMR-110}{37}\right)^2}$	Gaussian function	0.932	-	43	-	Various	Mudstone, siltstone, sandstone, shale, dolerite (hard rocks), granite, gneiss, mudstone, siltstone, sandstone, shale, and dolerite (soft rocks).
(15)	[19]	2013	$E_d = 10^{(RMR-16)/50}$	Power	0.64	-	420	7–92	Korea	Gneiss
(16)	[20]	2014	$E_d = 0.1627RMR - 5.0165$	Linear	0.6709	-	52	30–76	Iran	Sandy siltstone, mudstone, conglomerate, sandstone, dislocated rock mass, faulted rock mass, and shear zone.
(17)	[21]	2015	$E_d = 0.058e^{(0.0785RMR)}$	Exponential	0.97	-	4 Sites	-	Turkey	Basalt, tuffites, and diabases.
(18)	[22]	2013	$E_d = 9E - 7RMR^{3.868}$	Power	0.89	-	82	39–85	Iran	Grey-green schist, phyllite, dark grey to black limestone, and limy dolomite.

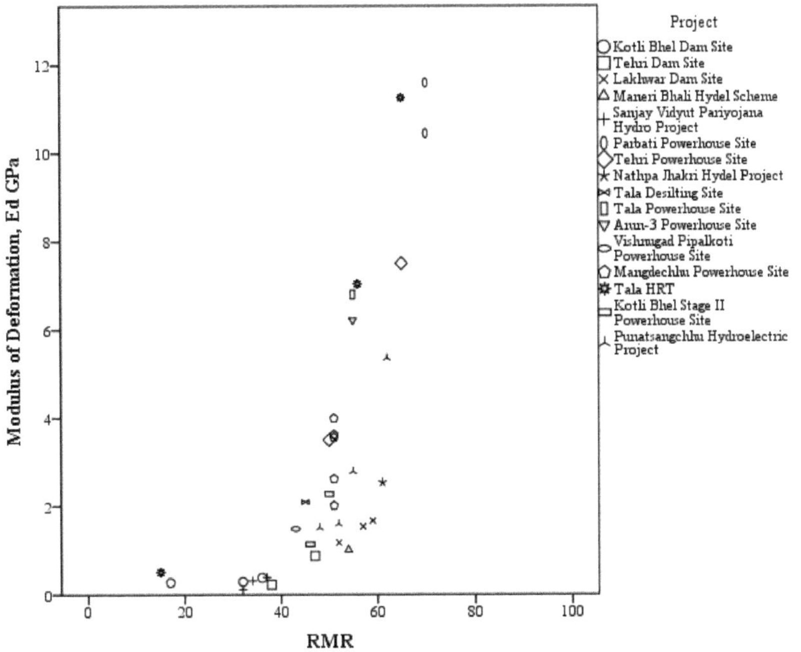

(**a**)

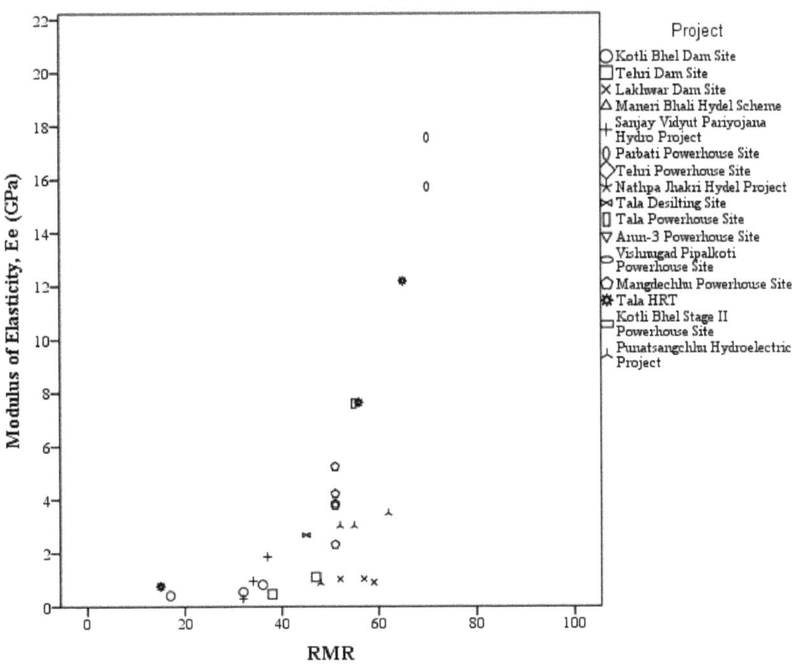

(**b**)

Figure 2. Plot between modulus of deformation and the RMR. (**a**) E_d vs. RMR. (**b**) E_e vs. RMR.

Statistical analysis was performed to establish a relationship between the RMR and E_d, and an equation to predict E_d from the RMR was proposed. The reliability and predictability of the proposed and the available equations were compared using statistical tools, and the reliable equation for the Himalayan region was presented. The equation was validated using the tested and estimated values in the 3D numerical model developed for Tala Hydroelectric Project, Bhutan. The instrumentation data were utilized for making comparisons with those of the modelling results.

3. Statistical Analysis

Statistical analyses using linear, logarithmic, cubic, and exponential functions were evaluated using Statistical Package for the Social Sciences (SPSS) software for the collected data and are presented in Figure 3 and Table 2. Figure 2a shows that the range of RMR values are from 15 to 70, and the range of E_d values are from 0.118 to 11.591 GPa. It is also observed that the cubic function given in Equation (19) has the highest value of the coefficient of regression (R^2), i.e., R^2 = 0.75 when compared to other functions, as shown in Figure 3.

$$E_d = 0.00011RMR^3 - 0.0083RMR^2 + 0.2RMR - 1.3 \quad (19)$$

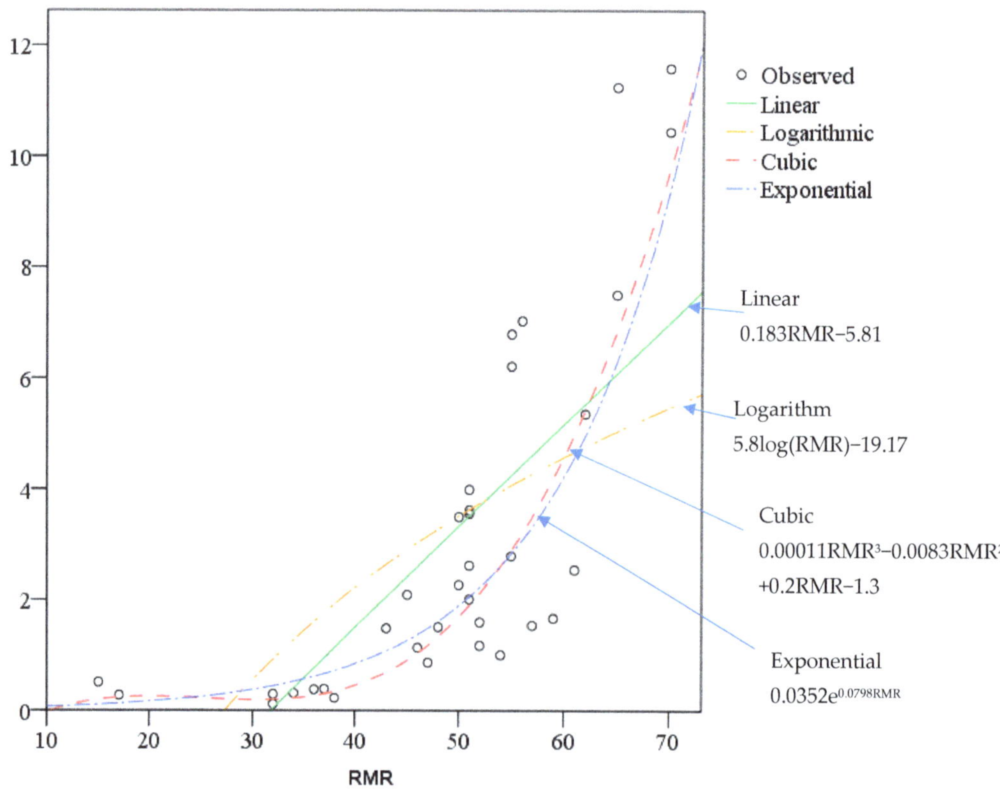

Figure 3. Relationship between modulus of deformation and RMR.

Table 2. Empirical equations for estimating E_d using RMR based on 35 test data.

S.No.	Type of Equation	Equation	Coefficient of Regression, R^2
1	Linear	0.183RMR − 5.81	0.53
2	Logarithmic	5.8log(RMR) − 19.17	0.37
3	Cubic	$0.00011RMR^3 - 0.0083RMR^2 + 0.2RMR - 1.3$	**0.75**
4	Exponential	$0.0352e^{0.0798RMR}$	0.708

To understand the prediction capacity, the root–mean–square error (RMSE) and variance accounted for (VAF) were calculated using Equations (20) and (21) for all the empirical equations discussed in Table 1, along with that of Equation (19). Root–mean–square error is defined as the standard deviation of the residuals. Residual is defined as the difference between the predicted and the actual values for each data point. In other words, residuals are nothing but prediction error. The RMSE is generally used as a measure in evaluating the performance of predictions and to check the efficiency of the model. The model is said to be accepted in regression analysis if the values of the RMSE and VAF are close to 0 and 100, respectively.

$$RMSE = \sqrt{\frac{1}{n}\sum_{i=1}^{n}(x - x')^2} \quad (20)$$

$$VAF = \left[1 - \frac{var(x - x')}{var(x)}\right]100 \quad (21)$$

The calculated RMSE and VAF values for Equation (19) are 1.70 and 74.33, respectively. Equation (19) was found to have a good prediction capacity compared to the other empirical equations listed in Table 1. The collected data could be further categorized based on the rock type, such as sedimentary and metamorphic rocks. The correlation between E_d and RMR values was made for rock types and the cubical function is shown in Figure 4.

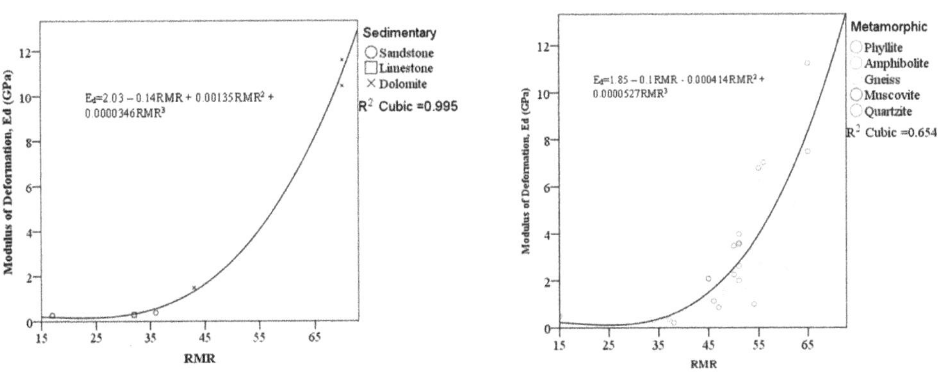

(**a**) (**b**)

Figure 4. Relationship between E_d and RMR for (**a**) sedimentary rocks and (**b**) metamorphic rocks.

The E_d values were calculated for the RMR values at in situ tested locations based on the empirical relations proposed by different authors and are shown in Figure 5. It is noticeable from Figure 5 that Equation (19) closely matches with that of the in situ tested-values curve. The empirical equations proposed by [15,21,22,30] are also in good comparison with that of the in situ tested value. The empirical equations proposed by [12,24,26,28,29] overestimated, and the remaining equations underestimated the E_d values.

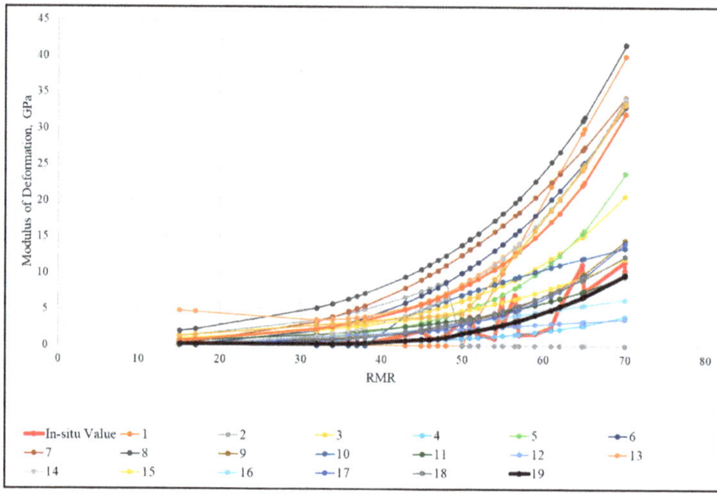

Figure 5. Estimated value of E_d based on empirical Equations (1)–(19).

4. Case Study—Tala Powerhouse Complex

The Tala Hydroelectric Project (1020 MW) is located on river Wangchhu, Western Bhutan [51,52]. The project consists of an underground machine-hall cavern, housing six units, each with a capacity of 175 MW. The machine-hall cavern (MHC) and the transformer-hall cavern (THC) dimensions are 206.4 m × 20.4 m × 44.5 m and 191 m × 16 m × 24.5 m, respectively. The rock pillar between the caverns is 40 m. The overburden ranges from 400 m to 500 m at the MHC and THC.

Hydrofracturing tests were carried out in the powerhouse cavern's exploratory drift to understand the stress field. The major principal stress is oriented in N50°W. The vertical stress of 10.865 MPa is calculated based on the overburden depth of 410 m. The ratio of maximum horizontal to vertical stress and minimum horizontal to vertical stress are 1.31 and 0.87, respectively [53,54]. The caverns are aligned in N37°W–S37°E direction across the strike of foliation [6].

4.1. Geology

The major lithology at the Tala Powerhouse complex consists of quartzite, phyllites, amphibolite schist, and phyllitic quartzite. The discontinuities were initially mapped in the exploratory drift (2 m × 2 m) driven in the machine-hall cavern along the crown level. The general foliation observed in the exploratory drift vary from N65° E–S65° W to N70° W–S70° E. The average foliation dip is 45.5°, and dip direction is N357°. Five sets of joints were observed in the exploratory drift in addition to the foliation. The rock quality index (Q) varied from 0.24 to 13.2 [53]. The representative value of RMR assessed in the caverns is 55.

4.2. 3D Numerical Modelling

Three-dimensional numerical modelling was carried out in this study using FLAC3D (Fast Lagrangian Analysis of Continua—three-dimensional) software. It utilizes an explicit finite-volume formulation for capturing models with complex behavior. The FLAC3D model, consisting of complex excavations of the machine-hall cavern, transformer-hall cavern, penstocks, bus ducts, and draft tubes considered in this study, is given in Figure 6. The in situ stress values obtained from the hydrofracturing tests were incorporated into the model before the start of the model simulation works.

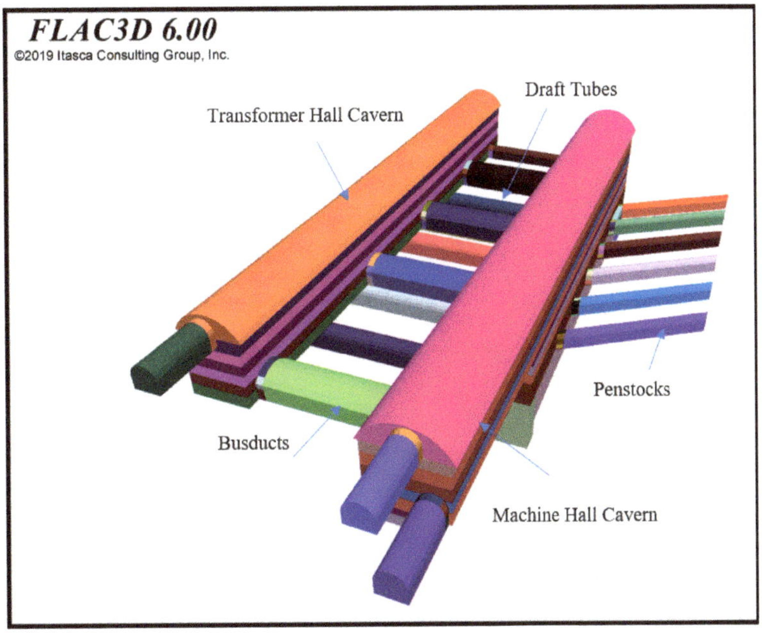

Figure 6. Three-dimensional view of the Tala Powerhouse complex, developed using FLAC3D.

4.3. Excavation Sequence and Support System

Initially, the machine-hall cavern's crown was excavated to the full width, followed by benching. The benching in the MHC and THC was taken up in 11 and 6 stages at the site. The bench heights in both caverns varied from 3 to 4 m. The excavation sequence adopted at the site was simulated in the 3D numerical model and is given in Figure 7. The support system installed at the site [55–58] and considered in the model in MHC and THC is shown in Table 3.

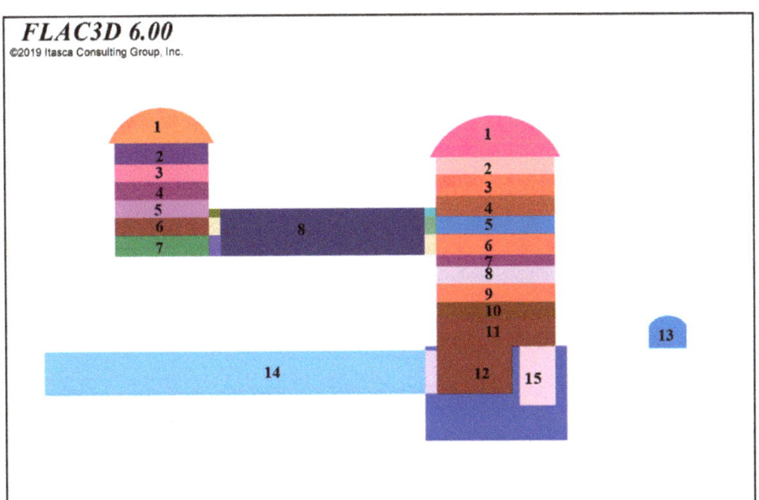

Figure 7. Excavation sequence simulated in 3D model.

Table 3. Support system considered in the model for MHC and THC.

Cavern	Support System
MHC–Crown	32 mm diameter, 8 m and 6 m long rock bolts at 1.5 m × 1.5 m pattern Steel-fiber-reinforced shotcrete (SFRS) of 100 mm thickness Steel ribs of ISMB 300 at 0.6 m spacing 32 mm/26.5 mm diameter, 12 m long Dywidag rock bolts at 1.5 m spacing
MHC–Walls	32 mm/26.5 mm diameter, 12 m long Dywidag rock bolts at 1.5 m spacing
THC–Crown	32 mm diameter, 8 m and 6 m long rock bolts at 3 m × 1.5 m pattern Steel-fiber-reinforced shotcrete (SFRS) of 100 mm thickness Steel ribs of ISMB 350 at 0.6 m spacing
THC–Walls	32 mm/26.5 mm diameter, 8 m long Dywidag rock bolts at 1.5 m spacing
MHC and THC Walls	Initial layer of shotcrete of 50 mm thickness Welded-wire mesh of 100 mm × 100 mm × 5 mm Final two shotcrete layers of 50 mm each

4.4. Material Properties

A linear, perfectly elastic–plastic constitutive model that follows the Mohr–Coulomb failure criterion was considered for the modelled rock mass [7] and postulated that the shear strengths of rocks comprise two parts—a constant cohesion and a normal stress-dependent frictional component. Two models were simulated in this study, considering the estimated E_d value of 2.89 GPa (Model A), based on the empirical Equation (19) for an RMR value of 55, and another model with an in situ tested (PLT), E_d value of 6.793 GPa [35,59] (Model B). Other material properties considered in the present analysis for both models are a density of 2650 kg/m^3, cohesion of 2.28 MPa, and friction angle of 28.3° [53].

4.5. Comparison of Modelling Results with Instrumentation Data

Various instruments were installed during different stages of excavation of the machine-hall cavern at the Tala Hydroelectric project, Bhutan. Displacements were measured using multipoint borehole extensometers (MPBX), reflective targets were used to measure the convergence of side walls using total station, and loads were measured by anchor load cells [6,60]. Wall convergence was measured using a total station that had an accuracy of 0.5 s. Reflective targets were installed opposite to each other on the walls of the machine-hall cavern for convergence measurements using the tie-distance method. The cavern walls convergence measured in the field was compared with the numerical modelling results for both Models A and B, respectively. The convergence was measured at the site for RD 15 m, 65 m, 110 m, and 150 m at EL 525 m, EL 520 m, and EL 515 m. Different benches were considered as the reference for a particular elevation based on the availability of the instrumentation data [6]. Bench 4 was taken as a reference for EL 525 m, bench 6 was taken as a reference for EL 520 m, and bench 7 was taken as a reference for EL 515 m. The measured and modelled convergence plots at EL 525 m, EL 520 m, and EL 515 m for two models, i.e., Model A with an E_d value of 2.89 GPa and Model B with an E_d value of 6.793 GPa, are shown in Figure 8.

Figure 8 shows that Model B is underpredicting the deformations in the powerhouse complex at all the locations, indicating that the in situ tested E_d value is on the higher side, enhancing the rock mass properties. Measured convergence matched well in Model A compared to Model B. Hence, the relation proposed in Equation (19) can be utilized to estimate the value of E_d. The displacement contours (in m) at RD 65 m after the complete excavation of the powerhouse complex for Models A and B are shown in Figure 9 (a) and (b), respectively.

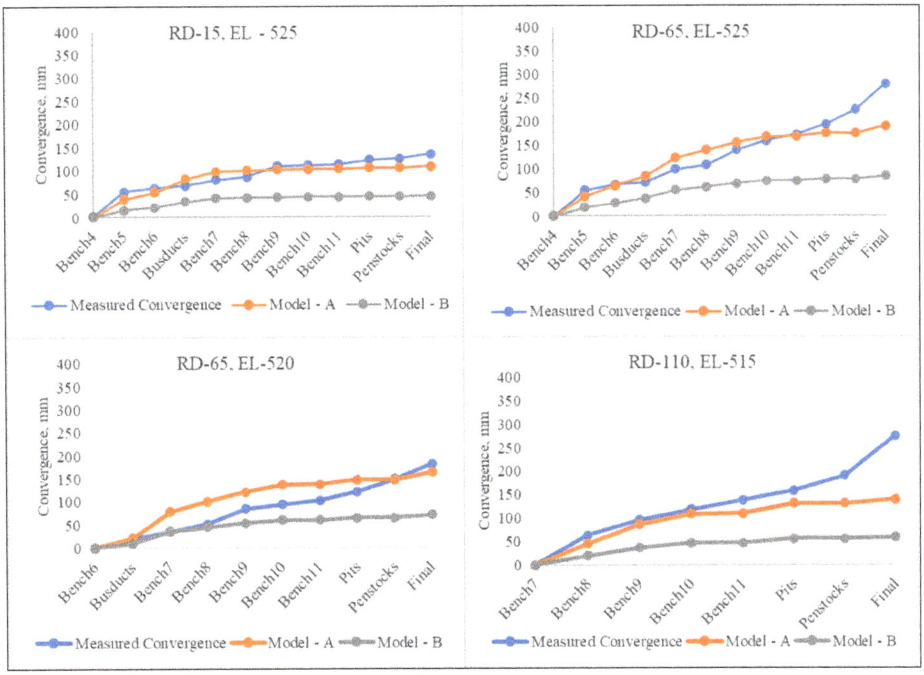

Figure 8. Plot between measured and modelled convergence at different elevations.

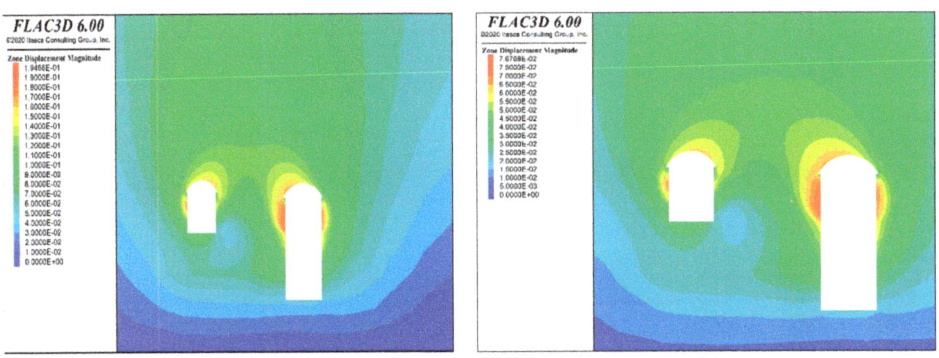

(**a**) (**b**)

Figure 9. Displacement contours at RD 65 after excavation. (**a**) Model A. (**b**) Model B.

5. Conclusions

This research has provided insight into the method for estimating the modulus of deformation using rock mass rating values. The conclusions derived from the current study are presented below:

- The review of various empirical models available for estimating E_d values indicates a considerable variation in the value of the deformation modulus for the Himalayan region. The empirical equations proposed by [14,20,21,29] are also in good comparison with the in situ tested value of E_d, while equations proposed by [11,23,25,27,28] overestimate, and the remaining equations underestimate E_d values.
- Based on the data obtained from 35 test locations, a predictive cubic equation (Equation (19)) could be developed, with R^2, RMSE, and VAF values of 0.75, 1.70, and 74.33,

respectively. These values indicate higher predictability and maximum accounted-for variance in E_d compared with other available correlations available in the literature.
- The 3D numerical modelling results show that the E_d value adopted based on the proposed Equation (19) (Model A) correlated well with that of the measured instrumentation data when compared with the value of E_d based on the in situ testing (Model B). Model B underpredicts the deformations in the powerhouse complex at all locations, indicating that the in situ tested E_d value is higher, enhancing the rock mass properties. Measured convergence matched well in Model A compared to Model B. Hence, the relation proposed in Equation (19) can be utilized to estimate the value of E_d.
- From the in situ tested data, the average ratio of E_e/E_d for the Himalayan region is 1.5.
- The proposed equation validates rock masses from the Himalayan region, with RMR values ranging from 15 to 70.

Author Contributions: Conceptualization, H.V.S.B.; methodology, H.V.S.B.; software, H.V.S.B. and S.R.N.; validation, H.V.S.B.; formal analysis, H.V.S.B.; investigation, H.V.S.B. and R.K.S.; resources, H.V.S.B., R.K.S. and S.R.N.; data curation, H.V.S.B.; writing—original draft preparation, H.V.S.B.; writing—review and editing, R.K.S. and S.R.N.; visualization, H.V.S.B.; supervision, R.K.S. and S.R.N. All authors have read and agreed to the published version of the manuscript.

Funding: This research received no external funding.

Institutional Review Board Statement: Not applicable.

Informed Consent Statement: Not Applicable.

Data Availability Statement: The data presented in this study are available on request from the corresponding author. The data are not publicly available due to the reason that they pertain to mega engineering projects.

Acknowledgments: The authors thank the project authorities and the director of the National Institute of Rock Mechanics (NIRM) for allowing them to conduct the studies. The support provided by the Department of Mining Engineering, Indian Institute of Technology (Indian School of Mines), Dhanbad, India, and BNV Siva Prasad and K Sudhakar, scientists, NIRM, is duly acknowledged. This forms part of the first author's doctoral research at IIT (ISM).

Conflicts of Interest: The authors declare no conflict of interest.

References

1. IHA (International Hydropower Association). *Hydropower Status Report: Sector Trends and Insights*; International Hydropower Association: London, UK, 2020.
2. CEA: HPMD (Central Electricity Authority: Hydro Project Monitoring Division). *Progress of On-Going Hydro Electric Projects, Quarterly Review No. 100, January–March 2020*; Central Electricity Authority: New Delhi, India, 2020.
3. Kansal, M.L.; Agarwal, S.S. Uncertainties-Based Potential Time and Cost Overrun Assessment While Planning a Hydropower Project. *ASCE-ASME J. Risk Uncertain. Eng. Syst. Part A Civ. Eng.* **2022**, *8*, 1–14. [CrossRef]
4. Siva Prasad, B.; Thapliyal, A.; Rabi, B.; Sripad, R. Delineation of Cavity in Downstream Surge Chamber at Punatsangchhu-II Hydroelectric Project, Bhutan. *J. Geol. Res.* **2019**, *1*, 5–11.
5. Najman, Y.; Clift, P.; Johnson, M.R.W.; Robertson, A.H.F. Early Stages of Foreland Basin Evolution in the Lesser Himalaya, N India. *Geol. Soc. Spec. Publ.* **1993**, *74*, 541–558. [CrossRef]
6. Naik, S.R. Studies on Stability Assessment of Large Caverns in Himalayan Region, National Institute of Technology Karnataka. Ph.D Thesis, National Institute of Technology Karnataka, Surathkal, India, 2017.
7. Brady, B.H.; Brown, E.T. *Rock Mechanics for Underground Mining*; Kluwer Academic Publishers: New York, NY, USA; Boston, MA, USA; Dordrecht, The Netherland; London, UK; Moscow, Russia, 1978; Volume 24, ISBN 140202116X.
8. Hoek, E.; Diederichs, M.S. Empirical Estimation of Rock Mass Modulus. *Int. J. Rock Mech. Min. Sci.* **2006**, *43*, 203–215. [CrossRef]
9. ISRM. *Commission on Terminology, Symbols and Graphic Representation*; International Society for Rock Mechanics (ISRM): Salzburg, Austria, 1975.
10. ISRM. *Suggested Method for Determing In Situ Deformability of Rock*; International Society for Rock Mechanics (ISRM): Salzburg, Austria, 1979; pp. 197–214.

11. Palmström, A.; Singh, R. The Deformation Modulus of Rock Masses-Comparisons between in Situ Tests and Indirect Estimates. *Tunn. Undergr. Sp. Technol.* **2001**, *16*, 115–131. [CrossRef]
12. Bieniawski, Z.T. Determining Rock Mass Deformability: Experience from Case Histories. *Int. J. Rock Mech. Min. Sci.* **1978**, *15*, 237–247. [CrossRef]
13. Aladejare, A.E.; Malachi Ozoji, T.; Adebayo Idris, M.; Lawal, A.I.; Onifade, M. Empirical Estimation of Rock Mass Deformation Modulus of Rocks: Comparison of Intact Rock Properties and Rock Mass Classifications as Inputs. *Arab. J. Geosci.* **2022**, *15*, 1033. [CrossRef]
14. Serafim, J.L.; Pereira, J.P. Considerations on the Geomechanical Classification of Bieniawski. In Proceedings of the International Symposium on Engineering Geology and Underground Openings, Lisbon, Portugal, 12–15 September 1983; pp. 1133–1144.
15. Chun, B.-S.; Lee, Y.; Jung, S. The Evaluation for Estimation Method of Deformation Modulus of Rock Mass Using RMR System. *J. Korean GEO-Environ. Soc.* **2006**, *7*, 25–32.
16. Isik, N.S.; Ulusay, R.; Doyuran, V. Deformation Modulus of Heavily Jointed-Sheared and Blocky Greywackes by Pressuremeter Tests: Numerical, Experimental and Empirical Assessments. *Eng. Geol.* **2008**, *101*, 269–282. [CrossRef]
17. Mohammadi, R.R. The Estimation of Rock Mass Deformation Modulus Using Regression and Artificial Neural Networks Analysis. *Sociology* **2010**, *35*, 1.
18. Shen, J.; Karakus, M.; Xu, C. A Comparative Study for Empirical Equations in Estimating Deformation Modulus of Rock Masses. *Tunn. Undergr. Sp. Technol.* **2012**, *32*, 245–250. [CrossRef]
19. Kang, S.S.; Kim, H.Y.; Jang, B.A. Correlation of in Situ Modulus of Deformation with Degree of Weathering, RMR and Q-System. *Environ. Earth Sci.* **2013**, *69*, 2671–2678. [CrossRef]
20. Nejati, H.R.; Ghazvinian, A.; Moosavi, S.A.; Sarfarazi, V. On the Use of the RMR System for Estimation of Rock Mass Deformation Modulus. *Bull. Eng. Geol. Environ.* **2014**, *73*, 531–540. [CrossRef]
21. Alemdag, S.; Gurocak, Z.; Gokceoglu, C. A Simple Regression Based Approach to Estimate Deformation Modulus of Rock Masses. *J. Afr. Earth Sci.* **2015**, *110*, 75–80. [CrossRef]
22. Khabbazi, A.; Ghafoori, M.; Lashkaripour, G.R.; Cheshomi, A. Estimation of the Rock Mass Deformation Modulus Using a Rock Classification System. *Geomech. Geoengin.* **2013**, *8*, 46–52. [CrossRef]
23. Jose, M.; Galeral, M.; A'lvarez, Z.B.T. Evalution of the Deformation Modulus of Rock Masses: Comparison of Pressuremeter and Dil-Atometer Tests with RMR Prediction. In Proceedings of the ISP5-Pressio International Symposium, Marne-la-Vallee, France, 22–24 August 2005.
24. Mehrotra, V.K. *Estimation of Engineering Parameters of Rock Mass*; University of Roorkee: Roorkee, India, 1992.
25. Kim, G. Revaluation of geomechanics classification of rock masses. In Proceedings of the Korean Geotechnical Society of Spring National Conference, Seoul, South Korea, 27 March 1993; pp. 33–40.
26. Jašarević, I.; Kovačević, M.S. Analyzing Applicability of Existing Classification for Hard Carbonate Rock in Mediterranean Area. In Proceedings of the ISRM International Symposium-EUROCK 1996, Turin, Italy, 16 September 1996; pp. 811–818.
27. Aydan, O.; Ulusay, R.; Kawamoto, T. Assessment of Rock Mass Strength for Underground Excavations. *Int. J. Rock Mech. Min. Sci. Geomech. Abstr.* **1997**, *34*, 705. [CrossRef]
28. Read, S.A.L.; Perrin, N.D.; Richards, L.R. Applicability of the Hoek-Brown Failure Criterion to New Zealand Greywacke Rocks. In Proceedings of the 9th ISRM Congress, Paris, France, 24 August 1999; pp. 655–660.
29. Diederichs, M.S.; Kaiser, P.K. Stability of Large Excavations in Laminated Hard Rock Masses: The Voussoir Analogue Revisited. *Int. J. Rock Mech. Min. Sci.* **1999**, *36*, 97–117. [CrossRef]
30. Gokceoglu, C.; Sonmez, H.; Kayabasi, A. Predicting the Deformation Moduli of Rock Masses. *Int. J. Rock Mech. Min. Sci.* **2003**, *40*, 701–710. [CrossRef]
31. Kayabasi, A.; Gokceoglu, C.; Ercanoglu, M. Estimating the Deformation Modulus of Rock Masses: A Comparative Study. *Int. J. Rock Mech. Min. Sci.* **2003**, *40*, 55–63. [CrossRef]
32. Bieniawski, Z.T. *Engineering Rock Mass Classifications: A Complete Manual for Engineers and Geologists in Mining, Civil, and Petroleum Engineering*; Wiley: Hoboken, NJ, USA, 1989.
33. Prasad, B.N.V.S.; Murthy, V.M.S.R.; Naik, S.R. Drillability Predictions in Aravalli and Himalayan Rocks—A Petro-Physico-Mechanical Approach. *Curr. Sci.* **2022**, *122*, 907–917. [CrossRef]
34. Siva Prasad, B.N.V.; Murthy, V.M.S.R.; Naik, S.R. Influence of Static and Dynamic Rock Properties on Drillability Prognosis for Mining and Tunnelling Projects. *Indian Geotech. J.* **2022**, *52*, 765–779. [CrossRef]
35. Siva Prasad, B.; Murthy, V.M.S.R.; Naik, S.R. Compendious Index for Drillability: A Rapid Tool to Assess Drill Penetration Rate and Bit Life for Rock Engineering Applications. *Bull. Eng. Geol. Environ.* **2023**, *82*, 1–20. [CrossRef]
36. Sekar Bellapu, H.V.; Sinha, R.K.; Naik, S.R. Estimation of Deformation Modulus of Rock Mass for an Underground Cavern Based on Back Analysis. *Lect. Notes Civ. Eng.* **2022**, *228*, 393–404. [CrossRef]
37. Bellapu, H.V.S.; Sinha, R.K.; Naik, S.R. Estimation of Modulus of Deformation by Different Methods for an Underground Cavern—A Case Study. *Indian Geotech. J.* **2022**, 1–7. [CrossRef]
38. Öge, İ.F. Determination of Deformation Modulus in a Weak Rock Mass by Using Menard Pressuremeter. *Int. J. Rock Mech. Min. Sci.* **2018**, *112*, 238–252. [CrossRef]
39. Zhang, B.; Mu, J.; Zheng, J.; Lv, Q.; Deng, J. A New Estimation Method and an Anisotropy Index for the Deformation Modulus of Jointed Rock Masses. *J. Rock Mech. Geotech. Eng.* **2022**, *14*, 153–168. [CrossRef]

40. Mehrotra, V.K.; Subhash, M. Use of In Situ Modulus for Classification of Rock Mass in Himalayas. In Proceedings of the International Symposium on Tunnelling for Water Resources and Power Projects, New Delhi, India, 19–23 January 1988; p. 402.
41. Agarwal, K.; Saran, S.; Jain, P.; Chandra, S. A Case Study w.r.t Geotechnical Aspects for the Pressure Tunnel of Sanjay Vidyut Pariyojana-Bhaba. In Proceedings of the International Symposium on Tunnelling for Water Resources and Power Projects, New Delhi, India, 19–23 January 1988; p. 402.
42. Sayeed, I.; Khanna, R. Geotechnical Investigations for Locating an Underground Powerhouse in Calcareous Rocks, Himachal Pradesh, India. In Proceedings of the Recent Advances in Rock Engineering (RARE), Bengaluru, India, 16–18 November 2016.
43. Singh, R. *Engineering in Rocks for Slopes, Foundations and Tunnels*; Ramamurthy, T., Ed.; PHI: Delhi, India, 2015.
44. Singh, R. Evaluation of Rock Mass Design Parameters for Tunneling at Tala Hydroelectric Project in Bhutan Himalayas. In Proceedings of the Seminar on Productivity and Speed in Tunnelling, Dehradun, India, 26–27 June 2003; pp. 41–54.
45. Sarwade, D.V.; Mishra, K.K.; Kapoor, V.K.; Kumar, N. Rock Mass Deformability: Empirical and Practical Approach. In Proceedings of the ISRM International Symposium—6th Asian Rock Mechanics Symposium, ARMS, New Delhi, India, 25–27 October 2010; pp. 23–27.
46. Naik, S.R.; Bhushan, R.; Bhusan, R.; Sekar, V.; Sudhakar, K. *3D Stress Analysis of Powerhouse Complex and Other Tunnels at PSP Tehri Project*; NIRM: Bengaluru, India, 2014.
47. Bellapu, H.V.S.; Bhusan, R.; Sudhakar, K.; Naik, S.R. *3D Numerical Model Studies for Stress Analysis of Underground Powerhouse Complex of Arun-3 H.E. Project (900 MW), Nepal*; NIRM: Bengaluru, India, 2020.
48. Vijay Sekar, B.; Bhusan, R.; Praveena, D.; Sudhakar, K.; Sivaprasad, B. *3D Numerical Modelling of Powerhouse Complex and Surge Tank for Vishnugad Pipalkoti Hydro Electric Project (444 MW), Uttarakhand*; NIRM: Bengaluru, India, 2021.
49. Naik, S.R.; Bhusan, R.; Sekar, B.H.V.; Sudhakar, K. *3D Stress Analysis of Underground Powerhouse Complex at Mangdechhu Hydroelectric Project, Bhutan*; NIRM: Bengaluru, India, 2018.
50. *National Institute of Rock Mechanics, Annual Report (2010–11)*; NIRM: Bengaluru, India, 2011.
51. Goyal, D.P.; Khazanchi, R.N. 1020 MW Tala Hydroelectric Project (Bhutan)–A Fast Construction Model. In *International Conference on Accelrated Construction of Hydropower Projects*; Central Board of Irrigation and Power: New Delhi, India; Tala Hydroelectric Project Authority: Gedu, Bhutan, 2003; pp. 26–27.
52. Jeyaseelan, R. Design of Large Caverns. In *International Conference on Accelrated Construction of Hydropower Projects*; Central Board of Irrigation and Power: New Delhi, India; Tala Hydroelectric Project Authority: Gedu, Bhutan, 2003; pp. 15–17.
53. Chopra, V.; Gupta, A. Roof Arch Failure of Machine Hall Cavern–Tala Hydroelectric Project. In *International Conference on Accelrated Construction of Hydropower Projects*; Central Board of Irrigation and Power: New Delhi, India; Tala Hydroelectric Project Authority: Gedu, Bhutan, 2003; pp. 29–31.
54. Bhasin, R.; Pabst, T.; Li, C. Stability Analysis of Surrounding Rock of a Large Hydropower Cavern in the Himalayas. In *Rock Characterisation, Modelling and Engineering Design Methods*; CRC Press: Boca Raton, FL, USA, 2013; pp. 659–664.
55. Singh, R.; Sharma, B.N.; Puri, P.K.; Gupta, M.; Goyal, D.P.; Chugh, I.K. Experience of Long Rock Bolts in Machine Hall Cavern at Tala Hydoelectric Project. In *International Conference on Accelrated Construction of Hydropower Projects*; Central Board of Irrigation and Power: New Delhi, India; Tala Hydroelectric Project Authority: Gedu, Bhutan, 2003; pp. 15–17.
56. Sharma, B.N.; Engineer, S. Experience of Dywidag Rock Bolts in Machine Hall Cavern at Tala Project. In *International Conference on Accelrated Construction of Hydropower Projects*; Central Board of Irrigation and Power: New Delhi, India; Tala Hydroelectric Project Authority: Gedu, Bhutan, 2003; Volume 9, pp. 68–74.
57. Singh, R. Quality Management of Powerhouse Cavern at Tala Hydoelectric Project in Bhutan Himalayas. In *International Conference on Accelrated Construction of Hydropower Projects*; Central Board of Irrigation and Power: New Delhi, India; Tala Hydroelectric Project Authority: Gedu, Bhutan, 2003; pp. 503–506.
58. Venkatesh, H.S.R.N.; Gupta, B.N.; Sharma, M.C.D.; Puri, P.K. Rock Mass Damage Control in Machine Hall Cavern of Tala Hydroelectric Project (Bhutan) by Near Field Vibration Monitoring. In *International Conference on Accelerated Construction of Hydropower Projects*; Central Board of Irrigation and Power: New Delhi, India; Tala Hydroelectric Project Authority: Gedu, Bhutan, 2003; Volume I, pp. 23–31.
59. Singh, R.; Dhawan, A.K. Experience of Deformability Measurement Using Goodman Jack. In Proceedings of the International Conference on Rock Engineering Techniques for Site Characterisation, ROCKSITE, Bangalore, India, 26 November 1999; pp. 29–36.
60. Singh, R. Instrumentation at Tala Hydroelectric Project in Bhutan. In *International Conference on Accelrated Construction of Hydropower Projects*; Central Board of Irrigation and Power: New Delhi, India; Tala Hydroelectric Project Authority: Gedu, Bhutan, 2003; pp. 42–66.

sustainability

MDPI

Article

Performance of Statistical and Intelligent Methods in Estimating Rock Compressive Strength

Xuesong Zhang [1,*], Farag M. A. Altalbawy [2,3], Tahani A. S. Gasmalla [4], Ali Hussein Demin Al-Khafaji [5], Amin Iraji [6], Rahmad B. Y. Syah [7] and Moncef L. Nehdi [8,*]

1 College of Pipeline and Civil Engineering, China University of Petroleum (East China), Qingdao 266580, China
2 Department of Chemistry, University College of Duba, University of Tabuk, Tabuk 71491, Saudi Arabia
3 National Institute of Laser Enhanced Sciences (NILES), University of Cairo, Giza 12613, Egypt
4 Department of Education, University College of Duba, University of Tabuk, Tabuk 71491, Saudi Arabia
5 Department of Laboratories, Techniques, Al-Mustaqbal University College, Babylon, Hillah 51001, Iraq
6 Engineering Faculty of Khoy, Urmia University of Technology, Urmia 5716693188, Iran
7 PUIN-Engineering Faculty, Universitas Medan Area, Medan 20223, Indonesia
8 Department of Civil Engineering, McMaster University, Hamilton, ON L8S 4M6, Canada
* Correspondence: snowpiner@upc.edu.cn (X.Z.); nehdim@mcmaster.ca (M.L.N.); Tel.: +905-525-9140 (ext. 23824) (M.L.N.)

Abstract: This research was conducted to forecast the uniaxial compressive strength (UCS) of rocks via the random forest, artificial neural network, Gaussian process regression, support vector machine, K-nearest neighbor, adaptive neuro-fuzzy inference system, simple regression, and multiple linear regression approaches. For this purpose, geo-mechanical and petrographic characteristics of sedimentary rocks in southern Iran were measured. The effect of petrography on geo-mechanical characteristics was assessed. The carbonate and sandstone samples were classified as mudstone to grainstone and calc-litharenite, respectively. Due to the shallow depth of the studied mines and the low amount of quartz minerals in the samples, the rock bursting phenomenon does not occur in these mines. To develop UCS predictor models, porosity, point load index, water absorption, P-wave velocity, and density were considered as inputs. Using variance accounted for, mean absolute percentage error, root-mean-square-error, determination coefficient (R^2), and performance index (PI), the efficiency of the methods was evaluated. Analysis of model criteria using multiple linear regression allowed for the development of a user-friendly equation, which proved to have adequate accuracy. All intelligent methods (with $R^2 > 90\%$) had excellent accuracy for estimating UCS. The percentage difference of the average of all six intelligent methods with the measured value was equal to +0.28%. By comparing the methods, the accuracy of the support vector machine with radial basis function in predicting UCS was ($R^2 = 0.99$ and PI = 1.92) and outperformed all the other methods investigated.

Keywords: UCS; intelligent and statistical methods; prediction; sedimentary rocks

Citation: Zhang, X.; Altalbawy, F.M.A.; Gasmalla, T.A.S.; Al-Khafaji, A.H.D.; Iraji, A.; Syah, R.B.Y.; Nehdi, M.L. Performance of Statistical and Intelligent Methods in Estimating Rock Compressive Strength. *Sustainability* **2023**, *15*, 5642. https://doi.org/10.3390/su15075642

Academic Editors: Mahdi Hasanipanah, Danial Jahed Armaghani, Jian Zhou and Jianjun Ma

Received: 27 December 2022
Revised: 26 February 2023
Accepted: 15 March 2023
Published: 23 March 2023

1. Introduction

Stability of slopes, prediction of drilling rate, classification of rock masses, and modeling of foundations require knowledge of the uniaxial compressive strength (UCS) of the rocks for designing projects [1–3]. Indirect determination of the UCS in places where the preparation of standard samples is difficult requires lots of time and is expensive. Hence, various researchers have predicted the UCS of the limestones and sandstones using statistical and intelligent methods [4–7]. Aladejare et al. [8] collected empirical relationships and models between UCS and other rock characteristics from previous studies. Several models were developed to estimate the rock UCS using Gaussian process regression (GPR) [9–13], feedforward multilayer perceptron artificial neural network (FMP-ANN) [14–19], random forest algorithm (RFA) [20–23], adaptive neuro-fuzzy inference system (ANFIS) [24–28], and multiple linear regression (MPLR) [3,4,7,8,25,29–31]. The results of Lawal et al.'s [9] study showed that the GPR method, with a correlation coefficient of almost 100%, is able to estimate the static and dynamic properties of sedimentary rocks. Moreover, a comparison of the RFA, MPLR, FMP-ANN methods in compressive strength estimation showed that FMP-ANN with the Levenberg–Marquardt algorithm has a higher accuracy than other methods [20]. The results of Matin et al.'s [22] study to select the effective variables using the random forest method showed that compression wave velocity is the most effective variable as an input for estimating compressive strength and the modulus of elasticity using predictive models. Hudaverdi [25] stated that the ANFIS method has a high efficiency in flyrock estimation with an average error of less than 8%. The results of the MPLR, ANFIS, and FMP-ANN methods in the UCS estimation showed that all three methods have a determination coefficient higher than 90%, while the ANFIS method has a better performance [26]. The comparison of the ANFIS, FMP-ANN and multiple regression methods by Yesiloglu-Gultekin and Gokceoglu [26] showed that the ANFIS method has higher accuracy for estimating compressive strength and the modulus of elasticity.

Mahmoodzadeh et al. [32] compared the K-nearest neighbor algorithm (KNNA), Gaussian process regression based on squared exponential kernel (GPR-SEK), support vector machine with radial basis function (SVR-RBF), and decision tree (DT) methods to forecast rock quality designation in a tunnel project and stated that the GPR-SEK method is more accurate than other methods. Xu et al. [33] forecasted the UCS of rock using intelligent technics. The SVM-RBF was used to predict UCS [34]. Rastegarnia et al. [19] used FMP-ANN and MPLR to predict the UCS of carbonates. They stated that FMP-ANN estimates the UCS more than the measured value. Trott et al. [35] used RFA to classify rock types. Barzegar et al. [36] predicted the UCS of travertine rocks using the SVM-RBF, FMP-ANN, and ANFIS methods and stated that the SVM-RBF showed higher accuracy than the other methods. Mohamad et al. [37] estimated the UCS of soft rocks using FMP-ANN and particle swarm optimization (PSO)-based ANN. Madhubabu et al. [6] used MPLR and FMP-ANN to estimate the UCS of the carbonate samples. Umrao et al. [24] used the ANFIS approach for estimating UCS based on density, porosity, and PWV. Moreover, using inteligent methods, Gül et al. [17] predicted the UCS of different rock types. Singh et al. [38] estimated the UCS of basalt samples via MPLR and ANFIS. Kaloop et al. [39] used GPR and multivariate adaptive regression splines (MARSs) to estimate rock UCS. They stated that the MARS showed higher acuracy than the GPR method. Some engineers and researchers are interested in simple empirical relationships using simple models such as simple and MPLR to estimate UCS. Therefore, simple empirical relationships are widely used to estimate rock UCS using statistical methods. Table 1 shows some of the relationships for estimating UCS by previous researchers.

This study was conducted to predict the UCS of sedimentary rocks based on porosity, point load index (PLI), density (D), water absorption by weight (WW), and P-wave velocity (PWV) using the FMP-ANN, GPR-SEK, KNNA, RFA, ANFIS, SVM, SR, and MPLR methods. Moreover, the types of kernel functions were investigated using the SVM method and the most accurate type of kernel function was introduced to estimate the UCS.

Sampling points, from 12 mines in the Bushehr province, south of Iran, are located between 50 and 52 degrees longitudes and 28 and 30 degrees latitudes. The mines are mainly travertine, limestone, and sandstone of the Aghajari and the Mishan formations.

Table 1. Relationships for estimating UCS by previous researchers.

Equation	Reference	Lithology
$UCS = 12.29PLI^{1.233}$	Teymen and Mengüç [40]	Various Rocks
$UCS = -37.82 + (0.017PWV)$	Salehin [41]	Sedimentary Rocks
$UCS = 0.043PWV - 136.8$	Aldeeky and Al Hattamleh [42]	Basalt Rocks
$UCS = 17.6PLI + 13.5$	Aliyu et al. [30]	Sedimentary Rocks
$UCS = 14.3PLI$	Aladejare [8]	Sedimentary Rocks
$UCS = 9.95PWV^{(1.21)}$	Kahraman [43]	Sedimentary rocks
$UCS = 0.034PWV - 86.36$	Wen et al. [7]	Limestone
$UCS = -5.10\phi + 110.79$	Edet [3]	Sandstone
$UCS = 0.025PWV - 8.619$	Azimian [29]	Limestone
$UCS = 6.6PWV^{1.6}$	Uyanık et al. [44]	Sedimentary rocks
$UCS = 22.18PWV - 30.32$	Selcuk and Nar [31]	Various Rocks
$UCS = 0.041PWV - 15.40$	Abdi and Khanlari [4]	Sandstones
$UCS = 2.304(PWV)^{2.43}$	Kılıç and Teyman [45]	Various Rocks
$UCS = 10 - 5D^{16.7}$	Aladejare [8]	Sedimentary rocks

2. Methodology

2.1. Laboratory Tests

Specimens with a diameter of 54 mm and a height to diameter ratio of 2 were prepared [46]. A wear device was used to parallel surfaces of specimens. Table 2 shows the methods used to measure geo-mechanical properties. Figure 1 displays some of the samples in laboratory tests.

Table 2. Methods used for performing tests.

Test	Standards and References	Descriptions
UCS	ISRM [47]	A constant loading rate of 0.7 MPa per second was applied to the samples. The amount of deformation was recorded using the corresponding gauge in the UCS test.
Point load index (PLI)	ASTM D5731 [48]	This test was done on irregular and cylindrical samples. Then the PLI was calculated.
Compressional wave velocity test	ASTM D2845 [49]	With a $\frac{1}{2}$ MHz frequency
Porosity (ϕ), density(D) and water absorption by weight (WW)	ISRM [47]	The total porosity (ϕ) of specimens was measured using the method of saturation and immersion way. Density was computed from the ratio of mass to sample volume.
Petrography	Folk [50], Dunham [51]	For classifying the samples using thin section images.

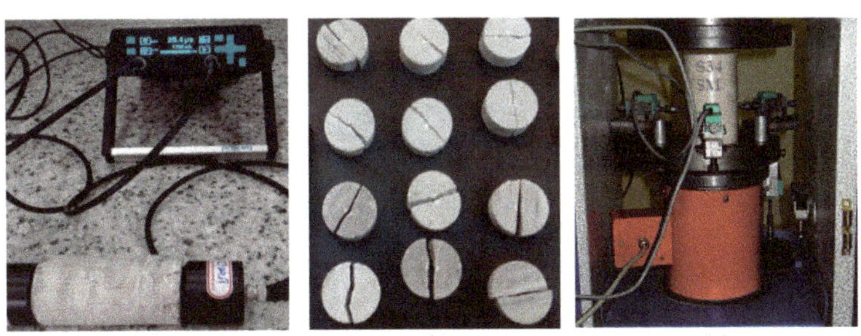

(a) (b) (c)

Figure 1. Example samples for PWV, PLI, and UCS tests: (**a**) device for measuring PWV, (**b**) example samples after PLI test, and (**c**) sample under UCS test.

2.2. Random Forest Algorithm (RFA)

The random forest method is one of the ensemble methods. In these methods, the model chosen for classification or regression is a combination of several models. Figure 2 shows the idea of the random forest algorithm. In this approach, each model issues its vote and the final result about the value is issued based on these votes [22,23,52]. The general principles of group training techniques are based on the assumption that their accuracy is higher than other training algorithms [53]. On the other hand, the accuracy of combined methods is higher than the accuracy of its components [53]. The RFA method has also been used in rock mechanics in recent years [20]. In the RFA method, the models used in the combined method, which are all of the decision tree type, form a forest. Each of the decision trees is made using a random selection of special attributes in each node to determine the branching. In other words, each tree is built based on the values of a random vector. These values have the identical scattering for all trees in the forest and are sampled independently. For classification, each tree issues its vote, and the final result is determined by the majority vote [54]. The number of trees and the number of chosen variables in each node are important parameters in the RFA [55]. In this method, by replacing the information every sampling time, some information is never sampled, and other data may be sampled several times. In other words, some input data for some trees will be out of the bag, that is, they will not participate in the creation of some trees. These data have the function of an internal validator for each tree, which is performed by estimating the out-of-bag error. If the out-of-bag data itself is predicted through trees, there will be an error for these predictions, and the average of these errors is called the out-of-bag error. This error indicates the influence of the unselected samples on the error rate of the final result of the random forest [56].

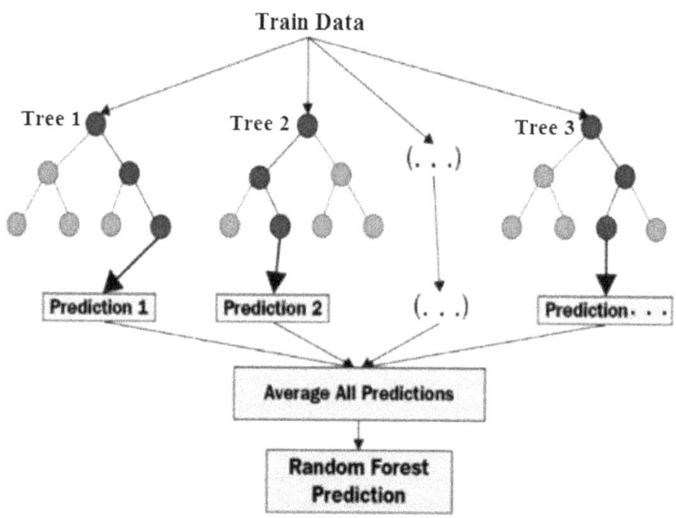

Figure 2. Concept of random forest algorithm.

2.3. Gaussian Process Regression Based on Squared Exponential Kernel (GPR-SEK)

Consider a d data set with n measurements: $d = \{(x_i, y_i)|i = 1, \ldots, n\}$, where x_i is the input vector with D dimension and y_i is the target output. This set, consisting of two components, input and output, will be denoted as measured points. To simplify the problem, the inputs of the collection are aggregated at $X = \{x_i, x_2, \ldots, x_n\}$ matrix and the outputs are also combined at $Y = \{y_i, y_2, \ldots, y_n\}$ matrix. Regression based on the data set d creates a new input x^* to arrive at the predicted distribution for the corresponding values of the measured y^* data. The Gaussian process (GP) is a group of accidental parameters, a restricted number of which are combined with Gaussian distributions (GDs) [57]. The GP is a generalization of GD. The GD is actually scattered between accidental parameters, while GP represents scattering between functions. The *f(x)* GP is described using the *m(x)* average and covariance functions according to Equations (1) and (2).

$$m(x) = E(f(x)) \tag{1}$$

$$c(x, x') = E(f(x) - m(x))(f(x') - m(x'))) \tag{2}$$

In relationships 1 and 2, $c(x, x')$ is the covariance or kernel function, which is computed at the x and x' points. The GP can be described as Equation (3).

$$f(x) \sim GP(m(x),\ \mathrm{c(x,x')}) \tag{3}$$

Usually, for simplification, the value of the average function is considered equal to zero [58]. In the *GP*, the correlation between the target and the input vector is based on Equation (4).

$$y_i = f(x_i) + \varepsilon \tag{4}$$

where $f(x_i)$ represents the arbitrary regression function and ε is the noise of the Gaussian function with zero mean and σ^2 variance (i.e., $\varepsilon \sim N(0,\ \sigma^2)$). Furthermore, it is supposed that $f = [f(x_1),\ \mathrm{f}(x_2),\ \ldots,\ \mathrm{f}(x_n)]^T$ has a performance according to the GP (i.e., $p(f|\mathrm{X}) = N(0,\ \mathrm{C})$). Here, C is the covariance matrix with the $c_{i,j} = c(x_i,\ \mathrm{x}_j)$ domains.

$$\mathrm{C}(\mathrm{X},\mathrm{X}) = \begin{pmatrix} c(x_1,x_1) & c(x_1,x_2) & \ldots & c(x_1,x_n) \\ c(x_2,x_1) & c(x_2,x_2) & \ldots & c(x_2,x_n) \\ \vdots & \vdots & \vdots & \vdots \\ c(x_n,x_1) & c(x_n,x_2) & \ldots & c(x_n,x_n) \end{pmatrix} \tag{5}$$

The $c_{i,j}$ is the covariance between the latent function values of $f(x_i)$ and $f(x_j)$. GP regression is used to calculate the predicted scattering for the *f** function values in the test points of $X^* = [x_1^*, x_2^* \ldots x_m^*]$. The distribution of y depends on the values of *f*, which is represented by an isotropic Gaussian as follows.

$$p\left(y \middle| f, x\right) = N(F, \sigma_n^2 I) \tag{6}$$

In relation (6), *I* is the identity matrix. According to the characteristics of the Gaussian function, the marginal distribution of *y* is determined as follows.

$$p(y|\mathrm{X}) = \int p\left(y|f,\mathrm{X}\right)p\left(f\middle|\mathrm{X}\right)df = N(0, \mathrm{C} + \sigma_n^2 I) \tag{7}$$

The integrated distribution of the observation data values, that is, the desired output, and the function values at the test points are written as follows [32].

$$\begin{bmatrix} y \\ f^* \end{bmatrix} \sim \mathrm{N}\left(0, \begin{bmatrix} \mathrm{C}(\mathrm{X},\mathrm{X}) + \sigma^2 I & \mathrm{C}(\mathrm{X},\mathrm{X}_*) \\ \mathrm{C}(\mathrm{X}_*,\mathrm{X}) & \mathrm{C}(\mathrm{X}_*,\mathrm{X}_*) \end{bmatrix}\right) \tag{8}$$

According to relation (3), and using standard rules to limit Gaussian, the following conditional distribution can be obtained.

$$p(f_*|\mathrm{X}, y, X_*) \sim N\left(\overline{f_*}, c(f_*)\right)) \tag{9}$$

$$\bar{f_*} = \mathrm{C}(\mathrm{X}_*,\mathrm{X})\left[\mathrm{C}(\mathrm{C},\mathrm{C}) + \sigma^2 I\right]^{-1} y \tag{10}$$

2.4. The SVM-RBF

To achieve the least error related to the test set, the SVM-RBF approach fits a linear line with epsilon (ε) thickness on the data [59]. In this method, a function such as $f(x) = m.x + b$ is used for forecasting, where m is weight vector and x and b are weights.

For minimizing weight vector and test error, this technique utilizes error functions for ignoring errors that are at a determined range from the real errors [60]. Hence, some deviation (derived from Equation (11)) from ε must be overlooked by including Equation (11) in Equation (12), which considers the ξ_i^- and ξ_i^+ deficiency parameters. Finally, the error values are optimized via Equation (12) using structural error minimization

$$|\xi|_\varepsilon = \begin{bmatrix} 0 & if\ |\xi| \leq \varepsilon \\ |\xi| - \varepsilon & otherwise \end{bmatrix} \tag{11}$$

$$\text{Minimize}: \{(\|m\|^2) * 1/2\} + \{(\sum_{i=1}^{N} (\xi_i^+ + \xi_i^-)) * C\}$$
$$\varepsilon_{\text{Constrains}}: \begin{bmatrix} m.x_i + b + \xi_i^+ - y_i \leq \varepsilon & i = 1, 2, \ldots, N \\ y_i - (b + m.x_i) \leq \xi_i^- + \varepsilon & i = 1, 2, \ldots, N \\ \xi_i^+ \geq 0,\ \xi_i^- \geq 0 & i = 1, 2, \ldots, N \end{bmatrix} \tag{12}$$

In Equation (12), $\{(\|m\|^2) * 1/2\}$ is the supervisory part, N is number of samples, ε is the allowable error, C is the complexity balance coefficient, and the ε values are the acceptable error range. As with the GPR method, various kernel functions are used in the SVM method [61]. Radial basis function (RBF), which is the most important kernel function, was used in the current research [62].

2.5. K Nearest Neighbor Algorithm (KNNA)

The KNNA is based on sample and performs classification based on K nearest neighbors. This method performs classification based on the similarity of the data. In fact, for each new test data, it calculates the K nearest neighbor distances and determines a label similar to the dominant label of this k neighbor for the desired point [63]. This method was introduced as a nonparametric method and does not make any assumption on the distribution of inputs. Therefore, it is extensively used in various fields [64].

In the KNNA classifier, an unknown value, is recognized by the similarity between known trained or labeled values based on the calculation of the distance between unknown values and labeled values. Then, K of the nearest values are selected as the basis for classification, and the unknown value (x test) is assigned to the class that has the most values among the closest values. For this purpose, three factors affect the KNNA classification: (1) the number of K of the neighbor and the changing of the value of K, which may amendment the classification results; (2) labeled dataset; therefore, adding or eliminating any value to the training samples affects the final results of the KNNA classifier; (3) the distance criterion. In KNNA, Euclidean distance is usually used as a distance criterion to measure the distance between two values [64,65]. This algorithm, as with the other algorithms used in this research, after examining the data in the program environment, divides the data into two parts, training data and test data, and builds the K nearest neighbor model and enters the training data into the model to train the model. Next, to determine the precision of the method, the test data is entered into the model for prediction and to evaluate the prediction accuracy in comparison with the labels of the test data [65,66].

2.6. ANFIS and FMP-ANN

The ANFIS and FMP-ANN methods have been widely introduced and described by previous researchers [18,25,67–70]. The transfer functions of neurons, membership functions, type of fuzzy inference system, and data training methods in these two methods are mentioned in the results section.

In the SVM-RBF, ANFIS, KNNA, *GPR-SEK*, and RFA methods, 30% and 70% of the whole data were used for the testing and training the models, respectively.

2.7. Performance Evaluation of Results

To appraise the methods, the correlation coefficient, the MAPE %, the RMSE, VAF, and the PI are defined in the form of Equations (13)–(16).

$$\text{MAPE} = \frac{1}{n}\sum_{i=1}^{n}\left|\frac{y-y'}{y}\right| * 100 \tag{13}$$

$$RMSE = \frac{1}{s^2 n}\sum_{i=1}^{n}(y-y')^2 \tag{14}$$

$$VAF = 100\left[1-\frac{var(y-y')}{var(y)}\right] \tag{15}$$

$$\text{PI} = \text{R}^2 + (\text{VAF}/100) - \text{RMSE} \tag{16}$$

In relationships 13 to 16, y is the value of the variable measured, y' is the predicted UCS, and n is the total data and s^2 is the sample variance. Equation (17) was used to normalize the data between -1 and 1.

$$X_i = 2\left(\frac{X - X_{min}}{X_{max} - X_{min}}\right) - 1 \tag{17}$$

In Equation (17), x is the measured variable, X_{min} is the minimum of the data, and X_{max} is the maximum of the data.

3. Results and Discussion

3.1. Geomechanical Properties of Samples

The maximum, minimum, and average engineering properties of 65 samples (37 samples of limestone, 11 samples of argillaceous limestone, and 17 samples of sandstone) are presented in Table 3. The average Es and UCS are 14.95 GPa and 37.54 MPa, respectively. Physical, mineralogical, and petrographic characteristics cause changes in the compressive strength of rocks [5,18,71]. Moreover, lithological properties such as the type of rock, the amount and type of minerals, the type of cement between the particles that comprise the rock and physical properties such as the amount of moisture, porosity, and water absorption have a significant effect on the compressive strength and, consequently, on the stability of mines [72,73]. As the amount of clay minerals increases, the resistance of the samples decreases [4,74]. Water absorption causes the swelling and instability of the mine wall in samples with a high percentage of clay minerals [75–77]. The number of joints changes the geomechanical properties and instability of the rocky slopes [78–80]. The engineering properties and stability of structures are affected by the amount of moisture [81].

Table 3. Laboratory results on sample.

Statistics \ Properties	Density (g/cm^3)	PLI (MPa)	Water Absorption (%)	Porosity (%)	UCS (MPa)	Es (GPa)	PWV (km/s)
Average	2.43	3.75	6.81	9.44	37.54	14.95	4.38
Std. Dev.	0.11	1.66	1.87	3.35	16.49	5.30	1.03
Kurtosis	0.13	(0.58)	(0.50)	(0.41)	(0.58)	(0.51)	(0.38)
Skewness	(0.42)	0.09	0.70	0.79	(0.71)	(0.62)	(0.78)
Min.	2.10	0.31	4.08	4.36	4.12	3.00	2.06
Max.	2.63	8.00	11.00	16.72	59.72	22.90	5.79
Specimens	65	65	65	65	65	65	65

In the sandstone samples of the present research, clay and gypsum were found. The cement of the samples is gypsum and calcite. The grains of these sandstones are semi-rounded to angular and have a moderate to poor grading. The study of the thin sections of the samples showed that the samples with higher clay content show lower resistance characteristics. Swelling clay minerals (such as montmorillonite) were not observed in the investigated samples. Generally, rock bursting occurs in deep mines and in quartz-rich rocks [12]. Because the depth of the studied mines is less than 50 m and the overburden stress is negligible, the risks of rock bursting have not been reported in them so far.

3.2. Petrographic Features

Texture has a special effect on the engineering properties of sedimentary rocks [82]. In carbonates, the rock texture is very different, but their mineralogy is not much different [83,84]. According to microscopic studies, the most basic mineral of limestone rocks was calcite. Mishan formation limestone rocks, with an early Miocene age, based on the Dunham [51] classification, were classified in the range of mudstone to grainstone. Moreover, argillaceous limestone samples of this formation were classified in the mudstone to packstone categories. Sandstone samples of the Aghajari formation with an upper miocene age were classified as calc-litharenite according to the Folk [50] classification. These rocks consist of carbonaceous rock fragments (26 to 75), volcanic gravel (2 to 35%), meta-morphic fragments (2 to 18%), feldspar (1 to 10%), dark minerals (1 to 8%), quartz (0 to 22%), and chert (2 to 11%).

3.3. Influence of Independent Variables on the UCS

Figure 3 shows the effect of variables on the UCS. There is a reasonable tendency among these characteristics. The UCS decreases with increasing WW and porosity. This Figure shows that porosity has the highest effect on the UCS. Numerous studies have reported linear relationships with high accuracy between the point load index (PLI) and UCS [8,30]. The results of the relationship between PVW and UCS show that PWV displays a high accuracy to estimate UCS (Figure 3).

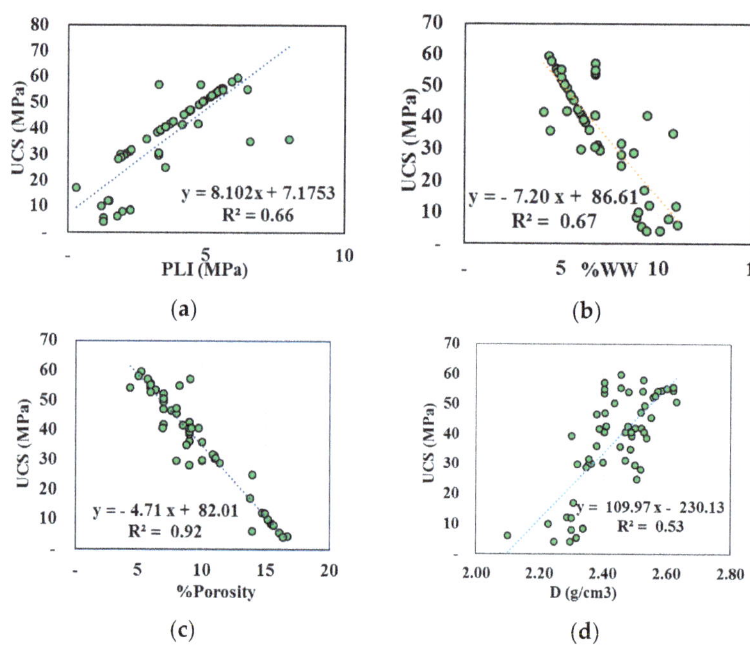

Figure 3. *Cont.*

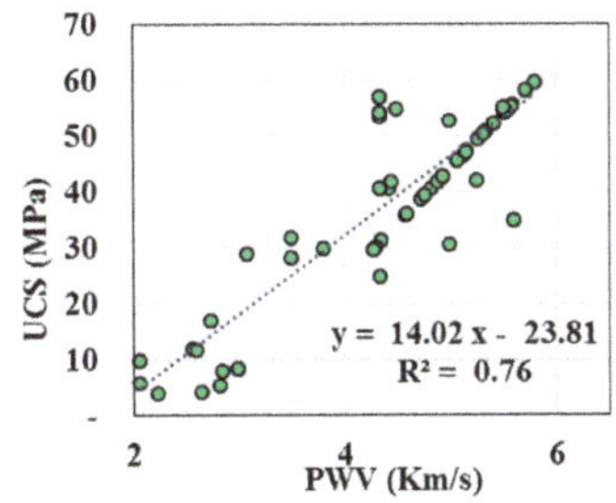

(e)

Figure 3. Effect of (**a**) point load index (PLI), (**b**) water absorption by weight (WW), (**c**) %porosity, (**d**) density (D), and (**e**) P−wave velocity (PWV) on the UCS.

3.4. Evaluation of Previous Emperical Relationships

For predicting the UCS of the rocks, some empirical relationships have been proposed (Table 1). In the current research, for each of the 65 samples of the present study, based on each of the empirical relationships in Table 1, UCS was predicted. Finally, the accuracy of the forecasted and actual UCS relationships were assessed.

Figure 4 displays the data scattering and the precision of correlation using PI and R^2. The results revealed that there is good compatibility between actual UCS and the estimated one using previous studies (Figure 4). A performance index (PI) was introduced by Yagiz et al. [85] for evaluating empirical equations and models. The value of this index is equal to two in the best case, and the lower it is, the lower the relationship performance. As can be seen, although the correlation coefficient is high, the performance index is negative, which indicates the poor performance of the previous researchers' relationships in estimating the UCS of the studied rocks (Figure 4). For this reason, various researchers have emphasized that empirical relationships should be determined for each region [85].

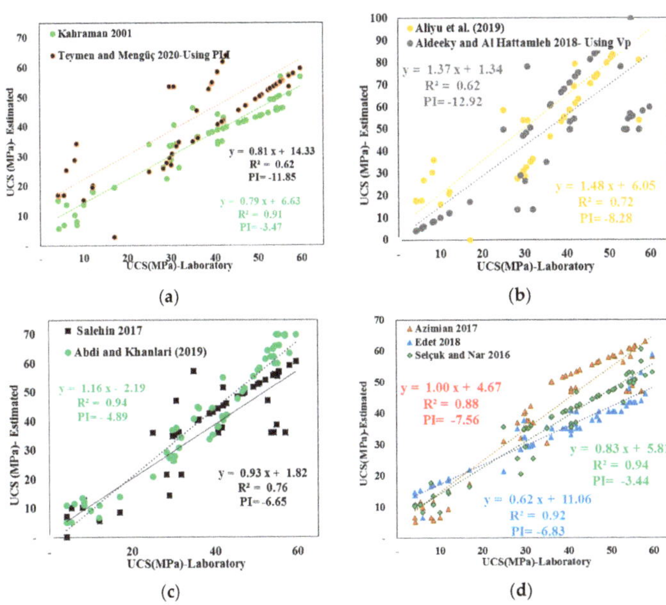

Figure 4. Measured UCS versus predicted UCS based on relationships of (**a**) Kahraman [43] and Teymen and Mengüç [40], (**b**) Aliyu et al. [30], Aldeeky and Al Hattamleh [42], (**c**) Salehin [41], Abdi and Khanlari [4], (**d**) Azimian [29], Edet [3], and Selcuk and Nar [31].

The type of rock, strength amount, method of conducting experiments, the test conditions (such as loading rate), and the petrography of the specimens of a specific study reveal the applicability of the proposed relationships for forecasting the UCS of rocks in other regions. For example, the average UCS of the samples of Edet [3] study was 32.22 MPa, and the average UCS of the present study is 37.54 MPa, which shows that the resistance of the Edet [3] study samples is lower than the resistance of the current research samples. As a result, the UCS values estimated from this researcher's relationship are mostly below the diagonal line (Figure 4d). The sample breaks faster and shows more resistance when the loading rate is increased.

3.5. Multiple Linear Regression (MPLR)

In the current work, MPLR analysis was performed using Minitab software (version 18). Equation (18) was developed to predict the UCS using this method. Various criteria to evaluate this relationship are presented below.

Various statistics (Tables 4 and 5) were used to evaluate relationship 18. The Durbin–Watson statistic DWS) and variance inflation factor (VIF) are used to evaluate the independence of errors and the correlation of independent variables, respectively [86]. The results showed that there is no problem in using relationships in terms of these two criteria because the DW is in the range of 1.5 to 2 and the VIF value is less than 10 (Table 5). Analysis of variance (ANOVA) results (Sig. < 0.00) show that the model has been well developed using MPLR. Sig. values (related to T-test) in Table 5 indicate the presence of variables in the multivariate regression output relationship. The constant value, density, and water absorption were removed from Equation (18) because the sig. value is more than 0.05.

Table 4. Multiple linear regression results to estimate UCS and Es.

Equation	R^2	RMSE (MPa)	MAPE%	VAF %	PI	DWS	ANOVA Results	Eq. No.
$UCS = 5.03PWV - 1.735\phi + 2.667PLI$	0.88	1.10	1.08	87.85	0.66	1.93	F-value = 79.37 p-value = 0.00	(18)

Table 5. Evaluation criteria of coefficients for relationship 18.

Term	Coefficients	T-Value	Significant Level (Sig.)	VIF (Variance Inflation Factor)
Constant	−32.1	−1.34	0.187	-
PWV	5.03	2.44	0.018	7.58
D	21.4	1.82	0.074	3.02
WW	0.281	0.35	0.728	3.81
ϕ	−1.735	−3.97	0.000	3.64
PLI	2.667	3.05	0.003	3.77

The normality of the error distribution is also one of the other criteria for evaluating empirical relationships. The normal distribution of errors related to the model provided by MPLR method shows that the proposed model can be used to estimate UCS (Figure 5).

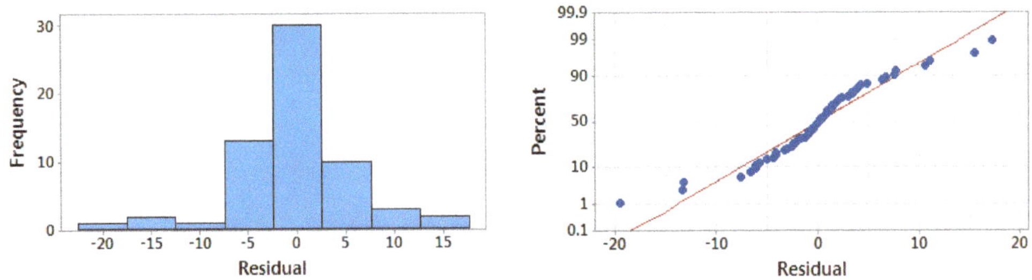

(a) (b)

Figure 5. Normal state of the residues of model 18 (**a**) histogram of residuals, and (**b**) normal probability plot.

3.6. The Results of Modeling Using RFA and GPR-SEK Methods

The RFA modeling was conducted using the R (version R4.2.1) software [54]. The GPR-SEK model was conducted using MATLAB software (MATLAB 2016b). In the RF method, the 10-fold cross-validation method was used to control the number of chosen parameters in each node of tree (m-try) and the number of trees (n-tree). According to this method, the number of 500 trees and five variables in each node has delivered the most satisfactory conditions for modeling. Therefore, these values were used for modeling purposes.

The random forest method works well for large amounts of data and has high accuracy. In the random forest method, because the amount of error decreases with the increase of trees, 500 trees were used to develop the model. Upon model execution, the results were evaluated using an out-of-bag (OOB) error estimation. The model was appraised by the test data, the results of which are presented in Figure 6. One of the advantages of the random forest algorithm is that it can determine the importance of variables in a problem. In this research, the significance of the inputs was achieved using the Gini significance index [54]. The results showed that porosity has higher importance than other parameters. In Figure 6, the error histogram, the graph of the measured, the forecasted UCS using the RFA method, and the GPR-SEK model are drawn. The GPR-SEK model was implemented based on the squared exponential kernel function. As can be seen in the figure, the results are close to the bisector line, and it can be said that the values have been predicted with good accuracy. Theoretically, if R^2 equals 100%, all the observed values will be similar to the fitted values and all the data points will be on the fitted line [87].

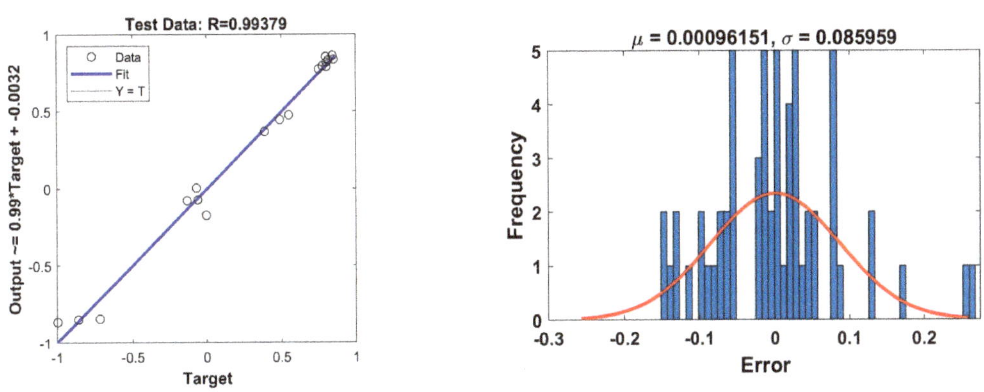

(a) (b)

Figure 6. *Cont.*

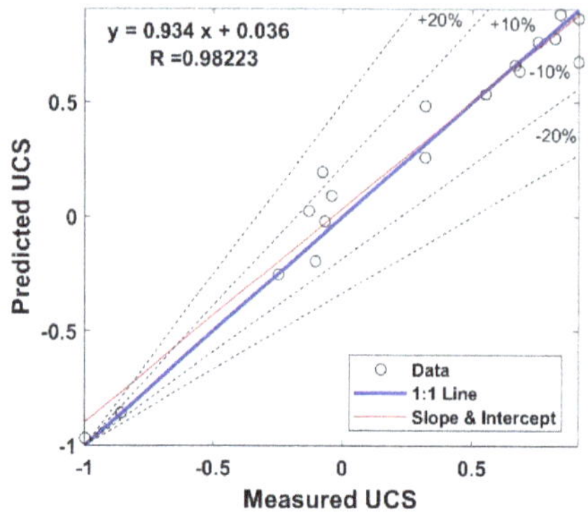

(**c**)

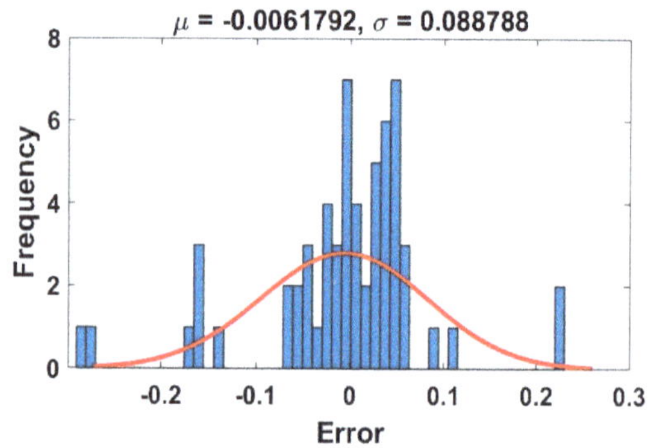

(**d**)

Figure 6. Results of (**a**) RFA for test data, (**b**) RFA for all data, (**c**) GPR-SEK for test data, and (**d**) GPR-SEK for all data.

3.7. The FMP-ANN Results

The FMP-ANN is widely used in engineering [88,89]. In the current study, for predicting UCS, various neurons in a hidden layer were investigated to develop optimal models. Based on equations proposed by previous researchers, the number of hidden layer neurons changes were determined (Table 6). The calculated number of hidden layer neurons changed from one to eleven according to Table 6. In this study, by checking this range using the FMP-ANN, this range was evaluated to achieve the ideal model architecture for forecasting UCS.

Table 6. Proposed equations by previous researchers to estimate the number of hidden layer neurons.

References	Neuron Numbers Calculated for This Study	Equations
Hecht-Nielsen [90]	≤3	$\leq 2 * N_i + 1$
Hush [91]	3	$3N_i$
Ripley [92]	3	$(N_i + N_O)/2$
Paola [93]	11	$\frac{2 + N_i * N_0 + 0.5N_0 * \left(N_0^2 + N_i\right) - 3}{(N_i + N_0)}$
Wang [94]	1	$2N_i/3$
Kaastra and Boyd [95]	2	$\sqrt{N_O * N_i}$
Kanellopoulos and Wilkinson [96]	1	$2N_i$

N_0 and N_i are the numbers of input and output neurons, respectively.

The used FMP-ANN method has a hidden layer with five inputs (PWV, point load index (PLI), porosity, density, and water absorption) and one output (UCS). Using MATLAB software, the Levenberg Marquardt (LM) training algorithm was used to train the network. The neuron transfer functions were the selected Sigmoid between the input layers and hidden layers and the Purelin between the hidden layers and output layers. In FMP-ANN modeling, the percentages of the validation, test, and training data in the present study were randomly selected as 15%, 15%, and 70% of the total data, respectively. The validation set is used to prevent overfitting, the training group is used to determine weights, and the test group is used to evaluate the FMP-ANN results [97–100]. The results showed that the third neuron is the most accurate neuron for estimating UCS. Figure 7 displays the optimal FMP-ANN chart achieved in the current research.

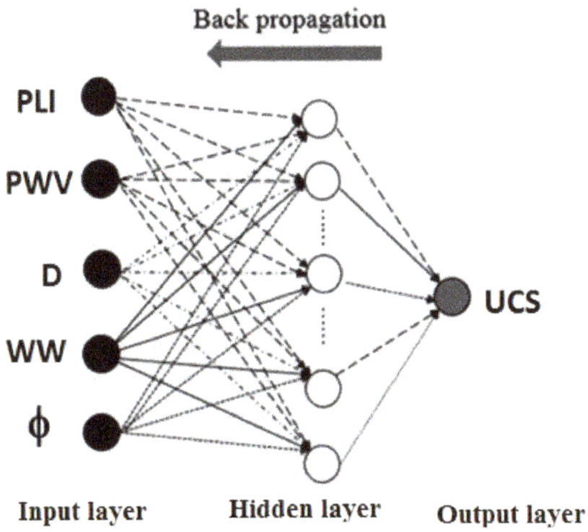

Figure 7. Used FMP-ANN structure.

Figure 8 shows the error variations in optimum results. The lowest error in epoch 4 was obtained for predicting the UCS (Figure 8). Moreover, in this research, the results of the FMP-ANN to estimate UCS have been compared with several methods. It was found that the accuracy of all methods was very high (the coefficient of determination is more than 97%).

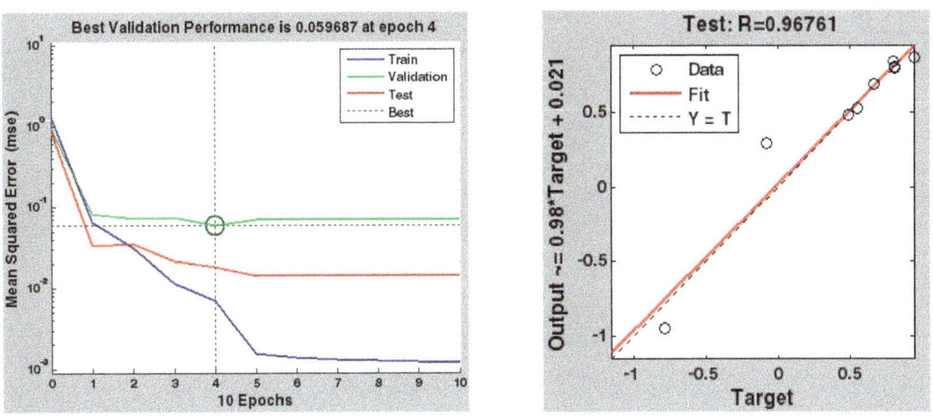

(**a**) (**b**)

Figure 8. The FMP−ANN results for the optimum model (**a**) error reduction trend and (**b**) corelation cofficient between measured and predicted UCS.

3.8. The KNNA Results

To apply the KNNA method to the data and determine the best K value, the KNNA was written in the form of a program in MATLAB software, which was run 310 times for K values from 1 to 30 programs; moreover, the amount of error was measured (Figure 9). Of the total data, 70% and 30% were used to train and test the model. The results showed that the lowest estimation error of the UCS was obtained at K = 2 (Figure 9). The error of this network for estimating the UCS with respect to the K values is equal to 0.11 (Figure 10). Figure 10 shows the KNNA results for estimating the UCS.

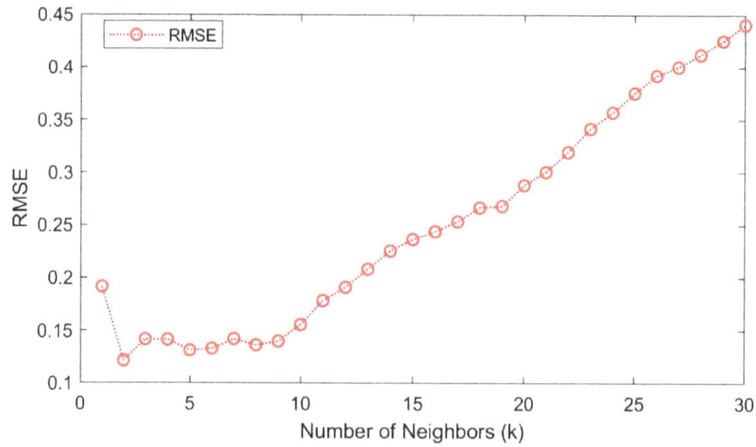

Figure 9. Obtained RMSE for forecasting UCS based on the KNNA for different values of K.

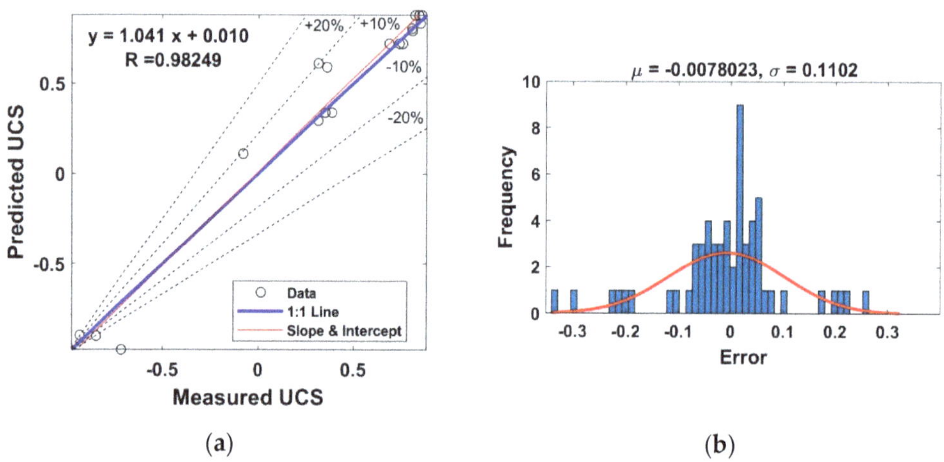

Figure 10. Accuracy of predicted UCS using KNNA: (**a**) correlation coefficient for test data and (**b**) error histogram for all data.

3.9. Results of SVM Method for Estimating UCS

The SVM algorithm uses a set of mathematical functions that are named kernels [101]. The most important kernel functions for solving engineering problems are listed in Table 7. Normally, three radial basis kernel functions (RBFs), polynomial of degree, d, and linear, are used in the support vector machine, and the use of each of these functions with different parameters in the estimation of rock strength may lead to different results [60,101]. Therefore, it is necessary to evaluate the efficiency and accuracy of each of these functions and to choose the appropriate kernel function in predicting resistance. These three kernel functions were also used in this research. It should be mentioned that the calculation process of SVM was performed based on coding in a MATLAB environment and that the parameters of the kernel functions were optimized using a trial and error process. The results of these investigations are presented in Table 8. It can be observed that, based on the statistical criteria, the accuracy of the kernel functions is as RBF> PK > LK. In this regard, Nguyen [102] investigated the performance of various kernel functions using the support vector machine method in estimating blast-induced ground vibration and stated that the radial basis function has the highest performance.

Table 7. The most important kernel functions for solving engineering problems [102].

Function	Description	Kernel Function Type
$k\left(x_i, x_j\right) = (x_i.x_j + 1)^d$	This kernel is widely used in image processing, where d is the degree of the polynomial.	Polynomial kernel (PK)
$k\left(x_i, x_j\right) = \exp(-\gamma\|x_i - x_j\|^2)$	This kernel is used for general purposes. It is used when there is no prior knowledge about the data. In $\gamma > 0$ condition, $\gamma = 1/2\sigma^2$ parameter is used.	Radial basis function (RBF)
$k\left(x_i, x_j\right) = x_i.x_j$	-	Linear kernel (LK)

Table 8. Evaluation of SVM model performance in UCS estimation using various kernel functions.

Kernel Function	Optimal Values of Parameters					Test Period				Train Period			
	ε	t	d	σ	c	RMSE	R^2	PI	MAPE	RMSE	R^2	PI	MAPE
PK	1.72	280.01	4	-	12.12	0.08	0.97	1.87	2.86	0.07	0.98	1.84	2.81
RBF	0.02	-	-	1.10	27	0.06	0.99	1.90	2.82	0.06	0.99	1.90	2.80
LK	0.45	-	-	-	0.90	0.09	0.96	1.83	2.84	0.09	0.97	1.81	

The error histogram and predicted and estimated UCS relationship with the optimal function (RBF function) are presented in Figure 11.

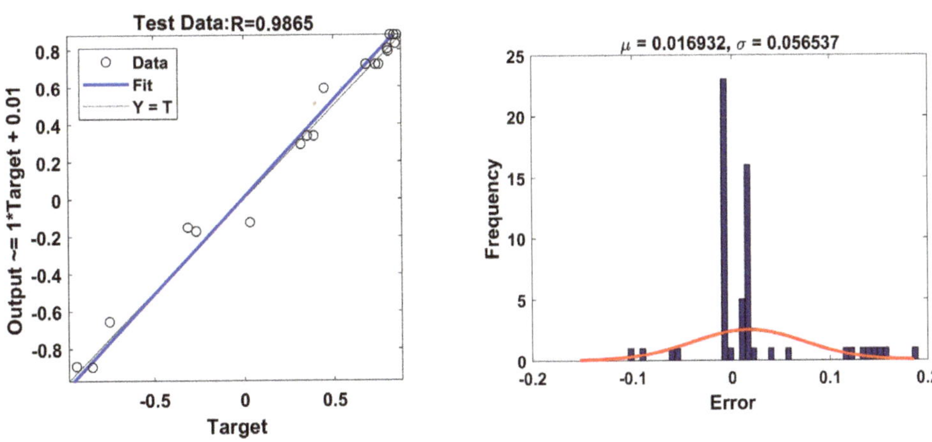

(a) (b)

Figure 11. Accuracy of predicted UCS using SVM-RBF: (**a**) correlation coefficient and (**b**) error histogram for all data.

3.10. Results of ANFIS Method for Estimating UCS

As with other intelligent methods, to test and train models using ANFIS, 30% and 70% of the whole data were used, respectively. The method of combining regression error propagation with least squares was used to train the model using the ANFIS. Table 9 shows the modeling features using the ANFIS method. A comparison of the performance of the methods for forecasting UCS based on different criteria has been reported in the next section.

Table 9. Modeling features using ANFIS.

FIS Generation Method	GENFIS4
Influence radius	0.60
Number of epochs	500
Error goal	0.00
Type	Sugeno
Rules	4
Number of membership functions (MFs)	6
Input MF type	Gauss MF
Output MF type	Linear

Figure 12 shows the error histogram and correlation coefficient of the ANFIS model in the test stage. This method, as with other used intelligent methods, has high accuracy in UCS estimation. The results of the intelligent models for estimating UCS from the test data performed better than the training data; therefore, it can be argued that overfitting did not occur.

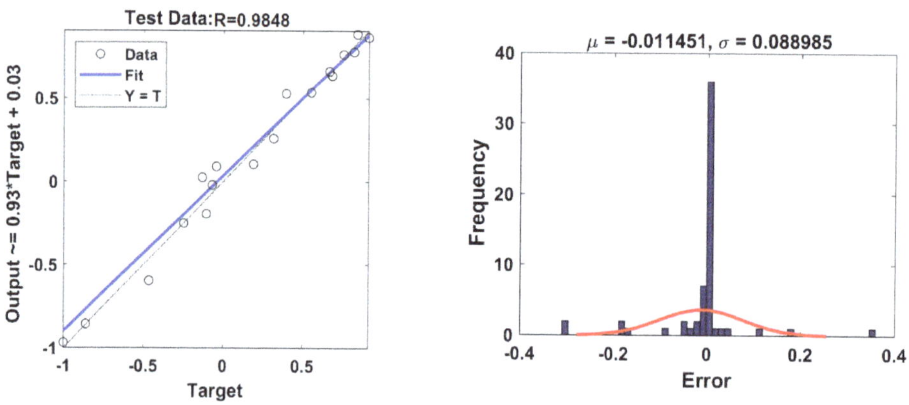

(**a**) (**b**)

Figure 12. Accuracy of predicted UCS using ANFIS: (**a**) correlation coefficient and (**b**) error histogram for all data.

3.11. *Evaluation of the Used Methods*

Table 10 and Figure 13 show the accuracy of the used methods for forecasting the UCS. According to the statistical criteria (i.e., R^2, MAPE%, RMSE, VAF, and PI), the SVM-RBF model displays greater precision than other models because the SVM uses the minimizing structural risk theorem and adapts the ability of the model to existing training data [103]. The number of input variables, number of samples, and training algorithm type also affect the accuracy of the methods [16,104,105]. Based on the correlation coefficient, all methods (R^2 > 90%) have excellent accuracy for estimating UCS.

Considering that all six intelligent methods showed very high accuracy in UCS estimation, the percentage difference of the average of all six intelligent methods with the measured value in the laboratory is equal to +0.28%. This amount of difference is less than 1% and indicates the high capability of intelligent methods for forecasting the UCS.

Table 10. Accuracy of approaches for predicting UCS.

APPROACHES	MAPE%	R^2	RMSE	VAF%	PI
RFA	9.27	0.98	0.09	97.63	1.87
SVM-RBF	2.83	0.99	0.06	98.96	1.92
ANFIS	2.98	0.98	0.09	97.86	1.87
KNNA	8.44	0.97	0.11	97.25	1.83
GPR-SEK	6.63	0.98	0.09	97.45	1.86
FMP-ANN	4.66	0.99	0.24	98.36	1.73

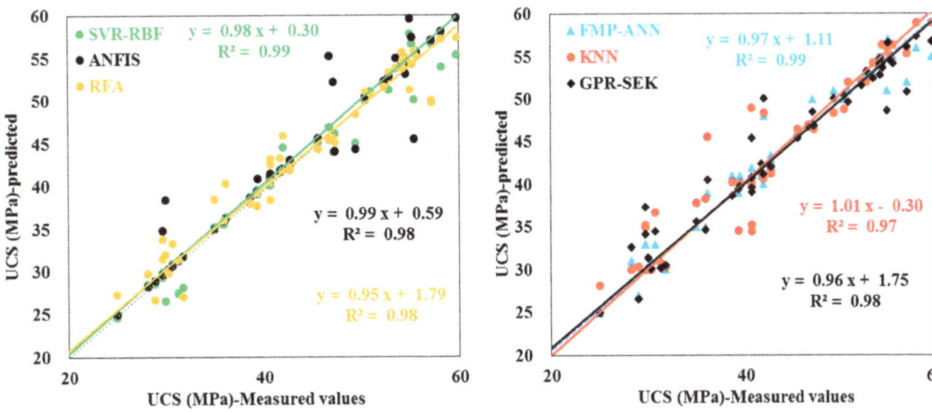

(a) (b)

Figure 13. Measured values versus predicted UCS using (**a**) SVR−RBF, ANFIS, and RFA methods, and (**b**) FMP−ANN, KNN, and GPR−SEK methods.

4. Conclusions

The UCS of rocks is a basic parameter necessary for assessing the construction of civil and mining structures, such as the stability of the mines and the bearing capacity of foundations. UCS estimation using core specimens is costly, difficult, and, in some cases, impossible. After assessing the geo-mechanical features of 55 samples of sandstone, limestone, and argillaceous limestone specimens, predictive models for estimating UCS were developed via intelligent and statistical approaches. The results showed that the carbonate and sandstone samples were classified as mudstone to grainstone and calc-litharenite, respectively. The PWV, WW, porosity, density, and PLI were considered as model inputs for predicting UCS. Statistical analysis allowed the development of equations with high accuracy to estimate UCS. Among the assessed linear, polynomial, and radial basis kernel functions, the accuracy of the other models was lower than that of SVM-RBF in forecasting UCS. The SVM-RBF model revealed that the R^2 and PI values were 0.99 and 1.92, respectively. The R^2 values of 98%, 98%, 97%, 98%, and 99% for forecasting the UCS were achieved using ANFIS, RFA, KNNA, GPR, and FMP-ANN, respectively. The number of samples and input variables had a significant impact on the performance of the methods. When the number of samples was small, the SVM method was more accurate. The percentage difference of the average of all six intelligent methods with the measured value was less than 1%, which indicates the superior capability of the intelligent methods in forecasting UCS.

Author Contributions: X.Z.: methodology, software, data curation. F.M.A.A.: Conceptualization, investigation. T.A.S.G.: writing—original draft preparation, methodology. A.H.D.A.-K.: analysis of results, validation. A.I.: performing field investigations and collecting samples R.B.Y.S.: writing—original draft preparation, resources. M.L.N.: supervision, project administration, funding acquisition. All authors have read and agreed to the published version of the manuscript.

Funding: This research received no external funding.

Institutional Review Board Statement: Not applicable.

Informed Consent Statement: The authors are fully aware and satisfied with the contents of the article.

Data Availability Statement: The data used in this study has been appropriately described in the manuscript.

Conflicts of Interest: The authors declare that they have no conflict of interest.

References

1. Yang, H.Q.; Zeng, Y.Y.; Lan, Y.F.; Zhou, X.P. Analysis of the excavation damaged zone around a tunnel accounting for geo-stress and unloading. *Int. J. Rock Mech. Min. Sci.* **2014**, *69*, 59–66. [CrossRef]
2. Yang, H.Q.; Xing, S.G.; Wang, Q.; Li, Z. Model test on the entrainment phenomenon and energy conversion mechanism of flow-like landslides. *Eng. Geol.* **2018**, *239*, 119–125. [CrossRef]
3. Edet, A. Correlation between Physico-mechanical Parameters and Geotechnical Evaluations of Some Sandstones along the Calabar/Odukpani–Ikom–Ogoja Highway Transect, Southeastern Nigeria. *Geotech. Geol. Eng.* **2018**, *36*, 135–149. [CrossRef]
4. Abdi, Y.; Khanlari, G.R. Estimation of mechanical properties of sandstones using P-wave velocity and Schmidt hardness. *New Find. Appl. Geol.* **2019**, *13*, 33–47.
5. Ajalloeian, R.; Mansouri, H.; Baradaran, E. Some carbonate rock texture effects on mechanical behavior, based on Koohrang tunnel data, Iran. *Bull. Eng. Geol. Environ.* **2017**, *76*, 295–307. [CrossRef]
6. Madhubabu, N.; Singh, P.K.; Kainthola, A.; Mahanta, B.; Tripathy, A.; Singh, T.N. Prediction of compressive strength and elastic modulus of carbonate rocks. *Measurement* **2016**, *88*, 202–213. [CrossRef]
7. Wen, L.; Luo, Z.Q.; Yang, S.J.; Guang, Y.Q.; Wang, W. Correlation of Geo-Mechanics Parameters with Uniaxial Compressive Strength and P-Wave Velocity on Dolomitic Limestone Using a Statistical Method. *Geotech. Geol. Eng.* **2018**, *37*, 1079–1094. [CrossRef]
8. Aladejare, A.E. Evaluation of empirical estimation of uniaxial compressive strength of rock using measurements from index and physical tests. *J. Rock Mech. Geotech. Eng.* **2020**, *12*, 256–268. [CrossRef]
9. Lawal, A.I.; Kwon, S.; Aladejare, A.E.; Oniyide, G.O. Prediction of the static and dynamic mechanical properties of sedimentary rock using soft computing methods. *Geomech. Eng.* **2022**, *28*, 313–324.
10. Lawal, A.I.; Olajuyi, S.I.; Kwon, S.; Aladejare, A.E.; Edo, T.M. Prediction of blast-induced ground vibration using GPR and blast-design parameters optimization based on novel grey-wolf optimization algorithm. *Acta Geophys.* **2021**, *69*, 1313–1324. [CrossRef]
11. Momeni, E.; Dowlatshahi, M.B.; Omidinasab, F.; Maizir, H.; Armaghani, D.J. Gaussian processregression technique to estimate the pile bearing capacity. *Arab. J. Sci. Eng.* **2020**, *45*, 8255–8267. [CrossRef]
12. Gao, W.; Karbasi, M.; Hasanipanah, M.; Zhang, X.; Guo, J. Developing GPR model for forecasting the rock fragmentation in surface mines. *Eng. Comput.* **2018**, *34*, 339–345. [CrossRef]
13. Dao, D.V.; Adeli, H.; Ly, H.B.; Le, L.M.; Le, V.M.; Le, T.T.; Pham, B.T. A sensitivity and robustness analysis of GPR and ANN for high-performance concrete compressive strength prediction using a Monte Carlo simulation. *Sustainability* **2020**, *12*, 830. [CrossRef]
14. Barham, W.S.; Rabab'ah Aldeeky, S.R.; Al Hattamleh, O.H. Mechanical and Physical Based Artificial Neural Network Models for the Prediction of the Unconfined Compressive Strength of Rock. *Geotech. Geol. Eng.* **2020**, *38*, 4779–4792. [CrossRef]
15. Armaghani, D.J.; Mamou, A.; Maraveas, C.; Roussis, P.C.; Siorikis, V.G.; Skentou, A.D.; Asteris, P.G. Predicting the unconfined compressive strength of granite using only two non-destructive test indexes. *Geomech. Eng.* **2021**, *25*, 317–330.
16. Kwak, N.S.; Ko, T.Y. Machine learning-based regression analysis for estimating Cerchar abrasivity index. *Geomech. Eng.* **2022**, *29*, 219–228.
17. Gül, E.; Ozdemir, E.; Sarıcı, D.E. Modeling uniaxial compressive strength of some rocks from turkey using soft computing techniques. *Measurement* **2021**, *171*, 108781. [CrossRef]
18. Alizadeh, S.M.; Iraji, A.; Tabasi, S.; Ahmed, A.A.A.; Motahari, M.R. Estimation of dynamic properties of sandstones based on index properties using artificial neural network and multivariate linear regression methods. *Acta Geophys.* **2022**, *70*, 225–242. [CrossRef]
19. Rastegarnia, A.; Teshniz, E.S.; Hosseini, S.; Shamsi, H.; Etemadifar, M. Estimation of punch strength index and static properties of sedimentary rocks using neural networks in south west of Iran. *Measurement* **2018**, *128*, 464–478. [CrossRef]

20. McElroy, P.D.; Bibang, H.; Emadi, H.; Kocoglu, Y.; Hussain, A.; Watson, M.C. Artificial neural network (ANN) approach to predict unconfined compressive strength (UCS) of oil and gas well cement reinforced with nanoparticles. *J. Nat. Gas Sci. Eng.* **2021**, *88*, 103816. [CrossRef]
21. Wang, M.; Wan, W.; Zhao, Y. Prediction of the uniaxial compressive strength of rocks from simple index tests using a random forest predictive model. *C. R. Méc.* **2020**, *348*, 3–32. [CrossRef]
22. Matin, S.S.; Farahzadi, L.; Makaremi, S.; Chelgani, S.C.; Sattari, G.H. Variable selection and prediction of uniaxial compressive strength and modulus of elasticity by random forest. *Appl. Soft Comput.* **2018**, *70*, 980–987. [CrossRef]
23. Gamal, H.; Alsaihati, A.; Elkatatny, S.; Haidary, S.; Abdulraheem, A. Rock strength prediction in real-time while drilling employing random forest and functional network techniques. *J. Energy Resour. Technol.* **2021**, *143*, 093004. [CrossRef]
24. Umrao, R.K.; Sharma, L.K.; Singh, R.; Singh, T.N. Determination of strength and modulus of elasticity of heterogenous sedimentary rocks: An ANFIS predictive technique. *Measurement* **2018**, *126*, 194–201. [CrossRef]
25. Hudaverdi, T. Prediction of flyrock throw distance in quarries by variable selection procedures and ANFIS modelling technique. *Environ. Earth Sci.* **2022**, *81*, 281. [CrossRef]
26. Yesiloglu-Gultekin, N.; Gokceoglu, C.; Sezer, E.A. Prediction of uniaxial compressive strength of granitic rocks by various nonlinear tools and comparison of their performances. *Int. J. Rock Mech. Min. Sci.* **2013**, *62*, 113–122. [CrossRef]
27. Yesiloglu-Gultekin, N.; Gokceoglu, C.A. Comparison Among Some Non-linear Prediction Tools on Indirect Determination of Uniaxial Compressive Strength and Modulus of Elasticity of Basalt. *J. Nondestruct. Eval.* **2022**, *41*, 10. [CrossRef]
28. Fathipour-Azar, H. Machine learning-assisted distinct element model calibration: ANFIS, SVM, GPR, and MARS approaches. *Acta Geotech.* **2022**, *17*, 1207–1217. [CrossRef]
29. Azimian, A. Application of statistical methods for predicting uniaxial compressive strength of limestone rocks using nondestructive tests. *Acta Geotech.* **2017**, *12*, 321–333. [CrossRef]
30. Aliyu, M.M.; Shang, J.; Murphy, W.; Lawrence, J.A.; Collier, R.; Kong, F.; Zhao, Z. Assessing the uniaxial compressive strength of extremely hard cryptocrystalline flint. *Int. J. Rock Mech. Min. Sci.* **2019**, *113*, 310–321. [CrossRef]
31. Selçuk, L.; Nar, A. Prediction of uniaxial compressive strength of intact rocks using ultrasonic pulse velocity and rebound-hammer number. *Q. J. Eng. Geol. Hydrogeol.* **2016**, *49*, 67–75. [CrossRef]
32. Mahmoodzadeh, A.; Mohammadi, M.; Ali, H.F.H.; Abdulhamid, S.N.; Ibrahim, H.H.; Noori, K.M.G. Dynamic prediction models of rock quality designation in tunneling projects. *Transp. Geotech.* **2021**, *27*, 100497.
33. Xu, C.; Amar, M.N.; Ghriga, M.A.; Ouaer, H.; Zhang, X.; Hasanipanah, M. Evolving support vector regression using Grey Wolf optimization; forecasting the geomechanical properties of rock. *Eng. Comput.* **2020**, *38*, 1819–1833. [CrossRef]
34. Ceryan, N. Application of support vector machines and relevance vector machines in predicting uniaxial compressive strength of volcanic rocks. *J. Afr. Earth Sci.* **2014**, *100*, 634–644. [CrossRef]
35. Trott, M.; Matthew, L.; Lindsay, H.; Daniel, L.M. Random forest rock type classification with integration of geochemical and photographic data. *Appl. Comput. Geosci.* **2022**, *15*, 100090. [CrossRef]
36. Barzegar, R.; Sattarpour, M.; Nikudel, M.R.; Moghaddam, A.A. Comparative evaluation of artificial intelligence models for prediction of uniaxial compressive strength of travertine rocks, case study: Azarshahr area, NW Iran. *Model. Earth Syst. Environ.* **2016**, *2*, 76. [CrossRef]
37. Mohamad, E.T.; Armaghani, D.J.; Momeni, E.; Abad, S. Prediction of the unconfined compressive strength of soft rocks: A PSO-based ANN approach. *Bull. Eng. Geol. Environ.* **2015**, *74*, 745–757. [CrossRef]
38. Singh, R.; Umrao, R.K.; Ahmad, M.; Ansari, M.K.; Sharma, L.K.; Singh, T.N. Prediction of geomechanical parameters using soft computing and multiple regression approach. *Measurement* **2017**, *99*, 108–119. [CrossRef]
39. Kaloop, M.R.; Bardhan, A.; Samui, P.; Hu, J.W.; Zarzoura, F. Computational intelligence approaches for estimating the unconfined compressive strength of rocks. *Arab. J. Geosci.* **2023**, *16*, 37. [CrossRef]
40. Teymen, A.; Mengüç, E.C. Comparative evaluation of different statistical tools for the prediction of uniaxial compressive strength of rocks. *Int. J. Min. Sci. Technol.* **2020**, *30*, 785–797. [CrossRef]
41. Salehin, S. Investigation into engineering parameters of marls from Seydoon dam in Iran. *J. Rock Mech. Geotech. Eng.* **2017**, *9*, 912–923. [CrossRef]
42. Aldeeky, H.; Al Hattamleh, O. Prediction of engineering properties of basalt rock in Jordan using ultrasonic pulse velocity test. *Geotech. Geol. Eng.* **2018**, *36*, 3511–3525. [CrossRef]
43. Kahraman, S. Evaluation of simple methods for assessing the uniaxial compressive strength of rock. *Int. J. Rock Mech. Min. Sci.* **2001**, *38*, 981–994. [CrossRef]
44. Uyanık, O.; Sabbağ, N.; Uyanık, N.A.; Öncü, Z. Prediction of mechanical and physical properties of some sedimentary rocks from ultrasonic velocities. *Bull. Eng. Geol. Environ.* **2019**, *78*, 6003–6016. [CrossRef]
45. Kılıç, A.; Teymen, A. Determination of mechanical properties of rocks using simple methods. *Bull. Eng. Geol. Environ.* **2008**, *67*, 237–244. [CrossRef]
46. *ASTM D2938-95*; Standard Test Method for Unconfined Compressive Strength of Intact Rock Core Specimens. ASTM International: West Conshohocken, PA, USA, 2002.
47. ISRM. *Rock Characterization Testing and Monitoring*; Brown, E.T., Ed.; ISRM Suggested Methods; Pergamon Press: Oxford, UK, 1981; Volume 211.

48. *ASTM D5731*; Standard Test Method for Determination of the Point Load Strength Index of Rock. ASTM International: West Conshohocken, PA, USA, 2002.
49. *ASTM D2845*; Test Methods for Ultra Violet Velocities Determination. ASTM International: West Conshohocken, PA, USA, 1983.
50. Folk, R.L. *Petrology of Sedimentary Rocks*; Hemphill: Austin, TX, USA, 1974; 600p.
51. Dunham, R.J. *Classification of Carbonate Rocks According to Depositional Textures*; American Association of Petroleum Geologists: Tulsa, OK, USA, 1962; pp. 108–121.
52. Zhou, J.; Huang, S.; Qiu, Y. Optimization of random forest through the use of MVO, GWO and MFO in evaluating the stability of underground entry-type excavations. *Tunn. Undergr. Space Technol.* **2022**, *124*, 104494. [CrossRef]
53. Mohri, M.; Rostamizadeh, A.; Talwalkar, A. Foundations of Machine Learning. In *Adaptive Computation and Machine Learning Series*; MIT Press: Cambridge, MA, USA, 2018.
54. Liaw, A.; Wiener, M. Classification and Regression by Random Forest. *R News* **2002**, *2*, 18–22.
55. Harris, J.; Grunsky, E.C. Predictive lithological mapping of Canada's North using Random Forest classification applied to geophysical and geochemical data. *Comput. Geosci.* **2015**, *80*, 9–25. [CrossRef]
56. James, G.; Witten, D.; Hastie, T.; Tibshirani, R. *An Introduction to Statistical Learning*; Springer: Berlin/Heidelberg, Germany, 2013; Volume 112, 607p.
57. Zhu, B.; Zhong, Q.; Chen, Y.; Liao, S.; Li, Z.; Shi, K.; Sotelo, M.A. A Novel Reconstruction Method for Temperature Distribution Measurement Based on Ultrasonic Tomography. *IEEE Trans. Ultrason. Ferroelectr. Freq. Control* **2022**, *69*, 2352–2370. [CrossRef]
58. Raja, M.N.A.; Jaffar, S.T.A.; Bardhan, A.; Shukla, S.K. Predicting and validating the load-settlement behavior of large-scale geosynthetic-reinforced soil abutments using hybrid intelligent modeling. *JRMGE* **2023**, *15*, 773–788.
59. Vapnik, V.N. *Statistical Learning Theory*; Wiley: New York, NY, USA, 1998; p. 736.
60. Fallah, M.; Pirali Zefrehei, A.R.; Hedayati, S.A.; Bagheri, T. Comparison of temporal and spatial patterns of water quality parameters in Anzali Wetland (southwest of the Caspian Sea) using Support vector machine model. *Casp. J. Environ. Sci.* **2021**, *19*, 95–104.
61. Kookalani, S.; Cheng, B. Structural Analysis of GFRP Elastic Gridshell Structures by Particle Swarm Optimization and Least Square Support Vector Machine Algorithms. *J. Civ. Eng. Mater. Appl.* **2021**, *5*, 12–23.
62. Yang, H.; Wang, Z.; Song, K. A new hybrid grey wolf optimizer-feature weighted-multiple kernel-support vector regression technique to predict TBM performance. *Eng. Comput.* **2020**, *38*, 2469–2485. [CrossRef]
63. Yang, H.; Song, K.; Zhou, J. Automated Recognition Model of Geomechanical Information Based on Operational Data of Tunneling Boring Machines. *Rock Mech. Rock Eng.* **2022**, *55*, 1499–1516. [CrossRef]
64. Duda, R.O.; Hart, P.E.; Stork, D.G. *Pattern Classification*; John Wiley & Sons: Hoboken, NJ, USA, 2012; Volume 3, pp. 731–739.
65. Tharwat, A.; Ghanem, A.M.; Hassanien, A.E. Three different classifiers for facial age estimation based on k-nearest neighbor. In Proceedings of the 2013 9th International Computer Engineering Conference (ICENCO), Giza, Egypt, 28–29 December 2013; IEEE: Piscataway, NJ, USA, 2013; pp. 55–60.
66. Aghighi, F.; Aghighi, H.; Ebadati, O.M. Evaluation of the efficiency of SVM and KNN Classification algorithms to extract urban effects from LiDAR cloud points. In *Second International Conference on Knowledge-Based Research in Computer Engineering & Information Technology, Tehran, Iran, 30 September 2016*; Majlisi University: Mobarakeh, Iran, 2017. (In Persian)
67. Saed, S.A.; Kamboozia, N.; Ziari, H.; Hofko, B. Experimental assessment and modeling of fracture and fatigue resistance of aged stone matrix asphalt (SMA) mixtures containing RAP materials and warm-mix additive using ANFIS method. *Mater. Struct.* **2021**, *54*, 225. [CrossRef]
68. Sobhani, B.; Safarianzengir, V. Monitoring and prediction of drought using TIBI fuzzy index in Iran. *Casp. J. Environ. Sci.* **2020**, *18*, 237–250.
69. Jang, J.S.R. ANFIS: Adaptive network based fuzzy inference system. *IEEE Trans. Syst. Man Cybern.* **1993**, *23*, 665–685. [CrossRef]
70. Moshahedi, A.; Mehranfar, N.A. Comprehensive Design for a Manufacturing System using Predictive Fuzzy Models. *J. Res. Sci. Eng. Technol.* **2021**, *9*, 1–23. [CrossRef]
71. Mokhberi, M.; Khademi, H. The use of stone columns to reduce the settlement of swelling soil using numerical modeling. *J. Civ. Eng. Mater. Appl.* **2017**, *1*, 45–60. [CrossRef]
72. Rastegarnia, A.; Lashkaripour, G.R.; Sharifi Teshnizi, E.; Ghafoori, M. Evaluation of engineering characteristics and estimation of dynamic properties of clay-bearing rocks. *Environ. Earth Sci.* **2021**, *80*, 621. [CrossRef]
73. Mikaeil, R.; Esmaeilzade, A.; Shaffiee Haghshenas, S. Investigation of the Relationship Between Schimazek's F-Abrasiveness Factor and Current Consumption in Rock Cutting Process. *J. Civ. Eng. Mater. Appl.* **2021**, *5*, 47–55.
74. Keykhah, H.; Dahan Zadeh, B. Stability Analysis of Upstream Slope of Earthen Dams Using the Finite Element method Against Sudden Change in the Water Surface of the Reservoir, Case Study: Ilam Earthen Dam in Ilam Province. *J. Civ. Eng. Mater. Appl.* **2018**, *2*, 24–30.
75. Taheri, S.; Ziad, H. Analysis and Comparison of Moisture Sensitivity and Mechanical Strength of Asphalt Mixtures Containing Additives and Carbon Reinforcement. *J. Civ. Eng. Mater. Appl.* **2021**, *5*, 1–8.
76. Sobhani, J.; Jafarpour, F.; Firozyar, F.; Pourkhorshidi, A.R. Simulated C3A Effects on the Chloride Binding in Portland Cement with NaCl and CaCl2 Cations. *J. Civ. Eng. Mater. Appl.* **2022**, *6*, 41–54.

77. Liu, B.; Yang, H.; Karekal, S. Effect of water content on argillization of mudstone during the tunneling process. *Rock Mech. Rock Eng.* **2020**, *53*, 799–813. [CrossRef]
78. Guo, Y.; Luo, L.; Wang, C. Research on Fault Activation and Its Influencing Factors on the Barrier Effect of Rock Mass Movement Induced by Mining. *Appl. Sci.* **2023**, *13*, 651. [CrossRef]
79. Yang, H.Q.; Li, Z.; Jie, T.Q.; Zhang, Z.Q. Effects of joints on the cutting behavior of disc cutter running on the jointed rock mass. *Tunn. Undergr. Space Technol.* **2018**, *81*, 112–120. [CrossRef]
80. Peng, J.; Xu, C.; Dai, B.; Sun, L.; Feng, J.; Huang, Q. Numerical Investigation of Brittleness Effect on Strength and Microcracking Behavior of Crystalline Rock. *Int. J. Geomech.* **2022**, *22*, 4022178. [CrossRef]
81. Ghavami, S.; Rajabi, M. Investigating the Influence of the Combination of Cement Kiln Dust and Fly Ash on Compaction and Strength Characteristics of High-Plasticity Clays. *J. Civ. Eng. Mater. Appl.* **2021**, *5*, 9–16.
82. Xiao, D.; Hu, Y.; Wang, Y.; Deng, H.; Zhang, J.; Tang, B.; Li, G. Wellbore cooling and heat energy utilization method for deep shale gas horizontal well drilling. *Appl. Therm. Eng.* **2022**, *213*, 118684. [CrossRef]
83. Rastegarnia, A.; Lashkaripour, G.R.; Ghafoori, M.; Farrokhad, S.S. Assessment of the engineering geological characteristics of the Bazoft dam site, SW Iran. *Q. J. Eng. Geol. Hydrogeol.* **2019**, *52*, 360–374. [CrossRef]
84. Kurtulus, C.; Bozkurt, A.; Endes, H. Physical and mechanical properties of serpentinized ultrabasic rocks in NW Turkey. *Pure Appl. Geophys.* **2012**, *169*, 1205–1215. [CrossRef]
85. Yagiz, S.; Sezer, E.A.; Gokceoglu, C. Artificial neural networks and nonlinear regression techniques to assess the influence of slake durability cycles on the prediction of uniaxial compressive strength and modulus of elasticity for carbonate rocks. *Int. J. Numer. Anal. Methods Géoméch.* **2012**, *36*, 1636–1650. [CrossRef]
86. Shirnezhad, Z.; Azma, A.; Foong, L.K.; Jahangir, A.; Rastegarnia, A. Assessment of water resources quality of a karstic aquifer in the Southwest of Iran. *Bull. Eng. Geol. Environ.* **2021**, *80*, 71–92. [CrossRef]
87. Shayesteh, A.; Ghasemisalehabadi, E.; Khordehbinan, M.W.; Rostami, T. Finite element method in statistical analysis of flexible pavement. *J. Mar. Sci. Technol.* **2017**, *25*, 15.
88. Rustamovich Sultanbekov, I.; Yurievna Myshkina, I.; Yurievna Gruditsyna, L. Development of an application for creation and learning of neural networks to utilize in environmental sciences. *Casp. J. Environ. Sci.* **2020**, *18*, 595–601.
89. Tabatabaei, M.; Salehpour Jam, A. Optimization of sediment rating curve coefficients using evolutionary algorithms and unsupervised artificial neural network. *Casp. J. Environ. Sci.* **2017**, *15*, 385–399.
90. Hecht-Nielsen, R. Kolmogorov's mapping neural network existence theorem. In Proceedings of the International Conference on Neural Networks, San Diego, CA, USA, 21–24 June 1987; IEEE Press: New York, NY, USA, 1987; Volume 3, pp. 11–14.
91. Hush, D. Classification with neural networks: A performance analysis. In Proceedings of the IEEE International Conference on Systems Engineering, Fairborn, OH, USA, 24–26 August 1989; IEEE: Piscataway, NJ, USA, 1989; pp. 277–280.
92. Ripley, B.D. *Statistical Aspects of Neural Networks*; Barndorff-Nielsen, O.E., Jensen, J.L., Kendall, W.S., Eds.; Networks and Chaos–Statistical and Probabilistic Aspects; Chapman and Hall: London, UK, 1993; pp. 40–123.
93. Paola, J.D. *Neural Network Classification of Multispectral Imagery*; The University of Arizona: Tucson, AZ, USA, 1994.
94. Wang, C. A Theory of Generalization in Learning Machines with Neural Application. Ph.D. Thesis, The University of Pennsylvania, Philadelphia, PA, USA, 1994.
95. Kaastra, I.; Boyd, M. Designing a neural network for forecasting financial and economic time series. *Neurocomputing* **1996**, *10*, 215–236. [CrossRef]
96. Kanellopoulos, I.; Wilkinson, G.G. Strategies and best practice for neural network image classification. *Int. J. Remote Sens.* **1997**, *18*, 711–725. [CrossRef]
97. Kavyanifar, B.; Tavakoli, B.; Torkaman, J.; Mohammad Taheri, A.; Ahmadi Orkomi, A. Coastal solid waste prediction by applying machine learning approaches (Case study: Noor, Mazandaran Province, Iran). *Casp. J. Environ. Sci.* **2020**, *18*, 227–236.
98. Rashidi Tazhan, O.; Pir Bavaghar, M.; Ghazanfari, H. Detecting pollarded stands in Northern Zagros forests, using artificial neural network classifier on multi-temporal lansat-8 (OLI) imageries (case study: Armarde, Baneh). *Casp. J. Environ. Sci.* **2019**, *17*, 83–96.
99. Dianati Tilaki, G.A.; Ahmadi Jolandan, M.; Gholami, V. Rangelands production modeling using an artificial neural network (ANN) and geographic information system (GIS) in Baladeh rangelands, North Iran. *Casp. J. Environ. Sci.* **2020**, *18*, 277–290.
100. Zhan, C.; Dai, Z.; Soltanian, M.R.; De Barros, F.P. Data-Worth Analysis for Heterogeneous Subsurface Structure Identification With a Stochastic Deep Learning Framework. *Water Resour. Res.* **2022**, *58*, e2022WR033241. [CrossRef]
101. Vapnik, V.; Chervonenkis, A. The necessary and sufficient conditions for consistency in the empirical risk minimization method. *Pattern Recognit. Image Anal.* **1991**, *1*, 283–305.
102. Nguyen, H. Support vector regression approach with different kernel functions for predicting blast-induced ground vibration: A case study in an open-pit coal mine of Vietnam. *SN Appl. Sci.* **2019**, *1*, 283. [CrossRef]
103. Al-Anazi, A.F.; Gates, I.D. Support vector regression to predict porosity and permeability: Effect of sample size. *Comput. Geosci.* **2012**, *39*, 64–76. [CrossRef]

104. Khajehzadeh, M.; Taha, M.R.; Eslami, M. Opposition-based firefly algorithm for earth slope stability evaluation. *China Ocean Eng.* **2014**, *28*, 713–724. [CrossRef]
105. Zhou, J.; Shen, X.; Qiu, Y.; Shi, X.; Khandelwal, M. Cross-correlation stacking-based microseismic source location using three metaheuristic optimization algorithms. *Tunn. Undergr. Space Technol.* **2022**, *126*, 104570. [CrossRef]

sustainability

MDPI

Article

Finite Element Analysis for the Mechanism of Stress Wave Propagation and Crack Extension Due to Blasting of a Frozen Rock Mass

Tingting Wang [1,2], Pingfeng Li [1,*], Chun'an Tang [3,4], Bingbing Zhang [1] and Jiang Yu [2]

1 Hongda Blasting Engineering Group Co., Ltd., Guangzhou 510623, China
2 School of Resources and Civil Engineering, Northeastern University, Shenyang 110819, China
3 State Key Laboratory of Coastal and Offshore Engineering, Dalian University of Technology, Dalian 116024, China
4 State Key Laboratory of Frozen Soil Engineering, Northwest Institute of Eco-Environment and Resources, Chinese Academy of Sciences, Lanzhou 730000, China
* Correspondence: hdbplpf@163.com

Abstract: The propagation mechanism of explosion stress waves in frozen rock mass is the main factor affecting the blasting efficiency and safety construction of strip mines in alpine cold regions. In order to study explosion stress wave propagation and crack extension in the blasting process of frozen rock mass with ice-filled cracks, $RFPA^{2D}$ is adopted to simulate the influence of the geometric parameters of ice-filled cracks (ice-filled crack thickness d, normal distance R from blasting hole to the ice-filled crack, and ice-filled crack angle α), loading intensity and loading rate on the explosion stress wave propagation effect and the damage range. The results show: The attenuation trend of explosion stress waves decreases gradually with an increase of thickness (e.g., In the case of R is 0.2 m, when d is 0.02 m, 0.04 m, and 0.08 m, the calculated attenuation factor of the minimum principal stress peak value is 7.128%, 18.056%, and 30.035%, respectively), and it decreases slightly with an increase of normal distance and ice-filled crack angle. The damage elements range of the ice-filled crack decreases when the ice-filled crack thickness and normal distance increases. The loading intensity and the loading rate have a significant influence on blasting hole fracture patterns. The ice-filled crack has a guiding effect on the growth of blasting cracks at the blasting hole. Nevertheless, the existence of ice-filled cracks inhibits the propagation of explosion stress waves in frozen rock mass.

Keywords: frozen rock mass blasting; ice-filled crack; explosion stress wave propagation; attenuation factor; numerical simulation

Citation: Wang, T.; Li, P.; Tang, C.; Zhang, B.; Yu, J. Finite Element Analysis for the Mechanism of Stress Wave Propagation and Crack Extension Due to Blasting of a Frozen Rock Mass. *Sustainability* **2023**, *15*, 4616. https://doi.org/10.3390/su15054616

Academic Editors: Mahdi Hasanipanah, Danial Jahed Armaghani and Jian Zhou

Received: 17 January 2023
Revised: 13 February 2023
Accepted: 1 March 2023
Published: 4 March 2023

1. Introduction

The design and disaster prevention of open pit blasting mining of mineral resources in cold regions have become key issues in the field of energy safety mining. There are a large number of naturally formed intermittent joint cracks, bedding and faults in rock mass. The existence of these structural planes affects the mechanical properties, vibration, permeability, energy transfer and other properties of rock mass. Affected by low temperature, the water in the primary fissures of open pit slopes becomes ice, which forms frozen rock mass [1–4]. The propagation and attenuation of explosion stress waves in frozen rock mass is slightly different from that in conventional rock mass, which affects the blasting effect and safety of strip mining in cold regions. Therefore, it is of great significance to study the explosion stress wave propagation and crack extension of ice-filled crack rock mass under explosion loading. This study will improve the efficiency of blasting energy utilization, blasting effect and disaster prevention of rock mass engineering in cold regions [5–7].

At present, many scholars have carried out a wealth of research on stress wave propagation in jointed rock masses. In terms of theoretical calculations, it is mainly divided into the discontinuous displacement method, the equivalent continuous medium method, and the continuous and discontinuous coupling method [8–10] to study the propagation characteristics of stress waves. The discontinuous displacement method is mainly used to analyze the stress wave propagation in a single crack or a group of parallel cracks. The fewer cracks there are, the better the analysis effect is. The equivalent continuous medium method can quickly calculate the propagation of stress waves in rock mass under a large number of cracks and uniform distribution. The continuous and discontinuous coupling method is used to analyze macroscopic joints and mesoscopic rock fissures in rock mass. Currently, this method focuses on the one-dimensional propagation law of stress waves [11,12].

In terms of physical tests, the separation Hopkinson pressure bar (SHPB) device has become the main research method to study the propagation of explosive stress waves in jointed rock masses [13–15]. Chen et al. [16] obtained the relationship between the transmission coefficient and the contact surface by the stress wave propagation experiment in artificial rock fractures. Kumar et al. [17] investigated the rate-dependent mechanical behavior of jointed rock with a non-persistent joint with different infill conditions under varying strain rates, i.e., 10^{-4} to 130 s^{-1} using an SHPB and static uniaxial compression test set-up. Certainly, it is a good method to study the propagation characteristics of explosion stress waves through the blasting simulation test. Luo et al. [18] used the dynamic caustics test system to study the penetration process of the main crack of the slotted hole and wing cracks of different angles. They drew the conclusion that the 90° pre-crack has a certain inhibitory effect on the reflected stretching wave. Ram et al. [19] studied the interaction between explosion waves and a structure by electric explosion technology.

The numerical simulation methods, in comparison with theoretical and experimental studies, provide easier and more economical conditions for studying stress wave propagation in jointed rock masses, especially for complex cases where theoretical and experimental solutions seem impossible.

The continuum-based method mainly contains the finite element method (FEM), XFEM (extended FEM), SPH (smoothed particle hydrodynamics), etc. [20–23]. Liang et al. [24] studied the dynamic fracture properties of rocks under different static stress conditions by $RFPA^{2D}$, and concluded that the crack propagation path became more discontinuous and rougher in a smaller-heterogeneity parameter case. Bendezu et al. [25] obtained the advantages and limitations of three methods (XFEM, the conventional finite element method (FEM) using a remeshing technique, and the element deletion method) that simulate the evolution of a rock fragmentation process. Based on the experimentally obtained mechanical properties, experienced peak pressure values inside the rock samples and blast-induced fracture patterns, Banadaki et al. [26] calibrated the Johnson-Holmquist model parameters in ANSYS Autodyn. Zhao et al. [27] analyzed the blasting-induced fracture propagation in coal masses by LS-DYNA, considering the dynamic compressive and tensile failure.

The discontinuum-based methods include the DEM (discrete element method) and the DDA (discontinuous deformation analysis). Yari et al. [28] studied the effect of the position of the joints relative to the blast hole on the blast wave propagation by 3D DEM models. Lak et al. [29] simulated the process of extension of blast-induced fractures in rock masses by the DEM, which considered fracture propagation from both the rock mass inherent fractures and newly induced cracks. Hajibagherpour et al. [30] simulated the mechanism of rock fragmentation due to blast-induced shock waves in a single blast hole by UDEC. Ning et al. [31] extended the DDA to model rock mass fracturing by coupling the rock mass failure process and the penetration effect of the explosion gas based on a generalized artificial joint concept.

Coupled or hybrid continuum-discontinuum-based methods include the FEM-SPH method [32], the DEM-SPH method [33], the MPM (material point method), the CDEM (continuum-discontinuum element method) [34], and the combined finite-discrete element method (FDEM) [35,36], etc. Trivino et al. [37] simulated blasting-induced crack initiation and propagation in a granitic outcrop using FDEM. Zhao et al. [38] studied the blasting effect disturbed by joint strength, joint stiffness, joint spacing, joint angle and other factors by CDEM.

The above-related research mainly focuses on the propagation characteristics of stress waves in jointed rock mass, while the explosive stress wave propagation in frozen rock masses at low temperature is slightly involved. Frozen rock mass with ice-filled cracks is very common in the mining process of mineral resources in cold regions, and its properties are different from that of conventional rock mass [39]. It is of practical value to study the propagation process of explosion stress waves in frozen rock mass with ice-filled cracks, the growth pattern of blasting cracks in frozen rock mass, and the attenuation of explosion stress waves after passing through ice, for the safe mining of strip mines in cold regions.

The aim of this study is to explore explosion stress wave propagation and crack extension in the blasting process of frozen rock mass with ice-filled cracks. The numerical model of frozen rock mass with ice-filled cracks is established by $RFPA^{2D}$ in Section 2. The influences of the geometrical parameters of ice-filled cracks (ice-filled crack thickness D, normal distance R from blasting hole to ice-filled crack, and ice-filled crack angle α), loading intensity and loading rate on the explosion stress wave propagation effect and the damage range are mainly analyzed in Section 3. This research can provide theoretical suggestions for improving the efficiency and disaster prevention of blasting engineering in cold regions.

2. The Principle of $RFPA^{2D}$

2.1. Overview of Mesoscopic

Rock failure process analysis (RFPA) is used to simulate the failure process of frozen rock mass with ice-filled cracks. However, we know that the rock is a heterogeneous material filled by the disorder of micro-structures, which plays a significant role on the mechanical properties of rock [40,41]. Therefore, rock heterogeneity should be considered and implemented in the numerical model. Rock heterogeneity can be well characterized by using the statistical method. In RFPA, the numerical testing sample is composed of elements with the same shape and size. It is assumed that the distribution of elemental mechanical parameters, including the strength, Poisson ratio, elastic modulus and density, can be depicted by the Weibull distribution function [42], as follows:

$$\phi(u) = \frac{m}{u_0}\left(\frac{u}{u_0}\right)^{m-1} \exp\left[-\left(\frac{u}{u_0}\right)^{m}\right] \tag{1}$$

where u is the mechanical and dynamic properties of elements, such as elastic modulus, strength, and density; u_0 is defined as the mean value of the element parameter; m is a shape parameter which is defined as the homogeneity index of the material; and $\varphi(u)$ is the distribution function of mechanical properties.

2.2. Evolutionary Damage Principle of RFPA Meso-Elements

$RFPA^{2D}$ uses the four-node iso-parametric element to describe the basic element. All of the elements are considered to be elastic and isotropic. The elastic damage constitutive method is adopted to elaborate the stress–strain relationship. The stress–strain curve of each element is considered as linearly elastic. When the damage threshold is reached, the maximum tensile stress criterion is used to judge the damage and failure of the element in tensile state, and the Mohr-Coulomb criterion is used to judge the damage and failure of the element in compressive and shear state. As the damage progresses, the elastic modulus of the element is gradually degraded. The modified elastic modulus can be expressed as follows:

$$E = (1 - D) \times E_0 \tag{2}$$

where E and E_0 are the elastic modulus after damage and the initial elastic modulus, respectively; and D is the damage variable.

Figure 1 shows the elastic damage constitutive relation of the element under uniaxial stress state. When shear failure occurs to the elements, the Mohr-Coulomb criterion is adopted:

$$F = \sigma_1 - \sigma_3 \frac{1+\sin\varphi}{1-\sin\varphi} \geq f_c \tag{3}$$

where σ_1, σ_3 and f_c are the maximum principal stress, the minimum principal stress and the uniaxial compressive strength, respectively; and ϕ is the friction angle.

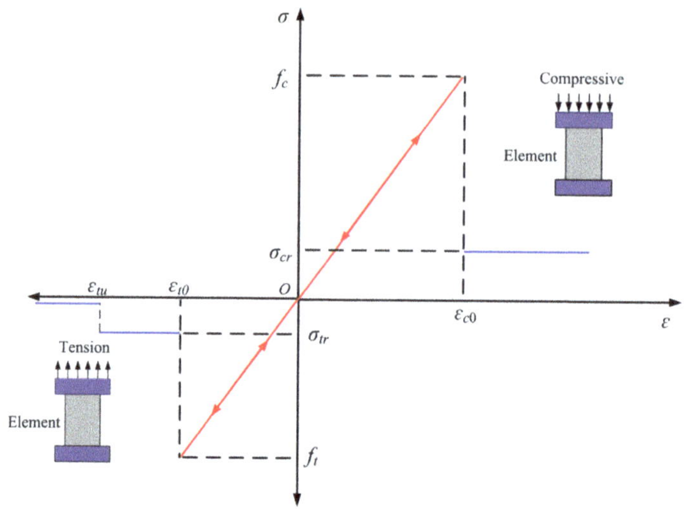

Figure 1. Elastic damage constitutive relation of the element under uniaxial stress state.

Damage variable D is introduced:

$$D = \begin{cases} 0 & \varepsilon \leq \varepsilon_{c0} \\ 1 - \frac{\sigma_{cr}}{\varepsilon E_0} & \varepsilon_{c0} \leq \varepsilon \end{cases} \tag{4}$$

where σ_{cr} is the compressive residual strength and ε_{c0} represents the maximum compressive strain.

When tensile failure occurs to the elements, the following is adopted:

$$\sigma_3 \leq -f_t \tag{5}$$

Damage variable D is introduced:

$$D = \begin{cases} 0 & \varepsilon \leq \varepsilon_{t0} \\ 1 - \frac{\sigma_{tr}}{\varepsilon E_0} & \varepsilon_{t0} < \varepsilon \leq \varepsilon_{tu} \\ 1 & \varepsilon > \varepsilon_{tu} \end{cases} \tag{6}$$

where f_t and σ_{tr} are the tensile strength of rock and the residual strength of tensile damage, respectively; ε_{t0} and ε_{tu} represent the maximum tensile strain and the ultimate tensile strain, respectively. When the maximum tensile strain of the rock element is reached, it loses carrying capacity.

2.3. RFPA Solution for Dynamic Finite Element Equations

The dynamic equilibrium equation for each node in motion can be expressed as [43]:

$$M\ddot{u} + C\dot{u} + Ku = Q \tag{7}$$

where $\ddot{u}$, $\dot{u}$, and u are the constant vectors of displacement, velocity and acceleration at t, respectively. K, M and C are the stiffness matrix, mass matrix and damping matrix of the system, respectively.

By substituting the equations relating velocity, acceleration and displacement in the Newmark method:

$$\ddot{u}_{t+\Delta t} = \frac{1}{\beta \Delta t^2}(u_{t+\Delta t} - u_t) - \frac{1}{\beta \Delta t}\dot{u}_t - \left(\frac{1}{2\beta} - 1\right)\ddot{u}_t \tag{8}$$

$$\ddot{u}_{t+\Delta t} = \frac{\gamma}{\beta \Delta t}(u_{t+\Delta t} - u_t) + \left(1 - \frac{1}{\beta}\right)\dot{u}_t - \left(\frac{\gamma}{2\beta} - 1\right)\ddot{u}_{t+\Delta t}\Delta t \tag{9}$$

By substituting Equations (8) and (9) into the dynamic equilibrium Equation (7):

$$\widehat{K}u_{t+\Delta t} = \widehat{Q}u_{t+\Delta t} \tag{10}$$

$$\widehat{K} = K + \frac{1}{\beta \Delta t^2}M + \frac{\gamma}{\beta \Delta t}C \tag{11}$$

$$\begin{aligned} \widehat{Q}_{t+\Delta t} = Q_{t+\Delta t} + M\left[\tfrac{1}{\beta \Delta t^2}(\alpha_{t+\Delta t} - \alpha_t) - \tfrac{1}{\beta \Delta t}\dot{\alpha}_t - \left(\tfrac{1}{2\beta} - 1\right)\ddot{\alpha}_t\right] \\ +C\left[\tfrac{\gamma}{\beta \Delta t}(\alpha_{t+\Delta t} - \alpha_t) + \left(1 - \tfrac{\gamma}{\beta}\right)\dot{\alpha}_t + \left(1 - \tfrac{\gamma}{2\beta}\right)\Delta t\ddot{\alpha}_t\right] \end{aligned} \tag{12}$$

where β and γ are the integration coefficients of the Newmark method. When $\gamma \geq 0.5$, $\beta \geq 0.25 \times (\gamma + 0.5)^2$, the Newmark method is unconditionally stable.

3. Numerical Simulation Blasting Process Analysis of Frozen Rock Mass with Ice-filled Cracks

3.1. Model Setup

$RFPA^{2D}$ has been widely used in the numerical simulation of stress wave propagation in conventional jointed rock masses [44,45]. Liang et al. [24] used $RFPA^{2D}$ to carry out the dynamic fracture characteristics of fractured rock under different static stress conditions and compared it with the experimental results of Yang et al. [46]. The simulation results showed that $RFPA^{2D}$ could well simulate crack propagation and stress wave attenuation under dynamic stress, which was in good agreement with the experimental results. In addition, some achievements have been made on the effect of ice-filled cracks on the propagation efficiency of rock explosion stress waves [47]. Therefore, explosion stress wave propagation and crack extension of frozen rock mass with ice-filled cracks during blasting is studied by $RFPA^{2D}$.

The blasting model of frozen rock mass is 4 m in length and 4 m in height. The mesh size is 0.01 m × 0.01 m. Single-hole blasting is adopted, and the blasting hole radius r is 0.05 m. The blasting load is simplified into triangular waves [24,45]. The loading case is shown in Figure 2. The total duration is 3×10^{-4} s, and the single-step loading time is 2×10^{-6} s.

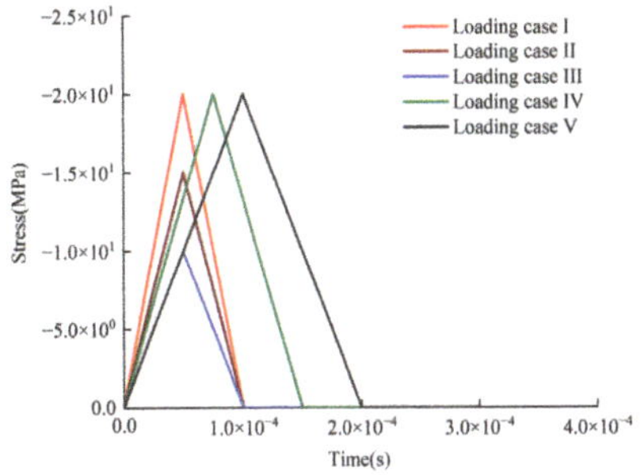

Figure 2. Explosion stress waves applied on the blasting hole.

The material parameters of the numerical simulation of rock and ice are shown in Table 1 [48]. Model I is used to study the influence of ice-filled crack thickness d, normal distance R from the ice-filled crack to the center of the blasting hole, loading intensity and loading rate on the propagation process of frozen rock mass explosion stress waves (Table 2). Model II is used to study the influences of the ice-filled crack angle α on the propagation process of frozen rock mass explosion stress waves (Table 3).

Table 1. Material parameters of the model.

	Elasticity Modulus (MPa)	m	Compressive Strength (MPa)	m	Poisson Ratio	Friction Angle	Density ($kg \times m^{-3}$)
Rock	32,000	5	147	5	0.3	30°	2600
Ice	6000	10	8	10	0.35	26.5°	917

Table 2. Calculation Model I.

		Explosion Model	Ice-Filled Crack Thickness d (m)		
			0.02	0.04	0.08
	0.2 m		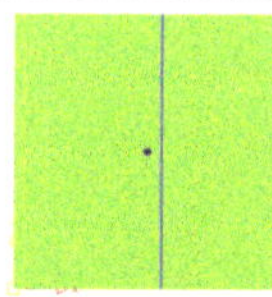	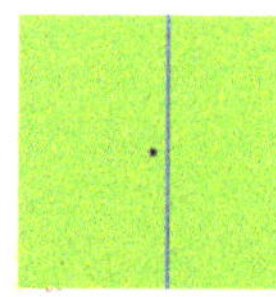	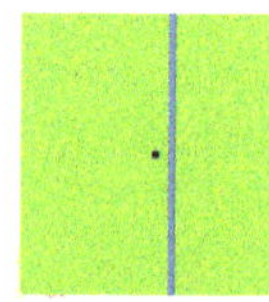
R	0.4 m	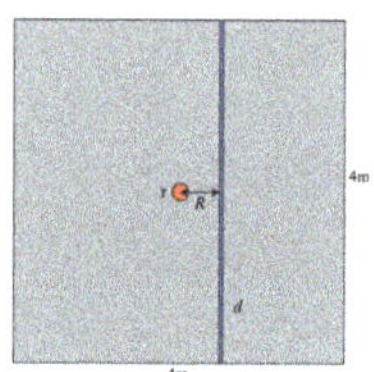	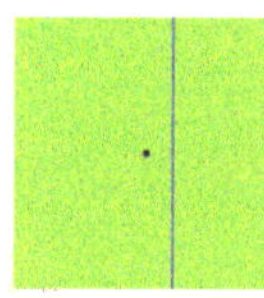	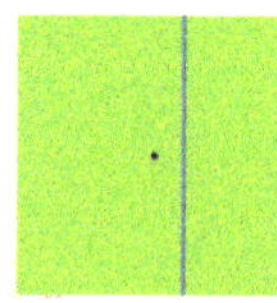	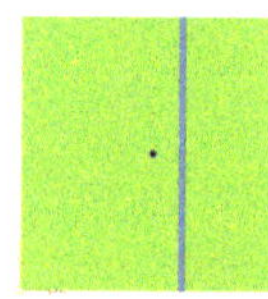
	0.8 m		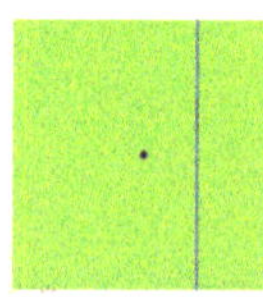	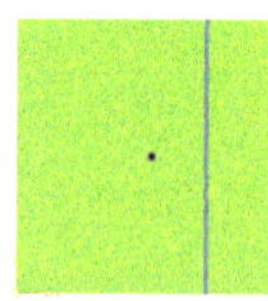	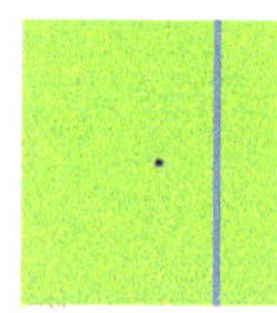

3.2. *Analysis of Blasting Failure Process of Intact Frozen Rock Mass*

3.2.1. Blasting Failure Process

Figure 3 shows the intact frozen rock mass explosion stress wave propagation and failure process. The cracks produced by blasting mainly occur near the blasting hole. The cracks show a uniform and divergent extension pattern. There is no ice-filled crack in the model, and explosion stress waves propagate equally on the left and right sides. At 0.54×10^{-4} s, damage elements begin to appear near the blasting hole; and at 1.48×10^{-4} s, macro blasting cracks are formed. After that, the macro cracks stop extending as the explosion stress wave has propagated out.

3.2.2. The Minimum Principal Stress at the Monitoring Point A

The monitoring point A is set (3.2 m, 2 m). In RFPA^{2D}, the pressure is positive, and the tensile stress is negative. Before 2.34×10^{-4} s, the minimum principal stress at the monitoring point A does not change, and the stress waves do not reach the monitoring point A. From 2.34×10^{-4} s to 3.04×10^{-4} s, the monitoring point is squeezed by stress waves, showing stress changes; and the peak value of the minimum principal stress is 0.576 MPa. After 3.04×10^{-4} s, the symbol of the minimum principal stress value changes, and tensile stress is generated at the monitoring point under explosion stress waves. (Figure 4).

Table 3. Calculation Model II.

Explosion Model		Ice-Filled Crack Angles α (°)		
		15°	30°	45°
R = 0.8 m d = 0.04 m	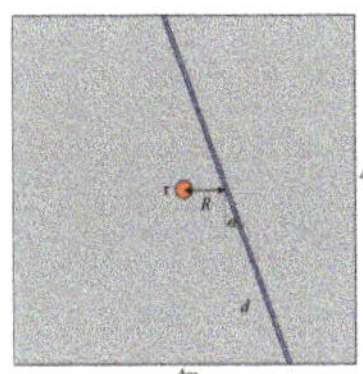	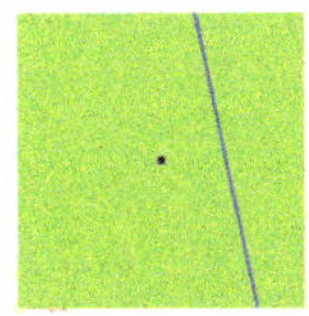	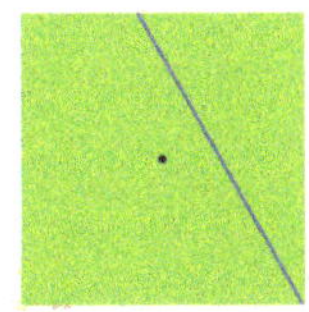	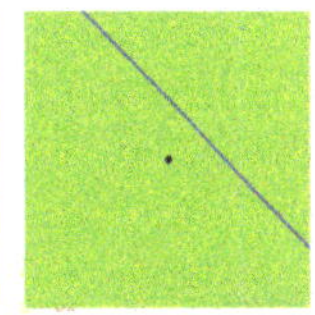
		60°		75°
		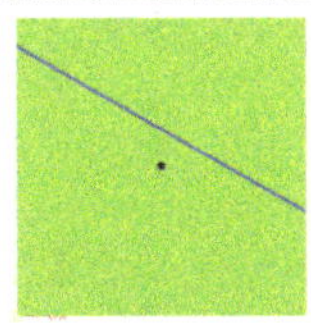		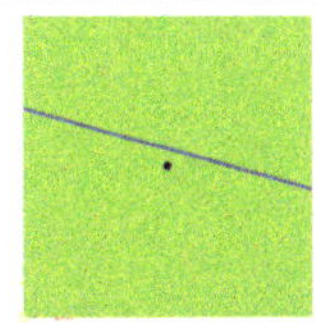

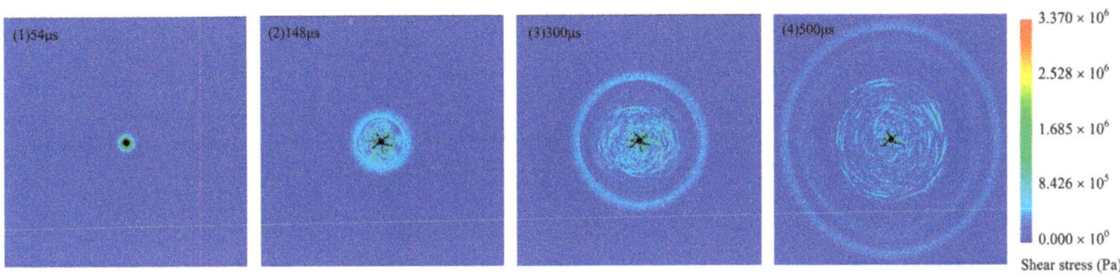

Figure 3. Intact frozen rock mass explosion stress wave propagation and failure process.

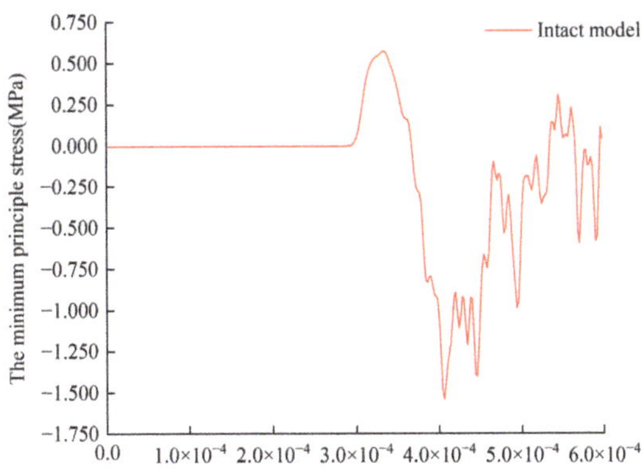

Figure 4. The minimum principal stress during blasting at intact frozen rock monitoring point.

3.3. Ice-Filled Crack Thickness d Influence on Blasting Effect of Frozen Rock Mass

3.3.1. Blasting Failure Process

By comparing the blasting effect of frozen rock with ice-filled cracks of 0.02 m, 0.04 m and 0.08 m thickness distributed in the same normal distance, the following observations are obtained. With the propagation of explosion stress waves, the numerical simulation results are the same as the blasting effect of intact frozen rock mass, and the blasting crack appears first at the blasting hole (Figure 5). Taking d = 0.02 m and R = 0.2 m frozen rock mass with ice-filled cracks as an example, at 0.54×10^{-4} s, damage elements occur near the blasting hole. At 1.48×10^{-4} s, the stress wave reaches the ice-filled crack. When the explosion stress wave propagates to the ice-filled crack, reflects and transmits. The reflected wave collides with the incident wave, which reduces both the energy of the incident wave and the explosion range. As the mechanical strength of ice is less than that of rock, damage elements gradually appear in the ice-filled crack first. At 3.00×10^{-4} s, the explosion stress wave continues to act on the ice-filled crack, and the range of damage elements intensifies. At 5×10^{-4} s, the stress waves have passed through the ice-filled crack, and the damage range remains stable.

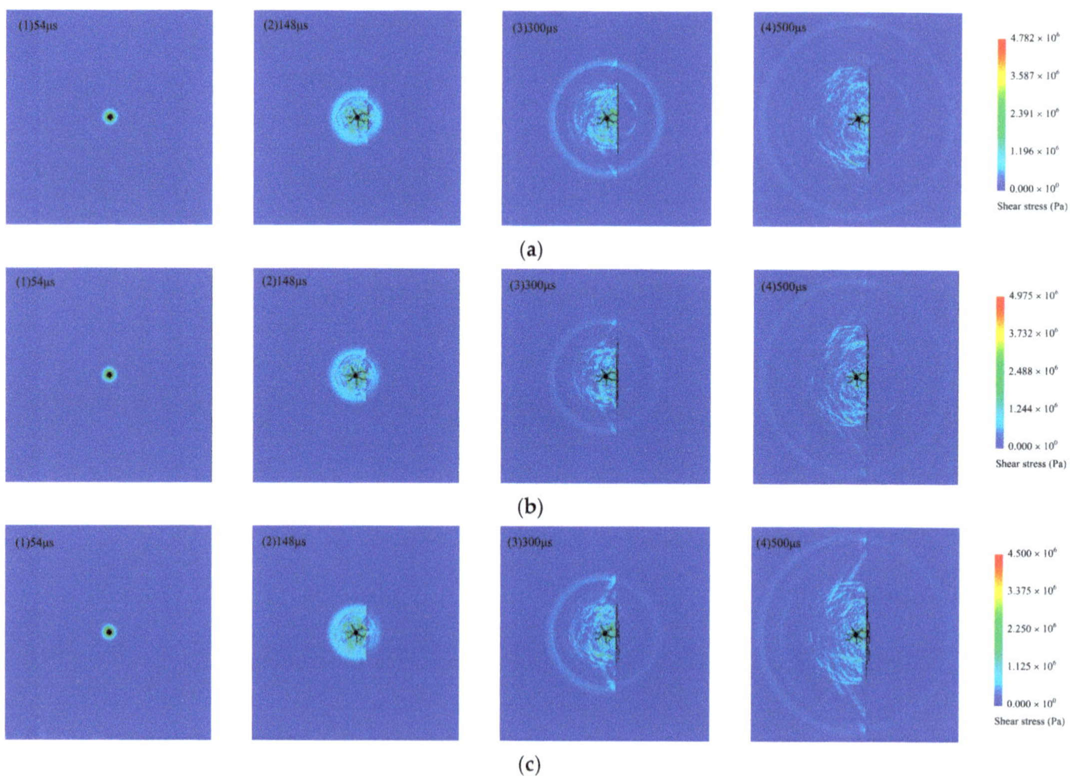

Figure 5. Explosion stress wave propagation and failure process of frozen rock mass with different ice-filled crack thicknesses (the normal distance R is 0.2 m). (**a1–a4**) d = 0.02 m; (**b1–b4**) d = 0.04 m; (**c1–c4**) d = 0.08 m.

By comparing Figure 5(a4,b4,c4), the damage elements range of the ice-filled crack decreases when the thickness of the ice-filled crack increases. By comparing the crack propagation of the non-ice side with the ice side, the crack propagation of the blasting is induced by the ice-filled crack.

3.3.2. The Minimum Principal Stress at the Monitoring Point A with Different Ice-Filled Crack Thicknesses

By comparing the minimum principal stress at the monitoring point A of ice-filled cracks with different thicknesses (Figure 6), the minimum principal stress presents a positive increase at first, then decreases to a negative fluctuation. The minimum principal stress amplitude is similar to that of the intact frozen rock mass. Taking R = 0.8 m frozen rock mass with ice-filled crack as an example.

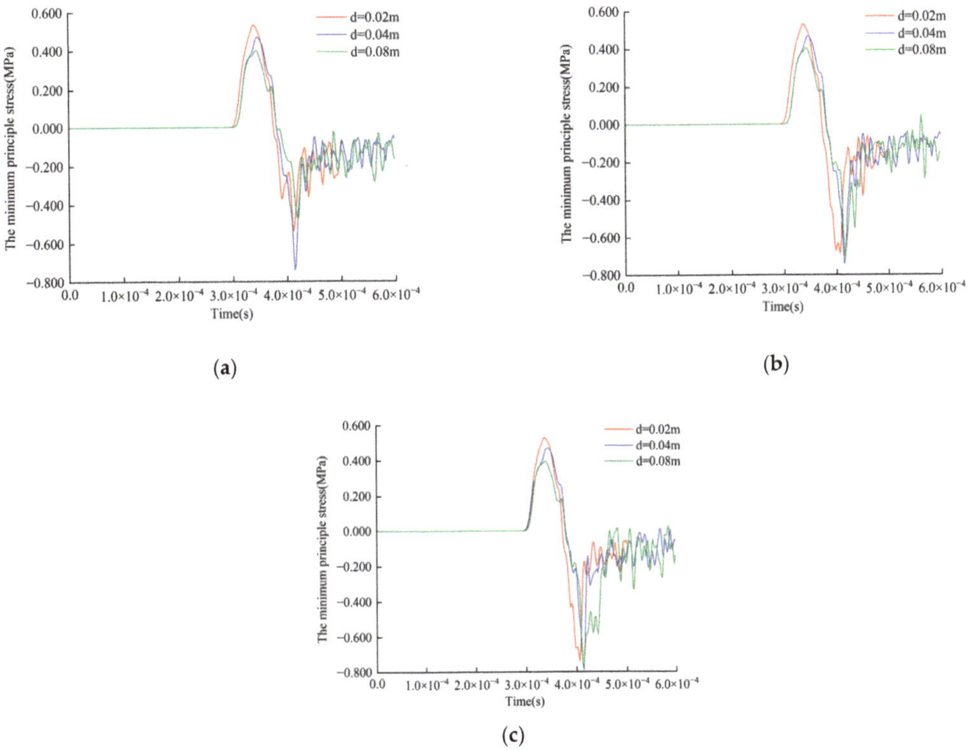

Figure 6. The minimum principal stress during blasting at frozen rock mass monitoring point with different ice-filled crack thicknesses. (**a**) R = 0.2 m; (**b**) R = 0.4 m; (**c**) R = 0.8 m.

When d = 0.02 m, the minimum principal stress is positive and the peak value is 0.528 MPa from 2.90×10^{-4} s to 3.72×10^{-4} s. The explosion stress waves produce extrusion effects on the monitoring point. After 3.72×10^{-4} s, the stress wave propagates through the monitoring point, tensile stress is generated on the monitoring point, and the minimum principal stress turns negative. When d = 0.04 m, the peak value of the minimum principal stress is 0.470 MPa from 2.92×10^{-4} s to 3.78×10^{-4} s. When d = 0.08 m, the peak value of the minimum principal stress is 0.395 MPa from 2.96×10^{-4} s to 3.76×10^{-4} s. The results show that the minimum principal stress decreases with an increase of ice-filled crack thickness.

At the monitoring point A, the peak value of the minimum principal stress of intact frozen rock mass is 0.576 MPa. The minimum principal stress peak value of frozen rock mass with ice-filled cracks is obviously smaller than that of intact frozen rock mass. It can be seen that the existence of ice-filled cracks inhibits the propagation of explosion stress waves in frozen rock mass.

3.4. Normal Distance R Influence on Blasting Effect of Frozen Rock Mass

3.4.1. Blasting Failure Process

Figure 7 shows the explosion stress wave propagation and failure process of frozen rock mass with different normal distance R. The crack pattern at the blasting hole is similar to that of the intact frozen rock mass, showing a uniform and divergent extension. When d = 0.02 m and 0.04 m, the damage elements ranges have little differences. When d = 0.08 m, the damage elements range have obvious differences. The smaller the normal distance R is, the closer the ice-filled crack is to the blasting hole, the more obvious the explosion stress wave effect is. The thicker the ice-filled crack is, the more apparent that it is affected by normal distance.

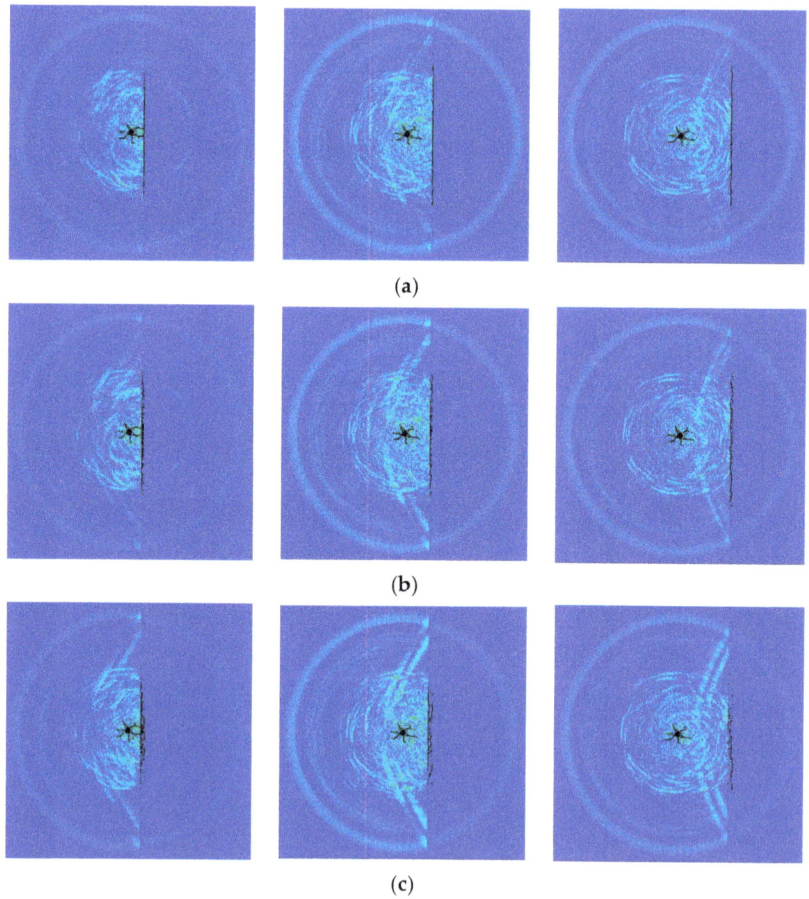

Figure 7. Explosion stress wave propagation and failure process of frozen rock mass with different normal distance R. (**a**) d = 0.02 m; (**b**) d = 0.04 m; (**c**) d = 0.08 m.

3.4.2. The Minimum Principal Stress at the Monitoring Point A with Different Normal Distances

Taking d = 0.04 m frozen rock mass with ice-filled crack as an example (Figure 6). When R = 0.2 m, the peak value of the minimum principal stress is 0.472 MPa from 2.98×10^{-4} s to 3.82×10^{-4} s. The explosion stress wave produces extrusion effects on the monitoring point. After 3.72×10^{-4} s, as the stress wave propagates through the monitoring point, tensile stress is generated on the monitoring point, so it is a negative value. When R = 0.4 m, the peak value of the minimum principal stress is 0.472 MPa from 2.98×10^{-4} s to 3.82×10^{-4} s. When R = 0.8 m, the peak value of the minimum principal stress is 0.470 MPa from 2.98×10^{-4} s to 3.80×10^{-4} s. When the normal distance between the ice-filled crack and the blasting hole increases, the reduction of the explosion stress wave propagation will decrease slightly.

3.5. Ice-Filled Crack Angle α Influence on Blasting Effect of Frozen Rock Mass

3.5.1. Blasting Failure Process

Figure 8 shows the explosion stress wave propagation and failure process of frozen rock mass with different ice-filled crack angles. Different angles affect the distance from the ice-filled crack to the blasting hole. When d = 0.04 m and R = 0.8 m, the larger the ice-filled crack angle is, the closer the ice-filled crack is to the blasting hole and the wider the damage range of the ice-filled crack.

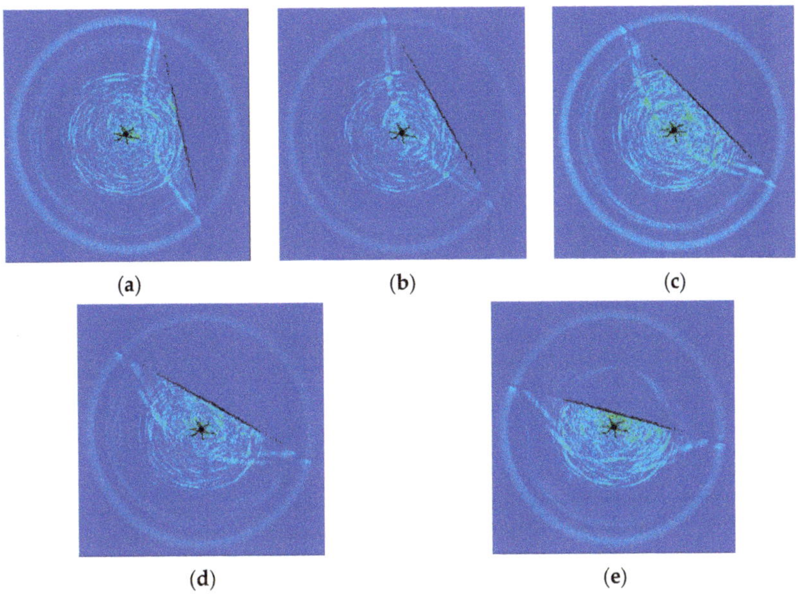

Figure 8. Damage and cracking after blasting of frozen rock mass with different angles. (**a**) 15°; (**b**) 30°; (**c**) 45°; (**d**) 60°; (**e**) 75°.

3.5.2. The Minimum Principal Stress at the Monitoring Point A with Different Loading Angles

Taking the frozen rock mass blasting process with the ice-filled crack angle 15° as an example. Before 2.92×10^{-4} s, the minimum principal stress does not change. From 2.92×10^{-4} s to 3.80×10^{-4} s, the peak value of the minimum principal stress is 0.462 MPa. After 3.80×10^{-4} s, the minimum principal stress turns negative. When the ice-filled crack angle is 30°, 45°, 60°, 75°, the peak value of the minimum principal stress is 0.464 MPa, 0.482 MPa, 0.475 MPa and 0.500 MPa, respectively (Figure 9). An increase of the ice-filled crack angle causes a slight decrease of the minimum principal stress peak value.

3.6. Loading Intensity Influence on Blasting Effect of Frozen Rock Mass

3.6.1. Blasting Failure Process

The mechanical properties of rock materials are affected by the loading peak value, which might influence the explosion stress wave propagation efficiency and blasting effect. Taking d = 0.04 m and R = 0.4 m frozen rock mass as an example to simulate the blasting process with different loading peak values. The failure patterns in loading cases I, II and III at selected times were compared in Figure 10.

With the propagation of the explosion stress wave, the blasting crack appears first at the blasting hole. At 0.54×10^{-4} s, the blasting hole begins to show damage elements. At 1.48×10^{-4} s, damage elements gradually appear in the ice-filled crack first. The blasting crack pattern at the blasting hole is obviously different due to different loading intensities.

At 3.00×10^{-4} s, the explosion stress wave continues to impact the ice-filled crack, and the range of damage elements intensifies. At 5×10^{-4} s, the stress waves have passed through the ice-filled crack and the damage range remains stable. By comparison with Figure 10(a4,b4,c4), as the loading intensity increases, the blasting crack pattern at the blasting hole becomes more apparent, and the damage elements range expands. Also, the effect of explosion stress wave propagation is different.

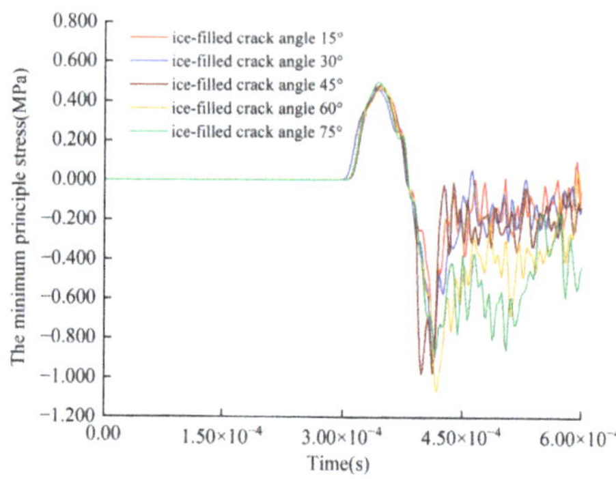

Figure 9. The minimum principal stress during blasting at frozen rock mass monitoring point with different ice-filled crack angles.

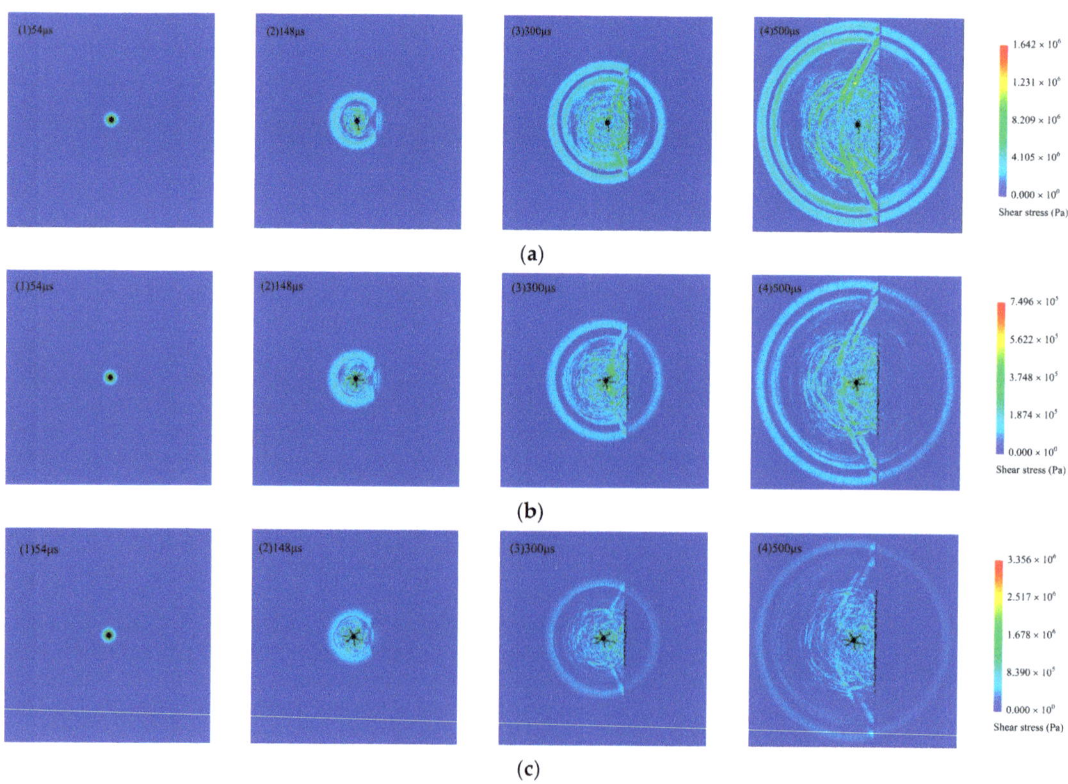

Figure 10. Damage and cracking after blasting of frozen rock mass with different loading intensities. (**a1–a4**) loading case III; (**b1–b4**) loading case II; (**c1–c4**) loading case I.

3.6.2. The Minimum Principal Stress at the Monitoring Point A with Different Loading Intensities

Taking the results of frozen rock mass with ice-filled crack blasting process when the loading intensity of the explosion stress wave is 15 MPa as an example. Before 2.96×10^{-4} s, the minimum principal stress has no change. From 2.96×10^{-4} s to 3.80×10^{-4} s, the peak value of the minimum principal stress is 0.362 MPa, and the explosion stress wave produces extrusion effects on the monitoring point. After 3.80×10^{-4} s, the minimum principal stress is negative. As the explosion stress wave propagates through the monitoring point, tensile stress is generated on the monitoring point. When the explosion stress wave loading intensity is 10 MPa, the minimum principal stress peak value is 0.233 MPa. When the explosion stress wave loading intensity is 20 MPa, the minimum principal stress peak value is 0.472 MPa (Figure 11). When the loading peak value of the explosion stress wave is larger, the effect of the explosion stress wave on the rock medium is more obvious.

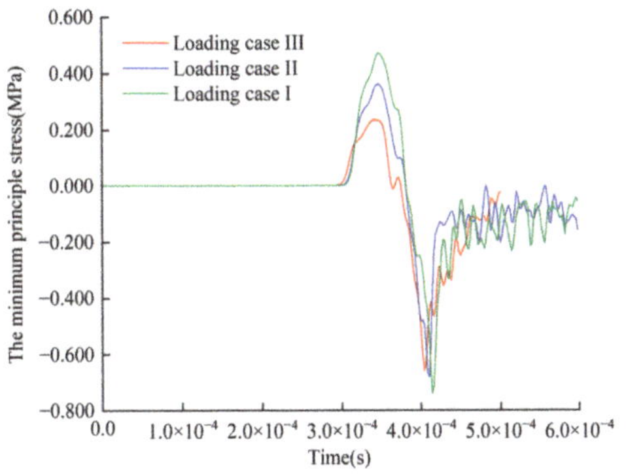

Figure 11. The minimum principal stress during blasting at frozen rock mass monitoring point with different loading intensities.

3.7. *Loading Rate Influence on Blasting Effect of Frozen Rock Mass*

3.7.1. Blasting Failure Process

Different blasting loading rates also have influence on the blasting effect. Taking d = 0.08 m and R = 0.4 m frozen rock mass with ice-filled crack as an example, the failure modes of loading cases I, IV and V at selected times are compared, as shown in Figure 12.

At 0.54×10^{-4} s, the blasting hole begins to produce damage elements. At 1.48×10^{-4} s, damage elements gradually appear in the ice-filled crack first. The blasting crack pattern at the blasting hole is different due to the different loading rates. At 3.00×10^{-4} s, the explosion stress wave continues to impact the ice-filled crack, and the range of damage elements intensifies. The blasting hole crack continues to expand. At 5×10^{-4} s, the stress waves have passed through the ice-filled crack and the damage range remains stable. By comparison with Figure 12(a4,b4,c4), as the loading rate decreases, the crack length at the blasting hole increases, and the ice-filled crack shows more damage elements.

3.7.2. The Minimum Principal Stress at the Monitoring Point A with Different Loading Rates

By comparing the minimum principal stress on the monitoring point with different loading rates in Figure 13, the following results are observed. Before 2.96×10^{-4} s, the minimum principal stress stays unchanged. From 2.96×10^{-4} s to 3.80×10^{-4} s, the peak value of the minimum principal stress is 0.362 MPa in loading case I. In loading cases IV and V, the variation of the minimum principal stress at the monitoring point A fluctuates dramatically. As the loading waveform is different, the minimum principal stress appears hysteresis, and the transmitted waves generated by the explosion stress waves through the ice-filled cracks are also different.

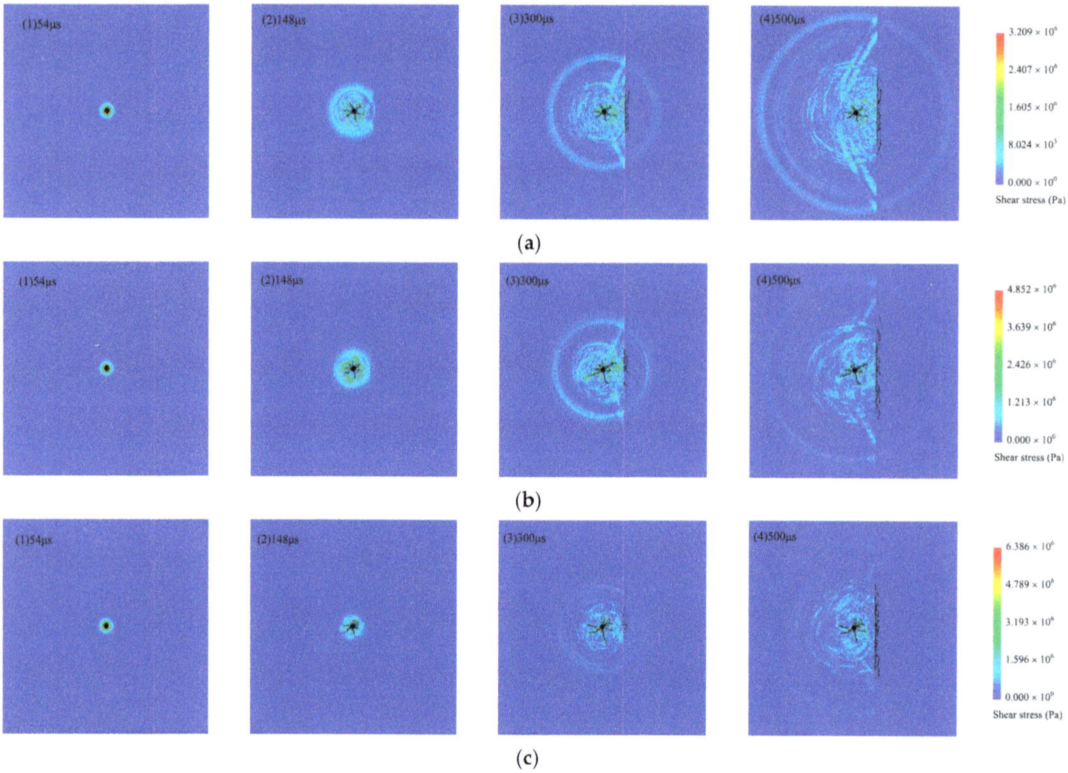

Figure 12. Damage and cracking after blasting of frozen rock mass with different loading rates. (**a1–a4**) Loading case I; (**b1–b4**) Loading case IV; (**c1–c4**) Loading case V.

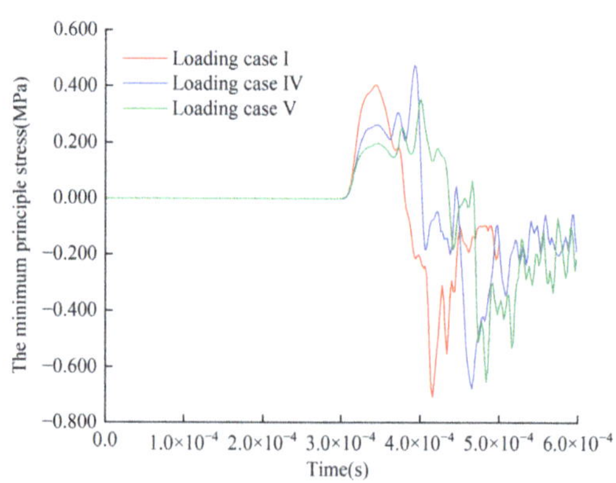

Figure 13. The minimum principal stress during blasting at ice-filled crack frozen rock mass monitoring point with different peak loading values.

3.8. Evaluation of the Explosion Stress Wave Attenuation Factor

In order to intuitively evaluate the effects of different ice-filled crack conditions on the explosion stress wave, the attenuation factor B of the explosion stress wave through frozen rock mass is calculated as follows,

$$B = \left| \frac{\beta_c - \beta_i}{\beta_i} \right| \times 100\% \quad (13)$$

where β_c is the peak value of the minimum principal stress at monitoring point of frozen rock mass under different conditions; and β_i is the minimum principal stress value at monitoring points of intact frozen rock mass, which is 0.576 MPa.

The attenuation trend of the explosion stress wave decreases gradually with an increase of thickness, and decreases slightly with an increase of direction, distance and ice-filled crack angle (Table 4).

Table 4. Explosion stress wave attenuation at monitoring points of frozen rock mass under different conditions.

No.	Width (m)	R (m)	Angle (°)	Loading Case	Attenuation (%)
1	0.02	0.2	0	I	7.128
2	0.02	0.4	0	I	7.292
3	0.02	0.8	0	I	8.333
4	0.04	0.2	0	I	18.056
5	0.04	0.4	0	I	18.056
6	0.04	0.8	0	I	18.403
7	0.08	0.2	0	I	30.035
8	0.08	0.4	0	I	29.689
9	0.08	0.8	0	I	31.424
10	0.04	0.8	15	I	19.792
11	0.04	0.8	30	I	19.444
12	0.04	0.8	45	I	16.319
13	0.04	0.8	60	I	17.535
14	0.04	0.8	75	I	13.194

4. Conclusions

In this work, the stress wave propagation and blasting crack extension mechanism of frozen rock mass with ice-filled cracks are analyzed by RFPA^{2D}. The findings facilitate blasting design and disaster prevention in cold region strip mining. Specifically, the explosion stress wave propagation effect and the damage range are obtained, which consider the geometrical parameters of ice-filled cracks (ice-filled crack thickness D, normal distance R from blasting hole to ice-filled crack, and ice-filled crack angle α), loading intensity and loading rate. The following conclusions can be drawn:

(1) The divergent crack extension pattern is always maintained at the blasting hole. Affected by explosion wave stress, the damage elements range of the ice-filled crack decreases when the ice-filled crack thickness and normal distance increase.
(2) The attenuation trend of explosion stress waves decreases with an increase of ice-filled crack thickness, and decreases slightly with an increase of normal distance and ice-filled crack angle.
(3) The loading intensity and the loading rate have a significant influence on blasting hole fracture patterns. The damage elements range of the ice-filled crack is enlarged when the loading intensity and the loading rate increase. The propagation effect of explosion stress waves is also different.
(4) The ice-filled crack has a guiding effect on the growth of the blasting crack at the blasting hole. Nevertheless, the existence of ice-filled cracks inhibits the propagation of explosion stress waves in frozen rock mass.

It should be noted that due to the lack of access to information of a real project, in this research, only numerical modelling has been done. It is better to combine these results with experiment and field validation to achieve a more comprehensive method to study frozen rock masses blasting process.

Author Contributions: Conceptualization, T.W. and P.L.; methodology, T.W., C.T. and P.L.; software, T.W., C.T. and J.Y.; validation, T.W., B.Z. and J.Y.; formal analysis, T.W.; investigation, T.W. and B.Z.; resources, T.W., C.T. and P.L.; data curation, T.W. and P.L.; writing—original draft preparation, T.W.; writing—review and editing, C.T. and P.L.; visualization, T.W.; supervision, C.T. and P.L.; project administration, C.T.; funding acquisition, C.T. and P.L. All authors have read and agreed to the published version of the manuscript.

Funding: This research was funded by the National Natural Science Foundation of China (Grant No. 42050201) and the Open Fund of State Key Laboratory of Frozen Soil Engineering (Grant No. SKLFSE202013).

Institutional Review Board Statement: Not applicable.

Informed Consent Statement: Informed consent was obtained from all subjects involved in the study.

Data Availability Statement: The data presented in this study are available on request from the corresponding author. The data are not publicly available due to privacy concerns.

Acknowledgments: Hongda Blasting Engineering Group Co., Ltd. are appreciated for their assistance.

Conflicts of Interest: The authors declare no conflict of interest.

References

1. Xia, C.C.; Lv, Z.T.; Li, Q.; Huang, J.H.; Bai, X.Y. Transversely Isotropic Frost Heave of Saturated Rock under Unidirectional Freezing Condition and Induced Frost Heaving Force in Cold Region Tunnels. *Cold Reg. Sci. Technol.* **2018**, *152*, 48–58. [CrossRef]
2. Liu, N.F.; Li, N.; Li, G.F.; Song, Z.P.; Wang, S.J. Method for Evaluating the Equivalent Thermal Conductivity of a Freezing Rock Mass Containing Systematic Fractures. *Rock Mech. Rock Eng.* **2022**, *55*, 7333–7355. [CrossRef]
3. Ma, D.D.; Xiang, H.S.; Ma, Q.Y.; Kaunda, E.E.; Huang, K.; Su, Q.Q.; Yao, Z.M. Dynamic damage constitutive model of frozen silty soil with prefabricated crack under uniaxial load. *J. Eng. Mech.* **2021**, *147*, 04021033. [CrossRef]
4. Liu, M.H.; Ma, W.; Niu, F.J.; Luo, J.; Yin, G.A. International Journal of Heat and Mass Transfer Thermal Performance of a Novel Crushed-Rock Embankment Structure for Expressway in Permafrost Regions. *Int. J. Heat Mass Transf.* **2018**, *127*, 1178–1188. [CrossRef]
5. Wan, W.K.; Li, C.C. Microscopic and Acoustic Interpretations of the Physics of Rock Burst and the Difference in Fracturing Patterns in Class I and Class II Rocks. *Rock Mech. Rock Eng.* **2022**, *55*, 6841–6862. [CrossRef]
6. Li, Y.Z.; Dai, F.; Wei, M.D.; Du, H.B. Numerical Investigation on Dynamic Fracture Behavior of Cracked Rocks under Mixed Mode I/II Loading. *Eng. Fract. Mech.* **2020**, *235*, 107176. [CrossRef]
7. Zhang, F.P.; Peng, J.Y.; Qiu, Z.G.; Chen, Q.K.; Li, Y.H.; Liu, J.P. Rock-like Brittle Material Fragmentation under Coupled Static Stress and Spherical Charge Explosion. *Eng. Geol.* **2017**, *220*, 266–273. [CrossRef]
8. Cai, J.G.; Zhao, J. Effects of Multiple Parallel Fractures on Apparent Attenuation of Stress Waves in Rock Masses. *Int. J. Rock Mech. Min. Sci.* **2000**, *37*, 661–682. [CrossRef]
9. Fan, L.F.; Ma, G.W.; Li, J.C. Nonlinear Viscoelastic Medium Equivalence for Stress Wave Propagation in a Jointed Rock Mass. *Int. J. Rock Mech. Min. Sci.* **2012**, *50*, 11–18. [CrossRef]
10. Yari, M.; Bagherpour, R.; Jamali, S.; Asadi, F. Selection of Most Proper Blasting Pattern in Mines Using Linear Assignment Method: Sungun Copper Mine. *Arch. Min. Sci.* **2015**, *60*, 375–386.
11. Lak, M.; Fatehi Marji, M.; Yarahmadi Bafghi, A.; Abdollahipour, A. Analytical and numerical modeling of rock blasting operations using a two-dimensional elasto-dynamic Green's function. *Int. J. Rock Mech. Min. Sci.* **2019**, *114*, 208–217. [CrossRef]
12. Pyrak-Nolte, L.J.; Morris, J.P. Single fractures under normal stress: The relation between fracture specific stiffness and fluid flow. *Int. J. Rock Mech. Min. Sci.* **2000**, *18*, 245–262. [CrossRef]
13. Li, J.C.; Ma, G.W. Experimental Study of Stress Wave Propagation across a Filled Rock Joint. *Int. J. Rock Mech.* **2009**, *46*, 471–478. [CrossRef]
14. Yari, M.; Monjezi, M.; Bagherpour, R. A novel investigation in blasting operation management using decision making methods. *Rud. Geol. Naft. Zb.* **2014**, *29*, 69–79.
15. Butt, H.S.U.; Xue, P.; Jiang, T.Z.; Wang, B. Parametric identification for material of viscoelastic SHPB from wave propagation data incorporating geometrical effects. *Int. J. Mech. Sci.* **2015**, *91*, 46–54. [CrossRef]

16. Chen, X.; Li, J.C.; Cai, M.F.; Zou, Y.; Zhao, J. A Further Study on Wave Propagation Across a Single Joint with Different Roughness. *Rock Mech. Rock Eng.* **2016**, *49*, 2701–2709. [CrossRef]
17. Kumar, S.; Tiwari, G.; Parameswaran, V.; Das, A. Rate-dependent mechanical behavior of jointed rock with an impersistent joint under different infill conditions. *J. Rock. Mech. Geotech.* **2022**, *14*, 1380–1389. [CrossRef]
18. Luo, H.H.; Yang, R.S.; Ma, X.M.; Zuo, J.J.; Zhang, Y.T.; Li, C.X. The Influence of Prefabricated Cracks at Different Angles on the Propagation Characteristics of Main Cracks in Slot Blasting. *J. Mater. Eng. Perform.* **2022**. [CrossRef]
19. Ram, O.; Sadot, O. Implementation of the exploding wire technique to study blastwave-structure interaction. *Exp. Fluids.* **2012**, *53*, 1335–1345. [CrossRef]
20. Gharehdash, S.; Barzegar, M.; Palymskiy, I.B.; Fomin, P.A. Blast induced fracture modelling using smoothed particle hydrodynamics. *Int. J. Impact Eng.* **2020**, *135*, 103235. [CrossRef]
21. Jeong, H.; Jeon, B.; Choi, S.; Jeon, S. Fracturing behavior around a blasthole in a brittle material under blasting loading. *Int. J. Impact Eng.* **2020**, *140*, 103562. [CrossRef]
22. Pramanik, R.; Deb, D. Implementation of Smoothed Particle Hydrodynamics for Detonation of Explosive with Application to Rock Fragmentation. *Rock Mech. Rock Eng.* **2015**, *48*, 1683–1698. [CrossRef]
23. Zhou, W.H.; Hu, C.Z.; Bao, J.; Zheng, J.J.; Liang, R. Numerical study on crack propagation and stress wave propagation during blasting of jointed rock mass. *Chin. J. Theor. Appl. Mech.* **2022**, *54*, 2501–2512. (In Chinese)
24. Liang, Z.Z.; Qian, X.K.; Zhang, Y.F.; Liao, Z.Y. Numerical Simulation of Dynamic Fracture Properties of Rocks under Different Static Stress Conditions. *J. Cent. South Univ.* **2022**, *29*, 624–644. [CrossRef]
25. Bendezu, M.; Romanel, C.; Roehl, D. Finite element analysis of blast-induced fracture propagation in hard rocks. *Comput. Struct.* **2017**, *182*, 1–13. [CrossRef]
26. Dehghan Banadaki, M.M.; Mohanty, B. Numerical simulation of stress wave induced fractures in rock. *Int. J. Impact Eng.* **2012**, *40–41*, 16–25. [CrossRef]
27. Zhao, J.J.; Zhang, Y.; Ranjith, P.G. Numerical Simulation of Blasting-Induced Fracture Expansion in Coal Masses. *Int. J. Rock Mech. Min. Sci.* **2017**, *100*, 28–39. [CrossRef]
28. Yari, M.; Ghadyani, D.; Jamali, S. Development of a 3D numerical model for simulating a blast wave propagation system considering me position of the blasting hole and in-situ discontinuities. *Rud. Geol. Naft. Zb.* **2022**, *37*, 67–78.
29. Lak, M.; Fatehi Marji, M.; Yarhamadi Bafghi, A.R.; Abdollahipour, A. Discrete element modeling of explosion-induced fracture extension in jointed rock masses. *J. Min. Environ.* **2019**, *10*, 125–138.
30. Hajibagherpour, A.R.; Mansouri, H.; Bahaaddini, M. Numerical modeling of the fractured zones around a blasthole. *Comput. Geotech.* **2020**, *123*, 103535. [CrossRef]
31. Ning, Y.J.; Yang, J.; Ma, G.W.; Chen, P.W. Modelling rock blasting considering explosion gas penetration using discontinuous deformation analysis. *Rock Mech. Rock Eng.* **2011**, *44*, 483–490. [CrossRef]
32. Johnson, G.R. Linking of Lagrangian particle methods to standard finite element methods for high velocity impact computations. *Nucl. Eng. Des.* **1994**, *150*, 265–274. [CrossRef]
33. Fakhimi, A.; Lanari, M. DEM-SPH simulation of rock blasting. *Comput. Geotech.* **2014**, *55*, 158–164. [CrossRef]
34. Yue, Z.W.; Zhou, J.; Feng, C.; Wang, X.; Peng, L.Z.; Cong, J.Y. Coupling of material point and continuum discontinuum element methods for simulating blast-induced fractures in rock. *Comput. Geotech.* **2022**, *144*, 104629. [CrossRef]
35. Wu, D.; Li, H.B.; Shao, Z.S.; Chen, S.H.; Zhou, C.H.; Liu, L.W. Effects of infilling materials on mechanical behaviors and cracking process of pre-cracked rock: Insights from a hybrid continuum-discontinuum method. *Eng. Fract. Mech.* **2021**, *253*, 107843. [CrossRef]
36. Yan, C.Z.; Xie, X.; Ren, Y.H.; Ke, W.H.; Wang, G. A FDEM-based 2D coupled thermal-hydro-mechanical model for multiphysical simulation of rock fracturing. *Int. J. Rock Mech. Min. Sci.* **2022**, *149*, 104964. [CrossRef]
37. Trivino, L.F.; Mohanty, B. Assessment of crack initiation and propagation in rock from explosion-induced stress waves and gas expansion by cross-hole seismometry and FEM-DEM method. *Int. J. Rock Mech. Min. Sci.* **2015**, *77*, 287–299. [CrossRef]
38. Zhao, A.P.; Feng, C.; Guo, R.K.; Li, S.H.; Jia, J.J. Effect of joints on blasting and stress wave propagation. *Chin. J. Rock Mech. Eng.* **2018**, *37*, 2027–2036. (In Chinese)
39. Bai, Y.; Shan, R.L.; Ju, Y.; Wu, Y.X.; Tong, X.; Han, T.Y.; Dou, H.Y. Experimental study on the strength, deformation and crack evolution behaviour of red sandstone samples containing two ice-filled fissures under triaxial compression. *Cold Reg. Sci. Technol.* **2020**, *174*, 103061. [CrossRef]
40. Wu, N.; Liang, Z.Z.; Li, Y.; Qian, X.K.; Gong, B. Effect of confining stress on representative elementary volume of jointed rock masses. *Geomech. Eng.* **2019**, *18*, 627–638.
41. Zhu, W.C.; Liu, J.; Yang, T.H.; Sheng, J.C.; Elsworth, D. Effects of local rock heterogeneities on the hydromechanics of fractured rocks using a digital-image-based technique. *Int. J. Rock Mech. Mining Sci.* **2006**, *43*, 1182–1199. [CrossRef]
42. Erarslan, N.; Liang, Z.Z.; Williams, D.J. Experimental and Numerical Studies on Determination of Indirect Tensile Strength of Rocks. *Rock Mech. Rock Eng.* **2012**, *45*, 739–751. [CrossRef]
43. Zhu, W.C.; Niu, L.L.; Li, S.H.; Xu, Z.H. Dynamic Brazilian Test of Rock under Intermediate Strain Rate: Pendulum Hammer-Driven SHPB Test and Numerical Simulation. *Rock Mech. Rock Eng.* **2015**, *48*, 1867–1881. [CrossRef]
44. Qian, X.K.; Liang, Z.Z.; Liao, Z.Y. A Three-Dimensional Numerical Investigation of Dynamic Fracture Characteristics of Rock Specimens with Preexisting Surface Flaws. *Adv. Civ. Eng.* **2018**, *2018*, 8027582. [CrossRef]

45. Qian, X.K.; Liang, Z.Z.; Liao, Z.Y.; Wang, K. Numerical investigation of dynamic fracture in rock specimens containing a pre-existing surface flaw with different dip angles. *Eng. Frac. Mech.* **2020**, *223*, 106675. [CrossRef]
46. Li, P.F.; Wang, T.T.; Tang, C.A.; Zhang, B.B. Propagation Characteristics of Explosion Stress Wave in Ice-filled Rock Mass. *Blasting* **2022**, *39*, 44–52. (In Chinese)
47. Yang, X.; Pu, C.J.; Tang, X.; Xiao, Z.X.; Guan, S.H.; Liao, T. Experimental study of effects of manual crack on blasting cracks propagation. *Blasting* **2014**, *31*, 26–31. (In Chinese)
48. Wang, T.T.; Li, P.F.; Tang, C.A.; Zhang, B.B.; Yu, J.; Geng, T. Tensile Characteristics and Fracture Mode of Frozen Fractured Rock Mass Based on Brazilian Splitting Test. *Appl. Sci.* **2022**, *12*, 11788. [CrossRef]

sustainability

MDPI

Article

Developing Two Hybrid Algorithms for Predicting the Elastic Modulus of Intact Rocks

Yuzhen Wang [1,2], Mohammad Rezaei [3,*], Rini Asnida Abdullah [4] and Mahdi Hasanipanah [4]

1 School of Water Conservancy and Civil Engineering, Zhengzhou University, Zhengzhou 450001, China
2 School of Civil Engineering, Henan Vocational College of Water Conservancy and Environment, Zhengzhou 450008, China
3 Department of Mining Engineering, Faculty of Engineering, University of Kurdistan, Sanandaj 66177-15175, Iran
4 Department of Geotechnics and Transportation, Faculty of Civil Engineering, Universiti Teknologi Malaysia, Skudai 81310, Johor Bahru, Malaysia
* Correspondence: m.rezaei@uok.ac.ir

Abstract: In the primary and final designs of projects related to rock mechanics and engineering geology, one of the key parameters that needs to be taken into account is the intact rock elastic modulus (E). To measure this parameter in a laboratory setting, core samples with high-quality and costly tools are required, which also makes for a time-consuming process. The aim of this study is to assess the effectiveness of two meta-heuristic-driven approaches to predicting E. The models proposed in this paper, which are based on integrated expert systems, hybridize the adaptive neuro-fuzzy inference system (ANFIS) with two optimization algorithms, i.e., the differential evolution (DE) and the firefly algorithm (FA). The performance quality of both ANFIS-DE and ANFIS-FA models was then evaluated by comparing them with ANFIS and neural network (NN) models. The ANFIS-DE and ANFIS-FA models were formed on the basis of the data collected from the Azad and Bakhtiari dam sites in Iran. After applying several statistical criteria, such as root mean square error (RMSE), the ANFIS-FA model was found superior to the ANFIS-DE, ANFIS, and NN models in terms of predicting the E value. Additionally, the sensitivity analysis results showed that the P-wave velocity further influenced E compared with the other independent variables.

Keywords: elastic modulus; ANFIS; differential evolution; firefly algorithm

Citation: Wang, Y.; Rezaei, M.; Abdullah, R.A.; Hasanipanah, M. Developing Two Hybrid Algorithms for Predicting the Elastic Modulus of Intact Rocks. *Sustainability* **2023**, *15*, 4230. https://doi.org/10.3390/su15054230

Academic Editor: Jianjun Ma

Received: 30 January 2023
Revised: 22 February 2023
Accepted: 23 February 2023
Published: 26 February 2023

1. Introduction

When planning most projects pertinent to geotechnical issues and rock engineering, it is of high importance to properly analyze how the intact rock behaves and carefully estimate its associated mechanical properties. The intact rock elastic modulus (E) has substantial effects on both the initial and final steps of designing geoscience-related projects, which include planning tunnels; designing blasting operations in rock materials; analyzing the constancy of rock slopes; and designing rock pillars, roads, dams, bridges, etc. Moreover, E is the most significant parameter applied to analyzing the stress-strain chart of rock specimens in a laboratory. E also plays an important role in analyzing the deformations and breakage of rocks surrounding underground excavation projects. As a result, inaccurate predictions of E can result in serious damages, leading to economic issues and severe safety problems due to the breakage probability during construction processes [1,2]. Thus, it is necessary to determine the E value quickly and accurately in order to correctly plan geo-engineering structures, accurately design mining- and civil engineering-related projects, and enhance the general safety level and effectiveness of operations at hand.

In general, the E value can be obtained using direct or indirect methods. The former are typically carried out within rock mechanics laboratories, where rock core specimens are subjected to experiments in a variety of conditions [1–3]. In contrast, indirect methods make use of predictive equations or models to estimate E. The direct methods have accuracy, but at the same time, they suffer from some drawbacks. First, it is not easy to provide the required specimens during the coring process with a high level of accuracy, especially in jointed, layered, and weakened rock structures. Second, it is both difficult and time consuming to prepare the core specimens with the appropriate geometry for the purpose of carrying out laboratory E tests. Such issues hinder the use of direct methods unless there is a high necessity [3–5].

Due to the above-mentioned challenges, various indirect methods have been introduced in the literature on the basis of predictive models/algorithms and equations to determine the E value of intact rocks. These methods have been normally configured based on arithmetical and smart intelligent models. Statistical models generally make use of simple or multiple regression models aiming to develop a number of empirical equations between the E value and effective mechanical and physical rock properties. The literature consists of numerous empirical equations formed based on analyzing petrology and mineralogy in addition to values estimated using the rock physical and mechanical properties, such as Schmidt hammer numbers [6], porosity of rock [7], slake durability of rock [8], and compressional/primary wave velocity [9].

In recent years, scholars have made several efforts to develop artificial intelligence (AI) models applicable to mining and rock engineering problems. Such efforts have resulted in a number of novel models proposed to estimate E and some other rock mechanical properties. These are mostly based on probable and intelligent methods, such as particle swarm optimization (PSO), fuzzy inference systems (FISs), genetic algorithms (GAs), Bayesian methods, adaptive neuro-fuzzy inference systems (ANFIS), tree models, extreme gradient-boosting (XGB), and artificial neural networks (ANNs), as well as their hybridized forms [10–16]. Sarkhani Benemaran et al. [17] employed an XGB model in combination with several optimization algorithms to predict the resilient modulus of flexible pavement foundations. They concluded the effectiveness of PSO-XGB models in this field. In another study, conducted by Shahani et al. [18], different AI models such as XGB, gradient-boosted tree regressors (GBTRs), Catboost, and light gradient-boosting machines were used to predict E. According to their results, the performance of GBTR was better than that of the other developed models. Recently, Tsang et al. [19] predicted the E values through some other models, i.e., extreme gradient-boosting trees, ANNs, random forests, and classification and regression trees. The results showed that the extreme gradient-boosting trees predicted the E value with the highest accuracy.

Such applications show that intelligent algorithms typically outperform the traditionally used statistical methods regarding E value prediction.

The present study is carried out to assess the potential of applying two hybrid evolutionary models to predict E. The proposed models are based on the integrated expert systems comprising ANFIS with two optimization algorithms, i.e., the firefly algorithm (FA) and differential evolution (DE). To check the effectiveness of FA and DE, the results of ANFIS-FA and ANFIS-DE are then compared with the ANFIS and NN results. The rest of this study is organized as follows. In Section 2, the source of the database is described. Then, the methodologies used in this paper and their implementations are explained in detail in Section 3. Finally, Sections 4 and 5 present the results/discussions and conclusions of this study, respectively.

2. Source of Database

An inclusive database is needed to be formed for E modeling by means of indirect intelligent methods. Such data were obtained through performing laboratory experiments on the core specimens provided from the excavation drilling processes carried out in two under-construction dam sites, namely the Bakhtiari and Azad dams located in Iran. The precise location of the Azad dam site is in the west of Iran, 40 km away from the western city of Sanandaj in the Kurdistan state. It is situated on the Sanandaj–Marivan cities road inside Kurdistan, with the $46^\circ32'57''$ to $35^\circ19'59''$ geographical coordinates of eastern and northern latitudes, respectively. The construction of this dam is currently in progress. It is mainly aimed at supplying electrical energy and producing power plant storage. The dimensions (length, height, and width) and water storage capacity of this dam are 595 m, 115 m, 11 m, and 260,000,000 m^3, respectively.

The Bakhtiari dam is located in the Zagros Highlands, 65 km southwest of the town of Dorud in the Lorestan state, and 70 km northeast of the town of Andimeshk in Khuzestan, Iran. The position of the dam site is at the 48°45′34.87″ to 32°57′23.58″ geographical coordinates of eastern and northern latitudes, respectively (Figure 1). The dam was built upon the Bakhtiari River, aiming to provide adequate water for many purposes, such as drinking, electrical power generation, flood control, and agricultural activities. The dam's body is at an elevation of 840 m. In addition, in the case of this dam, the peak elevation, crown width, crown length, and foundation width are 325 m, 10 m, 434 m, and 30 m, respectively [20]. The situations of both case studies (the Azad and Bakhtiari dams) on the Iran map are illustrated in Figure 1.

Figure 1. Situations of both case studies (the Azad and Bakhtiari Dams).

The Azad dam comprises a series of common structures, such as tailraces, higher deposits, different caverns, gurgitation storages, and access tunnels. Geologically, this dam is situated in Iran's famous formation, Sanandaj–Sirjan, with an alternation of schist, sandstone, limestone, and phylite rocks. The bedrock of the dam mainly comprises sandstone with a low degree of metamorphic, phyllite, and schist. Additionally, within the highland areas, lenses of limestone are also observable. From the stratigraphy point of view, rock outliers from the higher Cretaceous period to the present can be observed within the investigated region. Such rocks consist of four types from the past to the current session:

(1) metamorphic rock related to the Cretaceous period that includes a combination of clay and shale; (2) phyllite formation rocks related to the participation of the Cretaceous Paleocene periods, containing limestone and shale with sand; (3) rocks related to the participation of the Paleocene and Eocene periods, comprising sandstone, shale, limestone lenses, and volcanic rocks; and (4) formations related to the Quaternary period, consisting of shallow terraces and debris. From a tectonic viewpoint, the Sarvabad, Kargineh, and Satileh faults are situated 23, 4.5, and 32 km to the south, east, and northeast of the Azad dam, respectively [21]. The geological conditions and faults of the Azad dam site are shown in Figure 2.

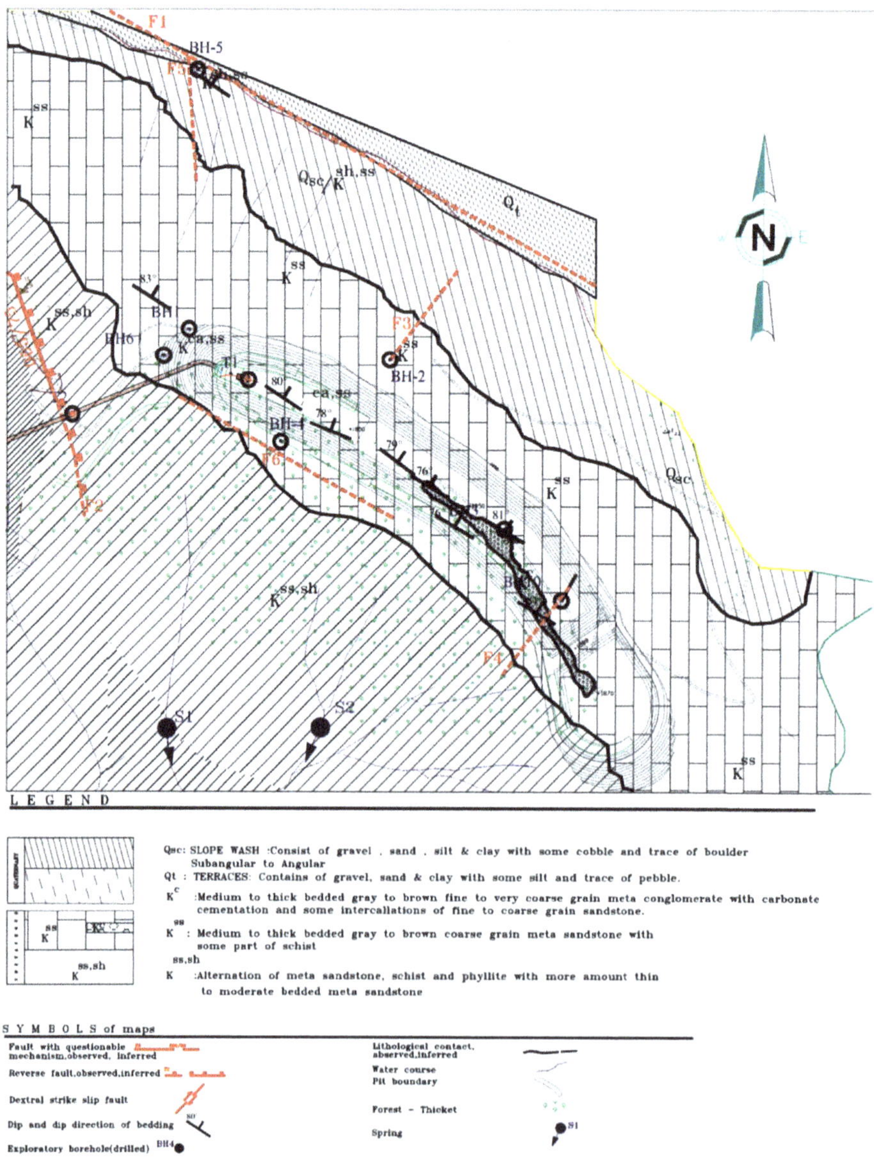

Figure 2. Geological conditions of the Azad dam site [15].

The Bakhtiari dam's bedrock is made of separate limestone and limestone combined with marl, which incorporates chert nodes. The limestone sections might be synthesized by a combined dolomite substance. From the perspective of geological structure, the dam area is positioned within the pleated Zagros, a portion of the tectono-sediment region of the Zagros. In the lowest northern portion, the area is restricted by pushed Zagros, while in the southwest, it is confined by the Khuzestan plain. With regard to the age of the compressed reservoirs of the area, they date back to between the Triassic and Pliocene eras, and then would have been wrinkled from the Plio-Pleistocene via the latest Alpine organic phase. A number of syncline and anticline sets have been created through such tectonism procedures. Primarily, the above arrangements have been identified by perpendicular axial levels related to the lots of pushed faults in the Zagros area. Additionally, key bed-rocks of the investigated area are made of limestone siliceous related to the current famous formation, Sarvak. This formation (Sarvak) belongs to the Bangestan collection of the middle period of the Cretaceous [20]. For a better review, the geological cross-section of the Bakhtiary dam and plant site is shown in Figure 3.

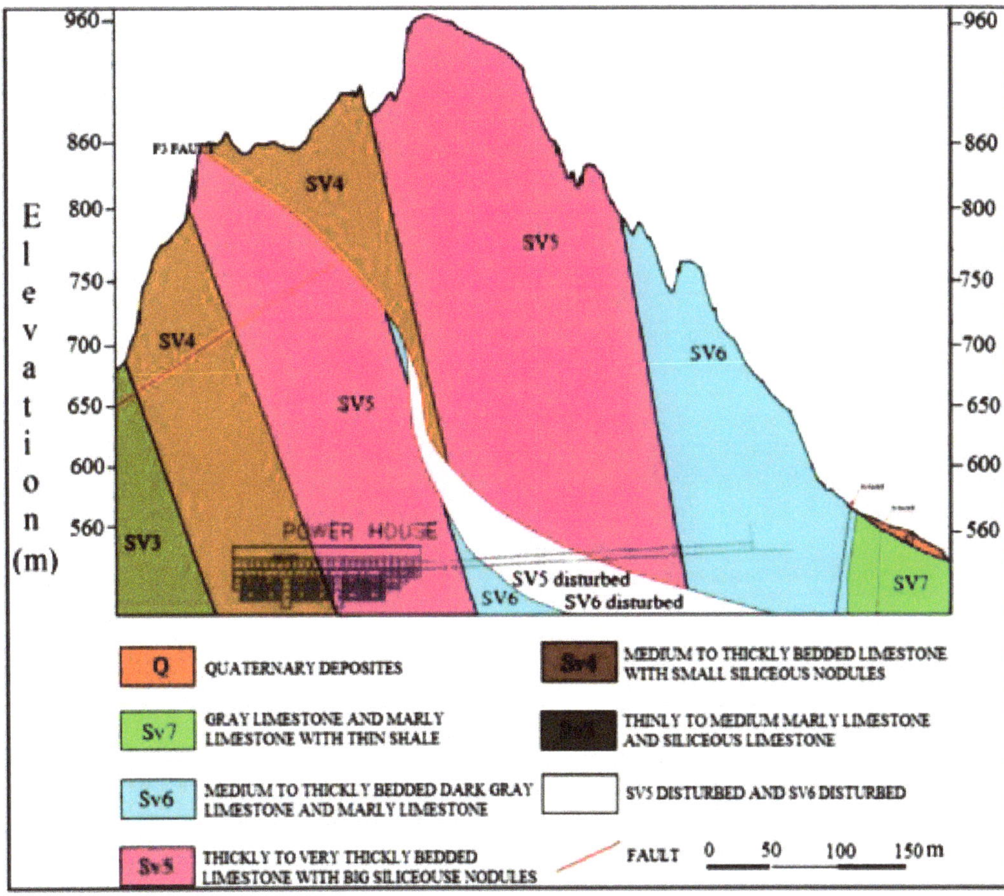

Figure 3. Geological cross-section of the Bakhtiary dam and plant site [15].

Database Description

To create inclusive datasets, adequate core samples with NX sizes, i.e., 54 mm in diameter, perpendicular cylindrical shapes, and ratios of height to diameter in the middle of

2:1 to 2.5:1 were used based on the process recommended by the ISRM [22,23]. The samples were arranged from the two dam sites introduced before. When the core specimens were prepared, their various characteristics, such as E value, porosity, density, and durability index, were measured in a laboratory. Furthermore, in the course of the coring operation, each sample's coring depth was recorded for the purpose of evaluating its impact on the rocks' geomechanical properties. The laboratory experiments in this study were carried out totally based on the ISRM and ASTM standard methods [22,23]. In this regard, a total of 50 test series were done successfully, and the outputs were documented in the cases of all variables noted above. As a result, 50 datasets were provided, aiming to construct the ANFIS-FA, ANFIS-DE, ANFIS, and NN models. Then, a sorting approach was adopted to divide the available database into training (constructing) and testing datasets. Roughly 20% of the database was determined as the testing dataset in order to be used later in the process of evaluating the models built in this paper.

Note that the ratio of 80 to 20 for training and testing groups has been widely suggested by many scholars, such as Ye et al. [24], Fang et al. [25], Nguyen et al. [26], and Zhou et al. [27]. Aside from that, we also tested the ratio of 70:30. Nevertheless, the 80:20 ratio had better performance; thus, this ratio was used in this study.

The statistical characteristics of all variables used in this study are shown in Table 1. For a better view, the frequency histogram of all input and output variables are depicted in Figures 4 and 5. For example, in Figure 4, regarding the depth of coring variables, 11, 21, 4, and 4 data were varied in the range of 0–50 m, 50–100 m, 100–150 m, and 150–250 m, respectively. In addition, Figure 6 illustrates the Pearson correlation plots for all variables.

Table 1. Modelling variables and the statistical characteristics datasets.

Statistical Characteristics	Variables													
	DC (m)		ρ (g/cm^3)		n (%)		DI (%)		ν		Vp (m/s)		E (GPa)	
	Train	Test	Train	Test	Train	Test	Train	Test	Train	Test	Train	Test	Train	Test
Mean	79.243	149.455	2.679	2.703	1.992	2.528	98.92	95.45	0.221	0.292	5.564	5.013	18.526	17.586
Standard Error	7.215	21.803	0.007	0.013	0.206	0.429	0.062	0.098	0.007	0.014	0.119	0.145	1.294	2.028
Standard Deviation	45.633	68.948	0.049	0.043	1.305	1.357	0.394	0.310	0.045	0.04	0.756	0.458	8.185	6.413
Sample Variance	2082.38	4753.87	0.002	0.002	1.704	1.842	0.156	0.096	0.002	0.002	0.572	0.210	67.009	41.134
Skewness	1.361	0.051	−1.127	−1.003	1.291	1.795	−0.403	−1.118	0.121	−0.677	−1.258	0.225	0.487	−0.484
Minimum	14	55.85	2.52	2.616	0.37	1.39	98.1	94.8	0.14	0.21	2.985	4.316	3	7.98
Maximum	213.4	248.5	2.74	2.75	5.81	5.81	99.6	95.8	0.3	0.34	6.652	5.82	42.8	25.17

DC: depth of coring, ρ: density, n: porosity, DI: durability, ν: Poisson ratio, Vp: P-wave velocity, E: elastic modulus.

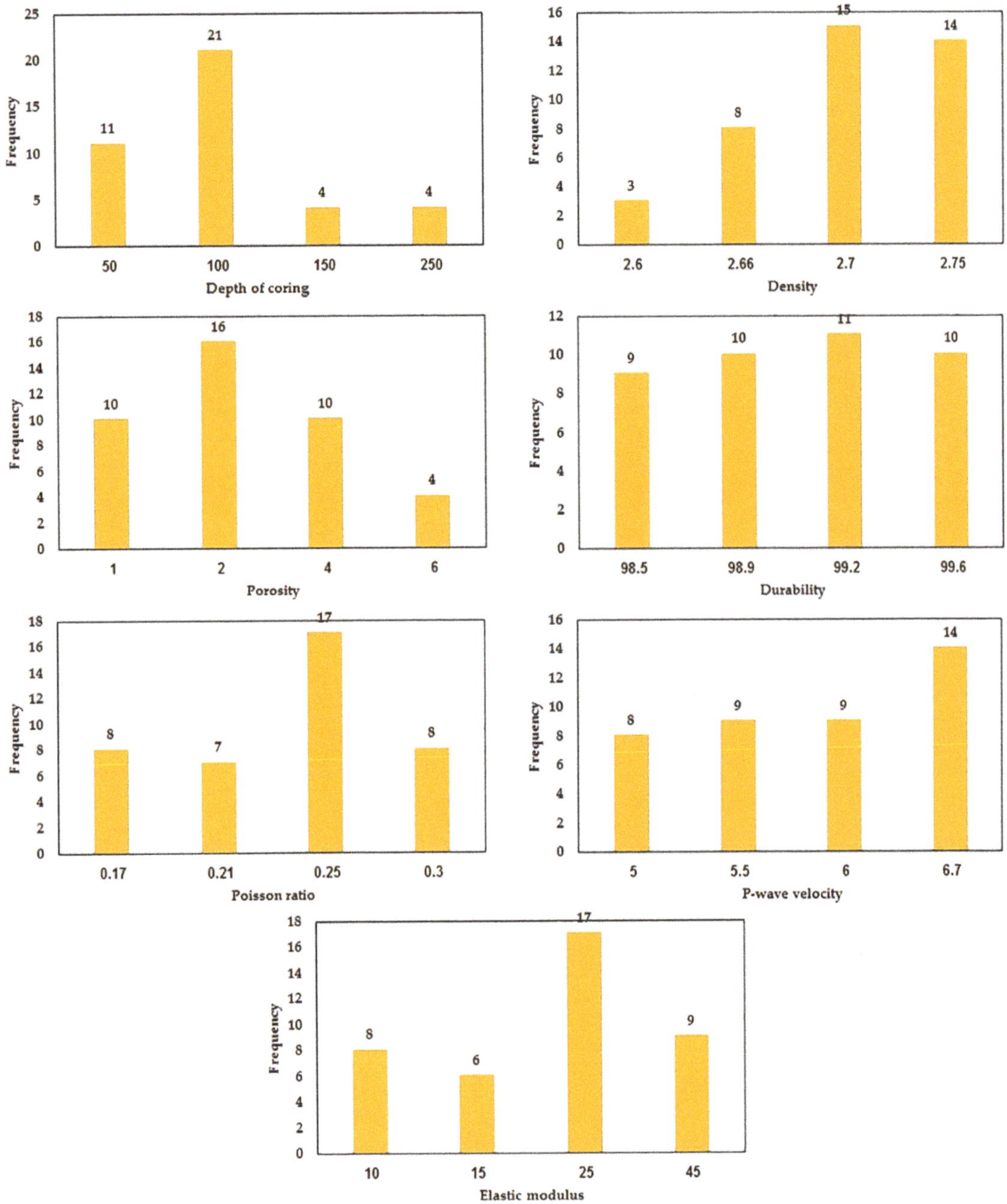

Figure 4. Frequency histogram of all variables for the training group.

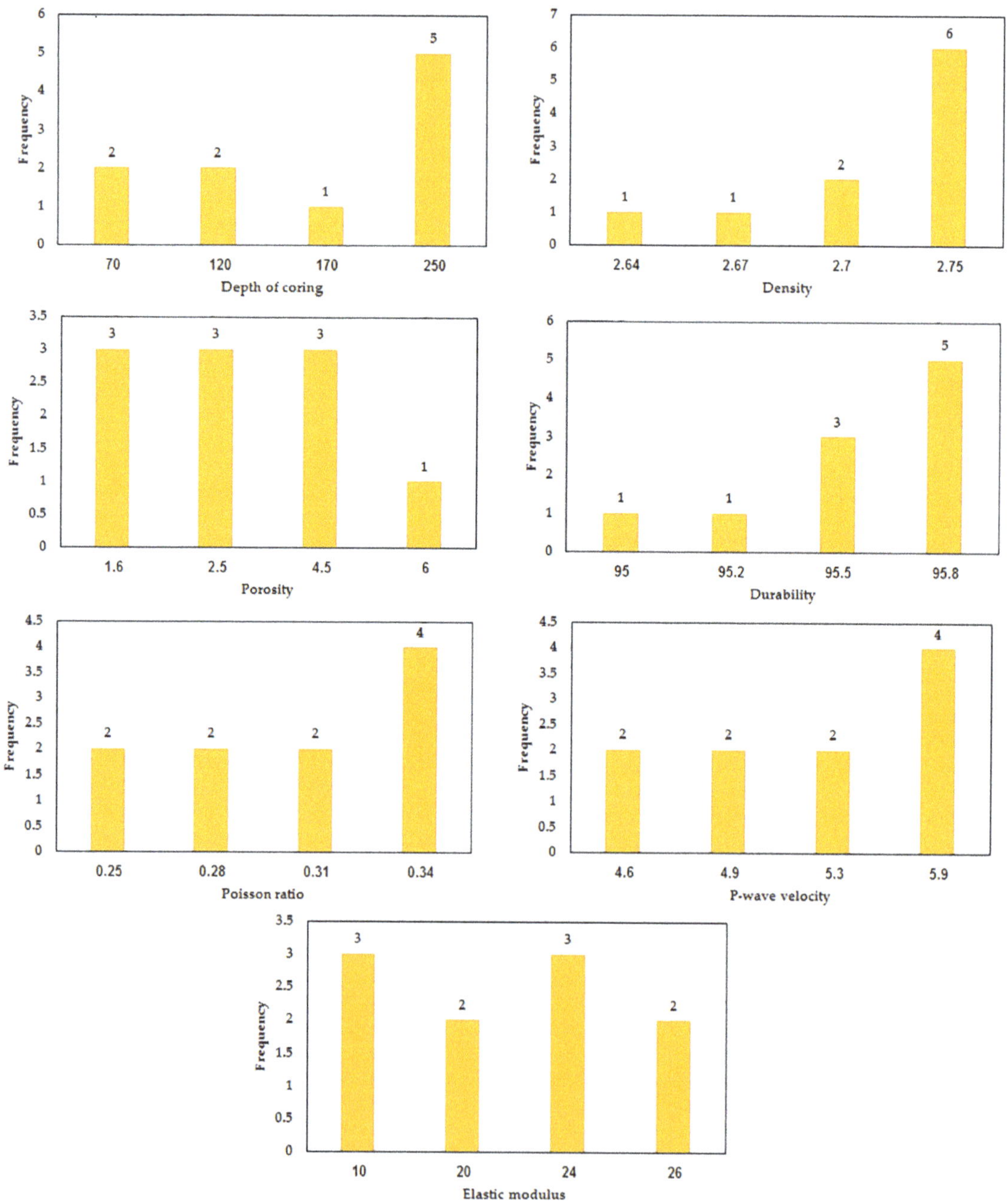

Figure 5. Frequency histogram of all variables for the testing group.

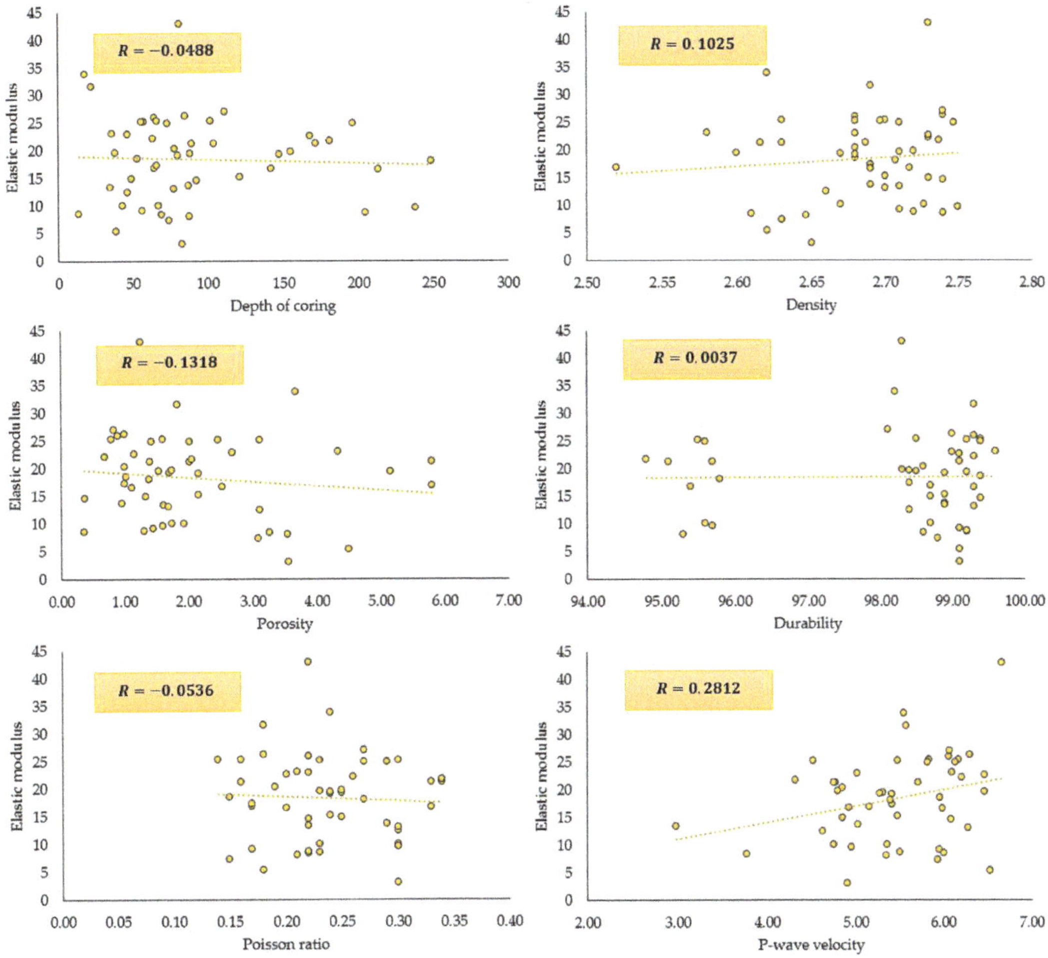

Figure 6. Pearson correlation plots for all input variables vs. the output (E).

3. Methodology

This section explains how ANFIS combined with the FA and DE algorithms is implemented. Additionally, the modeling process of the NN model is explained in this section. In the aforementioned models, of the total 50 datasets, 40 were used for the training phase and 10 were used for the testing phase. For a better overview, a schematic flowchart of the ANFIS-FA and ANFIS-DE proposed in this study is shown in Figure 7. It is worth mentioning that MATLAB@2018 was used to encode the proposed hybrid models.

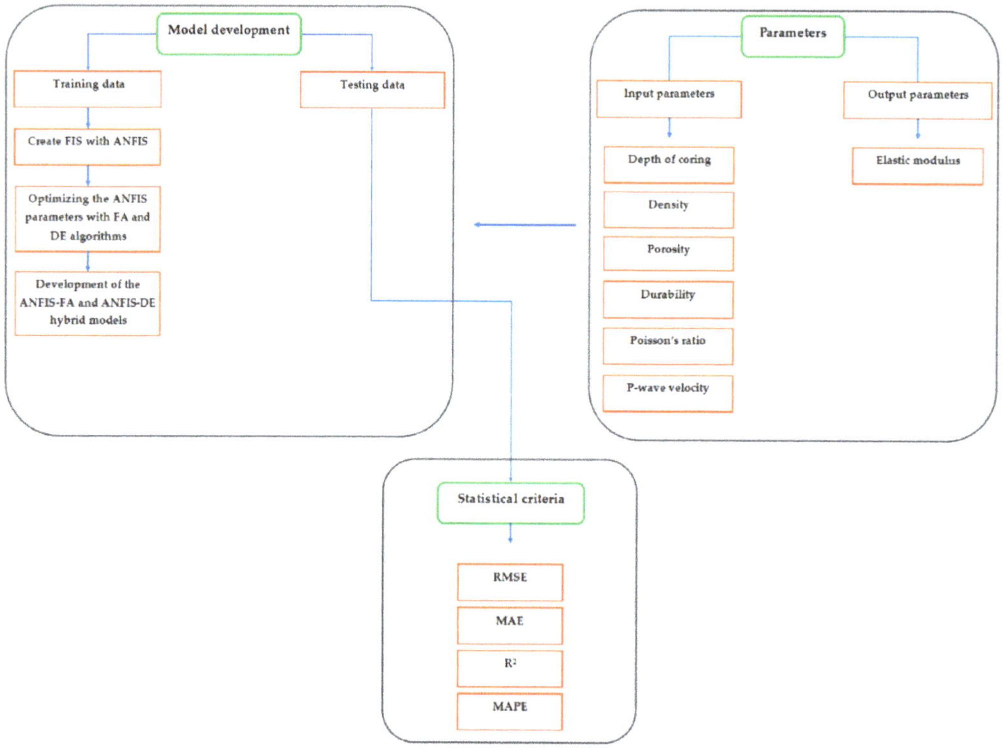

Figure 7. A schematic flowchart of the ANFIS-FA and ANFIS-DE proposed in this study.

3.1. *ANFIS Combined with DE*

Differential evolution (DE), which was originally proposed by Storn and Price [28], is an effective evolutionary algorithm that works on the basis of a global optimization approach. In general, DE offers three benefits: (1) a simple structure, (2) high-quality solutions achieved, and (3) easy implementation [29]. As a result, it is applied to a variety of conditions. In the present study, DE is used for the aim of minimizing the function of fitness using the amounts of optimized variable. By definition, the function of fitness refers to the root mean square error (RMSE) between the estimated and target datasets. DE, as an innovative algorithm, was implemented in order to adjust the functions of membership amounts of the ANFIS model and, consequently, enhance its overall prediction capability. Figure 4 illustrates the schematic presentation of the hybrid ANFIS-based DE algorithm.

To model ANFIS-DE, four parameters must be specified, namely the number of the iteration, crossover probability, mutation probability, and population size. For the selection of the optimal mutation probability, various values were examined, which can be seen in Table 2. The table also shows that by setting the mutation probability to 0.3, the optimum performance with the highest rank related to the testing phase (i.e., the maximum R^2 values) was attained. To obtain the best crossover probability, different values were examined (see Table 3). The table shows that the maximum R^2 values were attained when the crossover probability was fixed at 0.75. Different population sizes were also tested, as can be observed in Table 4. The table clearly demonstrates that when the population size was set to 250, the best result (the maximum R^2 values) was achieved. In these tests, the smallest amount of error was fixed at 1×10^{-5}, and the peak repetition was set to 500. Accordingly, the crossover probability, mutation probability, and scope of population were fixed to 0.75, 0.3, and 250, respectively. It is worth mentioning that the bolded amounts in Tables 2–4 are related to the best results (highest rank) obtained from the developed models.

Table 2. Selection of the most optimum mutation rate value in implementing the ANFIS-DE model.

Mutation Rate	Performance Criteria R^2		Rank
	Train	Test	
0.05	0.931	0.917	2
0.10	0.942	0.929	5
0.15	0.947	0.923	4
0.20	0.951	0.948	9
0.25	0.946	0.945	8
0.30	**0.958**	**0.952**	**10**
0.35	0.955	0.941	7
0.40	0.948	0.940	6
0.45	0.930	0.925	3
0.50	0.919	0.902	1

Table 3. Selection of the most optimum crossover value in implementing the ANFIS-DE model.

Crossover Rate	Performance Criteria R^2		Rank
	Train	Test	
0.60	0.947	0.932	2
0.65	0.944	0.938	3
0.70	0.953	0.946	5
0.75	**0.964**	**0.961**	**7**
0.80	0.960	0.952	6
0.85	0.943	0.940	4
0.90	0.932	0.927	1

Table 4. Selection of the most optimum population size value in implementing the ANFIS-DE model.

Population Size	Performance Criteria R^2		Rank
	Train	Test	
50	0.929	0.926	1
100	0.935	0.934	2
150	0.948	0.935	3
200	0.964	0.961	9
250	**0.976**	**0.970**	**10**
300	0.975	0.960	8
350	0.969	0.949	6
400	0.955	0.942	5
450	0.957	0.938	4
500	0.951	0.950	7

Figure 8 also shows the ANFIS-DE flowchart used in this study.

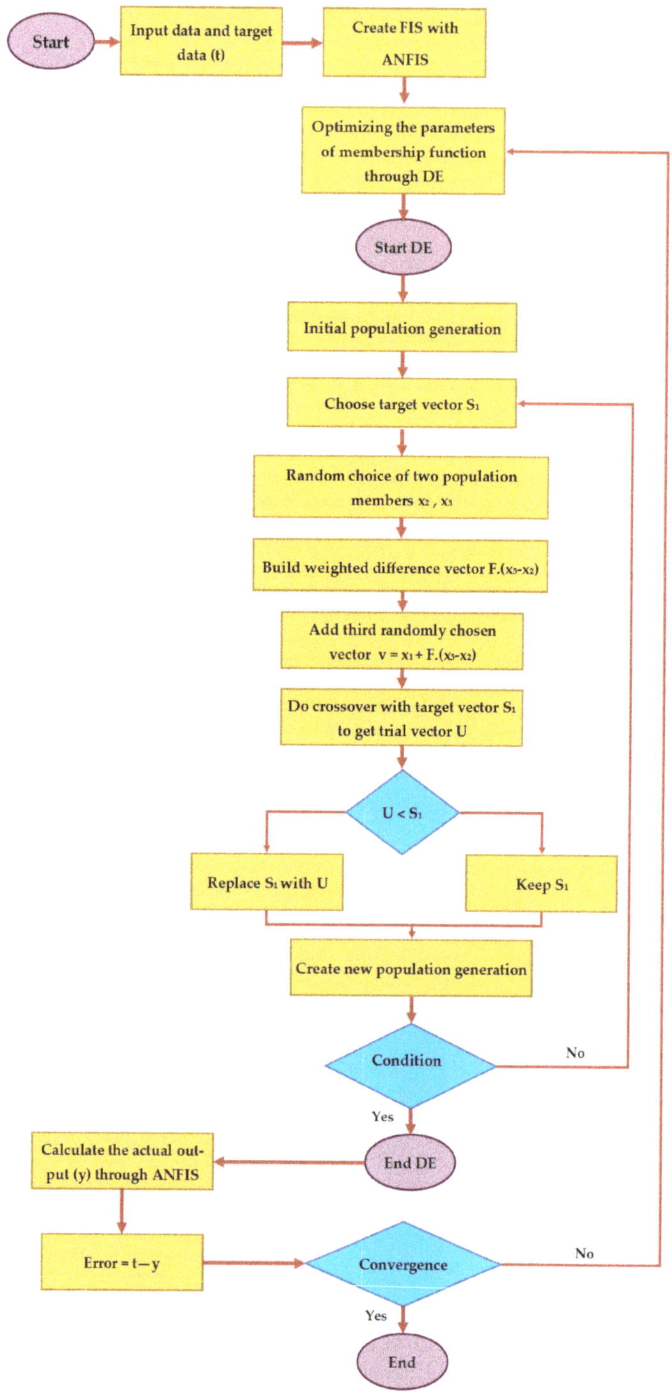

Figure 8. ANFIS-DE Flowchart.

3.2. ANFIS Combined with FA

This section introduces a hybridized model combining ANFIS and FA, called ANFIS-FA, with the objective of optimizing the premise parameters of ANFIS. To initiate the modeling process, there is a need to first determine the input-target variables/parameters. When the assembly of the input-target variables/parameters of the model is determined, it is time to determine the training samples. The reason for this is that a model should be capable of predicting the target parameter for those samples that have no effect on the model training process. As a result, the entire dataset was divided into two categories: training/construction and testing/examination samples. As previously mentioned, from among a total of 50 datasets, 20% (n = 10) were chosen in a random way and assigned to the testing group, while the remaining 80% (n = 40) were assigned to the training samples. In addition, it was required to initialize the ANFIS and FA parameters prior to ANFIS modeling in order to predict the target variables. With the use of a trial-and-error-based approach, the optimal amount of membership functions (MFs) was archived as 6. According to the literature [30–32], the α, β_0, γ, number of variables, and the population are the FA parameters. To improve the ANFIS performance, it was necessary to select the most appropriate values for the aforementioned FA parameters. Table 2 clearly shows that the number of variables is equal to six. As stated in the literature [31,32], in some cases, the value of 1 is suitable for the β_0 parameters. Therefore, in the modelling of ANFIS-FA, the value of β_0 was set to 1.

To select the most appropriate values for the α and γ parameters, various amounts of these parameters were examined, as given in Tables 5 and 6. Considering these tables, the most appropriate values (the highest R^2) for the α and γ parameters were obtained with $\alpha = 0.6$ and $\gamma = 1.5$. As a result, the values of 0.6 and 1.5 were used for the α and γ parameters in ANFIS-FA modelling. By setting the number of iterations to 1000, different values were also tested to select the most appropriate value for the population size, as shown in Table 7. The table shows that the best performance was attained with population = 200. Based on the above descriptions, the values of 6, 1, 0.6, 1.5 and 200 were set as the number of variables, β_0, α, γ, and population size, respectively. It is worth mentioning that the bolded amounts in Tables 5–7 are related to the best results (highest rank) obtained from the developed models. In this step, the most appropriate value of the number of iterations needed to be determined. According to the results, after the 15th iteration, no significant change was observed in the ANFIS-FA performance. In other words, after 15 iterations, the performance for different populations was constant. Accordingly, the number of iterations in ANFIS-FA modelling used in this study was set to 15. When the user-based defined parameters in the investigated models (FA and ANFIS) were determined, then the ANFIS training process was begun with the use of the training samples. To this end, the FA algorithm was used to optimize the primary part of the fuzzy If-Then rules, and the least-square method was applied to the optimization of the linear consequent fuzzy rules.

Additionally, the preliminary light strength corresponding to the primary generation was computed, and then each firefly's attractiveness level was measured. With the use of the movement equation, those fireflies that had a lower level of attraction were pushed toward the brighter firefly. Afterward, the light strength and individual firefly's position were updated, and the function of fitness was computed again. All steps involved in ANFIS-FA are displayed in Figure 9.

Table 5. Selection of the optimum α value in implementing the ANFIS-FA.

α	Performance Criteria R^2		Rank
	Train	Test	
0.10	0.945	0.933	1
0.20	0.947	0.941	4
0.30	0.954	0.943	5
0.40	0.951	0.948	6
0.50	0.958	0.955	8
0.60	**0.962**	**0.956**	**9**
0.70	0.951	0.949	7
0.80	0.944	0.938	2
0.90	0.941	0.940	3

Table 6. Selection of the optimum γ value in implementing the ANFIS-FA.

γ	Performance Criteria R^2		Rank
	Train	Test	
0.5	0.951	0.947	1
1	0.959	0.955	3
1.5	**0.973**	**0.968**	**6**
2	0.971	0.960	5
2.5	0.965	0.958	4
3	0.952	0.951	2

Table 7. Selection of the optimum population value in implementing the ANFIS-FA.

Population	Performance Criteria R^2		Rank
	Train	Test	
50	0.965	0.961	4
100	0.972	0.966	6
150	0.973	0.971	7
200	**0.979**	**0.975**	**8**
250	0.971	0.965	5
300	0.963	0.960	3
350	0.958	0.957	2
400	0.955	0.948	1

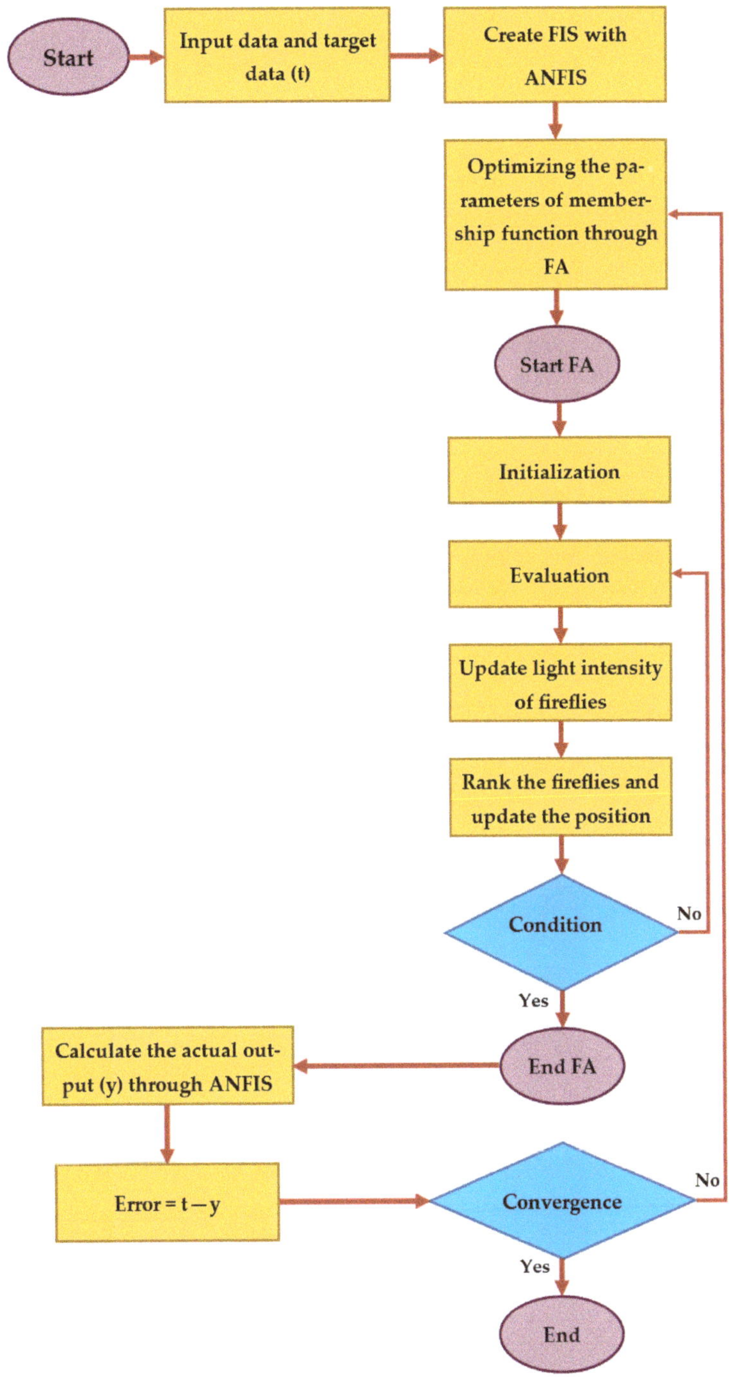

Figure 9. Implementing the ANFIS-FA used in this study.

3.3. Neural Network (NN)

Neural Networks (NNs), especially the Multi-Layer Perceptron (MLP), are widely used in prediction models applied to different engineering problems [31,32]. MLP, which is employed in the present study, contains three layers: input, hidden, and output layers. Therefore, as can be seen in Figure 6, the nodes that exist within the input layer correspond to DC, ρ, n, DI, v, and Vp, while those in the output layer correspond to E. Based on the trial-and-error approach, we considered the number of nodes within the hidden layer. The evaluation results showed that the existence of seven nodes within the hidden layer can result in a higher reliability. As can be observed in Figure 10, the hidden layer with seven nodes resulted in the optimal performance of NN (with the maximum R^2). It is worth mentioning that, to select the suitable number of nodes inside the hidden layer, different numbers were tested. As a result, the NN structure in the present research was built on the basis of six nodes within the first/input layer, seven nodes within the second/hidden layer, and one node within the last/output layer.

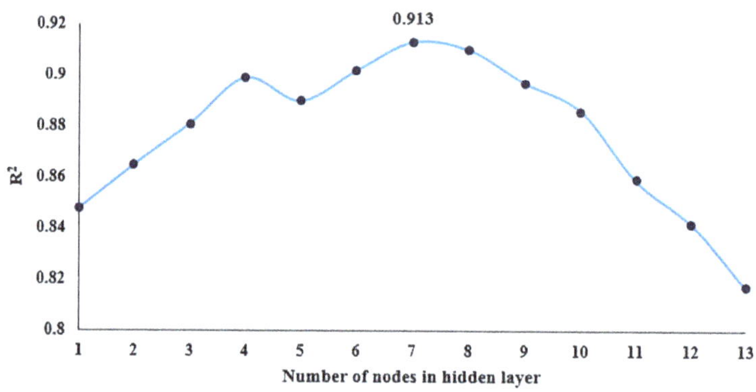

Figure 10. Use of various numbers of nodes in the second/hidden layer with their R^2 values.

4. Results and Discussion

The present study was aimed at examining the effectiveness of the FA and DE algorithms in optimizing ANFIS for the prediction of E. The results obtained from the proposed ANFIS-FA and ANFIS-DE models were compared to those of ANFIS and NN models. Here, the models' prediction capabilities were assessed regarding RMSE, mean of average percentage error (MAPE), mean of absolute error (MAE), variance account for (VAF), *A10-index*, and performance index (PI) [33–37], as presented in the following equations:

$$\text{MAE} = \frac{1}{n}\sum_{i=1}^{n}|A_i - P_i| \tag{1}$$

$$\text{RMSE} = \sqrt{\frac{\sum_{i=1}^{n}(A_i - P_i)^2}{n}} \tag{2}$$

$$\text{MAPE} = \left[\frac{1}{n}\sum_{i=1}^{n}\frac{|A_i - P_i|}{A_m}\right] \times 100 \tag{3}$$

$$\text{VAF} = \left[1 - \frac{var\ (A_i - P_i)}{var\ (A_i)}\right] \times 100 \tag{4}$$

$$A10 - index = \frac{m10}{n} \tag{5}$$

$$PI = \frac{1}{\overline{A_i}}\frac{RMSE}{R+1} \tag{6}$$

where n stands for the number of data (n = 50), and A_i, P_i, and $\overline{A_i}$ signify the actual, estimated, and average of actual E values, respectively. Additionally, m10 is the number of data with values of rate actual/predicted values (ranging from 0.9 to 1.1), and R in Equation (6) is the correlation coefficient. Table 8 presents the MAPE (%), MAE, RMSE, VAF(%), and *A10-Index* values attained by the developed models.

Table 8. Performance of the models to predict E by using six statistical criteria.

Model	Statistical Criteria												Total Rank	
	RMSE		MAE		MAPE (%)		VAF (%)		*A10-Index*		*PI*			
	Train; Rank	Test; Rank	Train; Rank	Test; Rank	Train; Rank	Test; Rank	Train; Rank	Test; Rank	Train; Rank	Test; Rank	Train; Rank	Test; Rank	Train	Test
ANFIS	2.337; 2	2.557; 2	2.258; 2	2.456; 2	10.173; 2	13.965; 2	93.274; 2	92.514; 2	0.35; 2	0.4; 2	0.064; 2	0.070; 2	12	12
NN	2.491; 1	2.781; 1	2.409; 1	2.639; 1	10.849; 1	15.006; 1	92.341; 1	90.985; 1	0.3; 1	0.4; 2	0.068; 1	0.076; 1	6	7
ANFIS-DE	1.447; 3	1.827; 3	1.384; 3	1.662; 3	6.236; 3	9.451; 3	97.436; 3	96.957; 3	0.75; 3	0.6; 3	0.039; 3	0.049; 3	18	18
ANFIS-FA	**0.909; 4**	**1.152; 4**	**0.865; 4**	**1.100; 4**	**3.899; 4**	**6.254; 4**	**98.962; 4**	**98.778; 4**	**0.925; 4**	**0.9; 4**	**0.024; 4**	**0.031; 4**	**24**	**24**

As can be observed in Table 8, the lowest MAPE (%), MAE, RMSE, and PI values were determined for the ANFIS-FA model as 6.254%, 1.1, 1.152, and 0.031, respectively. In addition, the highest VAF (%) and *A10-index* values were determined for the ANFIS-FA model as 98.778% and 0.9, respectively. These values were calculated for the ANFIS-DE model as 1.827, 1.662, 9.452%, 0.049, 96.957%, and 0.6, respectively; for the ANFIS model as 2.557, 2.456, 13.965%, 0.070, 92.514%, and 0.4, respectively; and for the NN model as 2.781, 2.639, 15.006%, 0.076, 90.985%, and 0.4, respectively. According to Table 8, the highest total rank values for both the training and testing groups were obtained by the ANFIS-FA model. It is worth mentioning that the bolded amounts in Table 8 are related to the best results (highest rank) obtained from the ANFIS-FA model. For a better overview, the predicted E values provided by all models in the testing phase are depicted in Figure 11. Additionally, Figure 12 shows the amount of error for each model related to the testing phase. According to these two figures, the prediction of E by the ANFIS-FA model is very accurate and closer to measured E values. In addition, Figures 13 and 14 demonstrate the scatter plots of actual versus estimated E values with the use of all predictive models. The figures show that the ANFIS-FA model obtained a greater value for the coefficients of determination (R^2). The R^2 values of 0.988, 0.970, 0.928, and 0.913 were obtained by the ANFIS-FA, ANFIS-DE, ANFIS, and NN models, respectively. Accordingly, FA was more effective in comparison with DE in regard to the ANFIS improvement. Furthermore, the absolute error of ANFIS-FA, ANFIS-DE, ANFIS, and NN models in predicting E for testing datasets (ten datasets) is depicted in Figure 15. According to this Figure, the orange-coloured line, which was obtained by the ANFIS-FA model, yields the lowest absolute error for all ten datasets. Moreover, the Taylor diagrams for both training and testing groups are shown in Figure 16. The results show that the ANFIS-FA has a stronger potential to predict E than the others. In this study, a sensitivity analysis was also performed. For this work, the effect of removing each input variable on E for the ANFIS-FA was calculated. In this regard, six new models based on the combination of input variables were constructed, as follows:

Model 1: inputs: all variables given in Table 1.
Model 2: inputs: all variables given in Table 1 except the depth of coring.
Model 3: inputs: all variables given in Table 1 except density.
Model 4: inputs: all variables given in Table 1 except porosity.
Model 5: inputs: all variables given in Table 1 except durability.
Model 6: inputs: all variables given in Table 1 except Poisson ratio.
Model 7: inputs: all variables given in Table 1 except P-wave velocity.

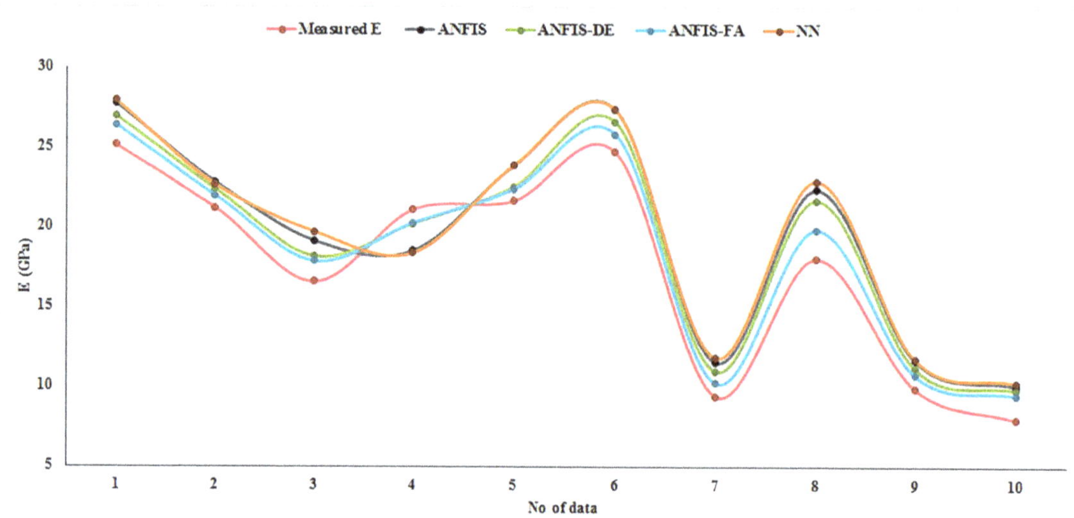

Figure 11. Comparing the values of E predicted by the four models.

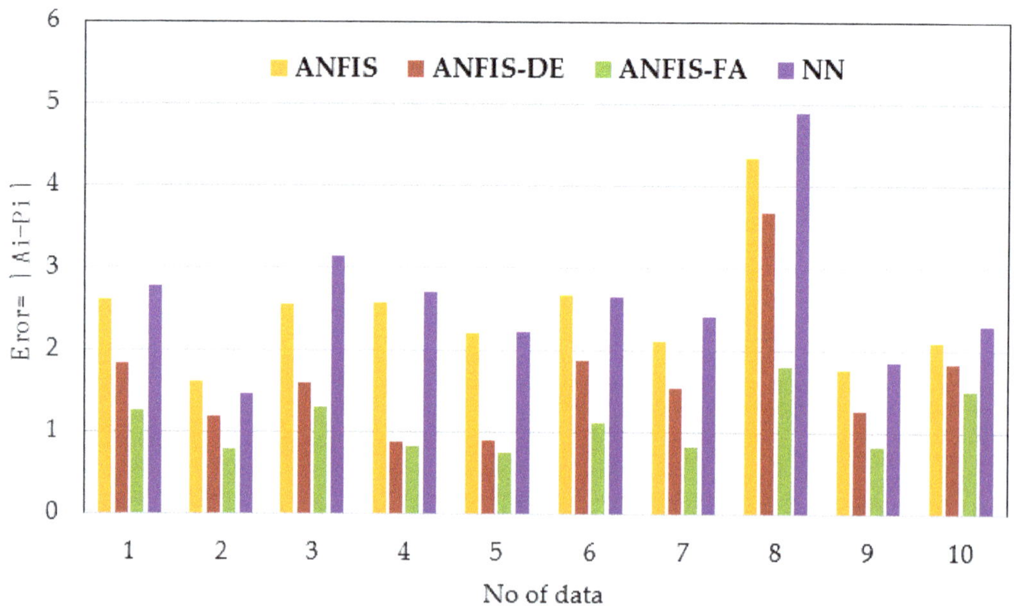

Figure 12. Amount of error for each model, as related to the testing phase.

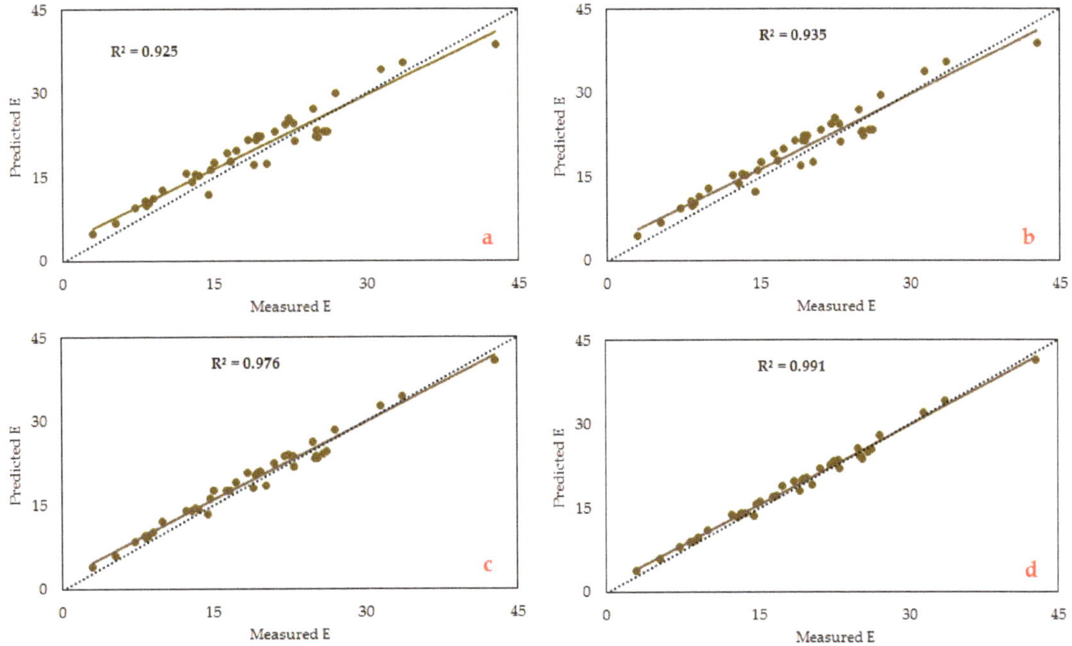

Figure 13. Comparison of the actual E value with those predicted, obtained by (**a**) NN, (**b**) ANFIS, (**c**) ANFIS-DE, and (**d**) ANFIS-FA for the training group.

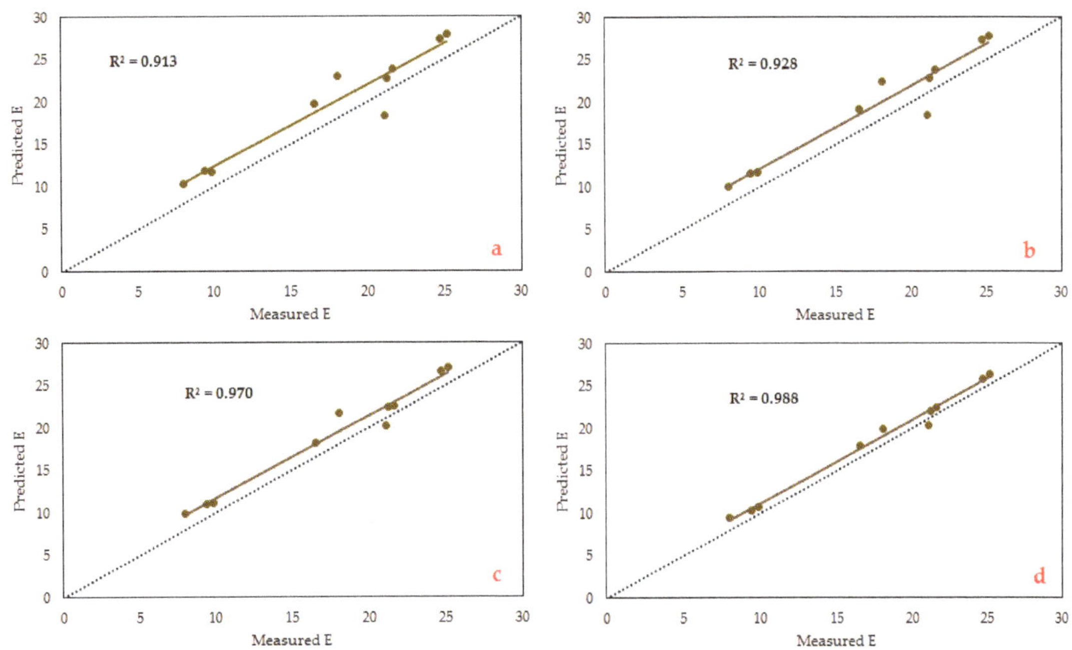

Figure 14. Comparison of the actual E value with those predicted, obtained by (**a**) NN, (**b**) ANFIS, (**c**) ANFIS-DE, and (**d**) ANFIS-FA for the testing group.

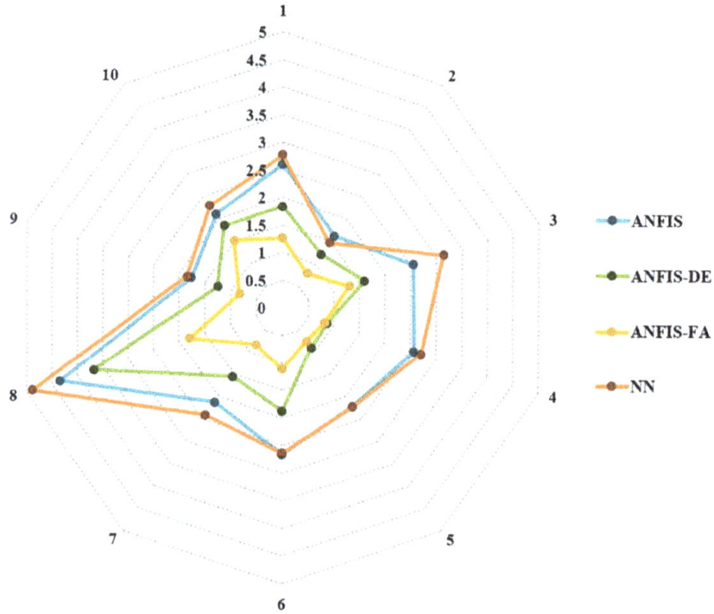

Figure 15. Absolute error values obtained by the models using the testing phase.

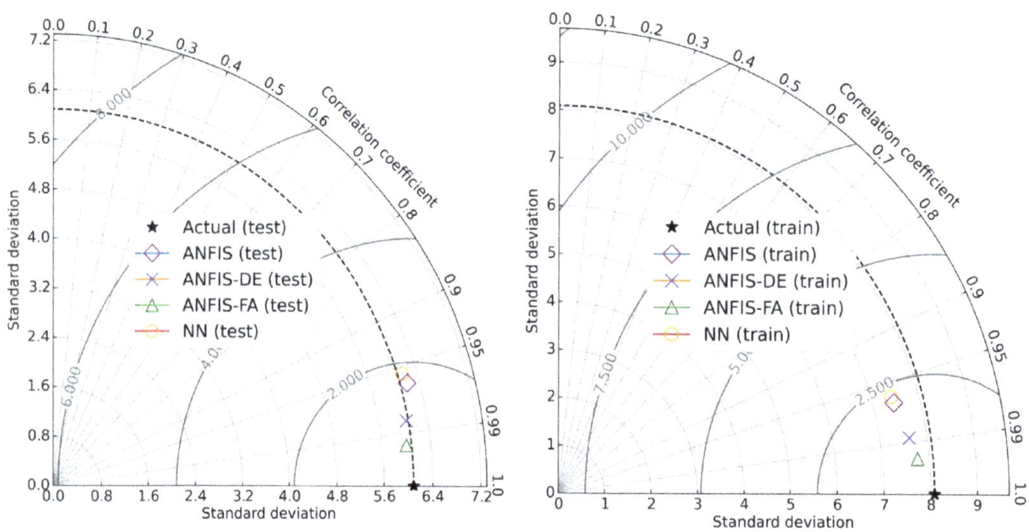

Figure 16. Taylor diagrams obtained by the predictive models.

The results of the above models are presented in Table 9, which shows that models 1 and 7 had the highest total rank (i.e., the best performance) and lowest total rank (i.e., the worst performance), respectively (Figure 17). Note that, the results of model 1 is bolded in Table 9. The results of presented in Table 9 indicated that once the P-wave velocity was removed from the modeling, the worst performance was obtained; thus, P-wave velocity can be determined as the most effective variable in the modeling.

Table 9. Performance of all seven ANFIS-FA models.

Model	Statistical Criteria								Total Rank	
	RMSE		MAE		MAPE (%)		VAF (%)			
	Train; Rank	Test; Rank	Train; Rank	Test; Rank	Train; Rank	Test; Rank	Train; Rank	Test; Rank	Train	Test
Model 1	**0.909; 7**	**1.152; 7**	**0.865; 7**	**1.100; 7**	**3.899; 7**	**6.254; 7**	**98.962; 7**	**98.778; 7**	**28**	**28**
Model 2	1.776; 3	2.306; 3	1.683; 4	2.204; 3	7.583; 4	12.532; 3	95.979; 3	95.848; 3	14	12
Model 3	1.570; 4	1.523; 6	1.508; 5	1.460; 6	6.790; 5	8.30; 6	96.910; 5	96.50; 5	19	23
Model 4	1.739; 5	2.056; 4	1.695; 3	2.022; 4	7.635; 4	11.497; 4	96.673; 4	96.254; 4	16	16
Model 5	1.856; 2	2.447; 2	1.808; 2	2.386; 2	8.146; 2	13.567; 2	95.899; 2	95.321; 2	8	8
Model 6	1.524; 6	1.917; 5	1.441; 6	1.838; 5	6.493; 6	10.451; 5	97.353; 6	96.808; 6	24	21
Model 7	2.279; 1	3.054; 1	2.167; 1	2.986; 1	9.761; 1	16.979; 1	93.408; 1	92.691; 1	4	4

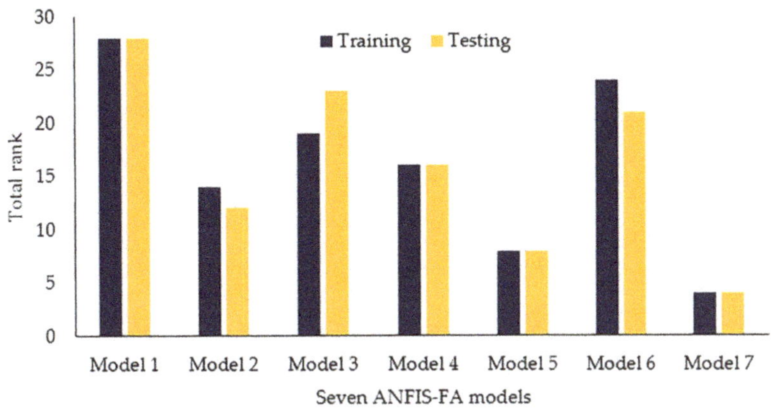

Figure 17. Total ranks of all seven constructed ANFIS-FA models to evaluate the sensitivity analysis.

5. Conclusions

The elastic modulus (E) is considered one of the most significant factors in the primary and ultimate plans of projects related to the geo-engineering field. As a result, it is highly necessary to predict E with a high accuracy level. This paper examined the use of two hybrid evolutionary models, namely ANFIS-FA and ANFIS-DE, to predict E. Additionally, the traditional ANFIS and NN models were developed for comparison aims. In total, 50 datasets were collected during the drilling process in the Azad and Bakhtiari under-construction dams in Iran. Out of the 50 datasets, 40 were used to construct the models, and the remaining datasets were used to test them. The input parameters considered in the construction of the models were porosity, density, depth of coring, Poisson's ratio, compressional/primary wave velocity, and durability, which were assigned as the input variables, whereas E was the output/target variable. Finally, some statistical indices were designed in order to demonstrate the capacity of the models in the prediction of E. According to the findings, the following results and remarks can be briefly listed:

1. The results demonstrated that ANFIS-FA was the most suitable model for the prediction of E in the cases studied. The ANFIS-DE, ANFIS, and NN models were identified as the next cases in this rank.
2. The FA and DE algorithms strongly improved the ANFIS performance in terms of predicting the E value. This confirms the effectiveness of FA and DE; accordingly,

these two algorithms can be effectively used to address other predicting problems in rock engineering fields.
3. The results of sensitivity analysis showed that the P-wave velocity was the most effective parameter on the intensity of E.
4. For future studies in this field, other evolutionary algorithms, e.g., the central force optimization, chicken swarm optimization, elephant search algorithm, and flower pollination algorithm, could be implemented to enhance the ANFIS performance.

Author Contributions: Conceptualization, M.R. and M.H.; methodology, Y.W. and M.H.; validation, Y.W., M.R., and M.H.; investigation, M.H. and M.R.; data curation, M.R.; writing—original draft preparation, Y.W., M.R., M.H., and R.A.A.; writing—review and editing, Y.W., M.R., M.H., and R.A.A.; supervision, M.R. and M.H. All authors have read and agreed to the published version of the manuscript.

Funding: This research was funded by the Project of Tackling Key Problems of Science and Technology in Henan Province (222102320164).

Institutional Review Board Statement: Not applicable.

Informed Consent Statement: Not applicable.

Data Availability Statement: The data used in this study may be available on request from the corresponding author.

Conflicts of Interest: The authors have no conflicts of interest to declare that are relevant to the content of this article.

Nomenclature

ANFIS	Adaptive neuro-fuzzy inference systems
ASTM	American Society for Testing and Materials
AI	Artificial intellginence
ANNs	Artificial neural networks
ρ	Density
DC	Depth of coring
DE	Differential evolution
DI	Durability
E	Elastic modulus
FA	Firefly algorithm
FIS	Fuzzy inference systems
GA	Genetic algorithm
ISRM	International Society for Rock Mechanics
MAE	Mean of absolute error
MAPE	Mean of average percentage error
MFs	Membership functions
MLP	Multi-layer perceptron
PSO	Particle swarm optimization
ν	Poisson ratio
n	Porosity
Vp	P-wave velocity
RMSE	Root mean square error
XGB	Extreme gradient boosting

References

1. Ocak, I.; Evren Seker, S. Estimation of Elastic Modulus of Intact Rocks by Artificial Neural Network. *Rock Mech. Rock Eng.* **2012**, *45*, 1047–1054. [CrossRef]
2. Jahed Armaghani, D.; Mohamad, E.T.; Momeni, E.; Monjezi, M.; Narayanasamy, M.S. Prediction of the strength and elasticity modulus of granite through an expert artificial neural network. *Arab. J. Geosci.* **2016**, *9*, 48. [CrossRef]
3. Sachpazis, C.I. Correlating Schmidt hardness with compressive strength and Young's modulus of carbonate rocks. *Int. Assoc. Eng. Geol. Bull.* **1990**, *42*, 75–83. [CrossRef]

4. Karakus, M.; Kumral, M.; Kilic, O. Predicting elastic properties of intact rocks from index tests using multiple regression modelling. *Int. J. Rock Mech. Min. Sci.* **2005**, *42*, 323–330. [CrossRef]
5. Sonmez, H.; Gekceoglu, C.; Nefeslioglu, H.A.; Kayabasi, A. Estimation of rock modulus for intact rocks with an artificial neural network and for rock masses with a new empirical equation. *Int. J. Rock Mech. Min. Sci.* **2006**, *43*, 224–235. [CrossRef]
6. Katz, O.; Reches, Z.; Roegiers, J.C. Evaluation of mechanical rock properties using Schmidt hammer. *Int. J. Rock Mech. Min. Sci.* **2000**, *37*, 723–728. [CrossRef]
7. Kilic, A.; Teymen, A. Determination of mechanical properties of rocks using simple methods. *Bull. Eng. Geol. Environ.* **2008**, *67*, 237–244. [CrossRef]
8. Yagiz, S. Correlation between slake durability and rock properties for some carbonate rocks. *Bull. Eng. Geol. Environ.* **2011**, *70*, 377–383. [CrossRef]
9. Nefeslioglu, H.A. Evaluation of geo-mechanical properties of very weak and weak rock materials by using non-destructive techniques: Ultrasonic pulse velocity measurements and reflectance spectroscopy. *Eng. Geol.* **2013**, *160*, 8–20. [CrossRef]
10. Singh, R.; Kainthola, A.; Singh, T.N. Estimation of elastic constant of rocks using an ANFIS approach. *Appl. Soft Comput.* **2012**, *12*, 40–45. [CrossRef]
11. Beiki, M.; Majdi, A.; Dadi Givshad, A. Application of genetic programming to predict the uniaxial compressive strength and elastic modulus of carbonate rocks. *Int. J. Rock Mech. Min. Sci.* **2013**, *63*, 159–169. [CrossRef]
12. Feng, X.; Jimenez, R. Bayesian prediction of elastic modulus of intact rocks using their uniaxial compressive strength. *Eng. Geol.* **2014**, *173*, 32–40. [CrossRef]
13. Ghasemi, E.; Kalhori, H.; Bagherpour, R.; Yagiz, S. Model tree approach for predicting uniaxial compressive strength and Young's modulus of carbonate rocks. *Bull. Eng. Geol. Environ.* **2018**, *77*, 331–343. [CrossRef]
14. Jahed Armaghani, D.; Mohamad, E.T.; Momeni, E.; Narayanasamy, M.S.; Amin, M.F.M. An adaptive neuro-fuzzy inference system for predicting unconfined compressive strength and Young's modulus: A study on Main Range granite. *Bull. Eng. Geol. Environ.* **2015**, *74*, 1301–1319. [CrossRef]
15. Rezaei, M. Indirect measurement of the elastic modulus of intact rocks using the Mamdani fuzzy inference system. *Measurement* **2018**, *129*, 319–331. [CrossRef]
16. Rezaei, M. Feasibility of novel techniques to predict the elastic modulus of rocks based on the laboratory data. *Int. J. Geotech. Eng.* **2020**, *14*, 25–34. [CrossRef]
17. Sarkhani Benemaran, R.; Esmaeili-Falak, M.; Javadi, A. Predicting resilient modulus of flexible pavement foundation using extreme gradient boosting based optimised models. *Int. J. Pavement Eng.* **2022**, 1–20. [CrossRef]
18. Shahani, N.M.; Zheng, X.; Guo, X.; Wei, X. Machine Learning-Based Intelligent Prediction of Elastic Modulus of Rocks at Thar Coalfield. *Sustainability* **2022**, *14*, 3689. [CrossRef]
19. Tsang, L.; He, B.; Rashid, A.S.A.; Jalil, A.T.; Sabri, M.M.S. Predicting the Young's Modulus of Rock Material Based on Petrographic and Rock Index Tests Using Boosting and Bagging Intelligence Techniques. *Appl. Sci.* **2022**, *12*, 10258. [CrossRef]
20. Asadizadeh, M.; Hossaini, M.F. Predicting rock mass deformation modulus by artificial intelligence approach based on dilatometer tests. *Arab J. Geosci.* **2016**, *9*, 96. [CrossRef]
21. Rezaei, M. Forecasting the stress concentration coefficient around the mined panel using soft computing methodology. *Eng. Comput.* **2019**, *35*, 451–466. [CrossRef]
22. ASTM. *Soil and Rock, Annual Book of ASTM Standards*; ASTM International: West Conshohocken, PA, USA, 2002; Volume 4.08.
23. ISRM. *Rock Characterization Testing and Monitoring, ISRM Suggested Methods*; Pergamon Press: Oxford, UK, 1981.
24. Ye, J.; Koopialipoor, M.; Zhou, J.; Armaghani, D.J.; He, X. A Novel Combination of Tree-Based Modeling and Monte Carlo Simulation for Assessing Risk Levels of Flyrock Induced by Mine Blasting. *Nat. Resour. Res.* **2021**, *30*, 225–243. [CrossRef]
25. Fang, Q.; Nguyen, H.; Bui, X.N.; Tran, Q.H. Estimation of Blast-Induced Air Overpressure in Quarry Mines Using Cubist-Based Genetic Algorithm. *Nat. Resour. Res.* **2020**, *29*, 593–607. [CrossRef]
26. Nguyen, H.; Bui, X.N.; Bui, H.B.; Mai, N.L. Correction to: A comparative study of artificial neural networks in predicting blast-induced air-blast overpressure at Deo Nai open-pit coal mine, Vietnam. *Neural. Comput. Appl.* **2021**, *33*, 10615. [CrossRef]
27. Zhou, J.; Li, C.; Koopialipoor, M.; Jahed Armaghani, D.; Pham, B.T. Development of a new methodology for estimating the amount of PPV in surface mines based on prediction and probabilistic models (GEP-MC). *Int. J. Min. Reclam. Environ.* **2021**, *35*, 48–68. [CrossRef]
28. Storn, R.; Price, K. Differential evolution–a simple and efficient heuristic for global optimization over continuous spaces. *J. Glob. Optim.* **1997**, *11*, 341–359. [CrossRef]
29. Bui, D.T.; Pradhan, B.; Nampak, H.; Bui, Q.T.; Tran, Q.A.; Nguyen, Q.P. Hybrid artificial intelligence approach based on neural fuzzy inference model and metaheuristic optimization for flood susceptibilitgy modeling in a high-frequency tropical cyclone area using GIS. *J. Hydrol.* **2016**, *540*, 317–330. [CrossRef]
30. Shang, Y.; Nguyen, H.; Bui, X.N.; Tran, Q.H.; Moayedi, H. A Novel Artificial Intelligence Approach to Predict Blast-Induced Ground Vibration in Open-Pit Mines Based on the Firefly Algorithm and Artificial Neural Network. *Nat. Resour. Res.* **2020**, *29*, 723–737. [CrossRef]
31. Cao, J.; Gao, J.; Nikafshan Rad, H.; Mohammed, A.S.; Hasanipanah, M.; Zhou, J. A novel systematic and evolved approach based on XGBoost-firefly algorithm to predict Young's modulus and unconfined compressive strength of rock. *Eng. Comput.* **2022**, *38*, 3829–3845. [CrossRef]

32. Zhou, J.; Li, C.; Arslan, C.H.; Hasanipanah, M.; Bakhshandeh Amnieh, H. Performance evaluation of hybrid FFA-ANFIS and GA-ANFIS models to predict particle size distribution of a muck-pile after blasting. *Eng. Comput.* **2021**, *37*, 265–274. [CrossRef]
33. Hasanipanah, M.; Monjezi, M.; Shahnazar, A.; Jahed Armaghanid, D.; Farazmand, A. Feasibility of indirect determination of blast induced ground vibration based on support vector machine. *Measurement* **2015**, *75*, 289–297. [CrossRef]
34. He, B.; Armaghani, D.J.; Lai, S.H. Assessment of tunnel blasting-induced overbreak: A novel metaheuristic-based random forest approach. *Tunn. Undergr. Space Technol.* **2023**, *133*, 104979. [CrossRef]
35. Ghanizadeh, A.R.; Ghanizadeh, A.; Asteris, P.G.; Fakharian, P.; Armaghani, D.J. Developing Bearing Capacity Model for Geogrid-Reinforced Stone Columns Improved Soft Clay utilizing MARS-EBS Hybrid Method. *Transp. Geotech.* **2022**, *38*, 100906. [CrossRef]
36. Ashrafian, A.; Taheri Amiri, M.J.; Rezaie-Balf, M.; Ozbakkaloglu, T.; Lotfi-Omran, O. Prediction of compressive strength and ultrasonic pulse velocity of fiber reinforced concrete incorporating nano silica using heuristic regression methods. *Constr. Build. Mater.* **2018**, *190*, 479–494. [CrossRef]
37. Armaghani, D.J.; Asteris, P.G. A comparative study of ANN and ANFIS models for the prediction of cement-based mortar materials compressive strength. *Neural Comput. Appl.* **2021**, *33*, 4501–4532. [CrossRef]

sustainability

MDPI

Article

Analyzing Geotechnical Characteristics of Soils in Erbil via GIS and ANNs

Zhvan Baqi Qader [1,*], Zuheir Karabash [2] and Ali Firat Cabalar [1]

1 Civil Engineering Department, University of Gaziantep, Gaziantep 27310, Turkey
2 Department of Dams and Water Resources Engineering, College of Engineering, University of Mosul, Mosul 41002, Iraq
* Correspondence: zhvan.b.qader@gmail.com

Abstract: The aim of this study is to analyze and model the geotechnical characteristics of soils in Erbil city using Geographic Information Systems (GIS) and Artificial Neural Networks (ANNs). The study used GIS to analyze the geotechnical properties of soils by collecting data from 102 boreholes in three different depth levels (1.5 m–3.5 m, 3.5 m–6.5 m and 6.5 m–9.5 m) to visualize and analyze soil characteristics such as fines content, moisture content, soil plasticity, shear strength parameters, compressibility, Standard penetration test (SPT), and bearing capacity. The paper also establishes the prediction of SPT-N value and bearing capacity based on geotechnical properties of soils using ANN methods and made correlations between SPT values and shear strength parameters with the bearing capacity of the soil. The results analyzed via GIS indicated that the soil classification was silty clay with a small amount of sandy gravel (CL) in most of the study area. According to the SPT–N values, most of the soils in Erbil City ranged between 33 and 50; a higher SPT value generally indicates denser and stronger soil. The value of the shear strength parameter for the maximum friction angle of the soil layers was found to be 36°, and the predominant cohesion was approximately 100 kPa. The compression index of soils ranged between 0.11 to 0.31. The results showed that the ANN models were able to accurately predict the geotechnical parameters of the soil types in the study area. In addition, the use of GIS and ANN techniques allowed for a comprehensive analysis of the geotechnical characteristics of the soils in Erbil, providing valuable information for future construction and development projects.

Keywords: Erbil; geotechnical characterization; GIS; ANN

Citation: Qader, Z.B.; Karabash, Z.; Cabalar, A.F. Analyzing Geotechnical Characteristics of Soils in Erbil via GIS and ANNs. *Sustainability* **2023**, *15*, 4030. https://doi.org/10.3390/su15054030

Academic Editors: Jian Zhou, Mahdi Hasanipanah and Danial Jahed Armaghani

Received: 12 January 2023
Revised: 14 February 2023
Accepted: 18 February 2023
Published: 22 February 2023

1. Introduction

One of the most important steps before constructing infrastructure is the geotechnical site investigation. It provides information on the site suitability for design criteria and possible construction problems such as time and resources. There are many methods for site investigations and in-situ tests, including pressure meter test, dilatometer test, SPT, cone penetration test (CPT), and plate load test [1]. For the construction of multistory buildings, highways, bridges, and industrial facilities, a soil survey is required to determine the type of soil, consistency, index properties, relative density, groundwater level, shear strength parameters, (SPT) value and bearing capacity [2]. It is necessary to know the bearing capacity of the soil layers for design, the choice of the foundation type, and the foundation depth for any superstructure [2–4]. A geotechnical investigation provides valuable information on the physical and mechanical properties of the soil and rock at a site, which is necessary for safe and durable engineering structures. The information collected from a geotechnical investigation is used to make informed decisions about the design and construction of the foundation, and to identify potential hazards, such as soil liquefaction or instability of slopes, that could compromise the safety of the structure. Therefore, to obtain the geotechnical parameters required for the calculation of the soil bearing capacity and settlement, in situ testing is required in addition to the collecting of disturbed and undisturbed specimens at different depths. Thus, several geotechnical experiments are conducted on these specimens to determine various parameters that are typically used to design the foundations [5]. Researchers have studied the reliability of SPT to determine the bearing capacity of soil [6–8]. Currently, the SPT test is used to evaluate the bearing capacity to design foundations [9,10].

GISs and ANNs could be used together in analyzing the geotechnical characteristics, and to predict the shear strength, settlement, and bearing capacity of the soil from the index properties of soils. The GISs provide an analytical function that is time–consuming for developing model entry data at different spatial scales [10,11]. A GIS is an organization of data that people interact with to integrate, analyze, and visualize data, to identify relationships, patterns and trends, and to resolve complicated issues; GIS has been used by many researchers to analyze various data [12–15]. ArcGIS was designed to capture data, store, update, process and present data, and to conduct analyses [16]. GIS can help to recognize possible challenges to the completion of the project early in the design process, which can help to avoid time losses for a construction project. Therefore, a GIS is a modular instrument that can be used to support geotechnical site assessments. It has been used to guide land preparation and to integrate field data with existing data [17].

ANN is one of the prevalent algorithms among researchers nowadays, specifically in geotechnical issues. ANN holds three significant advantages: first, the counting speed is high. Second, it has a strong fault-tolerant capability. Third, it is proficient in dealing with problems with complicated problem-solving rules [18]. The technique of utilizing ANN could be a suggestion for predictions, especially in cases where theoretical modeling does not give foreseen outcomes [19]. ANN aims to model the behavior of the nervous system in the human brain. ANN is an adequate solution for solving complex and nonlinear data modeling. Ref. [20] presents the estimation of standard penetration test values on cohesive soil using an artificial neural network without data normalization. Some previous studies investigated the assessment of geotechnical properties and determination of shear strength parameters by unitized ANN [21–23]. In the geotechnical domain, the development situations generally have multiple variables, making them challenging to model employing conventional mathematics [24].

In this study, the test results of 102 boreholes were gathered, categorized, and analyzed and modeled using ArcGIS10.7 software. These data were used to construct models using ANNs to predict the SPT values and bearing capacity of soils. The data represent the area of Erbil City in Iraq, covering a depth of 9.5 m below the ground surface. Data were collected from the Andrea Engineering Test Laboratory and the construction laboratory in Erbil. Therefore, for a geotechnical engineer, this information can be used to classify areas into zones according to GIS results. The purpose of this study is to analyze and model the geotechnical parameters such as the fines content, moisture content, soil plasticity, shear strength parameters, compressibility parameters, SPT–N values and soil bearing capacity via GIS to create a group of maps in different layers. In addition, the prediction of SPT-N value and bearing capacity based on geotechnical parameters were modeled by ANN methods. Correlations between shear strength parameters, SPT values and bearing capacities of soils were made by Minitab 17 programming.

By combining the capabilities of GIS and ANNs, researchers can develop models to analyze geotechnical issues at different spatial scales, producing results that are more accurate and efficient compared to manual methods In summary, GIS and ANNs complement each other by providing an analytical function that is efficient for developing model data entry at different spatial scales for geotechnical issues, allowing for more accurate predictions and a better understanding of the relationships between soil properties and other factors. To the best of our knowledge, this is the first study to cover mapping and modelling all the geotechnical characteristics of soil in Erbil city.

2. Study Area

Erbil is located in the northwestern region of Iraq. Geologically, it is in the low–folded belt of northern Iraq in the structural trough with a NW–SE axial trend, and within the foothill zone, which is part of the stable shelf tectonic unit of Iraq. Erbil City has an area of approximately 250 km^2 and GPS coordinates of 36°11′27.4″ N 44°00′33.7″ [25]. A location map of Erbil City is shown in Figure 1. From a geomorphological perspective, the area is flat with uncommon low–lying hills. In addition, Erbil City is stratigraphically covered by quaternary and Pleistocene deposits, which are dominated by clay, silt, and sand [26]. Erbil City is mainly covered by soils such as gravel and conglomerates with sand, clay, and silt. Conglomerates cover more than 80% of the study area [27].

Figure 1. Location of Erbil City, Iraq.

In the last decade, Erbil city had extensive development in the construction of railways, internal and ring roads. Therefore, collecting data, building a database, producing GIS maps for soil properties, and developing a model for soil behaviors and the bearing capacity of the foundations would be very useful for site engineers to make immediate decisions regarding the selection of project positions. Finally, the geotechnical properties at different depths were collected from the study area (Erbil city), analyzed and tabulated for the 102 boreholes. The locations of the boreholes were selected to ensure a uniform distribution throughout the study area. The borehole locations are shown in Figure 2.

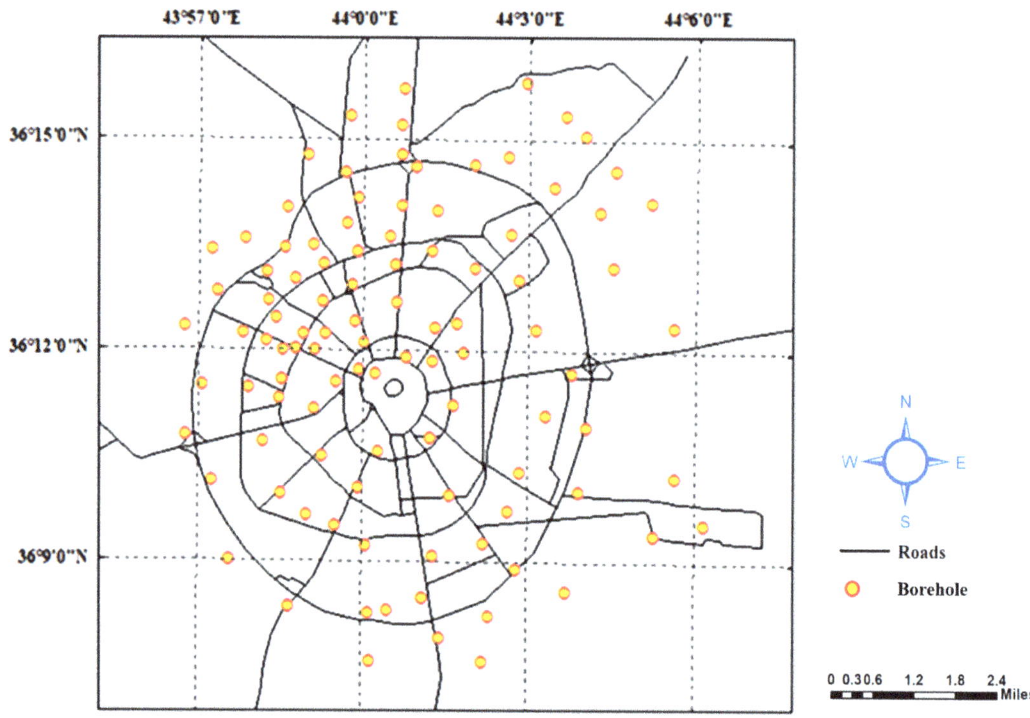

Figure 2. Borehole locations in the study area.

3. Methodology

The methodology of the study involved collecting data from the field in Erbil city center and analyzing it using both GIS and ANNs. The data was processed and analyzed using these tools to gain insights and conclusions about the study area. The flow chart represents in Figure 3 the methodology of this study as a tool to help readers understand the process used. A flow chart can show the different steps involved in integrating GIS, ANN, and lab analyses, making it easier for readers to follow and comprehend the study's methodology.

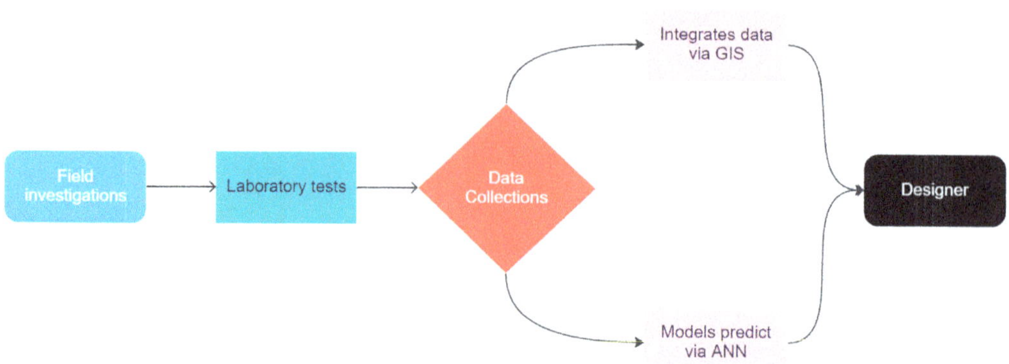

Figure 3. Flow chart showing in a simple way the methodology adopted in the present study.

3.1. Data Collection

In this study, the results of site investigations and numerous series of soil laboratory tests were collected from 102 borehole locations that cover the main part of the region. Soil investigation included drilling boreholes, taking both disturbed and undisturbed soil samples at 1–meter intervals from 1.0 m to 9.5 m depth, and various field tests. Laboratory work included a series of geotechnical tests to determine the soil index properties, sieve analysis, compressibility, settlement, and shear strength.

A sample of the data collected from the laboratory and field tests is presented in Table 1. This table includes some statistical information (e.g., min. values, max. values, average values, and standard deviation of the input and output data). All data of the study could find in Appendix A as a Table A1.

Table 1. Inputs and output of the present study.

No	LL%	PL%	PI%	WC%	c kN/m^2	ϕ	Fine Content	SPT-N Value kN/m^2	Q UL kN/m^2
	ASTM D 4318			ASTM D2216	ASTM 3080		ASTM D 6913	ASTM D1586	-
1	46	22	24	28.3	51	4	62.1	7	117
2	40	22	18	15.0	96	3	94.1	54	224
3	47	25	22	18.0	99	5	91.5	57	225
.	.	.	.	.	.	.	.	.	.
.	.	.	.	.	.	.	.	.	.
.	.	.	.	.	.	.	.	.	.
304	48	23	25	18.7	99	5	94.8	100	292
305	45	25	20	22.8	105	4	92.6	91	296
306	48	25	23	22.1	97	5	69.7	100	288
Min	0	0	0	12	29	3	42	5	74
Avarege	48.02	24.90	22.89	18.69	92.67	4.21	89.48	64.69	255.09
Max	71	36	37	29	136	6	100	100	375
SD *	7.51	3.67	5.59	2.87	19.55	0.62	11.65	25.93	60.98

SD * = Standard deviation.

3.2. Geographical Information Systems (GISs)

The results of the soil investigation and field and laboratory tests were employed to create a digital database for the study region. A database of geotechnical properties was used to provide the input values for the mapping software. In this study, the data were analyzed and presented as maps using ArcGIS (10.7) software. Deterministic methods (inverse distance weighting) were used to create maps and models of spatial data, which rely on probability and uncertainty. Deterministic methods use a fixed set of rules or algorithms to create maps and models, unlike geostatistical methods used for analyzing and modeling spatial data [11]. These methods are commonly used to analyze patterns, relationships, and trends in large, complex datasets.

Some of the most commonly used deterministic methods in ArcGIS include:

- Interpolation: This method is used to predict values at unsampled locations based on observed data. Interpolation methods in ArcGIS include inverse distance weighting, spline, and triangulated irregular network (TIN) interpolation.
- Buffering: This method is used to create a polygon around a feature that represents a specified distance. Buffers are commonly used in spatial analysis to identify areas that are within a certain distance of a feature of interest.
- Overlay: This method is used to combine two or more maps based on a set of rules or conditions. Overlays can be used to create a new map that shows the spatial relationships between features in the input maps.

- Reclassification: This method is used to change the values of a raster or vector layer based on a set of rules or conditions. Reclassification is often used to simplify complex data or to create new data layers based on existing data.
- Extraction: This method is used to select features from a map based on a set of conditions or rules. Extractions can be used to create new data layers that contain only the features that meet specific criteria [11].

The resulting digital maps illustrate the soil formation patterns, distribution, and geotechnical properties of soils at different depths. These maps simplify and help designers and site engineers make the right decisions in the construction of projects. The aim of drawing digital maps using the GIS method was to illustrate the bearing capacity of foundations at three different depths. In this study, the bearing capacity was estimated using two methods. The first Meyerhof method (1963) used shear strength parameters for (10 × 10 m). In the second method, standard penetration numbers were used for bearing capacity estimation [28].

3.3. Statistical Analysis

To make a correlation between the geotechnical properties, the data obtained from the soil investigation of all boreholes in the study area were correlated by MINITAB 17 software. For instance, correlations were made between the SPT values of soil strata with the shear strength parameters of soils and the ultimate bearing capacity. This method is used to find a correlation between the response (Y) and predictor (X) using regression analysis, which is an extensively used method for analyzing multifactor data.

3.4. Neural Network Model

The process for creating an artificial neural network is assumed by using the Matlab application. This study aims to make models by an artificial neural network. The network model developed was formed from data collection of geotechnical properties of soils in the study area. The ANN analysis result aims to predict SPT N-value and bearing capacity using the identical algorithm, the Back-propagation algorithm, and the same activation function. The network architecture was chosen using hidden layers and varying the number of neurons in the hidden layer. The relation number of neurons in the hidden layer is between 15 and 18 according to previous researches [29,30]. The network performance that has the smallest error and the correlation coefficient value that is proximate to 1 is most suitable for data predictions. Root Mean Squared Error (RMSE) is a commonly used evaluation metric in ANN models. RMSE measures the difference between the predicted and actual values, and it is expressed in the same units as the target variable. A low RMSE value indicates that the predictions are close to the actual values, while a high RMSE value displays that the predictions are far from the actual values. In ANN models, RMSE is used to evaluate the performance of the model and determine the quality of the predictions. A lower RMSE value shows a better fit between the predictions and the actual values and a more accurate model. In this research, the R^2, RMSE, and MAE values of the estimated and actual target parameters are computed in the implementation evaluation of regression models. The R^2, RMSE, and MAE represented mathematically as Equations (1)–(3):

$$R^2 = 1 - \frac{\sum_{I=1}^{N}\left(y_{mea} - y_{pre}\right)^2}{\sum_{I=1}^{N}\left(y_{mea} - y_{m}\right)^2} \tag{1}$$

$$RMSE = \sqrt{\frac{\sum_{I=1}^{N}\left(y_{mea} - y_{pre}\right)^2}{N}} \tag{2}$$

$$MAE = \frac{1}{N}\sum_{I=1}^{N}\left|y_{mea} - y_{pre}\right| \tag{3}$$

where y_{mea}, y_{pre}, and y_m represent the average of existing output, predicted output, and actual output, respectively. N denotes for all number of data. The degree of fitting is raised with R^2 immediacy to 1. RMSE and MAE are utilized to assess the model's prediction capability. For the RMSE and MAE, the prediction model will be more exact and its accuracy will be higher with a smallish value.

Figure 4 illustrates the structure of neural network models to predict SPT N value as output with two models: (a) using input as (LL%, PL%, PI%, WC, cohesion, ϕ, Fine content) and (b) using inputs as (LL%, PL%, PI%, WC, ϕ, Fine content). The structures of neural network models to predict ultimate bearing capacity are presented in Figure 5. The parameters were used as input in two models: (c) (LL%, PL%, PI%, WC, cohesion, ϕ, Fine content) and (d) (LL%, PL%, PI%, WC, ϕ, Fine content).

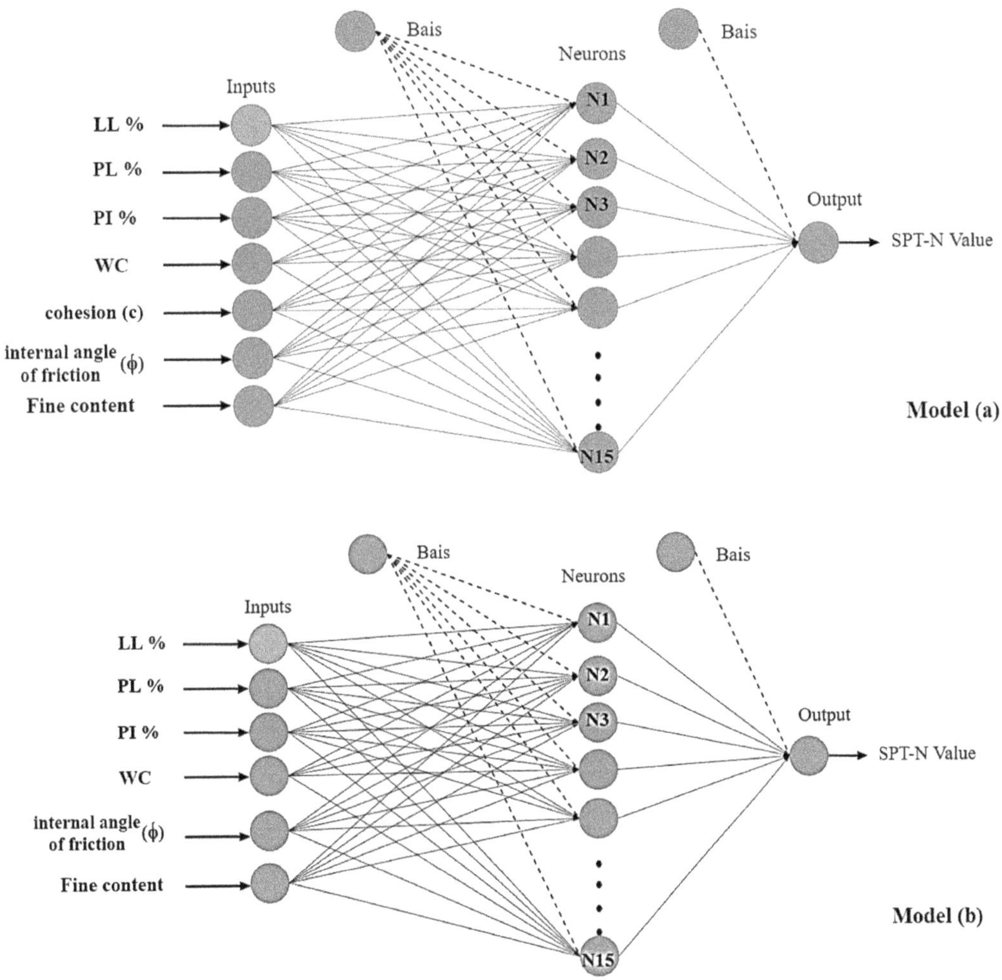

Figure 4. Structure of neural network models (a) and (b) to predict SPT-N value.

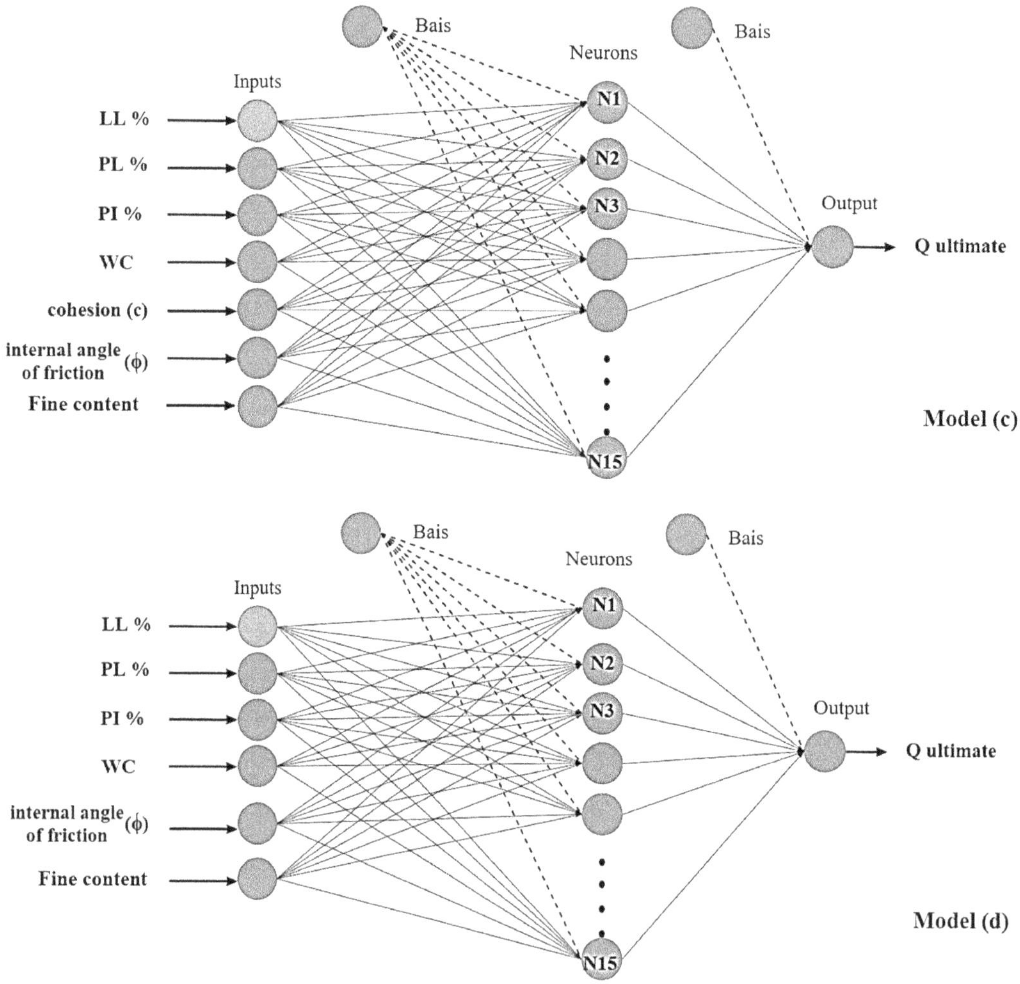

Figure 5. Structure of neural network models (c) and (d) to predict Q-ultimate.

4. Results and Discussions

4.1. Modeling of Soil Properties Using GIS Maps

GIS maps of the soil properties of the study area were produced. These maps included the distribution of the soil fines content, natural water content, liquid limit, plastic limit, cohesion, angle of internal friction, compression index, rebound index, SPT–N values, and bearing capacity of foundations at three different levels (1.5 m–3.5 m, 3.5 m–6.5 m, and 6.5 m–9.5 m). The soil characteristics at different depths were interpolated for the survey area to show the distribution of these properties in a clear way. In general, the soil characteristics in all the maps were divided into six major legends, each of which was represented by a unique color.

4.1.1. Fines Content Model

Fines content in soils is one of the most significant parameters that affect soil behaviors such as shear strength, compressibility, plasticity, and indirectly, the bearing capacity of foundations [31]. The results presented in Figure 6 indicate that the majority of the study area in Erbil city center has a high fines content. According to the figure, the fines content in most of the area is greater than 81%, meaning that the proportion of soil particles that are smaller than 0.075 mm in diameter is high. This high fine content will likely have significant effects on the soil's engineering properties, such as its shear strength, compressibility, and plasticity. The impact of soil properties on its ability to support loads, resist deformation, and transmit loads must be carefully considered in future construction and development projects in the area. The results from Figure 6 provide crucial information for engineers and planners in Erbil city center, emphasizing the need to consider the fines content of soil in decision-making for development projects. The compression level of soil becomes crucial when large particles are replaced by fine particles, and the impact of fines content is more pronounced when the soil is near saturation. This has been noted by other researchers in the field [31–33]. The relationship between natural water content and fines content is intertwined and cannot be evaluated separately. Sometimes, improving natural water content weakens the effect of fines content on soil shear strength, due to the sensitivity of fines to changes in natural water content. In dry conditions, fines do not significantly affect soil behavior due to the influence of suction [34,35].

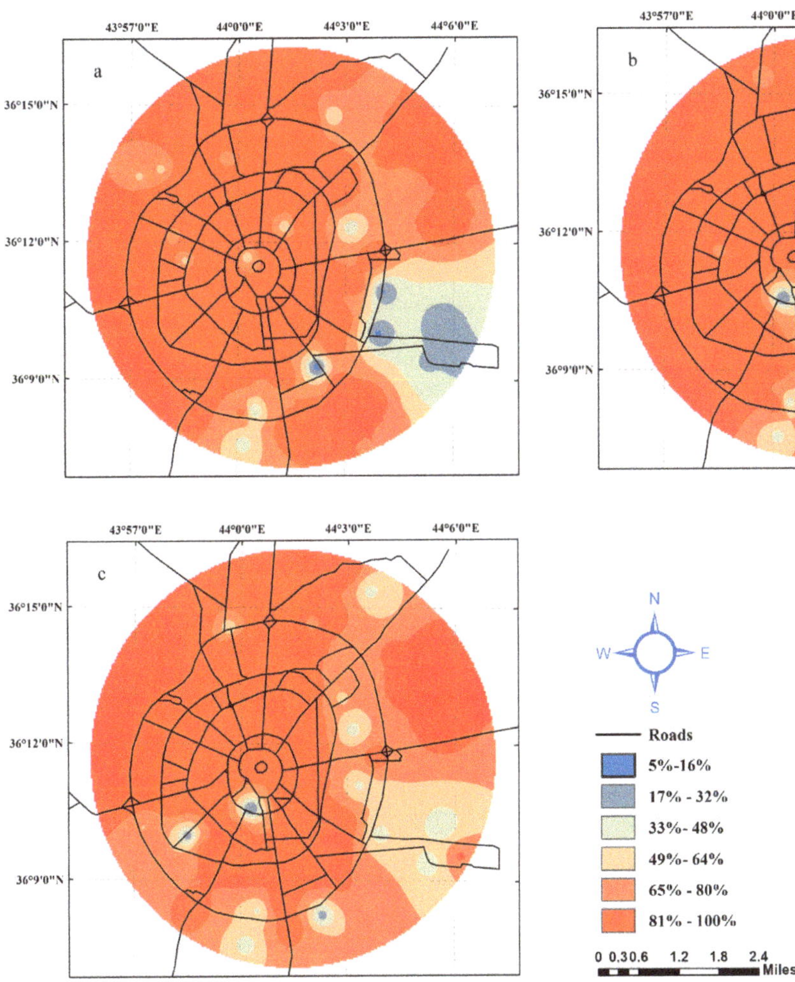

Figure 6. Fines content at depths (**a**) 1.5–3.5 m. (**b**) 3.5–6.5 m. (**c**) 6.5–9.5 m.

4.1.2. Atterberg Limits

Atterberg limits can be used to characterize soil behavior and classification, including the swelling potential of expansive soils, the consistency and plasticity of the soil. Two essential index properties can be obtained from the values of the natural water content: liquid limit and plastic limit [36]. Soil water content affects its consistency. A high water content in clay makes it behave like a liquid due to the reduction of attraction between clay particles caused by the excess water between the particles [37]. The liquid limit was used to determine the consistency of the fine-grained soil. This measure of soil consistency is useful for estimating soil consolidation properties and calculating the acceptable bearing capacity and settlement of foundations [38].

The variations in the soil liquid limit throughout the study area at the three different depths are illustrated in Figure 7. The results of the analysis of liquid limit values in the study area show that a significant portion of the soils have a liquid limit range between 40% and 52%. This range indicates the presence of low plastic clay, which is a type of soil that has low resistance to deformation when subjected to stress and is prone to collapsing. The presence of low plastic clay in high percentages in the study area is an important factor to consider in construction and development projects, as it may impact the stability and integrity of structures built on these soils. Additionally, the observation of soils with high liquid limit values (greater than 53%) in relatively central regions is also important. These central zones are considered critical points in the study area, as soils with high liquid limit values are prone to deformation and instability. These critical points need to be carefully evaluated and addressed in any future construction and development projects in the area to ensure the stability and integrity of the structures built on these soils. Numerous researchers have investigated the relationship between the liquid limit of soils and swelling potential. Some types of clay minerals with a high cation exchange capacity (CEC) suffer from expansion and an increase in the volume of the available water [38–41].

Plasticity is one of the most important features of clay, and the crystallinity of clay minerals is the primary source of this plasticity [39–41]. Soil is plastic when the water content is below the liquid and plastic limits. The plastic range, which is the difference between the two limit values, is called the plasticity index [42]. The plastic limit provides geotechnical engineers with indirect information about the activity, toughness index, and optimum moisture content of soils. Figure 8 shows the plastic limit variation in the study area at different depths. The analysis of the figure reveals that half of the study area has a plastic limit of soils that ranges between 19% and 24%, while the other half has a plastic limit ranging between 25% and 30%. This indicates that the soils in the study area have different levels of plasticity, which is a measure of the soil's ability to change shape without breaking under stress. The presence of soils with low plastic limits (Figure 8a) in small zones in the southeast direction is also an important observation. Soils with low plastic limits are less plastic and more brittle, making them more prone to cracking and failure under stress. These areas need to be carefully assessed and addressed in future construction projects in the area to ensure the stability and reliability of the structures built on these soils.

4.1.3. Natural Water Content Model

The natural water content can be considered a parameter that profoundly affects the geotechnical properties [43,44]. Figure 9 shows the variations in the natural water content in Erbil City. The figure presents the water content at depths of 1.5 m–3.5 m, 3.5 m–6.5 m, and 6.5 m–9.5 m. The results show that the natural water content in the study area is mostly found to range between 16% and 20% at the three levels of investigation depths. The natural water content of soils is an important factor to consider in construction and development projects, as it affects the soil's stability, bearing capacity, and compressibility. Soils with high natural water content are more susceptible to instability, while soils with low natural water content are more prone to drying and cracking.

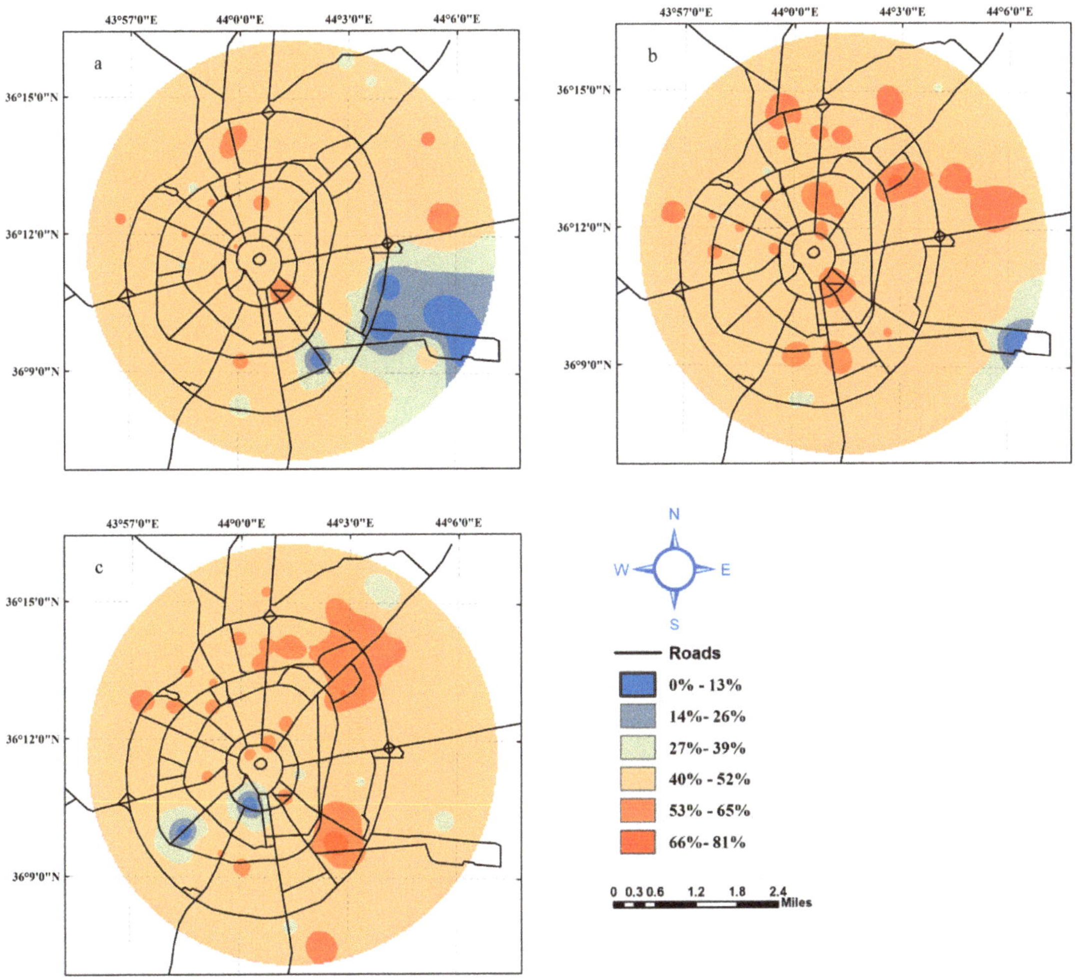

Figure 7. Liquid limit of soils at depths (**a**) 1.5–3.5 m. (**b**) 3.5–6.5 m. (**c**) 6.5–9.5 m.

This is consistent with the findings of many researchers working in this region [43,44]. The water table, during the time of exploration, was in very high depths below the natural ground level (NGL), meaning the groundwater was relatively shallow. The water table fluctuates seasonally, with an increase during spring. The soil above the water table affects its strength and compressibility, as more moisture results in decreased strength and increased compressibility. Saturated soil below the water table creates settling issues as the consolidation process reduces the natural water content under stable load.

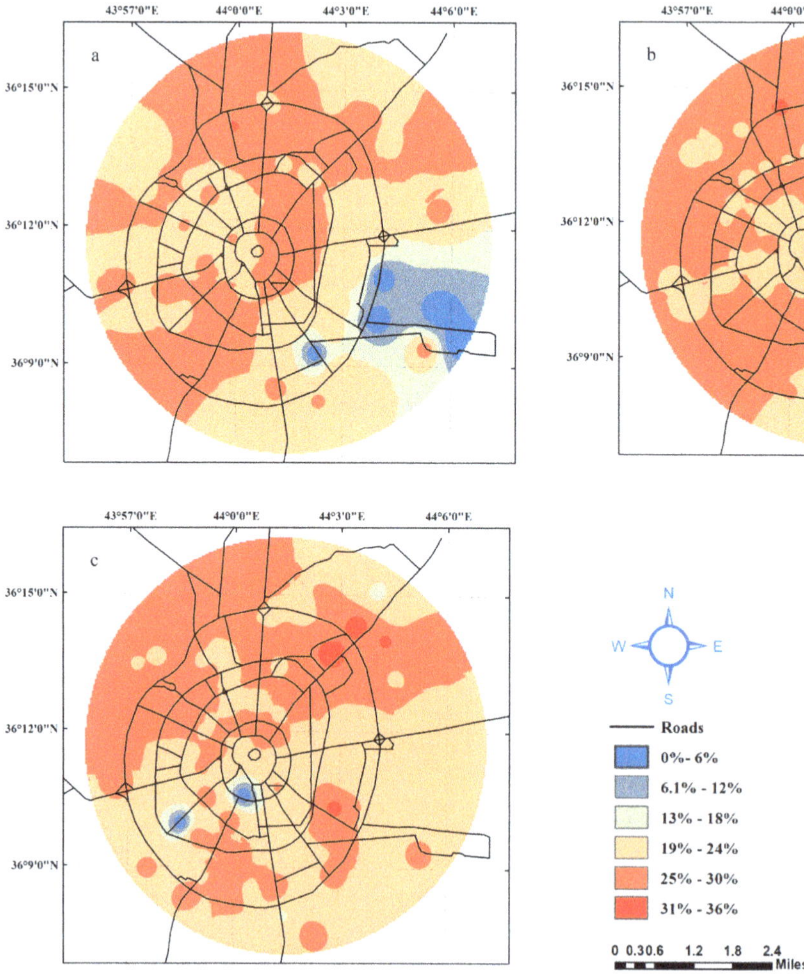

Figure 8. Plastic limit of soils at depths (**a**) 1.5–3.5 m. (**b**) 3.5–6.5 m. (**c**) 6.5–9.5 m.

4.1.4. Shear Strength Parameters

The shear strength of soils is a crucial parameter in many foundation engineering designs. It refers to the ability of soil to resist forces that cause slipping or sliding along a plane within the soil. The shear strength of soils is important in determining the bearing capacity of shallow and deep foundations, slope stability, tunnels, and lateral pressure on structures [28]. The soil's shear strength comes from its cohesive strength (c) and frictional strength, represented by the angle of internal friction (ϕ). Cohesive strength is due to the bonding force between soil grains and the binding material, while frictional strength arises from the friction, interlocking, and rolling of soil grains [44–47]. The strength of any soil decreases as the shear strain and expansion or contraction increase or decrease, respectively, with respect to the soil density due to applied loads. Shear strength parameters are widely utilized by different standard equations in the design of foundations, particularly in empirical equations. [48].

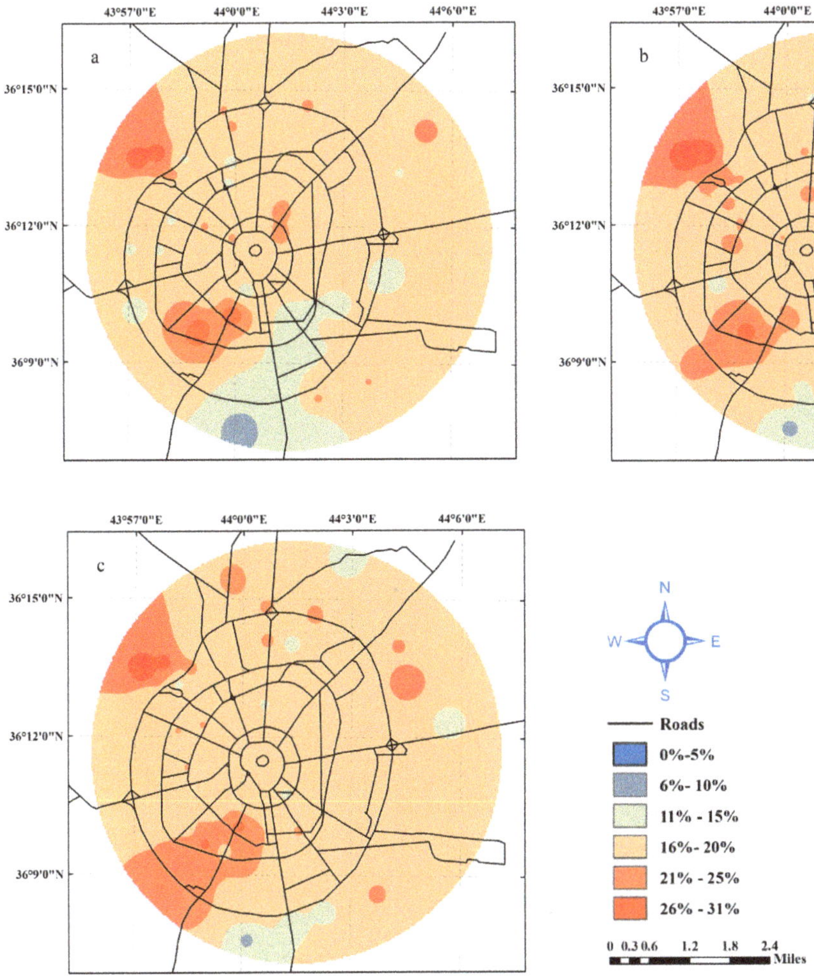

Figure 9. Natural water content of soils at depths (**a**) 1.5–3.5 m. (**b**) 3.5–6.5 m. (**c**) 6.5–9.5 m.

Figures 10 and 11 show the variations in the cohesion and angle of internal friction, respectively. The combination of these two parameters produces the shear strength of the soil, its variation throughout the study area, and the soil depth. Most of the soils in the study area at shallow depths (1.5 m–3.5 m) had cohesion values between 76 kPa and 100 kPa. However, there were relatively small regions with cohesion values greater than 100 kPa and less than 50 kPa. The area covered with soil had cohesion values greater than 100 kPa, which increased with the depth. The results of the figures are in agreement with the distribution of the fines content in the study area. The fines content of soils in the majority of the study area was found to be greater than 80%, and this high fine content is likely contributing to the high cohesion values in the area.

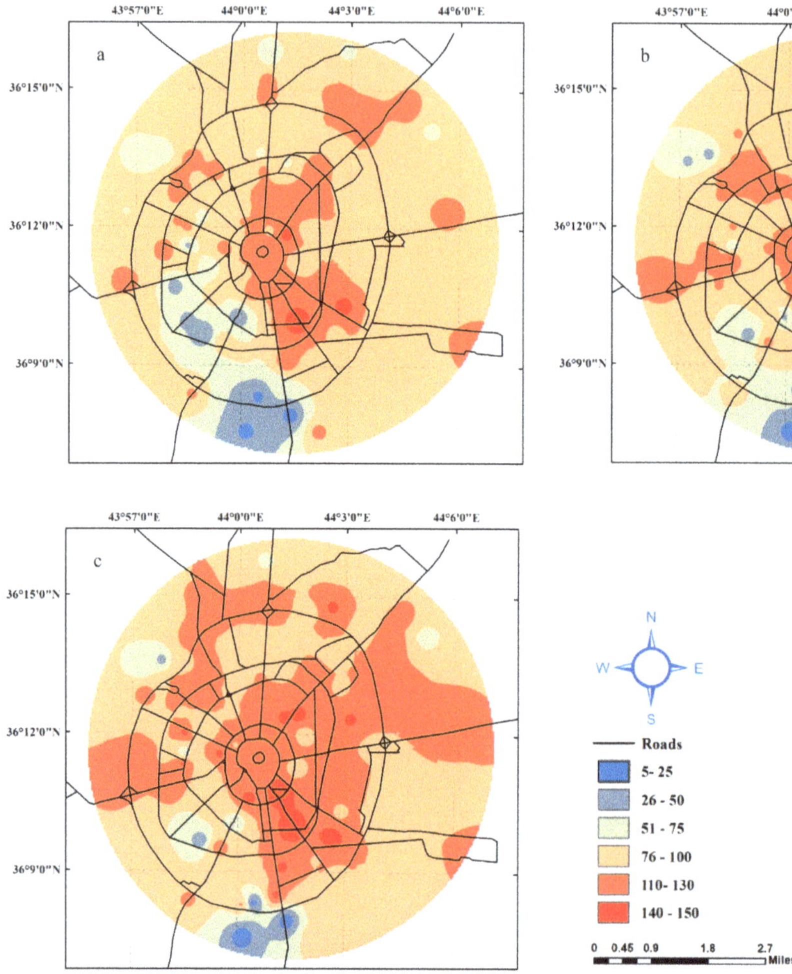

Figure 10. Cohesion of soils (kPa) at depths (**a**) 1.5–3.5 m. (**b**) 3.5–6.5 m. (**c**) 6.5–9.5 m.

Soils with high fines content produce higher cohesion values [49]. The angle of internal friction is a parameter of the soil shear strength and is employed in bearing capacity estimation, slope stability analyses, and estimation of soil lateral earth pressures. The soils in the study area were found to have an angle of internal friction between 2° and 6°, which was found to be similar at all three depths in this investigation. The east-south part of the study area had soils with an angle of internal friction ranging from 7° to 12°. These results indicate that the angle of friction values found in the study area are consistent with those found in similar studies in the region. The angle of internal friction is an important parameter in determining the shear strength of the soil, and these results suggest that the soils in the study area have moderate to low shear strength values. The results of the figures can be used to identify zones of high and low shear strength, which are important in determining the suitability of the soil for different types of structures. The information can also be used to design and construct structures that are appropriate for the soil conditions in the area [47,50].

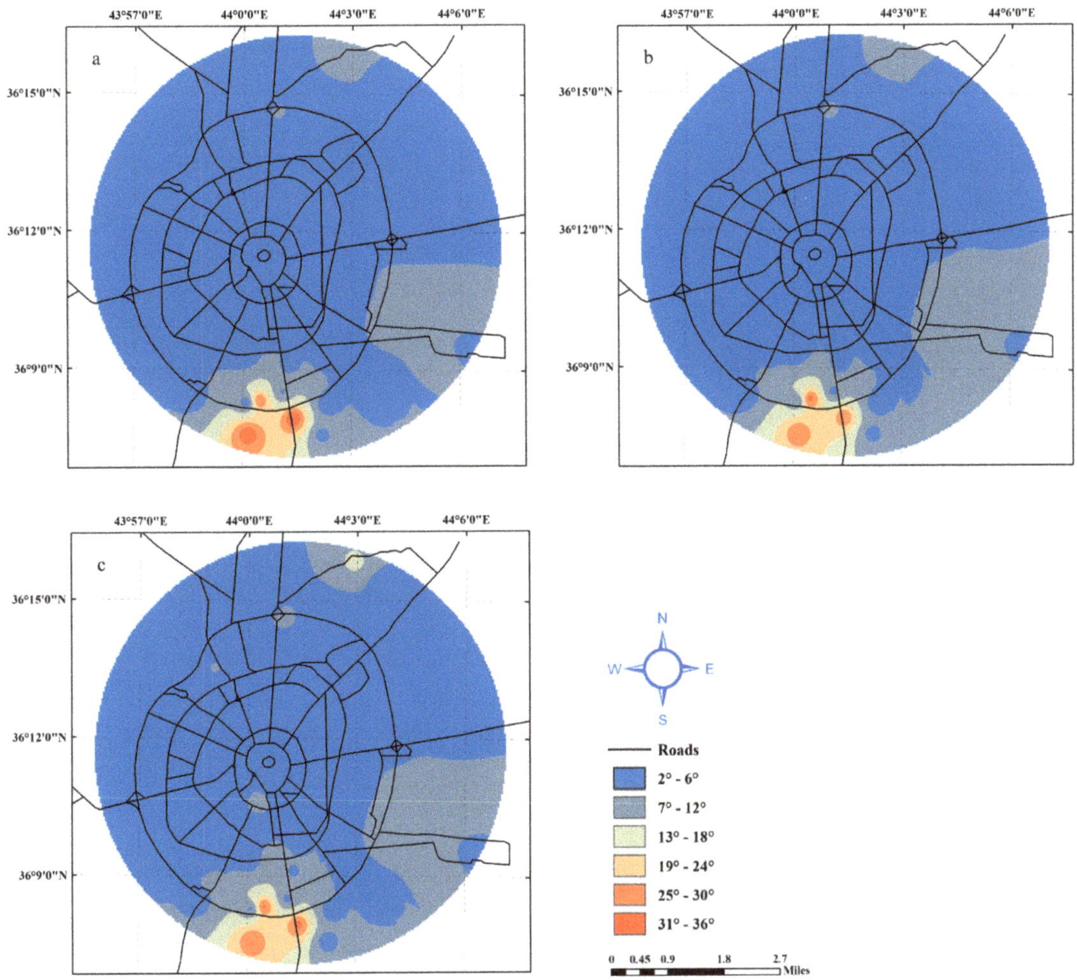

Figure 11. Angle of internal friction of soils at depths (**a**) 1.5–3.5 m. (**b**) 3.5–6.5 m. (**c**) 6.5–9.5 m.

4.1.5. Consolidation Parameter Model

The consolidation parameters (compression index and swelling index) of saturated clayey soil should be checked during the analysis and design of the foundations [51]. In this study, the distribution of the compression index and swelling index throughout the study area and their variation with the soil depth were investigated. Figure 12 shows that the compression index of soils in the study area, Erbil city center, has a range of values between 0.17 and 0.22, with a lower value found at greater depths. The compression index is an important factor in determining soil compression and consolidation. The results suggest that the shallow soil strata in the study area have high consolidation and settlement potential, while the settlement potential is expected to decrease with depth due to the variation in compression indices.

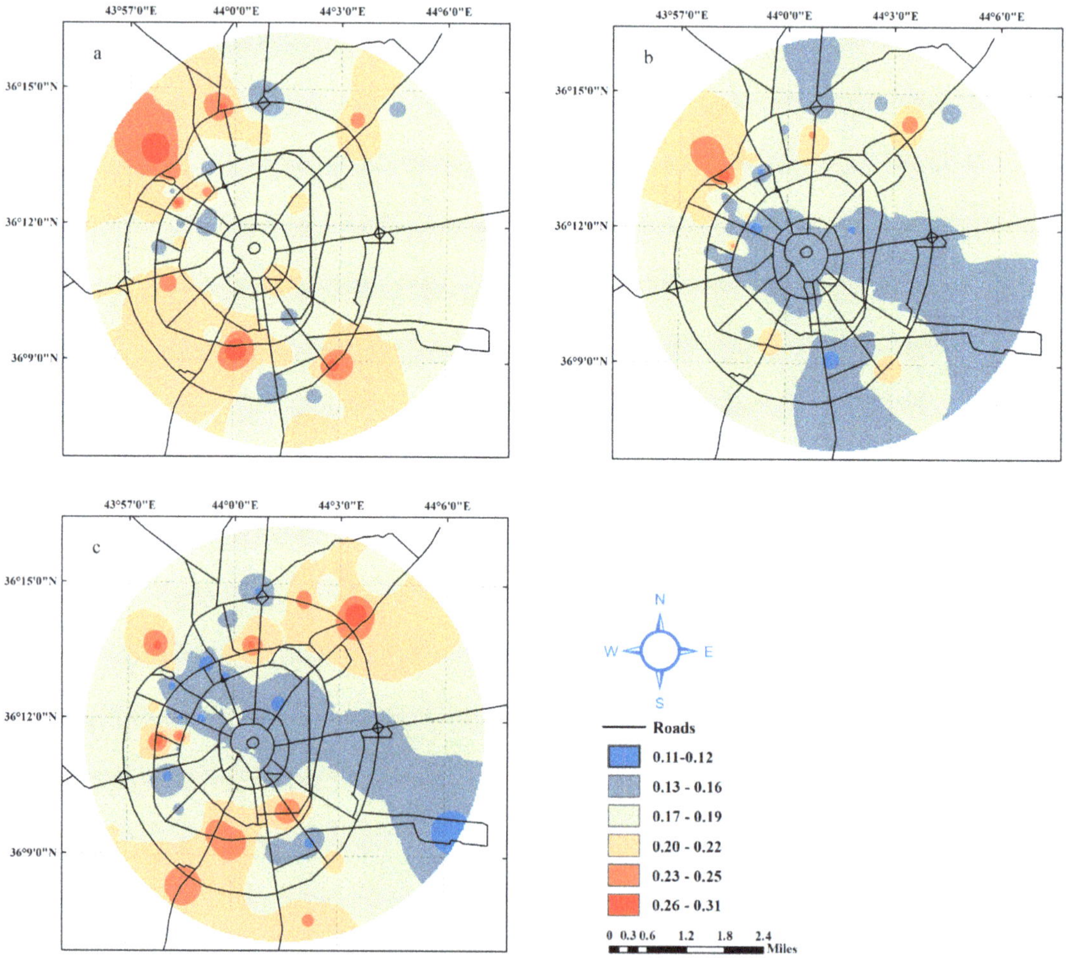

Figure 12. Compression index, Cc, of soils at depths (**a**) 1.5–3.5 m. (**b**) 3.5–6.5 m. (**c**) 6.5–9.5 m.

The rebound index is an important parameter in geotechnical engineering as it represents the soil's tendency to expand or contract under changes in moisture content. A high rebound index value designates that the soil is more susceptible to expansion and contraction, whereas a low value indicates that the soil is more stable. The results shown in Figure 13 suggest that the soil in the study area has moderate to low rebound index values in the range 0.015–0.078, indicating that the soil is relatively stable and less likely to expand or contract under changes in moisture content. This indicated that there was no soil swelling or shrinkage potential in the study area. The minimum values showed a consistent trend across all levels in the study area (Figure 13a), with the highest swelling index values located north of Erbil at the first level and in the south at the second and third levels. Most of the study area had moderate parameter values of 0.015–0.078. The consolidation parameters were consistent with previous studies [47,50,52].

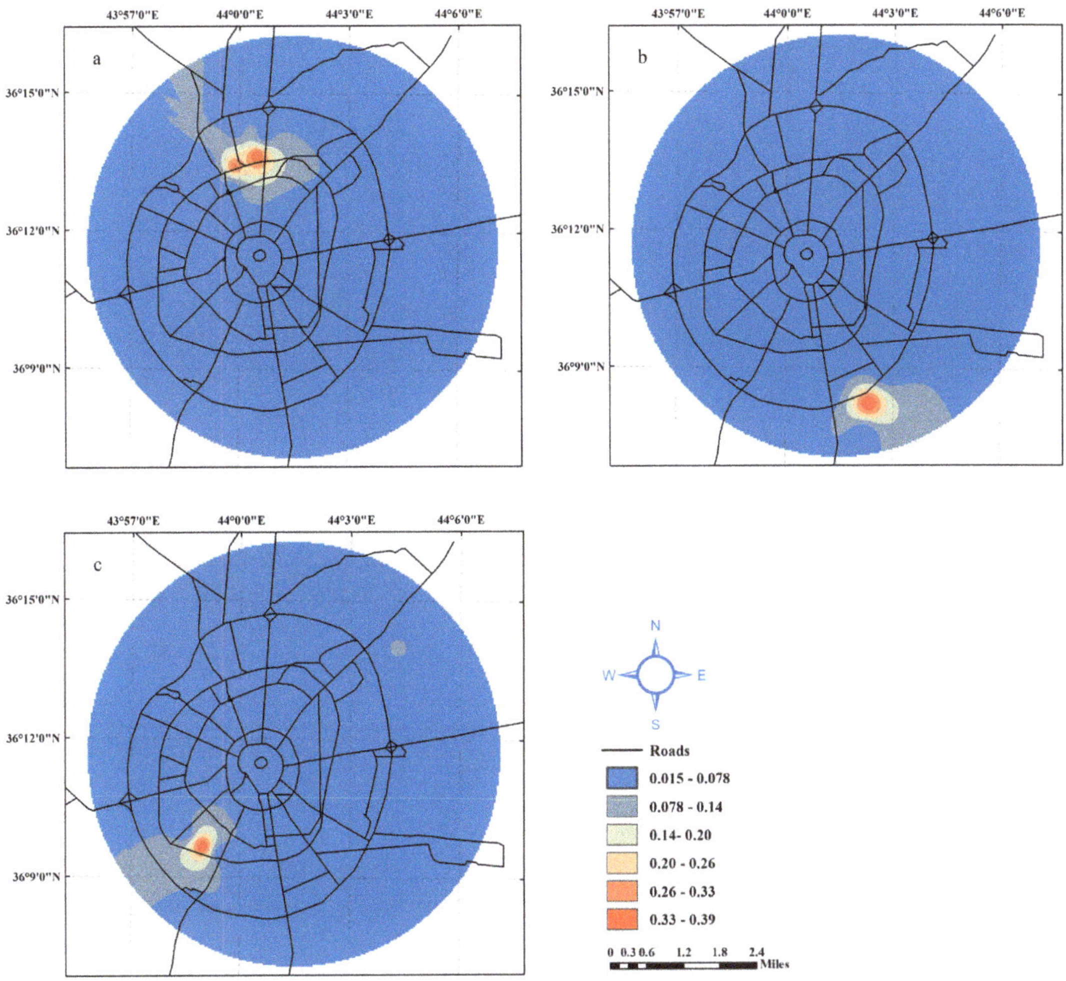

Figure 13. Rebound index, Cr, of soils at depths (**a**) 1.5–3.5 m. (**b**) 3.5–6.5 m. (**c**) 6.5–9.5 m.

4.1.6. Standard Penetration Test Model

The SPT is one of the many standard in situ tests used to identify soil type, stratigraphy, and relative strength measures during site investigations [53,54]. The eastern part of Erbil had higher SPT values, which is attributed to the higher unit weight of soils and the presence of stiffer and stronger soil layers as shown in Figure 14. The SPT values of soils at a depth of 6.5 m–9.5 m was mostly between 35 and 60, and the range increased from the west to the east of the study area. The increase in the SPT values with increasing depth is due to the influence of several factors. Soil type affects the SPT values as different soil types have different characteristics such as density, porosity, and strength, which all impact the SPT results. Unit weight, or the weight per unit volume of soil, is also a factor, as a higher unit weight results in higher SPT values. Confining pressure, the pressure applied to the soil from the surrounding material, also increases with depth, leading to higher SPT values. Overall, the results of SPT tests provide a good indication of the soil strength and its variation throughout the study area. To confirm the relationship between the SPT values and shear strength parameters (cohesion and angle of internal friction), data obtained from the boreholes were correlated. Figures 15 and 16 illustrate the correlation between the

SPT values and shear strength parameters for the soils within the study area. The figures evidently show that there is a good correlation between the SPT values, cohesion, and angle of internal friction. Generally, most values of SPT–N distribution from the center to the west of Erbil City were found to be between 17 and 48, while in the northeast and southeast, the values were higher than 50. As mentioned in various studies [55–57], the results of SPT–N showed that the soil in Erbil city is medium to very dense.

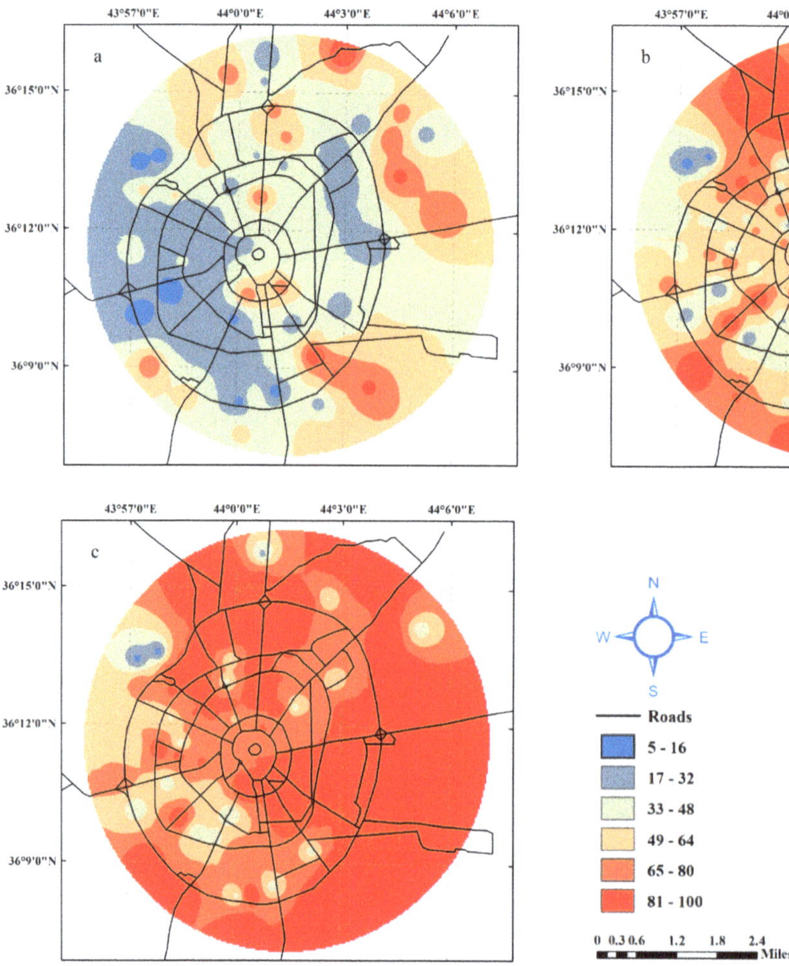

Figure 14. SPT-N values in site at depths (**a**) 1.5–3.5 m. (**b**) 3.5–6.5 m. (**c**) 6.5–9.5 m.

4.1.7. Bearing Capacity

The design of a footing depends on its soil's bearing capacity. Many methods for estimating soil bearing capacity exist, relying on factors such as soil shear strength, footing type, and SPT value [58]. In this study, the soil strata's ultimate bearing capacity was estimated using two methods, one based on shear strength parameters using Meyerhof's equation, and the other based on SPT-N values. The variation in the soil strata's ultimate bearing capacity based on Meyerhof's equation is shown in Figure 17.

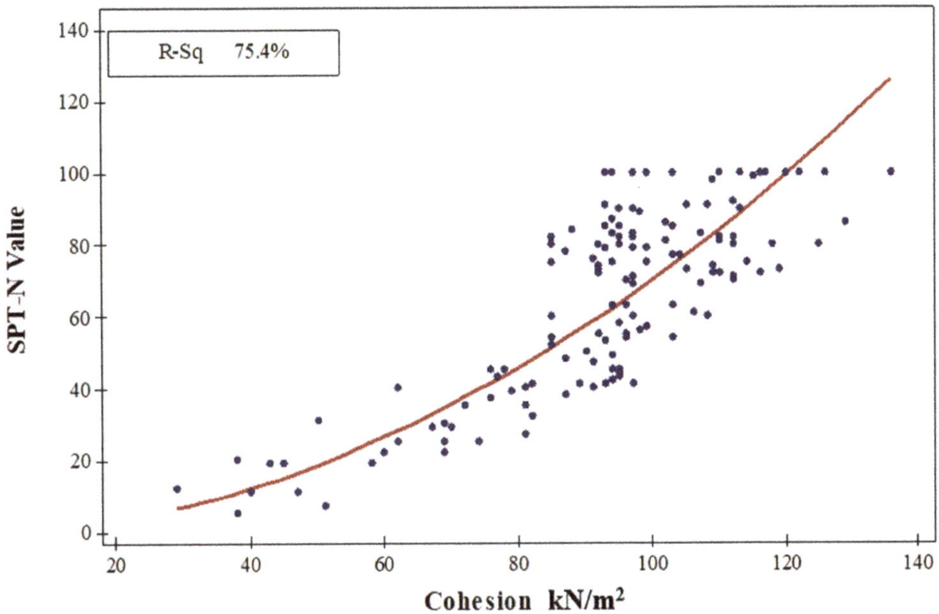

Figure 15. Correlation of SPT-N values with cohesion of soils in the study area.

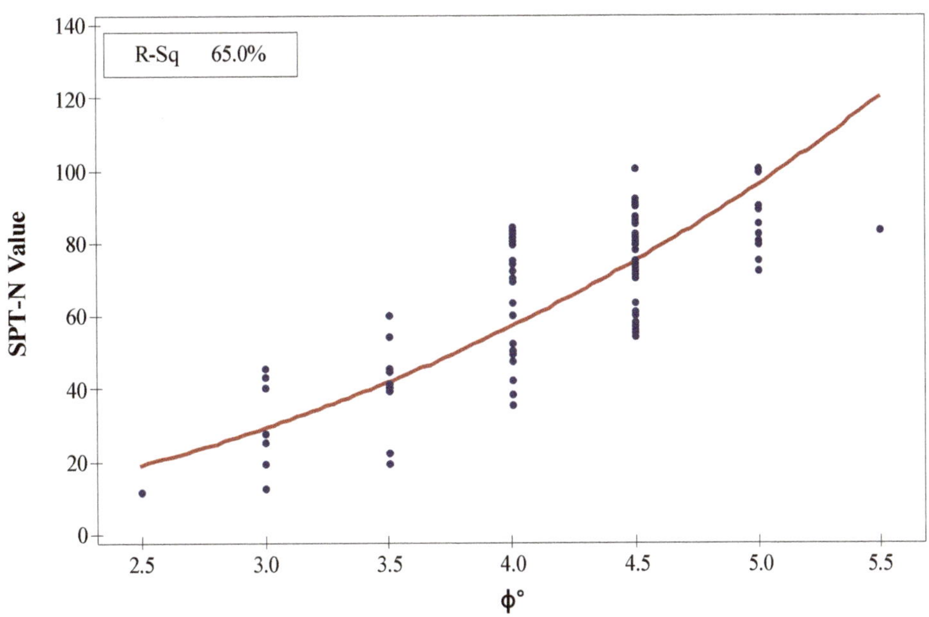

Figure 16. Correlation of SPT-N values with angle of internal friction (ϕ) of soils in the study.

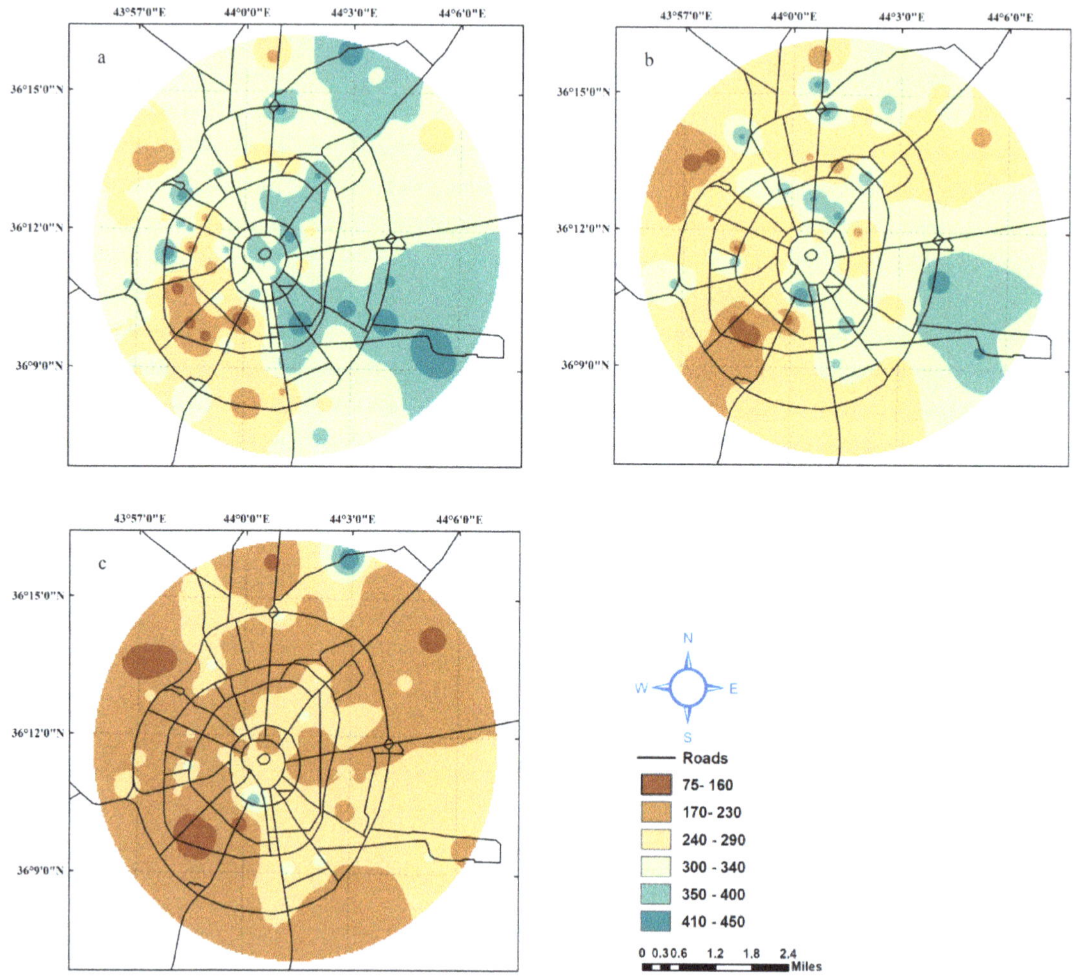

Figure 17. Ultimate bearing capacity according to Meyerhof's equation at depths (**a**) 1.5–3.5 m. (**b**) 3.5–6.5 m. (**c**) 6.5–9.5 m.

Figure 18 shows the ultimate bearing capacity of the soil strata in the study area at three different depths that were estimated from the SPT values. Changes in the ultimate bearing capacity were observed throughout the study area. The ultimate bearing capacity of the soil strata in the study area is a significant factor in the design of shallow and deep foundations. The results from Figure 18 indicate that the majority of the study area at shallow depths had an ultimate bearing capacity be-tween 170 kPa and 940 kPa. However, there were some small areas with lower ultimate bearing capacities. The ultimate bearing capacity increased with increasing depth, which can be attributed to the increase in soil confining pressure and soil unit weight. These findings are useful in determining the design and load-bearing capacity of foundations in the study area.

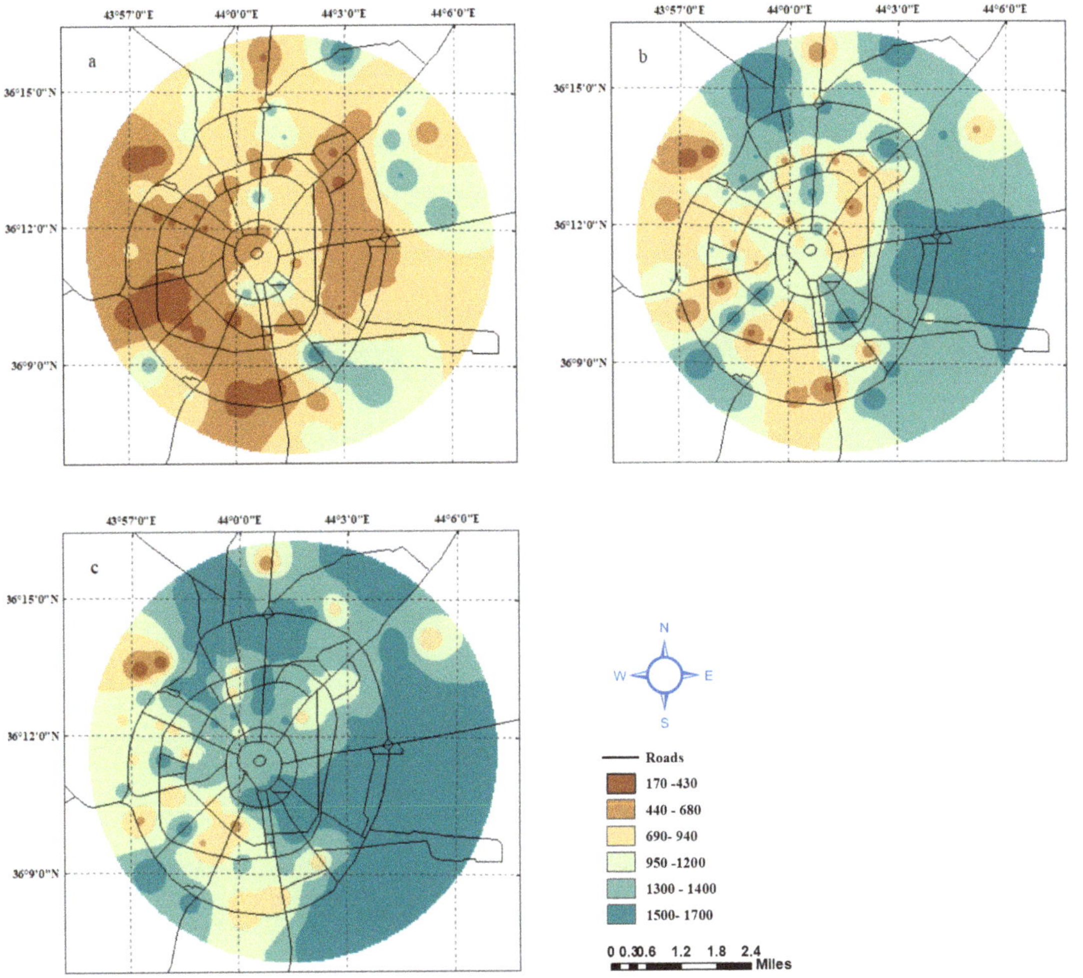

Figure 18. Ultimate bearing capacity from SPT-N value at depths (**a**) 1.5–3.5 m. (**b**) 3.5–6.5 m. (**c**) 6.5–9.5 m.

Figure 19 presents the correlation between the ultimate bearing capacity from Meyerhof's equation and the SPT values of the soils within the study area. This figure indicates a good correlation between the SPT values and the ultimate bearing capacity of the foundations. The correlation between the ultimate bearing capacity from the SPT values of the soil strata within the study area and soil cohesion (shear strength parameter). The ultimate bearing capacity from Meyerhof's equation and the ultimate bearing capacity from the SPT values of the soil strata within the study area were correlated, and the results are presented in Figure 20. The estimation of ultimate bearing capacity from the Standard Penetration Test (SPT) values is a relatively simple method compared to other methods that require more experimental tests and complex equations established on the shear strength parameters of the soil. The estimation of ultimate bearing capacity from SPT values is based on empirical correlations and has been widely used in the field of geotechnical engineering. The advantage of this method is its simplicity; however, it may not accurately reflect the actual soil conditions, which can be affected by factors such as soil type, stratigraphy, and loading conditions [59].

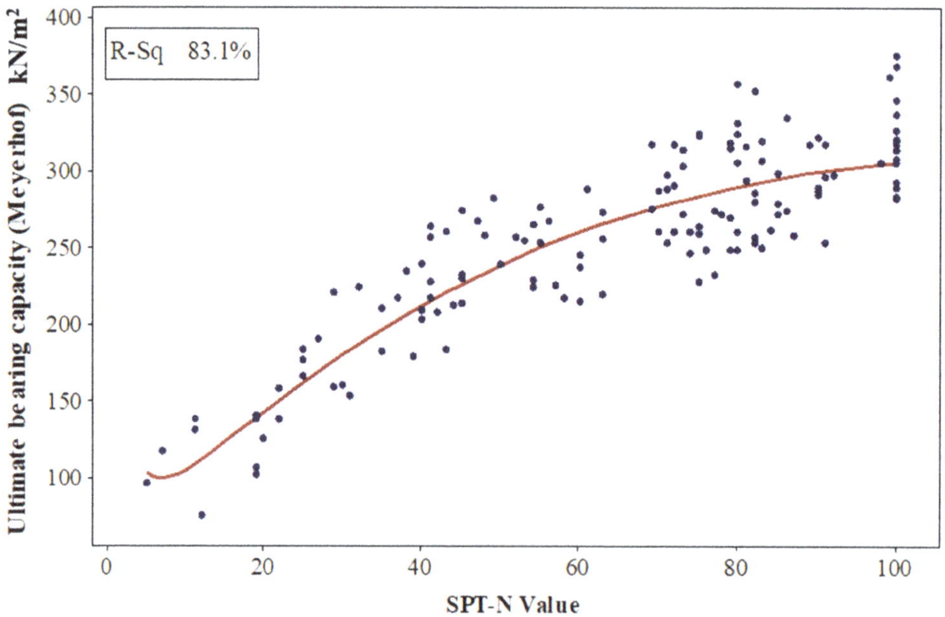

Figure 19. Correlation of ultimate bearing capacity (Meyerhof) with SPT-N values of soils in study area.

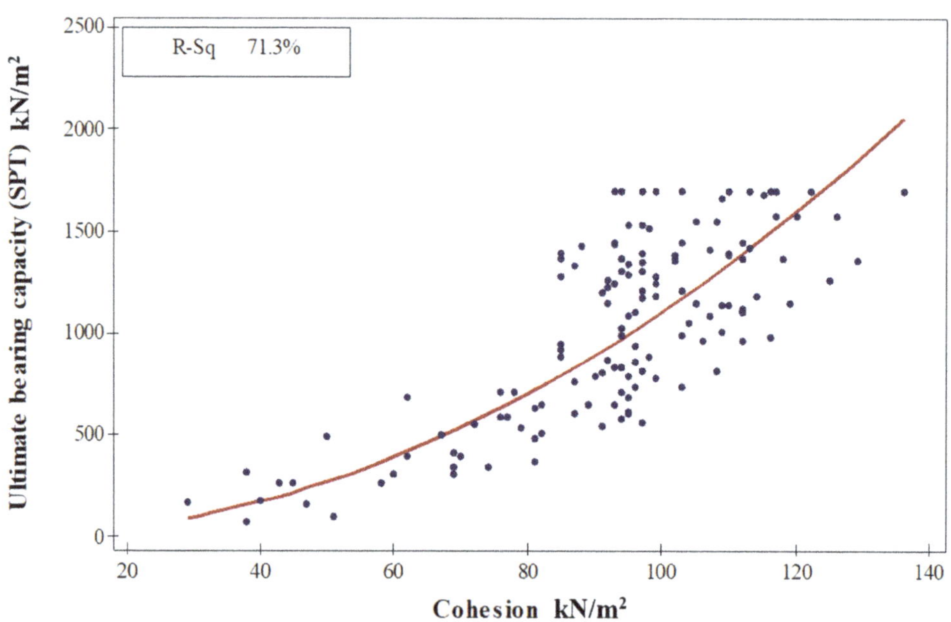

Figure 20. Correlation of ultimate bearing capacity (SPT-N) with soil cohesion of soils in study area.

4.2. Artificial Neural Network Models

ANN models are commonly used for regression and classification tasks, including prediction problems. ANN models consist of interconnected nodes, or artificial neurons, that process and transmit information. The nodes are organized into layers, and the connections between them can be adjusted during the training process to minimize the error between the predicted and actual outcomes. ANN models have been widely used in various fields due to their ability to handle complex relationships and make accurate predictions. The relationship between the predicted original SPT values and bearing capacity in the training data and test data was demonstrated in four models that have an acceptable R^2 and the smallest error. After input and output data are gathered and structured, training and test sets were established. 70% of the data were used for training and 15% used for testing and 15% for validation of the total data of boreholes. Predictions for SPT-N value from two models were developed.

4.2.1. Validation of Interpolations Based on Semivariograms

The transformed data's spatial autocorrelation is modeled using semivariograms/ covariance modeling for SPT-N values at depth 1.5–3.5 m, depicting the similarity decrease between data points as their distance increases. Binned values (red dots) are generated by grouping semivariograms/covariance points using square cells, while average points (blue crosses) are generated by binning empirical points in angular sectors. Binned points show local variation, while average values show smooth variation. A stable type model (dark blue line) is fitted to the empirical variogram for the measured data points.

At h = 0, the semivariogram should be 0. However, at an infinitesimally small separation distance, the semivariogram often exhibits a nugget effect, which is some value greater than zero. In this case, the nugget effect exists which is zero. The range is the distance at which the model levels out (5750.634). Locations closer than the range are spatially auto correlated, while farther locations are not. The partial sill is the sill minus the nugget (462.1716). The lag size is the distance class size (674.7548) with 12 lags. Semivariogram values are shown in Figure 21 with higher values in orange/red and lower values in blue/green.

Kriging estimates unknown spatial values. The search neighborhood step involves selecting nearby points with significant influence on the prediction location, determined by spatial auto-correlation. The method eliminates irrelevant points and weights nearby points using a search neighborhood of adjacent points, radius, and number of sectors to estimate values at the unknown location. Accurate neighborhood identification and selection of nearby points are crucial for successful kriging. As shown in Figure 22, five neighboring points are selected and a circle with four sectors is selected. The points highlighted in the data give an indicator of the weights associated with each point, and these weights are used to estimate the value at the unknown location, which is at the center of the crosshair.

4.2.2. Prediction for SPT-N Value

In Figures 23 and 24 the results of the ANN model (a) show a good agreement between the predicted and measured SPT-N values, with an R^2 of 0.92 for the training data and 0.81 for the testing data. The model uses Atterberg limit values, water content, cohesion, and internal friction as input variables and predicts the SPT value as its output. On the other hand, to predict SPT N value as output with (LL%, PL%, PI%, WC, ϕ, Fine content) as input, the result showed R^2 values of 0.90 and 0.8 for training and testing, respectively. As mentioned in previous studies [60], the predictions of the SPT values in model (a) and model (b) were conducted with more superficial R^2 values that give a significant agreement for using ANN modeling in geotechnical engineering that helps the engineering to be utilized in the design of infrastructures.

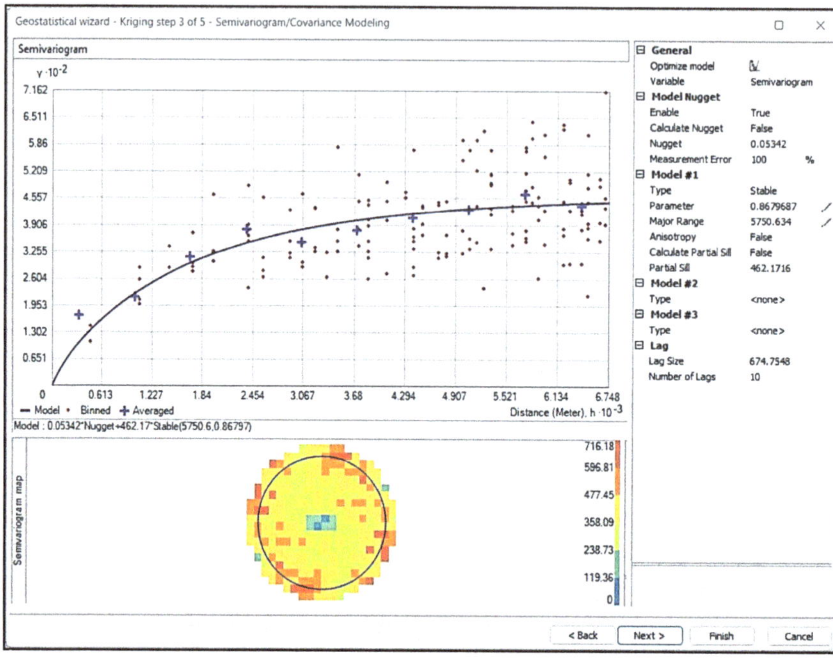

Figure 21. Semivariograms/Covariance Modeling.

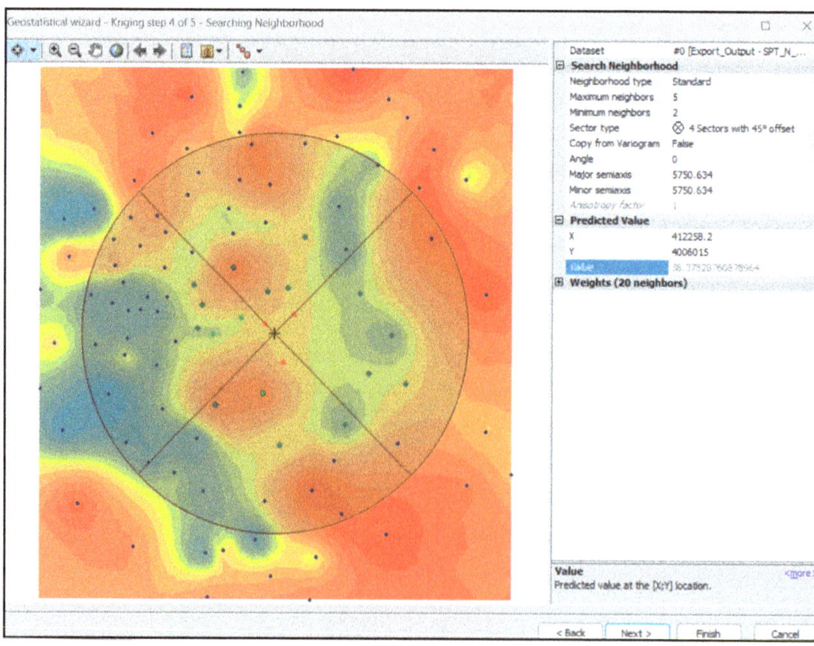

Figure 22. Searching Neighborhood Dialog Box of Ordinary Kriging.

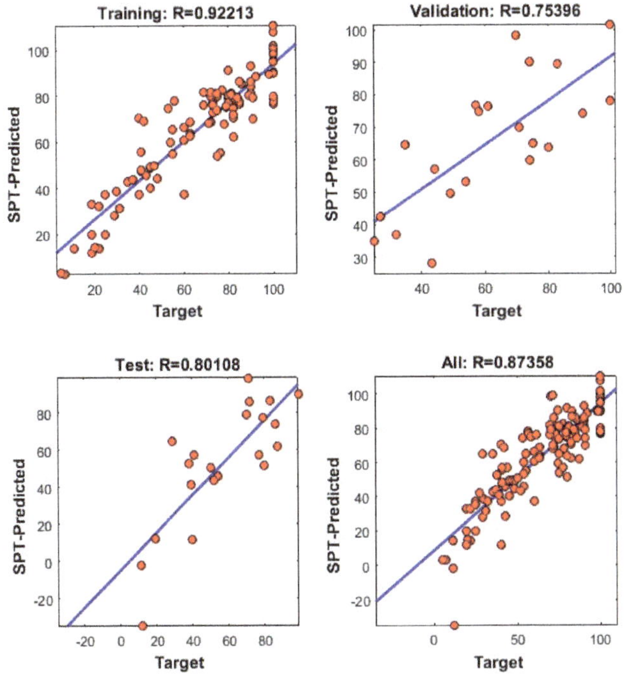

Figure 23. Model (a): relations between (LL%, PL%, PI%, WC, c, ϕ, Fine content) and SPT-N.

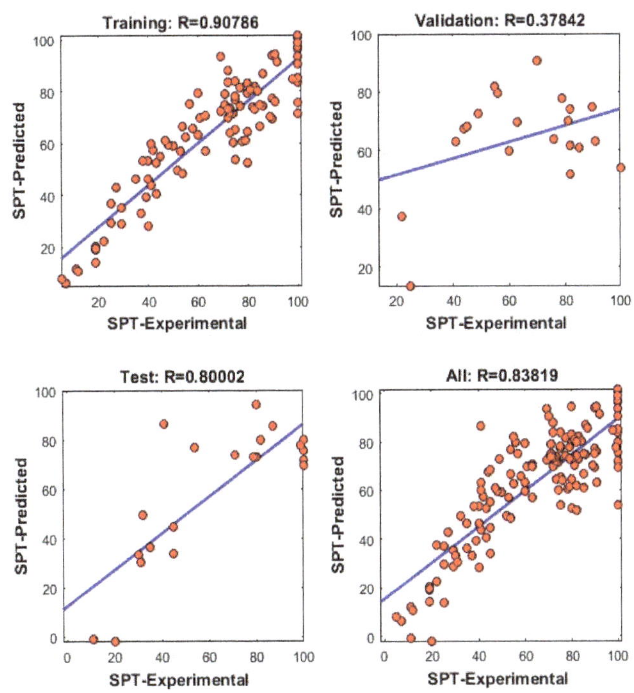

Figure 24. Model (b): relations between (LL%, PL%, PI%, WC, ϕ, Fine content) and SPT-N.

4.2.3. Prediction of Ultimate Bearing Capacity

The prediction of the ultimate bearing capacity of soil using ANNs is a machine learning approach that involves training a neural network model with a dataset of soil properties and corresponding ultimate bearing capacity values. The trained model can then be used to predict the ultimate bearing capacity of new soil samples based on their properties. Advantages of ANN prediction include the handling of non-linear relationships between soil properties and ultimate bearing capacity, the ability to incorporate complex relationships between soil properties and ultimate bearing capacity, and the ability to handle large datasets with many input variables.

To determine the best prediction results for the ultimate bearing capacity, two models were used. The inputs and outputs of the models are listed in Table 2 along with the value of R^2, which is a measure of the goodness of fit of the model to the data. R^2 ranges from 0 to 1, with a higher value indicating a better fit between the model and the data. The R^2 value provides an indication of how well the models are able to predict the ultimate bearing capacity based on the inputs.

Table 2. ANN models for Q-Ultimate prediction.

No	Model No.	Input	Output	Training	Validation	Testing	Adjust R^2
1	Model (c)	LL%, PL%, PI%, WC, c, ϕ, Fine content	Q-Ultimate	91.5	83.8	82	88.79
3	Model (d)	LL%, PL%, PI%, WC, ϕ, Fine content		73.97	34.8	86.98	70.8

The results of bearing capacity prediction using ANN modeling show similarities with the results of previous studies [60]. In Figures 25 and 26 the relationship between the predicted ultimate bearing capacity and the original ultimate bearing capacity on the training and testing data are illustrated.

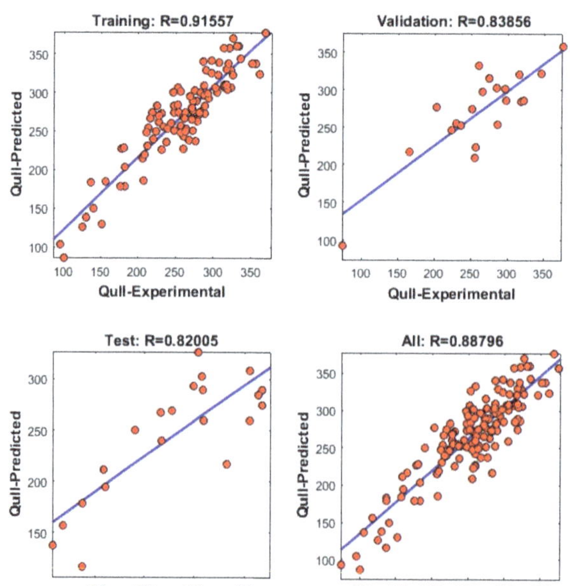

Figure 25. Model (c): relations between (LL%, PL%, PI%, WC, c, ϕ, Fine content) and Q ultimate.

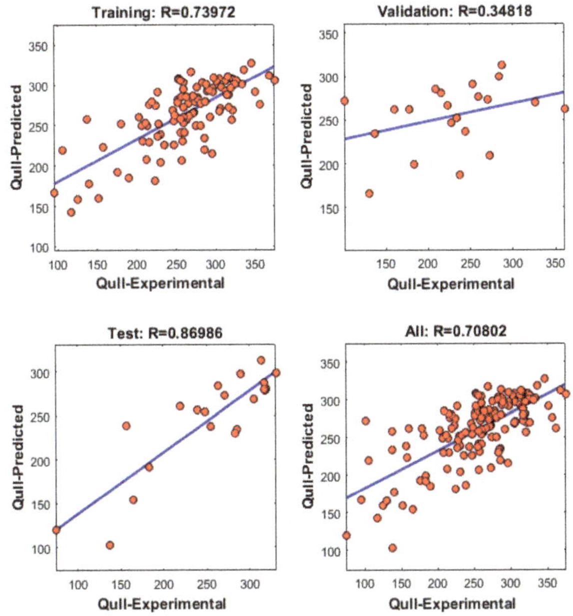

Figure 26. Model (d): relations between (LL%, PL%, PI%, WC, ϕ) and Q ultimate.

Model (c) has a high R^2 value, indicating a strong correlation between the predicted and actual values, and a low variance in the residuals. The high R^2 value for the training data (0.91) indicates that the model is able to fit the training data well, while the R^2 value of 0.82 for the testing data suggests that the model has good generalization ability and can predict unseen data with a certain degree of accuracy. In Model (d), the decrease in R^2 value is likely due to the absence of the cohesion value as an input. Cohesion is an important factor that affects the bearing capacity of soil, so its absence in the input could result in a decrease in the accuracy of the model. This highlights the importance of considering all relevant factors in the input variables of the model to improve its accuracy and predictability.

Soil cohesion is a measure of the shear strength of soil, which determines its ability to support loads. The results of research in geotechnical engineering have shown that soil cohesion is the most important factor in estimating soil bearing capacity. This is because it determines the resistance of soil to sliding or deformation under load, and is essential for ensuring the stability of structures built on the soil. Therefore, accurate determination of soil cohesion is crucial for safe and effective design of geotechnical structures.

4.2.4. Percentage Error of ANN Models

Figures 27 and 28 demonstrate the error percentage lines for model (a) and (b) outcomes referring to the difference between the predicted values and the actual values. It is a measure of the accuracy of the model's predictions. A lower error percentage indicates a more accurate model, while a higher error percentage means that the model is less accurate. The acceptable error percentage depends on the specific application and the acceptable level of error for that particular field. In some cases, a low error percentage, such as 1–2%, may be acceptable, while in others, a higher error percentage may be acceptable if the cost of a lower error is too high [61]. It is important to note that no model can be 100% accurate and some level of error is always present. The goal is to reduce the error to the lowest possible level while still making predictions that are useful and relevant.

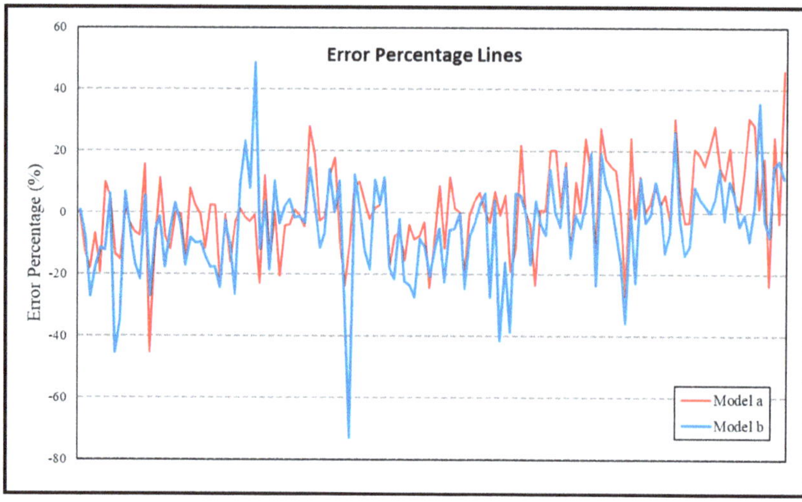

Figure 27. Error percentage lines of model (a) and (b) for prediction SPT-N value.

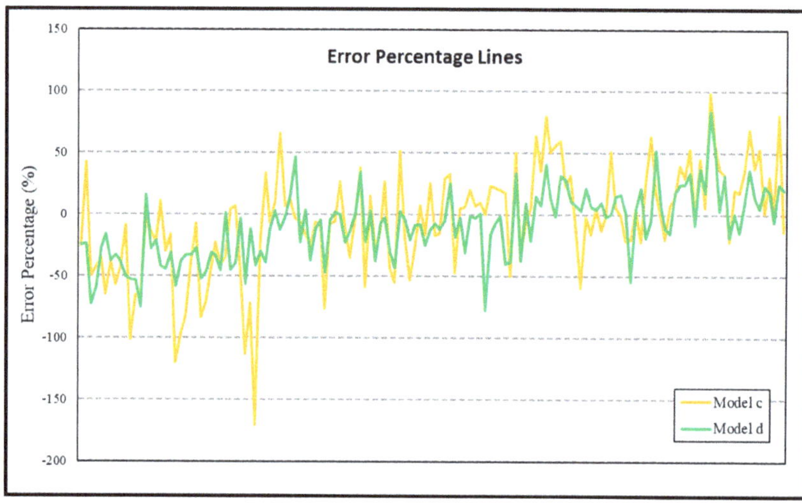

Figure 28. Error percentage lines of model (c) and (d) for prediction Q ultimate.

4.2.5. Analysis of Models

Sensitivity analysis is an important aspect of model evaluation, as it helps to assess the impact of individual model parameters on the final results. The absence of a sensitivity analysis in the study you are referring to could limit the understanding of the model's behavior and the confidence in the results obtained. Analysis of models is particularly important when using ANOVA, as the parameters of these models can have a significant impact on the spatial structure of the data. By conducting a sensitivity analysis, the researchers could determine the robustness of the model results. The sensitivity analysis to evaluate the effect of each model parameter based on the semivariograms shown in Table 3.

Table 3. Analysis of models and effect each parameter on the output of prediction.

Model a			Model b			Model c			Model d		
Parameters	F-Value	*p*-Value	Parameters	F-Value	*p*-Value	Parameters	F-Value	*p*-Value	Parameters	F-Value	*p*-Value
LL%	1.63	0.204	LL%	0.03	0.862	LL%	0.17	0.677	LL%	2.02	0.157
PL%	1.07	0.303	PL%	0.51	0.477	PL%	0.07	0.792	PL%	2.19	0.142
PI%	0.46	0.497	PI%	0.27	0.607	PI%	1.01	0.316	PI%	0.24	0.627
WC	2.52	0.115	WC	2.36	0.126	WC	1.12	0.291	WC	5.95	0.016
c	157.85	0.000	φ	0.75	0.387	c	320.21	0.000	φ	0.03	0.873
φ	10.29	0.002	No.200	46.73	0.000	φ	28.47	0.000	No.200	68.38	0.000
No.200	1.81	0.180				No.200	0.04	0.843			

A *p*-value less than 0.05 is commonly used as a threshold for determining statistical significance in hypothesis testing. In this context, a *p*-value less than 0.05 for a particular model parameter means that the results suggest a statistically significant relationship between the parameter and the outcome variable being modeled.

In this study, a *p*-value less than 0.05 for a particular model parameter would indicate that the parameter has a significant impact on the model results. This information can be useful in understanding the underlying relationships in the data and in guiding further analysis or interpretation of the results. However, it is important to consider other factors such as the sample size, the quality of the data, and the overall fit of the model in evaluating the reliability and robustness of the results.

5. Conclusions

This study focuses on developing maps for soil geotechnical properties that are widely utilized by geotechnical engineers in foundation design capacity. The main conclusions are as follows. An artificial neural network (NN) model was established for estimating SPT-N value and ultimate bearing capacity:

- GIS is an effective tool that can be used by engineers to analyze the preliminary exploration of geotechnical sites. Information from 102 boreholes, considering the main geotechnical properties, was collected, evaluated, and used as input data for GIS analysis.
- This information suggests that a significant portion of Erbil city has soil with a high proportion of fine-grained materials, such as clay and silt. High fines content can impact the soil's physical and engineering properties, such as its compressibility, permeability, and shear strength. The presence of high fines content can also increase the susceptibility of soil to swelling and shrinkage, which can lead to instability in structures built on or in the soil. The small zones in the southeast of the study area with lower fine contents may have different soil characteristics and may offer potential sites for structures that require more stable soil conditions. These findings are important for the design of infrastructure and buildings in the city.
- Atterberg limits in most of Erbil City were found to be between 40% and 52%, and 19% and 30% for the liquid and plastic limits, respectively. This indicates the high presence of low–plasticity clay and clayey silt. The results of the analysis of liquid limit and plastic limit values in the study area provide important information for engineers and planners in Erbil city center. They highlight the presence of low plastic clay in high percentages in the study area, as well as the need to carefully evaluate critical points with high liquid limit and plastic limit in future construction and development projects.
- Digital mapping of shear strength parameters showed that most soil strata at three different depths had an internal friction angle between 2° and 6°, and the cohesive strength ranged between 76 kPa and 130 kPa. The results of the cohesion values show that the soils in the study area at shallow depths have moderate to high cohesion values, and that the soils with high cohesion values tend to be located in areas with high fines content. However, the results of the angle of internal friction show that the soils in the study area have moderate to low shear strength values, with the soils in the east-south part of the area having slightly higher shear strength values. These findings are important in determining the suitability of the soil for different types of

structures and in designing and constructing structures that are appropriate for the soil conditions in the area.

- The soil in the study area mostly has a moderate compressibility and resilience, with a moderate to low amount of rebound. The compression index decreases with depth, suggesting that the soil becomes less compressible as one moves deeper into the ground. The rebound index indicates that the soil has a moderate to low ability to recover its original volume after being compressed. These findings provide valuable information for designing structures that are built on or into the soil in the study area.
- SPT values in the study area indicate moderate soil strength in the shallow strata, with a range of 17 to 48. As the depth of the soil strata increases, the SPT values increase and become higher, covering large parts of the study area. This suggests that the soil becomes stronger with increasing depth. The SPT is a widely used in-situ test for measuring soil strength, and these results provide valuable information for the design of foundations and other structures that are supported by the soil. The higher SPT values at greater depths indicate improved soil strength characteristics and can influence the design of these structures in terms of load-bearing capacity and stability.
- This conclusion suggests that the soil in Erbil City is not capable of supporting heavy loads without modification or special design measures. The ultimate bearing capacity is a measure of the maximum weight or load that a soil can support without failure. A value lower than 170 kPa indicates that the soil may not be suitable for supporting heavy structures, such as buildings and bridges, without additional treatment or specialized foundation design. Improving the soil, such as through compaction or stabilization, and utilizing special footing designs, such piles, can increase the soil's bearing capacity and ensure the stability and safety of structures built on the soil.
- At the preliminary design point, the completed digital geotechnical maps are vital. The designer could use the geotechnical parameters, consolidation characteristics and SPT as an effective visual display tool simply by using the digital values of these parameters for the proposed region, where the necessary decisions can be made.
- The correlation between the SPT values and shear strength parameters for the soils within the study area demonstrated a strong relationship between them.
- The results obtained from the models were compared with those measured from the field tests. It was found that predicted SPT-N values and Q-ultimate bearing capacity are quite close to the measured values. In order to check the prediction performance of the ANN model developed, several performance indices, such as R^2, MAPE, and RMSE were also calculated. The ANN model has shown good prediction performance based on the performance indices. Thus, the developed ANN model can be used to predict SPT-N and Q-ultimate bearing capacity from the soil parameters and borehole coordinates. The ANN model's implementation has also demonstrated that the neural network is a valuable tool to minimize the uncertainties encountered during geotechnical engineering projects. Therefore, using Artificial Neural Networks may provide new techniques and methodologies and minimize the potential inconsistency of correlations. ANN prediction is a useful tool for predicting the ultimate bearing capacity of soil, but it should be used in conjunction with other methods and validated with independent data to ensure accurate predictions.

Author Contributions: Conceptualization, K.Z., Z.B.Q. and A.F.C.; methodology, Z.B.Q.; software, Z.B.Q.; validation, Z.B.Q., A.F.C. and Z.K.; formal analysis, Z.B.Q.; investigation, Z.B.Q.; resources, Z.B.Q.; data curation, Z.B.Q.; writing—original draft preparation, Z.B.Q.; writing—review and editing, A.F.C., Z.B.Q. and Z.K.; visualization, Z.B.Q.; supervision, A.F.C. and Z.K.; project administration, A.F.C. All authors have read and agreed to the published version of the manuscript.

Funding: This research received no external funding.

Institutional Review Board Statement: Not applicable.

Informed Consent Statement: Not applicable.

Data Availability Statement: The data are available from the conforming author upon request.

Conflicts of Interest: The authors declare no conflict of interest.

Appendix A

The data of all boreholes are provided demonstrated in Table A1.

Table A1. Full data of all boreholes of study area.

BH NO	DEPTH m	X	Y	LL%	PL%	PI%	WC%	c (kN/m^2)	Φ (°)	Fine Content	SPT-N Value (kN/m^2)
1	1.5–3.5	406,851.5	4,009,570	46	22	24	28.3	51	4	62.1	7
2	1.5–3.5	405,972.9	4,009,284.2	40	22	18	15	96	3	94.1	54
3	1.5–3.5	407,487.6	4,007,947.1	47	25	22	18	99	5	91.5	57
4	1.5–3.5	406,121.1	4,008,204.7	40	22	18	15.0	96	3	94.1	54
5	1.5–3.5	407,443.2	4,008,702.7	50	27	23	20.8	109	5	92.9	41
6	1.5–3.5	407,941.4	4,009,347.7	50	28	22	13.9	108	4	91.8	45
7	1.5–3.5	408,217.9	4,008,512.23	35	20	19	20.0	96	4	96	41
8	1.5–3.5	408,997.4	4,008,917.8	39	21	18	16.3	109	5	89.4	45
9	1.5–3.5	408,743.4	4,009,398.1	47	25	22	18.0	99	5	91.5	57
10	1.5–3.5	408,026.1	4,010,406	45	24	21	16.6	96	5	95.5	63
11	1.5–3.5	409,617.8	4,011,321.6	46	27	19	20.2	99	5	90.7	41
12	1.5–3.5	408,588.1	4,011,780.2	45	25	20	19.0	97	4	91.6	60
13	1.5–3.5	409,762.1	4,012,805.4	48	26	22	19.7	94	5	81.3	75
14	1.5–3.5	411,184.2	4,011,787.2	49	23	26	19.6	108	4	96.3	33
15	1.5–3.5	411,184.3	4,012,540.7	47	25	22	15.6	105	5	97.4	30
16	1.5–3.5	411,247.7	4,013,522.9	48	26	22	18.4	58	3	92.7	19
17	1.5–3.5	409,969.9	4,010,643.8	58	31	27	20.5	91	4	88.9	40
18	1.5–3.5	409,662.9	4,009,977	56	30	26	18.8	94	4	69.1	42
19	1.5–3.5	409,915.1	4,009,232.2	45	24	21	13.6	77	3	95.9	29
20	1.5–3.5	410,862.1	4,009,607.35	44	25	19	18.3	69	5	94.5	30
21	1.5–3.5	411,994.6	4,009,216.9	42	23	19	19.2	70	4	98.2	29
22	1.5–3.5	411,166.1	4,010,438.9	50	26	24	17.9	95	5	92.2	58
23	1.5–3.5	412,126.9	4,010,308.3	51	29	29	19.8	95	5	96.2	70
24	1.5–3.5	411,550.5	4,011,441.2	44	23	21	14.8	95	7	100	80
25	1.5–3.5	413,143.3	4,011,488.9	44	25	19	20.5	97	4	93.5	41
26	1.5–3.5	414,064.7	4,011,692.8	41	23	18	17.4	110	4	53.6	37
27	1.5–3.5	415,317.3	4,010,885.4	50	26	24	15.9	116	5	68.5	23
28	1.5–3.5	414,149.3	4,009,671.4	51	27	24	19.1	94	4	87.3	19
29	1.5–3.5	414,362.5	4,008,450.6	52	24	28	19.7	90	4	86.4	17
30	1.5–3.5	416,171.1	4,012,249.6	45	24	21	15.3	98	6	79.8	43
31	1.5–3.5	415,609.8	4,012,761.8	38	21	17	16.0	95	5	70.6	45
32	1.5–3.5	414,549.1	4,013,637	38	21	17		70	12	97	100
33	1.5–3.5	417,922.5	4,010,462	53	27	26	21.6	69	3	92.6	25
34	1.5–3.5	416,961.1	4,011,319.3	49	23	26	15.1	112	5	86.7	71
35	1.5–3.5	416,542.1	4,010,233.5	43	23	20	17.8	91	3	89.3	79
36	1.5–3.5	416,909.4	4,008,754	44	24	20	14.8	92	4	66.2	85

Table A1. *Cont.*

BH NO	DEPTH m	X	Y	LL%	PL%	PI%	WC%	c (kN/m^2)	Φ (°)	Fine Content	SPT-N Value (kN/m^2)
37	1.5–3.5	406,775.3	4,007,107.5	50	26	24	18.3	82	4	92.1	18
38	1.5–3.5	405,227.4	4,007,274.3	53	27	26	17.3	74	3	96.4	25
39	1.5–3.5	407,403.9	4,006,885.3	40	22	18	13.6	105	5	63.8	19
40	1.5–3.5	407,715.1	4,007,488.5	43	22	21	19.1	90	4	93.2	20
41	1.5–3.5	408,445.4	4,007,069.4	45	24	21	15.8	58	5	95.9	19
42	1.5–3.5	408,736.7	4,005,101.9	44	25	19	17.6	60	4	95.5	22
43	1.5–3.5	408,248.5	4,006,694.7	43	24	19	19.9	69	4	92.1	22
44	1.5–3.5	407,872.4	4,006,657.22	56	25	31	15.6	116	5	95.5	28
45	1.5–3.5	409,311.9	4,005,786	46	24	22	17.7	112	4	97.4	32
46	1.5–3.5	408,775.59	4,006,631.24	46	22	24	21.1	85	4	95.8	19
47	1.5–3.5	409,973.4	4,006,133	53	25	28	20.9	87	5	93.6	31
48	1.5–3.5	409,048.64	4,007,069.39	46	24	22	16.6	89	4	96.2	29
49	1.5–3.5	409,874.19	4,007,406.1	46	24	22	17.7	92	5	94.1	37
50	1.5–3.5	408,985.14	4,007,907.6	54	26	28	15.3	99	4	94.1	39
51	1.5–3.5	411,015.2	4,007,902.7	55	24	31	15.3	112	5	95.2	82
52	1.5–3.5	418,521.8	4,007,163.4	57	25	32	19.6	104	4	92.9	77
53	1.5–3.5	409,785.25	4,008,333.1	48	22	26	14.4	97	4	95.5	22
54	1.5–3.5	410,998.6	4,008,878.3	49	26	23	18.8	105	4	92.2	32
55	1.5–3.5	410,109.1	4,006,834.44	41	19	22	16.8	101	4	87.3	41
56	1.5–3.5	410,419.8	4,006,001	45	21	24	16.3	123	5	52.3	28
57	1.5–3.5	412,049.02	4,007,215.97	52	24	28	21.8	109	5	51.8	44
58	1.5–3.5	411,263	4,006,447.5	51	27	24	19.4	96	4	94.2	33
59	1.5–3.5	412,664.18	4,007,328.4	49	25	24	19.4	124	4	92.6	38
60	1.5–3.5	411,990	4,006,315.2	44	25	19	21.7	136	5	95.7	39
61	1.5–3.5	413,164.9	4,008,779	43	23	20	18.9	119	5	96.6	41
62	1.5–3.5	412,846.74	4,006,559.81	49	26	23	17.4	79	4	93.5	39
63	1.5–3.5	412,536.5	4,005,174.2	48	27	21	18.0	81	4	93.5	35
64	1.5–3.5	415,747.4	4,005,960.69	48	23	25	16.8	96	4	96.4	21
65	1.5–3.5	414,814.8	4,007,156.9	40	21	19	17.9	97	5	39.7	31
66	1.5–3.5	415,071.9	4,004,866.9	47	24	23	19.0	95	5	90.8	34
67	1.5–3.5	417,971.23	4,001,685.01	48	27	21	17.4	88	8	30	61
68	1.5–3.5	418,519	4,003,222.9	0	0	0	17.7	85	10	18	37
69	1.5–3.5	415,926	4,002,878	0	0	0	19.0	83	9	14	50
70	1.5–3.5	416,167	4,004,572.3	0	0	0	11.0	80	8	15	36
71	1.5–3.5	419,289	4,001,977.6	0	0	0	18.6	114	5	25	40
72	1.5–3.5	411,924.27	4,004,297.59	57	25	32	16.4	116	5	94.7	72
73	1.5–3.5	410,482.29	4,003,953.63	48	23	25	13.4	109	5	91.4	74
74	1.5–3.5	412,426.98	4,002,802.69	45	21	24	12.1	144	4	95.6	23
75	1.5–3.5	414,365	4,003,408	44	21	23	13.7	135	4	94.6	20
76	1.5–3.5	411,975.73	4,001,202.95	47	21	26	13.8	118	5	93.2	39
77	1.5–3.5	414,022.9	4,002,387	45	22	23	15.4	95	4	94.7	44
78	1.5–3.5	411,697.91	4,000,107.57	41	27	14	11.9	55	6	77.8	8
79	1.5–3.5	410,200	3,999,706	35	23	12	12.6	47	3	83.7	11
80	1.5–3.5	413,372.73	4,001,528.38	0	0	0	13.7	98	5	8.4	100

Table A1. *Cont.*

BH NO	DEPTH m	X	Y	LL%	PL%	PI%	WC%	c (kN/m²)	Φ (°)	Fine Content	SPT-N Value (kN/m²)
81	1.5–3.5	414,253.79	4,000,845.76	47	22	25	17.5	86	5	91.7	81
82	1.5–3.5	415,598.5	4,000,259	45	22	24	20.1	89	3	93.5	83
83	1.5–3.5	413,510.13	3,999,595.57	47	25	22	20.5	81	3	95.6	27
84	1.5–3.5	412,160.93	3,999,051.88	41	22	19	12.7	12	36	100	45
85	1.5–3.5	410,732.9	3,999,785	40	21	19	10.8	15	30	33	47
86	1.5–3.5	410,253	3,998,445	43	23	20	5.7	19	35	42	53
87	1.5–3.5	413,309	3,998,392.5	41	21	20	12.3	108	5	81.6	60
88	1.5–3.5	410,136.16	4,001,490.72	55	26	29	17.3	81	3	96.2	22
89	1.5–3.5	409,296.11	4,002,026.51	46	24	22	22.1	96	5	96.8	31
90	1.5–3.5	408,932.6	4,003,851.9	50	24	26	14.8	90	3	97.3	29
91	1.5–3.5	408,515.58	4,002,310.93	44	30	14	31.1	25	2	90.1	18
92	1.5–3.5	409,939	4,003,009.6	44	28	16	29.2	29	3	89.2	12
93	1.5–3.5	407,821.05	4,002,866.56	42	21	21	23.2	45	4	91.6	19
94	1.5–3.5	407,298.5	4,004,255.62	49	26	23	16.7	38	3	89.8	5
95	1.5–3.5	405,937	4,003,224	43	23	20	13.3	90	4	93.3	7
96	1.5–3.5	407,774.75	4,005,380.1	38	21	17	16.9	98	4	96.6	39
97	1.5–3.5	407,844.33	4,005,870.11	39	22	17	18.9	43	4	56.7	19
98	1.5–3.5	406,914.85	4,005,671.15	45	23	22	14.5	121	6	94.6	23
99	1.5–3.5	405,242	4,004,431.7	46	25	21	15.6	110	4	98.1	25
100	1.5–3.5	405,684.54	4,005,717.45	43	21	22	14.8	95	4	94.3	41
101	1.5–3.5	408,025.12	3,999,907.5	46	24	22	15.7	103	4	96.2	54
102	1.5–3.5	406,400.4	4,001,124	42	26	16	15.5	94	4	96.1	75
103	3.5–6.5	406,851.5	4,009,570	49	23	26	29.4	40	4	83.5	11
104	3.5–6.5	405,972.9	4,009,284.2	48	23	25	29.1	45	5	87	9
105	3.5–6.5	407,487.6	4,007,947.1	50	24	26	19.2	109	5	95.1	72
106	3.5–6.5	406,121.1	4,008,204.7	46	26	20	19.4	85	4	97.9	60
107	3.5–6.5	407,443.2	4,008,702.7	52	26	26	23.2	112	4	94.9	70
108	3.5–6.5	407,941.4	4,009,347.7	52	29	23	14.8	103	4	93.7	77
109	3.5–6.5	408,217.9	4,008,512.23	40	22	18	20.5	113	5	100	90
110	3.5–6.5	408,997.4	4,008,917.8	36	23	13		105	5	98.4	73
111	3.5–6.5	408,743.4	4,009,398.1	44	24	20	19.3	102	5	95.1	86
112	3.5–6.5	408,026.1	4,010,406	42	23	19	15.5	102	6	94.1	80
113	3.5–6.5	409,617.8	4,011,321.6	64	33	31	19.5	89	4	80.5	97
114	3.5–6.5	408,588.1	4,011,780.2	48	25	23	20.1	93	5	89.1	99
115	3.5–6.5	409,762.1	4,012,805.4	46	25	21	17.2	98	5	72.5	100
116	3.5–6.5	411,184.2	4,011,787.2	45	24	21	14.3	104	4	95.7	50
117	3.5–6.5	411,184.3	4,012,540.7	49	28	21	18.2	132	6	92.9	62
118	3.5–6.5	411,247.7	4,013,522.9	45	27	18	19.0	62	4	100	25
119	3.5–6.5	409,969.9	4,010,643.8	51	27	24	18.7	93	5	94.3	91
120	3.5–6.5	409,662.9	4,009,977	54	29	25	19.3	95	5	94.2	89
121	3.5–6.5	409,915.1	4,009,232.2	47	23	24	17.5	87	5	94.6	33
122	3.5–6.5	410,862.1	4,009,607.35	46	22	24	20.7	82	4	98.2	41
123	3.5–6.5	411,994.6	4,009,216.9	45	21	23	17.0	72	5	97.5	35

Table A1. *Cont.*

BH NO	DEPTH m	X	Y	LL%	PL%	PI%	WC%	c (kN/m²)	Φ (°)	Fine Content	SPT-N Value (kN/m²)
124	3.5–6.5	411,166.1	4,010,438.9	54	25	29	18.6	91	4	97.8	76
125	3.5–6.5	412,126.9	4,010,308.3	55	26	29	19.3	93	4	92.1	79
126	3.5–6.5	411,550.5	4,011,441.2	46	22	24	14.8	97	8	100	100
127	3.5–6.5	413,143.3	4,011,488.9	46	25	21	21.1	99	4	95.2	75
128	3.5–6.5	414,064.7	4,011,692.8	60	32	28	21.2	147	5	86.6	50
129	3.5–6.5	415,317.3	4,010,885.4	50	28	22	20.2	98	5	91.9	56
130	3.5–6.5	414,149.3	4,009,671.4	42	23	19	19.4	95	5	41.6	100
131	3.5–6.5	414,362.5	4,008,450.6	71	36	35	19.7	95	5	59.7	43
132	3.5–6.5	416,171.1	4,012,249.6	38	21	17	16.8	95	5	82.6	82
133	3.5–6.5	415,609.8	4,012,761.8	44	25	19	16.1	97	6	85.8	83
134	3.5–6.5	414,549.1	4,013,637	42	23	19	10.4	75	11	100	100
135	3.5–6.5	417,922.5	4,010,462	48	26	22	20.2	76	5	96.1	37
136	3.5–6.5	416,961.1	4,011,319.3	48	23	25		114	5	69.1	75
137	3.5–6.5	416,542.1	4,010,233.5	45	26	19	21.6	94	5	94.9	87
138	3.5–6.5	416,909.4	4,008,754	54	32	22	19.6	94	4	89.4	83
139	3.5–6.5	406,775.3	4,007,107.5	53	28	25	19.3	78	3	97.9	45
140	3.5–6.5	405,227.4	4,007,274.3	54	28	26	16.8	82	4	98.6	32
141	3.5–6.5	407,403.9	4,006,885.3	38	21	17	17.3	98	5	92.5	39
142	3.5–6.5	407,715.1	4,007,488.5	47	23	24	21.5	92	5	96.1	55
143	3.5–6.5	408,445.4	4,007,069.4	44	23	21	19.3	81	3	96.3	40
144	3.5–6.5	408,736.7	4,005,101.9	46	23	23	19.2	87	4	93.8	38
145	3.5–6.5	408,248.5	4,006,694.7	52	26	26	22.3	93	5	94.6	53
146	3.5–6.5	407,872.4	4,006,657.22	57	26	31	18.7	106	5	95.6	61
147	3.5–6.5	409,311.9	4,005,786	55	26	29	17.6	103	4	96.9	63
148	3.5–6.5	408,775.59	4,006,631.24	47	23	25	15.8	96	5	92.2	70
149	3.5–6.5	409,973.4	4,006,133	43	21	22	20.5	107	4	87.9	69
150	3.5–6.5	409,048.64	4,007,069.39	46	23	23	17.3	90	4	94.6	50
151	3.5–6.5	409,874.19	4,007,406.1	45	22	23	17.5	92	5	94	73
152	3.5–6.5	408,985.14	4,007,907.6	56	30	26	18.2	110	5	93.3	72
153	3.5–6.5	411,015.2	4,007,902.7	63	28	35	21.8	102	6	94	91
154	3.5–6.5	418,521.8	4,007,163.4	56	24	32	16.6	129	5	96.1	86
155	3.5–6.5	409,785.25	4,008,333.1	51	27	24	18.2	118	6	96.4	40
156	3.5–6.5	410,998.6	4,008,878.3	49	26	23	18.7	99	5	96.3	100
157	3.5–6.5	410,109.1	4,006,834.44	46	19	27	17.2	98	5	90.3	30
158	3.5–6.5	410,419.8	4,006,001	48	20	28	18.2	126	5	87.2	50
159	3.5–6.5	412,049.02	4,007,215.97	54	28	26	17.8	134	6	96.4	63
160	3.5–6.5	411,263	4,006,447.5	55	24	31	19.5	99	5	94.2	79
161	3.5–6.5	412,664.18	4,007,328.4	46	26	20	17.2	118	4	96.6	22
162	3.5–6.5	411,990	4,006,315.2	46	25	21	17.1	121	5	96.5	38
163	3.5–6.5	413,164.9	4,008,779	47	23	24	16.3	115	6	69.1	36
164	3.5–6.5	412,846.74	4,006,559.81	45	24	21	23.8	76	4	94.7	45
165	3.5–6.5	412,536.5	4,005,174.2	47	26	21	18.9	89	5	98.1	41
166	3.5–6.5	415,747.4	4,005,960.69	45	21	24	20.4	126	5	98.4	100
167	3.5–6.5	414,814.8	4,007,156.9	44	22	22	20.4	120	5	96.4	100

Table A1. *Cont.*

BH NO	DEPTH m	X	Y	LL%	PL%	PI%	WC%	c (kN/m^2)	Φ (°)	Fine Content	SPT-N Value (kN/m^2)
168	3.5–6.5	415,071.9	4,004,866.9	46	23	23	18.9	117	5	90.8	100
169	3.5–6.5	417,971.23	4,001,685.01	45	25	20	17.4	91	9	39	76
170	3.5–6.5	418,519	4,003,222.9	45	24	21	17.7	95	11	64	90
171	3.5–6.5	415,926	4,002,878	41	23	19	19.0	89	11	39	68
172	3.5–6.5	416,167	4,004,572.3	40	20	24	18.0	76	9	55	99
173	3.5–6.5	419,289	4,001,977.6	0	0	0	18.6	112	5	78.8	92
174	3.5–6.5	411,924.27	4,004,297.59	62	26	36	16.6	125	5	68.3	80
175	3.5–6.5	410,482.29	4,003,953.63	49	23	26	16.6	115	6	17.6	69
176	3.5–6.5	412,426.98	4,002,802.69	48	25	23	18.1	140	5	39.9	100
177	3.5–6.5	414,365	4,003,408	39	21	18	17.5	92	4	98.1	80
178	3.5–6.5	411,975.73	4,001,202.95	61	26	35	16.9	137	5	97.2	72
179	3.5–6.5	414,022.9	4,002,387	53	26	27	16.6	119	5	63.6	73
180	3.5–6.5	411,697.91	4,000,107.57	44	23	21	20.3	98	5	50.4	20
181	3.5–6.5	410,200	3,999,706	37	25	12	17.8	60	6	94.2	37
182	3.5–6.5	413,372.73	4,001,528.38	42	21	21	16.3	94	6	87.5	29
183	3.5–6.5	414,253.79	4,000,845.76	46	22	24	20.1	92	4	76.5	91
184	3.5–6.5	408,932.6	4,000,259	48	23	25	19.2	94	4	94.7	100
185	3.5–6.5	413,510.13	3,999,595.57	48	24	24	17.4	85	3	88.6	97
186	3.5–6.5	412,160.93	3,999,051.88	39	21	18	14.1	38	29	100	100
187	3.5–6.5	410,732.9	3,999,785	38	20	18	10.8	8	32	41	100
188	3.5–6.5	410,253	3,998,445	41	20	21	8.7	13	29	44	100
189	3.5–6.5	413,309	3,998,392.5	46	24	22	13.4	94	5	61.1	63
190	3.5–6.5	410,136.16	4,001,490.72	58	27	31	18.4	93	5	99.1	41
191	3.5–6.5	409,296.11	4,002,026.51	50	26	24	21.0	108	4	93.5	39
192	3.5–6.5	409,707	4,003,851.9	51	28	23	15.0	87	3	96	48
193	3.5–6.5	408,515.58	4,002,310.93	50	28	22	28.2	38	4	95.3	20
194	3.5–6.5	409,939	4,003,009.6	41	24	17	25.4	40	5	78.4	29
195	3.5–6.5	407,821.05	4,002,866.56	45	26	19	24.7	51	4	87.9	100
196	3.5–6.5	407,298.5	4,004,255.62	45	23	22	12.2	102	5	93.4	20
197	3.5–6.5	405,937	4,003,224	41	23	18	15.0	87	3	89.8	26
198	3.5–6.5	407,774.75	4,005,380.1	48	23	25	19.9	128	6	91.4	81
199	3.5–6.5	407,844.33	4,005,870.11	38	26	12	23.7	50	4	66.7	31
200	3.5–6.5	406,914.85	4,005,671.15	55	26	29	17.4	112	5	98.7	71
201	3.5–6.5	405,242	4,004,431.7	48	23	25	17.2	125	5	93.8	70
202	3.5–6.5	405,684.54	4,005,717.45	50	27	23	16.2	94	3	98.1	45
203	3.5–6.5	408,025.12	3,999,907.5	48	26	22	17.7	81	3	99.1	81
204	3.5–6.5	406,400.4	4,001,124	49	26	23	21	85	4	98.8	100
205	6.5–9.5	406,851.5	4,009,570	47	23	24	26.9	41	3	83.5	10
206	6.5–9.5	405,972.9	4,009,284.2	51	24	27	30.4	51	3.5	93.8	12
207	6.5–9.5	407,487.6	4,007,947.1	56	30	26	19.1	107	4	93.5	83
208	6.5–9.5	406,121.1	4,008,204.7	54	28	26	16.4	112	5	95.7	80
209	6.5–9.5	407,443.2	4,008,702.7	49	27	22	13.6	105	4.5	94.8	96
210	6.5–9.5	407,941.4	4,009,347.7	55	26	29	21.7	97	4	97.5	82
211	6.5–9.5	408,217.9	4,008,512.23	42	28	0	18.3	117	5	100	100

Table A1. *Cont.*

BH NO	DEPTH m	X	Y	LL%	PL%	PI%	WC%	c (kN/m²)	Φ (°)	Fine Content	SPT-N Value (kN/m²)
212	6.5–9.5	408,997.4	4,008,917.8	54	26	28	18.5	110	5	98.4	82
213	6.5–9.5	408,743.4	4,009,398.1	42	23	19	17.4	108	6.5	97.3	95
214	6.5–9.5	408,026.1	4,010,406	41	28	13	17.4	110	5.5	91.7	82
215	6.5–9.5	409,617.8	4,011,321.6	46	25	21	18.4	122	5.5	52.1	100
216	6.5–9.5	408,588.1	4,011,780.2	42	25	17	19.1	103	5.5	86.3	100
217	6.5–9.5	409,762.1	4,012,805.4	47	26	21	21.4	99	5	90.9	100
218	6.5–9.5	411,184.2	4,011,787.2	47	26	21	21.9	118	4.5	91.7	80
219	6.5–9.5	411,184.3	4,012,540.7	53	25	28	18.3	103	4.5	90	85
220	6.5–9.5	411,247.7	4,013,522.9	47	28	0	19.7	67	4.5	100	29
221	6.5–9.5	409,969.9	4,010,643.8	56	30	26	18.8	93	4	90.7	85
222	6.5–9.5	409,662.9	4,009,977	39	21	18	18.2	95	4.5	82.4	90
223	6.5–9.5	409,915.1	4,009,232.2	44	22	22	16.2	91	4	91.3	47
224	6.5–9.5	410,862.1	4,009,607.35	55	28	27	17.4	85	4	96.1	80
225	6.5–9.5	411,994.6	4,009,216.9	48	23	25	16.1	87	4.5	98.5	78
226	6.5–9.5	411,166.1	4,010,438.9	53	25	28	21.7	92	4.5	90.1	100
227	6.5–9.5	412,126.9	4,010,308.3	56	26	30	12.9	91	4	96.6	99
228	6.5–9.5	411,550.5	4,011,441.2	47	21	26	15.6	108	8.5	100	100
229	6.5–9.5	413,143.3	4,011,488.9	41	21	20	21.7	95	3.5	76.9	90
230	6.5–9.5	414,064.7	4,011,692.8	59	28	31	16.6	130	5	93.8	47
231	6.5–9.5	415,317.3	4,010,885.4	64	33	31	18.7	92	4	64.2	74
232	6.5–9.5	414,149.3	4,009,671.4	56	33	23	18.1	92	4	73.4	72
233	6.5–9.5	414,362.5	4,008,450.6	66	29	37	18.2	96	4.5	61.6	55
234	6.5–9.5	416,171.1	4,012,249.6	32	15	17	16.1	94	4.5	52.4	100
235	6.5–9.5	415,609.8	4,012,761.8	36	20	16	18.2	93	4.5	42.3	100
236	6.5–9.5	414,549.1	4,013,637	44	21	23	11.3	78	14	100	100
237	6.5–9.5	417,922.5	4,010,462	43	25	18	17.1	62	3.5	94.9	40
238	6.5–9.5	416,961.1	4,011,319.3	46	22	24	17.2	110	4.5	74.3	81
239	6.5–9.5	416,542.1	4,010,233.5	55	31	24	20.7	116	4	88.7	100
240	6.5–9.5	416,909.4	4,008,754	48	24	24	22.7	110	5	86.2	100
241	6.5–9.5	406,775.3	4,007,107.5	52	27	25	17.8	85	4	98.4	52
242	6.5–9.5	405,227.4	4,007,274.3	51	26	25	18.2	85	4.5	94.3	54
243	6.5–9.5	407,403.9	4,006,885.3	45	25	20	20.5	103	4	94.4	45
244	6.5–9.5	407,715.1	4,007,488.5	49	26	23	20.1	97	4.5	96.1	71
245	6.5–9.5	408,445.4	4,007,069.4	48	26	22	20.7	85	4	93.1	82
246	6.5–9.5	408,736.7	4,005,101.9	54	25	29	16.6	83	3.5	94.1	83
247	6.5–9.5	408,248.5	4,006,694.7	50	27	23	19.2	97	5	93	79
248	6.5–9.5	407,872.4	4,006,657.22	51	25	26	19.4	138	4	93.3	60
249	6.5–9.5	409,311.9	4,005,786	49	26	23	18.2	130	4.5	97.5	70
250	6.5–9.5	408,775.59	4,006,631.24	48	23	25	17.6	97	5	94.9	69
251	6.5–9.5	409,973.4	4,006,133	44	24	20	17.7	99	5	92.2	75
252	6.5–9.5	409,048.64	4,007,069.39	48	24	24	18.4	97	4.5	90.4	90
253	6.5–9.5	409,874.19	4,007,406.1	46	23	23	15.8	98	5	93.3	89
254	6.5–9.5	408,985.14	4,007,907.6	56	25	31	18.8	108	4.5	96.7	91
255	6.5–9.5	411,015.2	4,007,902.7	45	23	22	14.5	115	5	93.2	99
256	6.5–9.5	418,521.8	4,007,163.4	48	23	25	13.5	113	4.5	93.8	100

Table A1. *Cont.*

BH NO	DEPTH m	X	Y	LL%	PL%	PI%	WC%	c (kN/m^2)	Φ (°)	Fine Content	SPT-N Value (kN/m^2)
257	6.5–9.5	409,785.25	4,008,333.1	46	24	22	18.8	95	3.5	94	43
258	6.5–9.5	410,998.6	4,008,878.3	49	27	22	18.6	103	4.5	92.9	100
259	6.5–9.5	410,109.1	4,006,834.44	42	28	14	17.2	108	5.5	92.1	59
260	6.5–9.5	410,419.8	4,006,001	55	23	32	17.1	120	4.5	91.2	72
261	6.5–9.5	412,049.02	4,007,215.97	55	24	31	15.4	130	5	82.3	89
262	6.5–9.5	411,263	4,006,447.5	55	26	29	18.4	102	4	92.7	81
263	6.5–9.5	412,664.18	4,007,328.4	48	25	23	16.4	133	5.5	96.1	43
264	6.5–9.5	411,990	4,006,315.2	49	26	23	16.5	120	5	89.7	71
265	6.5–9.5	413,164.9	4,008,779	45	27	18	20.2	123	5.5	86	69
266	6.5–9.5	412,846.74	4,006,559.81	45	24	21	19.6	85	4.5	94.7	75
267	6.5–9.5	412,536.5	4,005,174.2	37	18	19	20.2	88	4	98.0	84
268	6.5–9.5	415,747.4	4,005,960.69	46	22	24	18.4	89	4	91.2	100
269	6.5–9.5	414,814.8	4,007,156.9	43	22	21	16.3	128	5	42.3	87
270	6.5–9.5	415,071.9	4,004,866.9	38	21	17	16.3	122	4	45	100
271	6.5–9.5	417,971.23	4,001,685.01	47	25	22	17.2	97	9.5	42	100
272	6.5–9.5	418,519	4,003,222.9	37	21	16	16.8	90	9.5	36	100
273	6.5–9.5	415,926	4,002,878	44	23	21	17.0	93	10	42	100
274	6.5–9.5	416,167	4,004,572.3	43	23	20	20.1	80	9	60	100
275	6.5–9.5	419,289	4,001,977.6	44	24	20	17.7	109	4	81.4	98
276	6.5–9.5	411,924.27	4,004,297.59	56	25	31	14.4	142	5	96.3	100
277	6.5–9.5	410,482.29	4,003,953.63	0	0	0	15.6	131	7	18.1	100
278	6.5–9.5	412,426.98	4,002,802.69	43	23	20	20.5	136	4.5	93.1	100
279	6.5–9.5	414,365	4,003,408	58	32	26	17.2	97	5	93.7	100
280	6.5–9.5	411,975.73	4,001,202.95	51	24	27	18.1	129	5.5	97.6	72
281	6.5–9.5	414,022.9	4,002,387	81	26	55	15.3	140	5	95.1	79
282	6.5–9.5	411,697.91	4,000,107.57	51	28	23	17.5	131	4.5	94.2	40
283	6.5–9.5	410,200	3,999,706	48	28	22	16.8	103	5	61.4	45
284	6.5–9.5	413,372.73	4,001,528.38	45	21	24	17.2	99	6	93.7	42
285	6.5–9.5	414,253.79	4,000,845.76	49	26	23	18.6	96	4.5	82	100
286	6.5–9.5	415,598.5	4,000,259	41	23	18	21.0	89	4	76.5	100
287	6.5–9.5	413,510.13	3,999,595.57	43	23	20	13.9	87	3.5	26.6	93
288	6.5–9.5	412,160.93	3,999,051.88	36	17	19	13.2	6	32	100	80
289	6.5–9.5	410,732.9	3,999,785	43	21	22	17.0	16	31	45	100
290	6.5–9.5	410,253	3,998,445	42	22	20	9.0	7	30	41	100
291	6.5–9.5	413,309	3,998,392.5	59	26	33	15.7	94	3.5	70.8	100
292	6.5–9.5	410,136.16	4,001,490.72	55	26	29	21.8	97	5	95.9	79
293	6.5–9.5	409,296.11	4,002,026.51	53	25	28	19.2	115	4.5	95.2	42
294	6.5–9.5	408,932.6	4,003,851.9	50	28	22	17.5	94	4	96.8	49
295	6.5–9.5	408,515.58	4,002,310.93	45	28	17	25.9	33	4.5	87.9	38
296	6.5–9.5	409,939	4,003,009.6	43	28	15	26.4	55	4.5	93	33
297	6.5–9.5	407,821.05	4,002,866.56	0	0	0	19.7	52	3.5	25.8	100
298	6.5–9.5	407,298.5	4,004,255.62	41	23	18	15.2	119	4.5	94.7	40
299	6.5–9.5	405,937	4,003,224	46	24	22	19.3	93	4.5	63.7	38
300	6.5–9.5	407,774.75	4,005,380.1	43	21	22	20.5	130	6	94.2	89

Table A1. *Cont.*

BH NO	DEPTH m	X	Y	LL%	PL%	PI%	WC%	c (kN/m^2)	Φ (°)	Fine Content	SPT-N Value (kN/m^2)
301	6.5–9.5	407,844.33	4,005,870.11	43	23	20	19.8	49	3.5	98.9	38
302	6.5–9.5	406,914.85	4,005,671.15	48	23	25	18.7	99	4.5	94.8	100
303	6.5–9.5	405,242	4,004,431.7	50	26	24	15.5	115	5	96	73
304	6.5–9.5	405,684.54	4,005,717.45	48	23	25	18.7	99	5	94.8	100
305	6.5–9.5	408,025.12	3,999,907.5	45	25	20	22.8	105	4	92.6	91
306	6.5–9.5	406,400.4	4,001,124	48	25	23	22.1	97	5	69.7	100

References

1. Clayton, C.R.; Matthews, M.C.; Simons, N.E. *Site Investigation*; Granada: London, UK, 1982.
2. Roy, S.; Bhalla, S.K. Role of geotechnical properties of soil on civil engineering structures. *Resour. Environ.* **2017**, *7*, 103–109.
3. Skempton, A. Effective stress in soils, concrete and rocks. *Sel. Pap. Soil Mech.* **1984**, *1032*, 4–16.
4. Lupiezowiec, M. The application of c-φ reduction method to estimate the bearing capacity of subsoil. *ACEE Archit. Civ. Eng. Environ.* **2013**, *6*, 35–43.
5. Patel, A. *Geotechnical Investigations and Improvement of Ground Conditions*; Woodhead Publishing: Cambridge, UK, 2019.
6. Bozbey, I.; Togrol, E. Correlation of standard penetration test and pressuremeter data: A case study from Istanbul, Turkey. *Bull. Eng. Geol. Environ.* **2010**, *69*, 505–515. [CrossRef]
7. Yusof, N.Q.; Zabidi, H. Reliability of using standard penetration test (SPT) in predicting properties of soil. *J. Phys. Conf. Ser.* **2018**, *1082*, 012094. [CrossRef]
8. Cabalar, A.F.; Akbulut, N. *Evaluation of Actual and Estimated Hydraulic Conductivity of Sands with Different Gradation and Shape*; Springer Plus: Berlin/Heidelberg, Germany, 2016; Volume 5, pp. 1–6.
9. Player, R.S. Geographic information system (GIS) use in geotechnical engineering. In Proceedings of the GeoCongress 2006: Geotechnical Engineering in the Information Technology Age 2006, Atlanta, GA, USA, 26 February–1 March 2006; pp. 1–6.
10. Ahmed, C.; Mohammed, A.; Saboonchi, A. ArcGIS mapping, characterisations and modelling the physical and mechanical properties of the Sulaimani City soils, Kurdistan Region, Iraq. *Geomech. Geoeng.* **2022**, *17*, 384–397. [CrossRef]
11. Wan-Mohamad, W.N.; Abdul-Ghani, A.N. The use of geographic information system (GIS) for geotechnical data processing and presentation. *Procedia Eng.* **2011**, *20*, 397–406. [CrossRef]
12. Antoniou, A.A.; Papadimitriou, A.G.; Tsiambaos, G. A geographical information system managing geotechnical data for Athens (Greece) and its use for automated seismic microzonation. *Nat. Hazards* **2008**, *47*, 369–395. [CrossRef]
13. Cabalar, A.F.; Canbolat, A.; Akbulut, N.; Tercan, S.H.; Isik, H. Soil liquefaction potential in Kahramanmaras, Turkey. *Geomat. Nat. Hazards Risk* **2019**, *10*, 1822–1838. [CrossRef]
14. Mancini, F.; Stecchi, F.; Gabbianelli, G. GIS-based assessment of risk due to salt mining activities at Tuzla (Bosnia and Herzegovina). *Eng. Geol.* **2009**, *109*, 170–182. [CrossRef]
15. Naji, D.M.; Akin, M.K.; Cabalar, A.F. A comparative study on the VS30 and N30 based seismic site classification in Kahramanmaras, Turkey. *Adv. Civ. Eng.* **2020**, *2020*, 1–5. [CrossRef]
16. ESRI Environmental Systems Research Institute. 2005. Available online: https://www.esri.com/ (accessed on 10 February 2022).
17. Arnous, M.O. Geotechnical site investigations for possible urban extensions at Suez City, Egypt using GIS. *Arab. J. Geosci.* **2013**, *6*, 1349–1369. [CrossRef]
18. Chao, Z.; Ma, G.; Zhang, Y.; Zhu, Y.; Hu, H. The application of artificial neural network in geotechnical engineering. In *IOP Con-ference Series: Earth and Environmental Science*; IOP Publishing: Bristol, UK, 2018; Volume 189, p. 022054.
19. Sulewska, M.J. Applying artificial neural networks for analysis of geotechnical problems. *Comput. Assist. Methods Eng. Sci.* **2017**, *18*, 231–241.
20. Nugroho, S.A.; Fernando, H.; Suryanita, R. Estimation of standard penetration test value on cohesive soil using artificial neural network without data normalization. *Int. J. Artif. Intell. ISSN* **2022**, *2252*, 8938. [CrossRef]
21. Johora, F.T.; Hickey, C.J.; Yasarer, H. Predicting Geotechnical Parameters from Seismic Wave Velocity Using Artificial Neural Networks. *Appl. Sci.* **2022**, *12*, 12815. [CrossRef]
22. Mermerdaş, K.; Arbili, M.M. Explicit formulation of drying and autogenous shrinkage of concretes with binary and ternary blends of silica fume and fly ash. *Constr. Build. Mater.* **2015**, *94*, 371–379. [CrossRef]
23. Arbili, M.M.; Ghaffoori, F.K.; Mermerdaş, K. Statistical analysis of the performance of the soft computing based prediction model for shrinkage of concrete including mineral admixtures. *ZANCO J. Pure Appl. Sci* **2016**, *28*, 574–579.
24. Ghoreishi, B.; Khaleghi Esfahani, M.; Alizadeh Lushabi, N.; Amini, O.; Aghamolaie, I.; Hashim, N.A.; Alizadeh, S.M. Assessment of geotechnical properties and determination of shear strength parameters. *Geotech. Geol. Eng.* **2021**, *39*, 461–478. [CrossRef]
25. Hameed, H. Water Harvesting in Erbil Governorate, Kurdistan Region, Iraq: Detection of Suitable Sites Using Geographic in-Formation System and Remote Sensing. Master's Thesis, Lund University, Lund, Sweden, 2013.

26. Jassim, S.Z.; Goff, J.C. *Geology of Iraq: Dolin*; Prague and Moravian Museum: Brno, Czech Republic, 2006; p. 408.
27. Al-Sanjari, M.N.; Al-Tamimi, M.A. Interpretation of water quality parameters for Tigris River within Mosul City by using principal components analysis. *Tikrit J. Pure Sci.* **2009**, *14*, 68–74.
28. Das, B.M.; Sivakugan, N. *Principles of Foundation Engineering*; Cengage Learning: Belmont, CA, USA, 2018.
29. Fernando, H.; Nugroho, S.A.; Suryanita, R.; Kikumoto, M. Prediction of SPT value based on CPT data and soil properties using ANN with and without normalization. *Int. J. Artif. Intell. Res.* **2021**, *5*, 123–131. [CrossRef]
30. Chakraborty, A.; Goswami, D. Prediction of slope stability using multiple linear regression (MLR) and artificial neural network (ANN). *Arab. J. Geosci.* **2017**, *10*, 385. [CrossRef]
31. Duong, T.V.; Tang, A.M.; Cui, Y.J.; Trinh, V.N.; Dupla, J.C.; Calon, N.; Canou, J.; Robinet, A. Effects of fines and water contents on the mechanical behavior of interlayer soil in ancient railway sub-structure. *Soils Found.* **2013**, *53*, 868–878. [CrossRef]
32. Ebrahimi, A. Behavior or fouled ballast. In *Railway Track and Structures*; Simmons-Boardman Publishing Corporation: New York, NY, USA, 2011; Volume 107.
33. Kim, D.; Sagong, M.; Lee, Y. Effects of fine aggregate content on the mechanical properties of the compacted decomposed gra-nitic soils. *Constr. Build. Mater.* **2005**, *19*, 189–196. [CrossRef]
34. Inam, A.; Ishikawa, T.; Miura, S. Effect of principal stress axis rotation on cyclic plastic deformation characteristics of unsaturated base course material. *Soils Found.* **2012**, *52*, 465–480. [CrossRef]
35. Wang, Q.; Tang, A.M.; Cui, Y.J.; Delage, P.; Barnichon, J.D.; Ye, W.M. The effects of technological voids on the hydro-mechanical behaviour of compacted bentonite–sand mixture. *Soils Found.* **2013**, *53*, 232–245. [CrossRef]
36. Koester, J.P. The influence of test procedure on correlation of Atterberg limits with liquefaction in fine-grained soils. *ASTM Geotech. Test. J.* **1992**, *15*, 352–361.
37. Mitchell, J.K.; Soga, K. *Fundamentals of Soil Behavior*; John Wiley & Sons: New York, NY, USA, 2005.
38. Bogati, K. Ground Improvement by Jet Grouting Techniques. Bachelor's Thesis, HAMK University of Applied Sciences Construction Engineering, Amberg, Germany, 2019.
39. Carter, M.; Bentley, S.P. *Correlations of Soil Properties*; Pentech Press Publishers: London, UK, 1991.
40. Likos, W.J.; Song, X.; Xiao, M.; Cerato, A.; Lu, N. Fundamental challenges in unsaturated soil mechanics. *Geotech. Fundam. Addressing New World Chall.* **2019**, 209–236. [CrossRef]
41. Spagnoli, G.; Feinendegen, M. Relationship between measured plastic limit and plastic limit estimated from undrained shear strength, water content ratio and liquidity index. *Clay Miner.* **2017**, *52*, 509–519. [CrossRef]
42. Casagrande, A. Notes on the design of the liquid limit device. *Geotechnique* **1958**, *8*, 84–91. [CrossRef]
43. Gardner, C.M.; Robinson, D.; Blyth, K.; Cooper, J.D. Soil water content. In *Soil and Environmental Analysis*; CRC Press: Boca Raton, FL, USA, 2000; pp. 13–76.
44. Das, B.M.; Das, B.M. *Advanced Soil Mechanics*; Taylor & Francis: New York, NY, USA, 2008.
45. Wu, W.; Mhaimeed, A.S.; Al-Shafie, W.M.; Ziadat, F.; Dhehibi, B.; Nangia, V.; De Pauw, E. Mapping soil salinity changes using re-mote sensing in Central Iraq. *Geoderma Reg.* **2014**, *2*, 21–31. [CrossRef]
46. Khudhur, S.M.; Khudhur, N.S. Soil pollution assessment from industrial area of Erbil City. *J. Zankoi Sulaimani* **2015**, *17*, 225–238. [CrossRef]
47. Aldefae, A.H.; Mohammed, J.; Saleem, H.D. Digital maps of mechanical geotechnical parameters using GIS. *Cogent Eng.* **2020**, *7*, 1779563. [CrossRef]
48. Nam, S.; Gutierrez, M.; Diplas, P.; Petrie, J. Determination of the shear strength of unsaturated soils using the multistage direct shear test. *Eng. Geol.* **2011**, *122*, 272–280. [CrossRef]
49. Bapeer, G.B. The study of infiltration rate and atterberg limits of soils in Koi Sanjaq City, Erbil Governorate, Kurdistan Region, North Iraq. *Iraqi Bull. Geol. Min.* **2011**, *7*, 41–55.
50. Nareeman, B.J.; Fattah, M.Y. Effect of Soil Reinforcement on Shear Strength and Settlement of Cohesive Frictional Soil. *Geomate J.* **2012**, *3*, 308–313. [CrossRef]
51. Gunduz, Z.; Arman, H. Possible relationships between compression and recompression indices of a low-plasticity clayey soil. *Arab. J. Sci. Eng.* **2007**, *32*, 179.
52. Qurtas, S.S. Using groundwater levels and Specific Yield to Estimate the Recharge, South of Erbil, Kurdistan Region, Iraq. *Acad. J. Nawroz Univ.* **2018**, *7*, 191–196. [CrossRef]
53. Bowles, J. *Foundation Analysis and Design*; McGraw-Hill Book Company: New York, NY, USA, 1982; 816p.
54. Kumar, R.; Bhargava, K.; Choudhury, D. Estimation of engineering properties of soils from field SPT using random number generation. *INAE Lett.* **2016**, *1*, 77–84. [CrossRef]
55. Hasan, A.M.; Mawlood, Y.; Ahmed, A.A.; Ibrahim, H. Correlation of Shear Wave Velocity with SPT-N for a Tower-Building Site at Erbil City. *J. Duhok Univ.* **2020**, *23*, 235–245. [CrossRef]
56. Mahmoud, M.A. Reliability of using standard penetration test (SPT) in predicting properties of silty clay with sand soil. *Int. J. Civ. Struct. Eng.* **2013**, *3*, 545–556.
57. Myslivec, A.; Kysela, Z. *The Bearing Capacity of Building Foundations*; Elsevier: Amsterdam, The Netherlands, 2014.
58. Das, B.M.; Sivakugan, N. *Fundamentals of Geotechnical Engineering*; Cengage Learning: Belmont, CA, USA, 2016.
59. Dauji, S.; Rafi, A. Spatial interpolation of SPT with artificial neural network. *Eng. J.* **2021**, *25*, 109–120. [CrossRef]

60. Jasim, M.M.; Al-Khaddar, R.M.; Al-Rumaithi, A. Prediction of bearing capacity, angle of internal friction, cohesion, and plasticity index using ANN (case study of Baghdad, Iraq). *Int. J. Civ. Eng. Technol.* **2019**, *10*, 2670–2679.
61. İpek, E.; McKee, S.A.; Caruana, R.; de Supinski, B.R.; Schulz, M. Efficiently exploring architectural design spaces via predictive modeling. *ACM SIGOPS Oper. Syst. Rev.* **2006**, *40*, 195–206. [CrossRef]

sustainability

Article

Experimental Study on Shear Strength Parameters of Round Gravel Soils in Plateau Alluvial-Lacustrine Deposits and Its Application

Zhijun Kong [1,2], Yanhui Guo [1,*], Shilin Mao [3] and Wei Zhang [4]

1 Faculty of Public Safety and Emergency Management, Kunming University of Science and Technology, Kunming 650093, China
2 College of Civil and Architectural Engineering, Yunnan Agricultural University, Kunming 650201, China
3 Yunnan Construction Investment No. 6 Construction Co., Ltd., Yuxi 653199, China
4 Zhongxing Shuchuang (Yunnan) Technology Co., Ltd., Kunming 650011, China
* Correspondence: guoyanhui0818@kust.edu.cn; Tel.: +86-158-0881-4866

Abstract: The shear strength parameters of conglomerate soils are crucial to the stability analysis of foundation support when excavating and supporting ultra-deep foundation pits in the highland alluvial lacustrine layer. The difference in water content of conglomerate soils in different regions will directly affect the values of shear strength parameters. At the same time, more research on the shear strength of conglomerate soils under different water contents is required. In this study, a series of large-scale direct shear tests were carried out on the round gravel soil in the plateau alluvial-lacustrine deposit, and the round gravel soil's shear strength curves under natural and saturated conditions and water content were obtained. The influence of different water content on the shear strength characteristics of the round gravel soil was discussed, and the shear strength parameters of the round gravel soil with different water content were used in the numerical simulation of ultra-deep foundation pit excavation and support. The stress and deformation laws of the foundation pit support were analyzed. The results show that the peak strength of the round gravel soil in the natural water content state appears between 30% and 45% of the shear displacement, while the peak strength in the saturated water content state appears around 45–55% of the shear displacement. The shear strength tends to be stable or slightly weakened with the increase of the shear displacement. The angle of internal friction and cohesion of round gravel soil in the natural water content state is greater than those in the saturated water content state. The simulation of the foundation excavation support shows that the shear strength parameter of the round gravel soil influences the force deformation of the support structure. The higher the water content of the round gravel soil, the more the shear strength parameter affects the soil displacement. The research results can provide some reference for optimizing project design parameters.

Keywords: plateau alluvial-lacustrine deposits; round gravelly soil; large scale direct shear test; shear strength parameters; numerical simulation

Citation: Kong, Z.; Guo, Y.; Mao, S.; Zhang, W. Experimental Study on Shear Strength Parameters of Round Gravel Soils in Plateau Alluvial-Lacustrine Deposits and Its Application. *Sustainability* **2023**, *15*, 3954. https://doi.org/10.3390/su15053954

Academic Editors: Jian Zhou, Mahdi Hasanipanah and Danial Jahed Armaghani

Received: 29 December 2022
Revised: 18 February 2023
Accepted: 19 February 2023
Published: 22 February 2023

1. Introduction

Plateau alluvial strata are formed by the joint action of river alluvium and lake marsh sedimentation. Generally, they consist of rounded gravels with high foundation strength and are often used for urban construction sites [1–5]. Influenced by geological activities on the Yunnan-Guizhou Plateau, rounded gravel strata are widely distributed in Kunming [6] and buried at a shallow depth. Influenced by the formation conditions, time, and geographical area of round gravel soil, the maximum particle size of round gravel soil reaches 60–80 mm, and the poor grading, strong permeability, and interparticle clay composition lead to the strength of round gravel soil are larger than general soil [7–12]. To a certain extent, these characteristics make round gravel soils have strong engineering applications, leading scholars at home and abroad to focus more on engineering applications or theoretical studies. In contrast, research on shear strength parameters of round gravel soils has yet to receive more attention [13–15]. With the urban construction of Kunming city, more and more projects are built on round gravel strata. The shear strength and stress-strain characteristics of round gravel soils usually need to be considered in the design calculation of foundation pits. At the same time, there are few relevant studies on round gravel soils, resulting in most foundation pits in the design calculation can only be taken according to empirical values, which seriously affects the stability of foundation pit construction. Especially nowadays, foundation pit projects are developing in the direction of super large and deep. During design calculation, a slight deviation of soil shear strength parameters greatly impacts foundation pits' stability. According to statistics, the annual failure rate of foundation pits in China reaches more than 10–15%, which causes huge economic losses to China and threatens the safety of residents and urban construction, mainly due to the lack of experience in design and construction. Therefore, it is necessary to conduct a systematic study on the shear strength of round gravel soil parameters for systematic research to avoid the experience of errors leading to safety accidents in foundation pit construction.

With the continuous in-depth research on the physical and mechanical properties of coarse-grained soils by scholars at home and abroad, round gravel soils have also been widely studied. In terms of experimental shear properties of round gravel soil, Tang Kaishun et al. [16] conducted a large triaxial compression test on round gravel soil in the Nanning area under different compaction conditions and analyzed and compared the effect of compaction on shear strength parameters of round gravel soil under various levels of surrounding pressure. The results showed that the shear strength of round gravel soil was positively correlated with compaction and significantly improved compared with empirical engineering values. Ma Shaokun et al. [17,18] conducted a large drained dynamic triaxial test on round gravel soil in a saturated state. They analyzed the variation of cumulative strain, stress-strain hysteresis loop, and pore pressure of the round gravel soil under different relative compactness, dynamic stress amplitude, and the number of vibrations. Stark et al. [19] investigated the effect of particle shape on the internal friction angle of beach gravel soils by direct shear experiments on beach gravels. The results showed that elliptical gravels have a greater effect on the internal friction angle of beach gravel soils, and increasing the content of elliptical particles in gravel soils can significantly increase their internal friction angle magnitude. Enomoto et al. [20] investigated the strength, deformation properties, and small strain properties of undisturbed well-graded gravel soils by a series of medium and large triaxial and unconfined compression tests. The results showed that the shear modulus might be large when the small strain properties of gravel soils are determined by the dynamic method in the laboratory. The dynamic method converges to the static method test results when the wavelength is significantly larger than the mean diameter. Secondly, when gravel soil's dry density and homogeneity coefficient exceeded certain values, the dynamic and static shear modulus values of in-situ and remodeled soils increased with the increase of dry density. For the study of the intrinsic model of round gravel soils. Chen Chen et al. [21] proposed a modified Duncan-Chang intrinsic model for round gravel soil based on the unified disturbance degree function based on the disturbance state theory with the relative density Dr as the disturbance parameter. Saberi et al. [22] established a new elastic-plastic intrinsic model for gravel soil based on double surface plasticity and critical state geomechanics. To analyze the piling characteristics of round gravel, Liu Gang et al. [23] proposed a construction method of particle ellipsoid model for round gravel, established an ellipsoid model database, and derived the piling characteristics of round gravel through numerical piling tests and comparative analysis with the results of cylinder piling tests. As for the research on the engineering application of round gravel soil, Ou Xiaoduo et al. [24] used ABAQUS to simulate the deep foundation excavation of round gravel-mudstone strata and analyze the effect of deep foundation excavation on the double-row pile support structure. Ni Xiaorong et al. [25] studied the applicability of long-spiral bored piles in round gravel strata. They proposed a new construction process of secondary pressurized piles for construction problems.

In summary, scholars at home and abroad have studied the strength, deformation, and other properties of round gravel soil influencing factors and have achieved certain research results. However, most research focuses on the engineering application of round gravel formation. Little attention is paid to the influence of water on the mechanical properties of round gravel soil, and the research results need to be more comprehensive. Due to the special geographical location and formation conditions of alluvial plateau strata, the shear strength characteristics of round gravel soil in the formation are different from those in other regions. They are greatly affected by the change in moisture content. At the same time, scholars at home and abroad have paid less attention to the physical and mechanical characteristics of round gravel soil affected by moisture content. Therefore, based on the foundation pit project of the 14th water purification plant in Kunming, this study carried out a large-scale direct shear experiment on the round gravel soil of the alluvial plateau layer, obtained the shear strength parameters of round gravel, and analyzed the change law of the shear resistance characteristics of round gravel soil under different moisture cuts.

2. Experimental Study of Shear Property Parameters of Round Gravel Soil

2.1. Testing Instruments

The experiment uses a DHJ-30 type coarse-grained soil stacked ring shear experimental machine, see Figure 1. the experimental machine adopts the plate frame structure with a host size of 2000 × 800 × 1400 mm. The equipment shear box is square outside and round inside, the size of Φ300 × 300 mm, the maximum axial pressure is 300 kN (normal stress 4.3 MPa), the maximum horizontal thrust is 300 kN, force sensor resolution is 0.1 kN. It can realize stress and strain type shear, in which the shear moving speed is 0.001~5 mm/min.

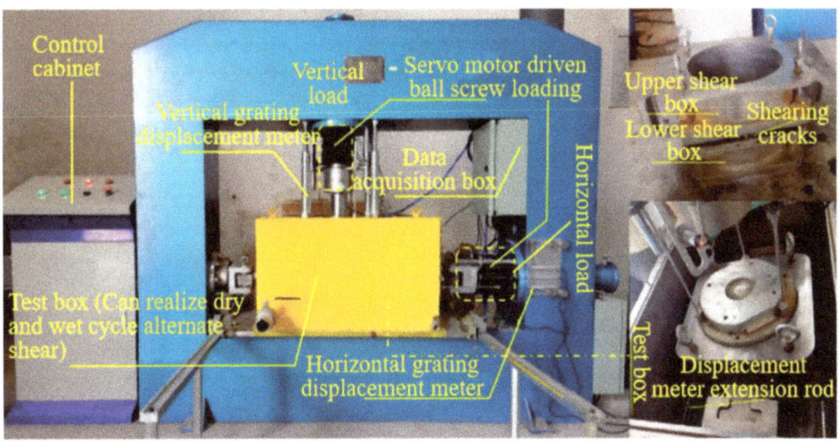

Figure 1. DHJ-30 indoor large direct shear experiment machine.

2.2. Experimental Soil Samples

2.2.1. Round Gravel Soil Particle Size Composition

The experimental soil sample was selected from the round gravel soil of the foundation pit project of fourteen water quality purification plants in Panlong District, Kunming, at a depth of 14 m. The lithology of the stratum is alluvial round gravel, dark gray, blue, saturated, slightly dense mainly, and locally medium dense, and the photo of the sampling point is shown in Figure 2.

(**a**) (**b**)

Figure 2. Photos of round gravel soil sampling points: (**a**) project site photos; (**b**) photo of sampling point.

Due to the disturbance of soil samples during sampling and transportation, the experimental process needs to reshape the soil samples, the specimen reshaping needs to sieve the retrieved soil samples, and the sieving process is executed according to the Geotechnical Experimental Methods Standard (GB/T50123-2019) [26]. According to the experimental requirements and equipment limitations, the experiments cannot be tested for super-size particles with particle sizes greater than 60 mm, so the soil sample gradation needs to be processed. Referring to the provisions of the geotechnical experimental method standard, when the content of super-size particles is greater than or equal to 5% of the total content, the equal mass substitution method is used. When the content of super-size particles is less than 5% of the total content, the rejection method is used. From the sieving experiment, it can be seen that the soil material of Kunming's fourteen water purification plants has less than 5% of soil particles larger than 60 mm in size. The rejection method is used to scale down the soil gradation on site, and the results of the sieving experiment are shown in Table 1.

Table 1. Experimental soil particle size composition rejection reduction treatment results.

Gradation Type	Percentage of Mass Smaller Than a Certain Particle Size/%						
	>60 mm	60~40 mm	40~20 mm	20~10 mm	10~5 mm	5~2 mm	<2 mm
Prototype gradation	0.48	7.39	25.61	21.62	13.73	11.01	20.16
Scaled gradation		7.43	25.73	21.72	13.80	11.06	20.26

Based on the results of particle gradation, the mass percentage of round gravel soil under different particle size grades is calculated, and the particle size grading curve is drawn. In this study, the cumulative curve of the particle size distribution of round gravelly soil after screening scale treatment is shown in Figure 3. It can be seen from Figure 3 that there are particles of all sizes of round gravelly soil in the 14th water purification plant in Kunming, indicating that there are fine particles in the coarse particles of Kunming No. 14 Water Purification Plant for filling.

2.2.2. Maximum Dry Density Experiment of Round Gravel Soil

Since the density of the experimental soil sample must be strictly controlled during the remolding process, it is necessary to measure the maximum dry density and the optimum moisture content of the round gravel soil, and the heavy compaction instrument is used for the experiment. During the experiment, take out particles larger than 40 mm and get their percentage P, and then compact the part of round gravel soil smaller than 40 mm. After the experiment, the maximum dry density and the best water content obtained from the experiment need to be corrected (applicable to the content of particles larger than 40 mm is less than 30%). Calculate the dry density of each point according to the experimental results, take the dry density as the ordinate and the water content as the abscissa, and draw the relationship curve between the dry density and the water content. The ordinate and abscissa of the peak point on the curve are the maximum dry density and the best water content, respectively. The results are shown in Table 2.

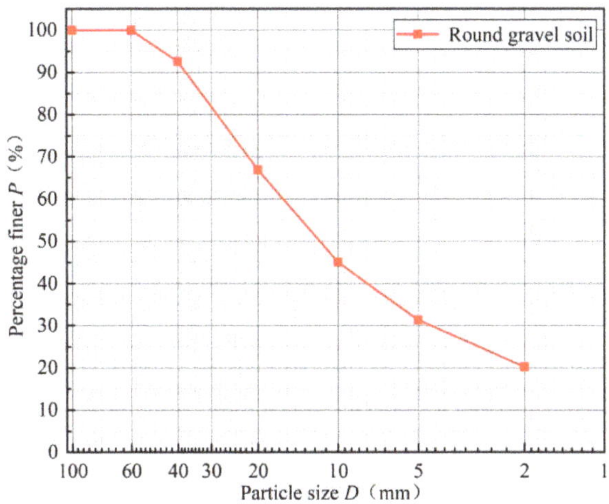

Figure 3. Particle size distribution curve of round gravel soil after scale reduction treatment.

Table 2. Results of round gravel soil compaction experiments.

Soil Sample Number	Design Water Content ω (%)	Weight of Cylinder and Soil (g)	Weight of Solid Barrel (g)	Combat Cylinder Volume (cm^3)	Wet Density ρ (g/cm^3)	After Experiment Water Content ω (%)
1	5	7653	3080	2159	2.12	8.5
2	7	7973	3080	2159	2.27	10.9
3	9	7913	3080	2159	2.24	12.2
4	11	7867	3080	2159	2.22	13.3
5	13	7833	3080	2159	2.20	15.1

The dry density and the water content curves are shown in Figure 4, which shows that the dry density increases with moisture content and then decreases. In contrast, the maximum dry unit weight and optimal moisture content correspond to the fitted curve's top, consistent with the dry density of fine-grained soil [27,28]. Wang et al. [29,30] found that the dry density of gravelly calcareous sand increased with increasing water content when the water content was greater than 8%. Although both are coarse-grained soils, the dry density shows a different variation pattern, mainly due to the fact that when the water content of gravelly calcareous sand exceeds 8%, the capillary suction between soil particles is weakened. Particle movement intensifies while particles' relative fragmentation rate increases, but it is always smaller than the relative fragmentation rate of dry calcareous gravelly sand, resulting in the increase of dry density with the increase of water content. In comparison, the round gravelly soil has already reached the liquefaction limit when the water content reaches 11%. If the water content increases, the specimen becomes liquefaction, and the dry density decreases.

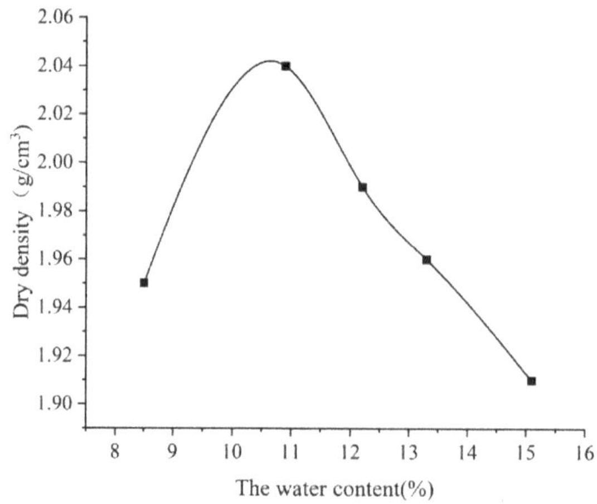

Figure 4. Dry density versus moisture content curve.

The results show that the optimum moisture content of round gravel soil is 11%, the maximum dry unit weight is 2.04 g/cm^3, and the corresponding wet density is 2.26 g/cm^3. Since the maximum dry density cannot be reached in the field for round gravel soil, in order to make the experimental specimens closer to the actual engineering site, the wet density corresponding to 0.87 of the maximum dry density is taken as 1.97 g/cm^3 for the indoor direct round gravel soil with reference to a large amount of relevant literature. The soil sample was prepared for the indoor direct shear experiment.

2.3. *Experimental Methods and Procedures*

2.3.1. Experimental Methods

The soil samples collected in the field were air-dried and sieved in the experiment. The soil material was weighed according to the particle gradation of round gravel soil after scale reduction and mixed evenly after spraying an appropriate amount of water according to the natural moisture content of 13.1%, divided into three parts equally, and loaded into the bogging bucket, enclosing for 24 h. Control the wet density of round gravel soil to 1.97 g/cm^3, the sample was loaded and compacted in layers, the inter-layer hair scraping treatment was required, and strictly control the filling density of the sample. For the direct shear experiment of round gravel soil under saturated conditions, after filling samples, water was added to the test chamber to cover the shear box, and the shear test shall be conducted after 24 h saturation. The experimental loading was strain-controlled, and the straight shear test was performed by fast shear, as shown in Figure 5.

2.3.2. Experimental Procedure

(1) Loading sample: according to the determined density, gradation, and moisture content of the filler, weigh the soil material in three parts, mix and blend, and load into the shear box in layers of compaction, each time loading to 1/3 of the total height of the shear box, until the control height, after completion, level the surface.
(2) Vertical loading: according to this experiment to determine the load level (low pressure: 100, 200, 300, 400 kPa) using servo motor control loading, stable pressure after observing the vertical displacement and event change curve until the stability standard control in the stability standard control at 0.002 mm/min.
(3) Horizontal shear: after the soil sample vertical loading stability, according to the same strain rate horizontal shear, the shear rate of 1 mm/min, while observing the experimental machine data acquisition system until the specimen damage. Experiment

until the soil sample horizontal shear displacement reaches 15% of the diameter of the specimen when the end of shear.

(4) The specimen damage determination: when the horizontal stress table readings fall, no longer rise or rise very little, the deformation change is large, that has been shear damage. If none of the above, when the shear deformation reaches 15% of the diameter of the shear box, stop the shear experiment. After the experiment, clear the soil on the shear box, analyze the shear surface characteristics, and take pictures.

Figure 5. Experimental steps for direct shear of round gravel soil: (**a**) sieving soil; (**b**) soil mixing and enclosing; (**c**) load soil sample into shear box; (**d**) lift the shear box into the test chamber; (**e**) install the sensor and start the experiment; (**f**) observe the shape of the cutting surface.

2.4. Test Results and Analysis

2.4.1. Shear Stress-Shear Displacement Curve Change Characteristics Analysis

Figure 6 shows the round gravel soil's fast shear stress-shear displacement curves in the natural water content state. As can be seen from the figure, the shear stress of the round gravel soil of the 14th water purification plant of Kunming City increases with the increase of shear displacement and then tends to be stable or decreases to a smaller extent, with the peak shear value appearing between 30% and 45% of the shear displacement. It can be seen that the curve jumps at some points, mainly since the initial density of the sample is small, and the shrinkage is continuous during shearing, resulting in the increase of compactness and strength and the increase of shear stress. Due to large interlocking particles of gravel in the sample, the overturning friction between the large particles of gravel requires a large horizontal thrust as the shearing proceeds, resulting in a rapid rise of shear stress. After the overturning, since the overturning friction resistance between coarse and fine grains is small, the shear stress decreases slightly while the shear plane is still in a tight state, and the shear stress continues to increase until reaching the peak. Low load specimen shear stress with the increase in shear displacement always maintains an increasing trend. With the increase in vertical load, this trend gradually weakened, combined with the end of the experiment shear surface. The difference in the distribution of large particles near the shear plane may cause this. The shear stress is high if there are many large particles in the shear plane.

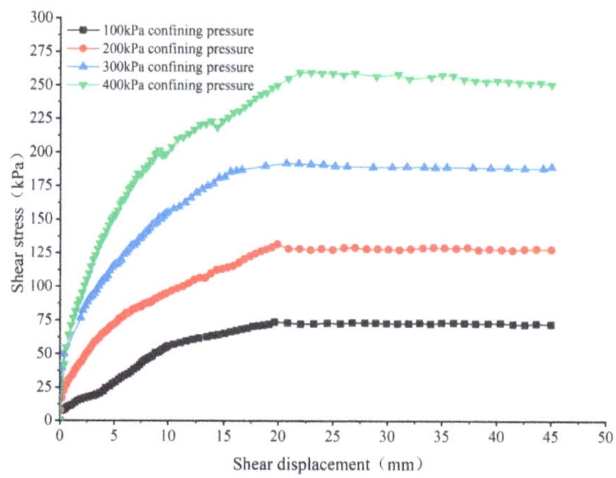

Figure 6. Shear displacement-shear stress curve for natural moisture content of round gravel soil.

In order to investigate the difference between the natural and saturated moisture content states of the round gravel soil, direct shear experiments were designed to be conducted under the saturated moisture content conditions of the round gravel soil. The shear displacement-shear stress relationship curve after shearing is shown in Figure 7. As seen in Figure 7, the shear stress of saturated gravel soil increases with the shear displacement and then stabilizes. The shear stress of saturated gravel soil is smaller than that of the natural state compared with the shear stress-shear displacement curve of the natural state. The shear surface morphology of round gravel soil specimens in a saturated water content state has no large particle fragmentation phenomenon. The reason is that the saturated state specimens with high water content, due to the lubricating effect of water molecules and the soil, are looser between the soil, resulting in reduced friction between soil particles; the soil particles directly overturn each other during shear.

2.4.2. Characterization of Shear Strength Parameters

According to the geotechnical test specification, the peak or stable value on the relationship curve between shear stress and horizontal displacement is taken as the shear strength. When there is no obvious peak, the shear stress at the horizontal displacement reaches 1/15~1/10 specimen diameter and is taken as the shear strength. Therefore, according to Figures 6 and 7, the peak shear stress is taken as the shear strength during this experiment. The relationship between the shear strength and the positive stress is linearly fitted by combining the Moore-Coulomb strength criterion. The results are shown in Figure 8.

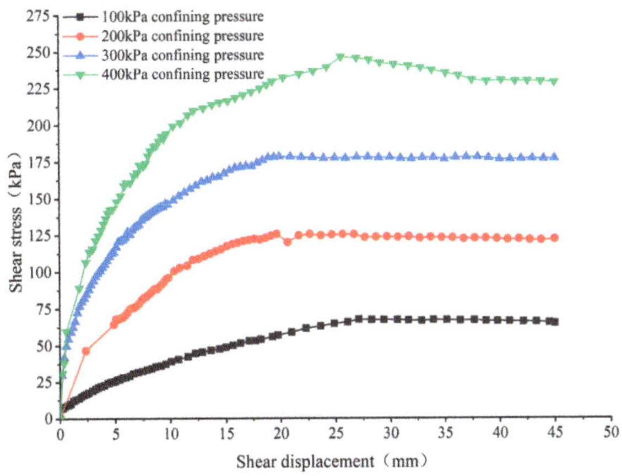

Figure 7. Shear displacement-shear stress curve of round gravel soil with saturated moisture content.

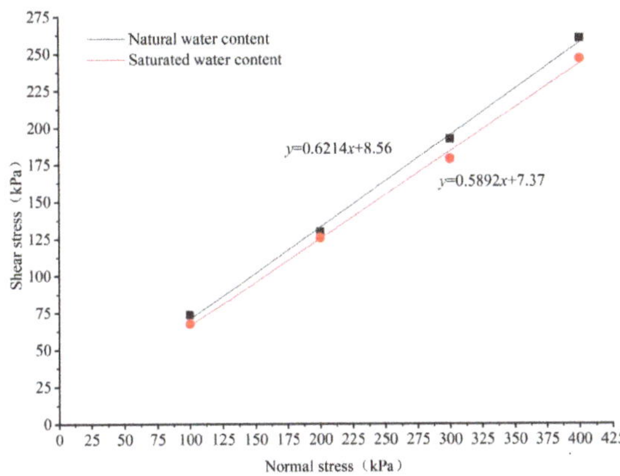

Figure 8. Fitted shear strength curves for natural and saturated moisture content of round gravel soils.

From Figure 8, it can be seen that the cohesion and internal friction angle of the saturated moisture content of the round gravel soil are smaller than those of the natural moisture content, which is since the water content of the saturated round gravel soil is larger than that of the natural state. The lubricating effect of the free water on the surface of the soil particles is enhanced, resulting in the weakening of the occlusal force between the particles and the weakening of the frictional effect, and the smaller horizontal thrust can make the soil produce larger shear displacement. The strength parameters of the round gravel soil in the natural and saturated state are shown in Table 3 from Figure 8.

Table 3. Shear strength parameters of round gravel soil.

	The Angle of Internal Friction φ (°)	Cohesive Forces c (kPa)
Natural moisture content of round gravel soil	31.9	8.56
Saturated moisture content of round gravel soil	30.5	7.37

3. Engineering Application Study on Shear Strength Parameters of Round Gravel Soil

3.1. Project Overview

The construction project of Kunming 14th, Water Purification Plant is a fully buried underground sewage treatment plant located in Panlong District, Kunming City, Yunnan Province, at the northern end of the Kunming Dianchi faulted basin. The shape of the basin is irregular, but generally it is longer from north to south and narrower from east to west, which is consistent with the structural trend. The ground elevation is between 1901.89 and 1907.57 m, and the maximum elevation difference is about 5.68 m. As shown in Figure 9, the stratigraphic structure within the depth of the pit survey mainly consists of the Quaternary artificial accumulation (Q4ml) layer: artificial fill; the Quaternary alluvium (Q4al + pl) layer: clay, round gravel, and powder; and the Quaternary marsh phase sediment (Q4h) layer: peaty soil and organic soil. The gravels are composed of chert, sandstone, and a small amount of basalt and quartz, rounded to subrounded, with good rounding, and filled with a small amount of silt and clay between grains, with poor cementation, and locally produced with thin laminated pebbles. Round gravel ③ has a grain size of 1.0–4.0 cm, a small amount of 5.0–6.0 cm, and a gravel content of 50–70%. Round gravel ④ has a grain size of 1.0–4.0 cm, with a small amount of 5.0–6.0 cm and a gravel content of 50–60%. Round gravel ⑤ grain size 0.5 cm~4.0 cm, a small amount of 5.0~6.0 cm, gravel content about 50~60%. Round gravel ⑥ Grain size 0.5 cm–3.0 cm, a small amount of 4.0–5.0 cm, gravel content about 50–65%. Round gravel ⑦ grain size 0.5 cm~2.5 cm, a small amount of 4.0~6.0 cm, gravel content of about 50~60%. Round gravel ⑧ grain size 0.5~2.0 cm, a small amount of 3.0~4.0 cm, gravel content about 55~65%. Regionally, it belongs to the Jinsha River system and Dianchi basin, and the surface water body is more developed. The groundwater type in the site is mainly upper stagnant water and diving, and the water level burial depth is between 0.8 and 1.5 m.

Figure 9. Site stratigraphic distribution.

The pit covers an area of 65,800 m^2, and the depth of pit support is 14~33 m. The pit is divided into four sections, one of which includes the intake pump room and aeration and sand sink. The modeling object is the intake pump room, with a design depth of 33 m, plan size of 32 m long and 25 m wide, perimeter length of 114 m, and area of 800 m^2. 1200 mm diaphragm wall + reinforced concrete internal bearing and anchor cable support form is used for the enclosure structure. The wall height is 43.75 m and 67.5 m, and the width is 5.10~6.00 m, the length of the single reinforcement cage is 44.25 m and 68 m, and the depth of continuous wall embedded in the subgrade is 22.8~42.1 m according to the geological condition. As Figures 10 and 11 show the section of the pit of the inlet pump room and the internal bearing structure, respectively.

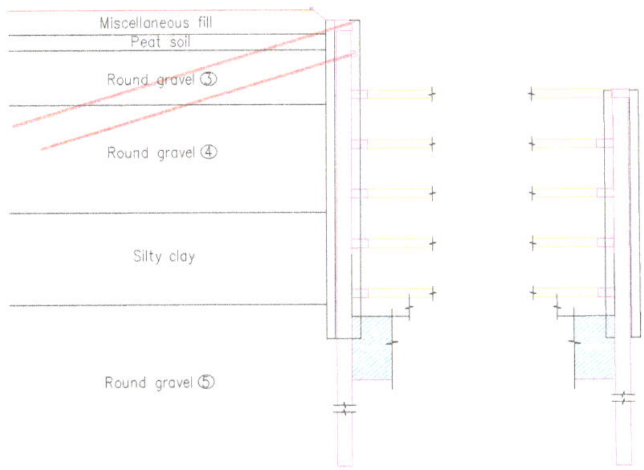

Figure 10. The section of the pit of the inlet pump room.

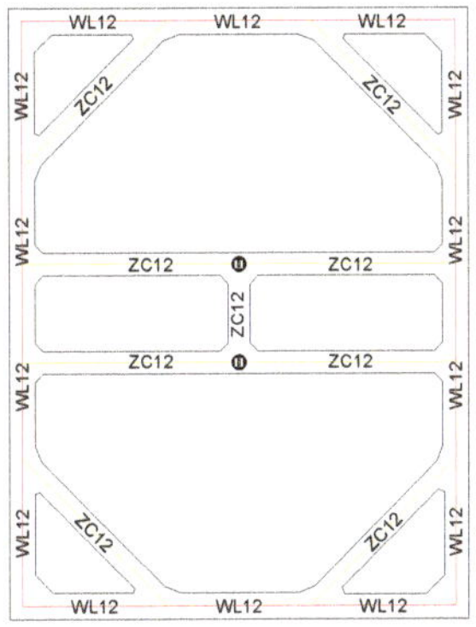

Figure 11. Structural drawing of internal bearing.

3.2. Pit Modeling and Parameter Selection

3.2.1. Computational Models

The width of the standard section of the foundation pit is 24.5 m, the length is 32.2 m, and the excavation depth is 29.4 m. the maximum length of the diaphragm wall is 70 m. The minimum length is 44 m. In order to reduce the influence of the simulated boundary conditions on the simulation process, the boundary is taken six times the excavation depth of the foundation pit. The model size is 180 m × 120 m × 120 m. as the construction depth of the diaphragm wall 2 is −8.7 m, the overall Foundation pit excavation to −8.7 m before starting into the pump room pit construction. Therefore, in this simulation pit excavation from −8.7 m to start, diaphragm wall 1 to the boundary to retaining wall simulation, calculation model as shown in Figure 12, simplification will not have a large impact on the accuracy of the calculation and the essence of the calculation. The model grid of the 3D finite element model of the foundation pit is mixed, the total number of cells of the overall foundation pit model is 77,202, and the total number of nodes is 47,899. The calculation uses displacement boundary conditions; in the left and right boundaries of the model, the displacement in the X direction is fixed, and in the front and back boundaries of the model, the displacement in the Y direction is fixed. Moreover, at the bottom of the model, the displacement in three directions is fixed. In the modeling process, the soil and diaphragm wall are simulated by solid units, the internal bearing and lattice columns are simulated by beam units, and implanted trusses simulate the prestressed anchor cables. Figure 13 shows the schematic diagram of the calculation model of the foundation support structure of the intake pump room. As shown in Figure 13, the diaphragm wall, internal bearing, and anchor cable together form the foundation support system. The three arrangements in the calculation model are shown in Figure 10.

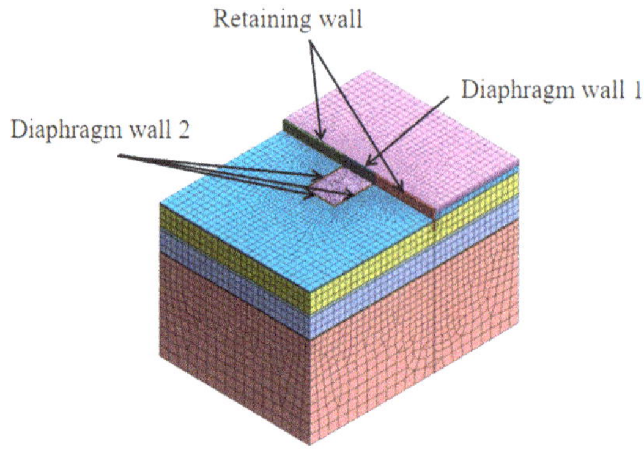

Figure 12. Calculation model for foundation pit excavation and support of intake pump house.

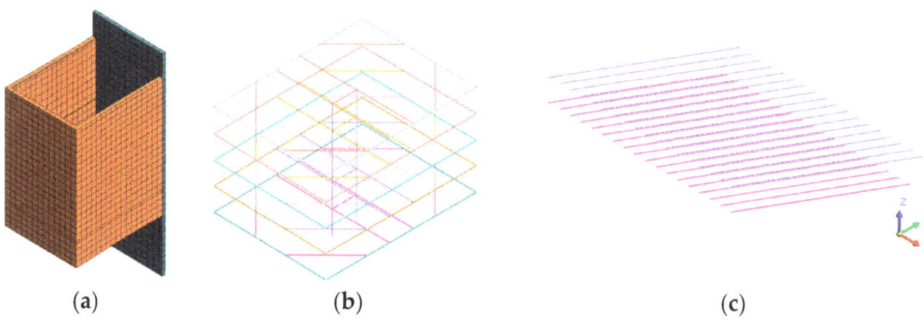

Figure 13. The schematic diagram of the calculation model of the foundation support structure of the intake pump room: (**a**) diaphragm wall; (**b**) interior bearing; (**c**) anchor cable.

3.2.2. Calculation Parameters Selection

Based on the large direct shear experiment of round gravel soil, the cohesion and internal friction angles in the natural and saturated states were determined to be 31.9°, 8.56 kPa, and 30.5°, 7.37 kPa, respectively, and the other soil parameters referred to in the modeling calculations were determined based on the Detailed Geotechnical Investigation Report of the Fourteenth Water Quality Purification Plant of Kunming City and related literature [31–33]. Since the support structures such as diaphragm walls, anchor cables, and internal bearings are in an elastic stress state, the soil in the model is an ideal elastoplastic medium. The soil model widely used in actual engineering is the Mohr-Coulomb elastoplastic model. Therefore, the Mohr-Coulomb principal model is used in the calculation. The mechanical calculation parameters of the soil are shown in Table 4, and the structural and mechanical parameters are shown in Table 5.

Table 4. Mechanical parameters of reference soil layers calculated by numerical simulation.

Number	Name of Soil Type	Volumetric Weight (kN/m^3)	Cohesive Forces (kPa)	The Angle of Internal Friction (°)	Poisson's Ratio	Elastic Modulus (MPa)
1	Miscellaneous fill	18.7	19.5	8.5	0.28	7
2	Peat soil	13.2	20	6	0.40	12.1
3	Round gravelly soil (natural)	19.4	8.56	31.9	0.46	196.67
4	Round gravelly soil (saturated)	19.4	7.37	30.5	0.46	196.67
5	Silty clay	19	40	12	0.30	16

Table 5. Structural mechanical calculation parameters.

Components	Elastic Modulus (MPa)	Poisson's Ratio	Volumetric Weight (kN/m^3)
Diaphragm wall	31,500	0.3	26
Anchor cable	195,000	0.3	78.5
Wai purlin	31,500	0.3	26
Interior bearing	31,500	0.3	26
Lattice column	31,500	0.3	26
Compaction grouting	25,000	0.3	26
Retaining wall	31,500	0.3	26

3.3. Results Analysis

3.3.1. Analysis of the Evolution Law of Foundation Pit and Surrounding Soil Displacement

In order to better analyze the surface settlement around the inlet pump room pit and the uplift of the pit bottom, the vertical displacement of the soil body is analyzed by selecting the short-side midline section of the diaphragm wall. As shown in Figure 14, the soil body is elevated 15 m below the pit bottom. It gradually decreases downward in a semicircle with the center of the pit bottom. The equilateral triangle centered on the lattice column has a small sinkage of 0.5 m from the center to the edge line. The soil in the middle of the two lattice columns has a large uplift. The maximum uplift reaches 8.68 cm under the saturation shear strength parameter and 7.95 cm under the natural shear strength parameter. The difference between the two reaches 7.3 mm, indicating that the shear strength parameter influences the soil's vertical displacement in excavating the foundation pit. Therefore, the shear strength parameter of the soil needs to be accurately determined in the calculation of the foundation pit design. Otherwise, it will easily lead to the excavation process of the foundation pit and the soil at the bottom of the pit will bulge too much and make the pit unstable. There is no sinking of the soil in a certain range around the excavation of the foundation pit, but there is a slight uplift. However, the uplift value is negligible. The soil outside the 5~15 m range of the diaphragm wall is affected by the excavation and sinks, and the sinking distance of the surrounding soil is close to the maximum settlement distance of 1.83 mm under the two parameters. The displacement is small and will not affect the surrounding buildings and roads.

Since the surface vertical displacement variation pattern under the conditions of two shear strength parameters of round gravel soil is consistent, the displacement variation curve under the conditions of natural shear strength parameters of round gravel soil is selected for analysis in this study, as shown in Figure 15. As can be seen from Figure 15, the soil at the edge of the foundation pit sinks after the end of excavation step 1. With the increased distance from the diaphragm wall, the sinking of the surrounding surface gradually decreases. The surrounding surface starts to rise outside the range of 40 m from the diaphragm wall. Furthermore, the soil around the foundation pit within 40 m is uplifted in the second step of excavation to the end. With the increase of excavation depth, the soil uplift around the foundation pit gradually increases. From the end of excavation step 2 to the end of the excavation, the uplift of the surrounding surface soil gradually decreases with the increase of distance from the diaphragm wall, and the decrease rate gradually increases with the increase of excavation depth. The main reason for the bulging of the soil around the foundation pit is that the foundation pit is excavated deeper, and the self-weight of the original soil is unloaded rapidly, so the bulging of the foundation is larger and drives the bulging of the surrounding surface soil.

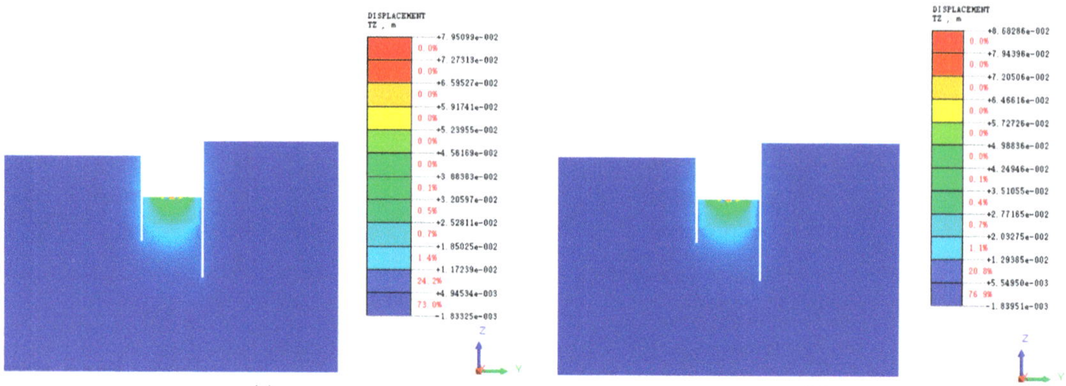

(a) (b)

Figure 14. Cloud map of vertical displacement of surrounding surface and soil at the bottom of the pit: (**a**) natural moisture content shear strength parameters; (**b**) saturated moisture content shear strength parameters.

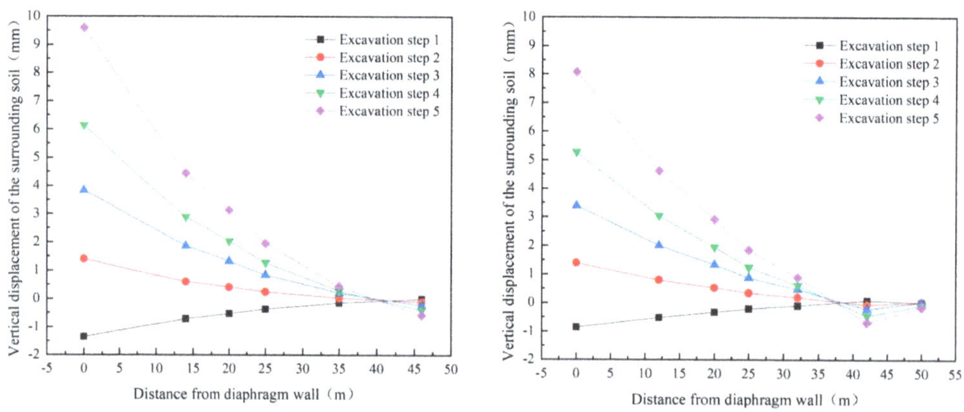

(a) (b)

Figure 15. Vertical displacement change curve of the ground surface around the foundation pit: (**a**) diaphragm wall 2 short side; (**b**) diaphragm wall 1 side.

Figures 16 and 17 show the displacement clouds of the surface and the soil at the bottom of the pit in X and Y directions around the end of the foundation pit excavation under natural and saturated shear strength parameters of round gravel soil, respectively. From Figure 16, it can be seen that after the end of the pit excavation, the soil's horizontal displacement at the pit bottom is locally larger, which is located on both sides of the lattice column, respectively. The maximum horizontal displacement is 15.7 mm, mainly due to the combined effect of the lattice column and excavation unloading. Affected by the reinforcement area of the pit bottom, the horizontal displacement of the pit bottom soil is slightly larger at the edge of the long side but smaller than the two sides of the lattice column. The horizontal displacement of the soil body under the saturated shear strength parameter of round gravel soil is slightly larger than that under natural conditions; the maximum is 17.7 mm.

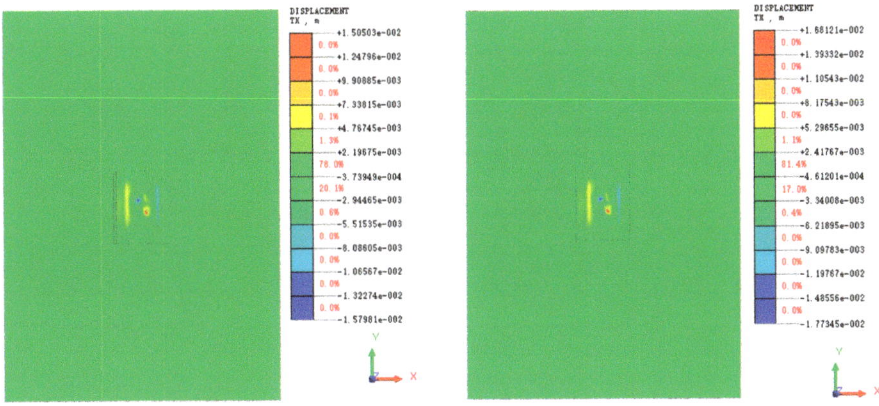

(**a**) (**b**)

Figure 16. X-direction displacement clouds of the surrounding surface and the soil at the bottom of the pit: (**a**) natural moisture content shear strength parameters; (**b**) saturated moisture content shear strength parameters.

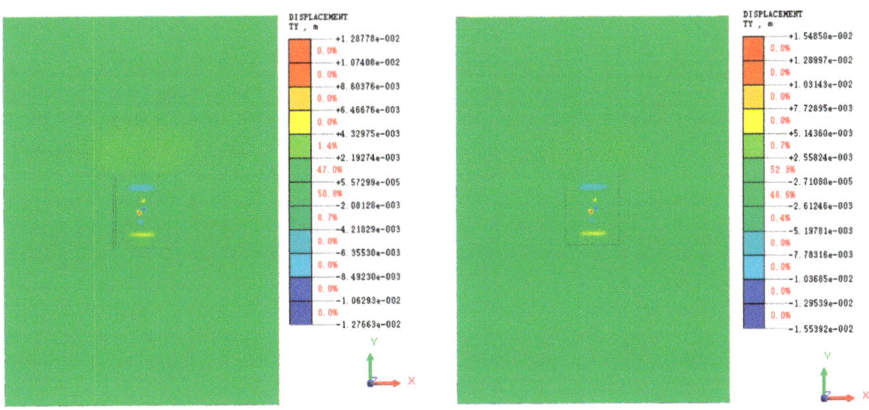

(**a**) (**b**)

Figure 17. Y-direction displacement clouds of the surrounding surface and the soil at the bottom of the pit: (**a**) natural moisture content shear strength parameters; (**b**) saturated moisture content shear strength parameters.

As shown in Figure 17, the maximum horizontal displacement of the subsoil of the foundation pit is located in the middle of the lattice column, and the horizontal displacement in the Y direction is slightly smaller than that in the X direction, which is mainly affected by the size and shape of the foundation pit. The horizontal displacement of the surrounding ground surface is very small and will not affect the surrounding buildings.

The horizontal displacement change curve under the natural shear strength parameter of round gravel soil is selected for analysis in this study, as shown in Figure 18. The midpoint of the foundation pit edge on the short side of diaphragm wall 1 and diaphragm wall 2 is taken as the monitoring point. As seen from the figure, during the first two excavation steps, the soil body on the side of diaphragm wall 1 is displaced into the pit. The horizontal displacement gradually decreases with the increase of burial depth. As the excavation progresses, the horizontal displacement gradually changes to the displacement outside the pit. Since the pit excavation unloads, the lower soil body extrudes the diaphragm wall to the pit so that the diaphragm wall above the initial excavation surface shows a trend of outward displacement, and the soil body is then displaced. Diaphragm wall 2 is displaced to the foundation pit's inner side, with the burial depth increasing gradually; the curve at −25 m appears to turn, mainly by the influence of stratigraphic changes.

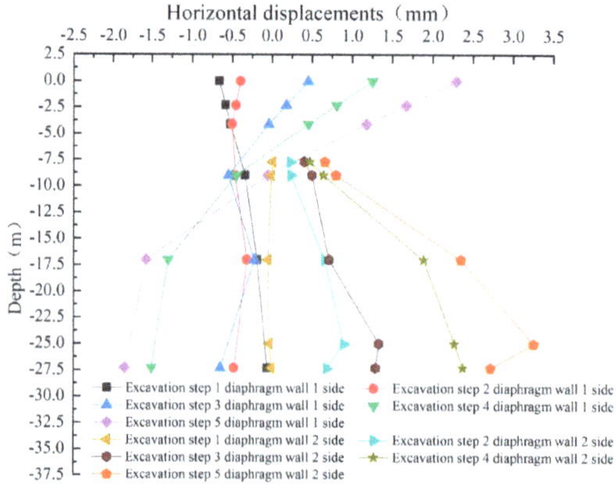

Figure 18. Horizontal displacement curve of soil around the foundation pit.

3.3.2. Analysis of Displacement Variation Law of Diaphragm Wall

The excavation process of the foundation pit will change the equilibrium state of the original soil, and the geotechnical body tends to an unstable state from the stable state in its natural state. As the main support structure, the underground diaphragm bears the role of water interception, seepage control, load bearing, and earth retaining and has a significant impact on maintaining the stability of the foundation pit. Excessive deformation of the diaphragm wall may cause wall damage or pit collapse. Therefore, the diaphragm wall displacement must be controlled within the specification requirements. As shown in Figures 19 and 20, the horizontal displacements of the diaphragm wall in the X and Y directions are shown. It can be seen from Figure 19 that when the round gravel stratum is in a state of natural and saturated moisture content, its shear strength parameters have little influence on the horizontal displacement of the diaphragm wall during the excavation of the foundation pit. Under the two-parameter conditions, the horizontal displacement of the diaphragm wall in the X direction is larger in the area below −29.4 m on the long side of the diaphragm wall. Under saturated conditions, the maximum horizontal displacement of the diaphragm wall in the X direction is 3.92 mm, which is far less than the standard warning value.

As shown in Figure 20, the horizontal displacement of the diaphragm wall along the Y-direction is larger in the range of the third inner support to the fifth inner support on the short side and the area below the pit bottom, with a mushroom-shaped distribution pattern. The maximum horizontal displacement in the Y-direction under the two parameters is 0.1 mm different, and it is also larger under the saturated condition, with a maximum displacement of 3.48 mm. it can be seen that the horizontal displacement under the two round gravel soil. The horizontal displacement under the shear strength parameter is close, and the displacement values are small and within the specification warning range, which indicates that the diaphragm wall better supports the foundation pit of the intake pump room.

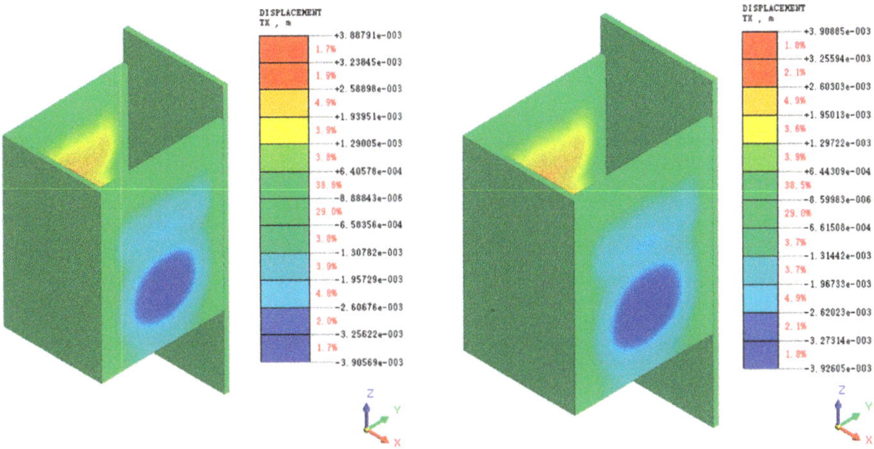

(**a**) (**b**)

Figure 19. Horizontal displacement cloud in X-direction of diaphragm wall: (**a**) natural moisture content shear strength parameters; (**b**) saturated moisture content shear strength parameters.

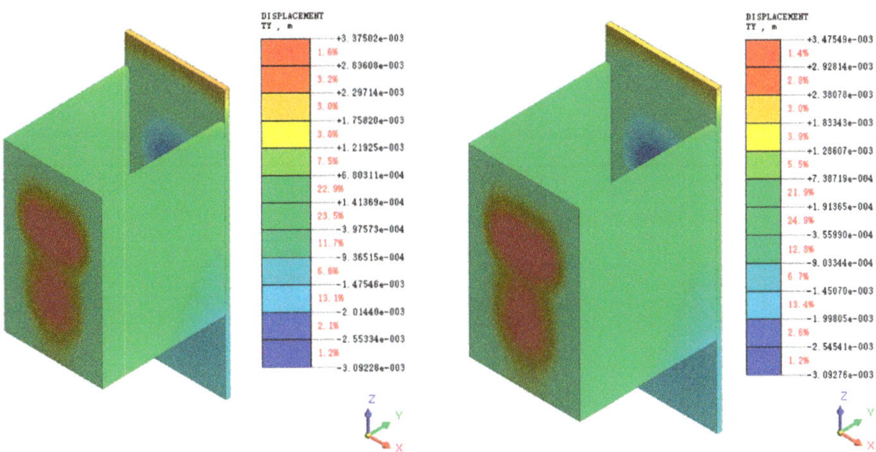

(**a**) (**b**)

Figure 20. Horizontal displacement cloud in Y-direction of diaphragm wall (**a**) natural moisture content shear strength parameters; (**b**) saturated moisture content shear strength parameters.

Select the midpoint of four diaphragm walls for analysis. Figure 21 shows the horizontal displacement curve of the diaphragm wall in X and Y under natural shear strength parameters of round gravel soil. As shown in Figure 21a, during the foundation pit excavation, the horizontal displacement of the long sides of diaphragm wall 2 at different burial depths is symmetrically distributed. Since the excavation depth of the first layer is shallow in the simulation process, the impact on the deep diaphragm wall is small, and the displacement decreases with the increase of the burial depth. During the excavation of the second layer to the fifth layer, it can be seen that the horizontal displacement of the diaphragm wall at the buried depth of 22.5 m has a big change trend. After excavating the second layer, the horizontal displacement of the diaphragm wall suddenly decreases from the buried depth of 22.5 m to the bottom of the pit. In contrast, the excavation of the third layer to the fifth layer continues to maintain an increasing or stable change trend, mainly due to each layer's excavation depth and the stratum's influence. From Figure 21b, it can be seen that the horizontal displacement of the top of wall 1 of the diaphragm wall gradually moves out of the pit during the excavation process, mainly due to the effect of the anchor cable in tension and the change of soil pressure behind the wall. It can be seen that with the increase of excavation depth, the horizontal displacement of the diaphragm wall gradually increases.

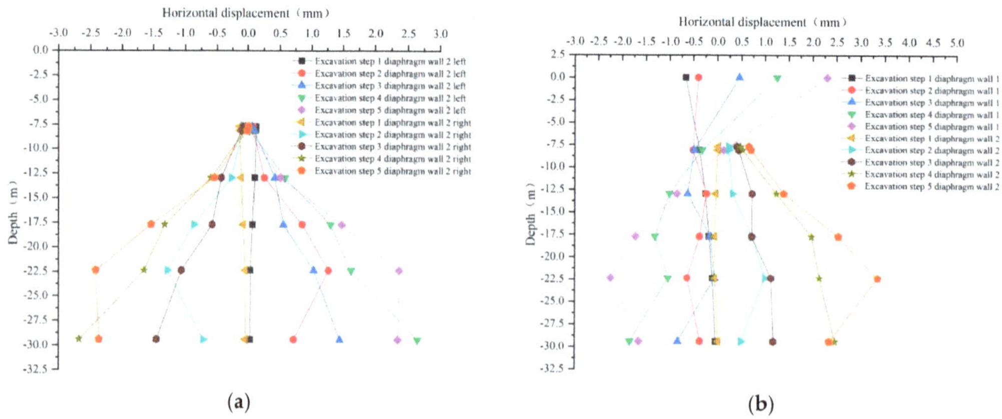

Figure 21. Horizontal displacement variation curve of diaphragm wall: (**a**) horizontal displacement in X direction; (**b**) horizontal displacement in Y direction.

Figure 22 shows the vertical displacement cloud of the diaphragm wall. As can be seen from the figure, the closer to the short side of the diaphragm wall 2, the greater the vertical displacement. The maximum vertical displacement is located in the wall near the fifth interior bearing on the short side of the diaphragm wall 2, with a maximum vertical displacement of 10 mm. It is larger under natural moisture content and shear strength parameter of round gravel soil, but the difference is only 0.1 mm. The top of the diaphragm wall in the foundation pit has a vertical displacement monitoring alarm value of 10–20 mm. However, the results show that the displacement of the top of the wall does not reach the alarm value, which shows that the diaphragm wall is stable.

3.3.3. Analysis of Displacement Variation Law of Diaphragm Wall

The diaphragm wall is a structure used to block the soil's lateral pressure on the pit's side wall, maintain the pit wall's stability, and ensure the shape of the pit excavation. Its stability is related to the smooth construction of the pit project and the safety of nearby buildings. If the pile structure is damaged or overstressed, it will affect the completion of the pit project and even threaten the safety of the engineering staff. Figures 23 and 24 show the maximum and minimum principal stresses of the diaphragm wall under the above two parameters, respectively. As can be seen from Figure 23, the minimum principal stresses in the diaphragm wall are mainly tensile stresses, and stress concentrations occur at the corners of the diaphragm wall between the third and fifth interior bearing. It can be seen that the inner side of the diaphragm wall is subject to compressive stress, the outer side of the wall is subject to tensile stress, and the tensile stress outside the wall is three times the compressive stress inside the wall. The maximum value of the minimum principal stress is 3.48 MPa. As shown in Figure 24, the maximum principal stress in the diaphragm wall is mainly compressive. The maximum compressive stress is located at the corner of two sides of the diaphragm wall between the third and fifth interior bearing. The compressive stress here is relatively concentrated, with a maximum of 6.1 MPa, which is 0.2 MPa higher than the calculated value under the natural shear strength parameter of round gravel soil. The main stress of the diaphragm wall has an obvious boundary at a depth of the foundation pit bottom, mainly since the excavation of the foundation pit leads to the redistribution of the soil stress behind the wall. However, the stress is small, indicating that the diaphragm wall has a good supporting effect and that the foundation pit is stable.

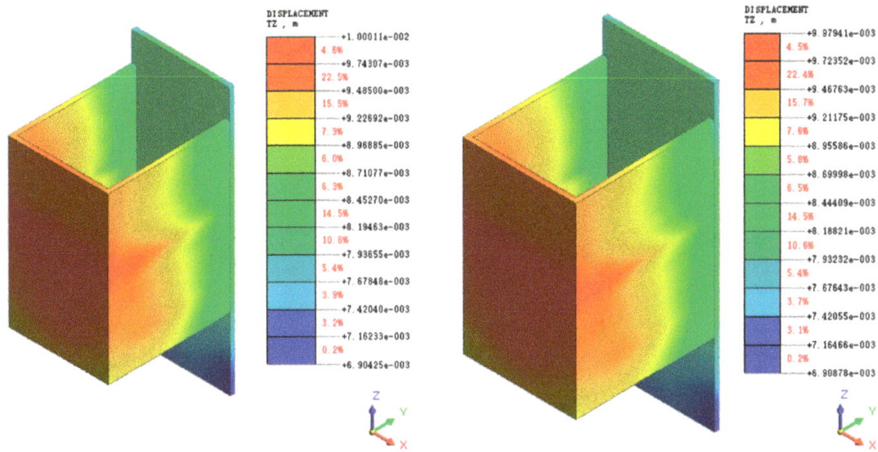

(a) (b)

Figure 22. Vertical displacement cloud of diaphragm wall: (**a**) natural moisture content shear strength parameters; (**b**) saturated moisture content shear strength parameters.

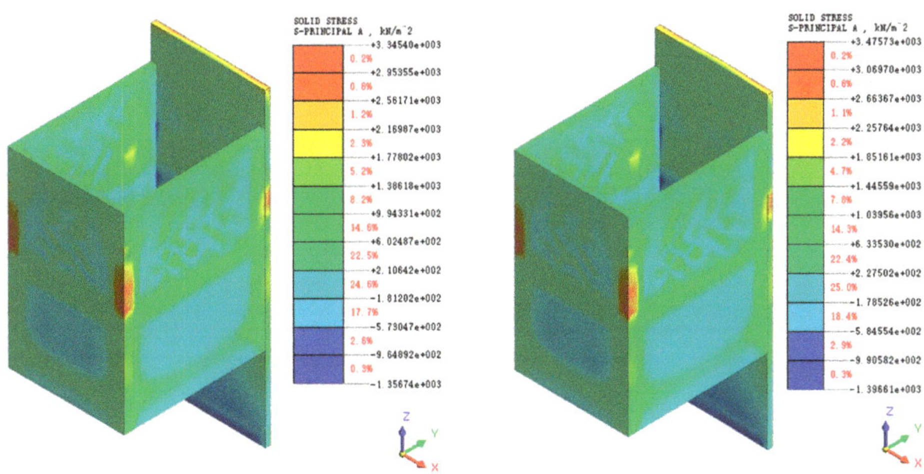

(a) (b)

Figure 23. Minimum principal stress cloud of diaphragm wall: (**a**) natural moisture content shear strength parameters; (**b**) saturated moisture content shear strength parameters.

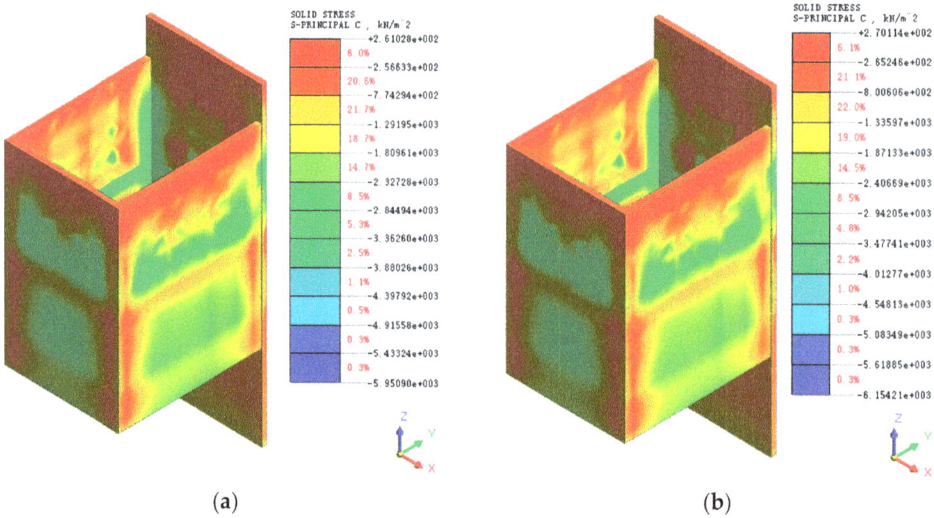

Figure 24. Maximum principal stress cloud of diaphragm wall: (**a**) natural moisture content shear strength parameters; (**b**) saturated moisture content shear strength parameters.

3.3.4. Analysis of the Variation Law of Interior Bearing Axial Force

As one of the main support structures of the intake pump room pit, the internal support bears the role of dispersing the force deformation of the underground diaphragm wall, and its force deformation significantly impacts the pit's stability. Figure 25 shows the axial force diagram of the interior bearing after the end of the pit excavation under the above two shear strength parameters. Figure 25 shows that the axial force of the diagonal brace in the fourth interior bearing is larger due to the large horizontal displacement of the diaphragm wall at that place after the excavation of the foundation pit is finished. As the excavation depth gradually increased, the maximum value of the interior bearing axial force gradually moved down to the fourth interior bearing since the horizontal displacement of the diaphragm wall gradually moved down to the fourth interior bearing. Under the natural shear strength parameter condition, the maximum axial force of the interior bearing from the first excavation to the bottom of the pit is 0.15 MN, 0.59 MN, 1.66 MN, 2.74 MN, 6.3 MN. The saturated shear strength parameters of round gravel soils under the maximum axial force of interior bearing are 0.15 MN, 0.65 MN, 1.85 MN, 2.95 MN, 6.53 MN. It can be seen that the maximum axial force of interior bearing axial force of the first four layers of soil excavation is small. The axial force of the interior bearing after excavating the last layer changes more. However, it is within the specification design range, indicating that the interior bearing has a good effect.

3.3.5. Anchor Cable Axial Force Variation Characteristics Analysis

The anchor cable can transfer the earth pressure on the support pile to the deep soil through the anchor cable's axial force, so the analysis of the axial force of the anchor cable is important for analyzing the foundation deformation. Figure 26 shows the axial force of prestressed anchor cable under natural and saturated shear strength parameters of round gravel soil. The maximum value of the anchor force is 262.9 kN in the free section of the second row of anchor ropes under saturated conditions and 261.9 kN under natural conditions, and the anchor force in the anchor section is gradually dispersed to the soil around the anchor end by the force transfer of mortar. Hence, the anchor force in the anchor section is smaller. Under the natural shear strength parameter condition, the maximum change of anchor cable axial force during excavation is 257 kN, 258 kN, 259 kN, 260 kN,

262 kN, and under the condition of saturation, the shear strength parameter is 258 kN, 259 kN, 260 kN, 261 kN, 263 kN. The reason is that this simulation mainly starts from the top of the wall of diaphragm wall 2, and the prestressed anchor cable only exists in diaphragm wall 1, which is higher than the simulation depth of the pit. Hence, the excavation simulation has little effect on the anchor cable force.

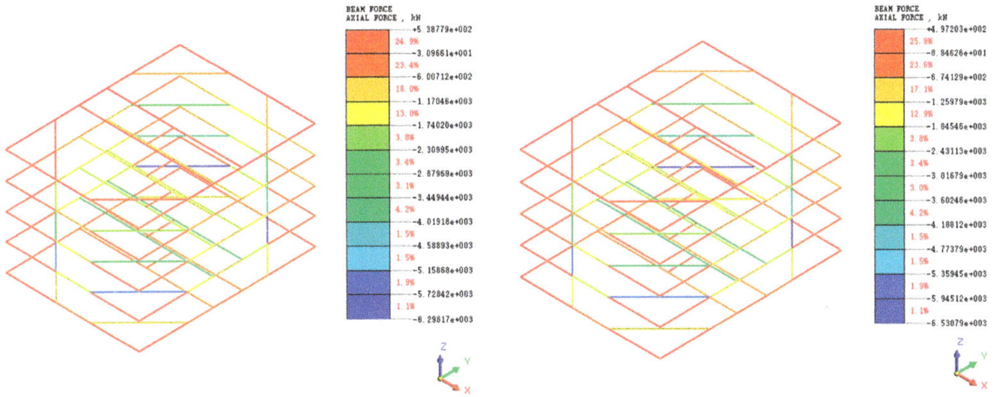

(**a**) (**b**)

Figure 25. Axial force cloud diagram of interior bearing system: (**a**) natural moisture content shear strength parameters; (**b**) saturated moisture content shear strength parameters.

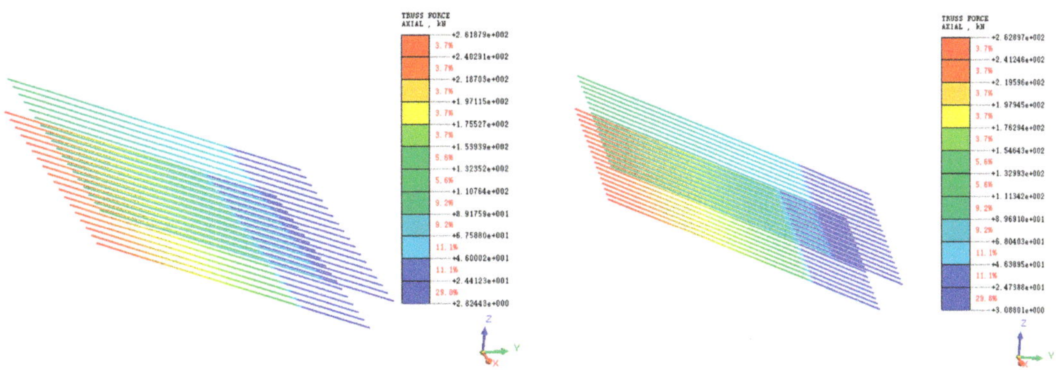

(**a**) (**b**)

Figure 26. Anchor cable axial force cloud: (**a**) natural moisture content shear strength parameters; (**b**) saturated moisture content shear strength parameters.

4. Conclusions

In this paper, based on a comprehensive analysis of the site geological conditions, the shear strength characteristics of round gravel soils and their parameter variation laws were analyzed by reshaping the soil samples retrieved from the site and conducting large-scale direct shear experiments on round gravel soils with natural and saturated moisture content. The research results were applied to the numerical simulation of foundation excavation to obtain the following conclusions.

(1) The shear strength characteristics of plateau alluvial-lacustrine alluvial round gravel soil under different water content conditions are studied and analyzed through large-scale direct shear tests. Under different water content conditions, the variation law

of shear stress-shear displacement of round gravel soil is slightly different. At low water content, with the increase of shear displacement, the particles of round gravel soil are sheared, and the shear surface contacts closely until shear failure, and the soil strength slightly increase. However, the shear resistance curve of round gravel soil with high water content gradually weakened with the increasing confining pressure.

(2) Large-scale direct shear experiments of round gravel soil show that the cohesion of round gravel soil in the natural state is 8.56 kPa, and the angle of internal friction is 31.9°. In the saturated state, the cohesion of round gravel soil is 7.37 kPa, and the angle of internal friction is 30.5°. With the increased water content, the round gravel soil's cohesive force and internal friction angle decrease significantly.

(3) The numerical simulation results of foundation pit excavation show that after the completion of construction, the pit bottom is subject to the joint influence of the reinforcement area and lattice columns. There is a large uplift, and the surrounding surface soil also shows a bulge within a certain range. Hence, further strengthening the monitoring and reinforcement of the surrounding structures is necessary. The stress deformation of the supporting structure is small, far less than the design value. With the increase of the conglomerate soil's internal friction angle and cohesion, the foundation support and soil deformation decreased, indicating that increasing the shear strength parameter of the conglomerate soil can effectively reduce the foundation deformation. The construction can be carried out by selecting conglomerate strata with different water content in practical engineering to achieve, reduce the project cost and improve the project economy.

Author Contributions: Conceptualization, Z.K. and Y.G.; software, Z.K.; formal analysis, Z.K.; investigation, Z.K.; resources, S.M. and W.Z.; writing—original draft preparation, Z.K.; writing—review and editing, Z.K., S.M., W.Z. and Y.G.; visualization, Z.K.; supervision, Y.G.; project administration, Y.G.; funding acquisition, S.M. and Y.G. All authors have read and agreed to the published version of the manuscript.

Funding: This research was funded by the scientific research fund project of Yunnan Provincial Department of Education, China (No. 2022J0065); key projects of analysis and testing fund of Kunming University of science and technology, China (No. 2021T20200145); China Postdoctoral Science Foundation Project (No. 2017M620433); General projects of Yunnan Fundamental Research Projects (No. 2018FB075); key technology research and pilot test of underground coal gasification, a major scientific and technological research project of CNPC (No. 2019e-25).

Data Availability Statement: The data involved in this study are given in full in the paper.

Conflicts of Interest: The authors declare no conflict of interest.

References

1. Tan, F.; Wu, S.; Huang, Z.F.; Chen, Z. Discussion on ground bearing capacity of the sandy pebble in the foundation of tall buildings in Chengdu area. *Build. Struct.* **2013**, *43*, 30–32, 83.
2. Hou, L.J.; Chen, X.C.; Chen, H.; Cui, C.L. Research on models of surface wave velocity method for determining bearing capacity of cobble soil foundation. *Rock Soil Mech.* **2008**, *29*, 2572–2576.
3. Soleimani, S.; Jiao, P.; Rajaei, S.; Forsati, R. A new approach for prediction of collapse settlement of sandy gravel soils. *Eng Comput.* **2018**, *34*, 15–24. [CrossRef]
4. Rücknagel, J.; Götze, P.; Hofmann, B.; Christen, O.; Marschall, K. The influence of soil gravel content on compaction behaviour and pre-compression stress. *Geoderma* **2013**, *209*, 226–232. [CrossRef]
5. Ghanizadeh, A.R.; Delaram, A.; Fakharian, P.; Armaghani, D.J. Developing Predictive Models of Collapse Settlement and Coefficient of Stress Release of Sandy-Gravel Soil via Evolutionary Polynomial Regression. *Appl. Sci.* **2022**, *12*, 9986. [CrossRef]
6. Jiang, J.C. Research on strong dewatering technology for deep foundation excavation in Kunming round gravel stratum. *Constr. Technol.* **2012**, *41*, 107–111.
7. Sun, T.; Chen, G.X.; Wang, B.H.; Li, X.J. Experimental research of effect of granule shape on shear modulus and damping ratio of gravel. *Chin. J. Rock Mech. Eng.* **2014**, *33*, 4211–4217.
8. Wang, Y.X.; Shao, S.J.; Wang, Z. Experimental study on mechanical behaviors and particle breakage of sandy gravel. *Chin. J. Rock Mech. Eng.* **2020**, *39*, 1287–1296.
9. Hara, T.; Kokusho, T.; Hiraoka, R. Undrained strength of gravelly soils with different particle gradations. *Mouth* **2004**, *277*, 1920.

10. Kuenza, K.; Towhata, I.; Orense, R.P.; Wassan, T.H. Undrained torsional shear tests on gravelly soils. *Landslides* **2004**, *1*, 185–194. [CrossRef]
11. Rollins, K.M.; Singh, M.; Roy, J. Simplified equations for shear-modulus degradation and damping of gravels. *J. Geotech. Geoenviron. Eng.* **2020**, *146*, 04020076. [CrossRef]
12. Hubler, J.F.; Athanasopoulos-Zekkos, A.; Zekkos, D. Monotonic and cyclic simple shear response of gravel-sand mixtures. *Soil Dyn. Earthq. Eng.* **2018**, *115*, 291–304. [CrossRef]
13. Liu, J.; Tang, Y.; Yi, L.; Peng, Y.C.; Zhou, Y.F. Creep Constitutive Model of Cobbly Soil and Its Engineering Application. *J. Yangtze River. Sci. Res. Inst.* **2022**, *39*, 107–112.
14. Tong, J.J.; Wang, M.N.; Yu, L.; Liu, D.G.; Xu, R. A study of the land subsidence around the deep foundation pit of the Chengdu subway station. *Hydrogeol. Eng. Geol.* **2015**, *42*, 97–101.
15. Penumadu, D.; Zhao, R. Triaxial compression behavior of sand and gravel using artificial neural networks (ANN). *Comput. Geotech.* **1999**, *24*, 207–230. [CrossRef]
16. Tang, K.S.; Xie, X.Y.; Yang, L. Research on mechanical characteristics of gravel soil based on large-scale triaxial tests. *Chin. J. Under Space Eng.* **2014**, *10*, 580–585.
17. Ma, S.K.; Wang, B.; Liu, Y.; Shao, Y.; Wang, H.G.; Wang, Y.L. Large-scale dynamic triaxial tests on saturated gravel soil in Nanning metro area. *Chin. J. Geotech. Eng.* **2019**, *41*, 168–174.
18. Ma, S.K.; Duan, Z.B.; Liu, Y.; Wang, B.; Shao, Y. Large triaxial test study on dynamic characteristics of round gravel. *Arab. J. Geosci.* **2020**, *13*, 1–9. [CrossRef]
19. Stark, N.; Hay, A.E.; Cheel, R.; Lake, C.B. The impact of particle shape on the angle of internal friction and the implications for sediment dynamics at a steep, mixed sand–gravel beach. *Earth Surf. Dynam.* **2014**, *2*, 469–480. [CrossRef]
20. Enomoto, T.; Qureshi, O.H.; Sato, T.; Koseki, J. Strength and deformation characteristics and small strain properties of undisturbed gravelly soils. *Soils Found.* **2013**, *53*, 951–965. [CrossRef]
21. Chen, C. Research on modified constitutive model of Shenyang circular-gravel based on disturbed state. *J. Henan Polytech. Univ. Nat. Sci.* **2017**, *36*, 125–131.
22. Saberi, M.; Annan, C.D.; Konrad, J.M. Constitutive modeling of gravelly soil–structure interface considering particle breakage. *J. Eng. Mech.* **2017**, *143*, 04017044. [CrossRef]
23. Liu, G.; Lu, R.; Zhao, M.Z.; Luo, Q.; Lv, C. Ellipsoid model based packing characteristics analysis of round gravels. *Rock. Soil Mech.* **2019**, *40*, 4371–4379.
24. Ou, X.D.; Huang, Z.Z.; Jiang, J.; Luo, F.Z.; Liang, Y.H. Influence of pit-in-pit excavation on double-row piles in composite stratum of round gravel and mudstone. *J. Yangtze River. Sci. Res. Inst.* **2022**, *39*, 78–85.
25. Ni, X.R.; Li, Z.L.; Wang, Y. Application of auger drilling secondary pressure fed technology into concrete piles in dry sand and gravel formations. *Constr. Technol.* **2015**, *44*, 134–136.
26. *GB/T50123-2019*; CSBTS (China State Bureau of Quality and Technical Supervision) Chinese Standard for Soil Test Method. CSBTS: Beijing, China, 2019.
27. Liu, L.L.; Sun, Q.C.; Wu, N.Y.; Liu, C.L.; Ning, F.L.; Cai, J.C. Fractal analyses of the shape factor in kozeny–carman equation for hydraulic permeability in hydrate-bearing sediments. *Fractals* **2021**, *29*, 2150217. [CrossRef]
28. Wei, R.C.; Liu, L.L.; Jia, C.; Zhao, H.L.; Dong, X.; Bu, Q.T.; Liu, C.L.; Wu, N.Y. Undrained Shear Properties of Shallow Clayey-Silty Sediments in the Shenhu Area of South China Sea. *Sustainability.* **2023**, *15*, 1175. [CrossRef]
29. Wang, X.Z.; Wang, X.; Shen, J.H.; Ding, H.Z.; Wen, D.S.; Zhu, C.Q.; Lv, S.Z. Foundation filling performance of calcareous soil on coral reefs in the South China Sea. *Appl. Ocean Res.* **2022**, *129*, 103386. [CrossRef]
30. Wang, X.; Shan, Y.; Cui, J.; Zhong, Y.; Shen, J.H.; Wang, X.Z.; Zhu, C.Q. Dilatancy of the foundation filling material of island-reefs in the South China Sea. *Constr. Build. Mater.* **2022**, *323*, 126524. [CrossRef]
31. Wang, Y.P.; Lu, Y.W.; Zhang, E.S.; Peng, Y.C.; Zuo, Y.Z.; Li, H.M. Comprehensive experimental study of strength and deformation characteristics and mechanical model parameters of sandy pebble soil. *J. Yangtze River. Sci. Res. Inst.* **2022**, *39*, 93–98.
32. Fragaszy, R.J.; Su, J.; Siddiqi, F.H.; Ho, C.L. Modeling strength of sandy gravel. *J. Geotech. Eng.* **1992**, *118*, 920–935. [CrossRef]
33. Guo, Y.H.; Yan, M.; Song, Q.; Yuan, G.; Fu, X.B. The influence of deep foundation pit excavation on the adjacent existing high pressure natural gas pipeline. *Chin. J. Under Space Eng.* **2021**, *17*, 840–847.

sustainability

Article

Mining Method Optimization of Difficult-to-Mine Complicated Orebody Using Pythagorean Fuzzy Sets and TOPSIS Method

Shuai Li [1], Qi Huang [1], Boyi Hu [1,*], Jilong Pan [1,*], Junyu Chen [2], Jianguo Yang [2], Xinghui Zhou [2], Xinmin Wang [1] and Haoxuan Yu [3]

1 School of Resources and Safety Engineering, Central South University, Changsha 410083, China
2 Zhejiang Suichang Gold Mine Co., Ltd., Lishui 323304, China
3 Civil Engineering Discipline, School of Engineering, Monash University, Malaysia Campus, Bandar Sunway 47500, Malaysia
* Correspondence: boicary@csu.edu.cn (B.H.); csujlpan@csu.edu.cn (J.P.)

Abstract: In Suichang gold mine, the altered rock type gold deposits were cut by faults and joint fissures, leading to complex resource endowment characteristics, large changes in occurrence, a serious complex of ore vein branches and great difficulty in mining. In order to select a suitable mining method for such a difficult and complicated orebody, a multi-factor and multi-index comprehensive evaluation system involving benefits, costs, safety and other aspects was constructed by using the Pythagorean fuzzy sets and TOPSIS method. Taking Suichang gold mine as an example, the weighted aggregation evaluation matrix was constructed, the closeness index of the four mining schemes were 0.8436, 0.3370, 0.4296 and 0.4334, and the mechanized upward horizontal layering method was determined as the optimal scheme. This method overcame the fuzzy comparison of economic and technical indicators directly, but converted them into corresponding fuzzy numbers to obtain accurate closeness index for optimization. The application of this method not only ensured a safe, efficient and environment-friendly mining effect, but also provided a reference for the optimization of the mining scheme of the severely branched composite orebody.

Keywords: mining method optimization; difficult-to-mine complicated orebody; multiple attribute decision making; Pythagorean fuzzy sets and TOPSIS method; mechanized upward horizontal layering method

Citation: Li, S.; Huang, Q.; Hu, B.; Pan, J.; Chen, J.; Yang, J.; Zhou, X.; Wang, X.; Yu, H. Mining Method Optimization of Difficult-to-Mine Complicated Orebody Using Pythagorean Fuzzy Sets and TOPSIS Method. *Sustainability* **2023**, *15*, 3692. https://doi.org/10.3390/su15043692

Academic Editor: Jianjun Ma

Received: 4 January 2023
Revised: 8 February 2023
Accepted: 13 February 2023
Published: 16 February 2023

1. Introduction

In the design process of new mines and reconstruction and upgrading of mines, the design of mining methods is an important part and is the core of the whole design work, which directly determines the subsequent technical personnel allocation, production organization management and industrial supporting facilities [1–3]. Choosing the most appropriate mining method to mine the deposit is very important to the safety, economy and environmental protection of mining operations and affects the benefits and long-term development of mining companies [4].

For a continuous orebody with regular shape, the most appropriate mining method can be selected by comparing the advantages and disadvantages and the economic and technical indicators. However, taking the difficult-to-mine complicated orebody (DCO) (see Figure 1) as example, it was greatly affected by faults and joint fissures, leading to obvious branching and compounding phenomenon. Since it is difficult to determine the optimal plan through traditional methods for DCO, more reliance is placed on the experience of design decision-makers and limited data, through the comprehensive comparison of various indicators. This is typical of Multiple Attribute Decision Making (MADM).

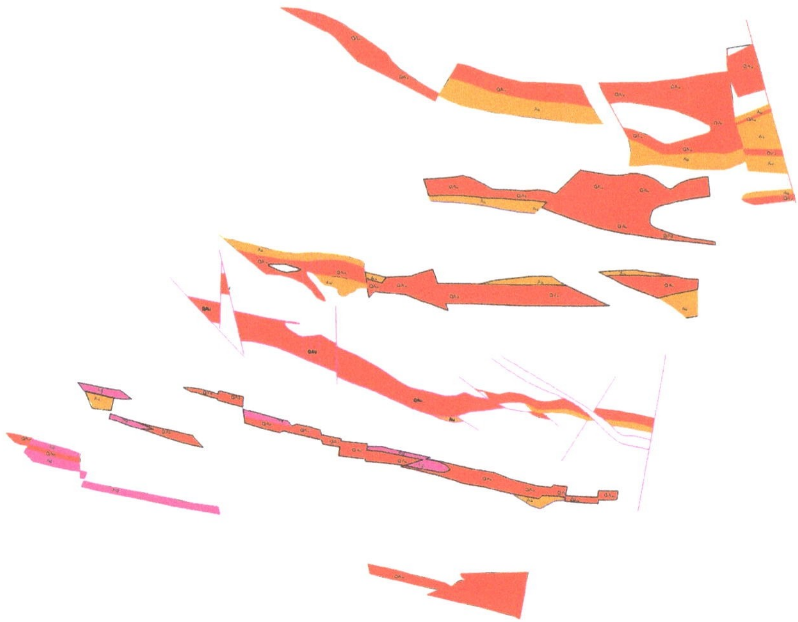

Figure 1. Plan of 140–260 m middle section of Suichang Gold Mine.

Multiple Attribute Decision Making refers to the decision making problem of selecting the best alternative or ranking under a condition of considering multiple indicators [5]. In order to solve such problems, researchers have developed a variety of multi criteria decision-making methods; typical representative methods include the Analytic Hierarchy Process [6], Entropy method [7], CRITIC method [8], TOPSIS method [9], GST (Grey Target Decision) method [10], DEA method [11], VIKOR method [12], Fuzzy comprehensive evaluation method [13], etc.

However, with increase in the complexity and scientific requirements of evaluation, a single decision-making method cannot guarantee optimal or accurate results [14,15]. Under this background, researchers have explored and developed mixed decision-making methods, for example, mixtures of Analytic Hierarchy Process and fuzzy, Entropy weight method and TOPSIS, Analytic Hierarchy Process, fuzzy set and VIKOR, fuzzy set and TOPSIS, etc.

In the field of mining engineering, many experts rely on these decision-making methods to optimize mining methods, for example, Karimnia [16] proposed the fuzzy analytical hierarchy process method, the most suitable method selected for Iran's Qapliq salt mine. Yavuz [17] used the AHP method and fuzzy multiple attribute decision-making method, respectively, in a lignite mine located in Istanbul, carried out sensitivity analysis of the two methods and concluded that the Room and pillar method with filling is the most appropriate method. Qinqiang Guo [18] used the mixed method of AHP to determine the index weight and TOPSIS to rank, selecting the most suitable mining method from the Soft Broken Complex Orebody, and achieved very good results, Iphar [19] is committed to developing a mobile application, integrating several decision-making methods, and the optimal mining method can be obtained by inputting the original parameters for reference by engineering researchers.

However, in view of the complexity of mining method selection, simple expert scoring cannot fully reflect the fuzzy information, and the cognitive differences between different experts are easy to cause distortion of results. In terms of sorting, different ranging methods may have different results. In this case, the development of fuzzy set theory provides a good idea for solving such problems [20]. Bajić [21] transformed the indicators into triangular fuzzy numbers, constructed a fuzzy decision matrix and a fuzzy performance matrix, used to select the optimal alternative, and verified this through sensitivity analysis. Memori [22] uses the TOPSIS method based on intuitionistic fuzzy sets, providing an accurate sustainable ranking of suppliers and a relevant solution for sustainable sourcing decisions that is validated through a real-world case study. Narayanamoorthy [23] selected the best scheme for the selection of industrial robots by using a combination of Interval-valued intuitionistic hesitant fuzzy entropy and VIKOR.

Pythagorean fuzzy sets (PFS) generalized by Yager [24] is a new method to deal with fuzzy problems. Its main contribution is to go beyond the limit that the sum of membership and no membership of fuzzy sets is less than 1 (see Figure 2). Compared with other fuzzy sets such as intuitionistic fuzzy sets, it can more fully and accurately represent uncertain information [25]. In view of this, this paper introduces a TOPSIS method based on PFS [26]. The framework of this PFS–TOPSIS method is illustrated (see Figure 3), which is used for mining method decision-making for the Suichang Gold Mine in Zhejiang Province, China, and has achieved good results in actual production.

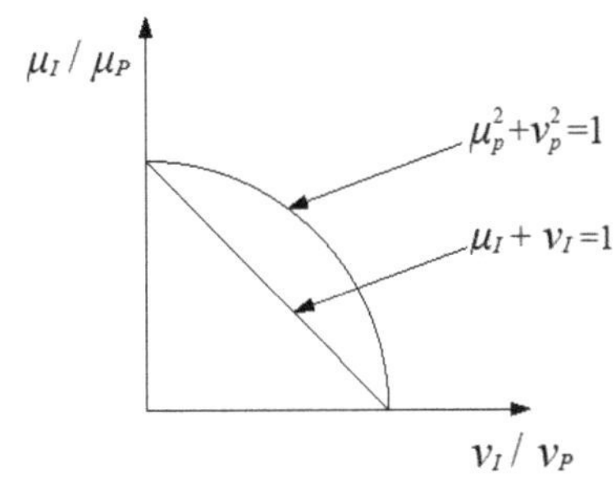

Figure 2. Spatial comparison between PFS and other fuzzy sets.

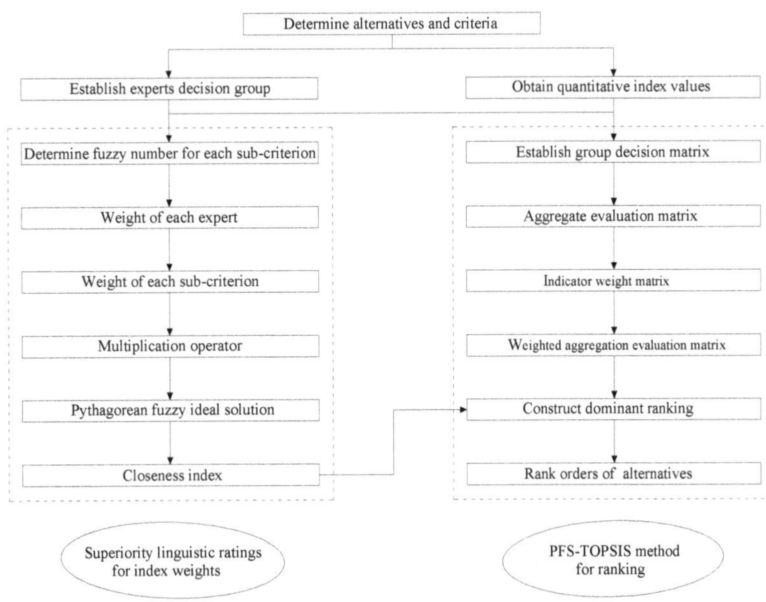

Figure 3. Framework of PFS-TOPSIS method.

2. Methods Introduction

2.1. Introduction of PFS Method

Definition 1. *Let X be a universe of discourse [27,28]. The PFS ξ on X is given by Equation (1).*

$$\xi = \{ [x, \mu_{\xi}(x), \nu_{\xi}(x) | x \epsilon X] \} \tag{1}$$

where the functions $\mu_{\xi}(x)$:X $\rightarrow$ [0, 1] and $\nu_{\xi}(x)$: X $\rightarrow$ [0, 1] define the degree of membership and the degree of non-membership of the element $x \in X$ to the set ξ, respectively, with the condition that $0 \leq (\mu_{\xi}(x))^2 + (\nu_{\xi}(x))^2 \leq 1, \forall x \in X$. $\pi_{\xi}(x) = \sqrt{1 - (\mu_{\xi}(x))^2 - (\nu_{\xi}(x))^2}$ is called the degree of indeterminacy of element $x \in X$. For convenience, they are called $(\mu_{\xi}(x), \nu_{\xi}(x))$ and a Pythagorean fuzzy number (PFN) denoted by $\xi = (\mu_{\xi}, \nu_{\xi})$ [29].

Definition 2. *For the collection $\xi_i = (\mu_{\xi_i} \nu_{\xi_i})$ (i = 1,2, … … n) of the of PFNs with the weight vector $w = (w_1\ w_2 \ldots\ldots w_n)$ of $\xi_i (i = 1,2, \ldots\ldots n)$ such that $\sum_{i=1}^{n} w_i = 1$, the Pythagorean fuzzy weighted averaging (PFWA) operator and the Pythagorean fuzzy weighted geometric (PFWG) operator can be defined as in Equations (2) and (3), respectively [30].*

$$\begin{aligned} PFWA_w(\xi_1, \xi_2, \cdots, \xi_n) &= w_1\xi_1 \oplus w_2\xi_2 \oplus \cdots \oplus w_n\xi_n \\ &= \left(\sqrt{1 - \prod_{i=1}^{n} \left(1 - (\mu_{\xi_i})^2\right)^{w_i}}, \prod_{i=1}^{n} (\nu_{\xi_i})^{w_i} \right), \end{aligned} \tag{2}$$

$$\begin{aligned} PFWG_w(\xi_1, \xi_2, \cdots, \xi_n) &= w_1\xi_1 \otimes w_2\xi_2 \otimes \cdots \otimes w_n\xi_n \\ &= \left(\prod_{i=1}^{n} (\mu_{\xi_i})^{w_i}, \sqrt{1 - \prod_{i=1}^{n} \left(1 - (v_{\xi_i})^2\right)^{w_i}} \right). \end{aligned} \tag{3}$$

Definition 3. *Let A and B be PFSs of X = {$x_1, x_2, \ldots x_n$} [27]. Then, the sum of A and B is defined as Equation (4).*

$$\begin{aligned} A \oplus B = \ & \{ \langle x, \sqrt{(\mu_A(x))^2 + (\mu_B(x))^2 - (\mu_A(x))^2(\mu_B(x))^2}, \\ & \nu_A(x)\nu_B(x) \rangle \mid x \in X \}, \end{aligned} \tag{4}$$

The product of A and B is defined as Equation (5).

$$\begin{aligned} A \otimes B = \ & \{ \langle x, \mu_A(x)\mu_B(x), \\ & \sqrt{(\nu_A(x))^2 + (\nu_B(x))^2 - (\nu_A(x))^2(\nu_B(x))^2} \rangle \mid x \in X \}. \end{aligned} \tag{5}$$

2.2. Introduction of TOPSIS Method

The TOPSIS method was developed by Hwang [31], first put forward in 1981, and is a method of ranking according to the closeness of a limited number of evaluation objects to the ideal target. After years of development, dozens of derivative methods have been created by combining various mathematical theories, and have been widely used in various fields such as economic, management, engineering, medicine, etc. Its core ideas and steps are as follows [32]:

(1) Quantification of evaluation indicators, converting natural language into numbers, and ensuring a certain distinction between good and bad, *D* is the evaluation objective and *X* is the evaluation index. The characteristic matrix is defined as Equation (6).

$$D = \begin{bmatrix} x_{11} & \cdots & x_{1j} & \cdots & x_{1n} \\ \vdots & & \vdots & & \vdots \\ x_{i1} & \cdots & x_{ij} & \cdots & x_{in} \\ \vdots & & \vdots & & \vdots \\ x_{m1} & \cdots & x_{mj} & \cdots & x_{mn} \end{bmatrix} = \begin{bmatrix} D_1(x_1) \\ \vdots \\ D_i(x_j) \\ \vdots \\ D_m(x_n) \end{bmatrix} = \left[X_1(x_1), \ldots X_j(x_i), \ldots, X_n(x_m)\right]. \quad (6)$$

(2) Normalize the characteristic matrix, obtain the normalized vector r_{ij} and establish the normalized matrix about the normalized vector. This is defined as Equation (7).

$$r_{ij} = \frac{x_{ij}}{\sqrt{\sum\limits_{i=1}^{m} x_{ij}^2}}, i = 1, 2, \ldots, m, j = 1, 2, \ldots, n. \quad (7)$$

(3) Normalize the value v_{ij} by calculating the weight; weight normalization matrix is defined as Equation (8).

$$v_{ij} = w_j r_{ij}, i = 1, 2, \ldots, m, j = 1, 2, \ldots, n. \quad (8)$$

(4) Determine ideal solution A^+ and anti-ideal solution A^-; in the ideal solution and anti-ideal solution, J_1 is the optimal value of profitability index set expressed on the i index; J_2 is the worst value of the i index of the loss index set. A^+ and A^- are defined as Equations (9) and (10).

$$A^+ = \left(max_i v_{ij} \middle| j \in J_1\right), \left(min_i v_{ij} \middle| j \in J_1\right), \middle| i = 1, 2, \ldots, m = v_1^+, v_2^+, \ldots, v_j^+, \ldots, v_n^+. \quad (9)$$

$$A^- = \left(min_i v_{ij} \middle| j \in J_1\right), \left(max_i v_{ij} \middle| j \in J_1\right), \middle| i = 1, 2, \ldots, m = v_1^-, v_2^-, \ldots, v_j^-, \ldots, v_n^-. \quad (10)$$

(5) Calculate the distance S^+ from the target to the ideal solution A^+ and the distance S^- from the target to the ideal solution A^-. The distances are defined as Equation (11).

$$S^+ = \sqrt{\sum\limits_{j=1}^{n} \left(V_{ij} - v_j^+\right)^2}, S^- = \sqrt{\sum\limits_{j=1}^{n} \left(V_{ij} - v_j^-\right)^2}, i = 1, 2, \ldots, m. \quad (11)$$

(6) Calculate the closeness index of the ideal solution. It is defined as Equation (12).

$$C_i^+ = \frac{S_i^-}{(S_i^+ + S_i^-)}, i = 1, 2, \ldots, m. \quad (12)$$

(7) Ranking according to the size of the ideal pasting progress.

2.3. Distance Measures and Similarity Measures for PFS

Distance measure for PFSs is a term that describes the difference between PFS. Let A and B be PFSs of $X = \{x_1, x_2, \ldots x_n\}$ with three parameters $\mu(x)$, $\nu(x)$ and $\pi(x)$. Here, some distance measures (DM) are presented for PFSs.

The normalized Hamming distance is defined as Equation (13).

$$d_{PFS}(A, B)_{nH} = \frac{1}{2n} \sum_{i=1}^{n} \left(\left| \mu_A{}^2(x_i) - \mu_B{}^2(x_i) \right| + \left| \nu_A{}^2(x_i) - \nu_B{}^2(x_i) \right| + \left| \pi_A{}^2(x_i) - \pi_B{}^2(x_i) \right| \right) \quad (13)$$

The normalized Euclidean distance is defined as Equation (14).

$$d_{PFS}(A, B)_{nE} = \sqrt{\frac{1}{2n} \sum_{i=1}^{n} \left((\mu_A{}^2(x_i) - \mu_B{}^2(x_i))^2 + (\nu_A{}^2(x_i) - \nu_B{}^2(x_i))^2 + (\pi_A{}^2(x_i) - \pi_B{}^2(x_i))^2 \right)} \quad (14)$$

The normalized Hausdorff distance is defined as Equation (15).

$$d_{PFS}(A,B)_{nHd} = \frac{1}{n}\sum_{i=1}^{n} \max\left[\mid \mu_A{}^2(x_i) - \mu_B{}^2(x_i) \mid, \mid \nu_A{}^2(x_i) - \nu_B{}^2(x_i) \mid\right] \tag{15}$$

For convenience, the above is called formula d_1, d_2 and d_3, and d_4 is defined as Equation (16).

$$d_4(A,B) = \frac{1}{n}\sum_{i=1}^{n} \frac{\mid \mu_A{}^2(x_i) - \mu_B{}^2(x_i) \mid + \mid \nu_A{}^2(x_i) - \nu_B{}^2(x_i) \mid}{\mu_A{}^2(x_i) + \mu_B{}^2(x_i) + \nu_A{}^2(x_i) + \nu_B{}^2(x_i)} \tag{16}$$

Like measure distance, similarity distance is also an important parameter between fuzzy sets. Let f be a monotone decreasing function. Then, the similarity measure between PFSs A and B can be defined as Equation (17).

$$s(A,B) = \frac{f(d(A,B)) - f(1)}{f(0) - f(1)} \tag{17}$$

By defining f, different similarity measures are obtained. Here, some simple methods are introduced. When $f(x) = 1 - x$, the similarity measure is defined as Equation (18).

$$s(A,B) = 1 - d(A,B) \tag{18}$$

When $f(x) = 1/(1 + x)$, the similarity measure is defined as Equation (19).

$$s(A,B) = \frac{1 - d(A,B)}{1 + d(A,B)} \tag{19}$$

2.4. PFS–TOPSIS Method for MADM

The Pythagorean fuzzy set has a broader value space than the traditional fuzzy set and can represent uncertain information in more detail. In addition, with better adaptability, the combination with other MADM methods has achieved many successful cases. TOPSIS is a classic evaluation or ranking method. Based on this, this section introduced a TOPSIS method based on the PFS. The detailed procedure is presented in the following:

Step 1: First, determine alternatives, criteria and experts, and also determine the corresponding transformation relationship between natural language and fuzzy numbers.

Step 2: Establish a group decision matrix scored by experts R = $(x_{(k)\ ij})_{l\times m}$, which can be defined as Equation (20).

$$x_{ij}^{(k)} = \left\{\left[\mu_{A_i}(C_j)^1, \nu_{A_i}(C_j)^1, \pi_{A_i}(C_j)^1\right]\cdots\left[\mu_{A_i}(C_j)^k, \nu_{A_i}(C_j)^k, \pi_{A_i}(C_j)^k\right]\cdots\left[\mu_{A_i}(C_j)^n, \nu_{A_i}(C_j)^n, \pi_{A_i}(C_j)^n\right]\right\}. \tag{20}$$

This represents PFS formed by n experts' evaluation of a certain index of a certain scheme. For convenience, $(\mu_{A_i}(C_j)^k, \nu_{A_i}(C_j)^k, \pi_{A_i}(C_j)^k)$ is represented by $(\mu_{ij}^k, \nu_{ij}^k, \pi_{ij}^k)$. Therefore, the group decision matrix is obtained as Equation (21).

$$R = \begin{array}{c} \\ A_1 \\ A_2 \\ \vdots \\ A_l \end{array} \begin{array}{c} \begin{array}{cccc} C_1 & C_2 & \cdots & C_m \end{array} \\ \begin{bmatrix} ((\mu_{11}^1,\nu_{11}^1,\pi_{11}^1)\cdots(\mu_{11}^n,\nu_{11}^n,\pi_{11}^n)) & ((\mu_{12}^1,\nu_{12}^1,\pi_{12}^1)\cdots(\mu_{12}^n,\nu_{12}^n,\pi_{12}^n)) & \cdots & ((\mu_{1m}^1,\nu_{1m}^1,\pi_{1m}^1)\cdots(\mu_{1m}^n,\nu_{1m}^n,\pi_{1m}^n)) \\ ((\mu_{21}^1,\nu_{21}^1,\pi_{21}^1)\cdots(\mu_{21}^n,\nu_{21}^n,\pi_{21}^n)) & ((\mu_{22}^1,\nu_{22}^1,\pi_{22}^1)\cdots(\mu_{22}^n,\nu_{22}^n,\pi_{22}^n)) & \cdots & ((\mu_{2m}^1,\nu_{2m}^1,\pi_{2m}^1)\cdots(\mu_{2m}^n,\nu_{2m}^n,\pi_{2m}^n)) \\ \vdots & \vdots & \ddots & \vdots \\ ((\mu_{l1}^1,\nu_{l1}^1,\pi_{l1}^1)\cdots(\mu_{l1}^n,\nu_{l1}^n,\pi_{l1}^n)) & ((\mu_{l2}^1,\nu_{l2}^1,\pi_{l2}^1)\cdots(\mu_{l2}^n,\nu_{l2}^n,\pi_{l2}^n)) & \cdots & ((\mu_{lm}^1,\nu_{lm}^1,\pi_{lm}^1)\cdots(\mu_{lm}^n,\nu_{lm}^n,\pi_{lm}^n)) \end{bmatrix} \end{array} \tag{21}$$

Step 3: Since the knowledge level, experience and focus of each expert are different, the importance of each expert is different, so the weight σ_k of each expert should be determined according to certain standards. At the same time, for the evaluation of the same indicator of the same scheme, the individual opinions of all experts need to be aggregated into a general evaluation view, i.e., transforming a PFS $x_{(k)_{ij}}$ into a Pythagorean fuzzy number $x_{ij} = (\mu_{A_i}(C_j), \nu_{A_i}(C_j), \pi_{A_i}(C_j))$. For convenience, this is expressed as $x_{ij} = (\mu_{ij}\ \nu_{ij}\ \pi_{ij})$; this transformation process is realized through the Python fuzzy aggregated averaging (PFWA) operator, which can be defined as Equation (22).

$$\begin{aligned} x_{ij} &= PFWA_\sigma\left(x_{ij}^1, x_{ij}^2, \cdots, x_{ij}^k, \cdots, x_{ij}^n\right) \\ &= \sigma_1 x_{ij}^1 \oplus \sigma_2 x_{ij}^2 \oplus \cdots \oplus \sigma_k x_{ij}^k \oplus \cdots \oplus \sigma_n x_{ij}^n \\ &= \left(\sqrt{1-\prod_{k=1}^n\left(1-\left(\mu_{ij}^k\right)^2\right)^{\sigma_k}}, \prod_{k=1}^n\left(\nu_{ij}^k\right)^{\sigma_k}, \right. \\ &\quad \left.\sqrt{\prod_{k=1}^n\left(1-\left(\mu_{ij}^k\right)^2\right)^{\sigma_k}-\left(\prod_{k=1}^n\left(\nu_{ij}^k\right)^{\sigma_k}\right)^2}\right). \end{aligned} \tag{22}$$

At the same time, the aggregation evaluation matrix can be obtained as Equation (23).

$$R_A = \begin{matrix} \\ A_1 \\ A_2 \\ \vdots \\ A_l \end{matrix} \begin{matrix} C_1 & C_2 & \cdots & C_m \\ \left[\begin{matrix} (\mu_{11}, \nu_{11}, \pi_{11}) & (\mu_{12} \cdot \nu_{12}, \pi_{12}) & \cdots & (\mu_{1m}, \nu_{1m}, \pi_{1m}) \\ (\mu_{21}, \nu_{21}, \pi_{21}) & (\mu_{22}, \nu_{22}, \pi_{22}) & \cdots & (\mu_{2m}, \nu_{2m}, \pi_{2m}) \\ \vdots & \vdots & \ddots & \vdots \\ (\mu_{l1}, \nu_{l1}, \pi_{l1}) & (\mu_{l2}, \nu_{l2}, \pi_{l2}) & \cdots & (\mu_{lm}, \nu_{lm}, \pi_{lm}) \end{matrix}\right] \end{matrix} \tag{23}$$

Step 4: All criteria may not have equal importance, so it is necessary to assign weight to indicators. This step is also determined by experts. Let $w_j^k = [\mu_j^k\ \nu_j^k\ \pi_j^k]$ be a PFN, which is used to indicate the evaluation and scoring of the j indicator by the k expert. Different experts' evaluation and scoring of an indicator also need to be aggregated into a Pythagorean fuzzy number $w_j = (\mu_j\ \nu_j\ \pi_j)$. This process is also implemented through the Python fuzzy aggregated averaging (PFWA) operator as in Equation (24).

$$\begin{aligned} w_j &= PFWA_\sigma\left(w_j^1, w_j^2, \cdots, w_j^k, \cdots w_j^n\right) \\ &= \sigma_1 w_j^1 \oplus \sigma_2 w_j^2 \oplus \cdots \oplus \sigma_k w_j^k \oplus \cdots \oplus \sigma_n w_j^n \\ &= \left(\sqrt{1-\prod_{k=1}^n\left(1-\left(\mu_j^k\right)^2\right)^{\sigma_k}}, \prod_{k=1}^n\left(\nu_j^k\right)^{\sigma_k}, \right. \\ &\quad \left.\sqrt{\prod_{k=1}^n(1-(\mu_j^k)^2)^{\sigma_k}-\left(\prod_{k=1}^n(\nu_j^k)^{\sigma_k}\right)^2}\right) \end{aligned} \tag{24}$$

At the same time, construct a weight matrix for all indicators W, $W= (w_1\ w_2 \ldots w_m)$.

Step 5: After the aggregation evaluation matrix and index weight matrix are obtained, the weighted aggregation Python fuzzy decision matrix (PFDM) can be obtained through the multiple operator $R_{WA} = (x_{Wij})_{l\times m}$, where $x_{Wij} = (\mu_{Wij}, \nu_{Wij}, \pi_{Wij})$, the multiplication operator, is as in Equation (25).

$$\begin{aligned} x_{Wij} &= x_{ij} \otimes w_j \\ &= \left(\mu_{ij}\cdot\mu_j, \sqrt{(\nu_{ij})^2+(\nu_j)^2-(\nu_{ij})^2\cdot(\nu_j)^2}, \right. \\ &\quad \left.\sqrt{1-(\mu_{ij}\cdot\mu_j)^2-(\nu_{ij})^2-(\nu_j)^2+(\nu_{ij})^2\cdot(\nu_j)^2}\right). \end{aligned} \tag{25}$$

The weighted aggregated PFDM can be constructed as in Equation (26).

$$R_{wA} = \begin{matrix} \\ A_1 \\ A_2 \\ \vdots \\ A_l \end{matrix} \begin{matrix} C_1 & C_2 & \cdots & C_m \\ \left[\begin{matrix} (\mu_{W11}, \nu_{W11}, \pi_{W11}) & (\mu_{W12}, \nu_{W12}, \pi_{W12}) & \cdots & (\mu_{W1m}, \nu_{W1m}, \pi_{W1m}) \\ (\mu_{W21}, \nu_{W21}, \pi_{W21}) & (\mu_{W22}, \nu_{W22}, \pi_{W22}) & \cdots & (\mu_{W2m}, \nu_{W2m}, \pi_{W2m}) \\ \vdots & \vdots & \ddots & \vdots \\ (\mu_{Wl1}, \nu_{Wl1}, \pi_{Wl1}) & (\mu_{Wl2}, \nu_{Wl2}, \pi_{Wl2}) & \cdots & (\mu_{Wlm}, \nu_{Wlm}, \pi_{Wlm}) \end{matrix}\right] \end{matrix} \tag{26}$$

Step 6: Let J_1 and J_2 be the collection of benefit-type criteria and cost-type criteria. The Pythagorean fuzzy positive ideal solution (PFPIS) A^+ and the Pythagorean fuzzy negative ideal solution (PFNIS) A^- are as in Equations (27)–(30).

$$A^+ = \{\langle C_j, \mu^+_{Wij}, \nu^+_{Wij}\rangle | C_j \in C, j = 1, 2, \cdots, m\}, \tag{27}$$

$$A^- = \{\langle C_j, \mu^-_{Wij}, \nu^-_{Wij}\rangle | C_j \in C, \ j = 1, 2, \cdots, m\}, \tag{28}$$

$$\mu^+_{Wij} = \begin{cases} \max\limits_{1\le i\le l} \mu_{Wij} \ if \ C_j \in J_1 \\ \min\limits_{1\le i\le l} \mu_{Wij} \ if \ C_j \in J_2 \end{cases}, \ \nu^+_{Wij} = \begin{cases} \min\limits_{1\le i\le l} \nu_{Wij} \ if \ C_j \in J_1 \\ \max\limits_{1\le i\le l} \nu_{Wij} \ if \ C_j \in J_2 \end{cases}, \tag{29}$$

$$\mu^-_{Wij} = \begin{cases} \min\limits_{1\le i\le l} \mu_{Wij} \ if \ C_j \in J_1 \\ \max\limits_{1\le i\le l} \mu_{Wij} \ if \ C_j \in J_2 \end{cases}, \ \nu^-_{Wij} = \begin{cases} \max\limits_{1\le i\le l} \nu_{Wij} \ if \ C_j \in J_1 \\ \min\limits_{1\le i\le l} \nu_{Wij} \ if \ C_j \in J_2 \end{cases} \tag{30}$$

Step 7: After obtaining PFPIS and PFNIS, the next step is to calculate the distance between each scheme and the optimal solution D (A_i, A^+) and the worst solution D (A_i, A^-). The normalized hamming distance formula is used. Then, the proximity between the alternatives and PFPIS is obtained and the calculation formula is as in Equation (31).

$$C(A_i) = \frac{D(A_i, A^-)}{D(A_i, A^+) + D(A_i, A^+)} \tag{31}$$

Step 8: According to the calculated closeness, rank each alternative from high to low and select the best one.

3. Results and Application

3.1. Background of Suichang Gold Mine

Suichang Gold Mine, the largest state-owned gold mining enterprise in Zhejiang Province (see Figure 4), the backbone of national gold system production and the first member of the Shanghai Gold Exchange, won the honorary titles of "National Green Mine", "National Excellent Mining Enterprise for Saving and Comprehensive Utilization of Mine Resources" and "National 4A Tourist Attraction" (see Figure 5). The mining rights area of Suichang Gold Mine is 2.3729 km^2 and the design production scale is 91,800 t/a. There are two gold and silver ore bodies in the main mining area, which are distributed in layers and veins, with an obvious branching compound phenomenon. The ore veins are 27~190 m long, with occurrence elevation of 125~317 m, dip angle of 35~85°, average thickness of 1~4 m and average grade of Au 15 g/t and Ag 400 g/t. The surrounding rock of the roof of the orebody in the middle section is relatively stable, while the roof of the orebody in the west section is controlled by the compressive torsional fracture, the surrounding rock is relatively broken and the joints are developed, often resulting in the collapse of the surrounding rock of the roof in the goaf [33].

At present, the mine faces the following three main technical problems.

(1) The stability of ore and rock in the altered zone is poor and mining technology is difficult. The endowment characteristics of altered rock type gold deposits are complex, the occurrence, grade and dip angle vary greatly and the ore veins intersect and branch seriously.
(2) The shrinkage method is not applicable to ore bodies with complex resource endowments such as large thickness changes and serious branching, the level of mechanized equipment is low, and the labor intensity of workers is high.
(3) The technology of replacing ore pillar with concrete is complex, with low labor efficiency and high cost.

Figure 4. Plan of the Suichang gold mine (accessed on 10 July 2022. https://www.fengyunditu.com/?ver=bd-wx-1604).

Figure 5. Construction of green mine at SuiChang gold mine: (a) exterior view of SuiChang National Mine Park; (b) the gold grottoes of the Dang Dynasty; (c) the gold grottoes of the Song Dynasty; (d) mining disaster size from the Ming Dynasty [29]. (cc by-sa 4.0).

3.2. Primary Selection of Mining Method

For the further development of the enterprise, Suichang Gold Mine decided to upgrade and reconstruct the mine, optimize mining methods, improve supporting facilities and working environment, and increase production capacity. After the preliminary analysis of technical conditions, investigation of engineering rock mechanics and investigation of mineability, four mining methods were preliminarily selected four mining methods. These are the mechanized upward horizontal layering method (MUH), general upward horizontal layering method (GUH), upward horizontal approach filling method (UHA) and shrinkage filling method (SFM).

MUH adopts trackless mechanized equipment such as drilling jumbos and scrapers for production, which can realize strong mining, strong extraction and strong filling, with large production capacity (see Figure 6).

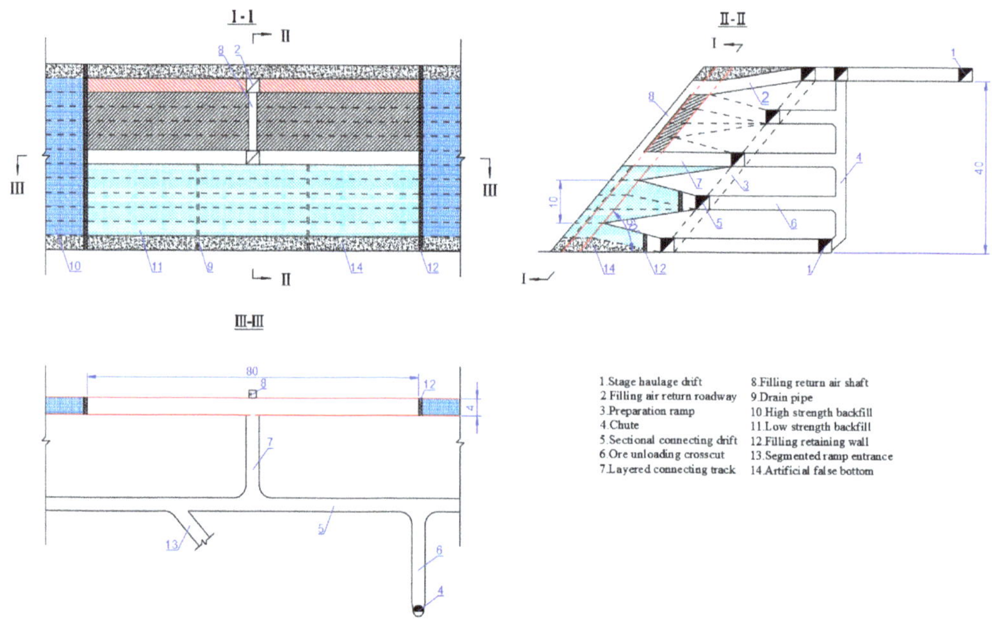

Figure 6. Mining method diagram of MUH.

GUH faces problems of small production capacity, low mechanization, high labor intensity, large amount of preparation work, large amount of reserved space pillar and bottom pillar and low ore drawing efficiency of its two electric harrows (see Figure 7).

UHA adopts trackless mechanized equipment such as the drilling jumbo and scraper, with large production capacity (see Figure 8).

SFM has low drilling efficiency, large amount of reserved space pillar and bottom pillar, difficulty of recovery, low recovery rate and large resource loss (see Figure 9).

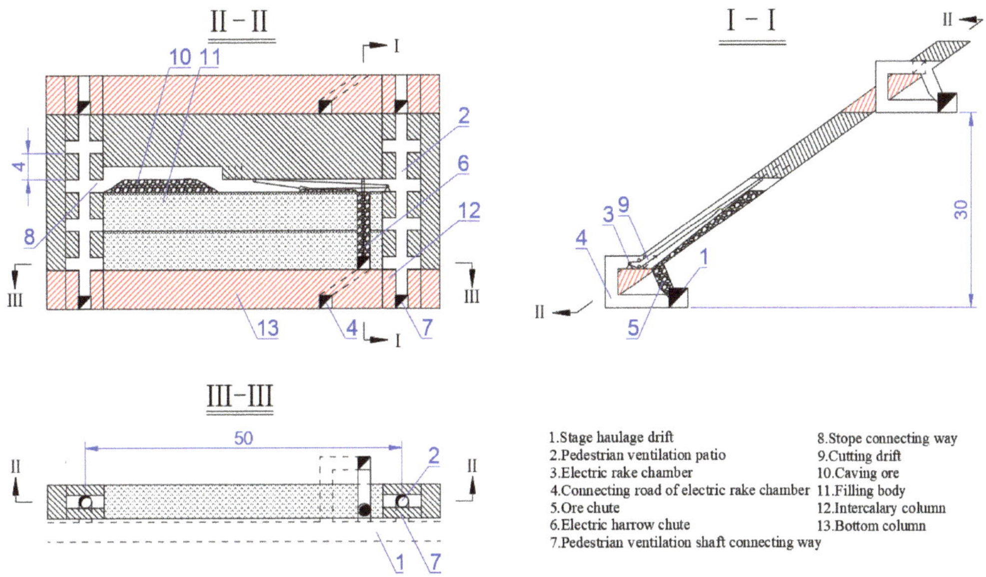

Figure 7. Mining method diagram of GUH.

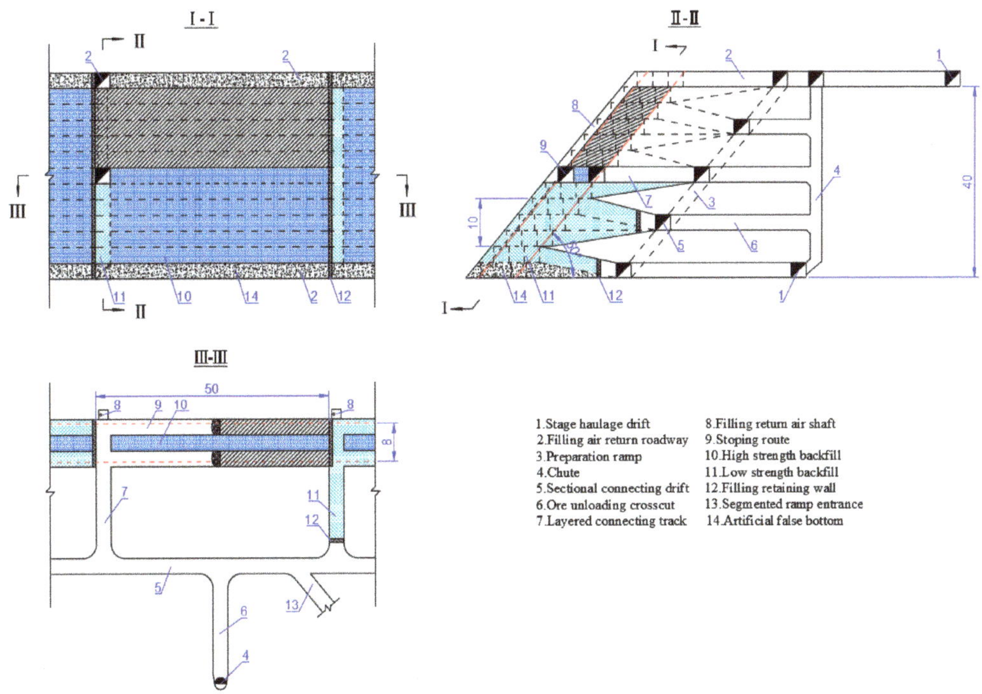

Figure 8. Mining method diagram of UHA.

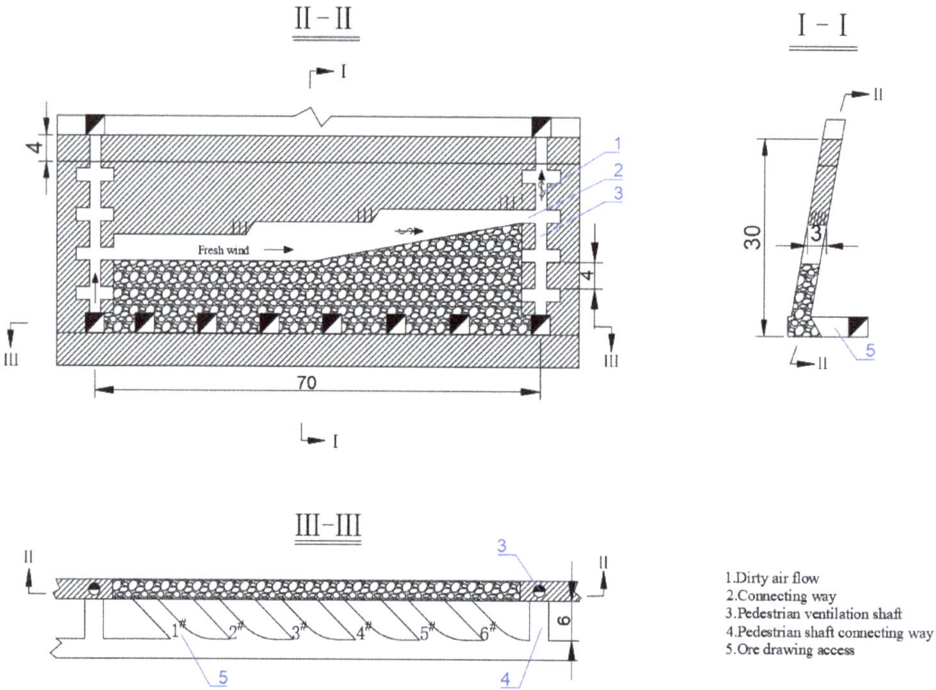

Figure 9. Mining method diagram of SFM.

3.3. Mining Method Optimization

The specific steps of the PFS–TOPSIS model for mining method optimization are as follows:

Step 1: In the optimization of mining methods, we selected three parties as scoring experts, namely the designer (E_1), the mining enterprise (E_2) and the operator (E_3). The four alternatives are MUHSM (A_1), GUHLM (A_2), UHAFM (A_3) and SFM (A_4). The indexes considered are ore recovery rate (C_1), stope production capacity (C_2), flexibility and adaptability (C_3), stope safety conditions (C_4), ore dilution rate (C_5), mining and cutting quantities (C_6), construction organization and labor intensity (C_7) and comprehensive total cost (C_8). Obviously, C_1–C_4 belongs to benefit index (J_1), and C_5–C_8 belongs to cost index (J_2). Next, the corresponding relationship between natural evaluation language and fuzzy number is defined. Table 1 defines the conversion criteria between the relative importance of indicators and PFN and Table 2 defines the conversion criteria between the relative superiority of the scheme and PFN.

Step 2: Three party experts will evaluate and score the superiority of the four schemes under each indicator, as shown in Table 3.

Table 1. The conversion criteria for the relative importance of indicators and PFN.

Linguistic Variables	PFNs
Very important (VI)	(0.90, 0.20, 0.39)
Important (I)	(0.75, 0.30, 0.59)
Medium (M)	(0.60, 0.50, 0.62)
Unimportant (U)	(0.45, 0.70, 0.55)
Very unimportant (VU)	(0.20, 0.90, 0.39)

Table 2. The conversion criteria for the relative superiority of the scheme and PFN.

Linguistic Variables	PFNs
Perfect (VI)	(1.00, 0.00, 0.00)
Very very good (VVG)	(0.90, 0.20, 0.39)
Very good (VG)	(0.80, 0.30, 0.52)
Good (G)	(0.70, 0.35, 0.62)
Medium (M)	(0.60, 0.50, 0.62)
Medium bad (MB)	(0.50, 0.60, 0.62)
Bad (B)	(0.40, 0.70, 0.59)
Very bad (VB)	(0.25, 0.80, 0.55)
Very very bad (VVB)	(0.10, 0.90, 0.42)

Table 3. Superiority evaluation result and fuzzy number.

Criteria	Alternatives	Expert		
		E_1	E_2	E_3
C_1	A_1	VG (0.80, 0.30, 0.52)	VG (0.80, 0.30, 0.52)	VVG (0.90, 0.20, 0.39)
	A_2	G (0.70, 0.35, 0.62)	M (0.60, 0.50, 0.62)	G (0.70, 0.35, 0.62)
	A_3	VVG (0.90, 0.20, 0.39)	VG (0.80, 0.30, 0.52)	VG (0.80, 0.30, 0.52)
	A_4	M (0.60, 0.50, 0.62)	G (0.70, 0.35, 0.62)	M (0.60, 0.50, 0.62)
C_2	A_1	VVG (0.90, 0.20, 0.39)	VG (0.80, 0.30, 0.52)	VG (0.80, 0.30, 0.52)
	A_2	M (0.60, 0.50, 0.62)	G (0.70, 0.35, 0.62)	M (0.60, 0.50, 0.62)
	A_3	G (0.70, 0.35, 0.62)	M (0.60, 0.50, 0.62)	M (0.60, 0.50, 0.62)
	A_4	VG (0.80, 0.30, 0.52)	G (0.70, 0.35, 0.62)	VG (0.80, 0.30, 0.52)
C_3	A_1	G (0.70, 0.35, 0.62)	G (0.70, 0.35, 0.62)	VG (0.80, 0.30, 0.52)
	A_2	G (0.70, 0.35, 0.62)	M (0.60, 0.50, 0.62)	G (0.70, 0.35, 0.62)
	A_3	VG (0.80, 0.30, 0.52)	VG (0.80, 0.30, 0.52)	G (0.70, 0.35, 0.62)
	A_4	M (0.60, 0.50, 0.62)	M (0.60, 0.50, 0.62)	M (0.60, 0.50, 0.62)
C_4	A_1	VG (0.80, 0.30, 0.52)	G (0.70, 0.35, 0.62)	VG (0.80, 0.30, 0.52)
	A_2	VG (0.80, 0.30, 0.52)	G (0.70, 0.35, 0.62)	G (0.70, 0.35, 0.62)
	A_3	G (0.70, 0.35, 0.62)	M (0.60, 0.50, 0.62)	M (0.60, 0.50, 0.62)
	A_4	G (0.70, 0.35, 0.62)	M (0.60, 0.50, 0.62)	M (0.60, 0.50, 0.62)
C_5	A_1	VB (0.25, 0.80, 0.55)	VB (0.25, 0.80, 0.55)	VB (0.25, 0.80, 0.55)
	A_2	M (0.60, 0.50, 0.62)	MB (0.50, 0.60, 0.62)	M (0.60, 0.50, 0.62)
	A_3	VB (0.25, 0.80, 0.55)	B (0.40, 0.70, 0.59)	VB (0.25, 0.80, 0.55)
	A_4	B (0.40, 0.70, 0.59)	VB (0.25, 0.80, 0.55)	B (0.40, 0.70, 0.59)
C_6	A_1	B (0.40, 0.70, 0.59)	MB (0.50, 0.60, 0.62)	MB (0.50, 0.60, 0.62)
	A_2	B (0.40, 0.70, 0.59)	B (0.40, 0.70, 0.59)	B (0.40, 0.70, 0.59)
	A_3	MB (0.50, 0.60, 0.62)	M (0.60, 0.50, 0.62)	MB (0.50, 0.60, 0.62)
	A_4	B (0.40, 0.70, 0.59)	VB (0.25, 0.80, 0.55)	B (0.40, 0.70, 0.59)
C_7	A_1	B (0.40, 0.70, 0.59)	B (0.40, 0.70, 0.59)	VB (0.25, 0.80, 0.55)
	A_2	M (0.60, 0.50, 0.62)	G (0.70, 0.35, 0.62)	G (0.70, 0.35, 0.62)
	A_3	MB (0.50, 0.60, 0.62)	MB (0.50, 0.60, 0.62)	M (0.60, 0.50, 0.62)
	A_4	M (0.60, 0.50, 0.62)	M (0.60, 0.50, 0.62)	G (0.70, 0.35, 0.62)
C_8	A_1	MB (0.50, 0.60, 0.62)	M (0.60, 0.50, 0.62)	MB (0.50, 0.60, 0.62)
	A_2	B (0.40, 0.70, 0.59)	B (0.40, 0.70, 0.59)	MB (0.50, 0.60, 0.62)
	A_3	M (0.60, 0.50, 0.62)	G (0.70, 0.35, 0.62)	M (0.60, 0.50, 0.62)
	A_4	B (0.40, 0.70, 0.59)	MB (0.50, 0.60, 0.62)	MB (0.50, 0.60, 0.62)

Step 3: Determine the importance of all experts, that is, assign a certain weight. Muhammad Akram proposed a method to determine the weight according to the three elements of PFN and established a corresponding relationship between natural language variables and weight. For convenience, this section directly refers to $\sigma_1 = 0.3252$, $\sigma_2 = 0.3754$ and $\sigma_3 = 0.2994$. The weighted aggregation of experts' scores is conducted through PFWA and the aggregation evaluation matrix is obtained, as shown in Table 4.

Table 4. Aggregate evaluation matrix.

	A_1	A_2	A_3	A_4
C_1	(0.838, 0.266, 0.476)	(0.667, 0.400, 0.629)	(0.841, 0.263, 0.473)	(0.642, 0.437, 0.630)
C_2	(0.841, 0.263, 0.473)	(0.642, 0.437, 0.630)	(0.637, 0.445, 0.629)	(0.768, 0.318, 0.556)
C_3	(0.735, 0.334, 0.590)	(0.667, 0.400, 0.629)	(0.775, 0.314, 0.548)	(0.600, 0.500, 0.624)
C_4	(0.768, 0.318, 0.556)	(0.738, 0.333, 0.587)	(0.637, 0.445, 0.629)	(0.637, 0.445, 0.629)
C_5	(0.250, 0.800, 0.545)	(0.566, 0.535, 0.627)	(0.317, 0.761, 0.566)	(0.353, 0.736, 0.578)
C_6	(0.471, 0.631, 0.616)	(0.400, 0.700, 0.591)	(0.542, 0.560, 0.627)	(0.353, 0.736, 0.578)
C_7	(0.363, 0.729, 0.580)	(0.672, 0.393, 0.628)	(0.534, 0.568, 0.626)	(0.634, 0.449, 0.630)
C_8	(0.542, 0.560, 0.627)	(0.434, 0.668, 0.604)	(0.642, 0.437, 0.630)	(0.471, 0.631, 0.616)

Step 4: Each expert will evaluate the superiority of various indicators, as shown in Table 5 for the indicator evaluation. The results will be weighted and aggregated into the indicator weight matrix as Equation (32).

$$W = \begin{bmatrix} (0.8695,\ 0.2258,\ 0.4393) \\ (0.8010,\ 0.3002,\ 0.5180) \\ (0.7140,\ 0.3496,\ 0.6066) \\ (0.9000,\ 0.2000,\ 0.3873) \\ (0.7709,\ 0.3545,\ 0.5292) \\ (0.8010,\ 0.3002,\ 0.5180) \\ (0.7838,\ 0.3137,\ 0.5360) \\ (0.7709,\ 0.3545,\ 0.5292) \end{bmatrix}^T \tag{32}$$

Table 5. Index evaluation.

Criteria	Experts		
	E_1	E_2	E_3
C_1	(0.90, 0.20, 0.39)	(0.90, 0.20, 0.39)	(0.75, 0.30, 0.59)
C_2	(0.75, 0.30, 0.59)	(0.90, 0.20, 0.39)	(0.60, 0.50, 0.62)
C_3	(0.75, 0.30, 0.59)	(0.75, 0.30, 0.59)	(0.60, 0.50, 0.62)
C_4	(0.90, 0.20, 0.39)	(0.90, 0.20, 0.39)	(0.90, 0.20, 0.39)
C_5	(0.60, 0.50, 0.62)	(0.90, 0.20, 0.39)	(0.60, 0.50, 0.62)
C_6	(0.75, 0.30, 0.59)	(0.90, 0.20, 0.39)	(0.60, 0.50, 0.62)
C_7	(0.60, 0.50, 0.62)	(0.75, 0.30, 0.59)	(0.90, 0.20, 0.39)
C_8	(0.60, 0.50, 0.62)	(0.90, 0.20, 0.39)	(0.60, 0.50, 0.62)

Step 5: After the aggregation evaluation matrix and index weight matrix are obtained, the weighted aggregation evaluation matrix can be obtained through the multiplication operator, as shown in Table 6.

Table 6. Weighted aggregate evaluation matrix.

	A_1	A_2	A_3	A_4
C_1	(0.729, 0.344, 0.592)	(0.580, 0.450, 0.679)	(0.731, 0.342, 0.590)	(0.558, 0.482, 0.676)
C_2	(0.674, 0.391, 0.627)	(0.514, 0.514, 0.687)	(0.510, 0.520, 0.685)	(0.615, 0.427, 0.663)
C_3	(0.525, 0.469, 0.710)	(0.476, 0.513, 0.714)	(0.553, 0.457, 0.697)	(0.428, 0.585, 0.689)
C_4	(0.691, 0.370, 0.621)	(0.664, 0.383, 0.642)	(0.573, 0.480, 0.664)	(0.573, 0.480, 0.664)
C_5	(0.193, 0.828, 0.526)	(0.436, 0.613, 0.659)	(0.244, 0.795, 0.555)	(0.272, 0.774, 0.572)
C_6	(0.377, 0.672, 0.637)	(0.320, 0.732, 0.601)	(0.434, 0.613, 0.660)	(0.283, 0.764, 0.580)
C_7	(0.285, 0.760, 0.584)	(0.527, 0.488, 0.696)	(0.419, 0.624, 0.660)	(0.497, 0.529, 0.688)
C_8	(0.418, 0.632, 0.653)	(0.335, 0.718, 0.610)	(0.495, 0.541, 0.680)	(0.363, 0.688, 0.628)

Step 6: The Pythagorean fuzzy positive ideal solution (PFPIS) A^+ and the Pythagorean fuzzy negative ideal solution (PFNIS) A^- are given as in Equations (33) and (34).

$$\begin{aligned} A^+ = \ & \{(0.731, 0.342, 0.590), (0.674, 0.391, 0.627), (0.553, 0.457, 0.697), \\ & (0.691, 0.370, 0.621), (0.193, 0.828, 0.526), (0.283, 0.764, 0.580), \\ & (0.285, 0.760, 0.584), (0.335, 0.718, 0.610)\} \end{aligned} \quad (33)$$

$$\begin{aligned} A^- = \ & \{(0.558, 0.482, 0.676), (0.510, 0.520, 0.685), (0.428, 0.585, 0.689), \\ & (0.573, 0.480, 0.664), (0.436, 0.613, 0.659), (0.434, 0.613, 0.660), \\ & (0.527, 0.488, 0.696), (0.495, 0.541, 0.680)\} \end{aligned} \quad (34)$$

Step 7: Calculate the distance between each scheme and the best solution and the worst solution and rank the schemes by calculating the closeness index to obtain the best scheme, as shown in Table 7.

Table 7. Optimal scheme ranking.

Alternatives	$D(A_i, A^+)$	$D(A_i, A^-)$	$C(A_i)$	Ranks
A_1	0.0352	0.1897	0.8436	1
A_2	0.1501	0.0763	0.3370	4
A_3	0.1270	0.0956	0.4296	3
A_4	0.1261	0.0964	0.4334	2

It can be seen from the above table that A_1 is the best choice, that is, the MUH is the best scheme along with mechanized mining (see Figure 10).

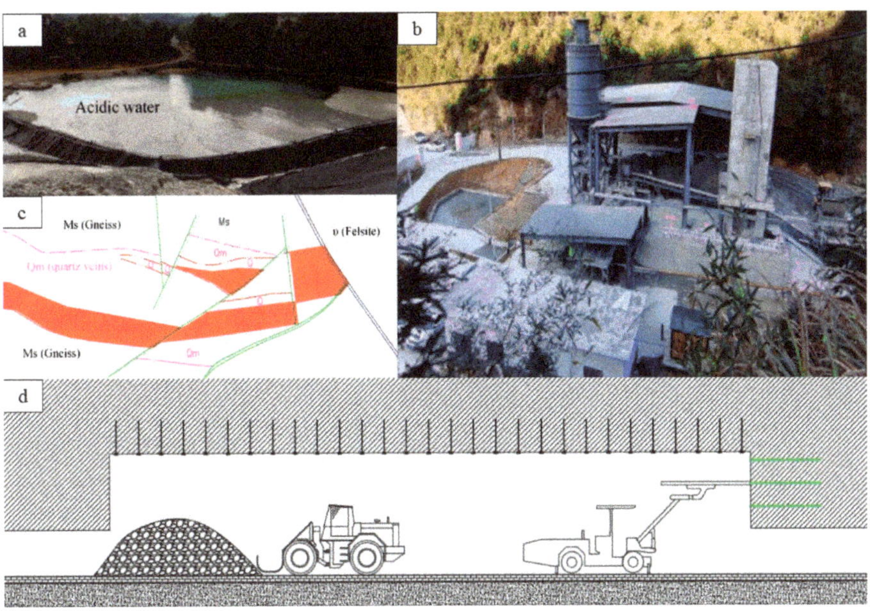

Figure 10. Mechanized mining in the Suichang gold mine: (**a**) filling material; (**b**) filling station; (**c**) resource survey; (**d**) mechanized mining [34]. (cc by-sa 4.0).

4. Discussion

The selection of mining method is very important and complex. In this paper, through a TOPSIS method based on PFS, the MUH is selected as the final scheme among the four mining methods suitable for Suichang Gold Mine. By combining the advantages of PFS which can fully represent fuzzy information with the advantages of TOPSIS ranking science, an ideal result is achieved.

However, due to the importance and particularity of mining method decision making, the above models and calculations cannot fully guarantee the scientific accuracy of the results. In fact, as TOPSIS methods that rank according to the proximity of good and bad solutions, the core influencing factors are the distance between the scheme and the positive and negative ideal solutions, as well as the ranking method. As mentioned above, there are many methods to measure the distance between PFS and each method has its own advantages and disadvantages and the most suitable application [35–38]. In terms of sorting, Hadi-Vencheh [39] believes that the traditional closeness index rank may not be able to produce an optimal alternative, being close to the PIS and far from the NIS. Consequently, they introduced the revised closeness index as Equation (35).

$$RC(A_i) = \frac{D(A_i, A^-)}{D_{\max}(A_i, A^-)} - \frac{D(A_i, A^+)}{D_{\min}(A_i, A^+)} \tag{35}$$

Mahanta [31] adopted a method of ranking by similarity in their article. It is defined in Equations (29) and (30).

$$S_r = \frac{S(D^i_+)}{S(D^i_+) + S(D^i_-)} \tag{36}$$

$$D^i_+ = D(A_i, A^+), D^i_- = D(A_i, A^-), S(D) = (1 - D)/(1 + D) \tag{37}$$

In view of this, the stability of the above results is analyzed through the four distance measures and three ranking methods mentioned above (see Figures 11–13), as shown in Tables 8 and 9, and the data in Figure 12 have been normalized.

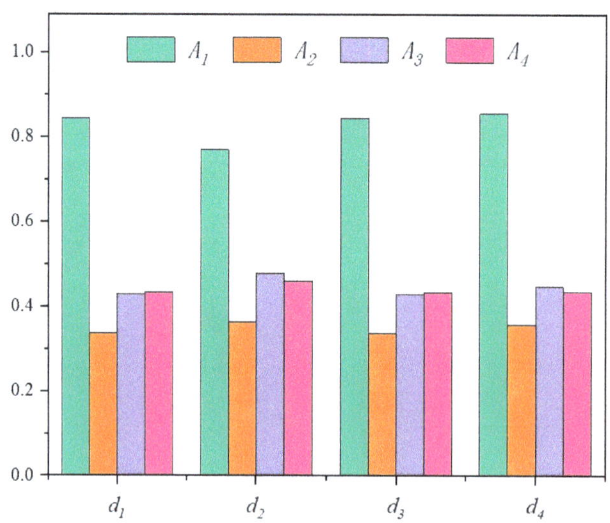

Figure 11. Closeness index rank based on different distance measures.

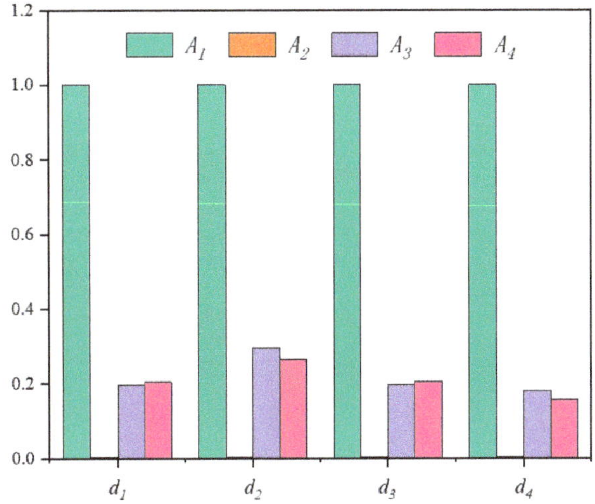

Figure 12. Revised closeness index rank based on different distance measures.

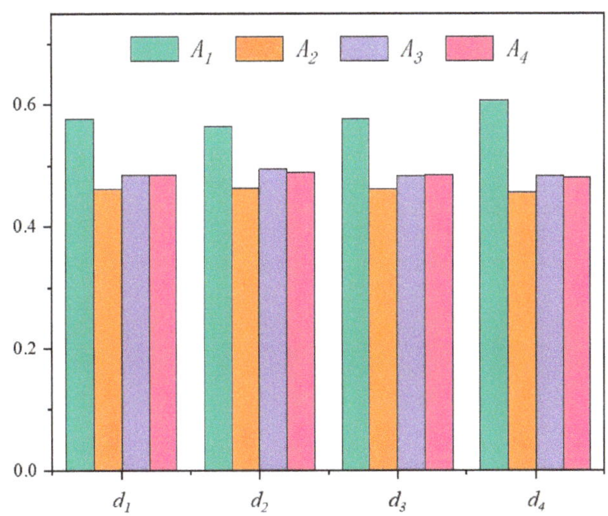

Figure 13. Relative similarity values rank based on different distance measures.

Table 8. Results of distance measures.

		A_1	A_2	A_3	A_4
d_1	$D(A,A^+)$	0.03517	0.15011	0.12696	0.12611
	$D(A,A^-)$	0.18968	0.07630	0.09560	0.09645
d_2	$D(A,A^+)$	0.05468	0.16732	0.13444	0.13656
	$D(A,A^-)$	0.18260	0.09535	0.12303	0.11595
d_3	$D(A,A^+)$	0.03516	0.15013	0.12699	0.12600
	$D(A,A^-)$	0.18969	0.07632	0.09548	0.09646
d_4	$D(A,A^+)$	0.04340	0.19304	0.16600	0.16947
	$D(A,A^-)$	0.25637	0.10775	0.13392	0.13089

Table 9. Results of ranking methods.

		A_1	A_2	A_3	A_4	Rank
$C(A_i)$	d_1	0.84359	0.33702	0.42955	0.43336	$A_1 > A_4 > A_3 > A_2$
	d_2	0.76956	0.36300	0.47783	0.45919	$A_1 > A_3 > A_4 > A_2$
	d_3	0.84361	0.33703	0.42920	0.43361	$A_1 > A_4 > A_3 > A_2$
	d_4	0.85523	0.35822	0.44652	0.43577	$A_1 > A_3 > A_4 > A_2$
$RC(A_i)$	d_1	0	−3.8661	−3.1061	−3.0775	$A_1 > A_4 > A_3 > A_2$
	d_2	0	−2.5379	−1.7851	−1.8625	$A_1 > A_3 > A_4 > A_2$
	d_3	0	−3.8669	−3.1078	−3.0747	$A_1 > A_4 > A_3 > A_2$
	d_4	0	−4.0280	−3.3029	−3.3947	$A_1 > A_3 > A_4 > A_2$
$S_r(A_i)$	d_1	0.57777	0.46267	0.48413	0.48499	$A_1 > A_4 > A_3 > A_2$
	d_2	0.56460	0.46343	0.49420	0.48953	$A_1 > A_3 > A_4 > A_2$
	d_3	0.57778	0.46267	0.48406	0.48505	$A_1 > A_4 > A_3 > A_2$
	d_4	0.60768	0.45645	0.48360	0.48027	$A_1 > A_3 > A_4 > A_2$

It can be seen from the analysis results that the best scheme obtained by changing the distance measures and ranking methods is still the MUH, so it can be considered that the results are accurate. In fact, in the actual production of Suichang Gold Mine, the application of this method has also achieved the ideal results of safety, efficiency and environmental protection.

Whether it can be considered that the model is universal and can be generalized in mining method optimization, the answer is obviously unknown. As is known, the factors that need to be considered in the optimization of mining methods are not completely fuzzy information, e.g., the recovery rate, cut ratio and other factors have certain empirical values. Therefore, the perfect solution is to build a corresponding transformation relationship between the exact value and the PFS and to build a mining method optimization model that combines fuzzy information with accurate data.

5. Conclusions

(1) Through the PFS–TOPSIS method, based on the selection of technical and economic mining methods, a comprehensive evaluation system with multiple factors and indicators was constructed and an accurate closeness index was obtained to optimize mining methods. This overcomes the uncertainty and unpredictability of the traditional optimization system and provides a reference for the mining of the difficult-to-mine complicated orebody.

(2) Taking Suichang Gold Mine as an example, according to the PFSTOPSIS method, a weighted aggregation evaluation matrix was constructed, and the closeness index of the four mining methods were calculated to be 0.8436, 0.3370, 0.4296 and 0.4334, respectively. The MUH has the highest closeness index, so this method was the best scheme.

(3) There were many ranging methods and ranking methods for PFS and only one method could not ensure the accuracy and scientific nature of the results. This paper mainly used the first ranging method, which was ranked by the traditional closeness index. Finally, it discussed the three methods of traditional closeness index, revised closeness index and relative similarity values for comprehensive ranking under the four distance measures. When using the first distance measure, the revised closeness index of the four mining methods was 0, −3.8661, −3.1061 and −3.0775, and the relative similarity values were 0.5777, 0.46267, 0.48413 and 0.48499. It was concluded that MUH was the best scheme, which not only verified the accuracy of the results, but also showed that PFS was applicable to the selection of mining methods.

Of course, there are still limitations to the mathematical approach used in this study, i.e., indicators with definite values need to be evaluated first in natural language and then converted to fuzzy sets. Even so, the method still performs well in problems with a large amount of fuzzy information, such as the selection of mining methods. Therefore, we advocate the application of such methods to more mines and encourage more and more researchers to test and optimize them in practice.

Author Contributions: Conceptualization, S.L.; resources, S.L.; writing—original draft preparation, Q.H.; writing—review and editing, Q.H., B.H. and J.P.; visualization, Q.H., B.H. and J.P.; supervision, J.C., J.Y. and. X.Z.; project administration, X.W. and H.Y.; funding acquisition, S.L. All authors have read and agreed to the published version of the manuscript.

Funding: Authors thank the financial support from the Natural Science Foundation of Hunan Province (Grant No. 2021JJ40745) and the National Natural Science Foundation of China (Grant No. 51804337).

Institutional Review Board Statement: Not applicable.

Informed Consent Statement: Not applicable.

Data Availability Statement: Not applicable.

Conflicts of Interest: The authors declare no conflict of interest.

References

1. Ma, D.; Duan, H.; Zhang, J.; Liu, X.; Li, Z. Numerical simulation of water-silt inrush hazard of fault rock: A three-phase flow model. *Rock Mech. Rock Eng.* **2022**, *55*, 5163–5182. [CrossRef]
2. Ma, D.; Duan, H.; Zhang, J. Solid grain migration on hydraulic properties of fault rocks in underground mining tunnel: Radial seepage experiments and verification of permeability prediction. *Tunn. Undergr. Space Technol.* **2022**, *126*, 104525. [CrossRef]
3. Ma, D.; Duan, H.; Zhang, J.; Bai, H. A state-of-the-art review on rock seepage mechanism of water inrush disaster in coal mines. *Int. J. Coal Sci. Technol.* **2022**, *9*, 50. [CrossRef]
4. Mijalkovski, S.; Peltechki, D.; Despodov, Z.; Mirakovski, D.; Adjiski, V.; Doneva, N. Methodology for underground mining method selection. *Min. Sci.* **2021**, *28*, 201–216. [CrossRef]
5. Ojha, M.; Singh, K.P.; Chakraborty, P.; Verma, S. A review of multi-objective optimisation and decision making using evolutionary algorithms. *Int. J. Bio-Inspir. Comput.* **2019**, *14*, 69–84. [CrossRef]
6. Saaty, T.L. What Is the Analytic Hierarchy Process? In *Mathematical Models for Decision Support*; Springer: Berlin/Heidelberg, Germany, 1988; pp. 109–121.
7. Jiskani, I.M.; Han, S.; Rehman, A.U.; Shahani, N.M. An Integrated Entropy Weight and Grey Clustering Method-Based Evaluation to Improve Safety in Mines. *Min. Metall. Explor.* **2021**, *38*, 1773–1787. [CrossRef]
8. Krishnan, A.R.; Kasim, M.M.; Hamid, R.; Ghazali, M.F. A Modified CRITIC Method to Estimate the Objective Weights of Decision Criteria. *Symmetry* **2021**, *13*, 973. [CrossRef]
9. Yang, W.G. Ingenious Solution for the Rank Reversal Problem of TOPSIS Method. *Math. Probl. Eng.* **2020**, *2020*, 9676518. [CrossRef]
10. Luo, D.; Wang, X. The multi-attribute grey target decision method for attribute value within three-parameter interval grey number. *Appl. Math. Model.* **2012**, *36*, 1957–1963. [CrossRef]
11. Omrani, H.; Alizadeh, A.; Naghizadeh, F. Incorporating decision makers' preferences into DEA and common weight DEA models based on the best-worst method (BWM). *Soft Comput.* **2020**, *24*, 3989–4002. [CrossRef]
12. Yang, W.G.; Wu, Y.J. A New Improvement Method to Avoid Rank Reversal in VIKOR. *IEEE Access* **2020**, *8*, 21261–21271. [CrossRef]
13. Zhang, J.; Li, H.; Liu, Y.; Feng, X. The improvement and application of fuzzy comprehensive evaluation method under the hybrid information. In Proceedings of the 25th Chinese Control and Decision Conference (CCDC), Guiyang, China, 25–27 May 2013; p. 20131612-1615.
14. Samanta, S.; Jana, D.K. A multi-item transportation problem with mode of transportation preference by MCDM method in interval type-2 fuzzy environment. *Neural Comput. Appl.* **2019**, *31*, 605–617. [CrossRef]
15. Bera, A.K.; Jana, D.K.; Banerjee, D.; Nandy, T. A two-phase multi-criteria fuzzy group decision making approach for supplierevaluation and order allocation considering multi-objective, multi-product and multi-period. *Ann. Data Sci.* **2021**, *8*, 577–601. [CrossRef]
16. Karimnia, H.; Bagloo, H. Optimum mining method selection using fuzzy analytical hierarchy process–Qapiliq salt mine, Iran. *Int. J. Min. Sci. Technol.* **2015**, *25*, 225–230. [CrossRef]
17. Yavuz, M. The application of the analytic hierarchy process (AHP) and Yager's method in underground mining method selection problem. *Int. J. Min. Reclam. Environ.* **2015**, *29*, 453–475. [CrossRef]
18. Guo, Q.Q.; Yu, H.X.; Dan, Z.Y.; Li, S. Mining Method Optimization of Gently Inclined and Soft Broken Complex Orebody Based on AHP and TOPSIS: Taking Miao-Ling Gold Mine of China as an Example. *Sustainability* **2021**, *13*, 12503. [CrossRef]
19. Iphar, M.; Alpay, S. A mobile application based on multi-criteria decision-making methods for underground mining method selection. *Int. J. Min. Reclam. Environ.* **2019**, *33*, 480–504. [CrossRef]

20. Atanassov, K.; Vassilev, P. Intuitionistic fuzzy sets and other fuzzy sets extensions representable by them. *J. Intel. Fuzzy Syst.* **2020**, *38*, 525–530. [CrossRef]
21. Bajic, S.; Bajic, D.; Gluscevic, B.; Vakanjac, V.R. Application of Fuzzy Analytic Hierarchy Process to Underground Mining Method Selection. *Symmetry* **2020**, *12*, 192. [CrossRef]
22. Memari, A.; Dargi, A.; Jokar, M.; Ahmad, R.; Abdul Rahim, A.R. Sustainable supplier selection: A multi-criteria intuitionistic fuzzy TOPSIS method. *J. Manuf. Syst.* **2019**, *50*, 9–24. [CrossRef]
23. Narayanamoorthy, S.; Geetha, S.; Rakkiyappan, R.; Joo, Y.H. Interval-valued intuitionistic hesitant fuzzy entropy based VIKOR method for industrial robots selection. *Expert Syst. Appl.* **2019**, *121*, 28–37. [CrossRef]
24. Yager, R.R. Pythagorean Membership Grades in Multicriteria Decision Making. *IEEE Trans. Fuzzy Syst.* **2014**, *22*, 958–965. [CrossRef]
25. Peng, X.D.; Selvachandran, G. Pythagorean fuzzy set: State of the art and future directions. *Artif. Intel. Rev.* **2019**, *52*, 1873–1927. [CrossRef]
26. Akram, M.; Dudek, W.A.; Ilyas, F. Group decision-making based on pythagorean fuzzy TOPSIS method. *Int. J. Intel. Syst.* **2019**, *34*, 1455–1475. [CrossRef]
27. Yager, R.R. Pythagorean Fuzzy Subsets. In Proceedings of the Joint World Congress of the International-Fuzzy-Systems-Association (IFSA)/Annual Meeting of the North-American-Fuzzy-Information-Processing-Society (NAFIPS), Edmonton, AB, Canada, 24–28 June 2013; p. 201357-61.
28. Yager, R.R.; Abbasov, A.M. Pythagorean Membership Grades, Complex Numbers, and Decision Making. *Int. J. Intel. Syst.* **2013**, *28*, 436–452. [CrossRef]
29. Zhang, X.L.; Xu, Z.S. Extension of TOPSIS to Multiple Criteria Decision Making with Pythagorean Fuzzy Sets. *Int. J. Intel. Syst.* **2014**, *29*, 1061–1078. [CrossRef]
30. Zhang, X.L. A Novel Approach Based on Similarity Measure for Pythagorean Fuzzy Multiple Criteria Group Decision Making. *Int. J. Intel. Syst.* **2016**, *31*, 593–611. [CrossRef]
31. Hwang, C.L.; Yoon, K. Methods for Multiple Attribute Decision Making. In *Multiple Attribute Decision Making*; Springer: Berlin/Heidelberg, Germany, 1981; pp. 58–191.
32. Wu, Z.; Li, Q.; Kong, D.; Chen, G.; Luo, D. The ANP-Fuzzy-TOPSIS model for the optimization of the scheme of large-section blasting. *Arab. J. Geosci.* **2020**, *13*, 1–9. [CrossRef]
33. *Feasibility Study Report on Safe, Efficient and Low Lean Filling Mining of High Grade Gold and Silver Resources in Soft Rock Strata*; Central South University: Changsha, China, 2021.
34. Yu, H.X.; Li, S.; Yu, L.F.; Wang, X. The Recent Progress China Has Made in Green Mine Construction, Part II: Typical Examples of Green Mines. *Int. J. Environ. Res. Public Health* **2022**, *19*, 8166. [CrossRef]
35. Ejegwa, P.A.; Awolola, J.A. Novel distance measures for Pythagorean fuzzy sets with applications to pattern recognition problems. *Granular Comput.* **2021**, *6*, 181–189. [CrossRef]
36. Mahanta, J.; Panda, S. Distance measure for Pythagorean fuzzy sets with varied applications. *Neural Comput. Appl.* **2021**, *33*, 17161–17171. [CrossRef]
37. Ejegwa, P.A. Distance and similarity measures for Pythagorean fuzzy sets. *Granular Comput.* **2020**, *5*, 225–238. [CrossRef]
38. Hussian, Z.; Yang, M.S. Distance and similarity measures of Pythagorean fuzzy sets based on the Hausdorff metric with application to fuzzy TOPSIS. *Int. J. Intel. Syst.* **2019**, *34*, 2633–2654. [CrossRef]
39. Hadi-Vencheh, A.; Mirjaberi, M. Fuzzy inferior ratio method for multiple attribute decision making problems. *Inform. Sci.* **2014**, *277*, 263–272. [CrossRef]

sustainability

MDPI

Article

Study on Physical and Mechanical Properties of High-Water Material Made by Seawater

Bangwen Lu [1,2], Changwu Liu [3,4,*], Jungang Guo [1,2] and Naiqi Feng [1,2]

1 Zhengzhou Institute of Multipurpose Utilization of Mineral Resources, Chinese Academy of Geological Sciences (CAGS), Zhengzhou 450006, China
2 China National Engineering Research Center for Utilization of Industrial Minerals, Zhengzhou 450006, China
3 College of Water Resource and Hydropower, Sichuan University, Chengdu 610065, China
4 State Key Laboratory of Hydraulics and Mountain River Engineering, Chengdu 610065, China
* Correspondence: liuchangwu@scu.edu.cn; Tel.: +86-0371-68632048

Abstract: In maritime engineering, marine-derived construction materials are seen as an efficient and cost-effective alternative. HWM is a novel inorganic cementitious material characterized by its high water content, rapid setting, and early strengthening. In this study, first, HWM was proposed to be produced from seawater and used in a maritime environment. Two groups of HWM samples with varied w/c ratios were prepared with fresh water and seawater, and their behavior was examined to assess the viability of HWM produced with seawater. The microstructures and chemical compositions were studied using SEM and XRD. Results indicated that as the w/c ratio increased from 3:1 to 6:1, the water content, density, and uniaxial compressive strength of HWM produced from seawater varied from 72.1% to 77.5%; 1.25 to 1.12 g/cm^3, and 1.47 MPa to 0.39 MPa, respectively, which is 2–10% lower, 0.8–2.2% higher, and 13–45% stronger than that from fresh water. The chemical composition of HWM mixed with seawater is predominantly composed of ettringite, C-S-H gel, aluminum ($Al(OH)_3$) glue, M-S-H gel, and $Mg(OH)_2$. SO_4^{2-} and Mg^{2+} in seawater participate in the hydration and hardening of HWM, resulting in an increase in the synthesis of ettringite and M-S-H gel, which makes the skeletal structure of HWM denser, hence increasing its strength. HWM derived from seawater retains excellent physical and mechanical properties. This work reveals the HWM-seawater interaction mechanism, elucidates the promising application prospect of HWM in maritime engineering, and paves the way to investigate its field performance.

Keywords: high-water material; seawater; water-cement ratio; microstructure; water content; strength

Citation: Lu, B.; Liu, C.; Guo, J.; Feng, N. Study on Physical and Mechanical Properties of High-Water Material Made by Seawater. *Sustainability* **2023**, *15*, 3334. https://doi.org/10.3390/su15043334

Academic Editors: Mahdi Hasanipanah, Danial Jahed Armaghani and Jian Zhou

Received: 23 November 2022
Revised: 17 January 2023
Accepted: 5 February 2023
Published: 11 February 2023

1. Introduction

In recent years, with the expansion of ocean exploitation, the construction of maritime infrastructure has grown at an unparalleled rate [1]. Cementing material has been widely used and chosen among manmade construction materials in maritime environments around the world due to its low cost and ease of construction [2]. Cementing material draws a matter of continuing concern due to a high environmental cost. It is reported that cement production is the third-largest producer of CO_2 in the world after transport and energy generation [3]. Sustainability for cementitious material has attracted widespread interest and has become a hot topic of research [4–6]. Recently, recycled cementing materials in cementing materials production have become more and more popular in terms of less consumption of natural materials and many environmental advantages of disposal and reusing of waste materials. The behavior of recycled cementing materials was investigated by a large number of researchers [7–13]. However, the long-term durability of recycled cementing materials, especially under various harsh environments, for example, in maritime environments, was reported to be obviously lower than conventional concrete [14]. Therefore, developing novel and sustainable cementing materials is still a scientific challenge for the sustainability of marine development.

In addition, cementitious materials are reported by previous research [15–17] to be susceptible to a range of physical and chemical breakdown processes in maritime environments. In addition, the use of standard construction materials (freshwater and river sand, etc.) on most islands and reefs in marine engineering is always constrained by time, transportation costs, and difficult geological conditions [17]. Therefore, how to obtain alternative and sustainable construction materials locally and economically is a significant challenge [18,19].

High-water material (HWM), also known as high-water back-filling material or high-water-content and quick-setting material, is a novel cementing material. HWM was first invented by Professor Henghu SUN from China University of Mining and Technology in 1989 and successfully used in the coal mining back-filling engineering practice [20]. HWM has been praised as green and sustainable cementing material with numerous excellent advantages, including high water content, good pumpability, rapid setting, high early strength, recrystallization recovery strength after the early failure of the stone body, a simple material production process, and low cost [21]. The chemical composition and mineral composition of the raw material of HWM were analyzed by Xie and Liu (2014) [22], and the hydrating and hardening mechanism of HWM was revealed by Xia et al. (2018) [23]. The basic physical and mechanical properties of HWM were tested by a large number of scholars, and the uniaxial and triaxial compressive strength and creep properties were measured as well. Xie et al. (2013) investigated the influence of curing time on the properties of HWM and revealed that the strength of HWM stones increased with the curing time [24]. Zhang et al. (2017) discussed the effects of water-cement ratios on the physical and mechanical characteristics of HWM [25]. Zhou et al. (2017) conducted an experimental study to investigate the failure characteristics of HWM under loading and divided the stress-strain curve of HWM, which had initial deformation, elastic, plastic deformation, and disruption four stages [26]. The research above proved that HWM is a promising cementing material with excellent physical and mechanical properties.

Initially, HWM was utilized mostly for back-filling, roadway support, etc., in underground coal mining [27–30]. Recent studies indicate that HWM has a promising application in maritime environments. HWM is lauded for its ability to "convert water into stone" because of its ultrahigh water-cement ratio and water content, which can exceed 10:1 and 95%, respectively, and its strength can exceed 5 MPa [20], which means less cementing materials are consumed, hence less CO_2 emission is produced. Given that water could be acquired locally and HWM slurry could be transported by pumping, HWM construction is practical, quick, and inexpensive in maritime environments [31].

Previous research has examined the behavior of HWM and demonstrated its usefulness in marine environments. Hou et al. (2012) recommended using HWM for the preservation and reinforcement of coral sand islands and reefs, as well as port reinforcement, and they measured the properties of HWM in marine environments [32]. The hydration mechanism, physical and mechanical properties, and port engineering application of HWM in maritime environments were examined [33]. He et al. (2014) discuss the environmental influences on the physical and mechanical properties of HWM cured in seawater [34]. However, there are still a number of scientific obstacles to implementing HWM in marine engineering. The majority of previous studies focused on the characteristics of HWM mixed with freshwater [34–37]. Given the shortage of fresh water in marine environments, HWM derived from marine sources (i.e., seawater) is viewed as an efficient and cost-effective alternative [38,39]. It is required to explore the viability of preparing HWM with seawater and the impact of the seawater environment on HWM's behavior.

In this study, the physical and mechanical characteristics and microstructure, and chemical composition of HWM produced with seawater were studied.

2. Materials and Methods

The raw materials of HWM used in this study consist of two parts: main material (A material, B material) and subsidiary material (A-A material, B-B material). Their chemical compositions are shown in Table 1. The main components of A material are bauxite. B material is composed of lime, gypsum, etc. The main components of A-A material are suspension agent, coagulant, and dispersant, including Na_2CO_3, $BaBiO_3$, etc. B-B material is composed of early strength agents, suspension dispersants, etc., including SiO_2, $CaSO_4$, etc. The proportion of ingredients employed in this study was A:A-A:B:B-B = 1:0.1:1:0.04, which follows the material formulation proposed by Sun and Song (1994) [20].

Table 1. Mineral components of high-water material used in this study.

ID	Materials	Main Compositions
A	Bauxite	$3CaO{\cdot}3Al_2O_3{\cdot}CaSO_4$, $2CaO{\cdot}SiO_2$ etc.
A-A	Additives	Na_2CO_3, $BaBiO_3$, etc.
B	Lime, gypsum, etc.	$CaSO_4$, $CaSO_4{\cdot}2H_2O$ $CaSO_4{\cdot}0.5H_2O$ etc.
B-B	Additives	SiO_2, $CaSO_4$, etc.

Figure 1 shows the preparation procedures for preparing HWM. It can be seen that A and A-A were mixed together first, and then enough water was added and stirred thoroughly to produce A seriflux. At the same time, B and B-B were mixed together, and then enough water was added and stirred thoroughly to produce B seriflux. A and B seriflux have a strong fluidity and keep good mobility as a liquid for more than 24 h. Therefore, A and B seriflux have excellent pumpability and can be transported by pipeline, which is convenient for construction in engineering practice. Then A and B seriflux were mixed together and stirred thoroughly to produce HWM stone. HWM has the characteristics of quick setting and early strength. HWM could be solidified in half an hour after mixing A and B seriflux together. The early strength of HWM stone could be over 2 MPa in 2 h, which is about 20% of the final strength, and could increase to more than 60–90% of final strength after 7 days, according to Sun and Song (1994) [20]. After that, the strength of HWM would increase flat. HWM stone should be cured in water for more than 28 days to get final strength. In addition, it was reported that the crystals of HWM could keep growing for a long term hence the mechanical strength of HWM could be recovered after the early failure of the stone body of HWM.

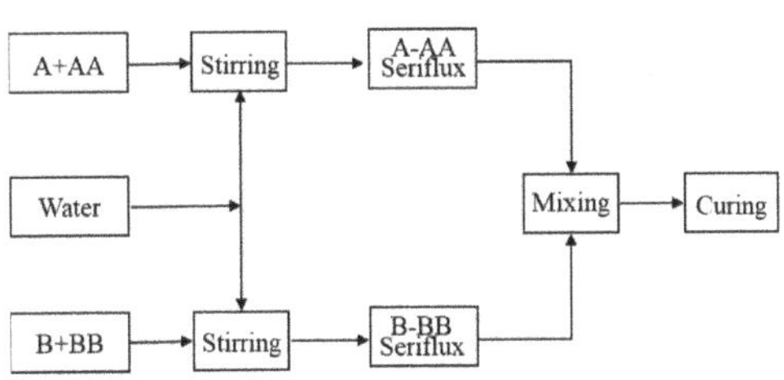

Figure 1. Procedures of preparing for HWM.

Table 2 shows the test plan in this study. The HWM samples were prepared and tested according to the Standard for the Test Method of Mechanical Properties of Ordinary Concrete (GB/T 50081-2002) [40] and Methods for Determination of Physical and Mechanical Properties of Coal and Rocks (GB/T 23561.1-2009) [41]. By varying the amount of water added, standard cylindrical HWM samples with a diameter of 50 mm and a height of 100 mm (50 mm × 100 mm) with different water-cement (w/c) ratios (i.e., w/c = 3:1, 4:1, 5:1, and 6:1) were produced. The HWM samples prepared with and cured in tap water were used as the control test group (Test ID: C). As a comparison, artificial seawater was used to prepare and cure samples of the seawater test group (Test ID: S). The artificial seawater used in this study was manmade in the laboratory by following ASTM D1141-98(2013) [42]. The temperature of the fresh water and seawater was about 18–22 °C.

Table 2. Specimen design and test plan.

Test ID	Test Group	Mixing and Curing Water	w/c	Parameters Investigated
C	Control test group	Tap water	3:1, 4:1, 5:1 and 6:1	Density, water content, strength, microstructure, chemical component etc.
S	Seawater test group	Artificial seawater		

Note: S3-2 means the 2nd HWM sample of the seawater test group with w/c = 3:1.

In accordance with the procedure proposed by He et al. (2014) [34], HWM was prepared and cured. The HWM samples were prepared in a standard cast iron mold (φ50 mm × 100 mm). As shown in Figure 1, the preparation procedures of preparing HWM in this study comprised weighing, mixing with water, combining A and B seriflux, injecting into the mold, and demolding. After demolding, the samples were then placed in tap water or seawater and cured for 28 days. All samples were put in an environment-controlled room at Sichuan University. The room temperature was kept at 26–30 °C, and the air humidity was 40–50%.

The samples were then utilized in the subsequent analysis. After analyzing the microstructure and chemical components with a scanning electron microscope (SEM) and X-rays, the fundamental physical characteristics, such as moisture content and density, were determined. MTS815.03 Electro-hydraulic Servo-controlled Rock Mechanics Testing System was used to test the mechanical properties of samples. Each group sample was subjected to the aforementioned tests three times, with the average value being the final test results. The above experiments were carried out in the State Key Laboratory of Hydraulics and Mountain River Engineering of Sichuan University, China.

3. Results

3.1. Density and Water Content

The density of HWM produced from water and seawater at various w/c ratios is depicted in Figure 2. It can be seen that as the w/c ratio grows, the average mass and bulk density of samples from two groups of HWM samples drop. As the w/c ratio increases from 3:1 to 6:1, the density of HWM produced from seawater decreases from 1.25 to 1.12 g/cm^3. The density steadily drops until it reaches the density of water (1 g/cm^3). However, the density of HWM prepared from seawater with the same w/c ratio is 0.8% to 2.2% more than that of HWM prepared from water.

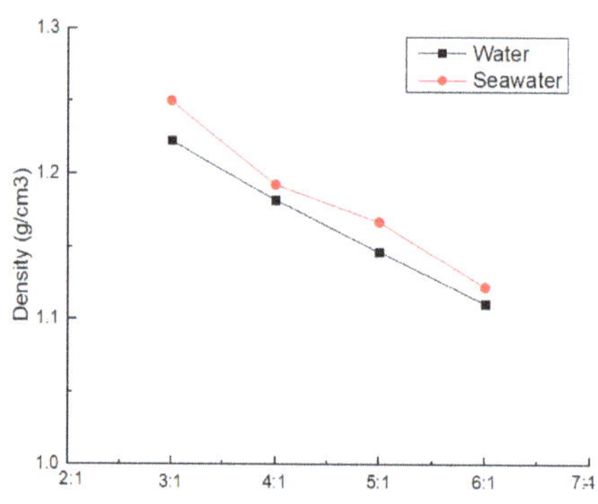

Figure 2. Density of HWM with various w/c ratios.

Figure 3 depicts the water content of HWM produced by mixing freshwater and seawater at various w/c ratios. As the w/c ratio increases, the moisture content of the two groups of HWM samples increases progressively. As the w/c ratio increases from 3:1 to 6:1, the water content of HWM produced from seawater increases from 72.1% to 77.5%. The HWM stone body with the same w/c ratios as the seawater test group had 2–10% less water than the control test group. Previous studies also found that when water-cement ratios increase, the density of HWM decreases while the water content of HWM increases, which is consistent with the findings based on the above test results.

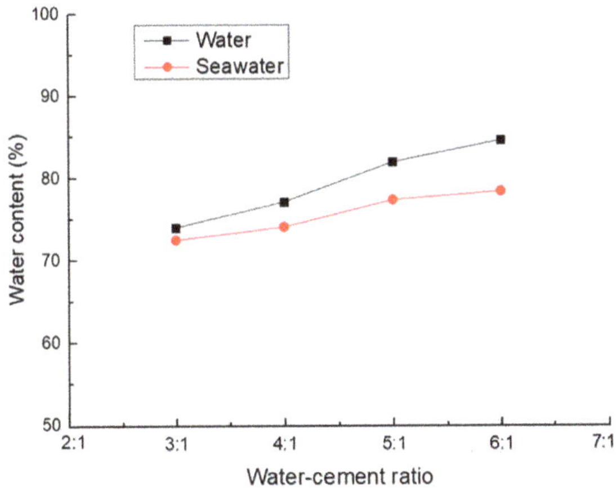

Figure 3. Water content of HWM with various w/c ratios.

3.2. Mechanical Characteristics

The stress-strain curves of S3 and S6 are shown in Figure 4.

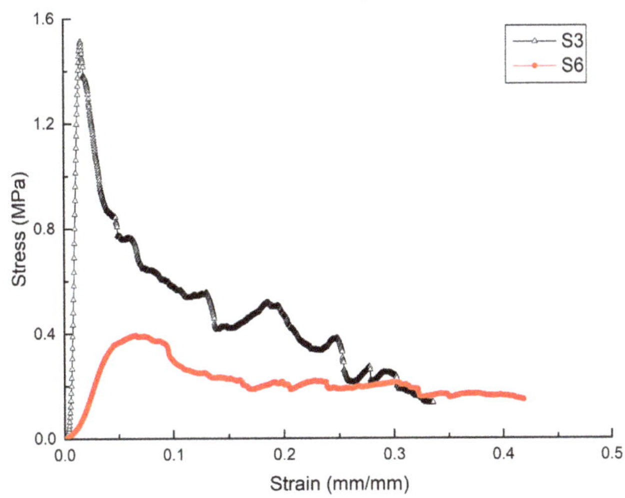

Figure 4. Typical strain-stress curves of HWM made by seawater.

In accordance with the rock mechanics classification standard, the stress-strain curve of a typical HWM can be classified into four stages: pore fracture compaction stage, elastic deformation to fracture development stage, unsteady fracture development stage, and post-peak failure stage. Among them, S3 has the shortest pore crack compaction stage, then enters the elastic deformation stage with a sharply rising stress-strain curve and an elastic modulus of 0.198 GPa. Afterward, S3 quickly reaches its maximum strength (i.e., 1.5 MPa), and the strain is 0.015%. As soon as it enters the post-peak phase, the residual stress is nearly nil.

Sample S6's stress-strain curve demonstrates a substantially longer pore fracture compaction stage and an unstable fracture development stage. In the elastic phase, its stress-strain curve is quite moderate, as the strain increases from 0.02 to 0.04 and the corresponding stress increases from 0.1089 MPa to 0.3276 MPa, while the linear phase elastic modulus is just 0.009 GPa. In the post-peak phase, stress slightly reduces as strain increases. The range of residual stress is 0.08–0.1 MPa, which corresponds to approximately 45–70 percent of the peak stress, while the range of strain is 0.12 to 0.20. It proved that HWM with higher water-cement ratios has better plasticity.

In the uniaxial compression test, Figure 5 depicts the typical failure images of the HWM produced from seawater with w/c = 3:1 (i.e., S3) and 6:1 (i.e., S6). Split failure can be seen to be the failure mode of S3. During loading, first, the stress of S3 sharply spiked while there was no vertical compression observed, after that the sample was split into many pieces in the internal axial direction, after which the stress fell sharply, and the sample could only hold a small axial strain, which is in accordance with the stress-strain curve depicted in Figure 4. At the same time, S6's mode of failure was a ductile failure. Under axial loading, the axial strain continued to grow, and the specimen was compressed vertically and extended laterally. As the axial compression increased, the upper portion of the specimen was initially crushed, and the shattered HWM blocks piled on the upper portion of the specimen and continued to support the load. The sample resembled a "compressed biscuit" in appearance. During this procedure, the material remained somewhat intact, and residual stress remained elevated. Figure 4 depicts the stress-strain curve for the material. A similar phenomenon was also observed in previous studies on the failure mode of HWM [43,44]. It was reported that HWM with higher w/c ratios owns better plasticity; therefore, it could generate a larger axial and horizontal strain under axial loading.

(**a**) w/c = 3:1 (**b**) w/c = 6:1

Figure 5. Compression failure mode of HWM made by seawater with various w/c.

During the uniaxial compression test, the surface of the S3 sample remained rather dry, and no water leaked out. While in S6, water precipitated continually, and as the strain increased, water bleeding accelerated. As seen in Figure 5, the bleed water was collected near the sample's base. Sun and Song (1994) proposed that the water in HWM could be divided into three parts, namely crystal water, absorbed water, and free water [20]. Most of the water in HWM is free water, and the amount of free water increases with the water-cement ratios. Free water is apt to be lost under loading. Therefore, a large amount of water was observed to be leaked out for HWM with higher water-cement ratios.

Two groups of HWM samples were then subjected to uniaxial compression tests to obtain the uniaxial compression strength. The loading rate for the tests was kept at 5 mm/min. Figure 6 depicts the sample strength of two distinct groups. It can be shown that as the w/c increased, the uniaxial compressive strength of the HWM samples in the two groups fell gradually. The sample strengths of the seawater test group with w/c = 3:1 and 6:1 were 1.47 MPa and 0.39 MPa, respectively, while those of the control test group were 1.30 MPa and 0.31 MPa, respectively. The HWM sample strengths of the seawater group were 13 to 60 percent higher than those of the freshwater group of HWM samples, respectively.

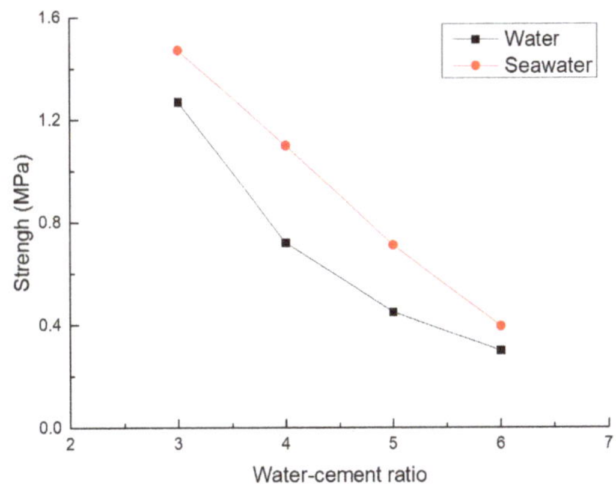

Figure 6. Strength of HWM made by seawater and water.

3.3. Microstructure and Chemical Compositions Analysis

Figure 7a–c show SEM images (×2000 times) of HWM samples of C3, S3, and S6, respectively.

(**a**) C3 (**b**) S3 (**c**) S6

Figure 7. SEM photos (×2000) of HWM made by water and seawater.

Figure 7a demonstrates that the hydration products of HWM produced by freshwater are dominated by ettringite crystals and have a needle-like, net-like, and predominantly rod-like structure. The crystals are staggered and interconnected to produce a dense network structure that serves as a framework and support. Filling the network structure with fibrous hydrated silica (C-S-H) gel and pom-shaped aluminum ($Al(OH)_3$) glue reduces the number of interior pores and increases the density. This explains why the water-cement ratio and water content of HWM are so high.

Comparing SEM images of HWM made by water and seawater with w/c = 3:1 (i.e., Figure 7a C3 and Figure 7b S3, respectively), the microstructures of both groups include a network structure composed of ettringite crystals, as well as fibrous hydrated silica (C-S-H) gel and pom-shaped aluminum ($Al(OH)_3$) glue within the network structure. However, there are more ettringite crystals within, and the network structure is significantly denser in HWM generated from seawater, which increases the network structure's strength in the HWM-hardened body. In addition to C-S-H gel and $Al(OH)_3$ glue, a significant amount of M-S-H gel and $Mg(OH)_2$ were found to be present in the network structure, which makes the structure denser. The strength of HWM with a stronger and denser microstructure is always greater. Consequently, the differences in the SEM images of HWM created using fresh water and seawater are consistent with the findings of Figure 6.

By comparing Figure 7b,c, the differences in the SEM images of HWM produced by seawater with a w/c ratio of 3:1 and 6:1 can be determined. It can be observed that HWM with w/c = 6:1 has significantly fewer ettringite crystals, as well as C-S-H gel, $Al(OH)_3$ glue, M-S-H gel, and $Mg(OH)_2$ inside the crystal network structure. HWM with a w/c ratio of 6:1 is able to absorb more water and has a larger water content than HWM with a w/c ratio of 3:1. This is because the network structure is significantly looser and has many more internal pores. In addition, the ettringite crystals of w/c = 6:1 are thinner, and the fibrous hydrated silicic acid gel and pom-shaped aluminum glue in the network structure are also significantly reduced, resulting in a structure with less compactness and more holes.

The C3 and S3 XRD scans are depicted in Figure 8. Ettringite (chemical formula: $3CaO{\bullet}Al_2O_3{\bullet}3CaSO_4{\bullet}32H_2O$), C-S-H, and $Al(OH)_3$ are the primary constituents of the stone body of the HWM, as seen. In addition to the aforementioned component, however, M-S-H gel and $Mg(OH)_2$ were also detected in the HWM of the seawater group.

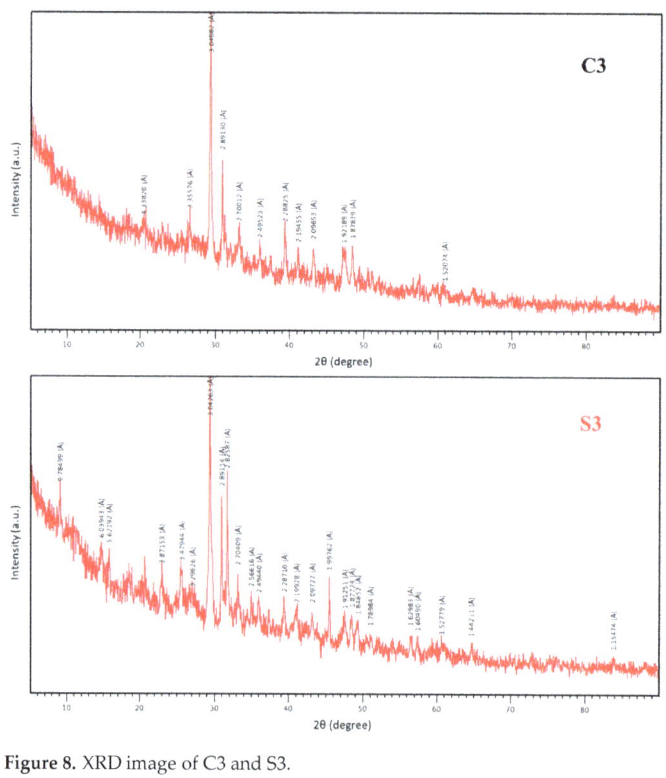

Figure 8. XRD image of C3 and S3.

4. Discussion

4.1. Effects of w/c on the Microscopic and Macroscopic Properties of HWM Made by Seawater

In comparison to fresh water, seawater has a substantial impact on the microstructure, chemical content, and physical and mechanical properties of HWM.

Zhang et al. (2017) analyzed the microstructure of HWM produced using potable water [25]. Figure 7 shows that compared to the SEM pictures of HWM made with fresh water, HWM made with seawater has a denser microstructure, a greater number of ettringite crystals, and a smaller porosity. This is compatible with the impacts of seawater on the macroscopic properties of HWM, specifically a decrease in water content, an increase in density, and an increase in strength.

Comparing the microstructure of HWM with different w/c ratios, the principal hydration products of HWM produced by seawater are the needle-like and prismatic Ettringite crystals, C-S-H and M-S-H gel, $Al(OH)_3$ glue, and $Mg(OH)_2$. Ettringite crystals in HWM with higher w/c ratios are narrower than those with lower w/c ratios.

In addition, when w/c ratios grow, fewer ettringite crystals are produced, much less fibrous hydrated silica gel and pom-shaped aluminum glue are filled in the network structure, and HWM has a looser microstructure and more interior pores. Comparing the macroscopic parameters of HWM with different w/c ratios, as illustrated in Figures 2–4 and 6, HWM formed from seawater with a greater w/c ratio has higher water content, lower density, and lower strength. Therefore, the microscopic and macroscopic properties of HWM in our investigation are compatible.

Additionally, w/c has a substantial impact on the failure modes of HWM. During the uniaxial compression test, hardly any water separated from samples with a w/c ratio of 3:1. The samples demonstrate brittle elasticity, and split failure is the predominant form of failure. The samples with a w/c ratio of 6:1 contain more water (78.4%). During the uniaxial compression test, there was water bleeding, and the rate of water bleeding increased as stress increased. The major mode of failure for samples with a w/c ratio of 6:1 is ductile failure. HWM with a greater w/c has excellent plasticity, whereas HWM with a lower w/c has excellent elasticity-brittleness. This may be the primary reason why failure modes of HWM with different w/c ratios vary.

He et al. (2014) studied the physical and mechanical qualities of HWM manufactured with fresh water but cured in seawater for 21 days. The results demonstrated that HWM cured in seawater could retain outstanding physical and mechanical properties [34]. Hou et al. (2012) conducted field testing with HWM derived from fresh water in maritime engineering, and the HWM demonstrated excellent performance [33]. In this investigation, it was determined that HWM might be prepared and cured in seawater while retaining its outstanding physical and mechanical qualities. HWM is, therefore, a promising building material for islands, reefs, and marine engineering.

4.2. Effects of Seawater on the Physical and Mechanical Properties of HWM

Xia et al. (2018) reported that the following reactions will occur after the mixing of two slurries during the hydrating and hardening process of HWM, as shown in Equations (1)–(4) [23]. During the above process, a large amount of $3(3CaO{\cdot}Al_2O_3{\cdot}CaSO_4{\cdot}12H_2O)$ (AFm) and $3CaO{\cdot}Al_2O_3{\cdot}3CaSO_4{\cdot}32H_2O$ (AFt) were produced.

$$3CaO{\cdot}Al_2O_3{\cdot}CaSO_4 + 2CaSO_4 + 38H_2O \rightarrow 3CaO{\cdot}Al_2O_3{\cdot}3CaSO_4{\cdot}32H_2O + 4Al(OH)_3 \quad (1)$$

$$3CaO{\cdot}Al_2O_3{\cdot}CaSO_4 + 18H_2O \rightarrow 3CaO{\cdot}Al_2O_3{\cdot}CaSO_4{\cdot}12H_2O + 4Al(OH)_3 \quad (2)$$

$$3Ca\,(OH)_2 + 3CaSO_4 + 2Al\,(OH)_3 + 2H_2O \rightarrow 3CaO{\cdot}Al_2O_3{\cdot}CaSO_4{\cdot}32H_2O \quad (3)$$

$$2CaO{\cdot}SiO_2 + nH_2O \rightarrow C\text{-}S\text{-}H + Ca(OH)_2 \quad (4)$$

It can be observed that a significant amount of the free water in the mixed slurry of A and B dissipates during the hydration reaction and converts into bound water of AFm and AFt, with AFt being the majority of the final reaction result (i.e., ettringite). During the hydration and hardening reaction of HWM, the formation of Ettringite crystals is largely reliant on the amount of water, with free water forming in HWM stones with a higher w/c ratio and less bound water forming in ettringite with a low w/c ratio. The ideal w/c ratio for HWM made using potable water is 6.86, and the HWM water content is 87.3% [24].

Seawater provides a high concentration of salt, with SO_4^{2-} and Mg^{2+} being the most influential ions on the hydration and hardening reactions of HWM.

According to reports [45,46], SO_4^{2-} reacts with Ca^{2+} to generate gypsum ($CaSO_4 \cdot 2H_2O$). On the one hand, gypsum can create ettringite (AFt) directly with tricalcium aluminate (C3A) in nail slurry. $CaSO_4$ must be present for AFm to react with gypsum dihydrate and form more ettringite (AFt). Equations (5)–(8) depict the primary chemical reaction equations for the procedure described above.

$$Ca(OH)_2 + Na_2SO_4 + 2H_2O \rightarrow CaSO_4 \cdot 2H_2O + 2NaOH \tag{5}$$

$$3(CaSO_4 \cdot 2H_2O) + 3CaO \cdot Al_2O_3 + 26H_2O \rightarrow 3CaO \cdot Al_2O_3 \cdot 3CaSO_4 \cdot 32H_2O\ (AFt) \tag{6}$$

$$\begin{gathered} 3CaO \cdot Al_2O_3 \cdot 3CaSO_4 \cdot 32H_2O + 3(CaSO_4 \cdot 2H_2O) \\ +4H_2O \rightarrow 3(3CaO \cdot Al_2O_3 \cdot CaSO_4 \cdot 12H_2O) \end{gathered} \tag{7}$$

$$\begin{gathered} 2(CaSO_4 \cdot 2H_2O) + 3CaO \cdot Al_2O_3 \cdot CaSO_4 \cdot 12H_2O + \\ 16H_2O \rightarrow 3CaO \cdot Al_2O_3 \cdot 3CaSO_4 \cdot 32H_2O \end{gathered} \tag{8}$$

As HWM is a novel cement-based material, its hydration and hardening processes are comparable to those of conventional cement-based materials, such as cement and concrete. When concrete is placed in seawater, ion exchange may occur, according to previous studies [47]. Ca^{2+} in C-S-H and $Ca(OH)_2$ would be replaced by Mg^{2+}, resulting in the formation of Mg-S-H gel and insoluble $Mg(OH)_2$. Equations (9)–(11) [46,48] depict the principal chemical reaction equations of the aforementioned procedure for the production of HWM from seawater. It can be observed that AFm combines with $MgSO_4$ to form more ettringite AFt, hence increasing the number of ettringite crystals in the body of the high-water material after it has been hardened. M-S-H gel and insoluble $Mg(OH)_2$ were used to strengthen the compactness of the hardened body of HWM by filling the network structure of Ettringite crystals. Therefore, it contributes to the enhancement of the tough body of HWM.

$$Ca(OH)_2 + MgSO_4 + 2H_2O \rightarrow CaSO_4 \cdot 2H_2O + Mg(OH)_2 \tag{9}$$

$$C\text{-}S\text{-}H + MgSO_4 + 2H_2O \rightarrow M\text{-}S\text{-}H + CaSO_4 \cdot 2H_2O \tag{10}$$

$$\begin{gathered} 4CaO \cdot Al_2O_3 \cdot 13H_2O + 3MgSO_4 + 2Ca(OH)_2 \\ +20H_2O \rightarrow 3CaO \cdot Al_2O_3 \cdot 3CaSO_4 \cdot 32H_2O + 3Mg(OH)_2 \end{gathered} \tag{11}$$

In conclusion, SO_4^{2-} and Mg^{2+} in seawater can enhance the synthesis of additional ettringite AFt in the hydration and hardening reaction of HWM, hence enhancing the strength of the hardened skeleton structure. In addition, M-S-H gel and insoluble $Mg(OH)_2$ were produced to fill the internal holes of the ettringite network structure. This improved the density of the hardening body of high-water material, hence enhancing the strength of seawater-based HWM.

5. Conclusions

In this study, HWM was first proposed to be produced from seawater and used in a maritime environment. Then the method of HWM produced from seawater was provided, and the physical and mechanical characteristics of HWM made from seawater were investigated, and the interaction mechanism between HWM and seawater was also discussed. The subsequent findings were reached:

(1) As the w/c ratio increases from 3:1 to 6:1, the water content, density, and uniaxial compressive strength of HWM produced from seawater varied from 72.1% to 77.5%; 1.25 to

1.12 g/cm^3, and 1.47 MPa to 0.39 MPa, respectively, which is 2–10% lower, 0.8–2.2% higher, and 13–45% stronger than that from fresh water. Compared to HWM samples made with fresh water, seawater test group samples show a 0.8–2.2% greater density, 2–10% lower water content, and 13–45% greater strength.

(2) The primary chemical components of HWM derived from seawater are Ettringite, C-S-H gel, aluminum ($Al(OH)_3$) glue, M-S-H gel, and $Mg(OH)_2$. Ettringite crystals stagger and link, forming a dense network structure that functions as a skeleton and support. The network structure was filled using C-S-H gel, aluminum ($Al(OH)_3$) glue, M-S-H gel, and $Mg(OH)_2$.

(3) The HWM-seawater interaction was revealed. SO_4^{2-} and Mg^{2+} in seawater contribute to the hydrating and hardening reaction of HWM, resulting in the production of additional Ettringite, M-S-H gel, and $Mg(OH)_2$, which makes the skeletal structure of HWM denser, hence enhancing its strength. HWM produced from seawater retains outstanding physical and mechanical characteristics.

(4) This study demonstrated that HWM produced from seawater possesses excellent physical and mechanical properties; consequently, HWM is a promising construction material for islands, reefs, and marine engineering, and it is suggested that additional field tests be conducted to verify the long-term behavior of HWM produced from seawater.

Author Contributions: Formal analysis, N.F.; Writing—original draft, B.L.; Writing—review & editing, C.L., J.G. and N.F. All authors have read and agreed to the published version of the manuscript.

Funding: This research was funded by the Geological Survey Program of China Geological Survey (Grants No. DD20190269, DD20221782).

Institutional Review Board Statement: Not applicable.

Informed Consent Statement: Not applicable.

Data Availability Statement: Not applicable.

Conflicts of Interest: The authors declare no conflict of interest.

References

1. To, W.-M.; Lee, P.K.C. China's Maritime Economic Development: A Review, the Future Trend, and Sustainability Implica-tions. *Sustainability* **2018**, *10*, 4844. [CrossRef]
2. Qu, F.; Li, W.; Dong, W.; Tam, V.W.; Yu, T. Durability deterioration of concrete under marine environment from material to structure: A critical review. *J. Build. Eng.* **2021**, *35*, 102074. [CrossRef]
3. Poudyal, L.; Adhikari, K. Environmental sustainability in cement industry: An integrated approach for green and economical cement production. *Resour. Environ. Sustain.* **2021**, *4*, 100024. [CrossRef]
4. Naqi, A.; Jang, J.G. Recent Progress in Green Cement Technology Utilizing Low-Carbon Emission Fuels and Raw Materials: A Review. *Sustainability* **2019**, *11*, 537. [CrossRef]
5. Mousavi, M.A.; Sadeghi-Nik, A.; Bahari, A.; Ashour, A.; Khayat, K.H. Cement Paste Modified by Nano-Montmorillonite and Carbon Nanotubes. *Mater. J.* **2022**, *119*, 173–185.
6. Mousavi, M.A.; Sadeghi-Nik, A.; Bahari, A.; Jin, C.; Ahmed, R.; Ozbakkaloglu, T.; de Brito, J. Strength optimization of cementitious composites reinforced by carbon nanotubes and Titania nanoparticles. *Constr. Build. Mater.* **2021**, *303*, 124510. [CrossRef]
7. Zeybek, Ö.; Özkılıç, Y.O.; Karalar, M.; Çelik, A.İ.; Qaidi, S.; Ahmad, J.; Burduhos-Nergis, D.D.; Burduhos-Nergis, D.P. Influence of Replacing Cement with Waste Glass on Mechanical Properties of Concrete. *Materials* **2022**, *15*, 7513. [CrossRef]
8. Karalar, M.; Bilir, T.; Çavuşlu, M.; Özkılıç, Y.O.; Sabri, M.M.S. Use of recycled coal bottom ash in reinforced concrete beams as replacement for aggregate. *Front. Mater.* **2022**, *9*, 1064604. [CrossRef]
9. Qaidi, S.; Najm, H.M.; Abed, S.M.; Özkılıç, Y.O.; Al Dughaishi, H.; Alosta, M.; Sabri, M.M.S.; Alkhatib, F.; Milad, A. Concrete Containing Waste Glass as an Environmentally Friendly Aggregate: A Review on Fresh and Mechanical Characteristics. *Materials* **2022**, *15*, 6222. [CrossRef]
10. Karalar, M.; Özkılıç, Y.O.; Aksoylu, C.; Sabri, M.M.S.; Beskopylny, A.N.; Stel'makh, S.A.; Shcherban', E.M. Flexural behavior of reinforced concrete beams using waste marble powder towards application of sustainable concrete. *Front. Mater.* **2022**, *9*, 1068791. [CrossRef]
11. Karalar, M.; Özkılıç, Y.O.; Deifalla, A.F.; Aksoylu, C.; Arslan, M.H.; Ahmad, M.; Sabri, M.M.S. Improvement in Bending Performance of Reinforced Concrete Beams Produced with Waste Lathe Scraps. *Sustainability* **2022**, *14*, 12660. [CrossRef]
12. Çelik, A.İ.; Özkılıç, Y.O.; Zeybek, Ö.; Özdöner, N.; Tayeh, B.A. Performance Assessment of Fiber-Reinforced Concrete Produced with Waste Lathe Fibers. *Sustainability* **2022**, *14*, 11817. [CrossRef]

13. Zeybek, Ö.; Özkılıç, Y.O.; Çelik, A.İ.; Deifalla, A.F.; Ahmad, M.; Sabri, M.M.S. Performance evaluation of fiber-reinforced concrete produced with steel fibers extracted from waste tire. *Front. Mater.* **2022**, *9*, 1057128. [CrossRef]
14. Wang, J.; Huang, T.; Liu, X.; Wu, P.; Guo, Z. Mechanical properties of recycled concrete in marine environment. *Sci. World J.* **2013**, *2013*, 728357. [CrossRef]
15. Liu, J.; Ju, B.; Xie, W.; Yu, H.; Xiao, H.; Dong, S.; Yang, W. Design and Evaluation of an Ultrahigh-Strength Coral Aggregate Concrete for Maritime and Reef Engineering. *Materials* **2021**, *14*, 5871. [CrossRef] [PubMed]
16. Cui, Y.; Jiang, J.; Fu, T.; Liu, S. Feasibility of using Waste Brine/Seawater and Sea Sand for the Production of Concrete: An Experimental Investigation from Mechanical Properties and Durability Perspectives. *Sustainability* **2022**, *14*, 13340. [CrossRef]
17. Yang, G.; Zhao, J.; Wang, Y. Durability properties of sustainable alkali-activated cementitious materials as marine engineering material: A review. *Mater. Today Sustain.* **2022**, *17*, 100099. [CrossRef]
18. Afzal, M.S.; Tahir, F.; Al-Ghamdi, S.G. Recommendations and Strategies to Mitigate Environmental Implications of Artificial Island Developments in the Gulf. *Sustainability* **2022**, *14*, 5027. [CrossRef]
19. Hayek, M.; Salgues, M.; Souche, J.-C.; Cunge, E.; Giraudel, C.; Paireau, O. Influence of the Intrinsic Characteristics of Cementitious Materials on Biofouling in the Marine Environment. *Sustainability* **2021**, *13*, 2625. [CrossRef]
20. Sun, H.H.; Song, C.Y. *Application of High-Water Solidifying Materials*; China University of Mining and Technology Press: Xuzhou, China, 1994.
21. Peng, M.X.; Jiang, J.H.; Zhang, X. Effect of composition on the performance and microstructures of mining high-water solidified materials. *Miner. Eng. Res.* **2011**, *26*, 56–59.
22. Xie, H.; Liu, C. Analysis on Raw Material of High-water and Quick-setting Material Stone. *Sci. Technol. Eng.* **2014**, *14*, 254–258.
23. Xia, J.; Zhao, W.; Liu, D.; Bai, J. Study on the hydration and hardening mechanism of the high-water rapid-setting material. *AEMCME* **2018**, *439*, 42–44. [CrossRef]
24. Xie, H.; Liu, C. Analysis on influence of moisture content on deformation characteristics of the high-water-content material stone. *J. Sichuan Univ. Sci. Eng. (Nat. Sci. Ed.)* **2013**, *45*, 1–6.
25. Zhang, Y.; Liu, C.; Xie, H.; Lu, B.W.; Wang, C.; Zou, J.L. Experimental research on influence of water cement ratio on mechanical characteristics of the high-water-content material. *Adv. Eng. Sci.* **2017**, *49*, 115–120.
26. Zhou, Q.; Xu, H.B.; Liu, J.H. Unstable failure characteristics and performance improvement of water-high filling materials. *J. China Coal Soc.* **2017**, *42*, 1123–1129. [CrossRef]
27. Zhang, F.T.; Wang, X.Y.; Bai, J.B.; Wang, G.Y.; Wu, B.W. Post-peak mechanical characteristics of the high-water material for backfilling the gob-side entry retaining: From experiment to field application. *Arab. J. Geosci.* **2020**, *13*, 386. [CrossRef]
28. Bai, E.; Guo, W.; Tan, Y.; Huang, G.; Guo, M.; Ma, Z. Roadway backfill mining with super-high-water material to protect surface buildings: A case study. *Appl. Sci.* **2019**, *10*, 107. [CrossRef]
29. Xie, S.; Wang, E.; Chen, D.; Li, H.; Jiang, Z.; Yang, H. Stability analysis and control technology of gob-side entry retaining with double roadways by filling with high-water material in gently inclined coal seam. *Int. J. Coal Sci. Technol.* **2022**, *9*, 1–18. [CrossRef]
30. Cao, W.; Wang, X.; Li, P.; Zhang, D.; Sun, C.; Qin, D. Wide Strip Backfill Mining for Surface Subsidence Control and Its Application in Critical Mining Conditions of a Coal Mine. *Sustainability* **2018**, *10*, 700. [CrossRef]
31. Li, S.; Yu, Z.; Yu, H.; Wang, X. The Recent Progress China Has Made in High-Concentration Backfill. *Sustainability* **2022**, *14*, 2758. [CrossRef]
32. Hou, L.T.; Tang, J.W.; Jiang, K.H. Basic characteristics of superabsorbent polymers and its application in port engineering. *Port Waterw. Eng.* **2007**, *5*, 54–56.
33. Hou, L.T. Applied experiments of superabsorbent polymers on repaired and strengthened calcareous sands island. *Concrete* **2012**, *12*, 122–124.
34. He, T.; Liu, C.W.; Ye, D.Y.; Wang, Y. Experimental study on environmental effects on the compressive strength of the high-water-content and quick-setting stone. *Adv. Eng. Sci.* **2014**, *46*, 86–91.
35. Wang, E.; Xie, S.; Chen, D.; Sun, Y.; Qi, P. Influence of different loading rates on the mechanics and failure characteristics of high-water-content material. *Geomech. Geoengin.* **2022**, *17*, 797–808. [CrossRef]
36. Xia, J.; Su, Q.; Liu, D. Optimal gypsum-lime content of high water material. *Mater. Lett.* **2018**, *215*, 284–287. [CrossRef]
37. Zuo, J.P.; Hong, Z.J.; Xiong, Z.Q.; Wang, C.; Song, H.Q. Influence of different W/C on the performances and hydration progress of dual liquid high water backfilling material. *Constr. Build. Mater.* **2018**, *190*, 910–917. [CrossRef]
38. Guan, G.; Wang, X.; He, J. Research Progress of Seawater Sea-Sand Concrete. *Bull. Chin. Ceram. Soc.* **2022**, *41*, 1483–1493.
39. Zhao, Y.; Hu, X.; Shi, C.; Zhang, Z.; Zhu, D. A review on seawater sea-sand concrete: Mixture proportion, hydration, microstructure and properties. *Constr. Build. Mater.* **2021**, *295*, 123602. [CrossRef]
40. *GB/T 50081*; Standard for the Test Method of Mechanical Properties of Ordinary Concrete. Standardization Administration of China: Beijing, China, 2002. (In Chinese)
41. *GB/T 23561.1*; Methods for Determination of Physical and Mechanical Properties of Coal and Rocks. Standardization Administration of China: Beijing, China, 2009. (In Chinese)
42. *ASTM D1141-98*; Standard Practice for the Preparation of Substitute Ocean Water. ASTM International: West Conshohocken, PA, USA, 2013. Available online: www.astm.org (accessed on 16 July 2021).
43. Feng, B.; Liu, C.W.; Xie, H.; Wu, M.S.; Jiang, Y. Experimental study on the size and the shape of high-water-content material that modified. *Adv. Eng. Sci.* **2017**, *49* (Suppl. S2), 121–127.

44. Wang, G.; Xiong, Z.Q.; Su, C.D. Analysis on size effect and failure characteristics of filling pillars with modified high-water material. *Coal Sci. Technol.* **2021**, *49*, 82–88.
45. Liao, Y.; Liu, C.; Kang, S. Effect of Mineral Admixtures on the Hydration of Sulphoaluminate Cement under Seawater Condition. *J. Chin. Ceram. Soc.* **2017**, *36*, 1333–1338.
46. Wang, C.L.; Liu, Z.P.; Zhang, T. Influence of Seawater and Curing Conditions on the Hydration Performance of Portland Cement. *Mater. Rep.* **2022**, *36*, 21010137-7.
47. Zhang, T.; Dieckmann, E.; Song, S.; Xie, J.; Yu, Z.; Cheeseman, C. Properties of magnesium silicate hydrate (MSH) cement mortars containing chicken feather fibres. *Constr. Build. Mater.* **2018**, *180*, 692–697. [CrossRef]
48. Song, Q.; Hu, Y.; Wang, Q.; Zhang, J.; Chen, Y. Research Development of Magnesium Silicate Hydrate Cement. *J. Chin. Ceram. Soc.* **2019**, *47*, 1642–1651.

sustainability

MDPI

Article

Estimating Flyrock Distance Induced Due to Mine Blasting by Extreme Learning Machine Coupled with an Equilibrium Optimizer

Ramesh Murlidhar Bhatawdekar [1], Radhikesh Kumar [2,*], Mohanad Muayad Sabri Sabri [3], Bishwajit Roy [4], Edy Tonnizam Mohamad [1], Deepak Kumar [5] and Sangki Kwon [6]

1 Centre of Tropical Geoengineering (GEOTROPIK), School of Civil Engineering, Faculty of Engineering, Universiti Teknologi Malaysia, Johor Bahru 81310, Malaysia
2 Department of Computer Science and Engineering, National Institute of Technology Patna, Ashok Raj Path, Patna 800005, India
3 Peter the Great St. Petersburg Polytechnic University, 195251 St. Petersburg, Russia
4 School of Computer Science, University of Petroleum and Energy Studies (UPES), Dehradun 248007, India
5 Department of Civil Engineering, National Institute of Technology Patna, Ashok Raj Path, Patna 800005, India
6 Department of Energy Resources Engineering, Inha University, Yong-Hyun Dong, Nam Ku, Incheon 402-751, Republic of Korea
* Correspondence: radhikesh.cs18@nitp.ac.in

Abstract: Blasting is essential for breaking hard rock in opencast mines and tunneling projects. It creates an adverse impact on flyrock. Thus, it is essential to forecast flyrock to minimize the environmental effects. The objective of this study is to forecast/estimate the amount of flyrock produced during blasting by applying three creative composite intelligent models: equilibrium optimizer-coupled extreme learning machine (EO-ELM), particle swarm optimization-based extreme learning machine (PSO-ELM), and particle swarm optimization-artificial neural network (PSO-ANN). To obtain a successful conclusion, we considered 114 blasting data parameters consisting of eight inputs (hole diameter, burden, stemming length, rock density, charge-per-meter, powder factor (PF), blastability index (BI), and weathering index), and one output parameter (flyrock distance). We then compared the results of different models using seven different performance indices. Every predictive model accomplished the results comparable with the measured values of flyrock. To show the effectiveness of the developed EO-ELM, the result from each model run 10-times is compared. The average result shows that the EO-ELM model in testing (R2 = 0.97, RMSE = 32.14, MAE = 19.78, MAPE = 20.37, NSE = 0.93, VAF = 93.97, A20 = 0.57) achieved a better performance as compared to the PSO-ANN model (R2 = 0.87, RMSE = 64.44, MAE = 36.02, MAPE = 29.96, NSE = 0.72, VAF = 74.72, A20 = 0.33) and PSO-ELM model (R2 = 0.88, RMSE = 48.55, MAE = 26.97, MAPE = 26.71, NSE = 0.84, VAF = 84.84, A20 = 0.51). Further, a non-parametric test is performed to assess the performance of these three models developed. It shows that the EO-ELM performed better in the prediction of flyrock compared to PSO-ELM and PSO-ANN. We did sensitivity analysis by introducing a new parameter, WI. Input parameters, PF and BI, showed the highest sensitivity with 0.98 each.

Keywords: flyrock; weathering index (WI); equilibrium optimizer (EO); particle swarm optimization (PSO); extreme learning machine (ELM); artificial neural network (ANN)

Citation: Bhatawdekar, R.M.; Kumar, R.; Sabri Sabri, M.M.; Roy, B.; Mohamad, E.T.; Kumar, D.; Kwon, S. Estimating Flyrock Distance Induced Due to Mine Blasting by Extreme Learning Machine Coupled with an Equilibrium Optimizer. *Sustainability* **2023**, *15*, 3265. https://doi.org/10.3390/su15043265

Academic Editors: Steve W. Lyon and Guang-Liang Feng

Received: 4 September 2022
Revised: 20 September 2022
Accepted: 17 October 2022
Published: 10 February 2023

1. Introduction

Surface mining is basically for breaking in situ rock, construction activities, and excavation; blasting is the most popular method throughout the globe [1–3]. Blasting is a system consisting of interaction between explosives and rock [4]. Blast design, properties of explosives, and rock mass are primarily the existing critical parameters for blasting and its performance [5,6]. Standard operating procedures are adopted for blast execution [7,8]. Desired blast fragmentation throw and shape muck pile affect the efficiency of loading equipment [9]. Blast fragmentation is also crucial for loading equipment and downstream operation of hauling and crushing [10,11]. Hence, these parameters are called favorable parameters [12,13], whereas flyrock, ground vibration, air overpressure, and dust affect the environment [7,14]. Therefore, these parameters are known as unfavorable parameters [8,15].

Flyrock is a geotechnical issue based on various rock mass properties. Various studies have been carried out to resolve geotechnical issues. Failure in geotechnical structure was studied with a multiscale work analysis approach [16]. An impressive micromechanical modeling (MM) framework was proposed by utilizing the discrete element method (DEM) and the micro-mechanical (MM) model [17]. The impact of flow direction vis-a-vis gravity direction on suffusion in geotechnical structures or slopes was resolved using the pioneering computational fluid dynamic-discrete element method (CFD-DEM) [18]. Further, the CFD-DEM model was helpful in investigating the particle shape effect and various levels of transmission at the macro- and micro-behavior levels during suffusion [19,20]. The CFD-DEM method played a significant role in the investigation of seepage in the underwater tunnel face [21]. The novel multi-scale approach, by deploying the smoothed particle hydrodynamics (SPH) method, was found to be efficient in computational analysis to understand granular collapse [22]. A possible solution for several engineering and industrial processes was found by developing DEM for irregular 3D particle shapes [23]. An algorithm was developed to accomplish 3D realistic stones of irregular geometries at random for specified samples with quantitative adjustable control [24]. In the convention of blasting safety criteria, defining frequency characteristics of blasting is of practical significance. A computational method associated with the wavelet frequency domain parameter or a main frequency band was proposed [25]. The innovative liquid carbon dioxide rock-breaking technology was found to be safer than explosive blasting. This technology needs further investigation to show that this technology is more efficient than traditional non-explosive techniques [26].

Flyrock has also been considered as an incidental or haphazard, exorbitant throw of rock pieces originating from blasting operations. Rock fragments from blasting may be thrown beyond the expected normal distance. This may result in a serious hazard to people working around the mines or severe danger to the property and machinery near the blasting site [27–29]. There are several causes of flyrock: usage of types of explosives, improper blast design, or explicit or uncertain conditions of the rock mass. Flyrock accidents are caused by poor security or blast management practices [30–32]. During the last decade, several researchers have developed many computational methods concurrently to forecast flyrock due to blasting [33].

Various researchers have pointed out that ANN, which is a branch of artificial intelligence (AI), is suitable for forecasting engineering problems [34,35]. For decades, many researchers had used ANN as a prediction method for flyrock distance [36–38]. However, ANN has some limitations, including slow learning speed and it falls into local minima due to the use of a gradient-based optimizer [39]. Furthermore, a specific ANN model for the prediction of flyrock is not available. On the other hand, the boundary condition of ANN models depends upon the variation in data sets [31]. Further, despite significant learning cycles on the same data set, there may be a marginal improvement in prediction performance. ANN models can easily find out the significance and sensitivity of input parameters [31]. As stated in the literature, various metaheuristic optimization algorithms (MOAs) could be deployed to forecast flyrock created by blasting due to the MOA's higher efficiency and to avoid the limitation of ANN. Some of the researchers compared ANN with other models, such as the ICA-coupled ANN model, which provided better performance as compared to ANN for the prediction of flyrock [40]. Furthermore, ANFIS showed superior performance as compared to ANN [41,42]. The PSO algorithm provided a powerful equation to predict flyrock due to blasting [43]. Gene expression programming (GEP) and the firefly algorithm (FA) were used to compare the results of flyrock prediction [44]. Further, hybrid algorithms were developed using optimization algorithms and ANN. MOAs provide a powerful ability to search for the best local solution from global optima. Therefore, MOA's estimated biases and weights of ANN can improve the prediction task of flyrock. The hybrid model, PSO-ANN, was developed for the prediction of flyrock [45]. The results of two different hybrid models, genetic algorithm-based ANN (GA-ANN) and recurrent fuzzy neural network (RFNN-GA), were compared with ANN to predict flyrock [46]. The results of three hybrid models, ICA-ANN, PSO-ANN, and GA-ANN, were compared to predict flyrock [47].

In recent years, one of the most exciting areas of study is machine learning (ML) [48]. In general, ML describes the ways of making predictions about and learning from data. As a subcategory of artificial intelligence, ML mainly aims at proposing and developing algorithms that can learn automatically from data.

Specifically, AI seeks to identify the objects existing in the neighboring areas and predict how the environment behaves in ways to make informed decisions. As a result, the ML techniques have a higher tendency to predict instead of estimate. For instance, it discusses the way researchers can make use of data obtained from an interferometry experiment to predict the interference pattern that would be seen under a variety of conditions. Furthermore, ML-based methods are mostly used to address high-dimensional problems of a higher complexity compared to the problems that may generally arise during a conventional statistics course [49]. The capability of generating and analyzing large data sets has dramatically increased during the last three decades. Such "big data" revolution has been prompted due to an exponential upsurge in computation capacities and memory, which is recognized as Moore's law. The ML models SVM [50,51] and ORLEM [52] were used by researchers for the prediction of flyrock due to blasting. In this research work, three hybrid models: PSO-ANN, PSO-ELM, and EO-ELM, are developed for the prediction of flyrock and their results are compared.

2. Models for the Prediction of Flyrock

Numerous researchers have proposed various approaches for flyrock prediction, which includes empirical, semi-empirical, and mathematical models.

2.1. Empirical Models for the Prediction of Flyrock

Various empirical models were developed by several researchers, mainly for blast production, which depend upon blast design and/or rock mass properties. Figure 1 shows a schematic diagram of flyrock, which may result in face burst, cratering, and rifting. Face burst results when geological discontinuities or planes of weakness exist. Cratering happens due to the escape of gases in the stemming zone due to back breaks or weak rock. Furthermore, it may be caused due to incorrect delay sequence (back rows firing first as compared to front rows). Rifting is due to stemming release with air pulse and associated with air blast. Inadequate stemming length and inappropriate stemming material are the causes of rifting [53].

Various empirical equations were developed by several researchers for flyrocks, which occur in numerous sizes and shapes. The prediction of maximum flyrock was recommended based on a factor of safety and hole diameter [54]. Further, the relationship between the ratio of stemming length to burden was established to maximum flyrock distance [55].

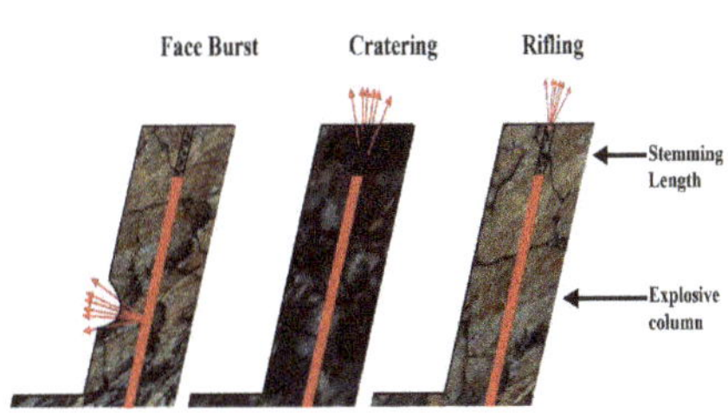

Figure 1. Schematic diagram of a mechanism of flyrock [56].

Researchers developed equations for flyrock prediction for the calculation of initial velocity based on the scaled burden method [57,58]. As per the equation, a charge per m and burden are the key parameters of face burst, as face burst increases with the increase in charge per m or decrease in burden. Furthermore, stemming length or charging per m are the key parameters of cratering. Flyrocks increase due to cratering, with the increase in charge per m or decrease in stemming length. Similarly, rifting depends upon a charge per m, burden, and drill hole angle. Flyrock increase, with the increase in charge per m or decrease in burden and rifting, is minimum in cases of vertical drill hole angle.

Flyrock prediction was developed based on blast design parameters (burden, stemming length, linear charge concentration, and specific charge) and rock mass properties (unconfined compressive strength and RQD) [58]. The empirical equation was established based on blast design parameters (stemming length, hole depth, burden, and spacing) and rock mass property (rock mass rating) for the prediction of flyrock [39]. These equations do not consider rock mass properties of tropically weathered rock. Various researchers have reported that empirical equations for the prediction of flyrock may be suitable for a particular site only and are not accurate [40,59].

2.2. *Mathematical Models for the Prediction of Flyrock*

Various researchers developed mathematical models for the prediction of flyrock. To estimate flyrock range, Lundborg [60] adopted a semi-empirical method to analyze the relationship between rock velocity and charge diameter. In terms of crater blasting in granite blocks, these authors introduced the relationship between the beginning velocity of the flyrock fragment, its size, and throw [61]. Two expressions were derived by Chiapetta (1983) in the case of distance that may be traveled by flyrock (Chiapetta RF, 1983). Furthermore, a relation was established by Roth (1979) to find out the flyrock travel range. According to this approach, all of the measurements were done on the flyrock range and the most important variable was to estimate the flyrock velocity at the beginning. Roth applied Gurney's proposed equation to compute the velocity at the beginning of the fragments thrown around through an explosion [61]. The limitation of mathematical equations is that their prediction is not accurate due to their being site-specific and having limited data input.

2.3. *Semi-Empirical Trajectory Physics-Based Models for the Prediction of Flyrock*

In semi-empirical trajectory physics-based models, the focus is on the beginning velocity of flyrock; therefore, they are most desired. One of the models developed by St. George and Gibson was modified by Little and Blair [62]. These models generally suffer from inexplicity in defining the velocity of detonation and the density of the explosive, which is applied to determine the blast-hole pressure and effects. The impact time applied to these equations is determined based on experimental observation instead of real monitoring.

2.4. *Artificial Intelligence Techniques*

Blasting is one of the major operations that causes several adverse environmental effects, such as generation of fines, ground vibrations, fumes, air blast, dust, and flyrock [44]. So, it is necessary to control these adverse effects while performing blasting operations [63]. During the last decade, various researchers have applied artificial intelligence (AI) techniques for the prediction, minimization, or optimization of these environmental effects. In recent years, ML is one of the trending methods. ML may be defined as computer algorithms that can improve automatically based on the nature of the signal or feedback provided to the learning system. It is divided into three categories, namely, reinforcement, unsupervised, and supervised learning. In "supervised learning", the model is trained based on provided inputs and their desired outputs and the goal is to learn the pattern by mapping input to output. In "Unsupervised learning", the model is trained based on provided inputs only where labels were not given so that it may find structure in the input on its own. In the case of "Reinforcement learning", the model gets across with a dynamic environment to perform a certain goal and after performing this goal it obtains feedback that is equivalent to rewards, which the model attempts to maximize. ANN is one of the most significant algorithms of ML. ANN is a parallelly distributed system that epitomizes the neural network of the human brain to create information processing models by composing different networks and connections [64]. It has the advantages of self-organizing, adaptive, and real-time learning features that enable it to overcome the defects of traditional logic-based AI in handling unstructured information and intuition [65]. During 1992–1997, Vladimir Vapnik with colleagues developed support-vector machine (SVM) models at AT&T Bell Laboratories. SVM is one of the models of supervised learning that have associated learning algorithms to analyze data for regression or classification.

Despite intended flexibility in singleton AI models, the previous studies indicate that these algorithms may fail to deliver expected results and may experience poor generalization capability. The reason for this is that they may become stuck in a locally optimal solution due to the use of stochastic selection or gradient-based learning algorithms of learning parameters [66,67]. To get the precise result and to perform the tasks adequately, a hybrid combination of soft computing techniques may be used as it uses data pre-processing techniques or metaheuristic optimization approaches to solve those problems [65,67]. Furthermore, the metaheuristic approach coupled with the ML model may enhance the ML model's performance as it may adequately reach the local solution from the global best solution [64,67].

The ANFIS model is a hybrid model developed with ANN and fuzzy interface system (FIS). Various researchers have used the ANFIS model for the prediction of flyrock due to blasting [68]. The results are highly promising, and comparative analysis suggests that the proposed modeling approach outperforms ANNs and other traditional time series models in terms of computational speed, forecast errors, efficiency, peak flow estimation, etc. It was observed that the ANFIS model preserves the potential of the ANN approach fully, and eases the model building process.

Zhou [69] utilized PSO and ANN techniques to minimize flyrock due to blasting [68]. PSO-ANN was found better as compared to the ANN model. ICA, GA, extreme learning machine (ELM), and biography-based optimization (BBO) have been applied for the prediction of various geotechnical issues by many researchers. Murlidhar [70] reviewed the ANN-GA [69], ANN-PSO [71], and ANN- ICA models. Each of these hybrid models provided better accuracy as compared to single models [42]. Murlidhar [72] have applied PSO-ELM, BBO-ELM, and ELM models to predict flyrock due to blasting [73].

ELM has insufficient generalization ability in dealing with the samples due to the random initialization of parameters, and PSO uses the individuals sharing information in the group to move the whole group to evolve from disorder to obtain the optimal solution [74]. PSO-ELM takes advantage of PSO to search for global optimal solutions and ELM to quickly deal with the nonlinear relationship [70]. In other words, PSO-ELM uses the PSO algorithm to optimize the input weight matrix and the hidden layer bias in ELM to obtain an optimal network [75–78]. Therefore, the PSO-ELM model provides better performance as compared to the singleton ELM model. Similarly, the BBO-ELM model provides better performance as compared to the singleton ELM model because the hybrid model provides the advantage of both BBO and ELM models [72]. Hence, in this paper, the authors decided to compare three hybrid models: PSO-ANN, PSO-ELM, and EO-ELM.

3. Background of Model

3.1. Extreme Learning Machine (ELM)

The ELM was introduced by Guang-Bin Huang (2004) [79] for feedforward neural networks. It consists of one hidden layer with multiple hidden nodes and their parameters do not require the tuning of input weights. In ELM, output weights of hidden nodes are generally learned only in one step that result in the learning of a linear model. ELM has a very high generalization capability and is considerably faster than the feedforward neural network (back-propagation algorithm (BP). This is because of the dependency among parameters of the different layers in a feedforward neural network. Due to these dependencies, one is obliged to adjust all of the parameters (weights and biases). So, several iterative learning steps are required to improve the learning performance of the feedforward network. For these reasons, ELM has been widely applied for classification, regression, sparse approximation compression, clustering, future learning, and many more.

Essence of ELM

The hidden layer's parameters do not require tuning of input weights. "Randomness" is one of the ways to implement ELM other than semi-randomness, which is considered in many traditional methods. In ELM, the hidden layer's mapping follows the rules of ridge regress theory [80], universal approximation, and neural network generalization theory [81]. Figure 2 shows the mapping of input space and feature space. ELM has the ability that may bridge the current gaps among linear systems, neural networks, matrix theories, SVM, random projection, Fourier series, and others.

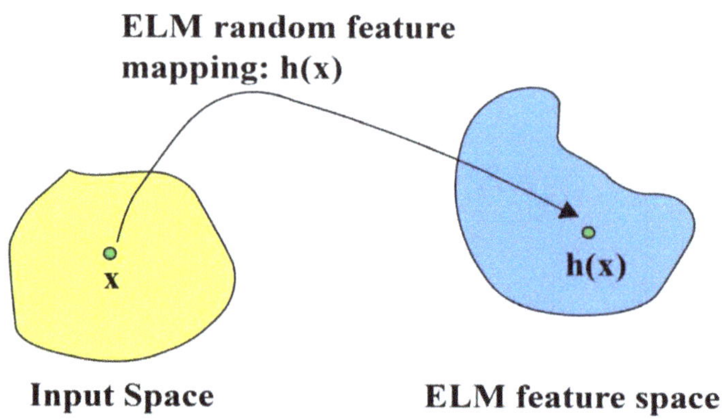

Figure 2. Mapping of input space and feature space.

The following steps need to be performed if a model consists of a particular number of hidden notes, node output function, and certain training set:

1. Randomly assign the hidden node's parameter
2. Calculate the output matrix of the hidden layer
3. Compute the output weights.

ELM is one of the ML algorithms which is free from tuning and works on the above-mentioned steps. It consists of hidden nodes with high importance and has a high-speed learning process.

3.2. Artificial Neural Network

The idea of an ANN is derived from the biological neural architecture where a very large number of biological neurons are interconnected through the links. It is an information processing model that is similar to the structure of neural networks present in the human brain in both structure and functions. It mimics the biological neural architecture in two ways: (i) the learning process is used by the network to acquire the knowledge from its surroundings and (ii) the acquired knowledge is stored by the synaptic weights (interconnection strength). An ANN consists of a network of interconnected processing units, which is capable of 'learning' to 'recognize' a complex input pattern, and predict the output pattern thereof. For this, the neural network is first 'trained' to analyze the input patterns and recognize the output that results from these inputs. The network is then able to recognize similarities in new input patterns and can predict the output. This property of a neural network makes it very useful for noisy (inexact) data to be interpolated and outputs predicted in terms of patterns that are already 'known' to it. This makes neural networks a ready replacement for older statistical techniques, such as linear regressing, multi-variable regression, and autocorrelation, etc. Even outputs that were previously not apparent to non-experts become recognizable, making the neural network a virtual expert. To solve different problems an ANN can be designed using the following three fundamental components:

- Transfer Function
- Network Architecture
- Learning Law.

3.3. *Equilibrium Optimization*

The Equilibrium Optimizer (EO) was developed based on a generic mass balance equation [82]. The EO algorithm is designed with a high level of search capability from exploratory and exploitative systems to randomly change solutions and prevent local minima. In the Equilibrium Optimization (EO) algorithm, the equilibrium state (optimal result) is finally achieved through equilibrium candidates or best so-far solutions through search agents randomly [83]. If EO is compared with many metaheuristic algorithms, EO starts the optimization process based on the initial population. The construction of the initial concentrations is based on several dimensions and particles in the search space with consistent random initialization. Particles with their concentrations are known as search agents, corresponding to particles and positions in the PSO algorithm. During the process of optimization in the initial period, there is no awareness and understanding about the equilibrium state. In the beginning, equilibrium candidates are identified by updating concentrations randomly to fit solutions. During the whole optimization process, based on different experimentation and case studies four best-so-far particles are identified and another particle is the concentration of arithmetical mean of the aforesaid four particles. Exploration capability is based on best-so-far particles while exploitation capability is based on the average value. In several engineering problems, the generation rate can be expressed as a function of time [84]. In the case of EO, selecting an appropriate generation rate as well as updating concentrations randomly enhances EO's exploratory performance during the initial irritation and exploitation search in the final iterations. Thus, EO supports from beginning to end the complete optimization process and avoids local minima. Exploration and exploitation processes are balanced to obtain adaptive values for monitoring parameters resulting in a significant reduction in the movement of particles. To study the efficiency and effectiveness of EO, quantitative and qualitative metrics were endorsed. The EO algorithm showed higher efficiency (i.e., shorter computational time or limited iterations) in achieving optimal or close to optimal solutions with specific or most of the problems examined. The EO algorithm is undoubtedly a better algorithm as compared to the metaheuristic algorithms, such as GA and PSO, or the recently developed algorithms, such as GWO, GA, and GSA. The performance of EO is statistically comparable with SHADE and LSHADE-SPACM algorithms.

3.4. *Particle Swarm Optimization (PSO)*

The particle swarm optimization algorithm was first introduced by Kennedy and Eberhart (1995) [85]. To start operation, this algorithm distributes a set of entities (particles, each of which stands for a feasible solution) randomly in the search space. An objective function is considered as the determining factor of the swarm's goal. The fitness of every entity/particle is determined by the value that is correspondent to the objective function. Figure 3 shows a standard flow chart of a PSO.

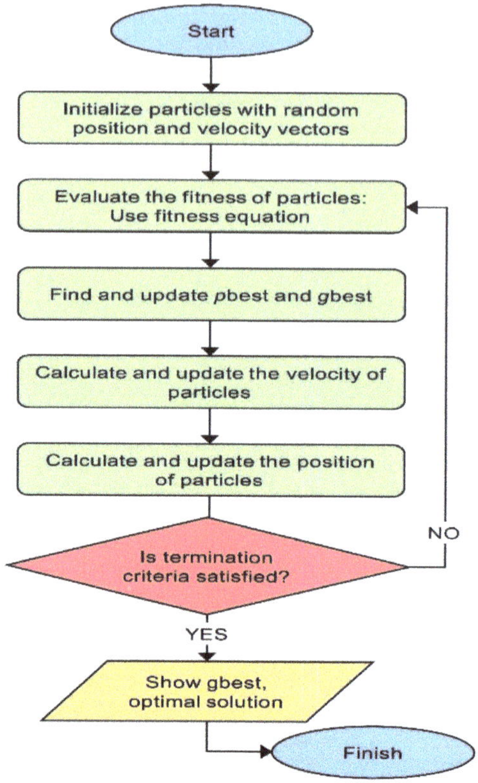

Figure 3. Standard flow chart of PSO [85].

The PSO algorithm uses a random or stochastically selected population. The first step in the process is to select a population of particles or solutions and iterate the solutions until an optimum is reached. Each particle is assigned a value and then updated as per the 'gbest' fitness for two outstanding values. The solutions higher than 'gbest' fitness values are chosen and the 'gbest' fitness factors are updated. Subsequently, wherever the fitness of the particle is better or higher than that of 'pbest', the corresponding parameters of 'pbest' are updated. The process would then enter the second phase, where the particles would be examined again.

3.5. Case Study and Data Collection

Basalt, granite, and limestone are common rocks for manufacturing aggregates in the construction industry of Thailand. Figure 4 shows potential rock resources of aggregates in different locations of Thailand. Small aggregate quarries produce less than 15,000 cubic meters per month. On the other hand, large aggregate quarries produce up to 150,000 cubic meters per month [86]. Vast quarries of limestone are used to supply limestone to manufacture Portland cement in factories situated 100 km from Thailand. The selected limestone quarry is an aggregate limestone quarry. Figure 5 shows photographs of an aggregate limestone quarry. Figure 6 shows the blasted muck pile of the face. During a blasting operation, flyrock is generated, which is concerning. In the selected quarry, the input parameters consisting of hole diameter, burden, stemming length, rock density, explosives charge per meter, powder factor, blast ability index, weathering index, and flyrock distance for 114 blasting events were collected. Figure 4 shows the location of the limestone quarry. Table 1 shows the details of all input parameters. The weathering index is site-specific.

Based on weathering index, the blast input parameters are decided. Hence, sensitivity analysis of parameters is compared with weathering index.

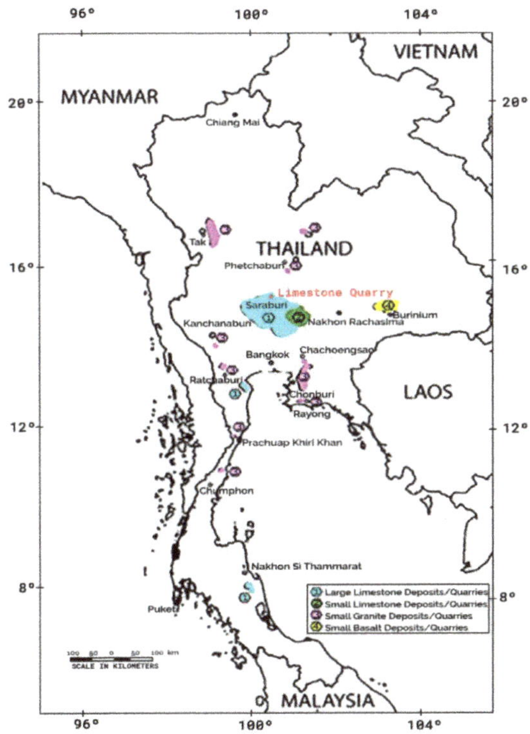

Figure 4. Location map of Thailand quarry [86].

Figure 5. Limestone quarry face at Thailand.

Figure 6. Blasted stone muck pile at quarry face.

Table 1. Input and output parameters.

Parameters	Hole Diameter	Burden	Stemming Length	Rock Density	Charge per M	Powder Factor	Blastability Index	Weathering Index	Flyrock
Symbol	D	B	ST	ρ	CPM	PF	BI	WI	FR
Unit	mm	m	m	Cum.t	kg/m	kg/cum	%	Ratio	m
Minimum	76	2.5	1.2	1.8	4.54	0.08	18.5	0.13	27
Quartile1	76	2.7	2	1.8	4.99	0.19	28.6	0.25	37
Average	90	3	2	2	7	0.30	43	0.76	81
Quartile3	102	3.6	2.95	2.5	8.99	0.40	54.6	0.88	82
Maximum	102	4.6	4	2.5	9.4	0.50	80.8	0.99	436

Weathering index (WI) is a new parameter introduced based on rock mass properties, such as water absorption (%), porosity (%), and point load index. Maximum values of water absorption and porosity are obtained for completely weathered granite. The maximum value of the point load index is obtained for fresh rock. At each blasting site, samples are collected, and each rock mass property is compared with the maximum value. The average of these ratios is known as the WI.

4. Model Development

4.1. Hybridization of PSO-ANN

The PSO algorithm was developed as a bird swarm simulation by Kennedy and Eberhart. Swarm intelligence is the capability of an individual bird to deal with the previous experiences of the whole swarm. In PSO, the decision-making process is essential, and it can be made based on the following:

- Personal experiences of individuals that give their best results.
- Experiences of other individuals that give the best results of the entire swarms.

Several researchers tried to enhance the generalization capabilities and performance of ANNs by using PSO algorithms because PSO is one of the robust global search methods that can enhance the performance capacity of ANN by adjusting its bias and weight of it. Furthermore, in the case of the local minimum, ANNs can increase the probability of convergence and, at that time, the PSO is likely to obtain the global minimum. Consequently, the developed PSO-ANN model acquires the search properties of both the ANN and PSO models. So, in the case of the PSO-ANN model, the PSO searches for the global minimum and then exploits towards the local solution, which can be employed by the ANN to find the best results in the search space [87].

In a learning procedure for PSO-ANN, it starts with random assignment of weights and biases of a group of random particles. After this step the PSO-ANN model is trained based on the assigned weights and biases and then, at each iteration, the error is calculated between the predicted and actual value. After that, the calculated error is reduced by changing the particle position. By changing the particle position the best solution is selected and accordingly a new error is achieved. This complete learning process continues unless or until the termination criteria are fulfilled.

4.2. Hybridization of PSO-ELM

Based on theory, Huang et al. [88] demonstrated the performance capability of ELM as a universal approximation, and the ability to engage many activation functions. Several researchers utilize ELM due to well-known features, such as fast learning capability and adequate ability to generalize, the same is deployed for prediction methods [79,89]. The generalizing ability of ELM is further enhanced by merging it with some other methods [90,91]. Several researchers from recent decades have successfully combined a nature-based/inspired algorithm to optimize the ELM model. Mohapatra et al. [92] designed a hybrid model consisting of the cuckoo search algorithm and ELM to classify medical data. The stability analysis of the photovoltaic interactive microgrid was carried out with a firefly algorithm by Satapathy et al. [93]. The evaluation of the aging degree of the insulated gate bipolar transistor was done with a whale optimization algorithm and ELM by Li et al. [94]. Figueiredo and Ludermir [95] studied the different topology of PSO—Global, Local, Von Neumann, Wheel, and Four Clusters—and showed, depending upon the problem, suitable topology, which need to be selected for the best PSO-ELM performance. Many researchers have used PSO-ELM for prediction in various engineering problems. PSO-ELM forecasting models were used to predict the regional groundwater depth [96]. The PSO-ELM approach was used for predicting landslide displacement interval [97]. The PSO-ELM model was deployed for predicting stabilized aggregate bases [98]. The PSO-ELM model was deployed to predict the vibration of the ground caused by the process of blasting [99]. Thus, from various research studies, an optimized version of ELM with other algorithms outperformed individual ELM accuracy levels in prediction jobs. ELM models generally get trapped in local minima because the initialization process is stochastic for the network input weights and hidden biases [100]. Various researchers have applied a combination of PSO and ELM to various areas reliably [101]. During the current study, to the best of our information/knowledge, the PSO-ELM model is developed for the first time to predict the flyrock caused by blasting. The flow chart of PSO-ELM is shown in Figure 7.

4.3. Hybridization of EO-ELM

This study proposes a new combination of hybrid ML models, called EO-ELM, where the EO optimizes the ELM learning parameters to find an optimal configuration of ELM for the prediction of flyrock. Here, the concentrations of EO are ELM learning parameters. The RMSE is considered an objective function for EO. The best equilibrium candidate found by EO is considered as the optimal configuration of ELM for prediction tasks.

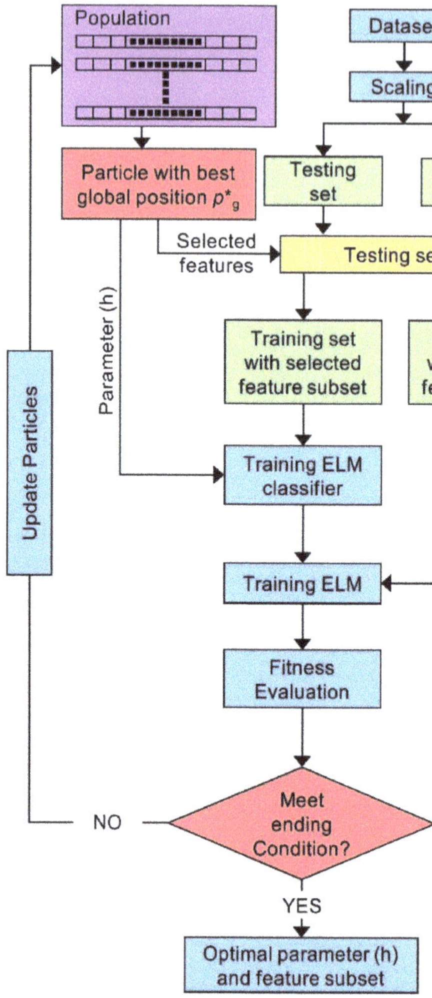

Figure 7. Flow chart of PSO–ELM.

In EO-ELM, initially, all particles do not know the solution space. The collaboration of five equilibrium candidates helps the concentration updating process of particles. At initial periods of iteration, the equilibrium candidates are diverse and the exponential term (F) produces large random numbers, which help particles to cover the entire solution space. Similarly, during the end period of iterations, particles are surrounded by equilibrium candidates which are in the optimum position with similar configurations. At these moments, the exponential term produces a lower value of random numbers, which helps fine-tune candidate solutions. The algorithm of EO-ELM is shown in the following Algorithm 1.

Algorithm 1: The algorithm of EO-ELM. Exponential term (*F*), λ is a turnover rate and defined as a random vector in between 0 and 1, a_2 is used to control the exploitation task. a_1 is used to control the exploration task, *sign*($\vec{r}$ −0.5) component consequences the direction of intensification and diversification of particles, *r* is defined as a random vector in between 0 and 1, generation rate (G), r_1, and r_2 denote the random values between 0 and 1. *GCP* is called generation rate control parameter

1. Select training and testing dataset
2. Begin ELM training
3. Set hidden units of ELM
4. Obtain the number of input weights and hidden biases
5. Initialize the populations (P)
6. Initialize the fitness of four equilibrium candidates
7. Assignment of EO parameters value (a_1 = 2, a_2 = 1, GP = 0.5)
8. for it = 1 to maximum iteration number do
9. for i = 1 to P do
10. Estimate the fitness of the ith particle
11. if fitness ($\vec{P}_i$) < fitness ($\vec{P}_{eq[1]}$)
12. Replace fitness ($\vec{P}_{eq[1]}$) with fitness ($\vec{P}_i$) and $\vec{P}_{eq[1]}$ with $\vec{P}_i$
13. elseif fitness ($\vec{P}_i$) < fitness ($\vec{P}_{eq[1]}$) & fitness ($\vec{P}_i$) < fitness ($\vec{P}_{eq[2]}$)
14. Replace fitness ($\vec{P}_{eq[2]}$) with fitness ($\vec{P}_i$) and $\vec{P}_{eq[2]}$ with $\vec{P}_i$
15. elseif fitness ($\vec{P}_i$) < fitness ($\vec{P}_{eq[1]}$) & fitness ($\vec{P}_i$) < fitness ($\vec{P}_{eq[2]}$) & fitness ($\vec{P}_i$) < fitness ($\vec{P}_{eq[3]}$)
16. Replace fitness ($\vec{P}_{eq[3]}$) with fitness ($\vec{P}_i$) and $\vec{P}_{eq[3]}$ with $\vec{P}_i$
17. elseif fitness ($\vec{P}_i$) < fitness ($\vec{P}_{eq[1]}$) & fitness ($\vec{P}_i$) < fitness ($\vec{P}_{eq[2]}$) & fitness ($\vec{P}_i$) < fitness ($\vec{P}_{eq[3]}$) & fitness ($\vec{P}_i$) < fitness ($\vec{P}_{eq[4]}$)
18. Replace fitness ($\vec{P}_{eq[4]}$) with fitness ($\vec{P}_i$) and $\vec{P}_{eq[4]}$ with $\vec{P}_i$
19. end if
20. end for
21. $\vec{P}_{mean} = (\vec{P}_{eq[1]} + \vec{P}_{eq[2]} + \vec{P}_{eq[3]} + \vec{P}_{eq[4]})/4$
22. $\vec{P}_{eq,pool} = \{\vec{P}_{eq[1]} + \vec{P}_{eq[2]} + \vec{P}_{eq[3]} + \vec{P}_{eq[4]} + \vec{P}_{eq[mean]}\}$ (Equilibrium pool)
23. Allocate $t = \left(1 - \frac{Iteration}{Max_{iteration}}\right)^{(a_2 \times \frac{Iteration}{Max_{iteration}})}$
24. for i = 1 to P do
25. Random generation of vectors $\vec{\lambda}$ and $\vec{r}$
26. Random selection of equilibrium candidate from equilibrium pool
27. Evaluate $\vec{F} = a_1 \times sign(\vec{r} - 0.5)[e^{-\vec{\lambda} t} - 1]$
28. Evaluate $\vec{GCP} = \{ \begin{matrix} 0.5 * r_1 r_2 \geq GP \\ 0 r_2 < GP \end{matrix}$
29. Evaluate $\vec{G}_0 = \vec{GCP} * (\vec{P}_{eq} - \vec{\lambda} \times \vec{P})$
30. Evaluate $\vec{G} = \vec{G}_0 \times \vec{F}$
31. $\vec{P} = \vec{P}_{eq} + (\vec{P} - \vec{P}_{eq}) \cdot \vec{F} + \frac{\vec{G}}{\vec{\lambda} V} * (1 - \vec{F})$ (Concentration update)
32. end for
33. end for
34. Set ELM optimal input weights and hidden biases using $\vec{P}_{eq[1]}$
35. Obtain output weights
36. ELM testing

4.4. Model Verification and Evaluation

One of the most important aspects of the model development process is model verification and evaluation, as it is necessary to understand the behavior of the model to check the evolution of the model towards acquiring the accurate result and to know whether the quality of the test model is excellent or not. To fulfil the desired, a training set is used to train the developed models and a different testing set is used to verify the model development. To evaluate the reliability of the developed model, seven different evaluation matrices, namely determination coefficient (R | 2), root mean square error (*RMSE*), variance account factor (*VAF*), mean absolute error (*MAE*), Nash–Sutcliffe efficiency (*NSE*), mean absolute percentage error (*MAPE*), and a-20 index (A20), were used to define the relation between the actual and predicted value. Out of these evaluation matrices, the *RMSE* shows the standard deviation of the error between the actual and predicted values. The *MAPE* shows the error value percentage with the original data; having 0% *MAPE* shows the perfect model. The *NSE* value is a normalized statistic and is used to measure the goodness of fit of the model. Similarly, in the case of *MAE*, the goodness of the model increases with a decrease in the value of *MAE*. Further, R2 indicates the correlation between the actual and predicted values. The closer the value is to 1, the more perfect the model (1). Similarly, if the value of A20 is closer to 1, this shows a perfect prediction model (2). VAF shows the ratio of error variance to the measured data variance. The calculation formulas for different evaluation matrices are as follows:

$$RMSE = \sqrt{\frac{1}{n}\sum_{i=1}^{n}(y - \overline{y})^2} \tag{1}$$

$$R^2 = \left(\frac{\sum_{i=1}^{q}\left(Y_{E_i} - Y_{\overline{E}_i}\right)\left(Y_{O_i} - Y_{\overline{O}_i}\right)}{\sqrt{\sum_{i=1}^{q}\left(Y_{E_i} - Y_{\overline{E}_i}\right)^2 \sum_{i=1}^{n}\left(Y_{O_i} - Y_{\overline{O}i}\right)^2}}\right)^2 \tag{2}$$

$$MAE = \frac{1}{n}\sum_{i=1}^{n}|(\hat{y}_i - y_i)| \tag{3}$$

$$MAPE = \frac{1}{n}\sum_{i=1}^{n}\left|\frac{y_i - \hat{y}_i}{y_i}\right| \times 100 \tag{4}$$

$$NSE = \left(1 - \frac{\sum_{i=1}^{n}\left(R_{O_i} - R_{E_i}\right)^2}{\sum_{i=1}^{n}\left(R_{O_i} - \overline{R_{O_i}}\right)^2}\right) \tag{5}$$

$$VAF\ (\%) = (1 - \frac{var(Y_{E_i} - Y_{O_i})}{var(Y_{E_i})}) \times 100 \tag{6}$$

$$A20 = \frac{m20}{M} \tag{7}$$

5. Results and Discussion

The objective of this study was to predict the flyrock distance. Therefore, crucial blast design parameters were selected as input parameters (hole diameter, burden, rock density, stemming length, charge per meter, powder factor, blastability index, and weathering index). After that, the PSO-ELM, PSO-ANN, and EO-ELM models were developed and used for the prediction of flyrock. When data are split into train and test sets, it always becomes a challenging task to develop a generalized data-driven model. This work used 80% and 20% data split ratios for train and test data, respectively. These split data are used to test and compare the performance of the developed models.

To prove the model's effectiveness with the split dataset, a 10-times average run of three models is checked and compared. An average run of optimization-coupled ML models is useful to check the randomness in problem-solving for the optimal parameter set of optimization algorithms. Table 2 shows the optimal parameter set of metaheuristic algorithms, which are initially set by heat and trial method.

Table 2. Optimal parameter values of metaheuristic algorithms for eight hidden neurons in ELM.

Model	Parameters	Value
EO-ELM	Maximum Iteration	500
	Size of Population	25
	a1	2.5
	a2	2.5
	GP	0.6
PSO-ELM	Maximum Iteration	500
	Size of Population	25
	C1	1
	C2	2
	W (inertia weight)	0.9
PSO-ANN	Maximum Iteration	500
	Size of Population	25
	C1	1
	C2	2
	W	0.98

During training, 500 iterations are set for each optimization-coupled ML model. The convergence plot of RMSE vs. iteration count is plotted and shown in Figure 8. Figure 8 shows faster (around 200 iterations) and better convergence ability of the EO-ELM compared to the PSO-ELM and PSO-ANN. The PSO-ELM and PSO-ANN become stuck in a local solution with premature convergence (Figure 8). Furthermore, the PSO-ANN may have an improper learning rate and overfitting issues due to the use of stochastic selection or gradient-based learning algorithms. As gradient-based learning algorithms intend to reach the minimum training error, but do not consider the magnitude of weights and only uses differentiable activation functions; due to this, they have less generalization performance [102]. In spite of this, other models use ELM, which is extremely fast and can train SLFNs using non-differentiable activation function to reach the solutions in a straightforward way without having issues, such as local minimum, improper learning rate, and overfitting. It tends to reach the smallest training error including the smallest norm of weights; due to this, it has the better generalization performance [102].

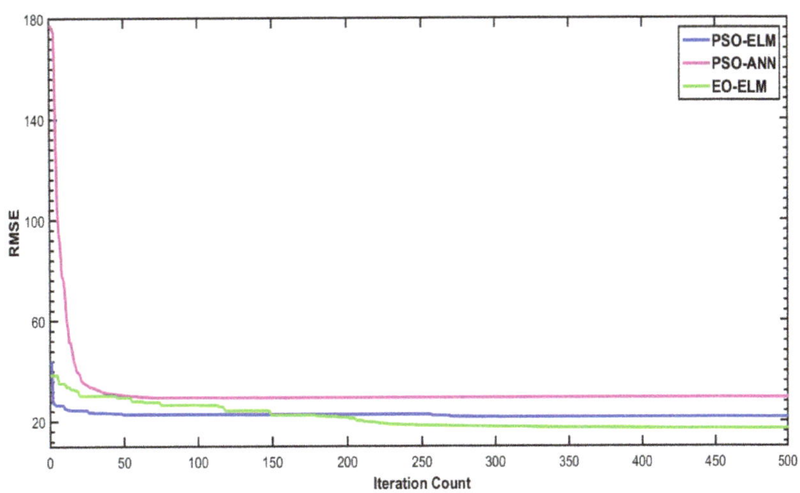

Figure 8. Convergence plot of three models (training period).

The training phase scatter diagrams for the EO-ELM, PSO-ANN, and PSO-ELM are shown in Figures 9–11, respectively. It is apparent from Figures 9–11 that the EO-ELM (Figure 9) predicts flyrock values more accurately compared to the PSO-ANN (Figure 10) and PSO-ELM (Figure 11). The testing phase scatter diagrams for the EO-ELM, PSO-ANN, and PSO-ELM are shown in Figures 12–14, respectively. It is evident that the EO-ELM (Figure 12) predicts the test flyrock data better compared to the PSO-ANN (Figure 13) and PSO-ELM (Figure 14). Table 3 shows linear equations of the predicted and measured values for the EO-ELM, PSO-ANN, and PSO-ELM for training and testing data, respectively.

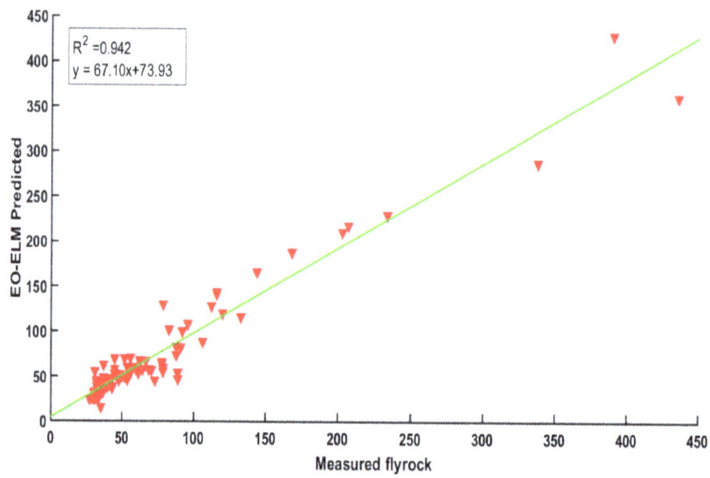

Figure 9. Comparison of predicted flyrock with the EO-ELM vs. measured flyrock.

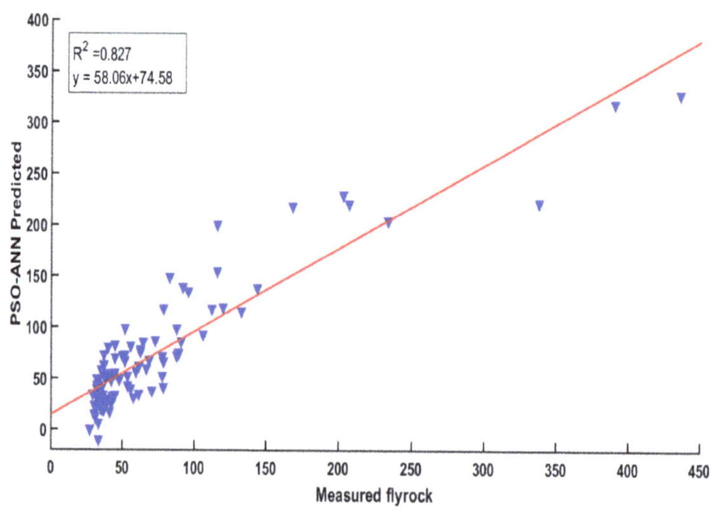

Figure 10. Measured flyrock and predicted values of flyrock with the PSO-ANN for training data.

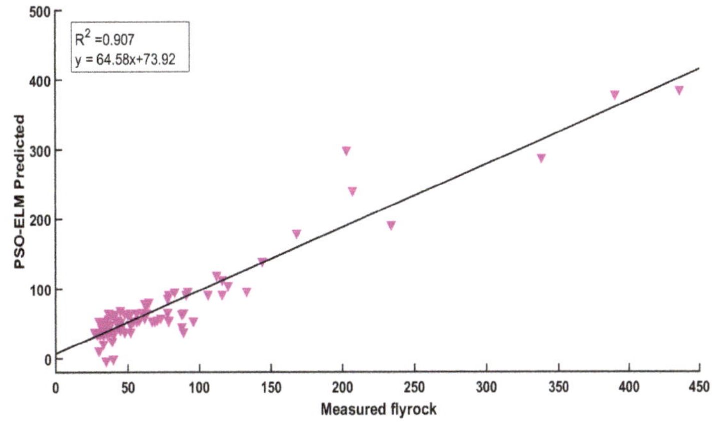

Figure 11. Measured flyrock and predicted values of flyrock with the PSO-ELM for training data.

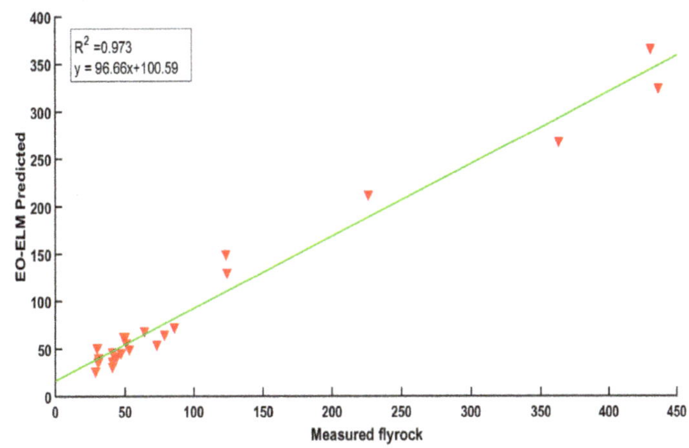

Figure 12. Comparison of measured flyrock vs. the EO-ELM prediction for testing data.

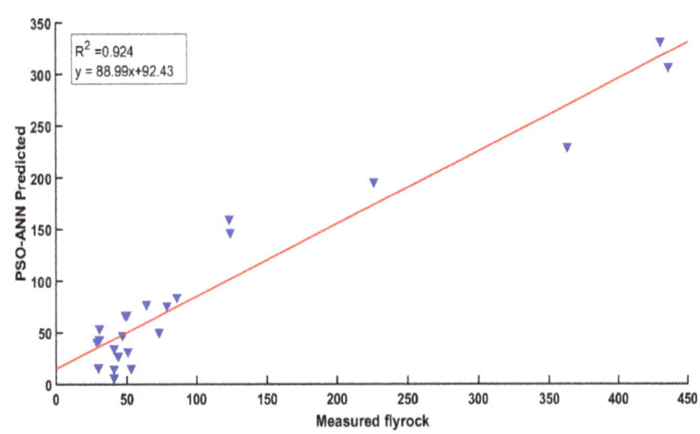

Figure 13. Comparison of measured flyrock vs. the PSO-ANN prediction for testing data.

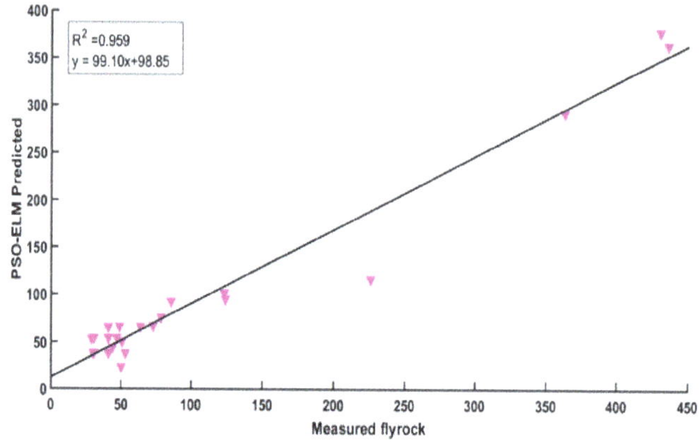

Figure 14. Comparison of measured flyrock vs. the PSO-ELM prediction for testing data.

Table 3. Linear equations for predicted and measured values for the EO-ELM, PSO-ANN, and PSO-ELM for training and testing data.

Model	Training Data	Testing Data
EO-ELM	67.10x + 73.93	96.66x + 100.59
PSO-ANN	58.06x + 74.58	88.9x + 92.43
PSO-ELM	64.58x + 73.92	99.10x + 98.85

Table 4 shows the better prediction efficiency of the EO-ELM in the training and testing period compared to the PSO-ANN and PSO-ELM in terms of seven matrices. In the testing period, the developed EO-ELM (R2 = 0.97, RMSE = 34.82, MAE = 20.3, MAPE = 17.60, NSE = 0.978, VAF = 97.88, A20 = 0.65) performed better compared to the PSO-ELM (R2 = 0.959, RMSE = 35.7 MAE = 23.53, MAPE = 21.84, NSE = 0.96, VAF = 95.79, A20 = 0.56) and PSO-ANN (R2 = 0.924, RMSE = 48.12, MAE = 31.68, MAPE = 24.25, NSE = 0.93, VAF = 92.89, A20 = 0.35). Furthermore, for better representation in terms of model deviations, the receiver operating characteristic (ROC) curve was drawn. It is evident that all of the models capture the good relationship in the prediction of flyrock during training (Figure 15) and minimum deviation was found for the EO-ELM followed by the PSO-ELM and PSO-ANN. Amongst the models, the EO-ELM shows comparatively lesser deviation during the training period. During the testing period, a similar pattern was observed, the EO-ELM shows the minimum deviations followed by the PSO-ELM and PSO-ANN (Figure 16).

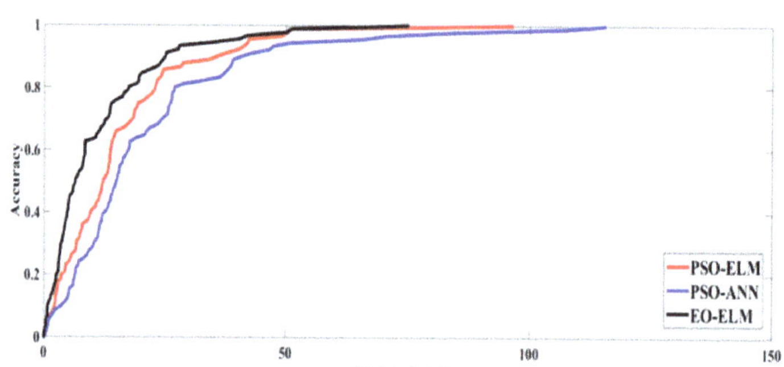

Figure 15. Diagram of REC for training dataset.

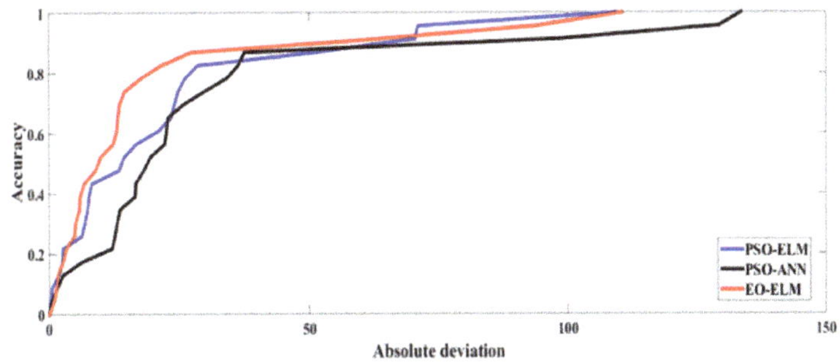

Figure 16. Diagram of REC for testing dataset.

Table 4. Comparison of three models in terms of seven matrices.

	Training Data Sets						
	R^2	**RMSE**	**MAE**	**MAPE**	**NSE**	**VAF**	**A20**
EO-ELM	0.942	17.02	11.26	21.20	0.946	94.62	0.53
PSO-ANN	0.827	29.5	21.07	27.04	0.821	82.19	0.43
PSO-ELM	0.907	21.56	15.64	24.27	0.900	90.08	0.45
	Testing Data Sets (Continued)						
	R^2	**RMSE**	**MAE**	**MAPE**	**NSE**	**VAF**	**A20**
EO-ELM	0.973	34.82	20.3	17.60	0.978	97.88	0.65
PSO-ANN	0.924	48.12	31.68	24.25	0.93	92.89	0.35
PSO-ELM	0.959	35.7	23.53	21.84	0.96	95.79	0.56

5.1. Average Performance of Models

Table 5 shows the average results for the 10-times run of three models. The EO-ELM shows the best average prediction performance compared to the PSO-ELM and PSO-ANN in terms of all matrices (Table 5). Figure 17a shows the most generalized performance of the EO-ELM (training phase) at each of the runs (10-times) compared to the PSO-ELM and PSO-ANN. Figure 17b shows that the EO-ELM has the best average convergence rate compared to the PSO-ELM and PSO-ANN.

Table 5. Comparison of average results for 1-times run of models.

	Training Data Sets						
	R^2	**RMSE**	**MAE**	**MAPE**	**NSE**	**VAF**	**A20**
EO-ELM	0.95	16.66	12.13	19.87	0.95	94.46	0.60
PSO-ANN	0.83	29.68	19.33	29.30	0.82	82.06	0.41
PSO-ELM	0.88	23.68	16.76	26.96	0.88	88.29	0.47
	Testing Data Sets (Continued)						
	R^2	**RMSE**	**MAE**	**MAPE**	**NSE**	**VAF**	**A20**
EO-ELM	0.97	32.14	19.78	20.37	0.93	93.97	0.57
PSO-ANN	0.87	64.44	36.02	29.96	0.72	74.72	0.33
PSO-ELM	0.88	48.55	26.97	26.71	0.84	84.84	0.51

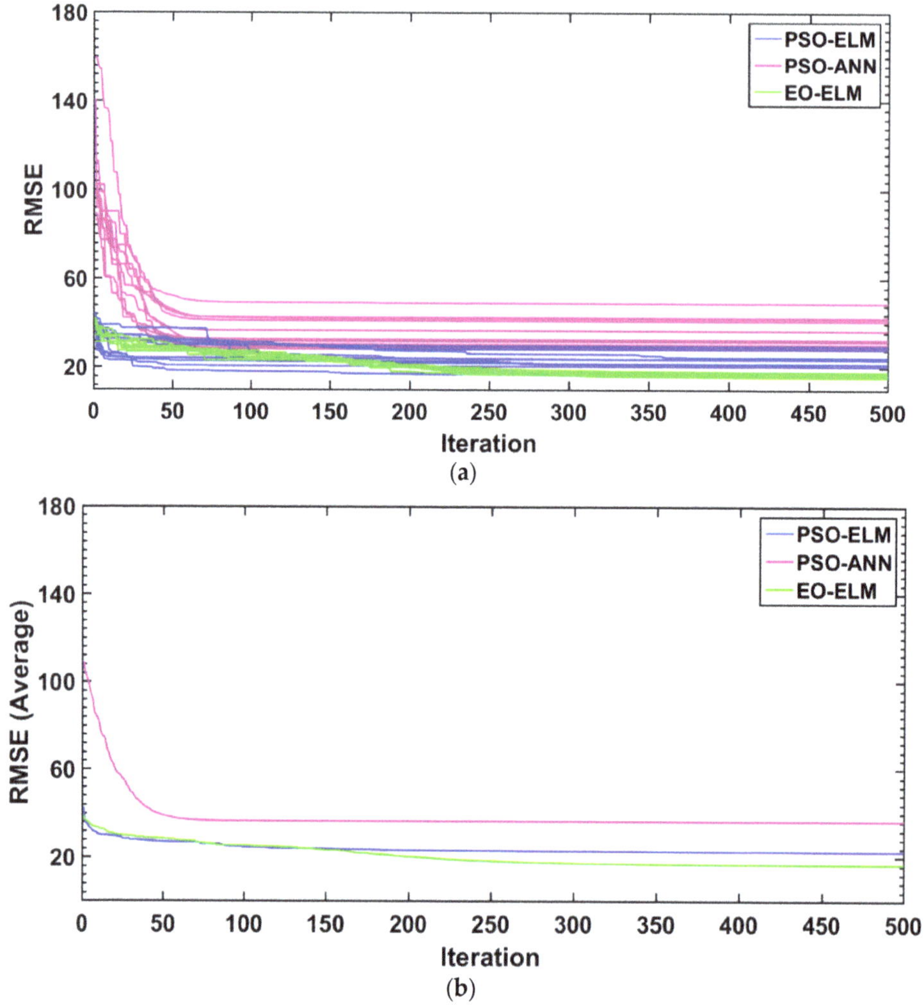

Figure 17. Convergence of the models for (**a**) 10 runs, and (**b**) average of 10 runs.

5.2. Anderson–Darling (A–D) Test

A non-parametric test called the A–D test was performed to assess the normality of all three models [68]. The p-values for the PSO-ELM, PSO-ANN, and EO-ELM models are less than the significance level of 0.05 (Table 6). Table 6 shows that the EO-ELM is the best performing model in estimating flyrock.

Table 6. Comparison of average results for 10-times run of models.

	Count	Mean	Median	SD	AD	p-Value
Actual	114	81.307	50.5	85.927	0	1
PSO-ELM	114	78.951	54.632	75.877	2.843	0.03244
PSO-ANN	114	78.183	55.035	70.484	2.308	0.00619
EO-ELM	114	79.311	53.492	76.092	0.8886	0.004215

5.3. Sensitivity Analysis

Sensitivity analysis of parameters, with respect to the weathering index, was carried out as shown in Figure 18. It showed the corresponding relationship of the parameter with respect to the measured flyrock distance based on the cosine amplitude. The application of this method was based on expressing all data pairs in a common *Z*-space. The following equation defines a data array Z based on data pairs of each input and output:

$$Z = \{z_1, z_2, z_3, z \ldots, z_i, z_n\}.Z = \{Z_1, Z_2, Z_3 \ldots, Z_i, \ldots, Z_n\} \tag{8}$$

whereas, z_i is a vector of length m in array Z, that as Equation (9):

$$z_i = \{z_{i1}, z_{i2}, z_{i3} \ldots, z_I\}.z_{i1} = \{Z_{i1}, Z_{i2}, Z_{i3} \ldots, Z_I\} \tag{9}$$

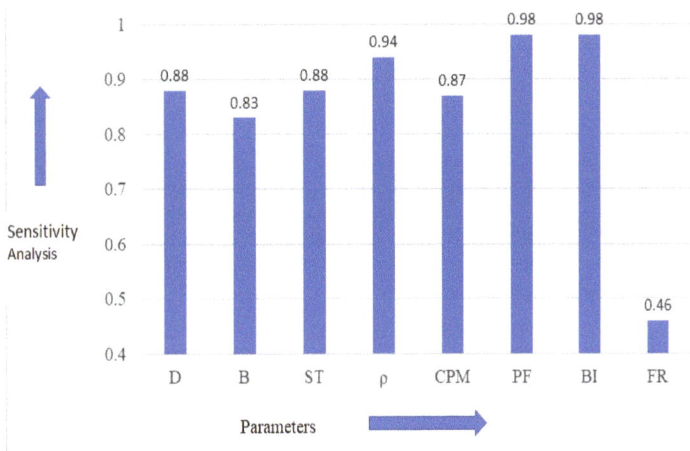

Figure 18. Sensitivity analysis of parameters with respect to weathering index.

Each point required *m*-coordinates for describing completely by training each of these data pairs in m dimensional space. The results were achieved as all of the points were in the spaced pair. The following equation shows the strength of relation (r_{ij}) between the data set Z_i and Z_j as Equation (10):

$$r_{ij} = \frac{\sum_{k=1}^{m} z_{ik} z_{jk}}{\sqrt{\sum_{k=1}^{m} z_{ik}^2 \sum_{k=1}^{m} z_k^2}} \tag{10}$$

The input parameters were selected, which were most sensitive, to apply to various prediction models and identify the best model suitable for comparing the predicted value and the measured value of flyrock distance.

6. Conclusions

This study uses three hybrid models: the EO–ELM, PSO–ANN, and PSO-ELM to predict flyrock. Out of these three hybrid models, the EO-ELM is proposed and the rest were used to validate the performance of the proposed model. The seven different matrices (R^2, RMSE, MAPE, NSE, MAE, VAF, and A20) are used for comparing the efficacy of the developed model. The developed EO-ELM model performed better compared to PSO-the ELM and PSO-ANN in predicting flyrock (Table 4). Further, all models were run 10-times and average results are shown in Table 5. It was observed that the EO-ELM model outperformed the PSO-ELM and PSO-ANN in average results. Furthermore, the 10-times run of the EO-ELM model showed better convergence capability (Figure 17) than the other two. Further, the A–D test showed that the EO-ELM model has better performance efficiency compared to the PSO-ELM and PSO-ANN. A sensitivity analysis was done introducing a new parameter, WI. The PF and BI showed the highest sensitivity with 0.98 each (Figure 18).

The limitation of this study is that this study was carried out for a particular limestone mine in Thailand. Therefore, the obtained results may not be suitable for other geological settings, i.e., in granite or other quarries. So, there is a need of further research by considering non controllable parameters of the blastability index, WI. On the other hand, by refining the controllable parameters, accuracy of the prediction of flyrock can be improved. There are a large number of mines near to the limestone mine under study and if large data sets are collected, further reliability of the prediction models can be improved through future research. The number of input parameters is eight. However, future studies of the prediction of flyrock can be done by limiting to five influential input parameters. The use of the latest technology, such as the video recording of flyrock with drones, will add value for future research. Furthermore, the wavelet frequency domain parameter or innovative liquid carbon dioxide rock-breaking technology are recent technologies used for blasting. Thus, there is a need to develop a new technology alternative to blasting.

Author Contributions: Conceptualization, E.T.M., R.M.B., D.K. and B.R.; methodology, R.M.B. and R.K.; software, R.K., D.K. and B.R.; formal analysis, R.M.B. and D.K.; resources, E.T.M. and R.M.B.; data curation, R.M.B.; writing—original draft, E.T.M., R.M.B., D.K., BR., M.M.S.S., S.K. and R.K.; writing—review and editing, R.MB., E.T.M., RK., M.M.S.S. and D.K.; supervision, E.T.M., D.K. and B.R.; funding acquisition, M.M.S.S. All authors have read and agreed to the published version of the manuscript.

Funding: The research is partially funded by the Ministry of Science and Higher Education of the Russian Federation under the strategic academic leadership program 'Priority 2030' (Agreement 075-15-2021-1333 dated 30 September 2021).

Institutional Review Board Statement: Not applicable.

Informed Consent Statement: Not applicable.

Data Availability Statement: The data presented in this study are available on request from the corresponding author.

Acknowledgments: Authors are thankful to Edy Toonizam Mohamad, Dean of Department of Civil Engineering, Universiti Teknologi Malaysia for encouragement for this research. Authors are also thankful to Vitaly Sergeev, Centre of Peter the Great St. Petersburg Polytechnic University, 195251 St. Petersburg, Russia for guidance and supervision of this paper.

Conflicts of Interest: The authors declare no conflict of interest.

References

1. Bhandari, S. Engineering Rock Blasting Operations. Available online: https://www.osti.gov/etdeweb/biblio/661808 (accessed on 3 September 2022).
2. Roy, P.P. *Rock Blasting: Effects and Operations*; IBH Publishing: New Delhi, India; Oxford, UK, 2005.
3. Mohamad, E.T.; Yi, C.S.; Murlidhar, B.R.; Saad, R. Effect of Geological Structure on Flyrock Prediction in Construction Blasting. *Geotech. Geol. Eng.* **2018**, *36*, 2217–2235. [CrossRef]
4. Li, H.B.; Zhao, J.; Li, J.R.; Liu, Y.Q.; Zhou, Q.C. Experimental studies on the strength of different rock types under dynamic compression. *Int. J. Rock Mech. Min. Sci.* **2004**, *41*, 68–73. [CrossRef]
5. Khandelwal, M.; Singh, T.N. Prediction of blast-induced ground vibration using artificial neural network. *Int. J. Rock Mech. Min. Sci.* **2009**, *46*, 1214–1222. [CrossRef]
6. Raina, A.K.; Murthy, V.M.S.R.; Soni, A.K. Estimating flyrock distance in bench blasting through blast induced pressure measurements in rock. *Int. J. Rock Mech. Min. Sci.* **2015**, *76*, 209–216. [CrossRef]
7. Bhatawdekar, R.M.; Edy, M.T.; Danial, J.A. Building information model for drilling and blasting for tropically weathered rock. *J. Mines Met. Fuels* **2019**, *67*, 494–500.
8. Mohamad, E.T.; Murlidhar, B.R.; Armaghani, D.J.; Saad, R.; Yi, C.S. Effect of Geological Structure and Blasting Practice in Fly Rock Accident at Johor, Malaysia. *J. Teknol.* **2016**, *78*. [CrossRef]

9. Sastry, V.R.; Chandar, K.R. Assessment of objective based blast performance: Ranking system. In *Measurement and Analysis of Blast Fragmentation*; CRC Press: New Delhi, India, 2009; pp. 13–17.
10. Kanchibotla, S.S.; Valery, W.; Morrell, S. Modelling fines in blast fragmentation and its impact on crushing and grinding. In *Explo '99—A Conference on Rock Breaking*; The Australasian Institute of Mining and Metallurgy: Kalgoorlie, Australia, 1999; pp. 137–144.
11. Armaghani, D.J. Rock fragmentation prediction through a new hybrid model based on imperial competitive algorithm and neural network. *Smart Constr. Res.* **2018**, *2*, 1–12. [CrossRef]
12. Thornton, D.; Kanchibotla, S.S.; Brunton, I. Modelling the Impact of Rockmass and Blast Design Variation on Blast Fragmentation. *Fragblast* **2002**, *6*, 169–188. [CrossRef]
13. Cunningham, C.V.B. The Kuz-Ram fragmentation model–20 years on. In *Brighton Conference Proceedings*; European Federation of Explosives Engineer: Brighton, UK, 2005; Volume 2005, pp. 201–210.
14. Venkatesh, H.S.; Bhatawdekar, R.M.; Adhikari, G.R.; Theresraj, A.I. Assessment and Mitigation of Ground Vibrations and Flyrock at a Limestone Quarry. 1999, pp. 145–152. Available online: https://www.academia.edu/34145884/ASSESSMENT_AND_MITIGATION_OF_GROUND_VIBRATIONS_AND_FLYROCK_AT_A_LIMESTONE_QUARRY (accessed on 3 September 2022).
15. Raina, A.K.; Chakraborty, A.K.; Choudhury, P.B.; Sinha, A. Flyrock danger zone demarcation in opencast mines: A risk based approach. *Bull. Eng. Geol. Environ.* **2011**, *70*, 163–172. [CrossRef]
16. Xiong, H.; Yin, Z.Y.; Nicot, F. A multiscale work-analysis approach for geotechnical structures. *Int. J. Numer. Anal. Methods Geomech.* **2019**, *43*, 1230–1250. [CrossRef]
17. Xiong, H.; Yin, Z.Y.; Nicot, F. Programming a micro-mechanical model of granular materials in Julia. *Adv. Eng. Softw.* **2020**, *145*, 102816. [CrossRef]
18. Xiong, H.; Yin, Z.Y.; Zhao, J.; Yang, Y. Investigating the effect of flow direction on suffusion and its impacts on gap-graded granular soils. *Acta Geotech.* **2021**, *16*, 399–419. [CrossRef]
19. Xiong, H.; Wu, H.; Bao, X.; Fei, J. Investigating effect of particle shape on suffusion by CFD-DEM modeling. *Constr. Build. Mater.* **2021**, *289*, 123043. [CrossRef]
20. Chen, F.; Xiong, H.; Wang, X.; Yin, Z. Transmission effect of eroded particles in suffusion using CFD-DEM coupling method. *Acta Geotech.* **2022**, 1–20. [CrossRef]
21. Fu, Y.; Zeng, D.; Xiong, H.; Li, X.; Chen, Y. Seepage effect on failure mechanisms of the underwater tunnel face via CFD-DEM coupling. *Comput. Geotech.* **2020**, *121*, 103449. [CrossRef]
22. Xiong, H.; Yin, Z.Y.; Nicot, F.; Wautier, A.; Marie, M.; Darve, F.; Veylon, G.; Philippe, P. A novel multi-scale large deformation approach for modelling of granular collapse. *Acta Geotech.* **2021**, *16*, 2371–2388. [CrossRef]
23. Wang, X.; Yin, Z.Y.; Xiong, H.; Su, D.; Feng, Y.T. A spherical-harmonic-based approach to discrete element modeling of 3D irregular particles. *Int. J. Numer. Methods Eng.* **2021**, *122*, 5626–5655. [CrossRef]
24. Wang, X.; Yin, Z.Y.; Zhang, J.Q.; Xiong, H.; Su, D. Three-dimensional reconstruction of realistic stone-based materials with controllable stone inclusion geometries. *Constr. Build. Mater.* **2021**, *305*, 124240. [CrossRef]
25. Chen, G.; Li, Q.-Y.; Li, D.-Q.; Wu, Z.-Y.; Liu, Y. Main frequency band of blast vibration signal based on wavelet packet transform. *Appl. Math. Model.* **2019**, *74*, 569–585. [CrossRef]
26. Li, Q.-Y.; Chen, G.; Luo, D.-Y.; Ma, H.-P.; Liu, Y. An experimental study of a novel liquid carbon dioxide rock-breaking technology. *Int. J. Rock Mech. Min. Sci.* **2020**, *128*, 104244. [CrossRef]
27. Hudaverdi, T.; Akyildiz, O. A new classification approach for prediction of flyrock throw in surface mines. *Bull. Eng. Geol. Environ.* **2019**, *78*, 177–187. [CrossRef]
28. Kecojevic, V.; Radomsky, M. Flyrock phenomena and area security in blasting-related accidents. *Saf. Sci.* **2005**, *43*, 739–750. [CrossRef]
29. Bajpayee, T.S.; Rehak, T.R.; Mowrey, G.L.; Ingram, D.K. Blasting injuries in surface mining with emphasis on flyrock and blast area security. *J. Saf. Res.* **2004**, *35*, 47–57. [CrossRef] [PubMed]
30. Adhikari, G.R. Studies on Flyrock at Limestone Quarries. *Rock Mech. Rock Eng.* **1999**, *32*, 291–301. [CrossRef]
31. Raina, A.K.; Murthy, V.M.S.R.; Soni, A.K. Flyrock in bench blasting: A comprehensive review. *Bull. Eng. Geol. Environ.* **2014**, *73*, 1199–1209. [CrossRef]
32. Hasanipanah, M.; Amnieh, H.B. A Fuzzy Rule-Based Approach to Address Uncertainty in Risk Assessment and Prediction of Blast-Induced Flyrock in a Quarry. *Nat. Resour. Res.* **2020**, *29*, 669–689. [CrossRef]
33. Bhatawdekari, R.M.; Danial, J.A.; Edy, T.M. A review of prediction of blast performance using computational techniques. In Proceedings of the ISERME 2018 International Symposium on Earth Resources Management & Environment, Thalawathugoda, Sri Lanka, 24 August 2018.
34. Sharma, L.K.; Vishal, V.; Singh, T.N. Developing novel models using neural networks and fuzzy systems for the prediction of strength of rocks from key geomechanical properties. *Measurement* **2017**, *102*, 158–169. [CrossRef]
35. Koohmishi, M. Assessment of strength of individual ballast aggregate by conducting point load test and establishment of classification method. *Int. J. Rock Mech. Min. Sci.* **2021**, *141*, 104711. [CrossRef]
36. Monjezi, M.; Bahrami, A.; Varjani, A.Y.; Sayadi, A.R. Prediction and controlling of flyrock in blasting operation using artificial neural network. *Arab. J. Geosci.* **2011**, *4*, 421–425. [CrossRef]
37. Ghasemi, E.; Amini, H.; Ataei, M.; Khalokakaei, R. Application of artificial intelligence techniques for predicting the flyrock distance caused by blasting operation. *Arab. J. Geosci.* **2014**, *7*, 193–202. [CrossRef]

38. Dehghani, H.; Ataee-Pour, M. Development of a model to predict peak particle velocity in a blasting operation. *Int. J. Rock Mech. Min. Sci.* **2011**, *48*, 51–58. [CrossRef]
39. Armaghani, D.J.; Mahdiyar, A.; Hasanipanah, M.; Faradonbeh, R.S.; Khandelwal, M.; Amnieh, H.B. Risk Assessment and Prediction of Flyrock Distance by Combined Multiple Regression Analysis and Monte Carlo Simulation of Quarry Blasting. *Rock Mech. Rock Eng.* **2016**, *49*, 3631–3641. [CrossRef]
40. Marto, A.; Hajihassani, M.; Armaghani, D.J.; Mohamad, E.T.; Makhtar, A.M. A Novel Approach for Blast-Induced Flyrock Prediction Based on Imperialist Competitive Algorithm and Artificial Neural Network. *Sci. World J.* **2014**. [CrossRef] [PubMed]
41. Jahed Armaghani, D.; Hajihassani, M.; Monjezi, M.; Mohamad, E.T.; Marto, A.; Moghaddam, M.R. Application of two intelligent systems in predicting environmental impacts of quarry blasting. *Arab. J. Geosci.* **2015**, *8*, 9647–9665. [CrossRef]
42. Trivedi, R.; Singh, T.N.; Gupta, N. Prediction of Blast-Induced Flyrock in Opencast Mines Using ANN and ANFIS. *Geotech. Geol. Eng.* **2015**, *33*, 875–891. [CrossRef]
43. Hasanipanah, M.; Armaghani, D.J.; Amnieh, H.B.; Majid, M.Z.A.; Tahir, M.M.D. Application of PSO to develop a powerful equation for prediction of flyrock due to blasting. *Neural Comput. Appl.* **2017**, *28*, 1043–1050. [CrossRef]
44. Faradonbeh, R.S.; Armaghani, D.J.; Amnieh, H.B.; Mohamad, E.T. Prediction and minimization of blast-induced flyrock using gene expression programming and firefly algorithm. *Neural Comput. Appl.* **2018**, *29*, 269–281. [CrossRef]
45. Kumar, N.; Mishra, B.; Bali, V. A Novel Approach for Blast-Induced Fly Rock Prediction Based on Particle Swarm Optimization and Artificial Neural Network. In *Proceedings of International Conference on Recent Advancement on Computer and Communication*; Springer: Berlin/Heidelberg, Germany, 2018; Volume 34, pp. 19–27. [CrossRef]
46. Rad, H.N.; Bakhshayeshi, I.; Jusoh, W.A.W.; Tahir, M.M.; Foong, L.K. Prediction of Flyrock in Mine Blasting: A New Computational Intelligence Approach. *Nat. Resour. Res.* **2020**, *29*, 609–623. [CrossRef]
47. Koopialipoor, M.; Fallah, A.; Armaghani, D.J.; Azizi, A.; Mohamad, E.T. Three hybrid intelligent models in estimating flyrock distance resulting from blasting. *Eng. Comput.* **2019**, *35*, 243–256. [CrossRef]
48. Kotsiantis, S.B.; Zaharakis, I.; Pintelas, P. Supervised machine learning: A review of classification techniques. *Emerg. Artif. Intell. Appl. Comput. Eng.* **2007**, *160*, 3–24.
49. Blum, A.L.; Langley, P. Selection of relevant features and examples in machine learning. *Artif. Intell.* **1997**, *97*, 245–271. [CrossRef]
50. Amini, H.; Gholami, R.; Monjezi, M.; Torabi, S.R.; Zadhesh, J. Evaluation of flyrock phenomenon due to blasting operation by support vector machine. *Neural Comput. Appl.* **2012**, *21*, 2077–2085. [CrossRef]
51. Rad, H.N.; Hasanipanah, M.; Rezaei, M.; Eghlim, A.L. Developing a least squares support vector machine for estimating the blast-induced flyrock. *Eng. Comput.* **2018**, *34*, 709–717. [CrossRef]
52. Lu, X.; Hasanipanah, M.; Brindhadevi, K.; Amnieh, H.B.; Khalafi, S. ORELM: A Novel Machine Learning Approach for Prediction of Flyrock in Mine Blasting. *Nat. Resour. Res.* **2020**, *29*, 641–654. [CrossRef]
53. Jahed Armaghani, D.; Tonnizam Mohamad, E.; Hajihassani, M.; Alavi Nezhad Khalil Abad, S.V.; Marto, A.; Moghaddam, M.R. Evaluation and prediction of flyrock resulting from blasting operations using empirical and computational methods. *Eng. Comput.* **2016**, *32*, 109–121. [CrossRef]
54. Ghasemi, E.; Sari, M.; Ataei, M. Development of an empirical model for predicting the effects of controllable blasting parameters on flyrock distance in surface mines. *Int. J. Rock Mech. Min. Sci.* **2012**, *52*, 163–170. [CrossRef]
55. Gupta, R.N. *Surface Blasting and Its Impact on Environment. Impact of Mining on Environment*; Ashish Publishing House: New Delhi, India, 1980; pp. 23–24.
56. Little, T.N. *Flyrock risk.* In *Proceedings EXPLO*; 2007; pp. 3–4.
57. Richards, A.; Moore, A. Flyrock control-by chance or design. In *Proceedings of the Annual Conference on Explosives and Blasting Technique*; International Society for Environmental Ethics (ISEE), 2004; Volume 1, pp. 335–348.
58. Trivedi, R.; Singh, T.N.; Raina, A.K. Prediction of blast-induced flyrock in Indian limestone mines using neural networks. *J. Rock Mech. Geotech. Eng.* **2014**, *6*, 447–454. [CrossRef]
59. Zhou, J.; Aghili, N.; Ghaleini, E.N.; Bui, D.T.; Tahir, M.M.; Koopialipoor, M. A Monte Carlo simulation approach for effective assessment of flyrock based on intelligent system of neural network. *Eng. Comput.* **2020**, *36*, 713–723. [CrossRef]
60. Lundborg, N.; Persson, A.; Ladegaard-Pedersen, A.; Holmberg, R. Keeping the lid on flyrock in open-pit blasting. *Eng. Min. J.* **1975**, *176*, 95–100.
61. Roth, J. *A Model for the Determination of Flyrock Range as a Function of Shot Conditions*; National Technical Information Service: Alexandria, VA, USA, 1979.
62. Stojadinović, S.; Pantović, R.; Žikić, M. Prediction of flyrock trajectories for forensic applications using ballistic flight equations. *Int. J. Rock Mech. Min. Sci.* **2011**, *48*, 1086–1094. [CrossRef]
63. Bui, X.-N.; Nguyen, H.; Le, H.-A.; Bui, H.-B.; Do, N.-H. Prediction of Blast-induced Air Over-pressure in Open-Pit Mine: Assessment of Different Artificial Intelligence Techniques. *Nat. Resour. Res.* **2020**, *29*, 571–591. [CrossRef]
64. Kasabov, N.; Scott, N.M.; Tu, E.; Marks, S.; Sengupta, N.; Capecci, E.; Othman, M.; Doborjeh, M.G.; Murli, N.; Hartono, R.; et al. Evolving spatio-temporal data machines based on the NeuCube neuromorphic framework: Design methodology and selected applications. *Neural Netw.* **2016**, *78*, 1–14. [CrossRef] [PubMed]
65. Hajihassani, M.; Armaghani, D.J.; Marto, A.; Mohamad, E.T. Ground vibration prediction in quarry blasting through an artificial neural network optimized by imperialist competitive algorithm. *Bull. Eng. Geol. Environ.* **2015**, *74*, 873–886. [CrossRef]

66. Roy, B.; Singh, M.P. An empirical-based rainfall-runoff modelling using optimization technique. *Int. J. River Basin Manag.* **2020**, *18*, 49–67. [CrossRef]
67. Roy, B.; Singh, M.P.; Singh, A. A novel approach for rainfall-runoff modelling using a biogeography-based optimization technique. *Int. J. River Basin Manag.* **2021**, *19*, 67–80. [CrossRef]
68. Kumar, R.; Singh, M.P.; Roy, B.; Shahid, A.H. A Comparative Assessment of Metaheuristic Optimized Extreme Learning Machine and Deep Neural Network in Multi-Step-Ahead Long-term Rainfall Prediction for All-Indian Regions. *Water Resour. Manag.* **2021**, *35*, 1927–1960. [CrossRef]
69. Zhou, J.; Koopialipoor, M.; Murlidhar, B.R.; Fatemi, S.A.; Tahir, M.M.; Armaghani, D.J.; Li, C. Use of Intelligent Methods to Design Effective Pattern Parameters of Mine Blasting to Minimize Flyrock Distance. *Nat. Resour. Res.* **2020**, *29*, 625–639. [CrossRef]
70. Murlidhar, B.R.; Armaghani, D.J.; Mohamad, E.T. Intelligence Prediction of Some Selected Environmental Issues of Blasting: A Review. *Open Constr. Build. Technol. J.* **2020**, *14*, 298–308. [CrossRef]
71. Armaghani, D.J.; Hasanipanah, M.; Mahdiyar, A.; Majid, M.Z.A.; Amnieh, H.B.; Tahir, M.M.D. Airblast prediction through a hybrid genetic algorithm-ANN model. *Neural Comput. Appl.* **2018**, *29*, 619–629. [CrossRef]
72. Murlidhar, B.R.; Kumar, D.; Armaghani, D.J.; Mohamad, E.T.; Roy, B.; Pham, B.T. A Novel Intelligent ELM-BBO Technique for Predicting Distance of Mine Blasting-Induced Flyrock. *Nat. Resour. Res.* **2020**, *29*, 4103–4120. [CrossRef]
73. Faradonbeh, R.S.; Hasanipanah, M.; Amnieh, H.B.; Armaghani, D.J.; Monjezi, M. Development of GP and GEP models to estimate an environmental issue induced by blasting operation. *Environ. Monit. Assess.* **2018**, *190*, 351. [CrossRef]
74. Anand, A.; Suganthi, L. Forecasting of Electricity Demand by Hybrid ANN-PSO Models. In *Deep Learning and Neural Networks: Concepts, Methodologies, Tools, and Applications*; IGI Global: Hershey, PA, USA, 2020; pp. 865–882. [CrossRef]
75. Armaghani, D.J.; Hajihassani, M.; Mohamad, E.T.; Marto, A.; Noorani, S.A. Blasting-induced flyrock and ground vibration prediction through an expert artificial neural network based on particle swarm optimization. *Arab. J. Geosci.* **2014**, *7*, 5383–5396. [CrossRef]
76. Cai, W.; Yang, J.; Yu, Y.; Song, Y.; Zhou, T.; Qin, J. PSO-ELM: A Hybrid Learning Model for Short-Term Traffic Flow Forecasting. *IEEE Access* **2020**, *8*, 6505–6514. [CrossRef]
77. Zeng, J.; Roy, B.; Kumar, D.; Mohammed, A.S.; Armaghani, D.J.; Zhou, J.; Mohamad, E.T. Proposing several hybrid PSO-extreme learning machine techniques to predict TBM performance. *Eng. Comput.* **2021**, 1–17. [CrossRef]
78. Kaloop, M.R.; Kumar, D.; Zarzoura, F.; Roy, B.; Hu, J.W. A wavelet—Particle swarm optimization—Extreme learning machine hybrid modeling for significant wave height prediction. *Ocean Eng.* **2020**, *213*, 107777. [CrossRef]
79. Cui, D.; Huang, G.-B.; Liu, T. ELM based smile detection using Distance Vector. *Pattern Recognit.* **2018**, *79*, 356–369. [CrossRef]
80. Hoerl, A.E.; Kennard, R.W. Ridge Regression: Biased Estimation for Nonorthogonal Problems. *Technometrics* **2000**, *42*, 80. [CrossRef]
81. Bartlett, P.L. The sample complexity of pattern classification with neural networks: The size of the weights is more important than the size of the network. *IEEE Trans. Inf. Theory* **1998**, *44*, 525–536. [CrossRef]
82. Faramarzi, A.; Heidarinejad, M.; Stephens, B.; Mirjalili, S. Equilibrium optimizer: A novel optimization algorithm. *Knowl. Based Syst.* **2020**, *191*, 105190. [CrossRef]
83. Nazaroff, W.W.; Alvarez-Cohen, L. *Environmental Engineering Science*; John Wiley & Sons: New York, NY, USA, 2001.
84. Guo, Z. Review of indoor emission source models. Part 1. Overview. *Environ. Pollut.* **2002**, *120*, 533–549. [CrossRef] [PubMed]
85. Kennedy, J.; Eberhart, R. Particle Swarm Optimization. In Proceedings of the ICNN'95—International Conference on Neural Networks, Perth, Australia, 27 November–1 December 1995; Volume 4, pp. 1942–1948. [CrossRef]
86. Tangchawal, S. *Planning and Evaluation for Quarries: Case Histories in Thailand*; IAEG2006: Nottingham, UK, 2006.
87. Jahed Armaghani, D.; Shoib, R.S.N.S.B.R.; Faizi, K.; Rashid, A.S.A. Developing a hybrid PSO–ANN model for estimating the ultimate bearing capacity of rock-socketed piles. *Neural Comput. Appl.* **2017**, *28*, 391–405. [CrossRef]
88. Huang, G.-B.; Chen, L.; Siew, C.-K. Universal Approximation Using Incremental Constructive Feedforward Networks with Random Hidden Nodes. *IEEE Trans. Neural Netw.* **2006**, *17*, 879–892. [CrossRef] [PubMed]
89. Zhu, H.; Tsang, E.C.C.; Zhu, J. Training an extreme learning machine by localized generalization error model. *Soft Comput.* **2018**, *22*, 3477–3485. [CrossRef]
90. Mohapatra, P.; Chakravarty, S.; Dash, P.K. An improved cuckoo search based extreme learning machine for medical data classification. *Swarm Evol. Comput.* **2015**, *24*, 25–49. [CrossRef]
91. Liu, H.; Mi, X.; Li, Y. An experimental investigation of three new hybrid wind speed forecasting models using multi-decomposing strategy and ELM algorithm. *Renew. Energy* **2018**, *123*, 694–705. [CrossRef]
92. Satapathy, P.; Dhar, S.; Dash, P.K. An evolutionary online sequential extreme learning machine for maximum power point tracking and control in multi-photovoltaic microgrid system. *Renew. Energy Focus* **2017**, *21*, 33–53. [CrossRef]
93. Li, L.-L.; Sun, J.; Tseng, M.-L.; Li, Z.-G. Extreme learning machine optimized by whale optimization algorithm using insulated gate bipolar transistor module aging degree evaluation. *Expert Syst. Appl.* **2019**, *127*, 58–67. [CrossRef]
94. Figueiredo, E.M.N.; Ludermir, T.B. Effect of the PSO Topologies on the Performance of the PSO-ELM. In Proceedings of the 2012 Brazilian Symposium on Neural Networks, Curitiba, Brazil, 20–25 October 2012; pp. 178–183. [CrossRef]
95. Liu, D.; Li, G.; Fu, Q.; Li, M.; Liu, C.; Faiz, M.A.; Khan, M.I.; Li, T.; Cui, S. Application of Particle Swarm Optimization and Extreme Learning Machine Forecasting Models for Regional Groundwater Depth Using Nonlinear Prediction Models as Preprocessor. *J. Hydrol. Eng.* **2018**, *23*. [CrossRef]

96. Wang, Y.; Tang, H.; Wen, T.; Ma, J. A hybrid intelligent approach for constructing landslide displacement prediction intervals. *Appl. Soft Comput.* **2019**, *81*, 105506. [CrossRef]
97. Kaloop, M.R.; Kumar, D.; Samui, P.; Gabr, A.R.; Hu, J.W.; Jin, X.; Roy, B. Particle Swarm Optimization Algorithm-Extreme Learning Machine (PSO-ELM) Model for Predicting Resilient Modulus of Stabilized Aggregate Bases. *Appl. Sci.* **2019**, *9*, 3221. [CrossRef]
98. Armaghani, D.J.; Kumar, D.; Samui, P.; Hasanipanah, M.; Roy, B. A novel approach for forecasting of ground vibrations resulting from blasting: Modified particle swarm optimization coupled extreme learning machine. *Eng. Comput.* **2021**, *37*, 3221–3235. [CrossRef]
99. Li, G.; Kumar, D.; Samui, P.; Nikafshan Rad, H.; Roy, B.; Hasanipanah, M. Developing a New Computational Intelligence Approach for Approximating the Blast-Induced Ground Vibration. *Appl. Sci.* **2020**, *10*, 434. [CrossRef]
100. Cao, J.; Lin, Z.; Huang, G.-B. Self-Adaptive Evolutionary Extreme Learning Machine. *Neural Process. Lett.* **2012**, *36*, 285–305. [CrossRef]
101. Chen, S.; Shang, Y.; Wu, M. Application of PSO-ELM in electronic system fault diagnosis. In Proceedings of the 2016 IEEE International Conference on Prognostics and Health Management (ICPHM), Ottawa, ON, Canada, 20–22 June 2016; pp. 1–5. [CrossRef]
102. Huang, G.-B.; Zhu, Q.-Y.; Siew, C.-K. Extreme learning machine: A new learning scheme of feedforward neural networks. In Proceedings of the 2004 IEEE International Joint Conference on Neural Networks (IEEE Cat. No. 04CH37541), Budapest, Hungary, 25–29 July 2004; pp. 985–990.

sustainability

Article

Stress and Deformation Characteristics of Tunnel Surrounding Rock under Alteration

Yapeng Chen [1], Tong Wu [1], Xiaoshi Yan [1], Shang Shi [2], Jianyong Li [2] and Jinyu Dong [2,*]

1 Beifang Investigation, Design & Research Co., Ltd., Tianjin 300222, China
2 College of Geosciences and Engineering, North China University of Water Resources and Electric Power, Zhengzhou 450046, China
* Correspondence: dongjy0552@126.com; Tel.: +86-139-0382-4025

Abstract: Based on a typical project in an altered rock area, this study carried out numerical simulations using the FLAC^{3D} software to calculate the changes in the stress field, deformation field, and plastic zone of the surrounding rock during the unsupported and supported excavation of a water transfer tunnel. The degree of alteration of the surrounding rock was considered as the base point. The following results were obtained: in the unsupported state, the tunnel surrounding rock was affected by different degrees of alteration, and compressive stress concentration appeared within a certain range at the bottom of the chamber. The value of all-directional stress decreased with the deepening of the degree of alteration, while the opposite was the case for the depth of influence. The displacement changes at the bottom and side walls of the chamber were large and increased significantly with the deepening of the degree of alteration; the displacement monitoring points distributed around the tunnel exhibited the same deformation trend. The plastic zone of the surrounding rock obviously expanded as the degree of alteration deepened. The stress, deformation field, and plastic zone of the tunnel surrounding rock were effectively controlled after the adoption of support measures. The results obtained by this study can be used as a reference for similar projects in altered rock areas.

Keywords: alteration; tunnel; numerical simulation; stress field; deformation field; plastic zone

Citation: Chen, Y.; Wu, T.; Yan, X.; Shi, S.; Li, J.; Dong, J. Stress and Deformation Characteristics of Tunnel Surrounding Rock under Alteration. *Sustainability* **2023**, *15*, 1161. https://doi.org/10.3390/su15021161

Academic Editors: Marc A. Rosen, Mahdi Hasanipanah, Danial Jahed Armaghani and Jian Zhou

Received: 13 September 2022
Revised: 24 December 2022
Accepted: 28 December 2022
Published: 7 January 2023

1. Introduction

To solve the problems of the uneven distribution of water resources and the contradiction between supply and demand, many large-scale water conservancy construction projects have been built around the world, among which water transmission tunnels are a key component with high importance. Extreme engineering geological conditions, such as significant faults, seasonal thawing in permafrost regions, and high-pressure water action, frequently cause the issue of surrounding rock deformation and damage during tunnel excavation [1–4]. Similarly, the stability of the surrounding rock is also significantly impacted by the complicated stress changes, rock extrusion, and deformation brought on by tunnel excavation. Dong et al. [5] exposed how tectonic stress forces affected the rock surrounding the tunnel's stress and deformation damage pattern. After conducting an excavation simulation under high ground stress for the underground chamber complex of Jinping I hydropower facility, Qian and Zhou [6] discovered that the rock body will manifest as two-dimensional band disintegration phenomenon and suggested corresponding support measures. After examining the primary stress rotation mechanism and the rock extrusion and deformation law during the excavation of high and deep buried tunnels, Cai et al. [7,8] pointed out that the three-dimensional spatial effect is more significant for the analysis of the stability of the surrounding rock. They proposed a three-dimensional orthotropic analysis and rock strength based on the GZZ strength criterion that can weaken the three-dimensional spatial effect and exert the rock body's own strength.

Additionally, as a particular class of engineering geological problems, alteration rocks have been exposed in several projects around the world, such as the Sanjiang orogenic section of the Yunnan–Tibet Railway in China [9], the Kerman Tunnel in the Urmia–Dokhtar Magmatic Arc (UDMA) in central Iran [10], multiple geothermal power plants in the Kuril–Kamchatka island arc, Russia [11], and the Wheal Martyn china pit in Southwest England [12]. Alteration rocks are a new class of rock that forms after diagenesis by hydrothermal erosion, tectonic dynamics, and secondary weathering that alters the elemental composition and structural features of the original rock to various degrees [13,14]. Researchers have examined the composition and characteristics of altered rocks in various geological contexts and discovered that some feldspar and mica minerals are mostly changed into clay minerals, sericite, chlorite, and chlorite after alteration [15–17]. The destruction of the original internal structure of the rocks caused by the alteration of tiny minerals increases the internal porosity of the rocks. The development of internal porosity as well as microcracks in the rock will produce continuity fracture damage after being disturbed by excavation [18,19]. Along with the long-term geological tectonic evolution and the effect of ground stress in the region, the distribution of alteration rocks is irregular, the internal structure is highly fragmented and the physical and mechanical properties are poor [20–23], which have a direct impact on site selection and the design and construction of projects.

The previous research system to examine the mechanical behavior of deeply buried tunnels from the viewpoints of intrinsic model and geological structure is reasonably well developed; the theoretical research on micro mineral analysis, physical and mechanical property changes, and alteration degree classification of altered rocks is also reasonably mature [24,25], but the pertinent engineering case studies are slightly lacking. Specifically in deeply buried tunnels, where the damage to the mechanical properties of rocks due to alteration is frequently beyond our original prediction, it has been common to observe construction challenges, schedule delays, and even safety threats brought on by improper support and poor response in the field. This is because there is little consideration of the alteration effect of rocks during engineering construction in altered rock areas. Therefore, there is scientific and practical value in analyzing the stress and deformation damage characteristics of tunnel excavation surrounding rocks under the effect of alteration for the design and construction of projects in altered rock areas. This study considered a water transfer tunnel with different degrees of alteration as an example. The FLAC^{3D} software was used to numerically simulate the excavation of a typical alteration tunnel section in its natural state and under two working conditions after support with the objective of analyzing the stress and deformation damage characteristics of the surrounding rock under different alteration degrees and elucidating the actual impact of alteration on the surrounding rock stability. The findings of this study provide the theoretical basis for optimizing the support scheme, and a reference for the design and construction of similar projects.

2. Engineering Geological Conditions

The water transmission tunnel is located in the central and western part of North Tianshan. This tunnel has a total length of 41.82 km, diameter of 5.3 m, and longitudinal slope of 1/564.8, which means that it is a deep and long buried tunnel. The study area has a complex geological structure with several northwest and northeast-trending compression–torsional faults, fold zones, and extrusion fracture zones. The tunnel's surrounding rocks belong to various lithologies, and mainly include Silurian, Devonian, Carboniferous sandstone, metamorphic sandstone, tuff, tuffaceous sandstone, and Hualixi-age granite. Among them, the granite section is 9.81 km long, dominated by diorite, granodiorite, and potassium granite, and was formed by the crustal movement of the Late Paleozoic and magmatic activity of the Hualixi period. After a long and complex tectonic–hydrothermal superposition modification, the granites in the study area have generally been affected by alteration, with the chloritization of black mica and clayification of feldspar minerals as the main alteration types.

Based on a field survey and field test, the degree of rock alteration can be divided into three categories: slight alteration, moderate alteration, and strong alteration. Slightly altered rocks have a relatively intact structure, produce a brittle hammering sound, and their rebound values are between 30 and 50; the cave blocks are less collapsed and larger in size. Moderately altered rocks have a partially broken structure, produce a muffled hammering sound, and their rebound values are between 10 and 30; the cave blocks are more collapsed and larger in size, and cavities can easily form at the top arch. The structure of strongly altered rocks is completely destroyed, with visible traces of the original structure, and these rocks can be crushed by hand; the rebound value is less than 10, and the cavern is in a large area of debris collapse, exhibiting obvious plastic deformation when wet. The alteration of the surrounding rock in the tunnel is widely distributed, and the construction process often leads to the large deformation of the surrounding rock, collapse, and other geological problems.

3. Numerical Computational Model

3.1. Computational Model

Based on site investigation and geological data, the geological model of the study area was generalized, and a three-dimensional (3D) numerical calculation tunnel model was established based on FLAC^{3D}. As shown in Figure 1, the water transmission tunnel model had an 8 m diameter, the height of the model was approximately 1313 m in the vertical direction (Z-axis), the calculated elevation at the bottom was 1000 m, and the highest elevation at the top surface was 2313 m. The model width was 200 m (X-axis direction), and 100 m was taken from each side of the tunnel's centerline. The extension of the model was 600 m (Y-axis direction), including 250 m for the slightly altered rock section (mileage K32 + 105–355), 200 m (mileage K32 + 355–555) for the moderately altered rock section, and 150 m (mileage K32 + 555–705) for the strongly altered rock section.

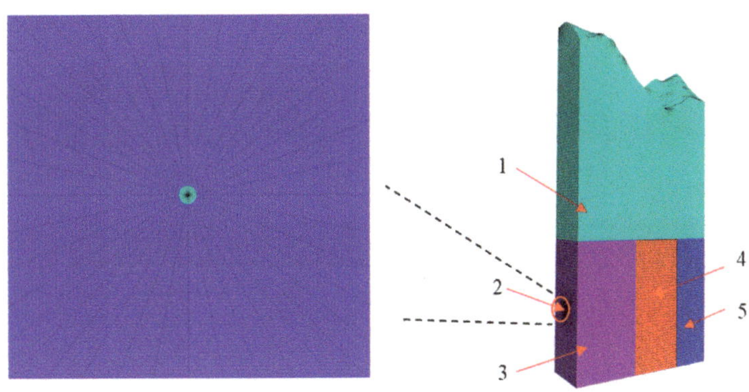

Figure 1. Calculation model diagram. 1. Low-resistivity overburden; 2. Water tunnel; 3. Slightly altered rock zone; 4. Moderately altered rock zone; 5. Strongly altered rock zone.

The geological model was simplified to four geotechnical materials, namely, the upper low resistance overburden and the lower three granites with different degrees of alteration. The tunnel passed through the lower altered rock layer, where the medium altered rock zone contained a fault. The geological model is shown in Figure 1. Since the model's overall size was too large, transitional meshing was employed to simplify the computation while guaranteeing that the mesh size close to the tunnel satisfied the requirements for calculation accuracy. The boundary conditions of the model were the X-directional displacement constraint along the tunnel extension boundary, the Y-directional displacement constraint along the vertical tunnel boundary, and the fixed constraint at the bottom boundary.

3.2. Application of Ground Stress and Selection of Calculation Parameters

To accurately obtain the tunnel ground stress field distribution, six sets of acoustic emission Kaiser effect ground stress tests and analyses were conducted on the borehole cores of the granite section, and the measured results are presented in Table 1. The test results reveal that the measured principal stress value of the borehole increased with the depth; the maximum principal compressive stress σ_1 was approximately horizontal, the dip angle a_1 was within $\pm$ 10°, and the dominant direction was north-northwest, which was more consistent with the direction of the regional tectonic stress field. The middle principal stress σ_2 had a gentle dip angle, and the dip angle a_2 varied within $\pm$ 28°; the minimum principal stress dip angle a_3 was above 64°.

Table 1. Earth stress test results.

Group	Depth (m)	Main Stress Values (MPa)			Main Stress Inclination (°)			Main Stress Direction (°)		
		σ_1	σ_2	σ_3	a_1	a_2	a_3	β_1	β_2	β_3
1	470–500	25.9	17.9	10.9	1.1	−25.4	64.6	18.5	−32.2	−73.9
2	580–610	28.7	18.4	13.9	9.6	−27.5	−60.6	13.8	−61.7	86.4
3	650–680	28.8	20.4	16.8	−2.6	−9.9	79.8	1.5	−70.8	−74.1
4	823–829	31.5	24.4	20.5	−8.6	23.6	64.7	11.8	−63.9	83.6
5	848–855	33.2	24.9	22.9	4.5	−16.2	−73.1	8.6	−86.8	83.6
6	883–886	35.2	26.5	23.9	2.7	6.8	82.6	5.2	64.7	63.5

Note: σ_1, σ_2, and σ_3 are the maximum principal compressive stress, intermediate principal stress, and minimum principal stress, respectively; a_1, a_2, and a_3 represent the angle (inclination) between the maximum principal compressive stress, intermediate principal stress, and minimum principal stress, and horizontal plane, respectively, with positive values representing the elevation angle and negative values representing the pitch angle; β_1, β_2, and β_3 are the angle between the projection of the three principal stresses on the oxy plane and x-axis, respectively, with positive values indicating counterclockwise rotation and negative values indicating clockwise rotation.

Based on the data in Table 1, the ground stress in the granite cave section can be projected, and the linear correlation equations between the maximum horizontal principal stress (σ_1), horizontal intermediate principal stress (σ_2), and minimum horizontal principal stress (σ_3) and the burial depth (H) can be derived as follows:

$$\sigma_1 = 0.0196H + 16.653, \quad R^2 = 0.9304$$

$$\sigma_2 = 0.0211H + 7.0488, \quad R^2 = 0.9688$$

$$\sigma_3 = 0.0303H - 3.3618, \quad R^2 = 0.9862$$

The principal stresses in the cavern line were projected according to the linear correlation equation, and the results are shown in Table 2. In the numerical calculation, σ_1, σ_2, and σ_3 were transformed along the X-direction (horizontal vertical tunnel axis direction), Y-direction (tunnel axis direction), and vertical direction to apply the Sxx, Syy, and Szz stresses to the model.

Table 2. Projected ground stress values in the calculated section of tunnel.

Main Lithology	Depth of Burial (m)			Density	Max. Main Stress σ_1 (MPa)			Intermediate Main Stress σ_2 (MPa)			Min. Principal Stress σ_3 (MPa)		
	Min.	Max.	Average	g/cm³	Min.	Max.	Average	Min.	Max.	Average	Min	Max	Average
monzonitic granite	594	1078	858	2.67	27.1	43.0	35.8	18.1	29.8	24.5	15.9	28.8	22.9

The elastic–plastic model and Mohr–Coulomb strength criterion were used in the calculation. The rock surrounding parameters were determined according to the standard of International Society of Rock Mechanics (ISRM, 2007) after testing at the Quality Inspection Center of Capital Construction Project of Haihe Water Conservancy Commission, Ministry of Water Resources of China and School of Earth Science and Engineering, North China

University of Water Resources and Electric Power, with reference to the Engineering Rock Quality Grading Standard (GBT50218-2014) and after considering the degree of rock alteration, as presented in Table 3.

Table 3. Proposed values of the main geological parameters of the tunnel envelope.

Surrounding Rock Type	Rock Density g/cm^3	Modulus of Elasticity *E* (GPa)	Deformation Modulus E_0 (GPa)	Poisson Ratio μ	Shearing Strength *C* (MPa)	φ (°)
Fresh	2.60~2.70	16~20	11~13	0.21~0.23	1.6~1.8	46~52
Slightly altered	2.45~2.60	12~15	9~11	0.25~0.28	1.0~1.2	38~44
Moderately altered	2.40~2.45	5~8	4~6	0.31~0.34	0.3~0.5	30~35
Strongly altered	2.30~2.40	0.2~1.0	0.1~0.4	0.36~0.38	0.05~0.10	22~26

According to the excavation design plan, after the tunnel excavation, the TBM shield was closed with synthetic coarse fiber concrete in time after the initial spraying. The slightly altered section adopted HW125 steel arch racks with a distance of 0.9 m; the moderately altered section adopted HW150 steel arch racks with a distance of 0.5 m; the strongly altered section adoptsedHW150 steel arch racks with a distance of 0.3 m. The longitudinal connection adopted Φ20 steel bars with a ring spacing of 1 m, and the top arch was equipped with Φ20 reinforcement rows within 150°. The support parameters are shown in Table 4, Table 5, and Table 6, respectively.

Table 4. Basic parameters of the anchor rods.

Diameter (mm)	Length (m)	Equivalent Elastic Modulus (GPa)	Tensile Strength (MPa)	Cement Slurry Stiffness (MPa)	Cement Slurry Cohesion (MPa)
22/25	2.0/2.5/3.5	200	360	15	0.8

Table 5. Basic parameters of the steel arch.

Elastic Modulus (GPa)	Poisson Ratio	Cross Sectional Area (cm^2)	Bulk Density (kN/m^3)	Y-Axis Moment of Inertia (10^4 cm^4)	Z-Axis Moment of Inertia (10^4 cm^4)
21	0.3	30.31	23.8	847	294

Table 6. Basic parameters of the concrete primary lining.

Concrete Grade	Equivalent Elastic Modulus (GPa)	Bulk Density (kN/m^3)	Poisson Ratio
C20	25.5	25	0.25

4. Analysis of the Calculation Results

For the in-depth investigation of the distribution and changes of the stress field and deformation field of the surrounding rock, typical sections were selected for detailed analysis in the surrounding rock with different degrees of alteration. Section K32 + 230 was selected in the slightly altered surrounding rock with a burial depth of 905 m; section K32 + 455 was selected in the moderately altered surrounding rock with a burial depth of 860 m; section K32 + 630 was selected in the strongly altered surrounding rock with a burial depth of 930 m.

4.1. Analysis of Stress Field

After the tunnel excavation, the horizontal maximum principal stress was approximately in line with the tunnel axis, ignoring its influence on the tunnel surrounding rock. The horizontal intermediate principal stress was approximately perpendicular to the tunnel axis, and the self-weight stress was the minimum principal stress. Provided that support measures were not used, under the influence of horizontal tectonic stress, a compressive stress concentration with a maximum value of 102.4 MPa appeared in the slightly altered surrounding rock section at the top and bottom 7.0 m of the chamber's depth, and gradually decreased to the initial stress state as the distance from the chamber increased, as shown in Figure 2a. Under the influence of chamber excavation and unloading, a stress reduction zone with a value of 54.6 MPa and influence depth of 16.0 m appeared in the surrounding rock on both sides and at the top of the chamber. Shear stress concentration appeared at the top and bottom corners of the cavern, and its maximum value was 29.5 MPa, as shown in Figure 2b. Compressive stress concentration with a maximum value of 87.36 MPa appeared in the medium alteration surrounding rock section 20 m at the top and bottom corners of the cavern, and gradually decreased to the initial stress state as the distance from the cavern chamber increased; the influence depth was 28.0 m, as shown in Figure 2c. A stress reduction zone appeared in the surrounding rock on both sides of the cavern chamber and at the top and bottom; its value was 42.3 MPa and its influence depth was 32.0 m. Shear stress concentration appeared at the top and bottom corner of the cavern, and its maximum value was 19.15 MPa, as shown in Figure 2d. In the strongly altered surrounding rock section, compressive stress concentration occurred at a depth of 13.0 m from the bottom of the cavern, and had the maximum value of 107.8 MPa, as shown in Figure 2e. In addition to the bottom of the cavern, a stress reduction zone with the minimum value of 49.04 MPa and small main stress influence depth of 35 m existed around the cavern. Shear stress concentration with an influence depth of 12.0 m existed at the top and bottom corners of the cavern. The maximum value was 19.7 MPa, as shown in Figure 2f. The depth of stress influence was significantly higher in the moderately altered and strongly altered surrounding rocks compared to the slightly altered surrounding rock.

After using support measures, the slightly altered surrounding rock section exhibited compressive stress concentration in the surrounding rock at the top and bottom corners of the cavern, and its maximum value was 102.1 MPa, as shown in Figure 3a. The stress reduction zone appeared at the bottom of the cavern, and its value was 59.9 MPa. Shear stress concentration appeared at the top and bottom corners of the cavern, and its maximum value was 31.5 MPa, as shown in Figure 3b. In the moderately altered surrounding rock section, compressive stress concentration appeared at the top and bottom corners of the cavern, and its maximum value was 101.6 MPa, as shown in Figure 3c. A stress reduction zone appeared at the side walls and bottom of the cavern, and its value was 61.4 MPa. Shear stress concentration appeared at the top and bottom corners of the cavern, and its maximum value was 24.9 MPa, as shown in Figure 3d. Compressive stress concentration appeared at the top and bottom corners of the cavern section, with a maximum value of 98.3 MPa, and gradually decreased to the initial stress state as the distance from the cavern increased, as shown in Figure 3e. A stress reduction zone with a minimum value of 61.06 MPa appeared in the side walls of the cavern. Shear stress concentration appeared at the top and bottom corners of the cavern, and its maximum value was 20.46 MPa, as shown in Figure 3f.

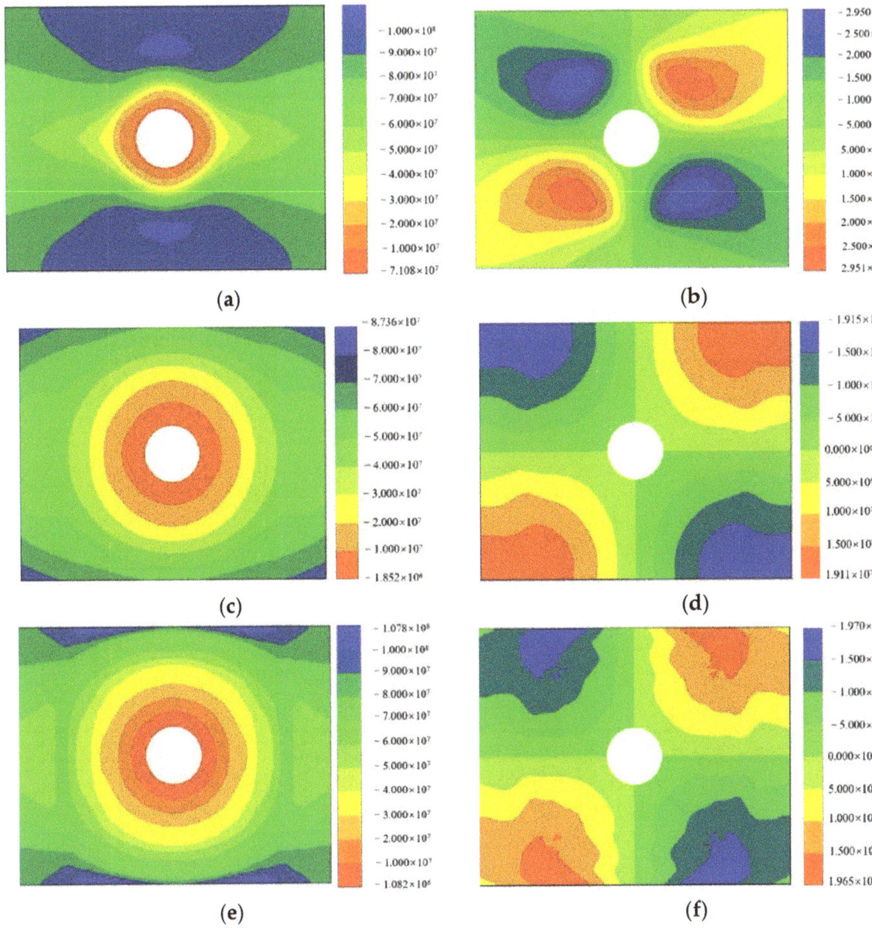

Figure 2. Main stress diagram of the tunnel excavation under unsupported conditions. (**a**) Middle main stress diagram of the slightly altered surrounding rock section. (**b**) Shear stress cloud diagram of the slightly altered surrounding rock section. (**c**) Middle main stress diagram of the moderately altered surrounding rock section. (**d**) Shear stress cloud diagram of the moderately altered surrounding rock section. (**e**) Middle main stress diagram of the strongly altered surrounding rock section. (**f**) Shear stress cloud diagram of the strongly altered surrounding rock section.

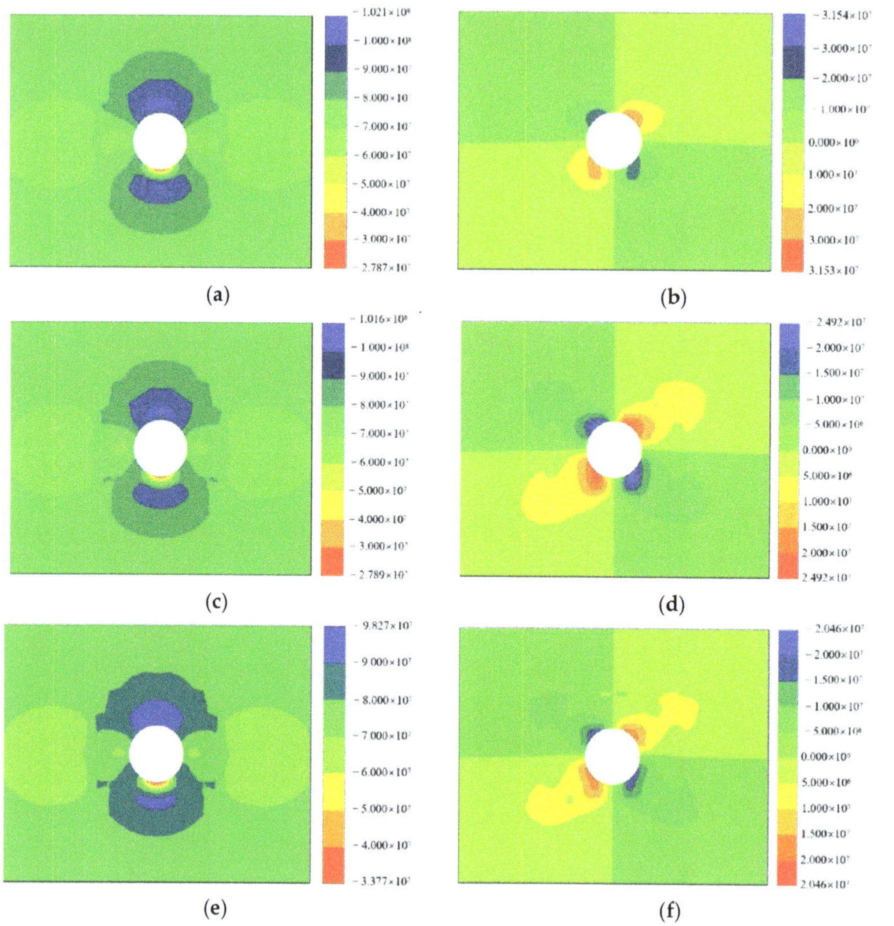

Figure 3. Main stress diagram of the excavation after the tunnel support. (**a**) Middle main stress diagram of the slightly altered surrounding rock section. (**b**) Shear stress cloud diagram of the slightly altered surrounding rock section. (**c**) Middle main stress diagram of the moderately altered surrounding rock section. (**d**) Shear stress cloud diagram of the moderately altered surrounding rock section. (**e**) Middle main stress diagram of the strongly altered surrounding rock section. (**f**) Shear stress cloud diagram of the strongly altered surrounding rock section.

The comparison between the stresses in the unsupported and supported cavern chambers is shown in Figure 4. As can be seen from the support, except for the strong alteration surrounding the rock section, owing to the occurrence of surrounding rock damage deformation and X-direction compressive stress reduction, the other stresses were increased by the surrounding rock owing to the restraining effect of the support structure.

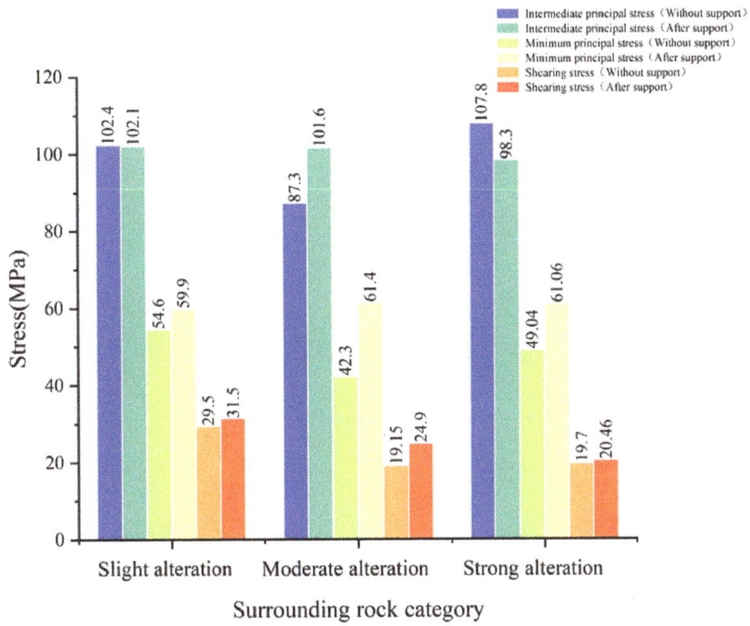

Figure 4. Comparison of surrounding rock stress in each profile before and after excavation and support of alteration rock tunnel.

4.2. Analysis of Surrounding Rock Displacement Field

In the natural state, the mountain displacement gradually decreases from top to bottom, with a stable trend in general, and the maximum displacement occurs in an area with large surface elevation and obvious surface undulation. The numerical simulation conducted by this study mainly considered the surrounding rock deformation caused by tunnel excavation. Therefore, the displacement in the natural state was considered to be zero, and the relative displacement of the surrounding rock after tunnel excavation was investigated on this basis.

In the tunnel excavation without support measures, the overall deformation of the surrounding rock after tunnel excavation was directed toward the cavern face; the top arch sank, the bottom slab rose, and the side walls moved inward. Figure 5 shows the displacement trend of each typical section of the surrounding rock under unsupported excavation. As can be seen, the maximum total displacement of the surrounding rock in the slightly altered section was 9.64 cm, and the displacement and impact depth of the top slab and sidewall were relatively large, while the displacement and impact depth of the bottom slab were relatively small. The deformation of the surrounding rock in the moderately altered section was larger than that in the slightly altered section, and the maximum total displacement was 166.4 cm. The maximum total displacement of the surrounding rock in the strongly altered section was 45.6 m, and the displacement at the base plate and sidewall was the most variable. In the displacement cloud distribution of each typical section (Figure 5), the area of the surrounding rock deformation caused by cavern excavation was pie shaped. Because of the large burial depth, the displacements of the surrounding rock in the vertical direction (Z-direction) and vertical tunnel direction (X-direction) were relatively large owing to the influence of self-weight stress and horizontal tectonic stress, while the displacement in the tunnel extension direction (Y-direction) was small. Owing to the influence of the degree of rock erosion, the displacement of the and moderately altered section (section 32 + 455, depth of 860 m) was much larger than that of the slightly altered section (section 32 + 230, depth of 905 m). Considering the discontinuity characteristics of the rock, it is concluded that large deformation damage occurs frequently. The strongly altered section (section 32 + 630, depth of 930 m) had actually been completely destroyed by large deformation.

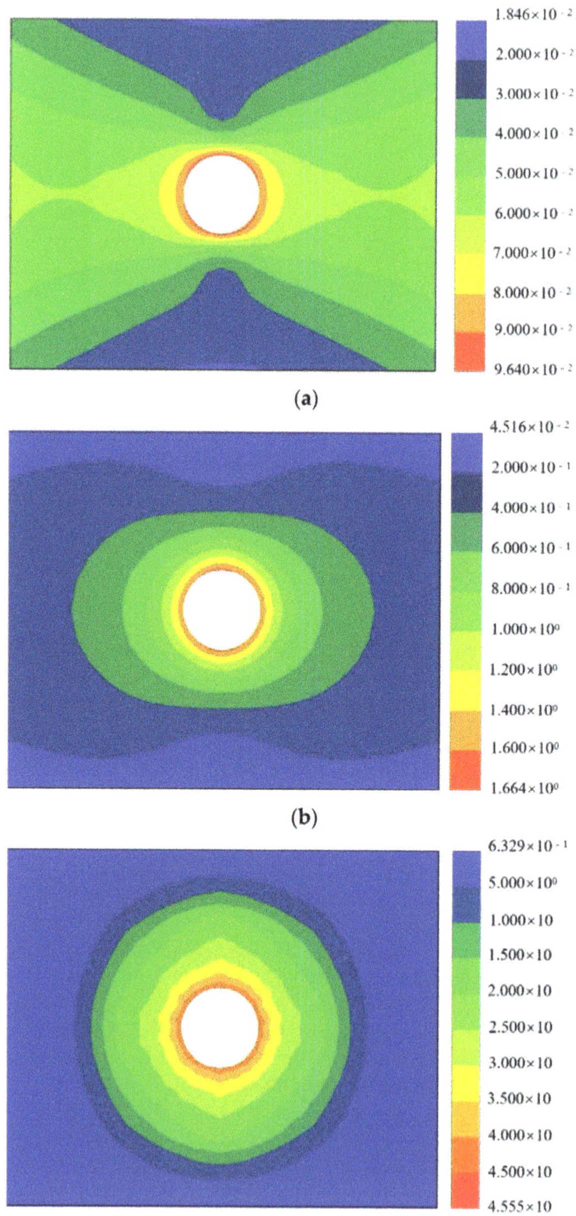

Figure 5. Displacement diagram of the tunnel excavation under unsupported conditions. (**a**) Total displacement diagram of the slightly altered surrounding rock section. (**b**) Total displacement diagram of the moderately altered surrounding rock section. (**c**) Total displacement diagram of the strongly altered surrounding rock section.

After adopting support measures, such as the anchors of a grouting reinforcement system, steel arch, shotcrete, and reinforcement mesh, the deformation area of the surrounding rock caused by tunnel excavation became butterfly shaped, and the deformation area of the surrounding rock was mainly concentrated at the side wall, whose displacement and influence depth were relatively large while the displacement and influence depth of the top arch and bottom slab were relatively small. Figure 6 shows the typical section displacement changes of each surrounding rock under supported excavation. As can been seen, the maximum total displacement of the slightly altered surrounding rock section was 1.02 cm, the maximum total displacement of the moderately altered surrounding rock section was 1.56 cm, and the maximum total displacement of the strongly altered surrounding rock section was 2.62 cm. The main reason for this is that the horizontal structural stress was the maximum main stress and the vertical stress was the minimum main stress in the tunnel. This led to the displacement of the surrounding rock, which was caused by the lateral expansion effect as the main contradiction, and the displacement of the top and bottom slab as the secondary contradiction, coupled with a high degree of fragmentation, extremely low mechanical strength, and the poor integrity of the strongly altered rock body, which led to the downward movement of the bottom slab under the influence of excavation disturbance and the constraint of the support effect during the excavation process (Figure 7a,c,e).

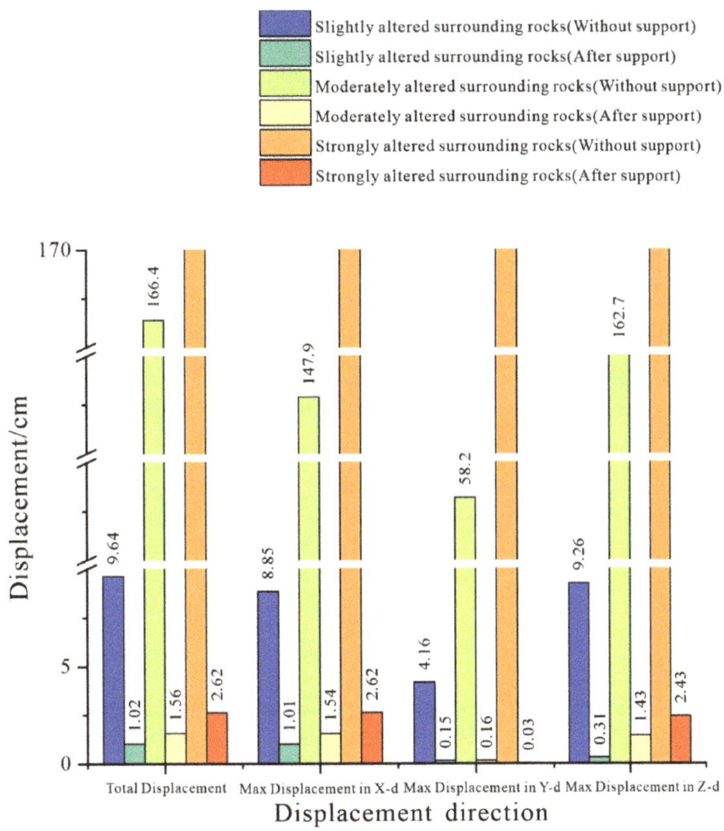

Figure 6. Comparison of the displacement in each profile after the excavation of the tunnel without support and after support.

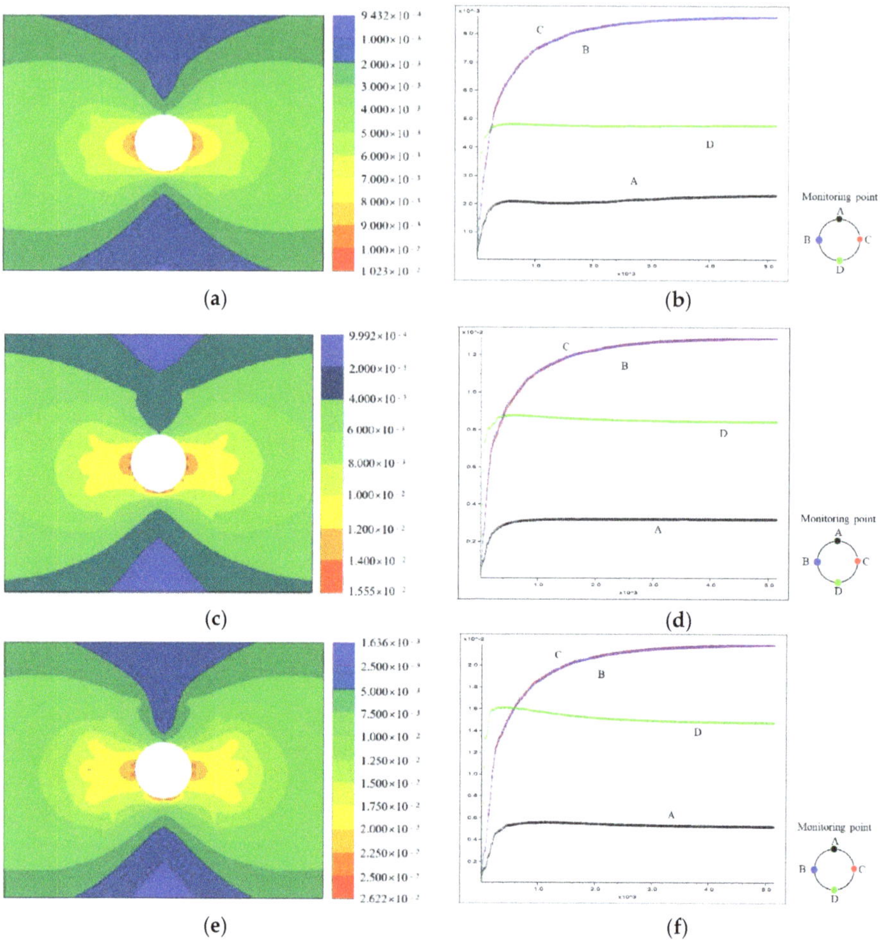

Figure 7. Excavation displacement map after tunnel support. (**a**) Total displacement diagram of the slightly altered surrounding rock section. (**b**) Displacement monitoring map of the slightly altered surrounding rock section. (**c**) Total displacement diagram of the moderately altered surrounding rock section. (**d**) Displacement monitoring map of the moderately altered surrounding rock section. (**e**) Total displacement diagram of the strongly altered surrounding rock section. (**f**) Displacement monitoring map of the strongly altered surrounding rock section.

In the calculation process, to monitor the deformation evolution process of each type of surrounding rock after tunnel excavation, displacement monitoring points were set at the top plate, both walls, and bottom plate of the cavern; the monitoring results are shown in Figure 7b,d,f. It can be seen that after the use of support measures, the displacement monitoring amount of all types of surrounding rocks in the cavern was significantly reduced, and the surrounding rocks of the tunnel exhibited the largest displacement at the cavern side walls, followed by the bottom slab and smallest top slab. Among them, the displacement of the top plate of the slightly altered surrounding rock was 0.23 cm, the maximum displacement of both walls was 0.86 cm, and the displacement of the bottom plate was 0.48 cm. The maximum displacement of the top plate of the moderately altered surrounding rock was 0.32 cm, the maximum displacement of both walls was 1.30 cm, and the maximum displacement of the bottom plate was 0.88 cm. The displacement of the top plate of the strongly altered surrounding rock was 0.56 cm, the maximum displacement of both walls was 2.19 cm, and the displacement of the bottom slab was 1.62 cm.

4.3. Analysis of the Plastic Zone of Surrounding Rock

The size and distribution characteristics of the plastic zone reflect the mechanical properties of the surrounding rock. Additionally, they characterize the actual size of the loosening zone of the tunnel surrounding rock after excavation and unloading, and the degree of disturbance to the surrounding rock in each excavation phase. After excavation, because of the increase in the large principal stress and decrease in small principal stress in the surrounding rock, the surrounding rock unit was in the form of compression–shear damage; the distribution of its plastic zone is shown in Figure 8.

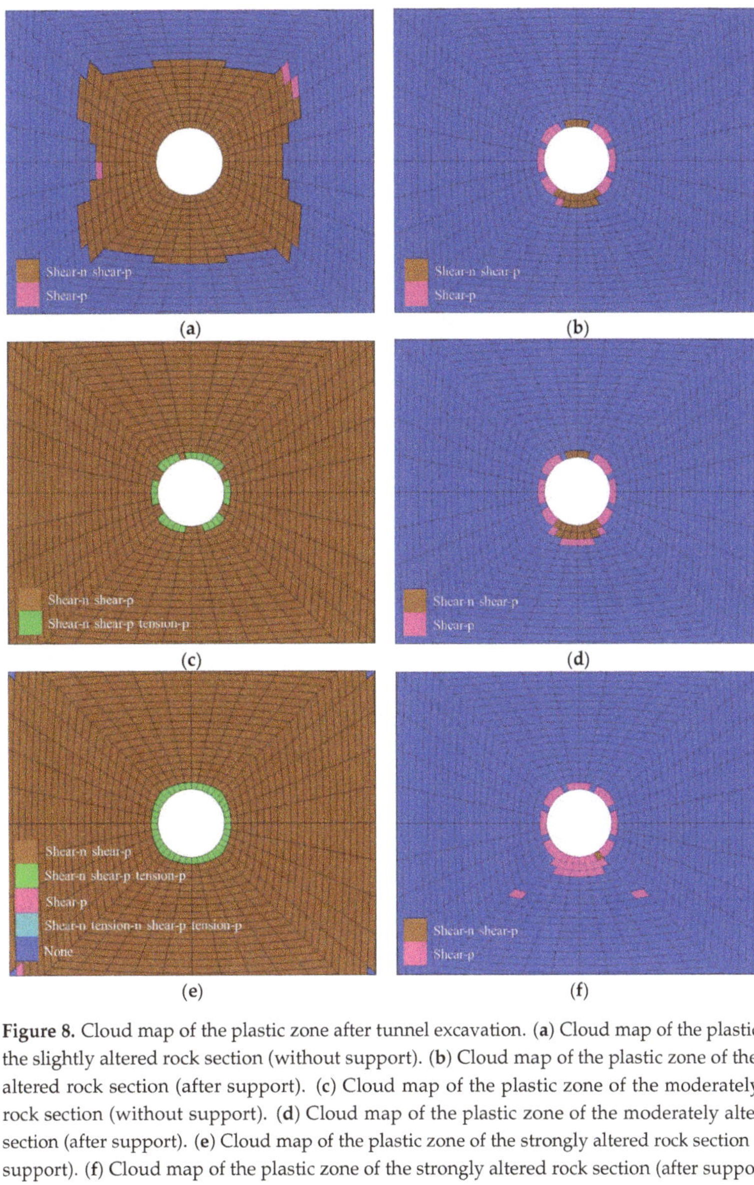

Figure 8. Cloud map of the plastic zone after tunnel excavation. (**a**) Cloud map of the plastic zone of the slightly altered rock section (without support). (**b**) Cloud map of the plastic zone of the slightly altered rock section (after support). (**c**) Cloud map of the plastic zone of the moderately altered rock section (without support). (**d**) Cloud map of the plastic zone of the moderately altered rock section (after support). (**e**) Cloud map of the plastic zone of the strongly altered rock section (without support). (**f**) Cloud map of the plastic zone of the strongly altered rock section (after support).

In the natural state, the damage depth of the plastic zone was approximately 8.0–10.0 m for slightly altered surrounding rock (section 32 + 230, depth of 905 m), approximately 20.0 m for moderately altered surrounding rock (section 32 + 455, depth of 860 m), and the strongly altered section of the perimeter rock had been completely destroyed and the depth of impact was too great, after tunnel excavation, as shown in Figure 8a,c,e.

After the use of support measures, the extent of the plastic zone in the surrounding rock decreased significantly compared with the unsupported condition, and the depth of damage of the plastic zone in the slightly altered surrounding rock (section 32 + 230, depth of 905 m) decreased to approximately 1.0–2.0 m. The depth of damage of the plastic zone in the moderately altered surrounding rock (section 32 + 455, depth of 860 m) decreased to approximately 1.0–3.0 m, and the depth of damage of the plastic zone in the strongly altered surrounding rock (section 32 + 630, depth of 930 m) decreased to approximately 0.6–1.0 m. The depth of damage in the plastic zone (section 32 + 630, depth of 930 m) decreased to approximately 1.0–3.0 m, as shown in Figure 8b,d,f.

4.4. *Recheck of the Surrounding Rock Deformation*

During the tunnel boring process, it was found that the deformation of the steel arch was mainly concentrated within 24 h after excavation, after which it gradually stabilized, and the final deformation was less different to the 24 h monitoring value. Therefore, 14 monitoring points with different degrees of alteration were randomly selected for each section of the excavation, and the above-mentioned support method was used to monitor and count the displacement changes of the steel arch after 24 h. The actual monitoring values are presented in Table 7 and Figure 9. Among them, the deformation of the steel arch in the slightly altered rock section ranged from 0.49 to 0.76 cm, and the average value was 0.62 cm. The deformation of the steel arch in the moderately altered rock section ranged from 0.57 to 1.29 cm, and the average value was 0.97 cm. The deformation of the steel arch in the strongly altered rock section ranged from 1.73 to 2.27 cm, and the average value was 1.93 cm.

After comparison with the numerical simulation results, it was found that the average value of the actual deformation of various altered rocks was basically consistent with the numerical simulation results, and the error was controlled within 1 cm, which validated the numerical simulation. The deformation of some moderately altered rocks was lower than the average value of deformation of slightly altered rocks, which indicates that moderately altered rocks still possessed a certain strength, and the rocks had good stability under supporting measures. The maximum deformation value of strongly altered rocks was 2.27 cm, which was significantly lower compared with unsupported excavation. Hence, the deformation of the surrounding rock was effectively controlled, which verified the reasonableness of the support measures.

Table 7. Comparison between the monitored and calculated values of the surrounding rock deformation in tunnel sections with different degrees of alteration.

The Degree of Surrounding Rock Alteration	The Actual Monitoring Range of Surrounding Rock Deformation (cm)	The Actual Monitoring Average Value of Surrounding Rock Deformation (cm)	The Numerical Simulation Calculation Value (cm)
Slightly altered	0.49~0.76	0.62	1.02
Moderately altered	0.57~1.29	0.97	1.56
Strongly altered	1.73~2.27	1.93	2.62

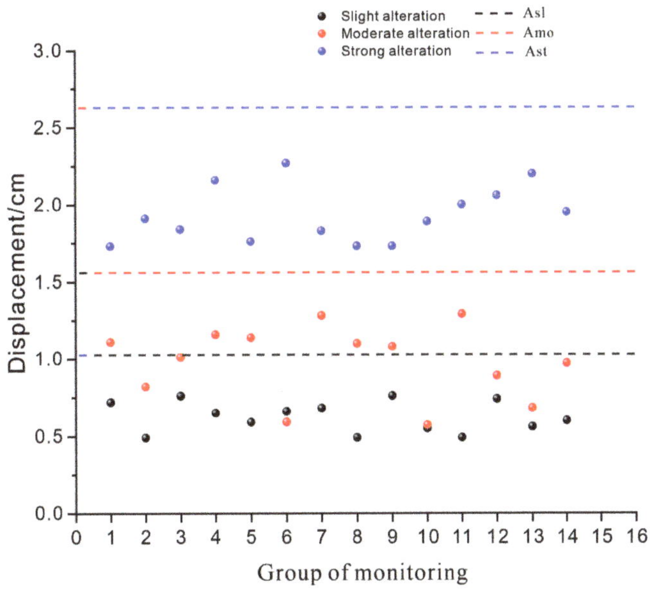

Figure 9. Actual monitoring value of tunnel surrounding rock deformation (Asl: calculated deformation value of the slightly altered surrounding rock; Amo: calculated deformation value of the moderately altered surrounding rock; Ast: calculated deformation value of the moderately altered surrounding rock.

5. Conclusions

This study carried out numerical simulations of a typical tunnel section excavated in the study area under unsupported and supported conditions using the FLAC^{3D} software. The stress field, displacement field, and plastic zone of a typical section of the surrounding rock were selected for calculation and comparison, and the results were analyzed. The following conclusions were drawn:

1. After tunnel excavation, under the action of tectonic stress, the initial horizontal tectonic stress field was much larger than the vertical self-gravity stress field, and the intermediate principal stress and minimum principal stress exerted greater influence on the surrounding rock. Stress redistribution in the mountain body led to the concentration of compressive and shear stresses within a certain depth of the cavern surrounding rock. Under the influence of horizontal tectonic stress, the displacement of the rock around the tunnel was maximum in the vertical tunnel direction (X-direction) followed by the vertical direction (Z-direction), and minimum in the tunnel extension direction (Y-direction). Based on the displacement monitoring curve of the surrounding rock, it is concluded that the displacement of the side wall of the cavern chamber was the largest, followed by the bottom plate and smallest top plate;
2. Under unsupported tunnel excavation, the surrounding rock underwent plastic damage at a certain depth range and large deformation damage was likely to occur owing to the discontinuous characteristics of the rock body. After implementing support measures, the stress in all directions of the tunnel surrounding rock increased owing to the restraint of the surrounding rock by the support structure, and the deformation of the surrounding rock was effectively controlled;
3. The comparison between the stress and displacement field results for the more strongly altered section and slightly altered section revealed that the deformation of the tunnel surrounding rock increased significantly with the deepening of alteration, and the alteration destroyed the original structure of the surrounding rock body, which led

to the reduction in its stiffness and strength, and thus increased the deformation of the surrounding rock such that destabilization damage occurred. After implementing support measures, the deformation of the tunnel decreased significantly. The actual monitoring value of the surrounding rock displacement was consistent with the simulation results, and the support scheme was reasonable, which provides a theoretical basis for the design and construction of similar projects.

Author Contributions: Conceptualization, Y.C.; methodology, J.D.; software, J.L.; validation, T.W., Y.C. and J.D.; formal analysis, S.S.; investigation, X.Y.; resources, Y.C.; data curation, S.S.; writing—original draft preparation, S.S.; writing—review and editing, J.D.; visualization, S.S.; supervision, J.D.; project administration, T.W.; funding acquisition, Y.C. All authors have read and agreed to the published version of the manuscript.

Funding: This research work was sponsored by the National Key Research and Development Project of China (Grant No. 2019YFC1509704), the National Natural Science Foundation of China (Grant Nos. U1704243, 41741019, 41977249 and 42090052), Henan Province Science and technology research project (Grant No. 192102310006).

Data Availability Statement: The data that supports the findings of this study is available from the corresponding author upon reasonable request.

Acknowledgments: We thank all project team members for their contributions to this study.

Conflicts of Interest: The authors declare no conflict of interest.

References

1. Wang, Y.; Jing, H.; Su, H.; Xie, J. Effect of a fault fracture zone on the stability of tunnel-surrounding rock. *Int. J. Geomech.* **2017**, *17*, 04016135. [CrossRef]
2. Fang, W.; Wang, G.; Wang, C. Effect of the Location of Fault Fracture Zones on the Stability of Symmetrical Submarine Tunnels. *Symmetry* **2021**, *13*, 1111. [CrossRef]
3. Shen, S.; Xia, C.; Huang, J.; Li, Y. Influence of seasonal melt layer depth on the stability of surrounding rock in permafrost regions based on the measurement. *Nat. Hazards* **2015**, *75*, 2545–2557. [CrossRef]
4. Yu, J.; Li, T.-B.; Zhang, J.-Z.; Cai, Y.-Y. Stress characteristics of surrounding rocks for inner water exosmosis in high-pressure hydraulic tunnels. *J. Cent. S. Univ.* **2014**, *21*, 2970–2976. [CrossRef]
5. Dong, J.Y.; Yang, J.H.; Zhou, J.J.; Yu, H. Influence of the Tectonic Stress Field on Stress and Deformation Failure Characteristics of a Tunnel's Surrounding Rock. *Mod. Tunn. Technol.* **2016**, *53*, 54–62. [CrossRef]
6. Qian, Q.; Zhou, X. Failure Behaviors and Rock Deformation During Excavation of Underground Cavern Group for Jinping I Hydropower Station. *Rock Mech. Rock Eng.* **2018**, *51*, 2639–2651. [CrossRef]
7. Cai, W.; Zhu, H.; Liang, W. Three-dimensional stress rotation and control mechanism of deep tunneling incorporating generalized Zhang–Zhu strength-based forward analysis. *Eng. Geol.* **2022**, *308*, 106806. [CrossRef]
8. Cai, W.; Zhu, H.; Liang, W. Three-dimensional tunnel face extrusion and reinforcement effects of underground excavations in deep rock masses. *Int. J. Rock Mech. Min. Sci.* **2022**, *150*, 104999. [CrossRef]
9. Zhang, Y.; Guo, C.; Qu, Y.; Song, Y. Engineering geological properties of altered rocks and implications for railway construction in the Sanjiang orogenic belt, Southwest China. *Bull. Eng. Geol. Environ.* **2011**, *70*, 143–152. [CrossRef]
10. Hashemnejad, A.; Aghda, S.M.F.; Talkhablou, M. Mechanized tunnelling in hydrothermally altered grounds: The effect of hydrothermal fluids on the rock behaviour in the central Iran. *Tunn. Undergr. Space Technol. Inc. Trenchless Technol. Res.* **2020**, *99*, 103340. [CrossRef]
11. Julia, F.; Vladimir, L.; Sergey, R.; David, Z. Effects of hydrothermal alterations on physical and mechanical properties of rocks in the Kuril–Kamchatka island arc. *Eng. Geol.* **2014**, *183*, 80–95. [CrossRef]
12. Coggan, J.; Stead, D.; Howe, J.; Faulks, C. Mineralogical controls on the engineering behavior of hydrothermally altered granites under uniaxial compression. *Eng. Geol.* **2013**, *160*, 89–102. [CrossRef]
13. Hu, S.X. *Petrography of Sympatric Alteration Rocks*; Geological Press: Beijing, China, 1980.
14. Bamba, T.; Togaried, K. *Wall Rock Alteration Ore Genesis*; The Society of Mining Geologists of Japan: Tokyo, Japan, 1988.
15. Tu, X.; Jian, B.; Wang, S.; Bian, H.; Wang, J.; Li, S. Swelling behavior induced by alteration in granite and its implications on underground excavation. *Tunn. Undergr. Space Technol.* **2005**, *20*, 378–389. [CrossRef]
16. Środoń, J.; Kuzmenkova, O.; Stanek, J.J.; Petit, S.; Beaufort, D.; Gilg, H.A.; Liivamägi, S.; Goryl, M.; Marynowski, L.; Szczerba, M. Hydrothermal alteration of the Ediacaran Volyn-Brest volcanics on the western margin of the east European Craton. *Precambrian Res.* **2019**, *325*, 217–235. [CrossRef]
17. Ren, M.; Wang, W.; Huang, Z.; Li, S.; Wu, Q.; Yu, H.; Yuan, G.; Sargent, P. Effect of alteration on the geochemistry and mechanical properties of granite from Pingjiang, Hunan Province, China. *Environ. Earth Sci.* **2022**, *81*, 60. [CrossRef]

18. Wang, X.; Asem, P.; Hu, C.; Labuz, J.F. Microcracking in tensile fracture of a brittle rock. *Eng. Fract. Mech.* **2021**, *251*, 107789. [CrossRef]
19. Wang, X.; Wang, E.; Liu, X.; Zhou, X. Failure mechanism of fractured rock and associated acoustic behaviors under different loading rates. *Eng. Fract. Mech.* **2021**, *247*, 107674. [CrossRef]
20. Del Potro, R.; Hürlimann, M. The decrease in the shear strength of volcanic materials with argillic hydrothermal alteration, insights from the summit region of Teide stratovolcano, Tenerife. *Eng. Geol.* **2008**, *104*, 135–143. [CrossRef]
21. Wang, H.; Xu, W.Y.; Shao, J. Experimental Researches on Hydro-Mechanical Properties of Altered Rock Under Confining Pressures. *Rock Mech. Rock Eng.* **2014**, *47*, 485–493. [CrossRef]
22. Pola, A.; Crosta, G.B.; Fusi, N.; Castellanza, R. General characterization of the mechanical behaviour of different volcanic rocks with respect to alteration. *Eng. Geol.* **2014**, *169*, 1–13. [CrossRef]
23. Callahan, O.A.; Eichhubl, P.; Olson, J.E.; Davatzes, N.C. Fracture mechanical properties of damaged and hydrothermally altered rocks, Dixie Valley-Stillwater Fault Zone, Nevada, USA. *J. Geophys. Res. Solid Earth* **2019**, *124*, 4069–4090. [CrossRef]
24. Khanlari, G.R.; Naseri, F. Investigation of physical deterioration of Malayer granitic rocks using a new weathering coefficient (Kr4). *Environ. Earth Sci.* **2016**, *75*, 414. [CrossRef]
25. Wyering, L.; Villeneuve, M.; Kennedy, B.; Gravley, D.; Siratovich, P. Using drilling and geological parameters to estimate rock strength in hydrothermally altered rock—A comparison of mechanical speciffc energy, R/N-W/D chart and Alteration Strength Index. *Geothermics* **2017**, *69*, 119–131. [CrossRef]

sustainability

MDPI

Article

The Infrared Radiation Characteristics of Sandstone Fracture Seepage under Coupled Stress-Hydro Effect

Ruoyu Cui [1,2], Kewang Cao [1,2,*], Xinci Li [3], Rana Muhammad Asad Khan [4,5], Naseer Muhammad Khan [6], Wei Liu [7], Qiangqiang Gao [7], Fagang Wang [7,8,9], Yuanzhong Yang [10], Jiangbo Quan [8,11] and Saad S. Alarifi [12]

1 School of Art, Anhui University of Finance and Economics, Bengbu 233030, China
2 State Key Laboratory for Geomechanics & Deep Underground Engineering, China University of Mining and Technology, Xuzhou 221116, China
3 School of Public Finance and Taxation, Zhongnan University of Economics and Law, Wuhan 430073, China
4 Department of Mining Engineering, Pak-Austria Fachhochschule Institute of Applied Sciences and Technology, Haripur 22620, Pakistan
5 Department of Civil and Environmental Engineering, Hanyang University, 222 Wangsimni-ro, Seongdong-gu, Seoul 04763, Republic of Korea
6 Department of Sustainable Advanced Geomechanical Engineering, Military College of Engineering, National University of Sciences and Technology, Risalpur 23200, Pakistan
7 Key Laboratory of Deep Coal Resource Mining, China University of Mining and Technology, Ministry of Education, Xuzhou 221116, China
8 Zijin Mining Group Co., Ltd., Longyan 364200, China
9 Zijin (Changsha) Engineering Technology Co., Ltd., Changsha 410114, China
10 Beijing Window Technology Co., Ltd., Beijing 100089, China
11 Luoyang Kunyu Mining Co., Ltd., Luoyang 471700, China
12 Department of Geology and Geophysics, College of Science, King Saud University, P.O. Box 2455, Riyadh 11451, Saudi Arabia
* Correspondence: tb18220001b0@cumt.edu.cn

Citation: Cui, R.; Cao, K.; Li, X.; Khan, R.M.A.; Khan, N.M.; Liu, W.; Gao, Q.; Wang, F.; Yang, Y.; Quan, J.; et al. The Infrared Radiation Characteristics of Sandstone Fracture Seepage under Coupled Stress-Hydro Effect. *Sustainability* **2022**, *14*, 16454. https://doi.org/10.3390/su142416454

Academic Editors: Mahdi Hasanipanah, Danial Jahed Armaghani and Jian Zhou

Received: 22 November 2022
Accepted: 5 December 2022
Published: 8 December 2022

Abstract: Effective monitoring of rock fracture and seepage is an important information means to ensure the safety of geotechnical engineering. Therefore, sandstone samples were subject to uniaxial compression under different hydraulic conditions in the presence of infrared radiation and observation. This study uses the multiple infrared radiation indexes (ΔAIRT, IRV, VDIIT) and image data to analyze the influence of coupled stress-hydro effect of infrared radiation change on sandstone surface. The main findings are: (1) The surface temperature of sandstone samples rises in the compaction and linear elastic stages, keeps stable or decreases in the fracture development stage, and rapidly decreases in the post-peak failure stage. (2) The samples with internal water pressure not more than 0.30 MPa, surface temperature and load curve at the compaction and linear elastic stage have a strong power function relationship, which a coefficient of determination is 0.8900. (3) The IRV curve appears as a pulse jump at the time of water seepage. After that, both the fracture development and the post-peak failure stages have stepped up. The VDIIT curve also appears to be a pulse jump at the time of water seepage, and obvious up and down fluctuations exist before water seepage and fracture. (4) Based on the Pauta Criterion, by analyzing the values of VDIIT during the experiment, the early warning threshold of sandstone fracture seepage is determined to be 0.00559. The research finding can provide an experimental and theoretical basis for the early warning of flood accidents in underground rock engineering.

Keywords: coupled stress-hydro effect; uniaxial loading; infrared radiation; warning threshold; non-destructive monitoring

1. Introduction

With the rapid development of social construction, many cities worldwide have taken the development and utilization of underground space as an important way to solve the population, resource and environmental crises and implement the concept of sustainable development [1–9]. However, the development of underground space and other geotechnical engineering is often restricted by groundwater [10–17]. Because water can cause changes in the physico-chemical properties of rocks, it is generally regarded as one of the most active and direct factors in geotechnical engineering disasters [18–25]. The rock in the engineering site is usually subject to the double action of water pressure and external force. Therefore, it is imperative to explore the dynamic law of rock fracture and seepage under the coupled stress-hydro effect and carry out the prediction research of water damage accidents to reduce the occurrence rate of geotechnical engineering accidents.

It has been found, that when the rock breaks, it usually releases electromagnetic energy [26–31], elastic energy [32–41], thermal energy [42–44], acoustic energy [45–51], and other kinds of energy [52]. Hence, generating a variety of disaster warning methods related to rock mass, such as infrared radiation method [53,54], electromagnetic radiation method [55,56], acoustic emission method [57,58], potential method [51,59] and microseismic method [60,61]. Among these, as a non-contact method, the infrared radiation method has the advantages of high accuracy [62], strong reliability [63], simple operation [64], and visualization [65], etc., which provides a convenient and accurate early warning method for rock fracture seepage monitoring under coupled stress-hydro effect.

Many scholars have recently studied the infrared radiation characteristics of rock failure and instability. Wu et al. [66] believe that there are abnormal changes in the infrared radiation image and temperature curve before rock failure. Lin et al. [67] found that the evolution law of average infrared radiation temperature is closely related to rock porosity, and the process of rock failure can be inferred from this. Huang et al. [68] found that the surface emissivity of loaded rock varies linearly with stress through experiments. Cao et al. [69] proposed a new index, "load-unload response ratio (LURR)" based on the rock failure characteristics under cyclic loading and unloading conditions. Zhang et al. [70] conducted some experiments on preflawed sandstone to investigate the infrared radiation characteristics during failure process and presented a new quantitative model based on Verhulst inverse function.

In addition to the above research on the infrared radiation characteristics of dry rocks, some scholars have conducted relevant experiments with water-bearing rocks. Cao et al. [71] carried out uniaxial loading tests of sandstone with different water content and thus proposed a quantitative analysis index of energy dissipation infrared radiation ratio, which was applied to predict and identify the failure of saturated rock. In addition, his team also found that rock saturation weakened its mechanical properties and amplified the changes in infrared radiation during the bearing process. According to the experimental results, they established a uniaxial loading constitutive model of rock based on infrared radiation [72]. Cai et al. [73] studied the infrared radiation characteristics of sandstone, granite, and marble with different water saturations during loading. They considered that the increment of infrared radiation has a great relationship with rock samples' water content and compressive strength. Shen et al. [74] proved through experiments that the maximum infrared radiation temperature of rock samples presents different characteristics at different stages of loading.

However, because many rock masses in the project site are under the simultaneous and continuous action of water pressure and external force, the above research cannot fully meet the requirement of real engineering application. Therefore, this paper innovatively designed the infrared radiation observation experiment of sandstone failure seepage under the coupled stress-hydro effect and established the quantitative characterization method of sandstone failure seepage through the infrared radiation response indicators, i.e., AIRT, IRV, and VDIIT. After that, the infrared radiation warning threshold of sandstone fracture and seepage was determined. The research results can provide an experimental and theoretical basis for the early warning of flood accidents in geotechnical engineering, e.g., tunnels, mines, and underground reservoirs.

2. Experimental Design

2.1. Experimental System and Equipment

The infrared radiation observation experiment for sandstone fracture and seepage under coupled stress-hydro effect, consists of a pressure control system, a hydraulic loading system, a digital camera system, and an infrared monitoring system. The influence of coupled stress-hydro effect on the infrared parameters of the sandstone surface is analysed through the collection of various parameters during the experiment; the initial position and damage characteristics of sandstone seepage fracture under various conditions are investigated. Figure 1 depicts the experiment system schematic diagram.

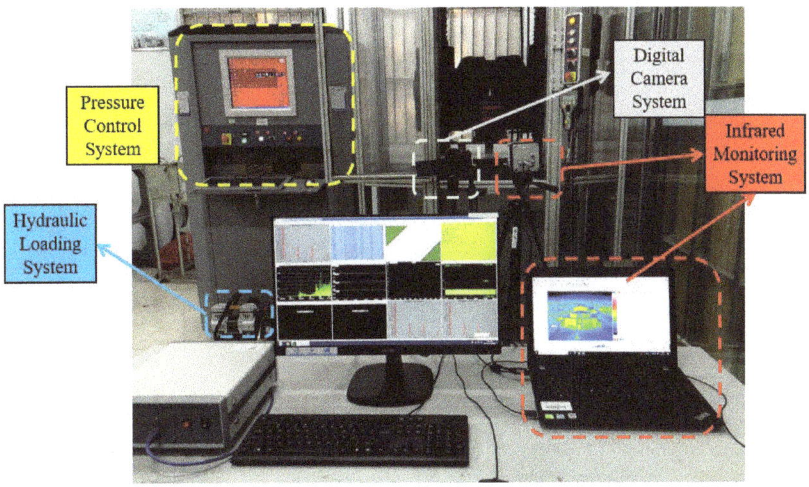

Figure 1. Experimental system diagram.

Pressure control system

This test uses the MTS C64.106 electro hydraulic servo universal testing equipment as the pressure control system, as shown in Figure 2. In Figure 1, the press may display the test parameter curve dynamically and record the axial stress, strain, displacement, and load in real-time. The sampling frequency is 1000 Hz, the maximum static load is 1000 kN. The loading, displacement, and deformation accuracy are within 0.5%.

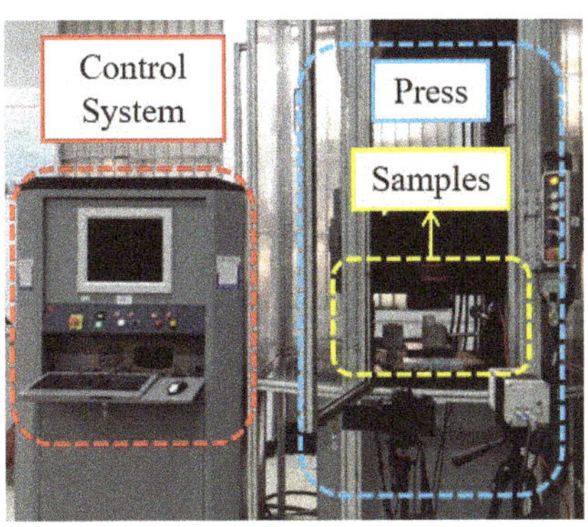

Figure 2. MTS electro-hydraulic servo universal testing machine.

Digital camera system

The Canon 600D SLR camera is used in this experiment to record the images. It has an 18 million pixel CMOS sensor, a digital 4 image processor, a 3-inch reversible LCD screen, full HD video recorder. The camera will record the entire experiment and be utilised to watch the sample fracture and water seepage process in the later stages of the experiment.

Hydraulic loading system

Pull a steel pipe out of the drilled sandstone, fill it with water, connect the pressure gauge, and link it to the air pump interface, before strong glue using to secure the perforated iron sheet to the top of the sandstone. An air pump compression technique is used to verify that the internal water pressure reaches and maintains a specific value during the test. The primary air compressor (Figure 3) characteristic parameters are listed in Table 1.

Figure 3. Air compressor.

Table 1. Main characteristic parameters of the air compressor.

Item	Parameter	Item	Parameter
Size	45 × 19 × 45 cm	Speed	2800 r/min
Matching power	980 W	Rated exhaust pressure	0.7 MPa
Overall weight	14 kg	Volume of air storage tank	8 L

Infrared observation system

An infrared thermal imager and its professional control system comprise the infrared radiation observation system, as shown in Figure 4. The infrared thermal imager is an uncooled infrared thermal imager with the type varioCAM HD head 880 from infra Tec, Dresden, Germany. The essential characteristics and performance of the infrared thermal imager are described in Table 2.

Figure 4. Infrared thermal imager.

Table 2. Main Characteristic Parameters of Infrared Thermal Imager System.

Item	Parameter
Thermal sensitivity	0.02 °C;
Thermal resolution	50 mK
Measuring band	7.5~14 μm
Lens	F1.0
Standard calibration range	−40~1200 °C
Pixel	2048 × 1536
High acquisition rate	240 Hz

2.2. Sample Preparation

The rock samples used by the authors were taken from the coal mine site and made by cutting a whole rock. First, the authors preliminarily screened the processed rock samples and eliminated the rock samples with surface cracks. Next, the authors used the U510 non-metallic ultrasonic detector to accurately measure the wave velocity of the sample. During the process, the probe should be kept in direct contact, and the probe should be in close contact with the sample through the couplant. At the same time, the authors eliminated the samples whose wave velocity deviated by more than 10% to ensure the maximum elimination of the dispersion of rock samples. The representative samples of cubic shape had dimensions 100 mm × 100 mm × 100 mm. To provide a space for water injection in the sandstone, drill a sandstone cylinder with a diameter of 50 mm and a depth of 65mm, at the centre of the sample surface with a drilling machine (see Figure 5a). After that, a steel pipe with an exterior diameter of 20 mm, an inner diameter of 18 mm, and a length of 200 mm (see Figure 5b) and an iron square piece of specification 80 mm, shall be used (see Figure 5c). Keep the drilled sandstone, steel pipe, and perforated iron sheet in a ventilated area for 24 h to allow the strong adhesive to set completely. Simultaneously, the small space at the three-part junction must be filled with strong glue and sealed to guarantee that the processed sample does not leak (see Figure 5d).

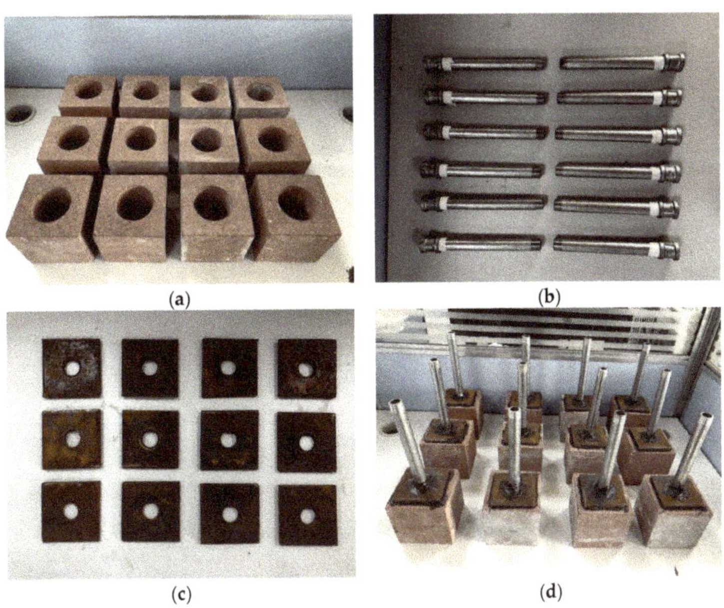

Figure 5. Sample manufacturing process. (**a**) Borehole sandstones; (**b**) Steel pipes; (**c**) Perforated iron plates; (**d**) Finished samples.

To ensure the accuracy of the experimental monitoring data, the flatness parameters of the rock sample surface are set as follows: the roughness of the rock sample surface is less than 0.1 mm, and there is no bulge and depression. The side of the rock sample shall be perpendicular to the upper and lower end faces, with a deviation of less than 0.05°. Rock samples are natural samples without special treatment such as drying or soaking. The rock sample shall be put into the laboratory one day before the experiment to ensure that the temperature of the rock sample is consistent with the temperature of the experimental environment. That is to prevent the accuracy of the infrared radiation response information from being disturbed by the heat transfer during the experiment.

The test is divided into 4 groups, with 3 samples in each group, 12 in total. Four different water pressures of 0 MPa, 0.15 MPa, 0.3 MPa and 0.45 MPa are, respectively, used to pressurize. The experimental samples grouping details are given in Table 3.

Table 3. Experimental grouping.

Water Pressure	Sample Number	Water Pressure	Sample Number	Water Pressure	Sample Number	Water Pressure	Sample Number
0 MPa	0–1 0–2 0–3	0.15 MPa	0.15–1 0.15–2 0.15–3	0.3 MPa	0.30–1 0.30–2 0.30–3	0.45 MPa	0.45–1 0.45–2 0.45–3

2.3. Experimental Process

Figure 1 depicts the pressure control system, water pressure loading system, digital camera system, and infrared observation system. The infrared observation instrument and digital camera are placed 1 m in front of the sample to enable observation and recording. The steel pipe is used to fill the interior chamber of the loaded sample with water sample, and the steel pipe is connected to the air compressor via the rubber pipe to assist later pressurisation. Place the loaded specimen on the presser pressure plate. Given the poor bearing capacity of the rock sample's cavity, a specification of 100 mm (length) × 35 mm (width) × 30 mm (height) metal cushion block is placed on the solid part of the rock sample and its center is aligned with the loading center, as illustrated in Figure 6. Simultaneously, the reference sample used for noise reduction must be put and kept on the loaded sample's side, and its height must be consistent with that of the loaded sample.

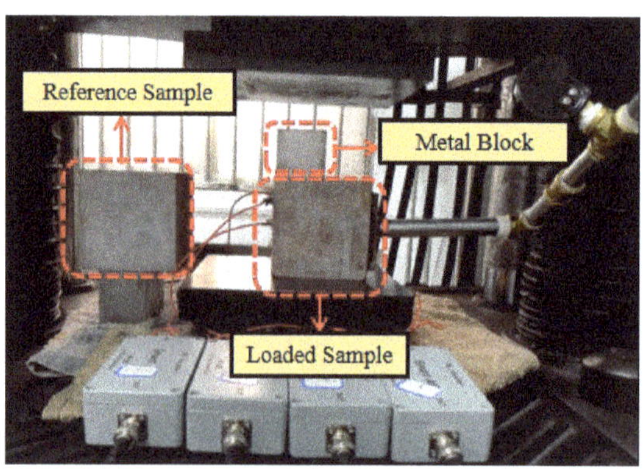

Figure 6. Details of samples.

After arranging the necessary equipment and samples, each experimenter is responsible for configuring the storage directory and other basic settings for the press, infrared thermal imager, and other equipment, with a 0.2 mm/min loading rate. After that, the air compressor increases the water pressure to the desired value, and the valve is closed to guarantee that the internal water pressure of the rock sample remains constant. After setup, the infrared thermal imager, digital camera, press, and other equipment will begin to operate in unison under the unified password. It is prohibited for laboratory staff to move around and close the laboratory windows, curtains, and all lighting sources that may create radiation interference during the information gathering procedure of the infrared thermal imager.

After the test, all equipment shall stop working at the same time, and the test personnel shall properly save the data of all equipment, and take photos of the fracture morphology of the rock sample. After cleaning the test bench, place the next sample to continue the experiment.

2.4. Experimental Data Processing Method

In the infrared radiation information collection system, the infrared thermal imager maps the rock samples' physical and structural changes during the experimental procedure to the infrared radiation temperature field. It shows in the form of infrared thermal images. When the difference between the background temperature and the temperature of the rock sample is large, the abnormal features of the infrared thermal picture are not visible (as shown in Figure 7), so the infrared radiation data of the rock sample must be extracted again in a small range.

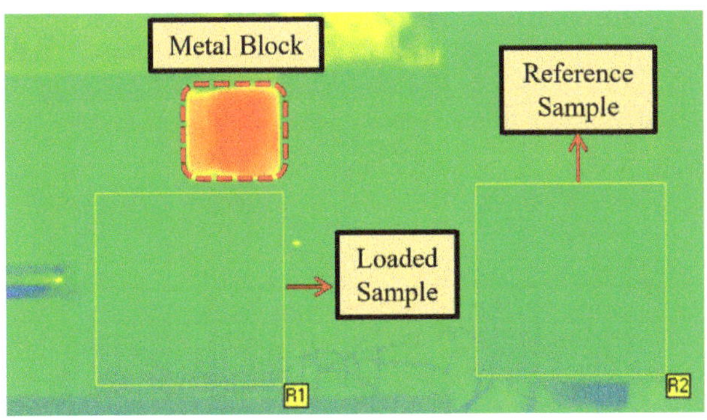

Figure 7. Infrared thermal image of rock sample.

A rectangular area (as illustrated in Figure 7) is constructed along the whole sample section in the infrared radiation acquisition system, and the infrared radiation response information of the rectangle area is then retrieved and preserved in the form of a series of two-dimensional matrices. The resampled infrared radiation data of frame P is a two-dimensional matrix with the following expression:

$$f_p(x,y) = \begin{bmatrix} f_p(1,1) & f_p(1,2) & \dots & f_p(1,L_y) \\ f_p(2,1) & f_p(2,2) & \dots & f_p(2,L_y) \\ \vdots & \vdots & \ddots & \vdots \\ f_p(L_x,1) & f_p(L_x,2) & \dots & f_p(L_x,L_y) \end{bmatrix} \tag{1}$$

where x represents the row number of the matrix $f_p(x,y)$ and y represents the column number; L_x and L_y are the maximum number of rows and columns of x and y, respectively.

According to the obtained temperature matrix, the following parameters can be calculated:

(1) Average infrared radiation temperature (AIRT) and ΔAverage infrared radiation temperature (ΔAIRT)

AIRT can directly reflect the bearing rock surface's overall infrared radiation field temperature. The average infrared radiation temperature (AIRT (p)) of the pth frame in the original infrared radiation thermal image sequence is expressed as:

$$\mathrm{AIRT}(p) = \frac{1}{L_x}\frac{1}{L_y}\sum_{x=1}^{L_x}\sum_{y=1}^{L_y} f_p(x,y) \quad (2)$$

Since the radiation interference of the loaded sample and the reference sample is almost synchronous in time and space, the AIRT of the loaded sample can be subtracted from the AIRT of the reference sample to obtain the denoised ΔAIRT. This can be calculated by using Equation (3):

$$\Delta\mathrm{AIRT}(p) = \mathrm{AIRT}(p) - \mathrm{AIRT}'(p) \quad (3)$$

(2) Infrared radiation variance (IRV)

The physical meaning of IRV is the changing trend of the dispersion degree of the temperature field in the original infrared radiation thermal image sequence diagram. The variance (IRV (p)) of the original infrared radiation thermal image sequence of the pth frame is expressed as:

$$\mathrm{IRV}(p) = \frac{1}{L_x}\frac{1}{L_y}\sum_{y=1}^{L_x}\sum_{x=1}^{L_y}\left[f_p(x,y) - \mathrm{AIRT}(p)\right]^2 \quad (4)$$

(3) Variance of differential infrared image temperature (VDIIT)

The physical meaning of VDIIT is the variation trend of the dispersion degree of the temperature field in the differential infrared radiation thermal image sequence diagram. The variance (VDIIT (P)) of the differential infrared radiation thermal image sequence of the pth frame is expressed as:

$$\mathrm{VDIIT}(p) = \frac{1}{L_x}\frac{1}{L_y}\sum_{y=1}^{L_x}\sum_{x=1}^{L_y}\left[\varphi_p(x,y) - \mathrm{AIRT}(p)\right]^2 \quad (5)$$

where $\varphi_p(x,y) = f_{p+1}(x,y) - f_p(x,y)$.

3. Experimental Results and Analysis

3.1. AIRT Response Characteristics of Sandstone Seepage

The loaded specimen ΔAIRT value has prominent change characteristics, mainly showing an upward-downward trend. According to the inflection point (breakpoint) of the corresponding load curve, the whole process can be divided into four stages: (I) compaction stage, (II) linear elastic stage, (III) fracture development stage, and (IV) post peak failure stage.

(1) sample 0–1

Figure 8 depicts the ΔAIRT-load curve for sample (0–1). At 0 s, the ΔAIRT was −0.147 °C. In the compaction stage, it showed a fluctuating upward trend. The sample enters the linear elastic stage at 340.17 s, at which the corresponding load is 54.27 kN, the ΔAIRT is −0.100 °C, which is 0.047 °C higher than that at the beginning. In the elastic stage, the ΔAIRT also showed an upward trend, and the temperature rise rate was almost the same as in the previous stage. Before and after 697 s, the load curve fluctuated, decreasing from 233.32 kN to 215.96 kN. At this time, the corresponding ΔAIRT was −0.043 °C, which indicates that the sample was at the end of the linear elastic stage, and the temperature increased by 0.057 °C compared with the initial stage. In the next fracture development stage, with the increase of the load curve, the ΔAIRT begins to decrease. At 796.19 s, the load curve has a peak value. After that, it rapidly decreases, which means the beginning of the post-peak failure stage. At this time, the corresponding ΔAIRT is −0.056 °C. At 802.80 s, the end of the experiment, the load decreased to 215.86 kN, and the ΔAIRT fell to −0.059 °C.

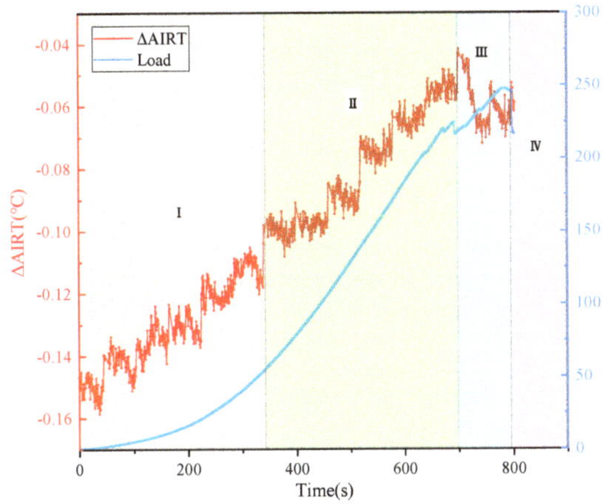

Figure 8. Experimental data of sample 0–1.

(2) sample 0–3

Figure 9 depicts the ΔAIRT-load curve for sample (0–3). The ΔAIRT at 0 s is 0.056 °C, it rises slowly at the compaction stage. In the range of 100.48 s to 120.35 s, ΔAIRT jumps from 0.063 °C to 0.081 °C, increasing by 0.018 °C. During this period, the load has not exceeded 2 kN, and the heat generation is caused by friction of particles in the rock sample. At 454.41 s, the bearing rock sample enters the linear elastic stage. At this time, ΔAIRT was 0.109 °C, which is 0.053 °C higher than the initial stage. In the linear elastic stage, ΔAIRT continued to rise. At 758 s, the load curve suddenly drops, and the drop amplitude reaches 13 kN. The ΔAIRT value was 0.134 °C, and the whole stage increased by 0.025 °C. The loaded rock sample thus enters the fracture development stage. In this stage, the load curve drops again at about 836 s, with a magnitude of 9 kN. At 919.91 s, the load curve reached the peak of 202.46 kN, followed by a sudden drop, which is a sign of the beginning of the post-peak failure stage. So far, the temperature drop of the whole fracture development stage is 0.057 °C. At the post-peak stage, the load curve and ΔAIRT both decreased rapidly, and the ΔAIRT fell to −0.041 °C.

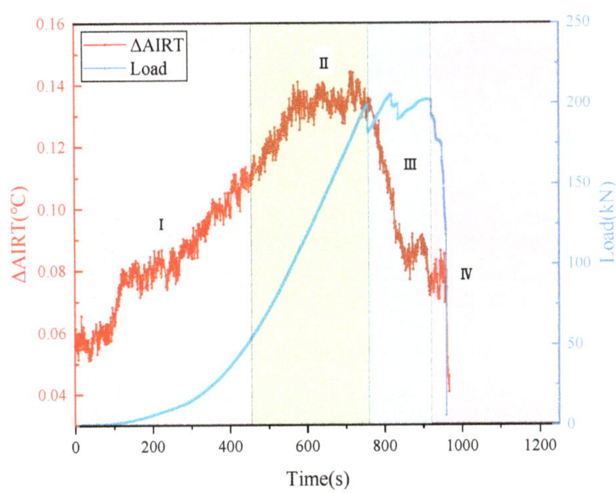

Figure 9. Experimental data of sample 0–3.

(3) sample 0.30–2

Figure 10 depicts the ΔAIRT-load curve for sample (0.30–2). The initial value of ΔAIRT was −0.671 °C. In the compaction stage (0 s~193.05 s), the load rises to 51 kN and ΔAIRT rises to −0.646 °C. In the linear elastic stage (193.05 s~524.76 s), the load curve and ΔAIRT showed a nearly linear rising trend, in which the load increased to 210.04 kN, ΔAIRT rose to −0.601 °C. During the whole fracture development stage (524.76 s~637.80 s), the load increased from 210.04 kN to 253.37 kN, which is also the peak load. During this period, the ΔAIRT was stable, and only slightly increased by 0.011 °C. After the peak of the load curve, the ΔAIRT and the load have a sudden drop trend, wherein ΔAIRT dropped to −0.773 °C, which overall decreased by 0.183 °C.

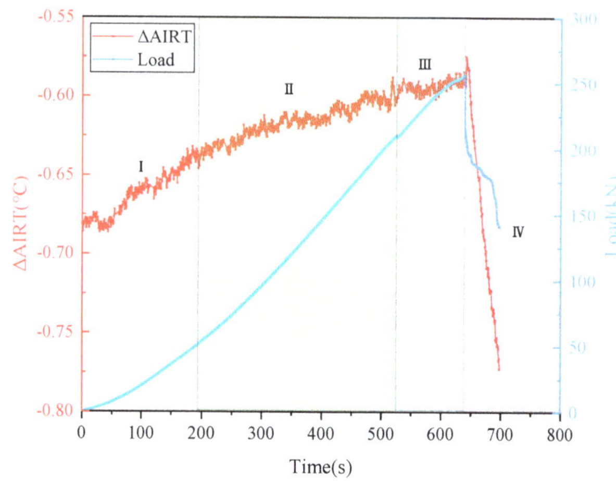

Figure 10. Experimental data of 0.30–2 sample.

Figure 11 shows the ΔAIRT's difference of the three samples at each stage. It revealed that in the compaction stage, the water body is confined to the interior of the rock sample. Its control effect on the ΔAIRT has not yet appeared, so the temperature change trend of the rock sample surface is dominated by the temperature rise caused by loading. In the linear elastic stage, the original fracture in the rock sample is gradually closed in the previous stage, the internal water body is difficult to seep out, so the surface temperature of the rock sample is still rising. In the fracture development stage, the internal cracks of the rock sample begin to grow, and gradually develop into macroscopic cracks, which is visible to the naked eye. The cracks begin to meet and penetrate, and the sample volume expands. In this stage, as the water begins to seep out along the developed fracture, its cooling effect on the rock sample surface begins to appear, in which 0–1 sample is cooled by 0.013 °C and 0–3 sample is cooled by 0.067 °C. Although the 0.30–2 sample still has a small temperature rise of 0.011 °C, the temperature rise trend has been significantly suppressed. In the post-peak failure stage, the load curve has decreased significantly since the macro fracture surface was formed. The water in the rock sample flows out in large quantities, resulting in a significant cooling effect. Among them, 0–1 sample is cooled by 0.003 °C, 0–3 sample is cooled by 0.118 °C, and 0.30–2 sample is cooled by 0.183 °C.

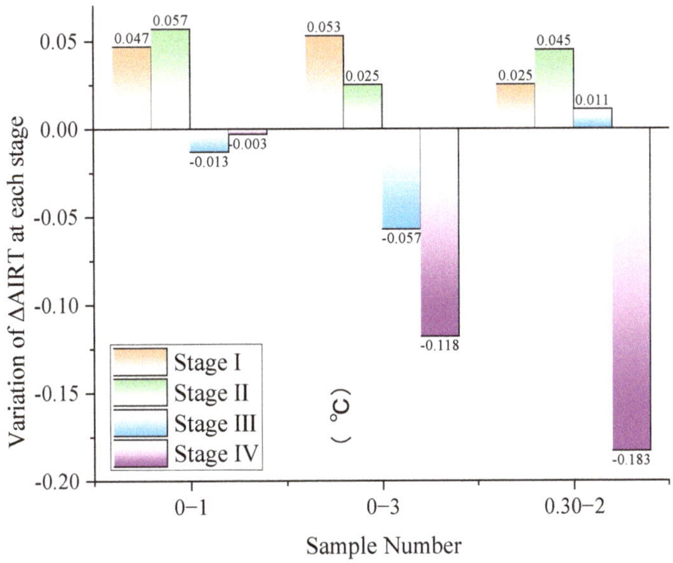

Figure 11. Temperature change in each stage.

In this experiment, when the internal water pressure is not more than 0.30 MPa, the water seeps out after the crack is developed, thus causing the cooling phenomenon. When the water pressure is 0.45 MPa, the internal water body seeps out in the linear elastic stage, thus, the ΔAIRT value changes from up to down. For rock samples with a water pressure of 0.45 MPa, the analysis of ΔAIRT changes will be given in combination with IRV and VDIIT.

3.2. The Functional Relationship between ΔAIRT and Load before Sandstone Seepage

By plotting the ΔAIRT-load curve of each rock sample, it can be found that when the internal water pressure is not more than 0.30 MPa, the load and ΔAIRT show a certain positive relationship in the compaction stage and the linear elastic stage, that is ΔAIRT increases with the increase of load. The 0–2 sample, 0–3 sample and 0.30–2 sample with typical experimental results are selected for analysis.

To further explore the functional relationship between ΔAIRT and load in this process, the ΔAIRT and load data of these rock samples in the compaction and linear elastic stages can be extracted. The function fitting can be performed according to the time parameters as shown in Figure 12. After several fitting times of, it was found that the trend of the power function model is consistent with the corresponding relationship point. The functional expression of ΔAIRT and load (L) is:

$$\Delta \mathrm{AIRT} = \mathrm{aL}^{\mathrm{b}} \tag{6}$$

where, ΔAIRT is the average infrared radiation temperature difference (°C), L is the axial load (kN), a and b are coefficients.

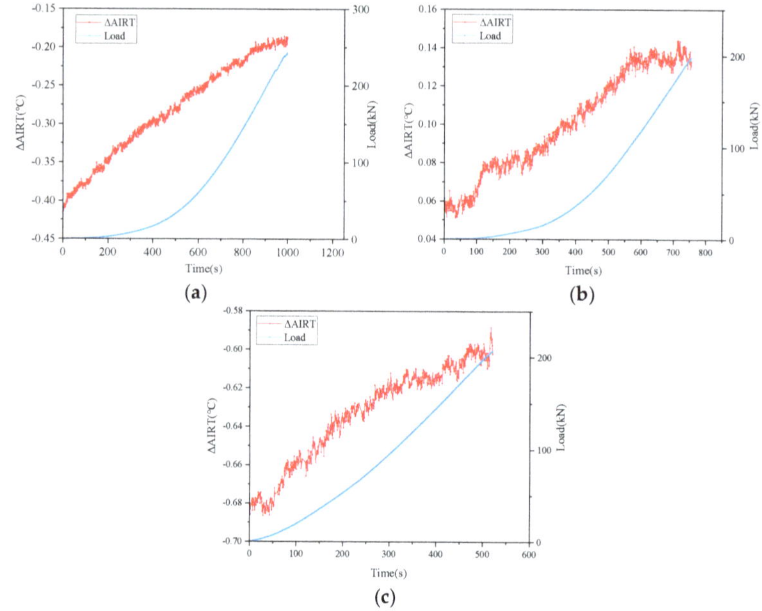

Figure 12. Data of each sample in the compaction and elastic stages. (**a**) 0–2 sample; (**b**) 0–3 sample; (**c**) 0.30–2 sample.

Figure 13 depicts a fitting curve of ΔAIRT-load of each sample. The maximum standard deviation of the fit curve for each sample is only 0.06429, and the minimum value of the function correlation coefficient is 0.8924, which indicates that the model selection and fit effect are ideal. The details about each sample are given in Table 4.

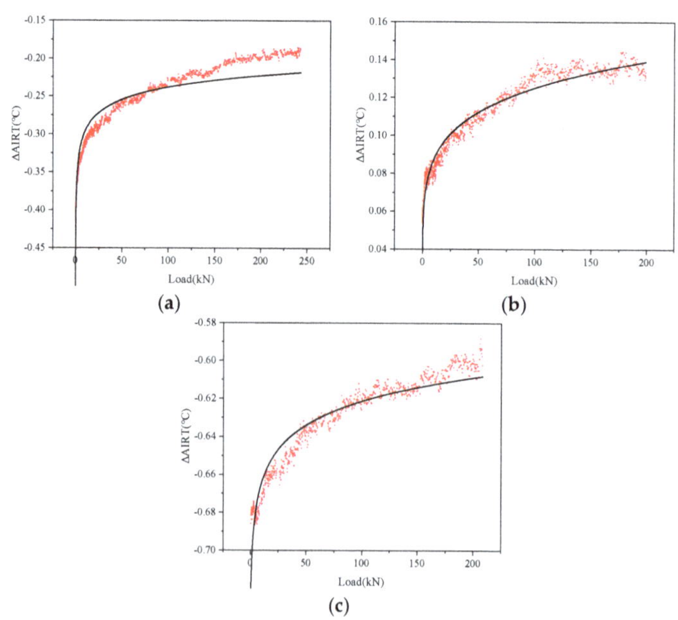

Figure 13. Fitting curve of each sample. (**a**) 0–2 sample; (**b**) 0–3 sample; (**c**) 0.30–2 sample.

Table 4. Function fitting parameters of ΔAIRT and load of some samples.

Sample	Function Correlation Coefficient	A	B	Standard Deviation
0–2	0.9063	−0.36895	−0.09505	6.429%
0–3	0.9623	0.06149	0.15406	2.705%
0.30–2	0.8924	−0.70989	−0.02894	2.517%

3.3. IRV Response Characteristics of Sandstone Seepage

Figure 14 depicts the ΔAIRT-load-IRV curve of 0.45–2 sample. It revealed that in the compaction stage (0 s−217.15 s), the ΔAIRT of 0.45–2 sample shows an upward trend, rising from −0.013 °C to 0.057 °C, increasing by 0.070 °C, and the corresponding load curve bends upward to 36.98 kN. As the water body inside the rock sample has not yet seeped out, the control effect on the surface temperature of the rock sample has not yet appeared. The temperature change trend is mainly dominated by the temperature rise caused by uniaxial loading, and the IRV has been stable between 0.013 and 0.016.

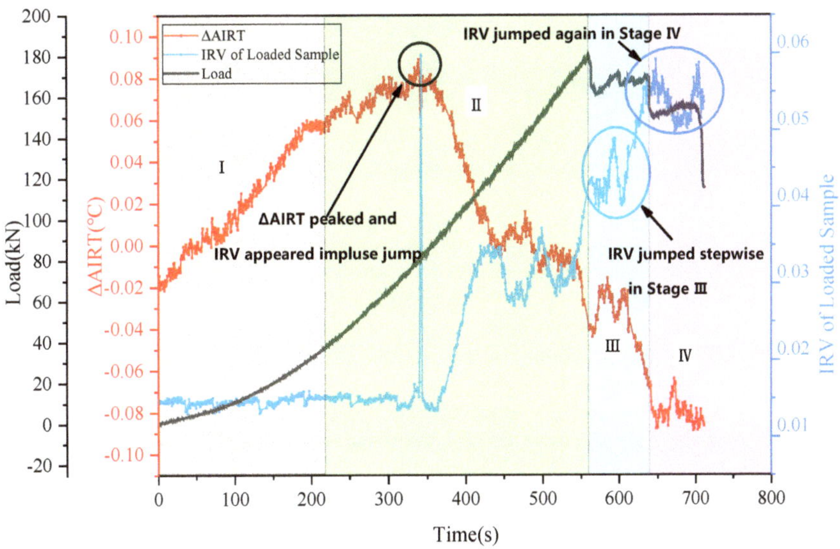

Figure 14. IRV of 0.45–2 sample and other related data.

Under 0.45 MPa water pressure, the water in the rock sample escapes and seeps through the original fissures in the rock and the micro pores between the particles around 340 s (see Figure 15). At the same time, the load value reached 80.41 kN, and remained in a linear rising state, while the AIRT and IRV curves exhibited significant changes. For example, at 339.47 s, the AIRT curve attained a peak value of 0.089 °C before beginning to fall. At 338.81 s, the IRV remained stable at 0.016, but at 342.78 s, the IRV pulse jumped at 0.060 before returning to 0.015 at 343.44 s. It can be seen that when the water pressure is strong (enough to cause a water seepage point and form a water seepage surface in a short time), the inflection point of ΔAIRT from rising to falling and the pulse type jump peak of IRV can be used as an early warning signal for water seepage (water inrush) of rock samples.

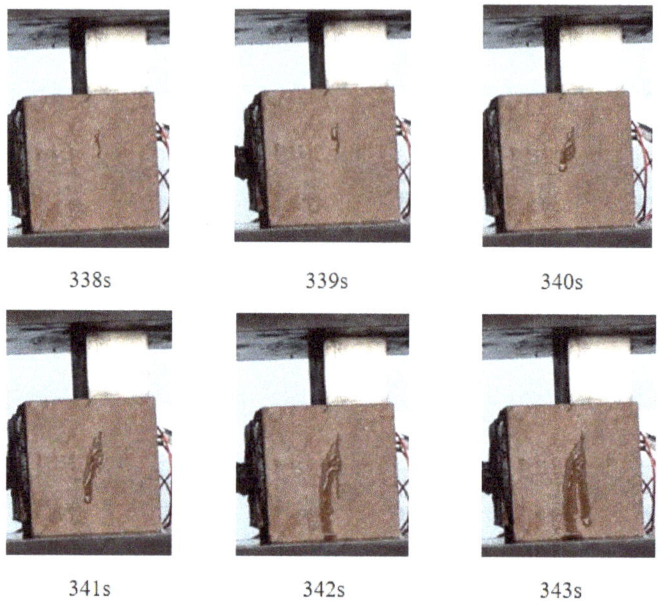

Figure 15. Water seepage diagram of 0.45–2 sample before and after 340 s.

After water seepage (water inrush) of rock samples, the ΔAIRT value drops rapidly. During the period from 339.47 s to 560.24 s, the ΔAIRT decreased from the peak value of 0.089 °C to −0.039 °C, and the cooling range was as high as 0.128 °C. While the IRV also increased from 0.015 to about 0.034 and maintained stable instability. During this period, the load curve has been in a linear rising state, reaching 179.46 kN at the end of the stage.

In the fracture development stage (560.24 s–638.80 s), IRV appears a step-type jump compared with the previous stage. In the later stage of the last linear elastic stage, IRV was once stable at around 0.034. Though the appearance of first peak of the load curve, the IRV curve rises rapidly, and remains constant at about 0.045 at this stage. This can be understood that the IRV response characteristics of the rock samples entering the fracture development stage. During this period, the load curve decreased from 179.46 kN to about 165.42 kN, and ΔAIRT was still in a downward trend.

In the post-peak failure stage (638.80 s–711.71 s), IRV continued to rise in stages. IRV, which was previously stable at 0.045 in the fracture development stage, rose rapidly to 0.055 and maintained a stable fluctuation trend. During this period, the load curve reached the peak of 168.61 kN and began to drop, while the ΔAIRT value was stable at −0.079 °C.

Figure 15 is an image recording of water seepage of the 0.45–2 sample before and after 340 s. The rock sample's observation surface is dry at 338 s, and a wet water point appears at the upper part of the rock sample, that is, the water seepage point. The seepage point is the starting point of the seepage process. After the water seepage, the wet area centered on the water seepage point began to expand. With time, it mainly expanded to the lower part of the rock sample and rapidly formed water droplets to slide down, which also affected the changes in ΔAIRT curve and IRV curve.

3.4. VDIIT Response Characteristics of Sandstone Seepage

Figure 16 is the ΔAIRT-load-VDIIT curve of the 0.45–2 sample. Figure 16 revealed the VDIIT of this rock sample is always around 0.0040 in the early stage of the experiment. However, between 250.20 s and 252.28 s, VDIIT jumped from 0.0040 to 0.0062, and then rapidly decreased to 0.0016. This point can be used as a precursor of water seepage of rock samples. After a development period, VDIIT suddenly increased to 0.0130 at 340.13 s, then jumped to 0.0408 at 342.78 s. This time point also corresponds to the inflection point of AIRT from rising to falling and the water seepage phenomenon, as shown in Figure 15.

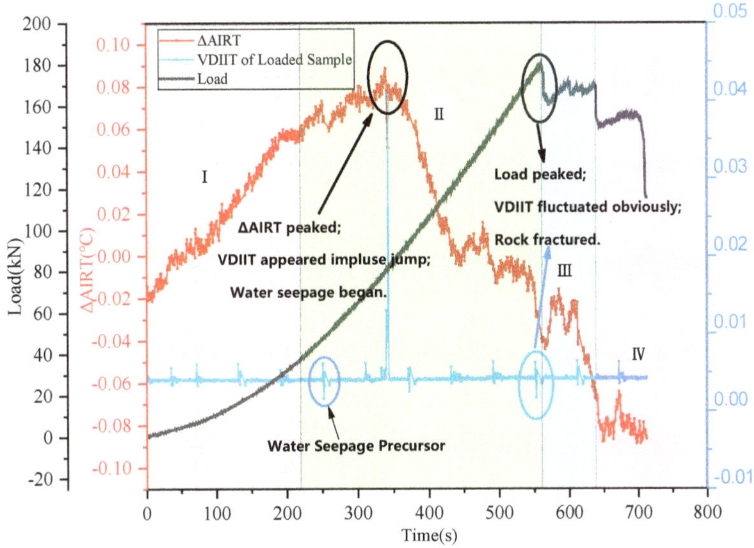

Figure 16. VDIIT of 0.45–2 sample and other related data.

The VDIIT also shows abnormal fluctuation when the rock sample breaks and water seep. At 551.27 s, VDIIT jumps from 0.0043 of the previous frame to 0.0063, then drops to 0.0017 at 552.60 s. It can be seen from the image records that this time point corresponds to the macro crack development and fracture water seepage process of the rock, as shown in Figure 17.

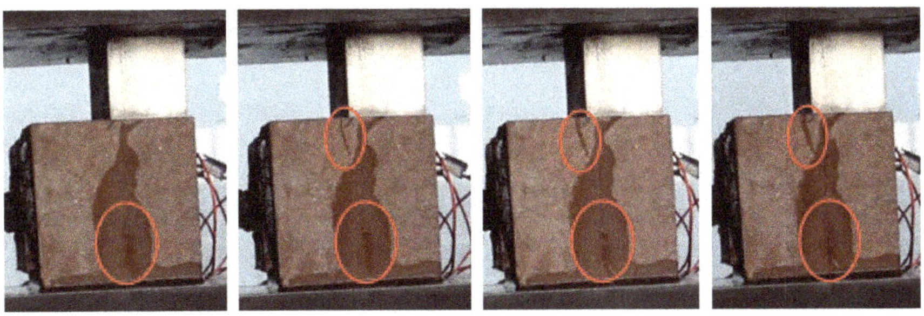

Figure 17. Macro crack development and fracture water seepage process of 0.45–2 sample before and after 552 s. The red circle represents the location of fracture development.

Figure 17 shows the macro crack development and fracture in the water seepage process of 0.45–2 sample before and after 552 s. At 551 s, longitudinal cracks appeared in the middle of the rock sample observation surface, and water flowed out at the lower part. At 552 s, a VDIIT fluctuated when longitudinal water seepage growth also appeared, at the left of the center upper part. Over time (555 s–560 s), the two seepage zones developed, expanded, and penetrated each other, forming a Y-shaped macro fracture. During this period, the ΔAIRT also rapidly decreased to 0.013 °C.

In order to reflect the mutation characteristics of VDIIT of the sample, the mutation threshold of VDIIT is determined based on the Pauta Criterion, and the discrimination criteria can be calculated using Equation (7).

$$|\varepsilon - \mu| > 3\sigma \tag{7}$$

where: ε is the VDIIT value, μ is the average value of VDIIT, σ is the standard deviation of VDIIT. The infrared radiation threshold is $\mu \pm 3\sigma$.

For example, the Sandstone sample 0.45–2 mutation threshold of VDIIT of the sandstone sample based on Pauta Criterion is 0.00798. It can be seen from Figure 16 that the VDIIT index based on the Pauta Criterion can effectively identify the VDIIT mutation in the process of water seepage (water inrush), which can be used as an early warning of disasters. Table 5 shows the mutation threshold statistics of all sandstone samples. The average value of the upper threshold of the mutation threshold is 0.00627, the maximum value is 0.00798, and the minimum value is 0.00559. When water seepage (water inrush) occurs in the rock, the difference in the infrared radiation temperature matrix increases, which makes VDIIT suddenly increase. Therefore, when identifying VDIIT mutations, it is only necessary to consider the upper threshold. In the application process, the minimum value of mutation threshold 0.00559 should be taken as the early warning threshold to ensure the accuracy of early warning.

Table 5. Water seepage (water inrush) threshold of sandstone samples.

Sample	Upper Threshold	Lower Threshold
0–1	0.00559	0.00367
0–2	0.00683	0.00455
0–3	0.00566	0.00368
0.15–1	0.00644	0.00401
0.15–2	0.00561	0.00383
0.15–3	0.00697	0.00487
0.30–1	0.00658	0.00434
0.30–2	0.00654	0.00427
0.30–3	0.00567	0.00362
0.45–1	0.00575	0.00376
0.45–2	0.00798	0.00048
0.45–3	0.00562	0.00360
Average value	0.00627	0.00372

4. Discussion

The rock, before uniaxial loading, contains a certain amount of pores, air and water. In the process of uniaxial loading, the internal stress of rock mass will increase, accompanied by pore compression, crack development, fracture and water seepage.The coupling effect of water pressure and stress leads to rock mass fracture, and the crack development will promote the seepage of water in the rock. The change of energy (ΔE) on the rock surface under coupled stress-hydro effect mainly includes the following five parts:

$$\Delta E = \Delta E_1 + \Delta E_2 + \Delta E_3 + \Delta E_4 + \Delta E_5 \tag{8}$$

ΔE_1 is energy that the gas escape process carries in the primary pore. It has been confirmed through laser Raman spectroscopy analysis technology that CH_4, CO_2, O_2 and other gases are in most rocks' pores [75]. Before the pores are damaged, the internal gas exists in the interior or surface of the pores in a free or adsorbed state. Under the action of external load, the pores are compressed or even destroyed, resulting in the escape of internal gas. In the process of gas escape, some energy will be taken away. Therefore, generally, $\Delta E_1 < 0$.

ΔE_2 is the energy generated by friction heat generation. In the rock's interior, the friction behavior will occur between the pores, fractures, joints and rock particles developed in all directions. Two factors affect the friction heat generation process: the positive pressure on the contact surface and the friction coefficient. When the friction coefficient is constant, the greater the normal stress on the contact surface, the greater the friction force, and the more work is done to overcome the friction force in the process of crack and particle sliding, thus causing the temperature of the contact surface to rise, so $\Delta E_2 > 0$.

ΔE_3 is the energy generated by the thermoelastic effect. The change of temperature rise of the loaded sample is in direct proportion to the change of stress, and the expression is:

$$\Delta T/T = K_0 \Delta\sigma \tag{9}$$

where: T is the absolute temperature of the solid unit; ΔT is the temperature change; $\Delta\sigma$ is the variation of the principal stress sum, and K_0 is the thermal elastic coefficient.

In the process of uniaxial loading, the principal stress increases with loading, so the sample temperature rises, $\Delta E_3 > 0$.

ΔE_4 is the heat generated by the expansion of original pores, fractures, and joints in the rock and the development of new fractures. With the increase of external load, internal pores and joints will shrink and close. With further loading, the pores will collapse, and the original fractures and joints will further expand, penetrate and merge, accompanied by new fractures. The increase of heat accompanies this process, so $\Delta E_4 > 0$.

ΔE_5 is the energy loss caused by the water body escaping. With the development, expansion, and penetration of rock fracture, the water in the sample begins to seep out. Water's specific heat capacity and thermal inertia are larger than rock's. The water has an evaporation effect, the temperature of water under the same conditions is lower than that of surrounding objects, and the water seepage part shows an obvious low-temperature area in the thermal image. With the increase of water seepage, the temperature of rock sample decreases continuously, so $\Delta E_5 < 0$.

The whole process of the experiment, includes the above five energy changes. With the difference in stress state, damage degree and other conditions of rock samples, the five changes work together to cause the rise and fall of rock sample temperature.

In the process of fracture and water seepage caused by rock mass compression, the seepage of water in the rock mass will lead to a decrease in infrared radiation temperature, while the thermal elastic effect, friction thermal effect, and crack propagation thermal effect will lead to the increase of infrared radiation temperature in the rock mass. Because the temperature drop of water seepage is higher than the temperature rise of rock fracture, the infrared radiation temperature of the rock will drop rapidly when the rock is near the fracture seepage. Infrared radiation has a strong sensitivity to water, which also provides the feasibility for monitoring the rock water seepage (water inrush) of underground engineering with infrared radiation.

Figures 15 and 17 show the whole process of sandstone before (fracture and water seepage) and after macro fracture development and water seepage, respectively. The microcracks on the surface of rock samples under uniaxial loading are mainly tensile cracks. For example, the microcracks on the surface of rock 0.45–2 sample first appear in the middle and late stages of loading, and gradually expand with the increase of stress, eventually forming large-scale fracture and water seepage. The authors believe this is related to the fact that the sandstone selected in this test is hard brittle sandstone. Hard rock has no obvious post-peak stage than soft rock samples such as mudstone. It is generally destroyed immediately after the peak strength, accompanied by sound. The hard rock stress-strain curves show a rapid decline after the peak stress. This type of rock strength is higher than soft rock. Most rock samples are axially split; the failure surface is nearly parallel to the axial tension failure. The specimen will not be damaged immediately after axial splitting, but also has a certain bearing capacity until a through tensile failure crack is formed in the rock. However, the water seepage in the rock sample impacts the rock's failure form; that is, a small number of rock samples appear shear microcracks on the surface in the middle and late stages of loading, and then the water seeps out along the shear microcracks. This is because the water seepage in the rock sample has a lubricating effect on the rock particles, reduces the friction force of particle crystals for friction sliding, and thus promotes the growth of primary cracks and the generation of new cracks (tensile cracks and shear cracks). The propagation of primary cracks can induce the generation of new cracks, and local damage is easy to occur, eventually leading to shear microcracks in a few rock samples. However, the seepage effect of water in the rock mass is affected by the rock mass microstructure. The rock microstructure in different areas of underground engineering is different, even in different regions of the same rock. In future research, the authors will further study the seepage effect of water in the rock mass and the corresponding infrared radiation characteristics, in combination with the microstructure characteristics of the rock mass to finally realize the monitoring and early warning of water seepage (water inrush) in underground engineering.

5. Conclusions

To explore the infrared radiation changes in the process of sandstone fracture and water seepage and determine the corresponding infrared radiation warning threshold, this study designed the infrared radiation observation experiment of sandstone failure seepage under the coupled stress-hydro effect and evaluated the corresponding relationship among load, AIRT, IRV, and VDIIT during the experiment. The following conclusions were drawn:

(1) During the experiment, the ΔAIRT of the sample mainly showed an upward-downward trend. Among them, the compaction stage and the linear elastic stage are both rising, the fracture development stage is stable or falling, and the post-peak failure stage has an obvious downward trend.

(2) For the sample with internal water pressure lower than 0.30 MPa, ΔAIRT at the compaction and linear elastic stage has a strong power function relationship with the load. The expression is $\Delta AIRT = aL^b$, where ΔAIRT is the difference between the infrared radiation average temperature of the loaded sample and the reference sample, L is the load, and a and b are the undetermined coefficients. The correlation coefficient of the function can reach above 0.8900, which has a strong reference value.

(3) The IRV curve has a pulse jump at the time of water seepage of the rock sample, and a step jump at the fracture development and the post-peak failure stages, respectively. The VDIIT curve has a pulse jump at the time of water seepage, and there is an obvious up and down fluctuation before water seepage and fracture.

(4) Based on the Pauta Criterion, a VDIIT mutation threshold for rock fracture seepage is proposed, which is 0.00559.

Author Contributions: R.C.—Conceptualization, writing-original draft and experiments; K.C.—Conceptualization, supervision and project administration; X.L.—Software and data curation; R.M.A.K.—Methodology; N.M.K.—Visulization; W.L.—Writing review&editing; Q.G.—Investigation and Data curation; F.W.—Experiments; Y.Y.—Investigation; J.Q.—Writing-review&editing; S.S.A.—Writing-review& editing. All authors have read and agreed to the published version of the manuscript.

Funding: This research was funded by the major university level scientific research project of Anhui University of Finance and Economics, "Research on the Connotative Characteristics and New Era Inheritance and Innovation of Northern Anhui Culture under the Yangtze River Delta Integration Strategy",(ACKYA20003). Also this research was supported by Researchers Supporting Project number (RSP2022R496), King Saud University, Riyadh, Saudi Arabia.

Data Availability Statement: The data that supports the findings of this study is available from the corresponding author upon reasonable request.

Acknowledgments: This research was funded by the major university level scientific research project of Anhui University of Finance and Economics, "Research on the Connotative Characteristics and New Era Inheritance and Innovation of Northern Anhui Culture under the Yangtze River Delta Integration Strategy",(ACKYA20003). Also this research was supported by Researchers Supporting

Project number (RSP2022R496), King Saud University, Riyadh, Saudi Arabia. Meanwhile, the authors gratefully acknowledge Yangyang Wang and Jiangtao Zhai for helpful suggestions.

Conflicts of Interest: The authors declare no conflict of interest.

References

1. Xie, H.; Zhao, J.W.; Zhou, H.W.; Ren, S.H.; Zhang, R.X. Secondary utilizations and perspectives of mined underground space. *Tunn. Undergr. Space Technol.* **2020**, *96*, 103129. [CrossRef]
2. Ma, T.; Liu, J.; Fu, J.; Wu, B. Drilling and completion technologies of coalbed methane exploitation: An overview. *Int. J. Coal Sci. Technol.* **2022**, *9*, 68. [CrossRef]
3. Armaghani, D.J.; Kumar, D.; Samui, P.; Hasanipanah, M.; Roy, B.J.E.W.C. A novel approach for forecasting of ground vibrations resulting from blasting: Modified particle swarm optimization coupled extreme learning machine. *Eng. Comput.* **2021**, *37*, 3221–3235. [CrossRef]
4. Vervoort, A. Various phases in surface movements linked to deep coal longwall mining: From start-up till the period after closure. *Int. J. Coal Sci. Technol.* **2021**, *8*, 412–426. [CrossRef]
5. Lian, X.; Hu, H.; Li, T.; Hu, D. Main geological and mining factors affecting ground cracks induced by underground coal mining in shanxi province, china. *Int. J. Coal Sci. Technol.* **2020**, *7*, 362–370. [CrossRef]
6. Wiatowski, M.; Kapusta, K.; Nowak, J.; Szyja, M.; Basa, W. An exsitu underground coal gasification experiment with a siderite interlayer: Course of the process, production gas, temperatures and energy efficiency. *Int. J. Coal Sci. Technol.* **2021**, *8*, 1447–1460. [CrossRef]
7. Qiao, X.; Chang, F. Underground location algorithm based on random forest and environmental factor compensation. *Int. J. Coal Sci. Technol.* **2021**, *8*, 1108–1117. [CrossRef]
8. Yi, M.; Wang, L.; Hao, C.; Liu, Q.; Wang, Z. Method for designing the optimal sealing depth in methane drainage boreholes to realize efficient drainage. *Int. J. Coal Sci. Technol.* **2021**, *8*, 1400–1410. [CrossRef]
9. Zhao, W.; Wang, K.; Zhang, R.; Dong, H.; Lou, Z.; An, F. Influence of combination forms of intact sub-layer and tectonically deformed sub-layer of coal on the gas drainage performance of boreholes: A numerical study. *Int. J. Coal Sci. Technol.* **2020**, *7*, 571–580. [CrossRef]
10. Yu, H.; Gui, H.; Zhao, H.; Wang, M.; Li, J.; Fang, H.; Jiang, Y.; Zhang, Y. Hydrochemical characteristics and water quality evaluation of shallow groundwater in suxian mining area, huaibei coalfield, china. *Int. J. Coal Sci. Technol.* **2020**, *7*, 825–835. [CrossRef]
11. Sun, K.; Fan, L.; Xia, Y.; Li, C.; Chen, J.; Gao, S.; Wu, B.; Peng, J.; Ji, Y. Impact of coal mining on groundwater of luohe formation in binchang mining area. *Int. J. Coal Sci. Technol.* **2020**, *8*, 88–102. [CrossRef]
12. Zhang, K.; Li, H.; Han, J.; Jiang, B.; Gao, J. Understanding of mineral change mechanisms in coal mine groundwater reservoir and their influences on effluent water quality: A experimental study. *Int. J. Coal Sci. Technol.* **2020**, *8*, 154–167. [CrossRef]
13. Armaghani, D.J.; Yagiz, S.; Mohamad, E.T.; Zhou, J. Prediction of TBM performance in fresh through weathered granite using empirical and statistical approaches. *Tunn. Undergr. Space Technol.* **2021**, *118*, 104183. [CrossRef]
14. Wang, G.Y.; You, G.; Shi, B.; Yu, J.; Tuck, M. Long-term land subsidence and strata compression in changzhou, china. *Eng. Geol.* **2009**, *104*, 109–118. [CrossRef]
15. Moon, J.; Fernandez, G. Effect of excavation-induced groundwater level drawdown on tunnel inflow in a jointed rock mass. *Eng. Geol.* **2010**, *110*, 33–42. [CrossRef]
16. Wu, Q.; Wang, M.; Wu, X. Investigations of groundwater bursting into coal mine seam floors from fault zones. *Int. J. Rock Mech. Min.* **2004**, *41*, 557–571. [CrossRef]
17. Liu, Z.; Wang, G.; Li, J.; Li, H.; Zhao, H.; Shi, H.; Lan, J. Water-immersion softening mechanism of coal rock mass based on split hopkinson pressure bar experiment. *Int. J. Coal Sci. Technol.* **2022**, *9*, 61. [CrossRef]
18. DeCuir, M.J.; Gupta, R.B.; Sastri, B. Beneficiation of coal using supercritical water and carbon dioxide extraction: Sulfur removal. *Int. J. Coal Sci. Technol.* **2020**, *8*, 717–726. [CrossRef]
19. Wang, J.; Yang, J.; Wu, F.; Hu, T.; Faisal, S.A. Analysis of fracture mechanism for surrounding rock hole based on water-filled blasting. *Int. J. Coal Sci. Technol.* **2020**, *7*, 704–713. [CrossRef]
20. Liu, A.; Liu, S.; Liu, P.; Wang, K. Water sorption on coal: Effects of oxygen-containing function groups and pore structure. *Int. J. Coal Sci. Technol.* **2021**, *8*, 983–1002. [CrossRef]
21. Hou, E.; Wen, Q.; Ye, Z.; Chen, W.; Wei, J. Height prediction of water-flowing fracture zone with a genetic-algorithm support-vector-machine method. *Int. J. Coal Sci. Technol.* **2020**, *7*, 740–751. [CrossRef]
22. Li, L.; Li, F.; Zhang, Y.; Yang, D.; Liu, X. Formation mechanism and height calculation of the caved zone and water-conducting fracture zone in solid backfill mining. *Int. J. Coal Sci. Technol.* **2020**, *7*, 208–215. [CrossRef]
23. Guo, P.; Gu, J.; Su, Y.; Wang, J.; Ding, Z. Effect of cyclic wetting–drying on tensile mechanical behavior and microstructure of clay-bearing sandstone. *Int. J. Coal Sci. Technol.* **2021**, *8*, 956–968. [CrossRef]
24. Qi, T.; Zhang, F.; Pei, X.; Feng, G.; Wei, H. Simulation research and application on response characteristics of detecting water-filled goaf by transient electromagnetic method. *Int. J. Coal Sci. Technol.* **2022**, *9*, 17. [CrossRef]

25. Ma, D.; Duan, H.; Zhang, J.; Bai, H. A state-of-the-art review on rock seepage mechanism of water inrush disaster in coal mines. *Int. J. Coal Sci. Technol.* **2022**, *9*, 50. [CrossRef]
26. Li, X.L.; Wang, E.Y.; Li, Z.H.; Liu, Z.T.; Song, D.Z.; Qiu, L.M. Rock burst monitoring by integrated microseismic and electromagnetic radiation methods. *Rock Mech. Rock Eng.* **2016**, *49*, 4393–4406. [CrossRef]
27. Di, Y.Y.; Wang, E.Y. Rock burst precursor electromagnetic radiation signal recognition method and early warning application based on recurrent neural networks. *Rock Mech. Rock Eng.* **2021**, *54*, 1449–1461. [CrossRef]
28. Xie, H.P.; Zhu, J.B.; Zhou, T.; Zhao, J. Novel three-dimensional rock dynamic tests using the true triaxial electromagnetic hopkinson bar system. *Rock Mech. Rock Eng.* **2021**, *54*, 2079–2086. [CrossRef]
29. Teimoori, K.; Cooper, R. Multiphysics study of microwave irradiation effects on rock breakage system. *Int. J. Rock Mech. Min. Sci.* **2021**, *140*, 104586. [CrossRef]
30. Song, D.Z.; Wang, E.Y.; Song, X.Y.; Jin, P.J.; Qiu, L.M. Changes in frequency of electromagnetic radiation from loaded coal rock. *Rock Mech. Rock Eng.* **2016**, *49*, 291–302. [CrossRef]
31. Liu, H.F.; Li, Z.L.; He, X.Q.; Wang, L.B.; Song, D.Z.; Tian, X.H.; Qiu, L.M.; Wang, W.X. Dynamic and static electrical characteristics of micro-surface of rocks by coupled use of atomic force microscope and micro-loading device. *Int. J. Rock Mech. Min. Sci.* **2021**, *148*, 104977. [CrossRef]
32. Asem, P.; Gardoni, P. Bayesian estimation of the normal and shear stiffness for rock sockets in weak sedimentary rocks. *Int. J. Rock Mech. Min. Sci.* **2019**, *124*, 104129. [CrossRef]
33. Momber, A.W. A refined model for solid particle rock erosion. *Rock Mech. Rock Eng.* **2016**, *49*, 467–475. [CrossRef]
34. Nikolenko, P.V.; Epshtein, S.A.; Shkuratnik, V.L.; Anufrenkova, P.S. Experimental study of coal fracture dynamics under the influence of cyclic freezing–thawing using shear elastic waves. *Int. J. Coal Sci. Technol.* **2020**, *8*, 562–574. [CrossRef]
35. Sun, C.; Li, G.; Gomah, M.E.; Xu, J.; Rong, H. Experimental investigation on the nanoindentation viscoelastic constitutive model of quartz and kaolinite in mudstone. *Int. J. Coal Sci. Technol.* **2021**, *8*, 925–937. [CrossRef]
36. Moore, J.R.; Geimer, P.R.; Finnegan, R.; Thorne, M.S. Use of seismic resonance measurements to determine the elastic modulus of freestanding rock masses. *Rock Mech. Rock Eng.* **2018**, *51*, 3937–3944. [CrossRef]
37. Wu, H.; Ju, Y.; Han, X.; Ren, Z.; Sun, Y.; Zhang, Y.; Han, T. Size effects in the uniaxial compressive properties of 3d printed models of rocks: An experimental investigation. *Int. J. Coal Sci. Technol.* **2022**, *9*, 83. [CrossRef]
38. Bobet, A. Lined circular tunnels in elastic transversely anisotropic rock at depth. *Rock Mech. Rock Eng.* **2011**, *44*, 149–167. [CrossRef]
39. Tang, X.H.; Zhang, Y.H.; Xu, J.J.; Rutqvist, J.; Hu, M.S.; Wang, Z.Z.; Liu, Q.S. Determining young's modulus of granite using accurate grain-based modeling with microscale rock mechanical experiments. *Int. J. Rock Mech. Min. Sci.* **2022**, *157*, 105167. [CrossRef]
40. Liu, H.H.; Rutqvist, J.; Birkholzer, J.T. Constitutive relationships for elastic deformation of clay rock: Data analysis. *Rock Mech. Rock Eng.* **2011**, *44*, 463–468. [CrossRef]
41. Du, F.; Ma, J.; Guo, X.; Wang, T.; Dong, X.; Li, J.; He, S.; Nuerjuma, D. Rockburst mechanism and the law of energy accumulation and release in mining roadway: A case study. *Int. J. Coal Sci. Technol.* **2022**, *9*, 67. [CrossRef]
42. Khan, N.M.; Ma, L.Q.; Cao, K.W.; Hussain, S.; Liu, W.; Xu, Y.J.; Yuan, Q.P.; Gu, J. Prediction of an early failure point using infrared radiation characteristics and energy evolution for sandstone with different water contents. *Bull. Eng. Geol. Environ.* **2021**, *80*, 6913–6936. [CrossRef]
43. Liu, L.Y.; Ji, H.G.; Elsworth, D.; Zhi, S.; Lv, X.F.; Wang, T. Dual-damage constitutive model to define thermal damage in rock. *Int. J. Rock Mech. Min. Sci.* **2020**, *126*, 104185. [CrossRef]
44. Vargas, E.D.; Velloso, R.Q.; Chavez, L.E.; Gusmao, L.; do Amaral, C.P. On the effect of thermally induced stresses in failures of some rock slopes in rio de janeiro, brazil. *Rock Mech. Rock Eng.* **2013**, *46*, 123–134. [CrossRef]
45. Liu, B.; Zhao, Y.; Zhang, C.; Zhou, J.; Li, Y.; Sun, Z. Characteristic strength and acoustic emission properties of weakly cemented sandstone at different depths under uniaxial compression. *Int. J. Coal Sci. Technol.* **2021**, *8*, 1288–1301. [CrossRef]
46. Dou, L.; Yang, K.; Chi, X. Fracture behavior and acoustic emission characteristics of sandstone samples with inclined precracks. *Int. J. Coal Sci. Technol.* **2020**, *8*, 77–87. [CrossRef]
47. Su, G.S.; Huang, J.H.; Xu, H.J.; Qin, Y.Z. Extracting acoustic emission features that precede hard rock instability with unsupervised learning. *Eng. Geol.* **2022**, *306*, 106761. [CrossRef]
48. Kong, X.G.; Wang, E.Y.; Li, S.G.; Lin, H.F.; Xiao, P.; Zhang, K.Z. Fractals and chaos characteristics of acoustic emission energy about gas-bearing coal during loaded failure. *Fractals* **2019**, *27*, 1950072. [CrossRef]
49. Chi, X.; Yang, K.; Wei, Z. Breaking and mining-induced stress evolution of overlying strata in the working face of a steeply dipping coal seam. *Int. J. Coal Sci. Technol.* **2021**, *8*, 614–625. [CrossRef]
50. Xue, D.; Lu, L.; Zhou, J.; Lu, L.; Liu, Y. Cluster modeling of the short-range correlation of acoustically emitted scattering signals. *Int. J. Coal Sci. Technol.* **2020**, *8*, 575–589. [CrossRef]
51. He, S.; Qin, M.; Qiu, L.; Song, D.; Zhang, X. Early warning of coal dynamic disaster by precursor of ae and emr "quiet period". *Int. J. Coal Sci. Technol.* **2022**, *9*, 46. [CrossRef]
52. Kong, X.G.; He, D.; Liu, X.F.; Wang, E.Y.; Li, S.G.; Liu, T.; Ji, P.F.; Deng, D.Y.; Yang, S.R. Strain characteristics and energy dissipation laws of gas-bearing coal during impact fracture process. *Energy* **2022**, *242*, 123028. [CrossRef]

53. Yao, Q.L.; Tang, C.J.; Xia, Z.; Liu, X.L.; Zhu, L.; Chong, Z.H.; Hui, X.D. Mechanisms of failure in coal samples from underground water reservoir. *Eng. Geol.* **2020**, *267*, 105494. [CrossRef]
54. Junique, T.; Vazquez, P.; Geraud, Y.; Thomachot-Schneider, C.; Sidibe, H. Microstructural evolution of granitic stones exposed to different thermal regimes analysed by infrared thermography. *Eng. Geol.* **2021**, *286*, 106057. [CrossRef]
55. He, X.Q.; Chen, W.X.; Nie, B.S.; Mitri, H. Electromagnetic emission theory and its application to dynamic phenomena in coal-rock. *Int. J. Rock Mech. Min.* **2011**, *48*, 1352–1358. [CrossRef]
56. Wu, R.; Zhang, P.; Kulatilake, P.H.S.W.; Luo, H.; He, Q. Stress and deformation analysis of gob-side pre-backfill driving procedure of longwall mining: A case study. *Int. J. Coal Sci.* **2021**, *8*, 1351–1370. [CrossRef]
57. Tang, J.; Zhang, X.; Sun, S.; Pan, Y.; Li, L. Evolution characteristics of precursor information of coal and gas outburst in deep rock cross-cut coal uncovering. *Int. J. Coal Sci. Technol.* **2022**, *9*, 5. [CrossRef]
58. Song, Z.; Ji, H.; Liu, Z.; Sun, L. Study on the critical stress threshold of weakly cemented sandstone damage based on the renormalization group method. *Int. J. Coal Sci. Technol.* **2020**, *7*, 693–703. [CrossRef]
59. Panthulu, T.V.; Krishnaiah, C.; Shirke, J.M. Detection of seepage paths in earth dams using self-potential and electrical resistivity methods. *Eng. Geol.* **2001**, *59*, 281–295. [CrossRef]
60. Zhou, J.R.; Wei, J.; Yang, T.H.; Zhu, W.C.; Li, L.C.; Zhang, P.H. Damage analysis of rock mass coupling joints, water and microseismicity. *Tunn. Undergr. Space Technol.* **2018**, *71*, 366–381. [CrossRef]
61. Liu, F.; Tang, C.A.; Zhang, Y.J.; Ma, T.H. Rockburst and microseismicity characteristics in the qinling water conveyance tunnel of the hanjiang-to-weihe river diversion project. *Int. J. Rock Mech. Min. Sci.* **2021**, *148*, 104973. [CrossRef]
62. Yang, H.Q.; Liu, B.L.; Karekal, S. Experimental investigation on infrared radiation features of fracturing process in jointed rock under concentrated load. *Int. J. Rock Mech. Min. Sci.* **2021**, *139*, 104619. [CrossRef]
63. Cao, K.W.; Ma, L.Q.; Zhang, D.S.; Lai, X.P.; Zhang, Z.T.; Khan, N.M. An experimental study of infrared radiation characteristics of sandstone in dilatancy process. *Int. J. Rock Mech. Min. Sci.* **2020**, *136*, 104503. [CrossRef]
64. Sun, X.M.; Xu, H.C.; He, M.C.; Zhang, F. Experimental investigation of the occurrence of rockburst in a rock specimen through infrared thermography and acoustic emission. *Int. J. Rock Mech. Min.* **2017**, *93*, 250–259. [CrossRef]
65. Liu, S.J.; Wu, L.X.; Wu, Y.H. Infrared radiation of rock at failure. *Int. J. Rock Mech. Min.* **2006**, *43*, 972–979. [CrossRef]
66. Wu, L.X.; Liu, S.J.; Wu, Y.H.; Wang, C.Y. Precursors for rock fracturing and failure—Part i: Irr image abnormalities. *Int. J. Rock Mech. Min.* **2006**, *43*, 473–482. [CrossRef]
67. Lin, Y.; Zheng, Z.B.; Zhou, K.P.; Gu, Z.Y.; Pan, Z. Research on the characteristics of infrared radiation and energy evolution law of red sandstone with different porosity during uniaxial compression. *Sustainability* **2022**, *14*, 7698. [CrossRef]
68. Huang, J.W.; Liu, S.J.; Gao, X.; Yang, Z.C.; Ni, Q.; Wu, L.X. Experimental study of the thermal infrared emissivity variation of loaded rock and its significance. *Remote Sens.* **2018**, *10*, 818. [CrossRef]
69. Cao, K.W.; Ma, L.Q.; Wu, Y.; Khan, N.M.; Spearing, A.J.S.; Hussain, S.; Yang, J. Cyclic fatigue characteristics of rock failure using infrared radiation as precursor to violent failure: Experimental insights from loading and unloading response. *Fatigue Fract. Eng. M* **2021**, *44*, 584–594. [CrossRef]
70. Zhang, K.; Liu, X.H.; Chen, Y.L.; Cheng, H.M. Quantitative description of infrared radiation characteristics of preflawed sandstone during fracturing process. *J. Rock Mech. Geotech.* **2021**, *13*, 131–142. [CrossRef]
71. Cao, K.; Ma, L.; Wu, Y.; Khan, N.M.; Yang, J. Using the characteristics of infrared radiation during the process of strain energy evolution in saturated rock as a precursor for violent failure. *Infrared Phys. Technol.* **2020**, *109*, 103406. [CrossRef]
72. Cao, K.W.; Ma, L.Q.; Wu, Y.; Spearing, A.J.S.; Khan, N.M.; Hussain, S.; Rehman, F.U. Statistical damage model for dry and saturated rock under uniaxial loading based on infrared radiation for possible stress prediction. *Eng. Fract. Mech.* **2022**, *260*, 108134. [CrossRef]
73. Cai, X.; Zhou, Z.L.; Tan, L.H.; Zang, H.Z.; Song, Z.Y. Water saturation effects on thermal infrared radiation features of rock materials during deformation and fracturing. *Rock Mech. Rock Eng.* **2020**, *53*, 4839–4856. [CrossRef]
74. Shen, R.X.; Li, H.R.; Wang, E.Y.; Chen, T.Q.; Li, T.X.; Tian, H.; Hou, Z.H. Infrared radiation characteristics and fracture precursor information extraction of loaded sandstone samples with varying moisture contents. *Int. J. Rock Mech. Min. Sci.* **2020**, *130*, 104344. [CrossRef]
75. Wu, L.X.; Liu, S.J.; Wu, Y.H.; Wang, C.Y. Precursors for rock fracturing and failure—Part ii: Irr t-curve abnormalities. *Int. J. Rock Mech. Min.* **2006**, *43*, 483–493. [CrossRef]

sustainability

MDPI

Article

Advanced Analysis of Collision-Induced Blast Fragmentation in V-Type Firing Pattern

Lalit Singh Chouhan [1,*], Avtar K. Raina [2], V. M. S. R. Murthy [1], Mohanad Muayad Sabri Sabri [3], Edy Tonnizam Mohamad [4] and Ramesh Murlidhar Bhatawdekar [5]

1 Department of Mining Engineering, Indian Institute of Technology (Indian School of Mines), Dhanbad 826 004, India
2 CSIR-Central Institute of Mining and Fuel Research, Nagpur 440001, India
3 Centre of Peter the Great St. Petersburg Polytechnic University, 195251 St. Petersburg, Russia
4 Centre of Tropical Geoengineering (GEOTROPIK), Institute of Smart Infrastructure and Innovative Construction (ISiiC), Universiti Teknologi Malaysia, Johor Bahru 81310, Malaysia
5 Centre of Tropical Geoengineering (GEOTROPIK), School of Civil Engineering, Faculty of Engineering, Universiti Teknologi Malaysia, Johor Bahru 81310, Malaysia
* Correspondence: lalit.2015dr1171@me.ism.ac.in

Abstract: The firing pattern of blastholes influences the geometric aspects of a blast design in terms of change in blasting burden and spacing. This in turn changes the effective stiffness of a blasthole and confinement of the explosive and aids in better fragmentation. However, during the blasting, the fragments tend to collide and further fragment the rock. In comparison with other patterns, the V-type firing pattern increases the chances of collision between the fragments during flight. The process is scantly documented and accordingly field experiments were conducted using three firing patterns, viz., line, diagonal, and V-type, in a mine with minor variation in rock factor and minor to moderate changes in blast design variables. Sixteen blast design variables such as burden, spacing, charge per hole, in-hole charge density, etc. along with firing pattern were considered for the analysis and fragmentation modeled with the help of surface response analysis and artificial neural networks. The analysis revealed that there is a significant influence of firing patterns on fragmentation. The V-type pattern showed significant reduction in fragment sizes that can be ascribed to in-flight collision processes. A surface response model was developed using advanced ANOVA and resulted in an adjusted R^2 and RMSE of 0.89, 0.025, respectively. Further, modeling with ANN was attempted that showed better results than ANOVA with R^2 and RMSE of 0.96 and 0.040 in training, and 0.884 and 0.049 in validation tests. Since, diagonal and V-type patterns have similar design parameters, the reduction in fragment size in the former pattern can be ascribed to the collision of rock fragments during their flight in blasting.

Keywords: blasting; V-type firing pattern; collision fragmentation; RSA; ANN

Citation: Chouhan, L.S.; Raina, A.K.; Murthy, V.M.S.R.; Sabri Sabri, M.M.; Mohamad, E.T.; Bhatawdekar, R.M. Advanced Analysis of Collision-Induced Blast Fragmentation in V-Type Firing Pattern. *Sustainability* **2022**, *14*, 15703. https://doi.org/10.3390/su142315703

Academic Editor: Kaihui Li

Received: 13 September 2022
Accepted: 14 November 2022
Published: 25 November 2022

1. Introduction

The objectives of blasting in mining are to ease the excavation operation by obtaining maximum yield through optimum fragment sizes while minimizing the adverse impacts of blasting such as blast-induced ground vibration, air overpressure, flyrock, and noise. To obtain the desired blast results, various blast design variables and factors, viz., burden, spacing, stemming length, type of explosive, powder factor, stiffness ratio, firing pattern, etc. are optimized as per the site-specific requirement. The end purpose of rock blasting in limestone mines, where investigations were carried out, is to produce the desired size feed for a crusher. Fragments produced by blasting should not only be small enough for economic loading of equipment, but should also pass easily through the crusher to realize equipment productivity [1]. To achieve the desired rock fragmentation by blasting, an effective way of determining the blast design variables should be selected [2]. There is ample evidence of the role of fragmentation on the overall mine–mill fragmentation system performance [3,4]. The efficiency of the system through fragmentation optimization is documented by several researchers [5–10] and can be achieved only when fragment sizes obtained from the blasting are measured. Therefore, fragmentation measurement of blasted muck piles is essential [11].

To reduce the production cost, the blast design should be revisited to match the cost of the mine–mill fragmentation system (MMFS) that includes unit operations such as drilling, blasting, loading, hauling, or conveying and crushing of primary or secondary nature [4]. In open pit mines, the cost of drilling and blasting operations covers approximately 15–20% of the total mining cost [12–14]. An increase in blasting cost reduces the cost of the downstream operations, viz., loading, hauling, crushing, and boulder breaking [3]. This requires the determination of an optimum fragment size range so that the cost of MMFS is optimized. If increasing the cost of blasting operations does not reduce the cost of successive operations, it will not impact the overall economics of the system.

Methods such as blast design evaluation or auditing lead to changes in main blast design variables such as burden, spacing, and stemming, hence, resulting in improved fragmentation and better system performance. One such method involves a change in firing sequence that can be considered as another critical blast design requirement for improving the rock fragmentation [15].

A proper design of the firing pattern, i.e., the delay required between hole to hole and row to row, plays a vital role not only in reducing fragmentation size but also helps to reduce ground vibration as well as back break. To maintain the continuous momentum for the inter-row displacement, a systematic release of the blasting energy is required which can be achieved with a proper burden. An improper delay in multi-row blast gives poor blast results, viz., poor rock fragmentation from the back rows, severe over/under break, large boulders from the collar region of the blasthole, and tight muck pile, etc. [16]. Different types of firing patterns, e.g., line, diagonal, and V-type are used in bench blasting. Each firing pattern has its own application and advantages [17,18]. The change in firing sequence from line firing to diagonal pattern changes the design geometries while blasting. This helps to reduce the blasted burden and increase the spacing and overall actual charge per unit mass of the blasthole. Thus, the effective stiffness and explosive confinement in the blasthole are modified significantly resulting in improved fragmentation particularly in the case of diagonal firing.

The V-type firing pattern has similar blasting design variables except those two limbs of a blast from the center fire towards each other. A distinction between the fragmentation in diagonal and V-type firing patterns should thus account for the fragmentation due to the collision of the fragments while in flight during blasting.

Although V-type and diagonal firing patterns provide a similar effective spacing to burden ratio, the V-type firing pattern is more suitable for achieving smaller fragmentation because it increases the opportunity for in-flight collision between broken rock fragments [15]. This particular characteristic of V-type firing is considered important to reduce fragment size and boulder occurrence within the blasted rock piles [17].

With this hypothesis studies have been conducted to document the reduction in blast fragmentation with V-type firing pattern. The primary focus of this paper is thus to evaluate the influence of in-flight collision between rock fragments on fragmentation and is probably the first of its kind study.

2. Influence of Firing Patterns on Fragmentation

As mentioned earlier, the firing patterns play a vital role in rock fragmentation size during blasting of the rock.

The firing pattern influences the rock fragmentation by following three ways, these are:

1. By changing burden and spacing during blasting which is also known as effective burden (B_e) and effective spacing (S_e) or blasting burden and blasting spacing;
2. Through possible in-flight collision between rock fragments during blasting and

3. a combination of the above two mechanisms.

There are many variations in firing patterns, but the most known firing patterns are:

1. Line firing pattern;
2. Diagonal firing pattern;
3. V-type firing pattern.

The objectives and effect of these patterns on rock fragmentation during blasting are summarized in Tables 1–5.

Table 1. Line firing pattern (a).

Firing Pattern	**Line Firing Pattern (Holes in Same Row Fired Simultaneously)**
Representative image	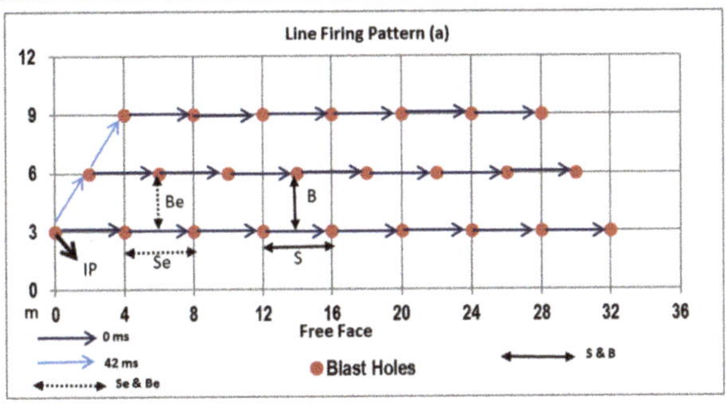
Main objectives	To achieve coarser fragmentation with lesser muck pile throw
Effect of firing pattern on fragmentation	$M_b = M_d$ where M_b is the ratio of S_e *to* B_e and M_d is the ratio of *S* to *B*. With this type of firing pattern, effective burden (B_e) is equal to the drill burden, which results in no advantage to the firing pattern on fragmentation. In addition, the movement of rock fragments is also in a single direction, with little possibility of inter collision of rock fragments during blasting. However, when rock fragments strike the ground, further fragmentation may take place which is governed by several variables, namely, discontinuities in the initial rock mass, their orientation at the time of impact, physicomechanical properties, incident angle, impact velocity, geometry and stiffness of the ground, and the presence of water [19].

Table 2. Line firing pattern (b).

Firing Pattern	**Line Firing Pattern (Holes in Same Row Fired Individually, But Firing of Successive Row Starts after Completion of the Preceding Row)**
Representative image	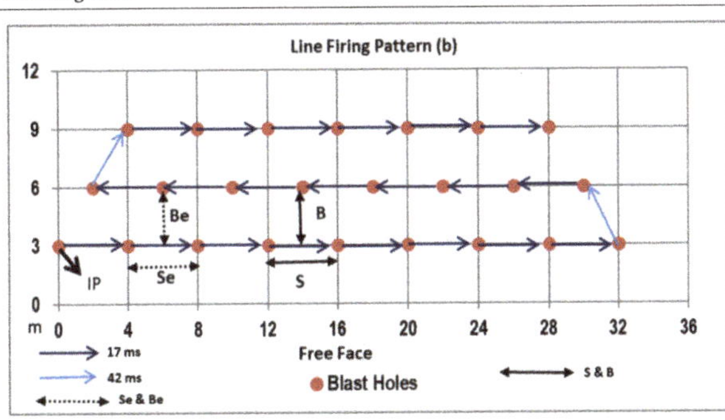
Main objectives	To achieve larger fragmentation with more muck pile throw.
Effect of firing pattern on fragmentation	$M_b = M_d$ Similar influence on rock fragmentation as provided by line firing pattern given in Table 1.

Table 3. Diagonal firing pattern (RHS).

Firing Pattern	Diagonal Firing Pattern (RHS)
Representative image	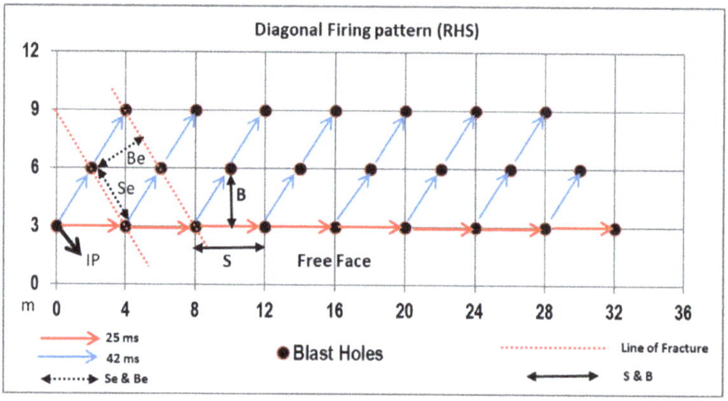
Main objectives	To achieve coarser and uniform fragmentation with medium muck pile throw.
Effect of firing pattern on fragmentation	$M_b < M_d$ Since effective burden (B_e) increases than the drill burden (B) during firing, a coarser fragmentation is achieved. Since movement of rock fragments is only in a single direction, this pattern also has little possibility of in-flight collision of rock fragments during blasting.

Table 4. Diagonal firing pattern (LHS).

Firing Pattern	Diagonal Firing Pattern (LHS)
Representative image	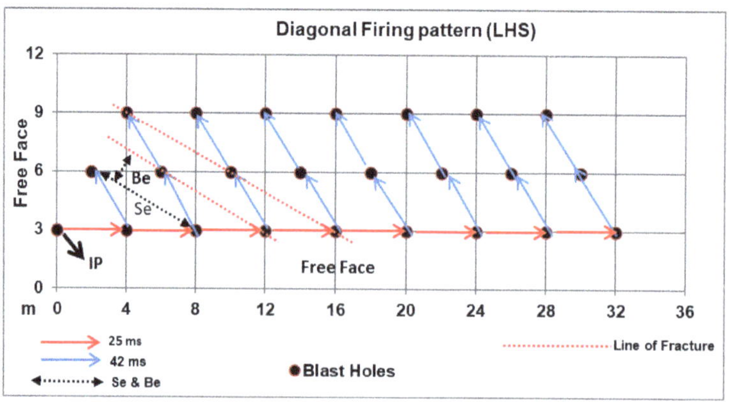
Main objectives	To achieve medium and uniform fragmentation with medium muck pile throw.
Effect of firing pattern on fragmentation	$M_b > M_d$ Since effective burden (B_e) decreases than the drill burden (B) during firing, a medium fragmentation is achieved. Since the movement of rock fragments is only in a single direction, this pattern also has little possibility of in-flight collision of rock fragments during blasting.

Table 5. V-type firing pattern.

Firing Pattern	**V-Type Firing Pattern**
Representative image	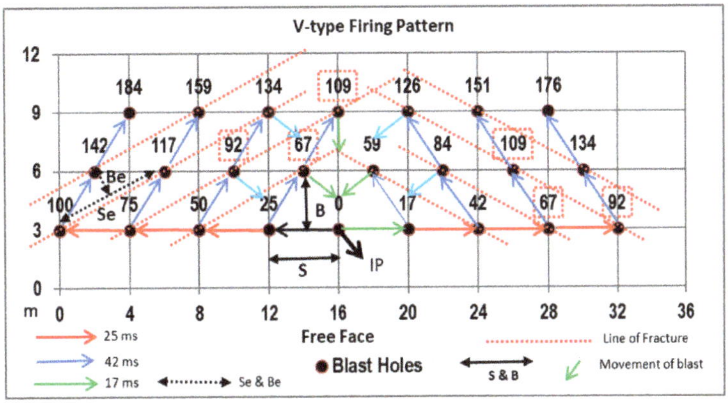
Main objectives	To achieve smaller and uniform fragmentation with optimum muck pile throw.
Effect of firing pattern on fragmentation	$M_b < M_d$ Since effective burden (B_e) decreases than the drill burden (B) during firing, a smaller fragmentation is achieved. In addition, it provides in-flight collision of rock fragments during blasting leading to further fragmentation.

$M_b = S_e/B_e$, $M_d = S/B$, S = drill spacing, B = drill burden, S_e = effective spacing during blasting, B_e = effective burden during blasting.

3. Data Compilation and Analysis

The study was conducted in a limestone mine in India. The deposit belongs to a hilly terrain of the Precambrian age of the Delhi Super Group. The annual production of the mines was 6 MTPa. Compressive strength of limestone varied from 80 to 110 MPa. The mine uses a blasthole diameter of 115 mm. ANFO was used as the explosive having a density of 800 kg/m^3 and the average VoD of the explosive measured through continuous in-hole method was 3700 m/s. The charging process of the blasthole is summarized in the following steps:

1. Create an air deck at the bottom of the hole with a wooden spacer having a length of 0.75 cm placed first;
2. Seal off the gap between the wooden spacer and the blasthole periphery, cut a primer cartridge, and put in the hole;
3. Lower the primer cartridge, attached with DTH of 250 ms, into the hole as a base charge;
4. Pour the prilled ANFO into the hole as a column charge;
5. Stem the blasthole with the help of drill cutting;
6. Connect blastholes of the first row with the help of TLD of 25 ms and the blasthole of the second and third row connected with the help of TLD of 42 ms as depicted in the figures given in Tables 1–5.

The blasts were initiated by a shock tube system with a delay sequencing of 17 ms, 25 ms, and 42 ms with a staggered drill hole pattern. The loading operations were performed by front end loader, shovel, and backhoe. The blast muck was transported by 55 MT rear dump trucks.

In order to achieve the objectives of this study, full scale blast trials were conducted in the mines by deploying line firing (L), diagonal firing (D), and V-type firing patterns (V). Other variables of the blast design varied over a narrow range and thus provided a means for comparing fragmentation in the above three firing patterns.

To assess the in-flight collision process between rock fragments during blasting in different firing pattern and its impacts, the following research methods were resorted to:

1. Determination of rock type, its strength and variation. Three main types of the rock formations are present in the area which were assigned three values for rock factor (*RF*);
2. Fragmentation analysis of blasted muck pile as explained in Table 6
3. Data analysis was carried out using Surface response analysis and artificial neural networking methods as explained in the following sections.

Table 6. Process of fragmentation assessment by the Fragalyst software.

Step	Image	Description
1		Images of muck pile captured with calibrator at different time intervals to cover the all sizes of fragments in the whole muck pile.
2		Import the image in Fragalyst software.
3		Calibrate the image by known dimensions of calibrator in software.
4.1		Edge Deduction process (Boundary)
4.2		Edge Deduction process (Super Impose)
4.3		Edge Deduction process (Segmentation)
5	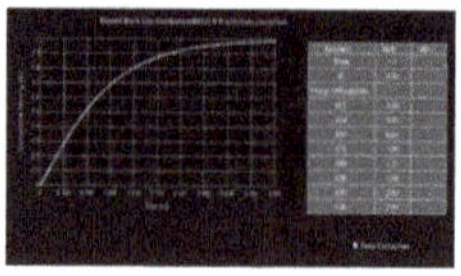	Fragmentation analysis—the data of Blasted Block Size Distribution (BBSD) i.e., a "3DR.file", is saved automatically in a folder in which images are kept.

Fragmentation Analysis of Blasted Muck Pile

A digital image analysis method using Fragalyst software was used for measurement of fragmentation in all the blasts monitored. The method requires muck pile images with a scale to calculate the size of fragments. The representative images of blasted muck pile, captured at different time intervals during the excavation operation to cover all the sizes of fragmentation in the whole muck pile, were thus taken. The process of fragmentation assessment by Fragalyst software is depicted in Table 6.

Ninety-two full-scale blasts were conducted while monitoring the blast design variables such as burden, spacing and stemming, bench height, specific charge, firing patterns, delay, and mean fragmentation, etc., with one free face availability. The statistics of the data generated are presented in Table 7. The difference in fragment sizes in the three types of firing patterns is evident from Figure 1d, wherein a reduction of around 26% in diagonal and around 45% in the case of the V-type firing pattern is registered in comparison with the fragmentation obtained in the line firing pattern.

Table 7. General statistics of the variables measured.

Statistics	Burden (B) (m)	Spacing (S) (m)	Stemming Length (l_s) (m)	Bench Height (H_b) (m)	Specific Charge (q) (kg/m^3)	Mean Fragmentation Size (k_{50}) (m^2)
Mean	2.82	3.86	2.81	9.08	0.48	0.32
Standard Error	0.01	0.02	0.04	0.08	0.00	0.01
Median	2.83	3.91	2.50	9.36	0.48	0.30
Mode	2.80	4.00	2.50	9.85	0.50	0.30
Standard Deviation	0.09	0.18	0.40	0.80	0.04	0.10
Sample Variance	0.01	0.03	0.16	0.63	0.00	0.01
Kurtosis	1.03	0.76	−1.21	1.52	−0.68	1.34
Skewness	-0.44	−1.35	0.43	−1.26	0.16	1.18
Range	0.49	0.65	1.59	3.85	0.16	0.42
Minimum	2.53	3.41	2.00	6.14	0.41	0.20
Maximum	3.03	4.06	3.59	9.99	0.57	0.62
Sum	259.28	354.91	258.49	835.68	44.27	29.19
Count	92.00	92.00	92.00	92.00	92.00	92.00

Representative images of fragmentation obtained from different firing patterns are provided in Figure 1a–c and the distribution of fragmentation in all the three firing patterns representing the average fragment sizes of all the blasts monitored are plotted in Figure 1d.

The distribution of some important blast design variables along with their ranges is given in Figure 2a–f.

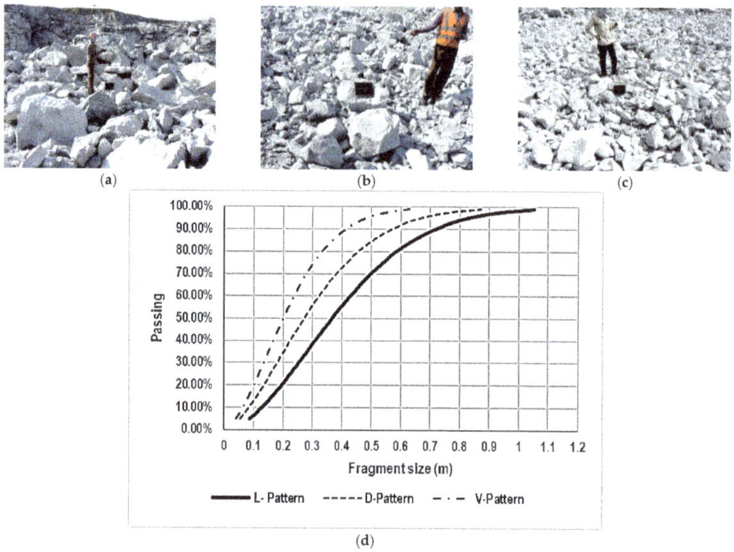

Figure 1. Influence of different firing patterns on rock fragmentation. (**a**) Rock fragmentation with line firing pattern (coarser size). (**b**) Rock fragmentation with diagonal firing pattern (medium size). (**c**) Rock fragmentation with V-type firing pattern (smaller size). (**d**) Average mean fragment size distributions of the three patterns for the test blasts.

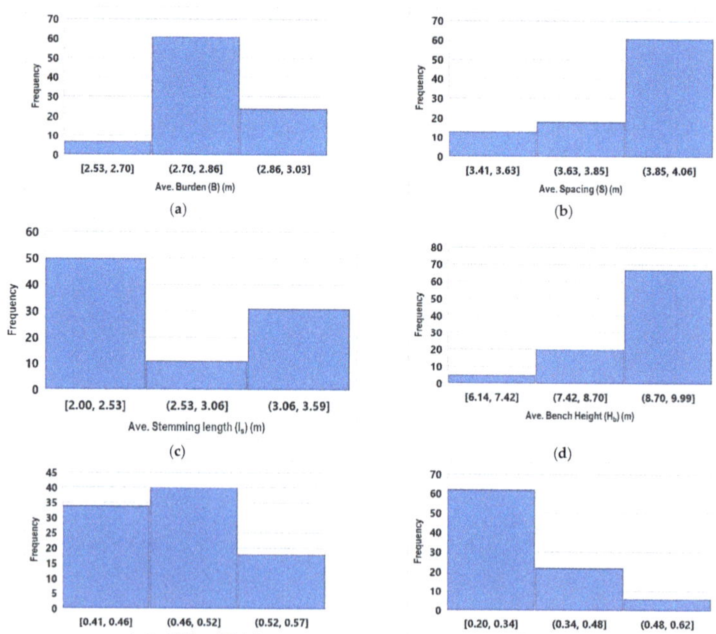

Figure 2. Histogram(s) of various blast variables: (**a**) burden, (**b**) spacing, (**c**) stemming, (**d**) bench height, (**e**) specific charge, and (**f**) mean fragmentation size, monitored during the study.

4. Model Development

4.1. Response Surface Analysis (RSA)

The impact of change in firing patterns on rock fragmentation during blasting can be evaluated properly if other variables of blasting are kept constant. However, it is important to note that in bench blasting there are variations in design pattern due to drilling, charging, and measuring errors. Moreover, there are conflicts in some factors and variables that make perfect modeling difficult. To develop an easy-to-use model for rock fragmentation prediction, response surface analysis (RSA) of the data was carried out while using the variables that most influence the fragmentation. The results of the RSA evaluation (Table 8) obtained through multivariate non-linear ANOVA method (Table 9), were finally used in developing the model. A back propagation algorithm being robust in nature was deployed to evaluate the variables over *p*-value, R^2, Akaike information criterion (AiCC) and Bayesian information criterion (BiCC), to eliminate insignificant and redundant terms in a quadratic model suggested by the initial analysis.

Table 8. Response surface design evaluation results.

Source	Sum of Squares	df	Mean Square	F-Value	*p*-Value
Model	0.5322	7	0.076	106.42	<0.0001
A-A	0.1431	2	0.0715	100.14	<0.0001
B-RF	0.0146	1	0.0146	20.4	<0.0001
C-ρ_{ee}	0.0161	1	0.0161	22.55	<0.0001
D-B × S	0.0146	1	0.0146	20.44	<0.0001
BC	0.0058	1	0.0058	8.12	0.0055
D^2	0.0124	1	0.0124	17.38	<0.0001
Residual	0.0579	81	0.0007		
Cor Total	0.59	88			

Table 9. ANOVA for the reduced RSA model.

Std. Dev.	0.0267	R^2	0.90
Mean	0.3087	Adjusted R^2	0.89
C.V. %	8.66	Predicted R^2	0.88
		Adequate Precision	38.70

The Model F-value of 106.42 implies that the model is significant. There is only a 0.01% chance that an F-value this large could occur due to noise and p-values less than 0.0500 indicating that the model terms are significant. In this case, A, B, C, D, BC, and D^2 (see Table 8 for terms) are significant model terms. Values of $p > 0.1000$ indicate the model terms are not significant. The modeling results are presented in Table 8.

The Predicted R^2 of 0.88 is in reasonable agreement with the Adjusted R^2 of 0.89, i.e., the difference is less than 0.2. Adequate precision, a measure of the signal to noise ratio, should be greater than four. In our case, the ratio of 38.70 indicates an adequate signal. This model can be used to navigate the design space. Several diagnostics were deployed (Figure 3) before accepting the final equation.

The plot of normal probability of externally studentized residuals follows a straight line (Figure 3a) indicating a proper transformation of the output, and that the residuals are within the expected ranges (Figure 3b) with no outliers. The Box-Cox plot for transformation (Figure 3c) confirms the transformation applied to the output, and all the data in the Cook's distance (Figure 3d) are quite well within the limits. The diagnosis thus points to the well-behaved structure of the design and analysis. The predicted vs. observed plot of the mean fragmentation size (k_{50}) shows that the prediction is quite significant with an adjusted

R^2 of 0.89 and predicted R^2 of 0.88. Accordingly, the final equation for prediction of mean fragmentation size (k_{50}) in terms of the independent variables is provided in Equation (1).

$$k_{50} = Int. + 0.21RF + 0.00204\rho_{ee} - 0.256(B \times S) + 0.0039(RF \times \rho_{ee}) + 0.134(B \times S)^2 \quad (1)$$

where *Int.* is the intercept and equals 0.471 for L, 0.389 for D, and 0.327 for the V-type firing patterns (for other symbols please see abbreviations at the end of this paper).

A comparison of mean fragmentation size predicted by the RSA model given in Equation (1) and measured value is shown in Figure 4 and confirms that the model can be used for mean fragmentation size (k_{50}) prediction.

To ascertain the surface response through the model and the interactions between the variables over space, several iterations were conducted while keeping two variables constant and varying one at a time. The results of such simulations are presented in Figure 5a–i.

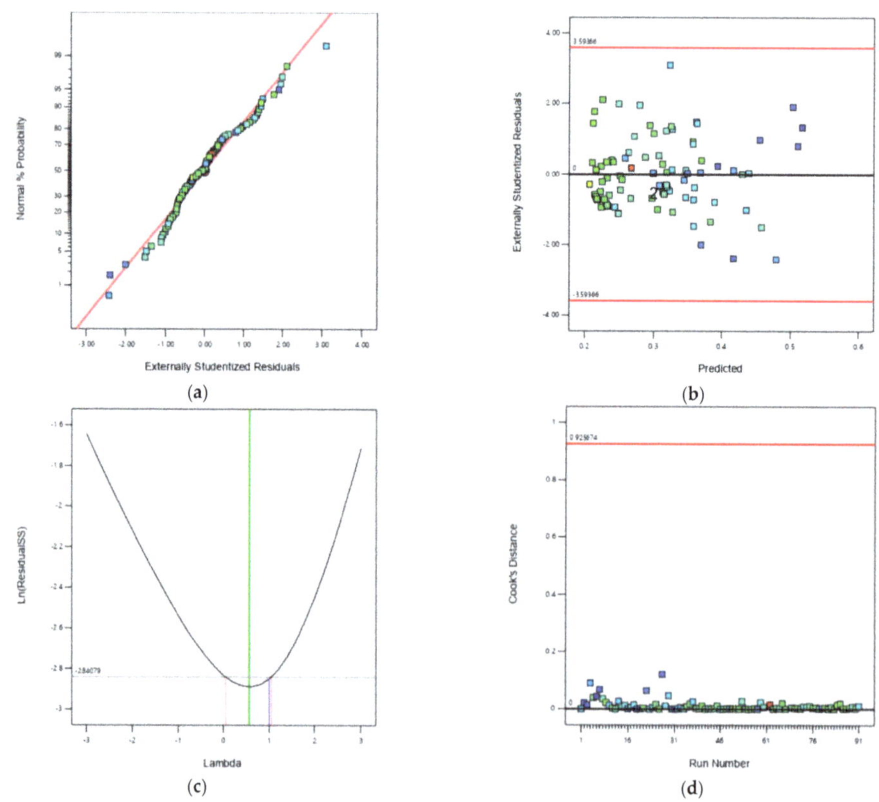

Figure 3. Diagnostic plots: (**a**) Externally Studentized Residuals vs Normal % Probability, (**b**) Externally Studentized Residuals vs. Predicted, (**c**) Box-Cox pot for power transformation, and (**d**) Cook's distance vs. Run Number.

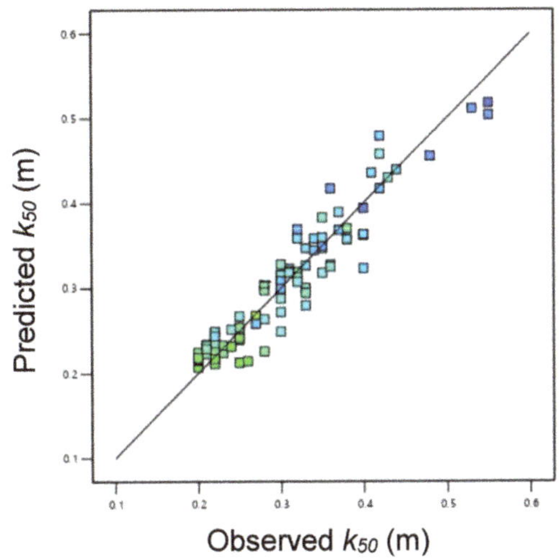

Figure 4. Predicted versus measured values of the mean fragmentation size (k_{50}) for the RSA model.

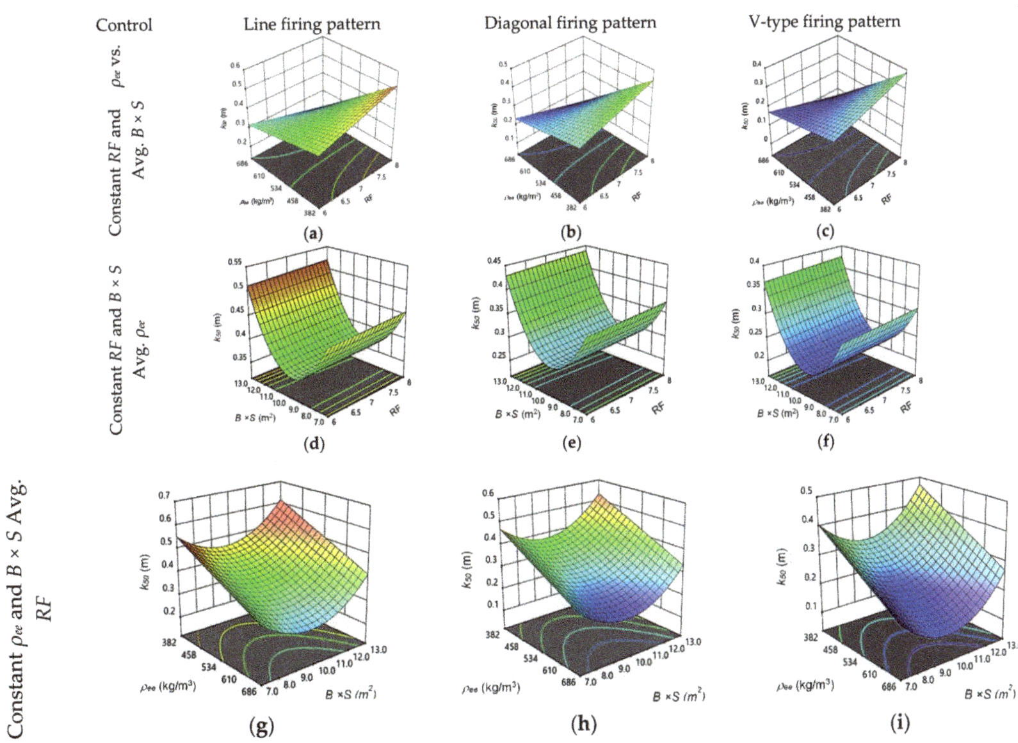

Figure 5. Interaction of different blast variables and response surface defined with the model for mean fragmentation size k_{50}, (**a–c**) relationship between RF and ρ_{ee} with k_{50} with average ($B \times S$), (**d–f**) influence of RF and ($B \times S$) on k_{50} with average value of ρ_{ee}, (**g–i**) relationship between ρ_{ee} and ($B \times S$) on k_{50} with average value of RF, for three different firing patterns tested.

Figure 5a–c show the relationship with *RF* and ρ_{ee} versus mean fragmentation size (k_{50}) with average ($B \times S$). Similar trends are observed in all three firing patterns. However, there is significant reduction in fragment size despite the constant variables at average ($B \times S$) in case of V-type of firing pattern.

Figure 5d–f shows the influence of RF and ($B \times S$) on the mean fragmentation size (k_{50}) with an average value of ρ_{ee} and have similar trends of fragment size (k_{50}) irrespective of the type of firing pattern. The figure confirms that the relationship of burden and spacing is not linear with fragmentation and an optimum value is evident for achieving the best possible fragment size. However, a significant change in fragment size is observed in the case of the V-type firing pattern.

The influence of ρ_{ee} and ($B \times S$) on the mean fragmentation size (k_{50}) with average value of RF is shown in Figure 5g–i. The trends in all the cases of firing patterns are similar, except the size of fragmentation that is varying over the three firing patterns tested. These figures also provide an optimum value of ($B \times S$) at which we can achieve smaller fragmentation with the same value of ρ_{ee}. There is a marked change in fragment size in the case of diagonal and V-type firing patterns. Distinct trends in the change in fragmentation with variation in ($B \times S$) and ρ_{ee} are, however, apparent from the figures.

4.2. Fragmentation Prediction Using Artificial Neural Network (ANN)

Artificial neural network (ANN) is a computational method consisting of several processing elements that receive inputs and deliver outputs based on their predefined activation functions. ANN consists of three layers, viz., input layer, the hidden layer, and the output layer. The input layer picks up the input signals and transfers them to the next layer and, finally, the output layer gives the prediction. The neural networks have to be trained with some training data to obtain a solution to a complex process output. The ANN and related methods have a capability to solve complicated problems, especially when the process and results are not fully understood [20]. The case is similar in blasting where the design variables present a complex relationship with rockmass, which in turn has several inconsistencies such as inhomogeneity and anisotropy.

Various algorithms have been suggested for training of the neural network, but the backpropagation algorithm is the most versatile and robust technique and provides the most efficient learning procedure for multy layer perceptron (MLP) networks. An experimental database including enough datasets is required to train the ANN model. Once the training process is completed, prediction can be made for a new input dataset.

Accordingly, to predict rock fragmentation by blasting, a back propagation ANN model was deployed for the data acquired and analyzed earlier by ANOVA. Several iterations were made to find the best possible network and hidden layers. The model that trained well and presented the best results is given in Figure 6. The plot of training progression thus obtained during the process is given in Figure 7.

In the above training process, the network is presented with a pair of patterns: an input pattern and the corresponding desired output pattern. The firing patterns can be treated as a string in the ANN training and therefore it is possible to estimate the mean fragment size from the trained network. The network can be queried for such results and hence compared.

Tables 10 and 11 show the input parameters and output parameters with their symbols and range, respectively, considered for developing the neural network. For introducing fragmentation to the network, an image analysis method, i.e., "Fragalyst" software was employed to determine muck pile size distribution. The process of fragmentation assessment by Fragalyst software is illustrated in Table 6. The 50% passing size (k_{50}) was chosen to determine the fragmentation quality. Out of a total of 92 datasets, 73 datasets were used to train the ANN model and 19 separate datasets (not used in training) were utilized for the purpose of validation thus representing the standard 80:20 ratio.

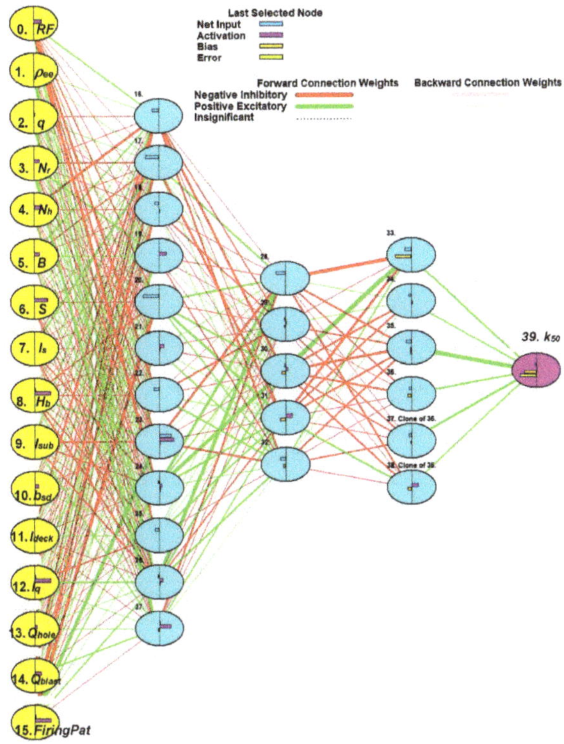

Figure 6. Suggested ANN topology for the study.

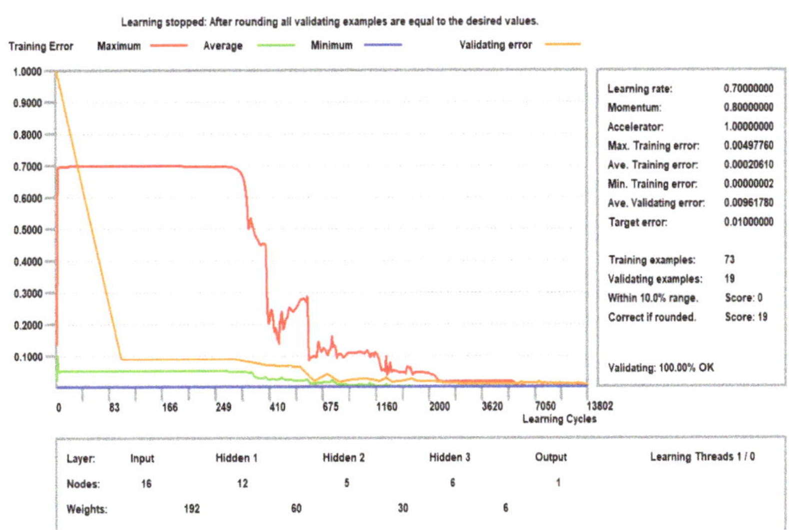

Figure 7. Neural network training and validation.

Table 10. Input parameters used for developing the neural network and their ranges.

Input Parameters	Symbols	Ranges
Burden (m)	B	2.53–3.02
Spacing (m)	S	3.41–4.06
Stemming length (m)	l_s	2.0–3.59
Bench Height (m)	H_b	6.13–9.98
Sub Drill (m)	l_{sub}	0.0–0.2
Air Decking Length (m)	l_{deck}	0.0–1.7
Number of holes	Nh	11–27
Number of rows	Nr	2–5
Charge length (m)	l_q	3.23–8
Charge/hole (kg)	Q_{hole}	29.75–61.87
Charge/blast (kg)	Q_{blast}	401.21–1246.95
Rock factor	RF	6–8
ρ_{ee} (kg/m^3)	ρ_{ee}	0.38–0.60
Specific Charge, kg/m^3)	q	0.41–0.57
firing pattern	$FPat$	L = 1, D = 2, V = 3
B × S	-	9.27–12.28

Table 11. Output parameter used for developing neural network and their ranges.

Output Parameters	Symbol	Range
Mean Fragmentation size (m)	k_{50}	0.20–0.62

The training results of the ANN model and the validation results are presented in Figure 8a,b respectively. The results of the analysis show R^2 and RMSE of 0.96 and 0.040 in the case of training and 0.884 and 0.049 in the case of validation tests. The results point to the fact that the ANN method can be well used for the prediction of fragmentation.

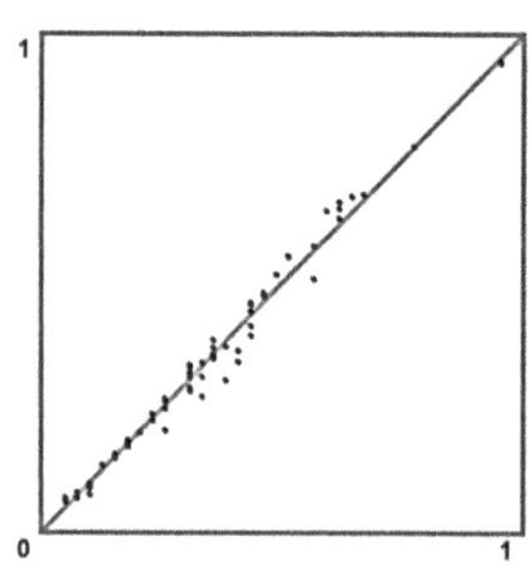

(a)

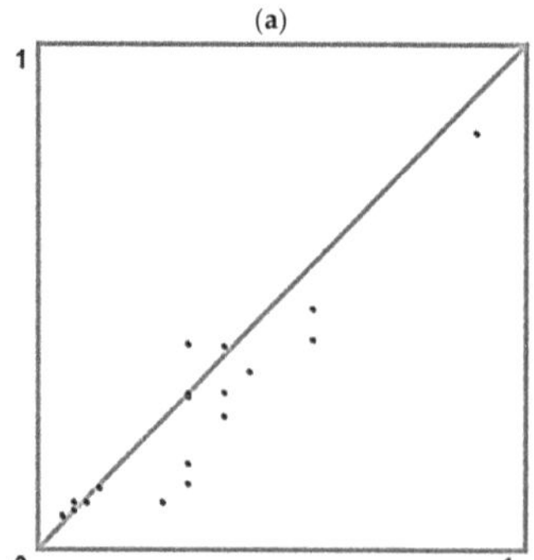

(b)

Figure 8. ANN training results (**a**), ANN-based validation results (**b**).

In addition to the above, the sum of the absolute weights of the connections from the input node to all the nodes in the first hidden layer defines the importance of the input variables. The relative importance can thus be worked out from such results. The results of such analysis are presented in Figure 9.

There are variables in the above list that can be grouped together or represent explosive distribution in a blasthole such as charge length, stemming length, decking length, and sub-drill length that has a significant contribution to fragmentation. This is probably the reason that some of the variables such as B and S assume less importance in ANN but have retained their importance in RSA when several such variables were combined in a single factor ρ_{ee}.

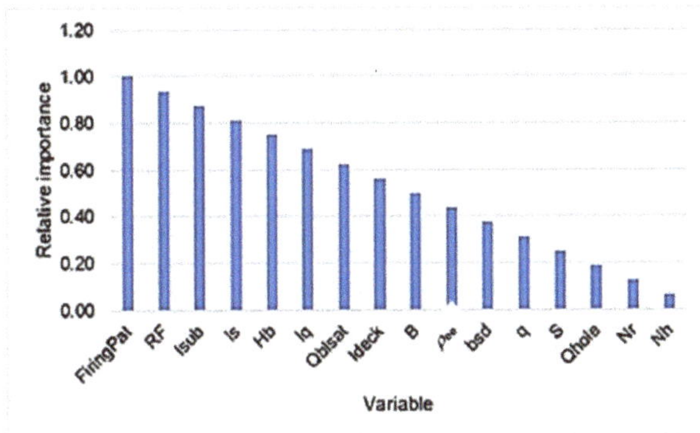

Figure 9. Importance of blast variables as determined by ANN.

5. Conclusions

A hypothesis that there is further breakage by the collision effect during flight of fragments in the blasting process was evaluated in this study. An experimental scheme that presents similar trends in design variables was adopted and comparison was made with the help of fragmentation measured in three types of firing patterns in blasting. The effect of firing patterns on fragmentation were evaluated with the help of 92 blasts in which 12 blasts were taken with a line firing pattern, 36 blasts with diagonal, and 44 blasts with a V-Type firing pattern in a limestone mine. The results acquired showed that there is a significant reduction in fragmentation in the case of the V-Type and diagonal firing patterns, respectively, and counts for 45% and 26% reduction in fragmentation in comparison to that of the line firing pattern and diagonal firing pattern. Since design variables are similar in the case of diagonal and V-type firing patterns, the reduction in fragmentation in the case of the latter pattern can be assigned to the impact of collision.

A surface response model was developed for prediction of the mean fragmentation size (k_{50}) that provided excellent results while using a rock factor with effective energy density in a blasthole and ($B \times S$) as modeling variables. The results of the analysis are provided in the form of models for the three types of firing patterns analyzed which showed significant R^2 and a strong agreement in adjusted and predicted R^2.

Further, the ANN method was deployed for assessing the predictability of the fragmentation using a back propagation algorithm and two hidden layers. The model trained well and validation tests yielded significant correlation between the predicted and observed values of mean fragment size of the blasts. Moreover, the importance of blast variables on rock fragmentation was evaluated with the help of ANN analysis in which the firing patterns and rock factor along with the charge distribution in the blasthole assumed higher significance.

Author Contributions: Conceptualization, L.S.C., A.K.R., V.M.S.R.M., E.T.M., and R.M.B.; methodology, A.K.R. and V.M.S.R.M.; software, L.S.C. and A.K.R.; formal analysis, A.K.R. and V.M.S.R.M.; resources, A.K.R. and V.M.S.R.M.; data curation, L.S.C.; writing—original draft, L.S.C., A.K.R., V.M.S.R.M., E.T.M., and R.M.B.; writing—review and editing, A.K.R., V.M.S.R.M., and M.M.S.S.; supervision, A.K.R.; funding acquisition, M.M.S.S. All authors have read and agreed to the published version of the manuscript.

Funding: This research received no external funding.

Institutional Review Board Statement: Not applicable.

Informed Consent Statement: Not applicable.

Data Availability Statement: The data presented in this study are available on request from the corresponding author.

Acknowledgments: The paper forms a part of the Ph.D. work of the first author. The authors are thankful to their respective employers for their permission to publish the findings. Help of several persons in conducting the study is duly acknowledged.

Conflicts of Interest: The authors declare that there is no conflict of interest to disclose.

Abbreviations

The following abbreviations are used in this manuscript:

Abbreviations	Explanations
H_b	Bench height (m)
B	Burden (m)
B_e	Effective burden (m),
S	Spacing (m)
S_e	Effective spacing (m)
M_b	Ratio of S_e to B_e
M_d	Ratio of S to B
l_s	Stemming length (m)
k_{50}	Mean fragment size (m)
d	Blasthole diameter(mm)
q	Specific charge (kg/m^3)
N_h	Number of holes
Nr	Number of rows
Q_{hole}	Explosive charge per hole (kg)
$FiringPat$	Firing pattern
TLD	Trunk Line Delay Detonator
DTH	Down The Line Delay Detonator
RHS	Right Hand Side (Connection of TLD leaning towards Right hand Side)
LHS	Left Hand Side (Connection of TLD leaning towards Left hand Side)
l_{sub}	Length of subgrade drilling (m)
l_{deck}	Length of decking (m)
l_q	Length of explosive charge in the hole (m)
Q_{blast}	Explosive charge per blast (kg)
RF	Rock factor
ρ_{ee}	Equivalent explosive charge density (kg/m^3), i.e., ratio of explosive per hole in kg to volume of charged section the blasthole where volume = $B \times S \times l_q$
bsd	Product of burden (B) and spacing (S) (m^2)
RSA	Response surface analysis
ANN	Artificial neural network
ANOVA	Analysis of variance
ANFO	Ammonium Nitrate Fuel oil

References

1. Božić, B. Control of fragmentation by blasting. *Rud. Geol. Naft. Zb* **1998**, *10*, 49–57.
2. Engin, I.C. A practical method of bench blasting design for desired fragmentation based on digital image processing technique and Kuz-Ram model. In Proceedings of the 9th International Symposium on Rock Fragmentation by Blasting, FRAGBLAST 9, Granada, Spain, 13–17 August 2009; pp. 257–263.
3. Mackenzie, A. Cost of explosives—Do you evaluate it properly? *Min. Congr. J.* **1966**, *52*, 32–41.
4. Hustrulid, W. *Blasting Principles for Open Pit Mining: Volume 1—General Design Concepts*; CRC Press: Boca Raton, FL, USA, 1999; Volume 1.
5. Torrealba-Vargas, J.; Esteban, R.; Roy, D.; Runnels, D. Mine to mill approach to optimize power consumption in a process plant operation by modelling and simulation. In Proceedings of the IMPC 2016 28th International Mineral Processing Congress, Quebec City, Canada, 11–15 September 2016.
6. Kim, K.M.; Kemeny, J. Site specific blasting model for mine-to-mill optimization. In Proceedings of the SME Annual Meeting and Exhibit and CMA 113th National Western Mining Conference, Denver, CO, USA, 28 February–2 March 2011; pp. 619–623.
7. Lewandowski, D.; Lipnicki, P. *Grinding Mill Process Optimization Algorithm*; IEEE: Toulouse, France, 2019.
8. Lam, M.; Jankovic, A.; Valery, W.; Kanchibotla, S. Maximising SAG mill throughput at Progera gold mine by optimizing blast fragmentation. In Proceedings of the International Conference on Autogenous and Semi-Autogenous Grinding Technology, Vancouver, BC, Canada, 25–28 September 2001; pp. 271–287.
9. Afum, B.O. Rock Fragmentation Evaluation towards Blast-To-Mill Concept of Blast Optimization in Hard Rock Mines. *J. Miner. Mater. Sci.* **2021**, *2*, 1–9. [CrossRef]
10. Zhang, Z.; Luukkanen, S. Feasibility and necessity of mine to mill optimization in mining industry. *Mater. Medica* **2021**, *2*, 63–66.
11. Siddiqui, F.; Shah, S.; Behan, M. Measurement of Size Distribution of Blasted Rock Using Digital Image Processing. *J. King Abdulaziz Univ. Sci.* **2009**, *20*, 81–93. [CrossRef]
12. Afeni, T.B. Optimization of drilling and blasting operations in an open pit mine-the SOMAIR experience. *Min. Sci. Technol.* **2009**, *19*, 736–739. [CrossRef]
13. da Gama, C.D.; Jimeno, C.L. Rock fragmentation control for blasting cost minimization and environmental impact abatement. In *Rock Fragmentation by Blasting*; CRC Press: Boca Raton, FL, USA, 1993; pp. 273–280.
14. Shim, H.J.; Ryu, D.W.; Chung, S.K.; Synn, J.H.; Song, J.J. Optimized blasting design for large-scale quarrying based on a 3-D spatial distribution of rock factor. *Int. J. Rock Mech. Min. Sci.* **2009**, *46*, 326–332. [CrossRef]
15. Chouhan, L.S.; Raina, A.K. Analysis of In-Flight Collision Process During V-Type Firing Pattern in Surface Blasting Using Simple Physics. *J. Inst. Eng. Ser. D* **2015**, *96*, 85–91. [CrossRef]
16. Choudhary, B.S. Firing Patterns and Its Effect on Muckpile Shape. *Int. J. Res. Eng. Technol.* **2013**, *2*, 32–45.
17. Konya, C.J.; Walter, E.J. Rock blasting and overbreak control (No. FHWA-HI-92-001; NHI-13211). *Security* **1991**, *132*, 430.
18. Jimeno, C.L.; Jimeno, E.L.; Carcedo, F.J.; de Ramiro, Y. Drilling and Blasting of Rocks. *Drill. Blasting Rocks* **1997**, 1–391. [CrossRef]
19. Luo, G.; Xiewen, H.; Yingjin, D.; Jiankang, F.; Xuefeng, M. A collision fragmentation model for predicting the distal reach of brittle fragmentable rock initiated from a cliff. *Bull. Eng. Geol. Environ.* **2019**, *78*, 579–592. [CrossRef]
20. Monjezi, M.; Bahrami, A.; Varjani, A.Y. Simultaneous prediction of fragmentation and flyrock in blasting operation using artificial neural networks. *Int. J. Rock Mech. Min. Sci.* **2010**, *47*, 476–480. [CrossRef]

sustainability

MDPI

Article

Spectrum Index for Estimating Ground Water Content Using Hyperspectral Information

Kicheol Lee [1], Ki Sung Kim [1], Jeongjun Park [2,*] and Gigwon Hong [3,*]

[1] Corporate Affiliated Research Institute, UCI Tech, 313, Inha-ro, Michuhol-gu, Incheon 22012, Korea
[2] Incheon Disaster Prevention Research Center, Incheon National University, 119 Academy-ro, Yeonsu-gu, Incheon 22012, Korea
[3] Department of Civil Engineering, Halla University, 28 Halladae-gil, Wonju-si 26404, Korea
* Correspondence: smearjun@hanmail.net (J.P.); g.hong@halla.ac.kr (G.H.); Tel.: +82-10-4722-2971 (J.P.); +82-10-9015-6263 (G.H.)

Abstract: Quality control considerably affects road stability and operability and is directly linked to the underlying ground compaction. The degree of compaction is largely determined by water content, which is typically measured at the actual construction site. However, conventional methods for measuring water content do not capture entire construction sites efficiently. Therefore, this study aimed to apply remote sensing of hyperspectral information to efficiently measure the groundwater content of large areas. A water content prediction equation was developed through an indoor experiment. The experimental samples comprised 0–40% (10% increase) of fine contents added to standard sand. As high water content is not required in road construction, 0–15% (1% increase) of water content was added. The test results were normalized, the internal and external environments were controlled for precise results, and a wavelength–reflection curve was derived for each test case. Data variability analyses were performed, and the appropriate wavelength for water content reflection, as well as reflectance, was determined and converted into a spectrum index. Finally, various fitting models were applied to the corresponding spectrum index for water content prediction. Reliable results were obtained with the reflectance corresponding to a wavelength of 720 nm applied as the spectrum index.

Keywords: spectrum index; water content; hyperspectral information

Citation: Lee, K.; Kim, K.S.; Park, J.; Hong, G. Spectrum Index for Estimating Ground Water Content Using Hyperspectral Information. *Sustainability* **2022**, *14*, 14318. https://doi.org/10.3390/su142114318

Academic Editors: Mahdi Hasanipanah, Danial Jahed Armaghani and Jian Zhou

Received: 19 September 2022
Accepted: 31 October 2022
Published: 2 November 2022

1. Introduction

The term sustainability refers to the ability to maintain a function and survive over time [1]. Several other definitions of the term have also been provided. For example, Brown [2] defined sustainability as the ability to meet the needs of future generations without reducing their opportunities, and the Brundtland Commission [3] defined it as technology to meet present needs without compromising the resources of future generations. In the field of engineering, especially geotechnical engineering, sustainability refers to the use of resources at a low cost while appropriately controlling harmful emissions. The term is divided into four concepts [4,5]: (1) robust design and construction, including social cost and inconvenience caused by construction; (2) the minimized use of resources and energy in the planning, design, construction, and maintenance of geotechnical facilities; (3) the use of methods and materials with a low impact on ecology and the environment; and (4) the reuse of geotechnical structures for minimizing waste.

The concepts of sustainability have been applied to the field of road construction, which is a subfield of geotechnical engineering. Corriere and Rizzo [6] defined "sustainable roads" as roads that can achieve basic design goals (compliance, safety, ease of mobility, maintenance, energy efficiency, transport capacity, etc.) during the construction, maintenance, and operation phases. Meanwhile, the sustainability of roads is mainly considered from the perspective of maintenance. Greenroads Foundation (USA) evaluated road sustainability using scores obtained from four-year tests on more than 50 types of roads The London Councils in the UK manages a "Highways Minor Works" toolkit to support the procurement of highway services, such as recycling, reducing transport distances, reducing overall lifetime costs, energy use, and CO_2 emissions [7,8]. In South Korea, development and research on smart roads are being actively conducted, which mainly requires automation equipment, datafication of information, and accurate quality control in a wide area [9,10].

In addition to the maintenance aspect, the practical implementation of "sustainable roads" is achieved through thorough, early-stage road excavation and pavement work. The compaction performed to secure the road bearing capacity not only ensures the durability of the asphalt or concrete road, but is also the determining process for road performance and quality, including the drivability of vehicles [11]. In general, the goal of compaction is 90% or more, which is calculated as the ratio of the dry unit weight on-site to the maximum dry unit weight. Furthermore, the water content used for the calculation is the ratio of soil water to soil weight, which affects the long-term stability of the subgrade, the quality of compaction, and the number of passes. It is essential to determine the groundwater content prior to performing actual compaction [12–20].

Groundwater content measurement is performed by the traditional methods of comparing the weight of an on-site sample with the weight of the sample after drying, using a scale; time domain reflectometry through the reverberation time of an electrical signal; and ground penetrating radar (GPR) through the intermittent measurement of water volume and dielectric constant [14–19]. However, the disadvantage of these methods is that they are time-consuming and labor-intensive in determining the water content distributed over the entire construction site by point measurement. Furthermore, the passive aspect of the data measurement process may vary the measured values depending on the skill level of the operator or increase the error range, which may cause reliability problems.

Accordingly, it is necessary to measure the water content in a range by a more reliable method rather than by the existing point measurement method. For example, remote hyperspectral sensing has been actively researched recently as a viable method. The field of remote hyperspectral sensing is broadly divided into spectroscopy, radiative transfer, imaging spectroscopy, and hyperspectral image processing, where spectral curves are derived through radiative transfer. A spectrum is a function of wavelength and indicates the distribution of reflectance; thus, the reflectance shown by the spectrum depends on the characteristics of the object [20]. In the construction industry, remote hyperspectral sensing usually involves the use of drones. Through this, orthographic images and hyperspectral information of a wide area are acquired and mainly used to classify mineral types, sizes, and qualities or to analyze vegetation distribution [21–27]. The photographed hyperspectral information indicates only the reflectance based on wavelength and is expressed as a spectrum index by substituting the equation corresponding to each property. The spectrum index is a value obtained by converting the spectral information (wavelength–reflection curve) obtained through a spectral experiment into a single value; this is equivalent to normalizing the necessary information. Thus, secondary processing of hyperspectral information is required to convert the photographed copy to suit the operator's intention.

Measuring groundwater content in ranges requires a conversion of the measured hyperspectral information into a spectrum index representing the water content. However, studies on the spectrum index related to water content have mainly focused on factors influencing water quality or moisture content in [25–29]. Through hyperspectral image analysis, Prošek et al. [28] classified the local waters, and Guo et al. [29] analyzed only the color change of the lake. On the other side, Ge et al. [30] verified the equation for calculating the spectrum index of various water contents to confirm soil aggregate structure and nutrient status. However, in general, many spectrum index models have a small R^2 value of calculated water content and measured water content. The most suitable spectrum index model still has a disadvantage: the target site does not reflect the low water content of agricultural land (measured water content of 10–30%).

The ground targeted in this study was a road construction site. In such a ground, compaction is usually carried out after filling, and the transported soil usually has a similar water content. That is, the water content measured at the surface represents the water content of the entire ground. In this ground, as work was performed on general sand, high water content was not recorded, but low water content was considered. Therefore, the application of the corresponding spectrum index model may show inappropriate results, necessitating the development of a spectrum index to indicate new water content. Thus, ground hyperspectral information was acquired from water content through a normalized and thoroughly controlled indoor experiment in this study. A subsequent spectrum index expressing water content through various combinations was obtained. The soil used in the experiment was prepared from 0 to 30% of fine particles (10% increments; particle size $\leq$ 0.075 mm) using standard sand, and it was used to simulate various types of soil while increasing water content by 1%. As the spectrum index must be used during image acquisition through drones, an indoor experimental system with the same measurement method was created, in which hyperspectral information was acquired for spectrum index calculation.

2. Methodology for Estimating Ground Water Content in Road Construction Site

The water content measurement process of this study is illustrated in Figure 1 and detailed as follows; (1) A hyperspectral camera is mounted on a drone, which is an unmanned aerial vehicle, and orthographic image and hyperspectral information of the ground for which water content is to be determined are acquired. In this case, the hyperspectral information is the pixel unit of the measurement area, which can be freely adjusted; (2) The acquired image is corrected through post-processing because there is shaking during the drone shooting process using the push-broom method. (3) Hyperspectral information existing in each pixel is converted into water content for a photographed copy where all processes have been completed. Here, hyperspectral information is the relationship between reflectance and wavelength; (4) A specific color is assigned to the water content converted for each pixel, and this is displayed on the map. That is, it implements a color-coded map (CCM).

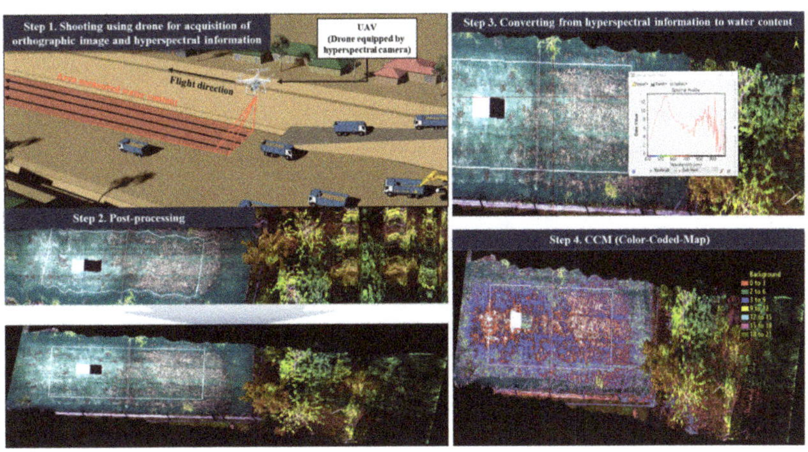

Figure 1. Process of color-coded map of water content in the road construction site.

This study aims to present a spectrum index that converts hyperspectral information into groundwater content at the fourth step illustrated in Figure 1. Thus, the spectrum index would refer to the groundwater content, which is a function of reflectance according to wavelength, as expressed in Equation (1). Here, w refers to water content, R_i refers to reflectance at a wavelength of i-nm, and i ranges from 400 nm to 1000 nm. Reflectance (R) is the ratio of the reflected energy to the total energy incident on the body, and it is expressed as a percentage. R is expressed through a complex process of reflection, absorption, and transmission of energy; it varies with wavelength and enables features to be identified on the body or surface to be measured [31].

$$\text{Spectrum index} = w = function(R_i) \quad (1)$$

Overall, we aim to present hyperspectral information measured with a hyperspectral camera as a function of Equation (1) (Figure 2), where, R_i in the function may represent one, two, or more points. Corresponding combinations and analyses are described in a later section.

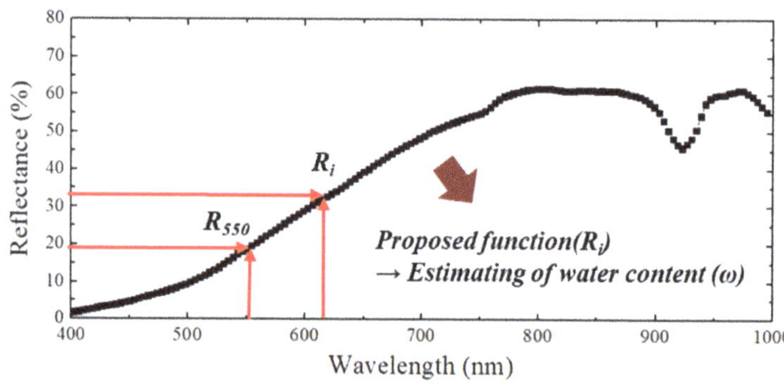

Figure 2. Methodology for estimating water content from hyperspectral information.

3. Laboratory Tests for Obtaining Hyperspectral Information

3.1. System for Obtaining Hyperspectral Information

An indoor experimental system (Figure 3a) was created to acquire hyperspectral information for determining the water content of the soil. The system consists of a hyperspectral camera (Micro HSI410shark, Coring, Seoul, Korea) capable of measuring wavelengths of 400–1000 nm at 2 nm intervals, a stage for push-broom scanning, and software to express the reflectance by wavelength in row data and graphs.

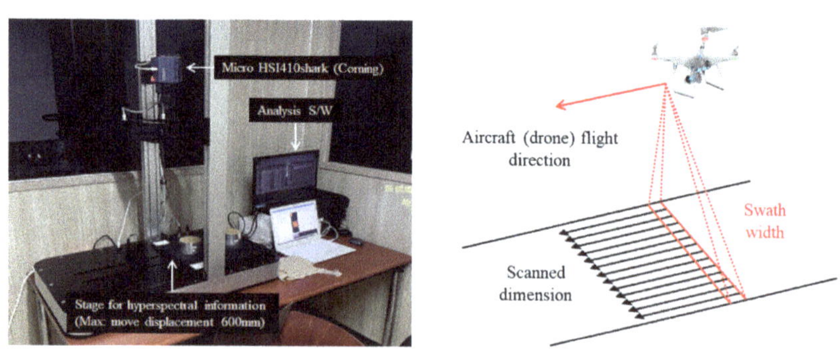

Figure 3. System setup: (**a**) laboratory conditions and; (**b**) Push-broom scanning.

The shooting method of a spectral camera is typically divided into staring or spectral scanning, which captures the entire scene in band-sequential format, and push-broom scanning, which generates a hyperspectral image in a line-by-line format [32]. The push-broom scanning method is a reliable method [33–35] mostly used for aviation photography using an unmanned aerial vehicle. The method also provides reasonable spatial resolution and high spectral resolution [36].

In push-broom scanning, a spectral camera is mounted to have a slit perpendicular to the moving direction of the drone (Figure 3b) to extract spectral information from a designated area of one pixel. The area of one pixel is determined by the sensing interval, focal length, and flight altitude. The drone captures all pixels corresponding to the orthogonal area while moving and then measures the frame of the line corresponding to the next pixel. The captured information is spectral information, including orthographic images.

In the actual field, push-broom scanning is applied as the drone moves, but in the indoor experiment, the movement of the sample located at the bottom was simulated. The simulation was intended to reduce errors due to changes in focus resulting from the camera movements and to omit the geometric correction step performed after actual on-site image acquisition.

3.2. Laboratory Test of Soil Sample

The ground measured in this study was a construction site mainly comprising sand composed of soil particles of various sizes, with differences in void ratio, compaction curve, and optimal water content depending on the particle size distribution. As the ground was composed of various sand types, differences were expected in the spectral information measured according to the water content. Therefore, it was necessary to acquire spectral information for various types of ground and convert it into water content. Thus, the derived water content should be constant regardless of the type of ground.

Therefore, for basic normalization, the base soil was set as standard sand. Standard sand is an aggregate used to improve the strength of cement, referring to poorly graded sand, which is granular soil with a particle size of 0.075–2.00 mm in accordance with [37]. Various ground simulations were performed with the addition of 10%, 20%, and 30% fine contents (particle size of 0.075 mm or less).

3.2.1. Sieve Test

The sieve test of standard sands with 0, 10, 20, and 30% fine contents was performed according to [38], and the results were as presented in Figure 4 and Table 1. The fine contents were added in relation to the total weight of the soil. As a result of the test, D_{10} was not measured at 20% and 30% of fine contents, but the pass rate for fine contents did not exceed 50%; D_{10} increased as it acted as a denominator in the coefficient of uniformity and the coefficient of curvature. Therefore, all samples used as a result of classification according to [39] could be classified as Poor Sand with uneven particle sizes.

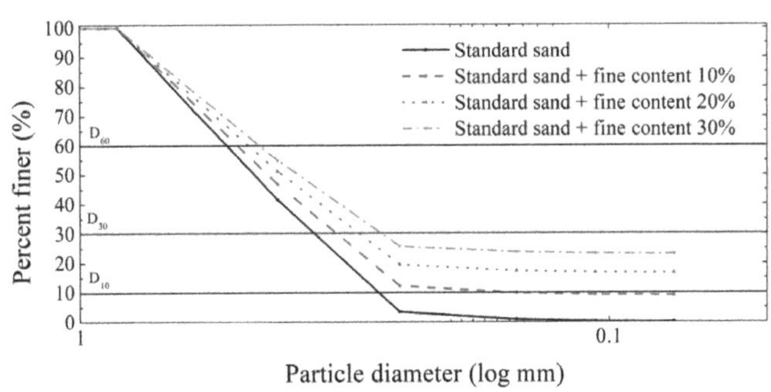

Figure 4. Particle size distribution curves of soil samples.

Table 1. Sieve analysis results of the used soil sample.

Fine Content in Standard Sand (%)	D_{10} [1] (mm)	D_{30} [2] (mm)	D_{60} [3] (mm)	Coefficient of Uniformity, C_u [4]	Coefficient of Curvature, C_c [5]	Percentage Passing No. 200 Sieve (%)	Soil Classification
0	0.274	0.363	0.530	1.934	0.907	0.06	SP
10	0.150	0.329	0.505	3.367	1.429	9.15	SP
20	-	0.300	0.482	-	-	16.72	SP
30	-	0.272	0.461	-	-	23.13	SP

[1] D_{10}: Particle diameter in percent finer of the soil corresponding to 10%; [2] D_{30}: Particle diameter in percent finer of the soil corresponding to 30%; [3] D_{60}: Particle diameter in percent finer of the soil corresponding to 60%; [4] C_u: Coefficient of uniformity that calculated by $C_u = D_{60}/D_{10}$; [5] C_c: Coefficient of curvature that calculated by $D_{30}{}^2/(D_{10}D_{60})$.

3.2.2. Standard Compaction Test

To produce soil samples with the same degree of compaction, it is necessary for us to know the maximum dry unit weight of each sample. Accordingly, the sieve test of standard sands with 0, 10, 20, and 30% fine contents was performed according to [40], and the results are presented in Figure 5. As the amount of fine content increased, the voids for the volume of the soil sample decreased, being filled with the fine content. Therefore, as shown in Figure 5, the optimum water content increased while the dry unit weight decreased with an increasing amount of fine content.

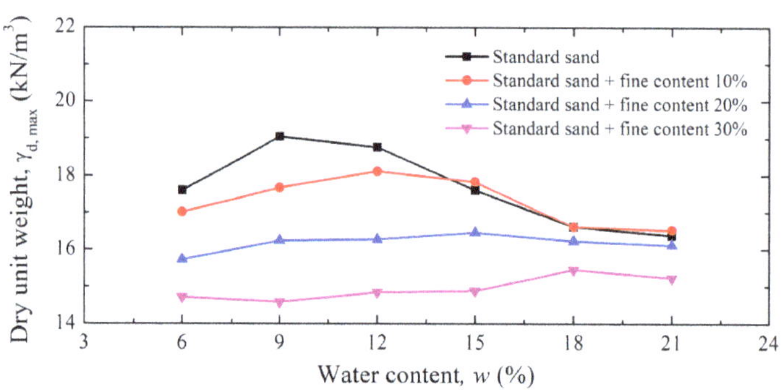

Figure 5. Compaction curves of soil samples.

3.2.3. Composition of Specimens

In general, the degree of compaction (ratio of on-site dry unit weight and maximum dry unit weight obtained through indoor experiments) is 95% at road construction sites. Thus, the soil sample specimens in this study were prepared with a compaction of 95%. Following the compaction curve in Figure 5, each sample was prepared with water content according to Figure 6a, followed by compaction as shown in Figure 6b. Here, the water content was set to 0–15%, making a total of 16 levels. Finally, specimens (Figure 6c) were obtained and placed on the specimen stage of the configured system (Figure 3a) to extract hyperspectral information.

The water content was the weight ratio of soil to water, and water as much as the water content set in this study was added to the weight of the sample. The volume of the experimental can (diameter = 10 cm, height = 6 cm) used for making the specimens was 471 cm^3, and the maximum weight of soil that the experimental can could contain was 914 g (standard sand), 870 g (standard sand + fine content 10%), 790 g (standard sand + fine content 20%), and 742 g (standard sand + fine content 30%), according to the maximum dry unit weights presented in Figure 5.

(a) (b) (c)

Figure 6. Composition of specimens: (**a**) Mixing of a soil sample with standard sand and water; (**b**) Compaction of the specimen in a circular petri; (**c**) Specimens of soil samples with fine and water contents.

Whenever the water content increased by 1%, 9.1 g, 8.7 g, 7.9 g, and 7.4 g of water was added. The composition of specimens was processed precisely, and when the water content was measured again after obtaining the spectral information, the water content was found to be the same as in the initial state. Hence, the measured spectral information reflected the water content of specific specimens accurately.

3.3. Hyperspectral Information of Soil Sample

Figure 7 illustrates the hyperspectral information (relationship between wavelength and reflectance) of soil samples according to the water content measured through the system. In all the indoor experimental results, the reflectance according to wavelength showed a similar trend. The reflectance increased non-linearly as the wavelength increased, showing a rapid increase at approximately 750 nm. The maximum reflectance was measured at 800 nm and gradually decreased to the vicinity of the wavelength of 920 nm. Subsequently, the reflectance exhibited non-linear behavior; for example, it increased again. According to the characteristics of each wavelength band, the reflectance increased in the visible-rays region (400–800 nm), and then it decreased and increased non-linearly in the wavelength band of infrared rays (800–1000 nm). In addition, as the water content increased, the reflectance according to the wavelength decreased. This may have been due to absorption occurring more than reflection as the amount of water increased.

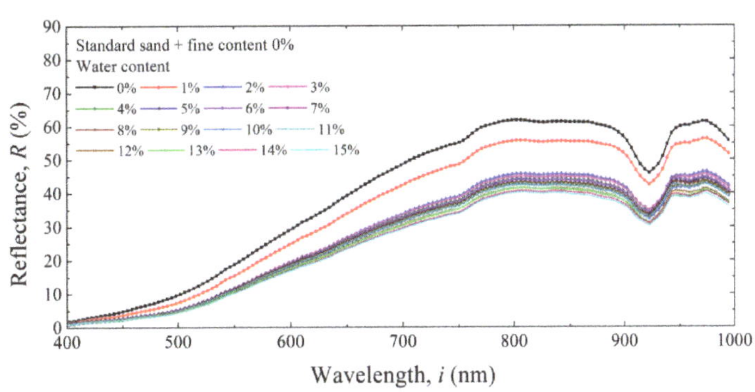

(**a**)

Figure 7. *Cont.*

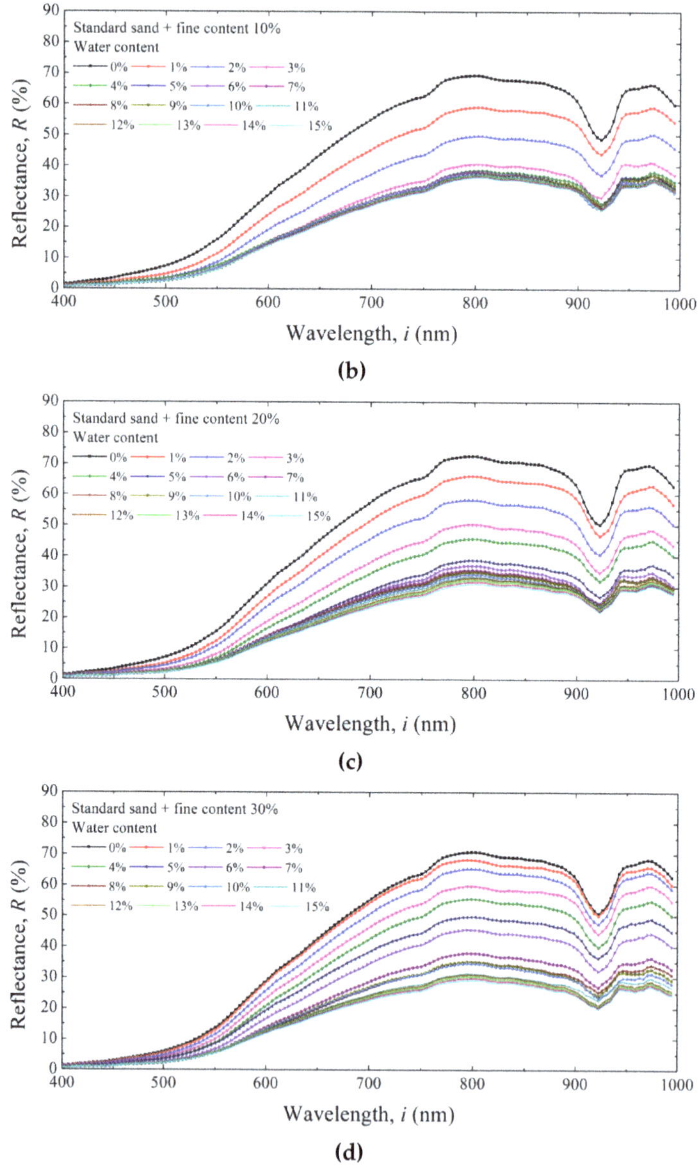

Figure 7. Relationship between wavelength and reflectance at fine contents of (**a**) 0%, (**b**) 10%, (**c**) 20%, and (**d**) 30%.

4. Estimation of the Spectrum Index for Water Content Prediction

4.1. Variability Analysis of Hyperspectral Information

For water content prediction, spectral information measured through experiments should be converted into a single-value spectrum index, which should be inserted into a water content prediction equation. Processing spectral information is time-consuming as numerous row data are collected from the row data obtained with a hyperspectral camera; this phenomenon is due to the relationship between wavelength and reflectance, as shown in Figure 7. However, as water content measurement is performed on the day of construction in actual road construction sites, with frequent changes made according to various conditions (rainfall, humidity, and temperature conditions), it is essential to minimize the processing time. Accordingly, a significant point (reflectance at a specific wavelength) should be extracted, which should be converted into a spectrum index.

The conditions under which a specific wavelength was selected are illustrated in Figures 8 and 9. The hyperspectral information processed was the ratio of fine content (0–30%, 10% increment) and water content (0–15%, 1% increment), and the wavelength–reflection curves were 64 in total.

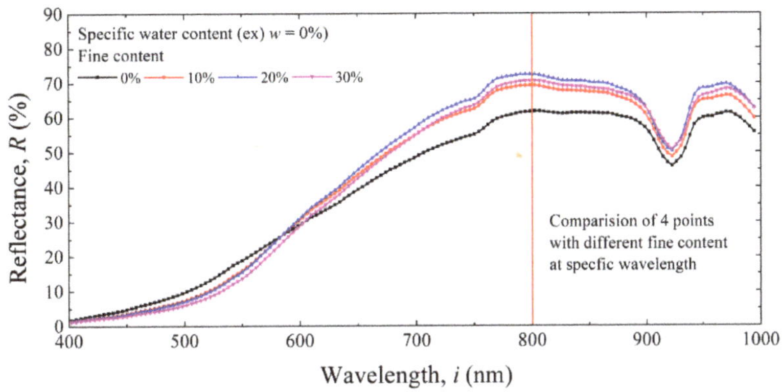

Figure 8. Selection of wavelength–reflection point for minimizing fine content effect.

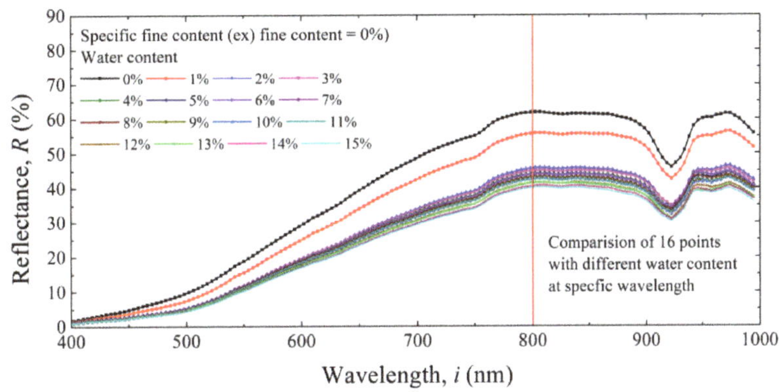

Figure 9. Selection of wavelength–reflection point for maximizing water content effect.

The first wavelength selection condition is that the difference in reflectance at a specific wavelength should be small regardless of the content of fine particles at a specific water content amount. If the difference between reflectance is large according to the content of fine particles, a different spectrum index and equation of water content prediction should be selected according to each ground condition. However, this step is practically impossible as it requires setting the ground conditions individually in a large construction site. Therefore, it is necessary to calculate a specific wavelength band with little difference in reflectance according to the change in fine content; this wavelength should show reflectance with slight variability (Figure 8).

The second condition is that there should be a clear difference in reflectance among water contents (Figure 9). If the difference in reflectance at a specific wavelength is not large, there is a possibility that the difference in water content may change rapidly even with a small change in reflectance. Hence, the error would be substantial, having a large impact on the final product, CCM. Therefore, it is necessary to determine a specific wavelength band with a large difference in reflectance according to a change in water content, which is equivalent to finding a point with a large variability.

Data variability can be evaluated by the coefficient of variation (COV), as in Equation (2). Here, COV is the ratio of standard deviation (SD) to mean. In general, COV is excellent at less than 10%, good from 10% to 20%, acceptable from 20–30%, and not acceptable beyond 30%.

$$COV(\%) = (SD/Mean) \times 100(\%) \quad (2)$$

4.1.1. Effects of Fine Contents

Figure 10 presents the COV (Dot in figure) of four points of fine content (0%, 10%, 20%, and 30%) and the average (Red line in figure) of all data according to the wavelength at a specific water content. In terms of average COV, the wavelength with the maximum COV was 29.03% at 500 nm, and the wavelength with the minimum COV was 720 nm, which was 10.23%. In the wavelength range of approximately 400–600 nm, the COV was high; in the range of 600–880 nm, it showed a smooth parabolic shape with a value of 10.23–12.40%. At wavelengths above 880 nm, the COV was approximately steady with slight fluctuations. Regarding the first condition for selecting the wavelength to be applied to the spectrum index, a wavelength with small fine content variability should be selected. For the optimal condition, it is appropriate to use the reflection of a wavelength of 720 nm although it also appears appropriate to use the reflectance of a wavelength of 600 nm or more, as the COV difference from 600 nm to 880 nm was approximately 2%.

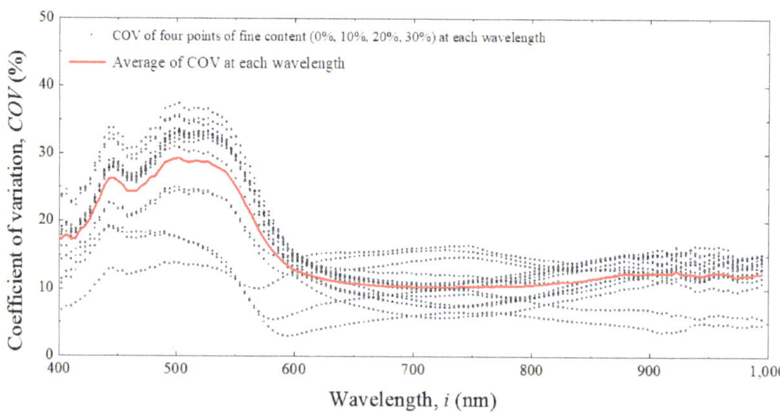

Figure 10. Analysis of using COV to select an appropriate wavelength for minimizing fine content effect.

4.1.2. Effects of Water Contents

Figure 11 illustrates the COV of reflectance by water content according to fine content amount as well as the average according to wavelength. Regarding the second condition for selecting the wavelength to be applied to the spectrum index, a wavelength with high water content variability should be selected. Therefore, to maximize the water content effect at the wavelength of 600–880 nm, a wavelength with a high COV was selected. Within that range, a minimum COV of 24.31% at 810 nm was measured as well as a maximum COV of 27.60% at 600 nm.

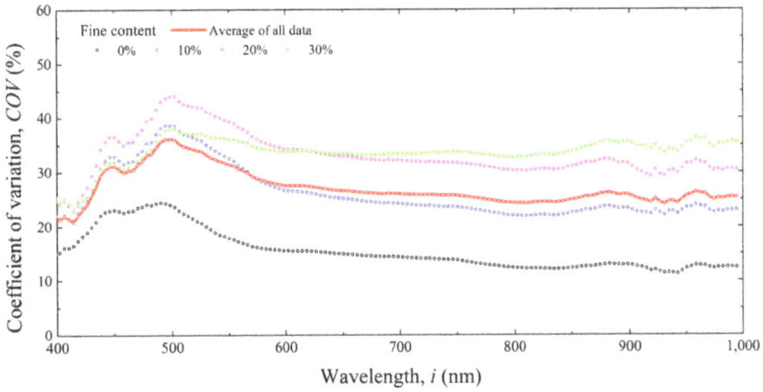

Figure 11. Analysis of COV for selecting appropriate wavelength for maximizing water content effect.

4.2. Spectrum Index Reflected by Selected Wavelength and Reflection

Following the variability analyses, a specific wavelength for spectrum index was selected as illustrated in Figure 12: (1) 720 nm wavelength and (2) 600–880 nm wavelength. The wavelength of 720 nm is the point with the least fine content effect. As water content had a high variability, it is most appropriate to use the wavelength from a single perspective. In this case, the spectrum index refers to a reflectance at 720 nm. We express the corresponding spectrum index as R_{720}. The variability analysis showed that COV exhibited a similar trend in the wavelength band of 600–880 nm. Therefore, all reflectance in the wavelength band of 600–880 nm are considered, and the spectrum index is expressed as an integral. In this paper, the integral is expressed as $I_{600–820}$ (integral from 600 nm to 820 nm of wavelength), and it refers to the area between the wavelength–reflectance curve and the x-axis (range of 600–820 nm).

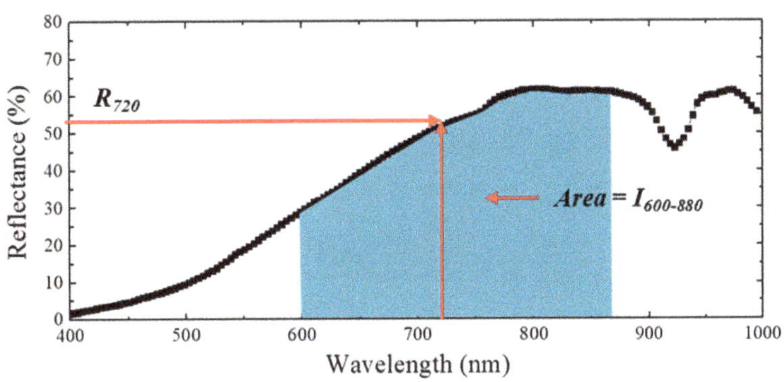

Figure 12. Selection of wavelength and reflection for spectrum index.

4.3. Equation for Predicting Water Content Using Spectrum Index

Data fitting was performed, as shown in Figure 13, to determine the water content prediction formula using the appropriate spectral index. The fine particle content was not classified separately in all the data here; R_{720} and $I_{600–880}$ were plotted on the x-axis against water content on the y-axis. The total number of data was 64. Figure 13 shows that the water content gradually decreased as the spectrum index increased, but the relationship was nonlinear. Therefore, it is necessary to derive a non-linear equation for the relationship.

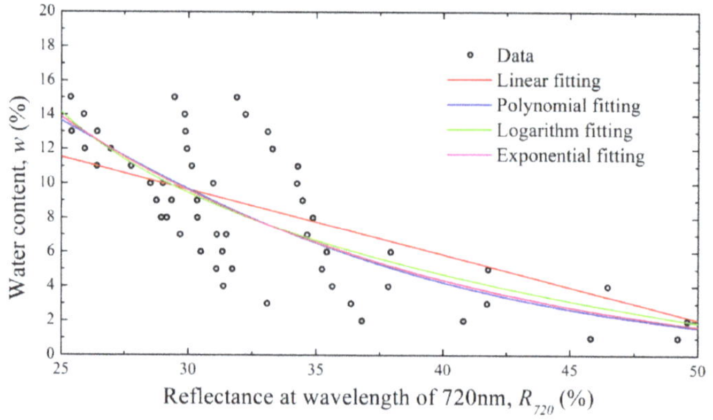

(**a**)

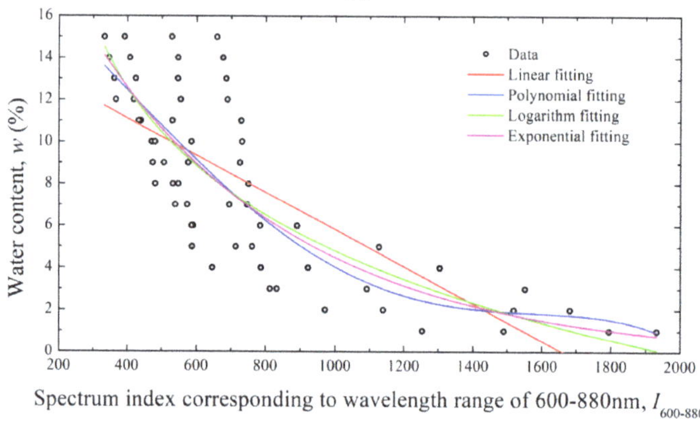

(**b**)

Figure 13. Spectrum index with water contents; (**a**) R_{720} and (**b**) $I_{600–820}$.

Various fitting equations, including linear, polynomial, logarithmic, and exponential equations, were considered. Table 2 presents the equations and their corresponding R^2 values. After fitting, R^2 was low for $I_{600–880}$ (using the integral area) compared with that of R_{720} (calculated as a single point). Therefore, it was appropriate to select R_{720} as a spectrum index; an exponential fitting model with a high correlation coefficient was selected as the equation for water content prediction.

Table 2. Results of the fitting.

Index	Fitting Model	Equation	R^2
R_{720}	Linear	$w = -0.379R_{720} + 21.021$	0.636
	Polynomial	$w = -8.38 \times 10^{-6}{R_{720}}^4 + 0.0012{R_{720}}^3 - 0.0462{R_{720}}^2 - 0.2631R_{720}122(0\ \text{index} + 33.7973)$	0.687
	Logarithm	$w = 25.767 - 7.004\ln(R_{720} - 19.738)$	0.695
	Exponential	$w = -1.172 + 79.648\exp(-0.0666R_{720})$	0.697
$I_{600–880}$	Linear	$w = -0.00883I_{600–880} + 14.658$	0.579
	Polynomial	$w = -5.37 \times 10^{-12}{I_{600–880}}^4 + 2.159 \times 10^{-8}{I_{600–880}}^3 -7.782 \times 10^{-5}{I_{600–880}}^2 + 0.0479I_{600–880}122(0\ \text{index} + 10.0868)$	0.637
	Logarithm	$w = 48.239 - 6.432\ln(I_{600–880} - 144.708)$	0.643
	Exponential	$w = -0.2228 + 25.032\exp(-0.00167I_{600–880})$	0.645

4.4. Comparison of the Literature with Proposed Spectrum Index Method

To verify the suitability of the water content prediction equation proposed in this study, a comparison with the existing theoretical equations was performed. Eleven of the thirty high-R^2 prediction equations investigated by Ge et al. [30] are presented in Table 3. Because the existing equations target only the spectrum index, a separate fitting should be performed for the water content prediction equation. According to [30], a linear fitting was performed. Therefore, to obtain the equation for predicting water content, the spectral information from this study was substituted into the spectrum index, and the equations were obtained individually through linear fitting.

Table 3. Comparison of bias factors of each inflection point.

Spectrum Index		Equation for Water Content Prediction	Ref.
Sort	Equation		
mNDVI705	$(R_{750} - R_{705})/(R_{740} + R_{705} + 2R_{445})$	$w = -105.01\text{mNDVI705} + 13.40$	[41]
NDVI	$(R_{800} - R_{680})/(R_{800} + R_{680})$	$w = 161.11\text{NDVI} - 20.57$	[42]
NDCI	$(R_{762} - R_{527})/(R_{762} + R_{527})$	$w = 8.04\text{NDCI} + 1.44$	[41]
NDVI705	$(R_{750} - R_{705})/(R_{750} + R_{705})$	$w = -120.54\text{NDV705I} + 14.53$	[43]
RVI	R_{800}/R_{680}	$\omega = 55.28\text{RVI} - 71.13$	[43]
NDRE	$(R_{750} - R_{705})/(R_{750} + R_{705})$	$w = -120.54\text{NDRE} + 14.53$	[44]
GNDVI	$(R_{750} - R_{550})/(R_{750} + R_{550})$	$w = 3.79\text{GNDVI} + 5.10$	[45]
OSAVI	$[(1 + 0.16)(R_{800} - R_{670})]/(R_{800} + R_{670} + 0.16)$	$w = 115.04\text{OSAVI} - 19.52$	[42]
VOG1	R_{740}/R_{720}	$w = -114.33\text{VOG1} + 127.58$	[46]
VOG2	$(R_{734} - R_{747})/(R_{715} - R_{726})$	$w = 2.71\text{VOG2} + 5.66$	[46]
VOG3	$(R_{734} - R_{747})/(R_{715} + R_{720})$	$w = 577.11\text{VOG3} + 14.58$	[46]

Figure 14 illustrates the results of the water content prediction equation proposed in this study and the measured and predicted water contents according to the existing theoretical formulas. Overall, in the existing equations, no significant change was observed in the predicted water content as the measured water content increased. In other words, the R^2 value was distributed from 0.002 to 0.122, indicating an extremely low correlation. Therefore, when the spectral information obtained in this study was substituted into the existing spectrum index, a considerable error was obtained, demonstrating that the water content prediction equation using the proposed R_{720} is appropriate.

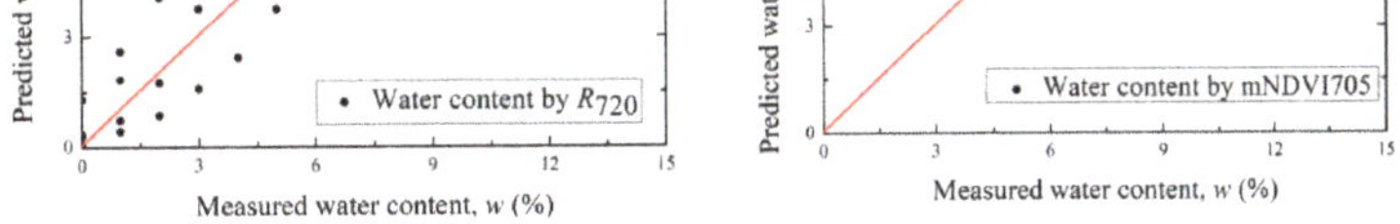

(a) (b)

Figure 14. *Cont.*

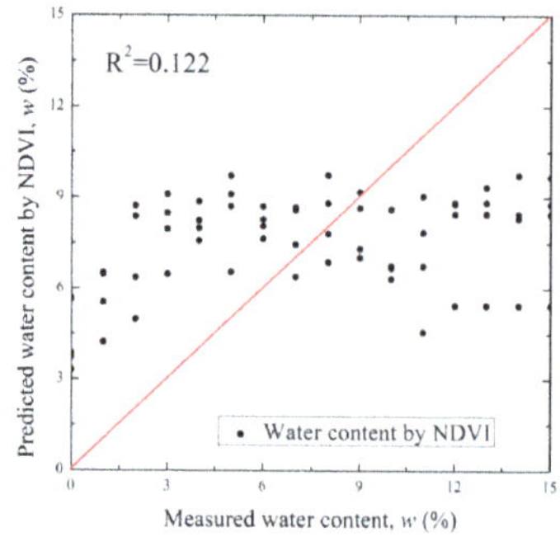

(c)

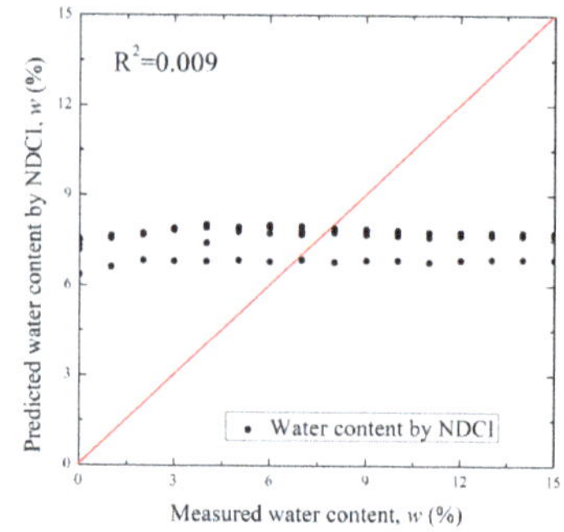

(d)

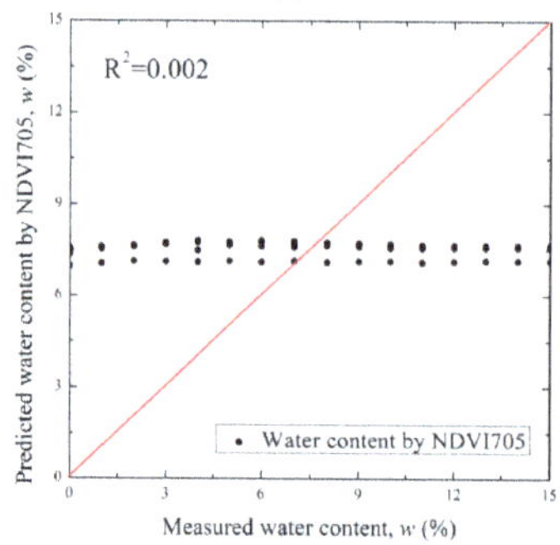

(e)

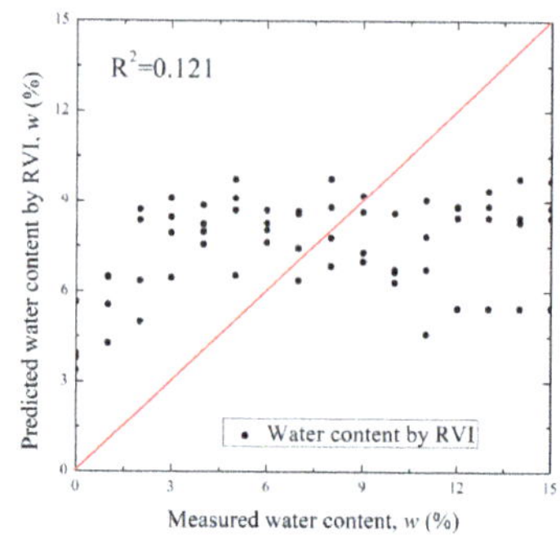

(f)

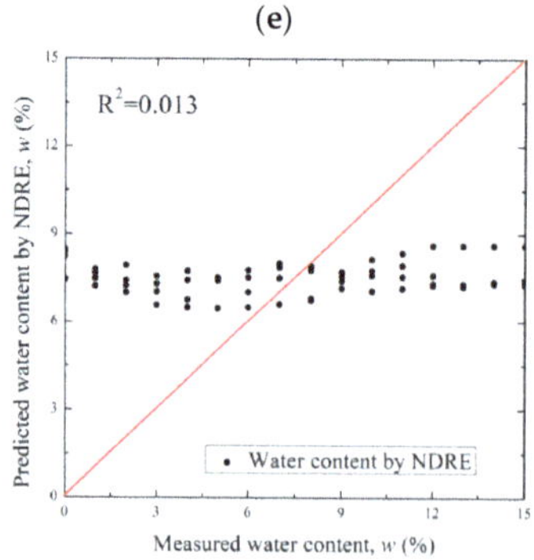

(g)

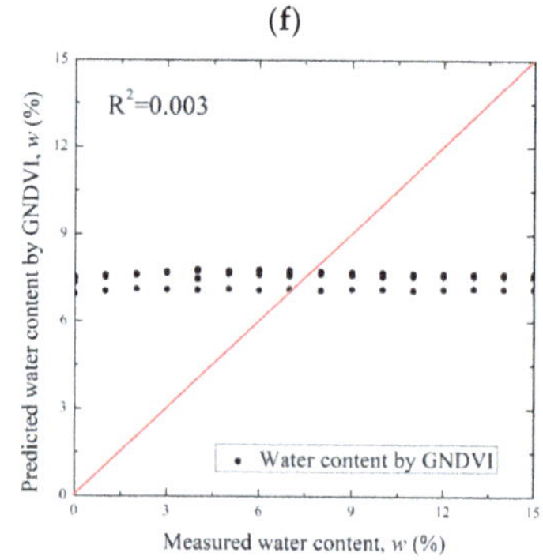

(h)

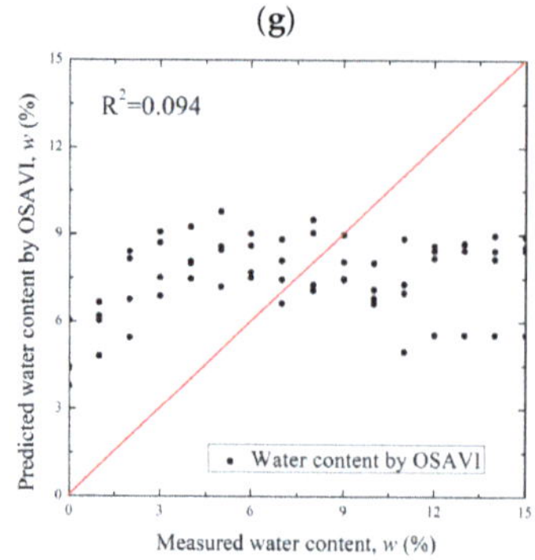

(i)

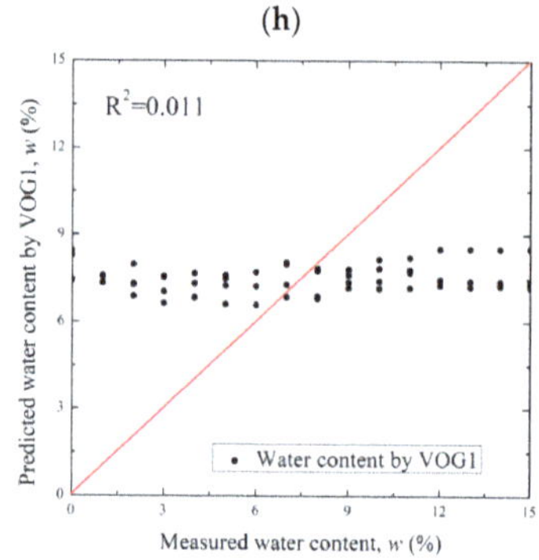

(j)

Figure 14. *Cont.*

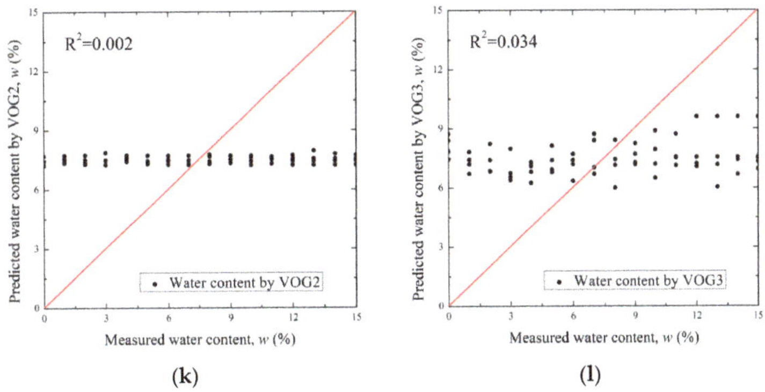

Figure 14. Comparison of predicted and measured water contents by (**a**) R_{720}, (**b**) mNDVI705, (**c**) NDVI, (**d**) NDCI, (**e**) NDVI705, (**f**) RVI, (**g**) NDRE, (**h**) GNDVI, (**i**) OSAVI, (**j**) VOG1, (**k**) VOG2, and (**l**) VOG3.

5. Conclusions

In this study, groundwater content was measured for determining road bearing capacity and for road quality control during road earthworks and pavement construction toward achieving "sustainable roads." The existing water content measurement method is the point measurement method; however, we aimed to acquire the water content of a wide area at once, necessitating the use of hyperspectral information. Hence, hyperspectral information was obtained through many indoor experiments, and a water content prediction formula was proposed. The conclusions drawn from this study are as follows.

1. In this study, sophisticated specimens were created by adding fine contents to standard sand, and hyperspectral information was obtained according to water content through precise laboratory tests. For hyperspectral information, a spectrum index was selected through various correlation analyses, and an equation to convert the spectrum index to water content was proposed.
2. The suitable wavelength for calculating the spectrum index was 600–880 nm, as determined through variability analysis based on the water content and fine contents. The variability analysis results showed that no difference existed in the results of the equation for water content prediction even when a single wavelength within the range was selected. When the integral value of reflectance was used at 600–880 nm, R^2 was rather low. This phenomenon was the result of the overlapping variability of wavelength and reflectance. Even when the R^2 of the corresponding index was measured, it was not appropriate as it increased the time for calculating the spectrum index.
3. The available equation for the prediction of the groundwater content is when the reflectance at a wavelength of 720 nm is applied to the exponential model. As a result of the linear regression analysis according to the measured and predicted water content, R^2 was measured to be the highest, which means that it is most suitable for representing the water content in the ground. In terms of spectral range, 720 nm is deep red light.
4. The correlation (R^2: 0.009–0.122) when the existing spectrum index for water content prediction was substituted into the hyperspectral information obtained in this study was measured to be very low. Even when the existing equation was substituted into the hyperspectral information obtained by Ge et al. [26], the R^2 ranged from 0.052 to 0.398, indicating that the reliability of the existing formula was low. Therefore, the R^2 (0.7067) of our proposed equation for water content prediction according to R_{720} was large and reliable. This is because the existing method calculated the water content in a linear line through a simple linear regression analysis of the spectrum index.
5. The disadvantage of this study is that the proposed equation was derived without going through an actual field test. Thus, in the field, errors may occur depending

on actual variables, such as weather, temperature, humidity, and the skill level of the drone operator. Therefore, it is necessary to test the accuracy and reliability of the equation derived from this study in the field, and the equation must be modified through additional data acquisition.

Author Contributions: Conceptualization, K.L., K.S.K., J.P. and G.H.; Data curation, K.L., J.P. and G.H.; Funding acquisition, K.S.K.; Formal analysis, K.L., K.S.K. and J.P.; Investigation, K.L., K.S.K., J.P. and G.H.; Methodology, G.H.; Project administration, J.P. and G.H.; Resources, K.S.K. and G.H.; Supervision, J.P.; Validation, K.L., J.P. and G.H.; Visualization, K.L.; Writing—original draft preparation, K.L. and K.S.K.; Writing—review and editing, K.S.K., J.P. and G.H. All authors have read and agreed to the published version of the manuscript.

Funding: This research was funded by the Ministry of Land, Korea Agency for Infrastructure Technology Advancement (KAIA), grant number 22SMIP-A157182-03.

Institutional Review Board Statement: Not applicable.

Informed Consent Statement: Not applicable.

Data Availability Statement: Not applicable.

Conflicts of Interest: The authors declare no conflict of interest.

References

1. Basu, D.; Misra, A.; Puppala, A.J. Sustainability and geotechnical engineering: Perspectives and review. *Can. Geotech. J.* **2015**, *52*, 96–113. [CrossRef]
2. Brown, R.L. *Building a Sustainable Society*; W.W. Norton: New York, NY, USA, 1981.
3. Brundtland, G.H. *Our Common Future: Report of the World Commission on Environment and Development*; Oxford University Press: Oxford, UK, 1987.
4. Socolow, R.; Andrews, C.; Berkhout, F.; Thomas, V. *Industrial Ecology and Global Change*; Cambridge University Press: Cambridge, UK, 1997; pp. 23–41.
5. Kibert, C.J. *Sustainable Construction*; John Wiley and Sons Inc.: Hoboken, NJ, USA, 2008.
6. Corriere, F.; Rizzo, A. Sustainability in road design: A methodological proposal for the drafting of guideline. *Procedia-Soc. Behav. Sci.* **2012**, *53*, 39–48. [CrossRef]
7. Muench, S.T.; Anderson, J.; Bevan, T. Greenroads: A Sustainability Rating System for Roadways. *Int. J. Pavement Res. Technol.* **2010**, *3*, 270–279.
8. Muench, S.T.; Anderson, J.L.; Hatfield, J.P.; Koester, J.R.; Söderlund, M.; Weiland, C. *Greenroads Manual v1. 5*; University of Washington: Washington, DC, USA, 2011.
9. Jeon, D.H.; Cho, J.Y.; Jhun, J.P.; Ahn, J.H.; Jeong, S.; Jeong, S.Y.; Kumar, A.; Ryu, C.H.; Hwang, W.; Park, H.; et al. A lever-type piezoelectric energy harvester with deformation-guiding mechanism for electric vehicle charging station on smart road. *Energy* **2021**, *218*, 119540. [CrossRef]
10. Kim, T.W.; Ryu, I.; Lee, H.; Jang, J.A. Dynamic Spatial Area Design for Transportation Management at Smart Road Lighting Platform System in Korea. In Proceedings of the 2021 International Conference on Information and Communication Technology Convergence (ICTC), Jeju, Korea, 20–22 October 2021.
11. Ma, Y.; Luan, Y.C.; Zhang, W.G.; Zhang, Y.Q. Numerical simulation of intelligent compaction for subgrade construction. *J. Cent. South Univ.* **2020**, *27*, 2173–2184. [CrossRef]
12. Ahmed, H.A. Electrical Resistivity Method for Water Content Characterization of Unsaturated Clay Soil. Doctoral Theses, Durham University, Durham, UK, 2014.
13. Roodposhti, H.R.; Hafizi, M.K.; Kermani, M.R.S.; Nik, M.R.G. Electrical resistivity method for water content and compaction evaluation, a laboratory test on construction material. *J. Appl. Geophys.* **2019**, *168*, 49–58. [CrossRef]
14. Abdullah, N.H.H.; Kuan, N.W.; Ibrahim, A.; Ismail, B.N.; Majid, M.R.A.; Ramli, R.; Mansor, N.S. Determination of soil water content using time domain reflectometer (TDR) for clayey soil. In Proceedings of the AIP Conference Proceedings, Maharashtra, India, 5–6 July 2018.
15. Wen, M.M.; Liu, G.; Horton, R.; Noborio, K. An in situ probe-spacing-correction thermo-TDR sensor to measure soil water content accurately. *Eur. J. Soil Sci.* **2018**, *69*, 1030–1034. [CrossRef]
16. Peng, W.; Lu, Y.; Xie, X.; Ren, T.; Horton, R. An improved thermo-TDR technique for monitoring soil thermal properties, water content, bulk density, and porosity. *Vadose Zone J.* **2019**, *18*, 1–9. [CrossRef]
17. Ercoli, M.; Di Matteo, L.; Pauselli, C.; Mancinelli, P.; Frapiccini, S.; Talegalli, L.; Cannata, A. Integrated GPR and laboratory water content measures of sandy soils: From laboratory to field scale. *Constr. Build. Mater.* **2018**, *159*, 734–744. [CrossRef]
18. Klotzsche, A.; Jonard, F.; Looms, M.C.; van der Kruk, J.; Huisman, J.A. Measuring soil water content with ground penetrating radar: A decade of progress. *Vadose Zone J.* **2018**, *17*, 1–9. [CrossRef]

19. Zhou, L.; Yu, D.; Wang, Z.; Wang, X. Soil water content estimation using high-frequency ground penetrating radar. *Water* **2019**, *11*, 1036. [CrossRef]
20. Eismann, M.T.; Stocker, A.D.; Nasrabadi, N.M. Automated hyperspectral cueing for civilian search and rescue. *Proc. IEEE* **2009**, *97*, 1031–1055. [CrossRef]
21. Van Der Meer, F.D. *Imaging Spectrometry-Basic Principles and Prospective Applications*; Kluwer Academic Publishers: Dordrecht, The Netherlands, 2003.
22. Bassani, C.; Cavalli, R.M.; Cavalcante, F.; Cuomo, V.; Palombo, A.; Pascucci, S.; Pignatti, S. Deterioration status of asbestos-cement roofing sheets assessed by analyzing hyperspectral data. *Remote Sens. Environ.* **2007**, *109*, 361–378. [CrossRef]
23. Smith, M.L.; Ollinger, S.V.; Martin, M.E.; Aber, J.D.; Hallett, R.A.; Goodale, C.L. Direct estimation of aboveground forest productivity through hyperspectral remote sensing of canopy nitrogen. *Ecol. Appl.* **2002**, *12*, 1286–1302. [CrossRef]
24. Kokaly, R.F.; Asner, G.P.; Ollinger, S.V.; Martin, M.E.; Wessman, C.A. Characterizing canopy biochemistry from imaging spectroscopy and its application to ecosystem studies. *Remote Sens. Environ.* **2009**, *113*, S78–S91. [CrossRef]
25. Zhang, F.; Zhou, G.S. Research progress on monitoring vegetation water content by using hyperspectral remote sensing. *Chin. J. Plant Ecol.* **2018**, *42*, 517.
26. Zhang, F.; Zhou, G. Estimation of vegetation water content using hyperspectral vegetation indices: A comparison of crop water indicators in response to water stress treatments for summer maize. *BMC Ecol.* **2019**, *19*, 1–12. [CrossRef]
27. Kovar, M.; Brestic, M.; Sytar, O.; Barek, V.; Hauptvogel, P.; Zivcak, M. Evaluation of hyperspectral reflectance parameters to assess the leaf water content in soybean. *Water* **2019**, *11*, 443. [CrossRef]
28. Prošek, J.; Gdulová, K.; Barták, V.; Vojar, J.; Solský, M.; Rocchini, D.; Moudrý, V. Integration of hyperspectral and LiDAR data for mapping small water bodies. *Int. J. Appl. Earth Obs. Geoinf.* **2020**, *92*, 102181. [CrossRef]
29. Guo, Y.; Bi, Q.; Li, Y.; Du, C.; Huang, J.; Chen, W.; Shi, L.; Ji, G. Sparse Representing Denoising of Hyperspectral Data for Water Color Remote Sensing. *Appl. Sci.* **2022**, *12*, 7501. [CrossRef]
30. Ge, X.; Wang, J.; Ding, J.; Cao, X.; Zhang, Z.; Liu, J.; Li, X. Combining UAV-based hyperspectral imagery and machine learning algorithms for soil moisture content monitoring. *PeerJ* **2019**, *6926*, 1–27. [CrossRef] [PubMed]
31. Jain, S.K.; Singh, V.P. *Water Resources Systems Planning and Management*; Elsevier: Amsterdam, The Netherland, 2003.
32. Lu, G.; Fei, B. Medical hyperspectral imaging: A review. *J. Biomed. Opt.* **2014**, *19*, 010901. [CrossRef] [PubMed]
33. Angel, Y.; Turner, D.; Parkes, S.; Malbeteau, Y.; Lucieer, A.; McCabe, M.F. Automated georectification and mosaicking of UAV-based hyperspectral imagery from push-broom sensors. *Remote Sens.* **2020**, *12*, 34. [CrossRef]
34. Jurado, J.M.; Pádua, L.; Hruška, J.; Feito, F.R.; Sousa, J.J. An Efficient Method for Generating UAV-Based Hyperspectral Mosaics Using Push-Broom Sensors. *IEEE J. Sel. Top. Appl. Earth Obs. Remote Sens.* **2021**, *14*, 6515–6531.
35. Yi, L.; Chen, J.M.; Zhang, G.; Xu, X.; Ming, X.; Guo, W. Seamless Mosaicking of UAV-Based Push-Broom Hyperspectral Images for Environment Monitoring. *Remote Sens.* **2021**, *13*, 4720. [CrossRef]
36. Ortega, S.; Guerra, R.; Diaz, M.; Fabelo, H.; López, S.; Callico, G.M.; Sarmiento, R. Hyperspectral push-broom microscope development and characterization. *IEEE Access* **2019**, *7*, 122473–122491. [CrossRef]
37. *ISO 679*; Cement-Test Methods-Determination of Strength. International Organization for Standardization: Geneva, Switzerland, 2009.
38. *ASTM D422*; Standard Test Method for Particle-Size Analysis of Soils. American Society for Testing of Materials: West Conshohocken, PA, USA, 2016.
39. *ASTM D2487*; Standard Practice for Classification of Soils for Engineering Purposes (Unified Soil Classification System). American Society for Testing of Materials: West Conshohocken, PA, USA, 2017.
40. *ASTM D698*; Standard Test Methods for Laboratory Compaction Characteristics of Soil Using Standard Effort (12,400 ft-lbf/ft3 (600 kN-m/m^3)). American Society for Testing of Materials: West Conshohocken, PA, USA, 2017.
41. Liang, L.; Di, L.; Zhang, L.; Deng, M.; Qin, Z.; Zhao, S.; Lin, H. Estimation of crop LAI using hyperspectral vegetation indices and a hybrid inversion method. *Remote Sens. Environ.* **2015**, *165*, 123–134. [CrossRef]
42. Haboudane, D.; Miller, J.R.; Pattey, E.; Zarco-Tejada, P.J.; Strachan, I.B. Hyperspectral vegetation indices and novel algorithms for predicting green LAI of crop canopies: Modeling and validation in the context of precision agriculture. *Remote Sens. Environ.* **2004**, *90*, 337–352. [CrossRef]
43. Sims, D.A.; Gamon, J.A. Relationships between leaf pigment content and spectral reflectance across a wide range of species, leaf structures and developmental stages. *Remote Sens. Environ.* **2002**, *81*, 337–354. [CrossRef]
44. Broge, N.H.; Leblanc, E. Comparing prediction power and stability of broadband and hyperspectral vegetation indices for estimation of green leaf area index and canopy chlorophyll density. *Remote Sens. Environ.* **2001**, *76*, 156–172. [CrossRef]
45. Yao, X.; Wang, N.; Liu, Y.; Cheng, T.; Tian, Y.; Chen, Q.; Zhu, Y. Estimation of wheat LAI at middle to high levels using unmanned aerial vehicle narrowband multispectral imagery. *Remote Sens.* **2017**, *9*, 1304. [CrossRef]
46. Vogelmann, J.E.; Rock, B.N.; Moss, D.M. Red edge spectral measurements from sugar maple leaves. *Int. J. Remote Sens.* **1993**, *14*, 1563–1575. [CrossRef]

sustainability

MDPI

Article

Assessment of the Blasting Efficiency of a Long and Large-Diameter Uncharged Hole Boring Method in Tunnel Blasting Using 3D Numerical Analysis

Min-Seong Kim [1], Chang-Yong Kim [1], Myung-Kyu Song [2] and Sean Seungwon Lee [2,*]

[1] Department of Geotechnical Engineering Research, Korea Institute of Civil Engineering and Building Technology, 283 Goyang-daero, Goyang-si 10223, Korea
[2] Department of Earth Resources and Environmental Engineering, Hanyang University, 222 Wangsimni-ro, Seoul 04763, Korea
* Correspondence: seanlee@hanyang.ac.kr

Abstract: Cut blasting is one of the most essential processes to reduce blast-induced vibration in tunnel blasting. The long and large-diameter uncharged hole boring (LLB) method is an example of one of the cut blasting methods, which utilizes large-diameter uncharged holes drilled in the tunnel face. In this study, blasting simulations were performed to analyze its blasting mechanism, and the LLB method and the traditional burn-cut method were simulated to compare their blasting efficiency. A 3D numerical analysis using LS-DYNA code, a highly non-linear transient dynamic finite element analysis using explicit time integration, was used to simulate the blasting process, and a Johnson–Holmquist constitutive material model, which is optimal for simulating brittle materials under dynamic conditions, was used to simulate the rock behavior under blasting. The modified LLB method showed a 3.75-fold increase in the advance per round compared to the burn-cut method, due to the increased formation of long and large-diameter uncharged holes compared to blast holes. This modified LLB method used 30% less explosives, so its failure range was approximately 1.25 times less than that of the burn-cut method, but its advance was approximately 4 times larger than the burn-cut method, which was similar to the original LLB method. This confirmed that the modified LLB method is significantly more efficient in terms of increased blasting efficiency (particularly the advance per round) as well as reduced blast-induced vibration, compared to the traditional cut blasting method.

Keywords: tunnel excavation; cut blasting method; LLB method; uncharged hole; numerical simulation

Citation: Kim, M.-S.; Kim, C.-Y.; Song, M.-K.; Lee, S.S. Assessment of the Blasting Efficiency of a Long and Large-Diameter Uncharged Hole Boring Method in Tunnel Blasting Using 3D Numerical Analysis. *Sustainability* **2022**, *14*, 13347. https://doi.org/10.3390/su142013347

Academic Editors: Jian Zhou, Mahdi Hasanipanah and Danial Jahed Armaghani

Received: 14 September 2022
Accepted: 10 October 2022
Published: 17 October 2022

1. Introduction

A drill and blast method is the most common rock excavation method in mining and civil engineering. However, this method causes environmental pollution and hazards such as noise, vibration, and flying rocks generated by the blasting, and blast-induced vibration can damage nearby structures in urban areas, when it exceeds specific values [1,2]. In tunnel construction, the advance rate is one of the key factors because it is directly related to the overall construction period and influences the economic feasibility of a project. Therefore, the ultimate goal of blast excavation in the new Austrian tunneling method (NATM) is to maximize the advance per round, while meeting the allowable criteria for blast-induced vibration. It is well known that blast-induced vibration is the most significant factor that affects the surrounding structures [3]. In the rock fragmentation caused by blasting, it has been found that only 20–30% of the energy is used in the breakage of the surrounding rocks, while the residual energies are dissipated in the forms of vibration, noise, and flying rocks [4,5]; thus, the current blasting techniques have inevitable limitations [6]. When the vibration exceeds a critical value, internal cracks in the rock mass can be induced, potentially harming nearby structures [7]. Given the recent rise in environmental concerns, complaints about blast-induced vibration and noise continue to increase, and regulations are becoming commensurately stricter [8].

Several parameters affect blast-induced vibration [9], and the free face is one of the most efficient factors in maximizing the blasting efficiency, as well as reducing vibration. Unlike open-pit mines, tunnel structures generally have only one free face, so the cut blasting method is necessary to create artificial additional free faces to improve the blasting efficiency. V-cuts and burn-cuts, which are the traditional cut blasting methods, are widely used in tunnel blasting. Using the traditional cut method, more explosives are needed to increase the advance, but there are limits to increasing the amount of explosives, while keeping to the allowable limit of vibration.

A long and large-diameter uncharged hole boring (LLB) method is an advanced cut blasting method that involves boring large-diameter uncharged holes. This method has been reported to have significantly less blast-induced vibration than traditional cut methods [10]. The high-performance LLB machine generally drills 50 m with a 382-mm hammer bit at a time, considering the overall working process and drilling time. Thus, this method has the great advantage of creating large-diameter uncharged holes (free faces) that are deeper than regular blast holes, thus significantly increasing the advance per round compared to the traditional cut method [11].

However, the cut methods in tunnel blasting that utilize large-diameter uncharged holes (over 350 mm) have not yet been fully analyzed in terms of failure mechanisms and blasting efficiency. In addition, in some tunnel construction sites where the LLB method was used, the advance per round was similar to, or even less than, the traditional cut methods. In the actual tunnel construction site, if explosive material detonates inside the rock mass, there is a limit to investigating the failure mechanism of the LLB method because an instantaneous explosion reaction occurs. To investigate the failure mechanism of blasting more accurately, experimental studies should be conducted, but several risks and limitations are involved because they are very expensive and time-consuming [12]. Instead, numerical approaches have been widely used to overcome the many limitations of experimental research [13–15].

In this study, a series of numerical analyses were carried out to investigate the failure mechanism of tunnel blasting with the LLB method. The LS-DYNA software, which is a finite element software that can handle dynamic and non-linear problems, was used to simulate the tunnel blasting. The traditional burn-cut method and the LLB method, as employed to create uncharged holes for reducing blast-induced vibration, were modeled, to compare their blasting mechanisms and efficiency, especially the advance per round. Additional analysis was performed to compare the blasting efficiency when a reduced amount of explosives was used with the LLB method.

2. Long and Large-Diameter Uncharged Hole Boring Method

2.1. Introduction to the LLB Method

The LLB method is an advanced cut blasting method to minimize the vibration generated by blasting in a NATM tunnel. This method creates long and large-diameter (382-mm) uncharged holes at a cut area using a high-performance boring machine, as shown in Figure 1. The LLB hole serves to release the confining stresses of the rock mass before blasting. When an explosive detonates in the blast hole, the adjacent uncharged hole can not only provide a larger space for moving breaking rocks, but can also change the stress distribution in the rocks around the uncharged hole, which is called the "uncharged hole effect" [16].

One of the outstanding characteristics of this method is that it typically drills 50 m at a time, and can efficiently drill up to 65 m without any problems. Thus, additional free faces are formed beyond the depth of the blast hole, which theoretically has the advantage of increasing the advance per round. In addition, one to three holes are generally drilled in the LLB method, depending on what is applicable in the site conditions. Unlike the conventional cut blasting methods, such as V-cut and burn-cut, this method is applicable for both weak and hard rock masses. The field application of the LLB machine and a tunnel face with completed drilling are shown in Figure 2 [10].

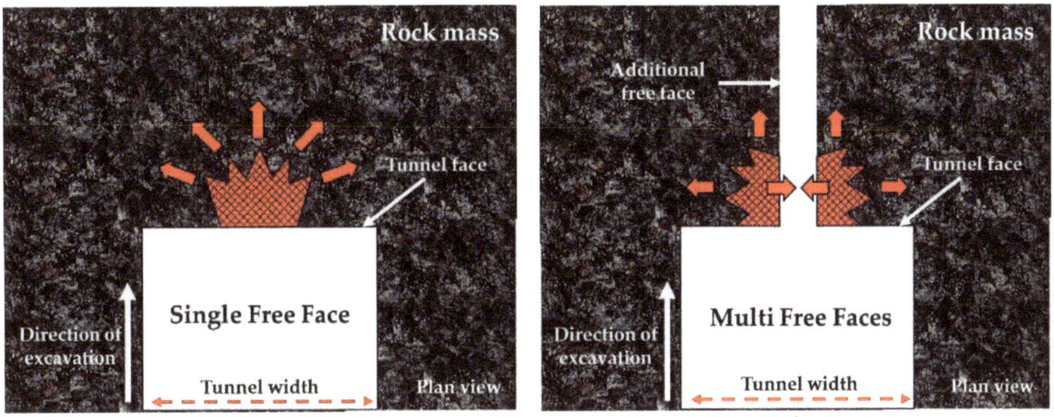

(a) (b)

Figure 1. Comparison of general blasting and controlled blasting using the LLB method. (**a**) General blasting; (**b**) controlled blasting using the LLB method.

(a) (b)

Figure 2. Images of field application of the LLB machine and a tunnel face with completed drilling. (**a**) Field application of the LLB machine; (**b**) tunnel face with completed drilling.

2.2. A Mechanism of Tensile Fracturing by Long and Large-Diameter Uncharged Holes

When an explosive detonates in a blast hole, gaseous detonation products fill the blast hole with high pressure and temperature. This high-pressure detonation gas is immediately applied to the surface of a blast hole and generates radial compressive stress, which is generally greater than the strength of the rock [17,18]. The compressive stress wave generated by blasting turns into a tensile stress wave when it reaches a free face, and the reflected tensile wave returns through the rock mass [19–21]. This behavior is generally called the Hopkinson effect, and the principle of this theory is illustrated in Figure 3 [22].

It is well known that the dynamic tensile strength of rock is much lower than the dynamic compressive strength [23,24]. If the tension stress waves exceed the dynamic tensile strength of the rock, cracks will be generated gradually, and this tensile stress leads to greater damage to the rock mass in the vicinity of the free face [16]. Therefore, when a propagated compressive stress wave from the detonated explosive reaches an uncharged hole, it turns into a tensile stress wave; thus, the rock is effectively crushed around the uncharged hole. A tunnel structure generally has only one free face, which is the tunnel face. To effectively blast in tunneling, adding free faces is the most vital factor for initiating or maximizing the Hopkinson effect.

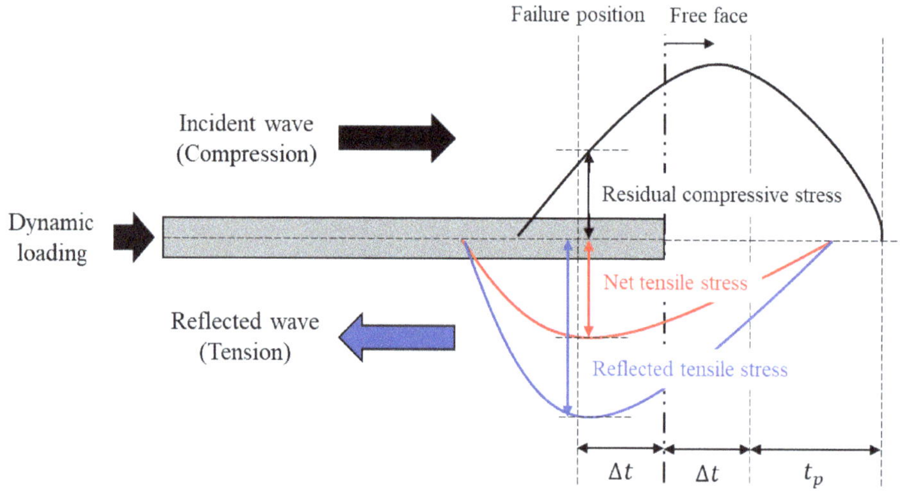

Figure 3. Principle of the Hopkinson effect.

3. Numerical Analysis

3.1. Analysis Model

This study used the LS-DYNA software, which is widely used for blasting simulations, to simulate dynamic and non-linear problems. The arbitrary Lagrangian–Eulerian (ALE) formulation was employed to investigate blast-induced pressure and its interaction with various materials [25–27]. The ALE formulation contains both Lagrangian and Eulerian formulations to utilize the advantages of each approach [28]; it allows the mesh to move independently from the material flow, and each element in the mesh can contain mixtures of different materials [29].

The burn-cut method and the LLB method were modeled to investigate their effects on the blasting efficiency, including the advance per round in tunnel blasting. Both analysis models included the rock, explosives, and stemming in the charge hole, as well as uncharged holes. The rock was modeled using the Lagrangian method, with a geometry of 1500 mm × 1500 mm × 1500 mm to focus on the cut area, as shown in Figure 4. The diameter of the blast holes was 50 mm and they contained 300 mm of explosives and 700 mm of stemming, which were modeled with the ALE method. The diameters of the uncharged holes for the burn-cut and the LLB method were 102 mm and 382 mm, respectively. The spacing between the blast hole and the uncharged hole for the burn-cut was 200–300 mm and 400 mm for the LLB method. The geometries were modeled based on actual design cases from subway tunnel construction sites in South Korea. In both cases, the explosives were set to detonate at the same time, and the boundary conditions were set as non-reflecting boundaries for all sides, except the tunnel face.

3.2. JH-2 Constitutive Model for Rock Material

A Johnson–Holmquist (JH-1) constitutive material model was initially proposed for studying the behavior of metals such as copper and nickel under large strain, high strain rate, and high-pressure conditions. Based on the JH-1 model, an improved material model, named as the JH-2 constitutive model, was suggested to describe the mechanical behavior under dynamic conditions, considering the softening property of brittle materials. The JH-2 model is widely used in LS-DYNA to simulate the dynamic behavior (blasting) of rocks. This improved model represented the strength and damage of material as functions of the representative variables and the damage evolution within the material was considered. In addition, this model considers the pressure, strain-rate dependent strength, damage, and fracture of materials [30–32]. Granite is one of the most brittle materials and a widely distributed underground rock material in South Korea. Therefore, granite properties were used here to simulate the tunnel blasting, as shown in Table 1 [33]. A general overview of the model is illustrated in Figure 5.

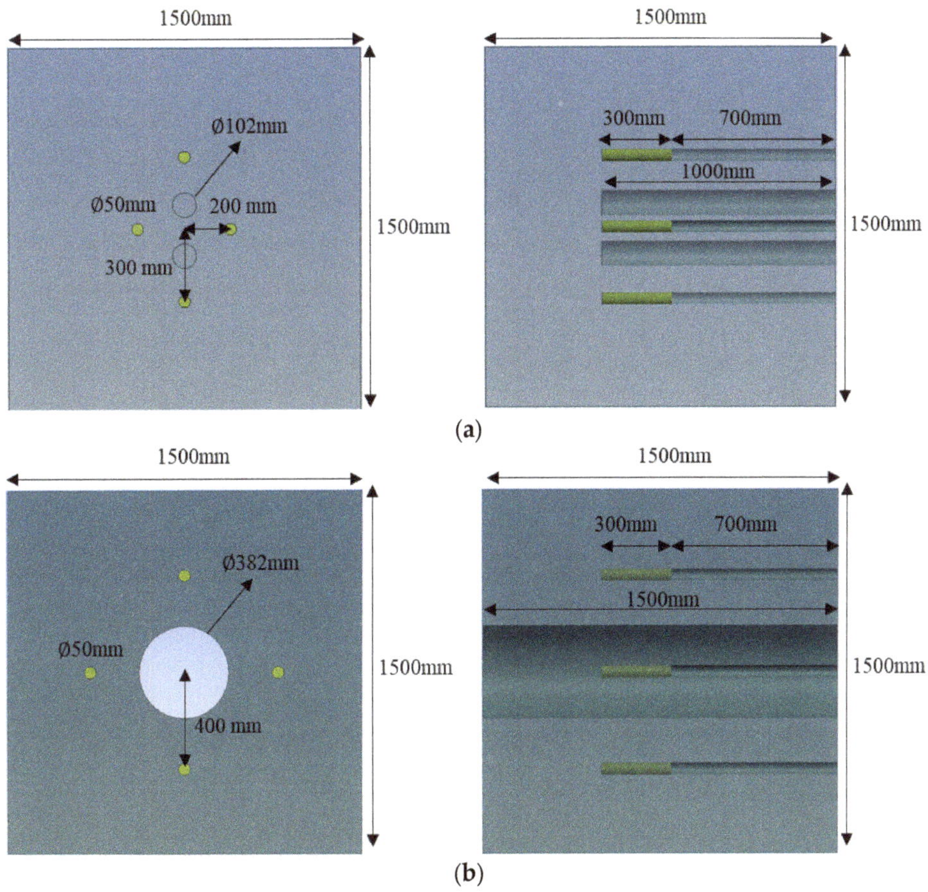

Figure 4. Geometry of numerical simulations for the burn-cut and the LLB methods. (**a**) Burn-cut method; (**b**) LLB method.

Table 1. Input parameter for JH-2 model in LS-DYNA.

Parameter	Value	Parameter	Value
Density (kg/m^3)	2560	Maximum normalized fractured strength	0.160
Shear modulus (GPa)	11.606	Hugoniot elastic limit (GPa)	4.500
Intact normalized strength parameter A	1.248	Pressure component at the Hugoniot elastic limit (GPa)	2.930
Fractured normalized strength parameter B	0.680	Fraction of elastic energy loss	1.000
Strength parameter C	0.005	Parameter for a plastic strain to fracture D_1	0.008
Fractured strength parameter M	0.830	Parameter for a plastic strain to fracture D_2	0.435
Intact strength parameter N	0.676	First pressure coefficient K1 (GPa)	10.720
Reference strain rate	1.000	Second pressure coefficient K2 (GPa)	−386
Maximum tensile strength (GPa)	0.015	Elastic constant K3 (GPa)	12,800

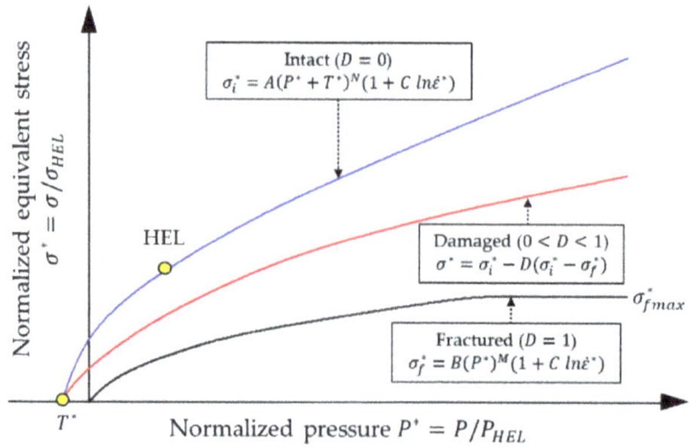

Figure 5. Strength model for the JH-2 constitutive model.

The JH-2 model includes the following three types of strength: intact state, damaged state, and fractured state, as shown in Figure 5. The three states have their own strength equations, which present the relationship between the normalized equivalent stress and the normalized pressure, expressed as

$$\sigma^* = \sigma_i^* - D(\sigma_i^* - \sigma_f^*) = \sigma/\sigma_{HEL} \tag{1}$$

where σ_i^* is the normalized intact equivalent stress, σ_f^* is the normalized fracture stress, D is the damage coefficient ($0 \leq D \leq 1$), σ is the actual equivalent stress calculated by the von Mises stress formula, and σ_{HEL} is the equivalent stress at the Hugoniot elastic limit (HEL). The HEL stands for the net compressive stress at which a one-dimensional shock wave with uniaxial strain exceeds the elastic limit of the material. The brittle material begins to soften when the damage begins to accumulate, and this process can be expressed by Equation (1). Although the softening does not continue when the material is completely damaged ($D = 1$), it allows for gradual softening of the brittle material under increasing plastic strain. The normalized intact strength and fracture strength are given by

$$\sigma_i^* = A(P^* + T^*)^N \left(1 + C \times \ln \dot{\varepsilon}^*\right) \tag{2}$$

$$\sigma_f^* = B(P^*)^M \left(1 + C \times \ln \dot{\varepsilon}^*\right) \tag{3}$$

where A, B, C, M, and N are all constants. The normalized pressure is $P^* = P/P_{HEL}$, where P is the actual pressure and P^* is the pressure at the HEL. The normalized maximum tensile hydrostatic pressure is $T^* = T/P_{HEL}$, where T is the maximum tensile hydrostatic pressure. The strain is $\dot{\varepsilon}^* = \dot{\varepsilon}/\dot{\varepsilon}_0$, where $\dot{\varepsilon}$ is the actual strain; $\dot{\varepsilon}_0 = 1\ \text{s}^{-1}$ is the reference strain rate.

The damage is mainly accumulated due to the generation of fractures, and the damage graph is shown in Figure 6a. The damage in the JH-2 model can be expressed as follows:

$$D = \sum \Delta\varepsilon^p/\varepsilon_f^p = \sum \Delta\varepsilon^p/[D_1(P^* + T^*)^{D_2}] \tag{4}$$

where $\Delta\varepsilon^p$ is the plastic strain during a cycle of integration, ε_f^p is the plastic strain to the fracture under constant pressure P, and D_1 and D_2 are the damage factors for ε_f^p. The equation of state (EOS) for the JH-2 constitutive model, as shown in Figure 6b, presents the relationship between hydrostatic pressure and volumetric strain, which consists of a pure elastic stage and a plastic damage stage. The hydrostatic pressure before the fracture and after the damage starts to accumulate can be expressed as follows:

$$P = K_1\mu + K_2\mu^2 + K_3\mu^3 \cdots (D = 0) \tag{5}$$

$$P = K_1\mu + K_2\mu^2 + K_3\mu^3 + \Delta P \cdots (0 < D < 1) \quad (6)$$

where K_1, K_2, and K_3 are constants (K_1 is the bulk modulus), and $\mu = \rho/\rho_0 - 1$ for the current density ρ and the initial density ρ_0. When the fracture occurs in the material, which is caused by the bulking energy, the incremental pressure ΔP is added. In this analysis model, erosion was set to occur when the tensile stress exceeds the maximum tensile strength of the rock mass, using blasting to simulate the rock failure.

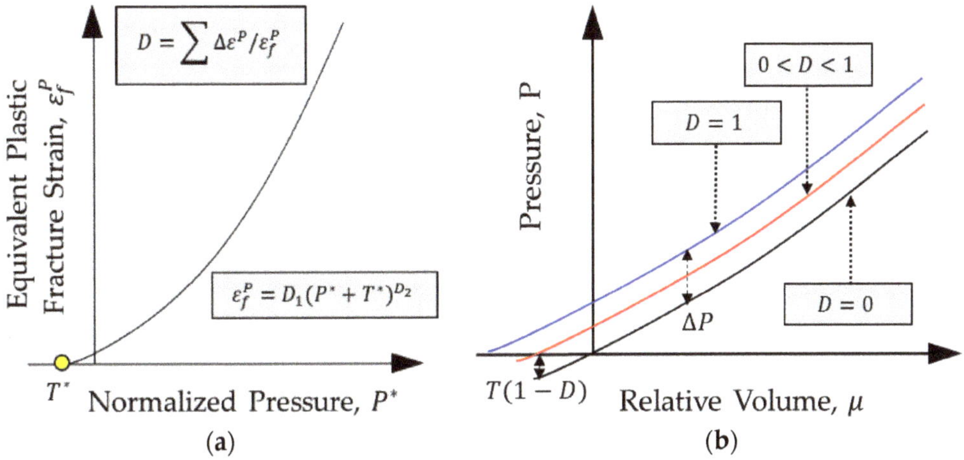

Figure 6. Damage and EOS models for the JH-2 constitutive model. (**a**) Damage model; (**b**) EOS model.

3.3. Explosive and Stemming Material Models

To model the pressure generated by the expansion of the detonation product, a Jones-Wilkins–Lee (JWL) equation of state (EOS) was used [34]. The JWL EOS is widely used in explosives modeling for describing the relationship of pressure–volume–energy for detonation products. In this study, JWL EOS was used to simulate the blasting process, and this equation can be expressed as follows:

$$P = A\left(1 - \frac{\omega}{R_1 V}\right)^{-R_1 V} + B\left(1 - \frac{\omega}{R_2 V}\right)^{-R_2 V} + \frac{\omega}{V}E_0 \quad (7)$$

where P is pressure, E and V are the detonation energy per unit volume and the relative volume, and A, B, R_1, R_2, and ω are the EOS coefficients, respectively. A high explosive burn (HEB) material card was used to simulate the detonation of the explosives. The input values for JWL EOS and HEB for an emulsion material are provided in Table 2 [35].

Table 2. Input parameter for JWL EOS in LS-DYNA.

JWL	Parameter	A (GPa)	B (GPa)	R_1	R_2	ω	E_0 (GPa/m^3/m^3)	V_0 (m^3/m^3)
	Value	276	8.44	5.215	2.112	0.501	3.868	1.0
HEB	Parameter	RO (kg/m^3)			D (m/s)		P_{cj} (GPa)	
	Value	1180			5122		9.530	

where RO, D, and P_{cj} are the density, detonation velocity, and Chapman–Jouguet pressure, respectively.

Stemming is a material used to prevent the release of detonation gases by filling the remaining areas in the blast hole, and it helps the pressure generated by the explosives to crush the rock efficiently. Here, the stemming material was modeled using the FHWA_SOIL material model developed by the U.S. Federal Highway Administration (FHWA) [36]. This model is an isotropic material with damage, and it is effective at modeling the behavior of soil under the consideration of strain-softening, kinematic hardening, strain rate effects, element deletion, excess pore water effects, and stability with no soil confinement; the input parameters are listed in Table 3 [37–39].

Table 3. Input parameter for FHWA_SOIL model in LS-DYNA.

Parameter	Value	Parameter	Value
Density (kg/m^3)	2350	Eccentricity parameter	0.700
Specific gravity	2.650	Moisture content	6.200
Density of water (kg/m^3)	1000	Skeleton bulk modulus (MPa)	0.153
Viscoplasticity parameter V_n	1.100	Minimum internal friction angle (radians)	0.063
Viscoplasticity parameter γ_r	0.0	Volumetric strain at initial damage threshold	0.001
Maximum number of plasticity iterations	10.00	Void formation energy	10.00
Bulk modulus (MPa)	15.30	Strain hardening, percent of φ_{max} where non-linear effects start	10.00
Shear modulus (MPa)	19.50	Pore water effects on bulk modulus PWD1	0.0
Peak shear strength angle (radians)	0.420	Pore water effects on effective pressure PWD2	0.0
Cohesion (MPa)	0.011	Strain hardening, amount of non-linear effects	10.00

The uncharged holes were modeled using the NULL material card, and the void was modeled using the LINEAR_POLYNOMIAL EOS material card, which is given by

$$P = C_0 + C_1\mu + C_2\mu^2 + C_3\mu^3 + \left(C_4 + C_5\mu + C_6\mu^2\right)E \tag{8}$$

where C_0, C_1, C_2, C_3, C_4, C_5, and C_6 are constants, and $\mu = \rho/\rho_0 - 1$, which is the ratio of the current density to the initial density. The detailed parameters for air are given in Table 4.

Table 4. Material model and EOS parameter for air.

ρ (kg/m^3)	C_0	C_1	C_2	C_3	C_4	C_5	C_6	E_0 (MPa)
1.29	0.0	0.0	0.0	0.0	0.4	0.4	0.0	0.25

3.4. Results of Numerical Analyses

Figure 7 displays the results of numerical analyses for the burn-cut and LLB methods over time, and shows side views of the analysis models in the wireframe element mode for observing the state of fracturing around the blast holes inside the rock mass. The number of elements of the burn-cut and LLB methods was 6,800,190 and 6,816,395, respectively.

When the explosives detonate after 0 s, the high-temperature and high-pressure gases are generated by the four blast holes; thus, compressive fractures are generated near the blast holes, and the explosion energy is propagated in the form of a compressive stress wave to the rock mass [40,41]. The propagated compressive stress wave reaches the uncharged holes before it reaches the tunnel face because of the shorter burden. The compressive stress wave first reaches the surface of the uncharged holes, and then is reflected as a tensile wave that begins to crush the rock around the uncharged holes at 0.2 ms in both cases. At 0.2 ms, in the burn-cut method, the spacing between the explosives and the uncharged hole is shorter than that of the LLB method; thus, the crushed zone around the uncharged hole was wider than in the LLB method. Then, the crushed area near the uncharged holes gradually increases. At the same time, the continuously propagated stress wave finally reaches the tunnel face and the rock is crushed, due to the reflected tensile stress at 0.8 and 1.0 ms, respectively. The crushed zones continuously increase due to the high pressure and the reflected tensile stress generated by the uncharged holes and the tunnel face.

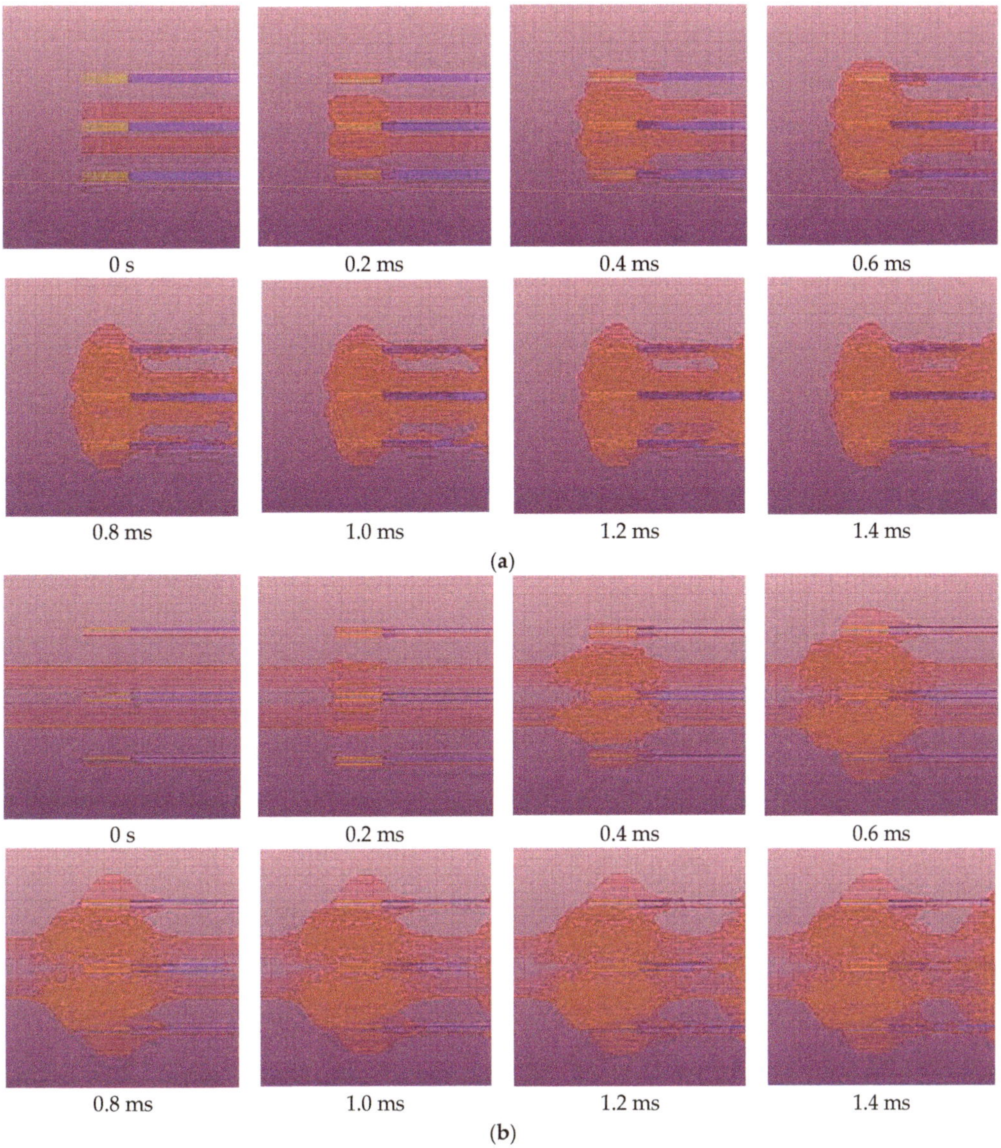

Figure 7. Results of the numerical analyses of the burn-cut and LLB methods. (**a**) Burn-cut method; (**b**) LLB method.

4. Discussion

4.1. Effect of Increasing the Advance Rate in the Burn-Cut and LLB Methods

Figure 8 summarizes the numerical analysis results at the last step (at 1.4 ms) to investigate the influence of the advance rate in the burn-cut and LLB methods. In the case of the Burn-cut method, the rock was crushed up to 0.09 m from the end line of the explosives, whereas the rock was crushed up to 0.37 m in the LLB method, which was more than four times the advance rate. It is believed that the large-diameter uncharged hole that was longer than the ordinary blast holes contributed significantly to the generation of more extensive tensile failure. In particular, in the burn-cut method, the blast-induced stress was used more to crush rocks between the blast holes and the tunnel face than to increase the advance. On the other hand, blast-induced stress in the LLB method was used more to increase the advance. Therefore, the long and large-diameter uncharged hole is believed to contribute greatly to increasing the advance.

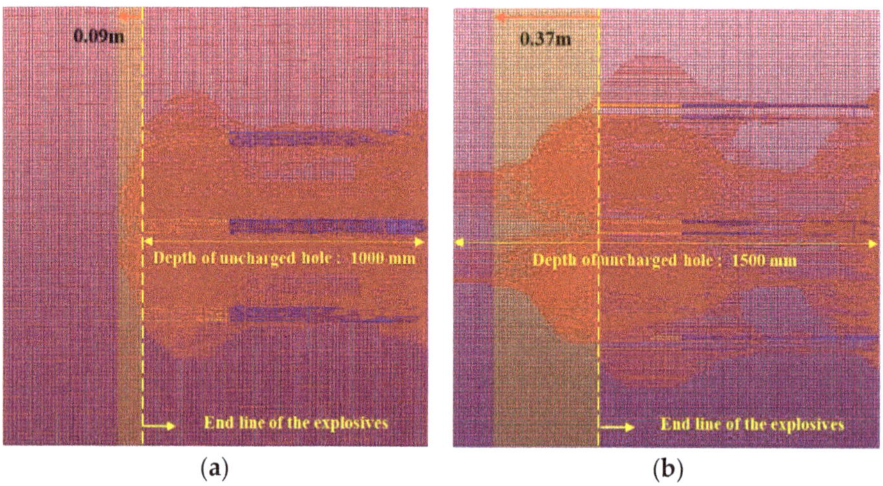

Figure 8. Comparison of the advance between the burn-cut and LLB methods. (**a**) Burn-cut method; (**b**) LLB method.

When explosives detonate near the cut area in the first stage, the long and large-diameter uncharged hole provides a larger space for crushed rock to move into, thus reducing the confining pressures of the surrounding rocks. Efficient moving of these crushed rocks creates an additional free face and reduces interference between the crushed rocks in the second stage. Considering that the entire blasting process uses the sequential blasting technique, which has a detonating delay time in the actual blasting, this method is significantly beneficial in that the long and large-diameter uncharged hole provides a larger space to move the crushed rocks into. In addition, the formed long and large-diameter uncharged hole before blasting provides an opportunity to reduce the amount of explosives; thus, the advance rate can be increased and the blast-induced vibration can be reduced.

4.2. *Effect of Increasing the Advance in the Burn-Cut and LLB Methods*

Additional analysis of the LLB method was carried out to investigate the blasting efficiency with the large-diameter uncharged hole with the same depth as the blast holes (a shorter depth than in the original LLB method). Figure 9 summarizes the results of the numerical analysis for the original LLB method and the modified LLB with a shorter uncharged hole. The number of elements of the modified LLB method was 6,719,248. The results show that the advance in the modified LLB method with its shorter uncharged hole was 0.11 m, which was approximately 3.36 times less than in the original LLB method. However, the advance in the modified LLB method was approximately 1.22 times longer than in the burn-cut method, as shown in Figure 8a. This suggests that the advance increased due to the long and large-diameter uncharged hole that caused a more extensive range of tensile failure compared to the burn-cut method. Given that the wider and longer uncharged hole significantly contributed to increasing the blasting efficiency as well as the advance, the LLB method has pronounced advantages compared to the traditional burn-cut method.

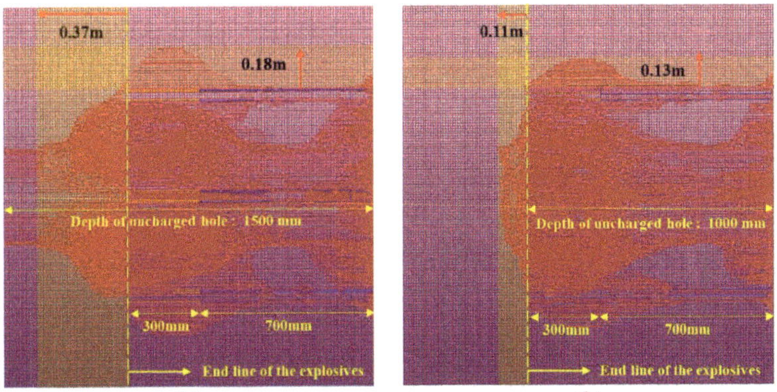

(a) (b)

Figure 9. Comparison of blasting efficiency depending on the length of the large-diameter uncharged hole. (**a**) Original LLB method; (**b**) modified LLB method with reduced uncharged hole length.

4.3. Effect of Reducing the Amount of Explosives

Additional numerical analysis was performed to estimate the effect of reducing the amount of explosives when applying the LLB method. The total length of the explosives was reduced by 30%, from 300 to 210 mm, and the results are illustrated in Figure 10. The number of elements of the modified LLB method was 6,801,550. The crushed zone around the blast hole was about 2.25 times smaller than with the original LLB design, but the advance was similar, 0.37 mm compared to 0.36 mm with the original LLB design. Although the overall failure range decreased as the amount of explosives was decreased, the tensile failure caused by the long and large-diameter uncharged hole had a significant influence. Therefore, in sequential blasting, the modified LLB method with a reduced amount of explosives in the cut area can be expected to have similar blasting efficiency as the original LLB method.

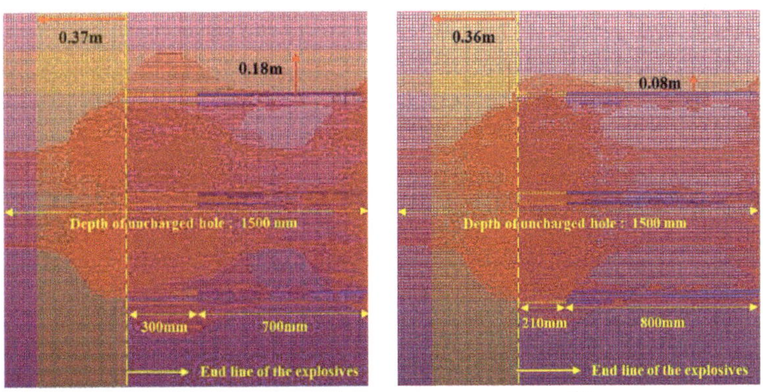

(a) (b)

Figure 10. Comparison of failure zones depending on the amount of explosives in the LLB methods. (**a**) Total explosives length of 300 mm (original LLB method); (**b**) total explosives length of 210 mm (modified LLB method).

Figure 11 shows the numerical analysis results of blasting with the burn-cut method and the modified LLB method with a reduced amount of explosives. As mentioned above, although the total influence range near the explosives was decreased due to the reduction in the amount of explosives in the modified LLB method, its advance was four times longer than that of the burn-cut method, as shown in Figure 11. Even in terms of blasting design, it is estimated that it is desirable to reduce the amount of explosives compared to the existing design, due to the large-diameter uncharged hole formed before blasting. Thus, the modified LLB method showed pronounced advantages in reducing blast-induced vibration by reducing the amount of explosives in the cut area, as well as increasing the advance.

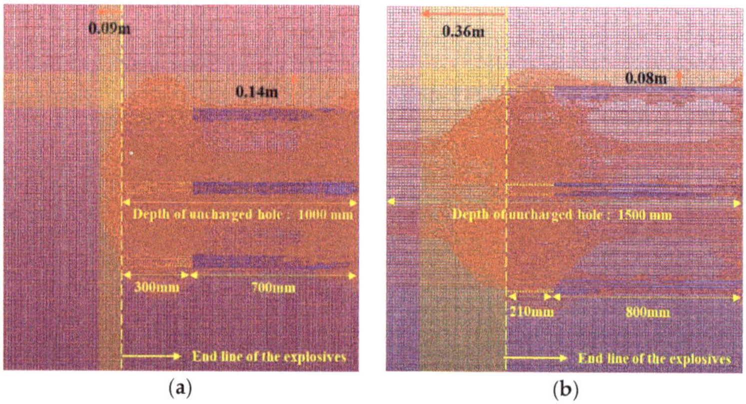

Figure 11. Comparison of range of failure in the burn-cut and the modified LLB methods. (**a**) Burn-cut method; (**b**) design of reduced explosives in the modified LLB method.

4.4. Comparison of Each Analysis Model using Eroding Elements

Figure 12 presents the total deleted eroding volume by detonation in the original LLB method, the burn-cut method, and the modified LLB method with reduced explosives. The total number of eroding elements can be used to compare the extent of the failure zones according to each method, and the number of deleted elements for each method was 101,953, 86,622, and 81,753, respectively. The original LLB method showed the widest range of the crushing zone, which was about 1.18 times larger than in the burn-cut method. In the modified LLB method with 30% reduced explosives, the total number of deleted eroding elements decreased by about 1.25 times compared to the original LLB, and was about 1.06 times lower than the burn-cut method. The modified LLB method showed similar failure patterns compared to the original LLB method, but the number of deleted eroding elements around the blast holes decreased as the amount of explosives decreased. The main effect of the burn-cut method was the reduction in the failure zone between the blast holes and the tunnel face, rather than contributing to increasing the advance.

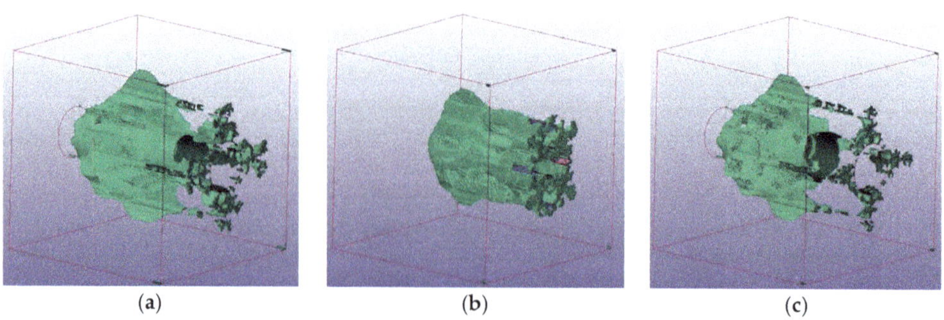

Figure 12. Comparison of deleted elements for each analysis model. (**a**) Original LLB method; (**b**) burn-cut method; (**c**) modified LLB method with reduced explosives.

In summary, the reduction in explosives in the modified LLB method reduced the overall size of the failure zones, but had a substantial advantage in that it contributed significantly to increasing the advance rate. Therefore, the LLB method clearly had better crushing efficiency than the traditional burn-cut method and, even when the amount of explosives in the cut area was reduced, this method had similar blasting efficiency compared to the traditional burn-cut method. Therefore, the modified LLB method can be considered an excellent alternative blasting method in that it not only reduces blast-induced vibration by reducing the amount of explosives needed in the cut area, but also increases the advance rate. However, since this is the result of the analysis using computer simulation, numerous field tests should be carried out for validation of the results.

5. Conclusions

In this study, numerical analysis using LS-DYNA software was performed to investigate the blast mechanism and efficiency by applying an advanced LLB cut blasting method that can effectively reduce blast-induced vibration. The Johnson–Holmquist (JH-2) constitutive model, which was developed for brittle materials subjected to dynamic conditions, was used to model the tunnel rock material, and the explosives and stemming materials were modeled using relevant emulsion and soil properties, respectively. The original LLB method was compared to the traditional burn-cut method as well as the modified LLB method, which shortened the depth of its large-diameter uncharged hole to match that of the blast holes and reduced the amount of explosives by 30%.

The numerical analysis confirmed that the new LLB method not only contributed to increasing the rock failure range in the cut area by forming an approximately 3.75 times larger uncharged hole than in the conventional burn-cut method, but also increased the advance rate by about 3.36 times by generating more tensile failure in the excavation direction. The modified LLB method used 30% less explosives and produced about 1.25 times fewer deleted eroding elements than the original LLB method (and 1.06 times fewer than the burn-cut method). In the traditional burn-cut method, which uses relatively small-diameter uncharged holes, explosives are used in crushing rocks more finely instead of contributing to increasing the advance rate. In contrast, the modified LLB method, with its long and large-diameter uncharged hole, generated more tensile failure in the excavation direction, so its advance rate increased four-fold compared to the burn-cut method. The long and large-diameter uncharged hole, which was already formed before blasting, reduces the initial confining stress of rock mass and can efficiently utilize the free face effect, thereby reducing the amount of explosives. Therefore, the modified LLB method, with its reduced amount of explosives, is significantly beneficial as an alternative method for reducing blast-induced vibration and increasing blasting efficiency, particularly in terms of advance per round, compared to the traditional burn-cut method. If a multi-LLB method that uses several long and large-diameter uncharged holes instead of one hole is considered, the blasting efficiency would increase significantly. Therefore, based on the numerical analysis, this method can be considered as an economical and eco-friendly blasting method for reducing the overall construction period and addressing environmental concerns.

Author Contributions: Conceptualization, S.S.L.; Data curation, M.-S.K.; Formal analysis, M.-S.K.; Funding acquisition, S.S.L.; Investigation, S.S.L.; Methodology, S.S.L.; Project administration, S.S.L.; Resources, M.-K.S.; Software, M.-S.K.; Supervision, S.S.L.; Validation, C.-Y.K.; Visualization, M.-S.K.; Roles/Writing—original draft, M.-S.K.; Writing—review and editing, S.S.L. All authors have read and agreed to the published version of the manuscript.

Funding: This work was supported by a Korea Agency for Infrastructure Technology Advancement (KAIA) grant funded by the Ministry of Land, Infrastructure, and Transport (grant 22UUTI-C157786-03).

Institutional Review Board Statement: Not applicable.

Informed Consent Statement: Not applicable.

Data Availability Statement: Some or all data that support the findings of this study are available from the corresponding author upon reasonable request.

Conflicts of Interest: The authors declare that they have no known competing financial interest or personal relationship that could have appeared to influenced the work reported in this paper.

References

1. New, B.M. Ground vibration caused by construction works. *Tunn. Undergr. Space Technol.* **1990**, *5*, 179–190. [CrossRef]
2. Berta, G. Blasting-induced vibration in tunneling. *Tunn. Undergr. Space Technol.* **1994**, *9*, 175–187. [CrossRef]
3. Singh, P.K.; Vogt, W. Effect of direction of initiation on ground vibrations. *Int. J. Min. Reclam.* **1998**, *12*, 75–78. [CrossRef]
4. Singh, T.N.; Singh, V. An intelligent approach to prediction and control ground vibration in mines. *Geotech. Geol. Earthq. Eng.* **2005**, *23*, 249–262. [CrossRef]
5. Lu, C.P.; Dou, L.M.; Wu, X.R.; Xie, Y.S. Case study of blast-induced shock wave propagation in coal and rock. *Int. J. Rock Mech. Min. Sci.* **2010**, *47*, 1046–1054. [CrossRef]
6. Kuzu, C.; Guclu, E. The problem of human response to blast induced vibrations in tunnel construction and mitigation of vibration effects using cautious blasting in half-face blasting rounds. *Tunn. Undergr. Space Technol.* **2009**, *24*, 53–61. [CrossRef]
7. Ainalis, D.; Kaufmann, O.; Tshibangu, J.P.; Verlinden, O.; Kouroussis, G. Modelling the source of blasting for the numerical simulation of blast-induced ground vibrations: A review. *Rock Mech. Rock Eng.* **2017**, *50*, 171–193. [CrossRef]
8. Mesec, J.; Kovac, I.; Soldo, B. Estimation of particle velocity based on blast event measurements at different rock units. *Soil Dyn. Earthq. Eng.* **2010**, *30*, 1004–1009. [CrossRef]
9. Yan, Y.; Hou, X.M.; Fei, H.L. Review of predicting the blast-induced ground vibrations to reduce impacts on ambient urban communities. *J. Clean Prod.* **2020**, *260*, 121135–121155. [CrossRef]
10. Kim, M.S.; Lee, S.S. The efficiency of large hole boring (MSP) method in the reduction of blast-induced vibration. *Appl. Sci.* **2021**, *11*, 1814. [CrossRef]
11. Beak, J.H.; Beak, S.H.; Han, D.H.; Won, A.R.; Kim, C.S. A study on the design of PLHBM. *Explos. Blasting.* **2012**, *30*, 66–76.
12. Ma, G.W.; An, X.M. Numerical simulation of blasting-induced rock fractures. *Int. J. Rock Mech. Min. Sci.* **2008**, *45*, 966–975. [CrossRef]
13. Qui, X.Y.; Hao, Y.; Shi, X.; Hao, H.; Zhang, S.; Gou, Y. Numerical simulation of stress wave interaction in short-delay blasting with a single free surface. *PLoS ONE.* **2018**, *13*, e0204166. [CrossRef]
14. Jeong, H.Y.; Jeon, B.K.; Choi, S.B.; Jeon, S.W. Fracturing behavior around a blasthole in a brittle material under blasting loading. *Int. J. Impact Eng.* **2020**, *140*, 103562–103576. [CrossRef]
15. Zhang, H.; Li, T.C.; Du, Y.T.; Zhu, Q.W.; Zhang, X.T. Theoretical and numerical investigation of deep-hole cut blasting based on cavity cutting and fragment throwing. *Tunn. Undergr. Space Technol.* **2021**, *111*, 103854–103871. [CrossRef]
16. Yang, Y.Z.; Shao, Z.S.; Mi, J.F.; Xiong, X.F. Effect of adjacent hole on the blast-induced stress concentration in rock blasting. *Adv. Civ. Eng.* **2018**, *3*, 5172878. [CrossRef]
17. Kutter, H.K.; Fairhurst, C. On the fracture process in blasting. *Int. J. Rock Mech. Min. Sci. Geomech. Abstr.* **1971**, *8*, 189–202. [CrossRef]
18. Esen, S.; Onederra, L.; Bilgin, H. Modelling the size of the crushed zone around a blasthole. *Int. J. Rock Mech. Min. Sci.* **2003**, *40*, 485–495. [CrossRef]
19. Hopkinson, B. A method of measuring the pressure produced in the detonation of high, explosives or by the impact of bullets. *Philos. Trans. R. Soc. Lond. Ser. A* **1913**, *89*, 411–413. [CrossRef]
20. Kolsky, H. *Stress Waves in Solids*, 1st ed.; Courier Corporation: Oxford, UK, 1953.
21. Chai, S.B.; Li, J.C.; Zhang, Q.B.; Li, H.B.; Li, N.N. Stress wave propagation across a rock mass with two non-parallel joints. *Rock Mech. Rock Eng.* **2016**, *49*, 4023–4032. [CrossRef]
22. Oh, S.W.; Choi, B.H.; Min, G.J.; Jung, Y.B.; Cho, S.H. An experimental study on the dynamic increase factor and strain rate dependency of the tensile strength of rock materials. *Explos. Blasting* **2021**, *39*, 10–21.
23. Cho, S.H.; Ogata, Y.; Kaneko, K. Strain-rate dependency of the dynamic tensile strength of rock. *Int. J. Rock Mech. Min. Sci.* **2003**, *40*, 763–777. [CrossRef]
24. Li, X.H.; Zhu, Z.M.; Wang, M.; Shu, Y.; Deng, S.; Xiao, D.J. Influence of blasting load directions on tunnel stability in fractured rock mass. *J. Rock Mech. Geotech. Eng.* **2022**, *14*, 346–365. [CrossRef]
25. Yang, J.H.; Cai, J.Y.; Yao, C.; Li, P.; Jiang, Q.H.; Zhou, C.B. Comparative study of tunnel blast-induced vibration on tunnel surfaces and inside surrounding rock. *Rock Mech. Rock Eng.* **2019**, *52*, 4747–4761. [CrossRef]
26. Hu, Y.G.; Yang, Z.W.; Huang, S.L.; Lu, W.B.; Zhao, G. A new safety control method of blasting excavation in high rock slope with joints. *Rock Mech. Rock Eng.* **2020**, *53*, 3015–3029. [CrossRef]
27. Yin, Y.; Sun, Q.; Zou, B.P.; Mu, Q.Y. Numerical study on an innovative shaped charge approach of rock blasting and the timing sequence effect in microsecond magnitude. *Rock Mech. Rock Eng.* **2021**, *54*, 4523–4542. [CrossRef]
28. Souli, M.; Bouamoul, A.; Nguyen-Dang, T.V. Ale formulation with explosive mass scaling for blast loading: Experimental and numerical investigation. *Comput. Model Eng. Sci.* **2012**, *86*, 469–486. [CrossRef]

29. Olovsson, L.; Souli, M.; Do, I. LS-DYNA—ALE Capabilities (Arbitrary-Lagrangian-Eulerian) Fluid-Structure Interaction Modeling. Livermore Software Technology Corporation. 2003. Available online: https://ftp.lstc.com/anonymous/outgoing/jday/aletutorial-278p.pdf (accessed on 10 March 2022).
30. Johnson, G.R.; Holmquist, T.J. Response of boron carbide subjected to large strains, high strain rates, and high pressures. *J. Appl. Phys.* **1999**, *85*, 8060–8073. [CrossRef]
31. Holmquist, T.J.; Johnson, G.R. Response of silicon carbide to high velocity impact. *J. Appl. Phys.* **2002**, *91*, 5858–5866. [CrossRef]
32. Wang, J.X.; Yin, Y.; Esmaieli, K. Numerical simulations of rock blasting damage based on laboratory-scale experiments. *J. Geophys. Eng.* **2018**, *15*, 2399–2417. [CrossRef]
33. Wang, J.X.; Yin, Y.; Lu, C.W. Johnson–Holmquist-II(JH-2) constitutive model for rock materials: Parameter determination and application in tunnel smooth blasting. *Appl. Sci.* **2018**, *8*, 1675. [CrossRef]
34. Lee, E.L.; Horning, H.C.; Kury, J.W. *A Diabatic Expansion of High Explosives Detonation Products*; TID4500-UCRL 50422; Lawrence Livermore National Laboratory, University of California: Livermore, CA, USA, 1968.
35. Yi, C.P. *Improved Blasting Results with Precise Initiation—Numerical Simulation of Sublevel Caving Blasting*; Report; Swedish Blasting Research Centre: Stockholm, Sweden, 2013; Volume 3.
36. Lewis, B.A. *Manual for LS-DYNA Soil Material Model 147*; Federal Highway Administration: McLean, VA, USA, 2004.
37. Koneshwaran, S.; Thambiratnam, D.P.; Gallage, C. Blast response of segmented bored tunnel using coupled SPH-FE method. *Structure* **2015**, *2*, 58–71. [CrossRef]
38. Livermore Software Technology. *LS-DYNA Keyword User's Manual—Volume II R12*; Livermore Software Technology: Livermore, CA, USA, 2020.
39. Baranowski, P.; Mazurkiewicz, L.; Malachowski, J.; Pytlik, M. Experimental testing and numerical simulations of blast-induced fracture of dolomite rock. *Meccanica* **2020**, *55*, 2338–2352. [CrossRef]
40. Cho, S.H.; Kaneko, K. Rock Fragmentation Control in Blasting. *Mater. Trans.* **2004**, *45*, 1722–1730. [CrossRef]
41. Shadabfar, M.; Gokdemir, C.; Zhou, M.L.; Kordestani, H.; Muho, E.V. Estimation of Damage Induced by Single-Hole Rock Blasting: A Review on Analytical, Numerical, and Experimental Solutions. *Energies* **2020**, *14*, 29. [CrossRef]

sustainability

MDPI

Article

Estimation of Active Earth Pressure for Narrow Unsaturated Backfills Considering Soil Arching Effect and Interlayer Shear Stress

Daxi Fu [1,2], Minghui Yang [3], Bo Deng [4,*] and Hutao Gong [2]

1. School of Civil Engineering, Beijing Jiaotong University, Beijing 100044, China
2. Henan Provincial Transportation Planning and Design Institute Co., Ltd., Zhengzhou 450015, China
3. School of Architecture and Civil Engineering, Xiamen University, Xiamen 361005, China
4. College of Civil Engineering, University of South China, Hengyang 421001, China

* Correspondence: parl_d@126.com

Abstract: Affected by climatic conditions such as rainfall, evaporation and air temperature, most of the backfill soil is in an unsaturated state, and the decrease in matric suction leads to the failure of the retaining structure. In view of this, this study takes the vertical rigid retaining wall with narrow unsaturated backfill as the research object, assuming that the backfill behind the wall forms a circular soil arch and considering the interlayer shear stress; thus, the analytical solution of the active earth pressure of narrow unsaturated soil is derived based on the thin layer element method. The reliability of this method is verified by comparing with the experimental and existing theoretical results. A parameter analysis demonstrates that with the increase in the interface friction angle of the moving wall–soil, the average shear stress coefficient of zone I and zone II increases gradually, but with the increase in the interface friction angle of the fixed wall–soil, the average shear stress coefficient of zone I decreases; with the increase in effective internal friction angle and effective cohesion, the active earth pressure decreases and the tension crack depth increases; with the increase in the interface friction angle, the active earth pressure in the upper part of the retaining wall increases slightly, while the active earth pressure in the lower part decreases obviously; with the increase in matric suction, the active earth pressure first decreases rapidly and then increases gradually, and the tension crack depth first increases and then decreases, but the distribution pattern of the horizontal active earth pressure remains unchanged; the active earth pressure decreases with the decrease in the aspect ratio, and when the aspect ratio is smaller, the attenuation is more obvious; until the aspect ratio reaches a certain value, the active earth pressure is basically unchanged.

Keywords: active earth pressure; unsaturated narrow backfill; interlayer shear stress; arching effect; thin-layer element method

Citation: Fu, D.; Yang, M.; Deng, B.; Gong, H. Estimation of Active Earth Pressure for Narrow Unsaturated Backfills Considering Soil Arching Effect and Interlayer Shear Stress. *Sustainability* **2022**, *14*, 12699. https://doi.org/10.3390/su141912699

Academic Editors: Mahdi Hasanipanah, Danial Jahed Armaghani and Jian Zhou

Received: 13 September 2022
Accepted: 27 September 2022
Published: 6 October 2022

1. Introduction

The lateral pressure of the retaining wall is a key factor affecting the safety and stability of the retaining wall. An accurate estimation of the earth pressure behind a retaining wall is the premise of rational design. Coulomb's and Rankine's earth pressure theory have been used by engineering designers until now because of their simple form and clear concept. Both of them assume that the backfill behind the wall is semi-infinite and is fully saturated or dry. Furthermore, the earth pressure obtained by the two theories is linearly distributed along the depth. However, it is proven that the active earth pressure distribution behind the wall is nonlinear due to the rough wall–soil interfaces [1–5].

With the rapid development of urban construction, underground buildings are becoming more and more intensive, so that the backfills behind many retaining walls cannot meet the conditions of semi-infinite soil. The adjacent underground structures prevent the sliding surface in the backfill from fully developing to the backfill surface, resulting in a large difference between the earth pressure behind the wall and that in the semi-infinite soil case. Several studies have demonstrated that the active earth pressure acting on the retaining wall decreases with the narrowing of the backfill behind the wall [3,5–12]. In addition, some scholars have discussed the shape of the slip surface in the soil under the condition of narrow backfill. A simple but effective method is to assume that the linear sliding surface is truncated by the adjacent wall to form a trapezoidal sliding wedge [5,13]. Greco [14] believed that the sliding surface will reflect and form a multi fold line failure mode after it develops to the adjacent wall. Yang et al. [15] conducted a series of model tests and observed that the slip surfaces are curvilinear planes developed from the heel of the retaining wall to the crest of the backfill.

On the other hand, after the soil arch theory was put forward by Janssen [16], the relevant earth pressure research was carried out by many geotechnical scholars. Handy [17] defined the active soil arch as the trajectory of small principal stress, assumed that the trajectory of the soil arch between two parallel retaining walls is catenary, and established the expression of the earth pressure distribution behind the wall by the slice method. Paik and Salgado [18] assumed that the soil arch is a circular arc, and the sliding surface of the backfill is the Rankine sliding surface, but the deflection of the principal stress at the sliding surface is not considered. Goel [19] assumed the soil arch as a parabola and solved the theoretical expression of active earth pressure in the form of polar coordinates in the case of the linear failure mode and parabola failure mode. Many subsequent studies have proven that the soil arching effect cannot be ignored in the calculation of earth pressure [9,20–22].

Additionally, due to the complex natural geological environment, various retaining structures are inevitably backfilled by unsaturated soil. In this case, the traditional earth pressure theories treat backfills as completely saturated or dry soils. Nevertheless, many engineering cases demonstrate that economic and safe retaining structure design can only be carried out if the lateral pressure change caused by the change of matric suction in unsaturated backfill is reasonably considered [23–25]. The following work has been carried out by researchers: Pufahl et al. [26] obtained the Rankine earth pressure solution of unsaturated soil based on the strength theory of two stress variables [27]. Liang et al. [28] combined the strength theory of unsaturated soil with the Coulomb earth pressure theory, and obtained the unified solution of Coulomb earth pressure in unsaturated soil. Since then, the limit equilibrium method has been widely used to solve the earth pressure of unsaturated soil [24,29,30]. Moreover, the slip line theory [31,32] and the upper bound method of limit analysis [33–35] have also been extended to the calculation of earth pressure in unsaturated soils. However, most unsaturated earth pressure theories still assume that the backfill is in semi-infinite soil conditions when analyzing the finite fill behind the wall, which is seriously inconsistent with the actual situation, resulting in too conservative calculation results. In addition, two ideas are generally adopted for the theoretical derivation of unsaturated earth pressure at present. One is based on the traditional Coulomb earth pressure method, and the overall sliding wedge is used to solve the problem, but the distribution of earth pressure cannot be obtained. The other method is to use the horizontal thin-layer slicing method, but most theories do not consider the comprehensive influence of the soil arching effect and interlayer shear stress in the analysis, resulting in a large deviation between the predicted results and the actual situation.

Based on the above analysis, this study takes the narrow unsaturated backfill as the analysis object, assumes that the trajectory of small principal stress is an arc and the soil slip surface behind the wall is a Coulomb slip surface, and the Mohr stress circle is used to calculate the average vertical stress and shear stress of the horizontal interlayer; on this basis, the expression of active earth pressure, its resultant force and the height of the action point is given by the horizontal thin layer analysis method, and it is compared with the model test and other theoretical results in the relevant literature. At the same time, through parameter analysis, the effects of fill aspect ratio, cohesion, internal friction angle, matrix suction and wall–soil interface friction angle on active earth pressure, its resultant force and the height of the action point are studied.

2. Shear Strength for Unsaturated Soils

In order to more clearly describe the characteristic that the shear strength of unsaturated soil increases with the increase in matric suction, Fredlund et al. [27] introduced a new material variable φ^b related to matric suction, and proposed a shear strength theory of two independent stress state variables for calculating the shear strength of unsaturated soil. As shown in Figure 1, in the three-dimensional space of two stress variables and shear stress, the shear strength envelope is a plane. The expression of shear strength is given by:

$$\tau_f = (\sigma - u_a)\tan\varphi' + c' + (u_a - u_w)\tan\varphi^b \tag{1}$$

where σ is the total normal stress; u_a and u_w are the pore air pressure and the pore water pressure; $(\sigma - u_a)$ represents the net normal stress on the failure plane; φ' is the effective internal friction angle; c' is the effective cohesion; $(u_a - u_w)$ represents the matric suction; φ^b is matric suction angle, which indicates the rate of increase in the soil strength with respect to an increase in matric suction.

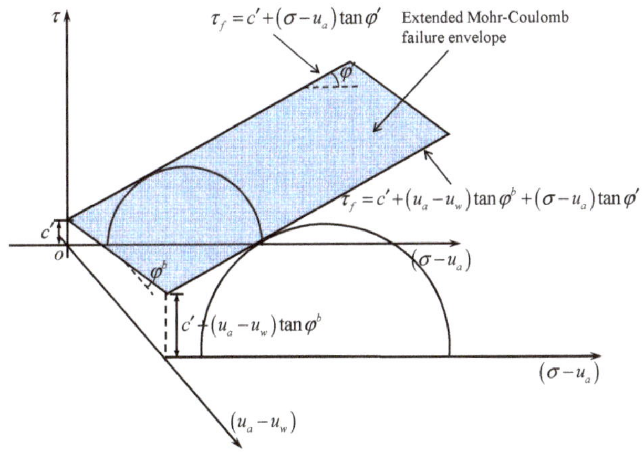

Figure 1. Extended Mohr-Coulomb failure envelope of unsaturated soils.

Subsequent studies [36,37] found that the ϕ^b is not a fixed value. With the change of matric suction, φ^b will take different values. Based on the microscopic analysis of unsaturated soil, Vanapalli et al. [38] suggested that the SWCC model be used to correlate the relationship between ϕ^b and volumetric water content of soil, which has the following form:

$$\tan\varphi^b = \left(\frac{\theta - \theta_r}{\theta_s - \theta_r}\right)\tan\varphi' \tag{2}$$

where θ is the volumetric water content; θ_r is the residual volumetric water content; and θ_s is the saturated volumetric water content. Using the SWCC model proposed by Van Genuchten [39], the matric suction angle corresponding to different matric suction can be obtained from the following formula:

$$\tan\varphi^b = \frac{\tan\varphi'}{\{1 + [\alpha(u_a - u_w)]^n\}^{1-1/n}} \tag{3}$$

where α and n are SWCC fitting parameters; parameter α is approximate to the inverse of air entry suction; parameter n reflects the pore size distribution.

Through the above analysis, when the value of matric suction is determined, the shear strength provided by the matric suction can be calculated. The intercept of the shear strength envelope in the corresponding net stress and shear stress plane coordinate system is the sum of the shear strength provided by the effective cohesion and matric suction, as shown in Figure 1. In the subsequent calculation, we will express the contribution of matric suction to shear strength through total cohesion:

$$c_t = c' + (u_a - u_w) \tan \varphi^b \tag{4}$$

3. Lateral Pressure Coefficient and Shear Stress Coefficient

Terzaghi [40] defines the phenomenon of earth pressure transfer from the yield zone to the adjacent static zone as a soil arching effect, and the existence of the soil arching effect changes the distribution of the earth pressure. The basic idea of calculating the earth pressure by soil arching effect is to obtain the active lateral earth pressure coefficient by assuming the geometric shape of the principal stress arch, and then solve the average vertical pressure and earth pressure of the soil layer according to the differential element force balance equation. In the calculation of the lateral earth pressure, the circular earth arch is widely used and the earth pressure distribution is in good agreement with the test results [18,19,41]. However, few studies have considered the influence of adjacent underground buildings on the soil arch.

As shown in Figure 2, in the case of narrow backfill, assuming that the backs of two rigid walls are vertical, and the backfill surface is horizontal, it is considered that there is an overload q0 on the surface of the backfill, and the backfill of the wall is in an unsaturated state. Suppose the wall height is H, the soil weight is γ, the soil cohesion is c', the soil friction angle is φ', the friction angle between the retaining wall and the soil is δ_1, and the friction between the existing building and the soil is angle is δ_2. The active slip surface intersects with the adjacent underground buildings to form a trapezoidal soil wedge, the Coulomb slip surface is assumed to be the slip surface under this working condition, and the inclination angle β of the Coulomb-type slip surface passing through the wall heel is:

$$\beta = \arctan\left(\sqrt{\tan^2 \varphi' + \frac{\tan \varphi'}{\tan(\varphi' + \delta_1)}} + \tan \varphi'\right) \tag{5}$$

Different lateral earth pressure coefficients and interlayer shear stress coefficients will be obtained when the soil arching effect is analyzed in the upper rectangular region (zone I) and the lower triangular region (zone II). In this section, the lateral earth pressure coefficient and the interlayer shear stress coefficient of each region are solved first, and the influence of the wall soil friction angle on them is discussed. Then, in the next section, the distribution of earth pressure, resultant force and their action points in each region are solved according to the horizontal thin layer element method.

Considering the total cohesion of unsaturated soils, the coordinate system $\sigma' o \tau$ is obtained by translating the coordinate system $\sigma o \tau$ to the left by $c_t \cot\phi'$, as shown in Figure 3. Next, the lateral earth pressure coefficient and interlayer shear stress coefficient will be calculated by analyzing the soil arching effect in different areas.

Zone I:

From the stress Mohr's circle in Figure 3, the angles θ_A and θ_B between the large principal stress and the horizontal direction at points A and B in Figure 2 can be expressed as:

$$\theta_A = \frac{\pi}{2} - \frac{1}{2}\arcsin\left(\frac{\sin \delta_1}{\sin \varphi'}\right) + \frac{\delta_1}{2} \tag{6}$$

$$\theta_B = \frac{\pi}{2} - \frac{1}{2}\arcsin\left(\frac{\sin \delta_2}{\sin \varphi'}\right) + \frac{\delta_2}{2} \tag{7}$$

where δ_1 and δ_2 are the interface friction angle of the moving wall and the fixed wall.

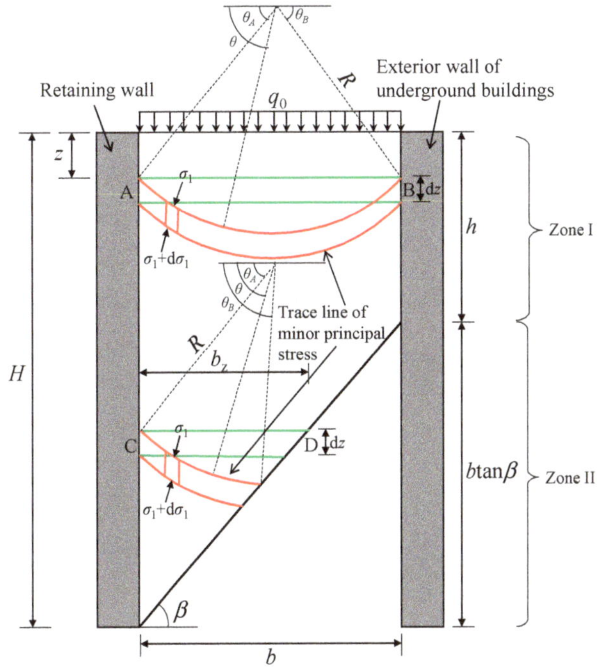

Figure 2. Narrow backfill behind retaining wall and its trajectory of minor principal stress.

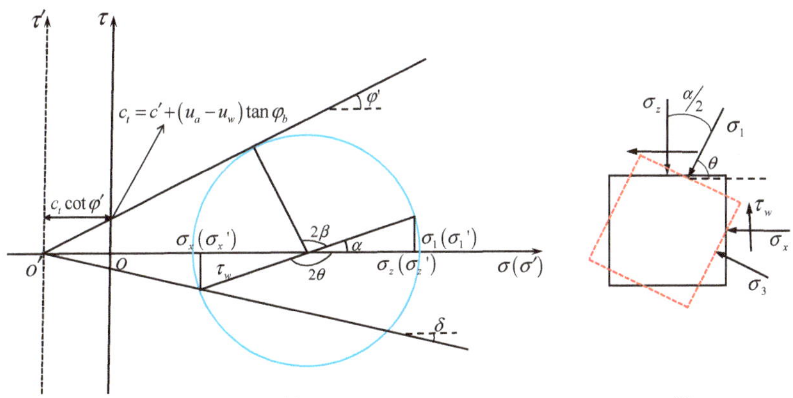

(a) (b)

Figure 3. Stress state behind retaining wall. (**a**) Mohr stress circle; (**b**) strss state.

The width of the horizontal thin layer element in zone I can be calculated as:

$$b = R(\cos\theta_A + \cos\theta_B) \tag{8}$$

where R is the radius of circular soil arch.

In the new coordinate system, the horizontal stress, vertical stress and shear stress at any point in the soil arch can be expressed as:

$$\sigma_x{}' = \sigma_1{}'\left(\cos^2\theta + K_a \sin^2\theta\right) \tag{9}$$

$$\sigma_z{}' = \sigma_1{}'\left(\sin^2\theta + K_a\cos^2\theta\right) \tag{10}$$

$$\tau = \sigma_1{}'(1 - K_a)\sin\theta\cos\theta \tag{11}$$

where σ'_1 is the major principal stress of the new coordinate system; K_a is the Rankine active earth pressure coefficient, which can be calculated by $K_a = \frac{1-\sin\varphi'}{1+\sin\varphi'}$.

Handy [17] and Paik [18] defined the lateral active earth pressure coefficient as the ratio of the horizontal stress behind the retaining wall to the average value of the vertical stress on the soil arch trajectory. Similarly, we define the interlayer shear stress coefficient as the ratio of the average shear stress on the soil arch trajectory to the average vertical stress on the soil arch trajectory.

The average shear stress and the average vertical stress on the soil arch trajectory can be obtained by the integral method:

$$\overline{\sigma}_{z1}{}' = \frac{\int_{\theta_A}^{\pi-\theta_B}\sigma_1{}'(\sin^2\theta + K_a\cos^2\theta)R\sin\theta d\theta}{b} = \sigma_1{}'\left(1 + \frac{(K_a-1)(\cos^3\theta_A + \cos^3\theta_B)}{3(\cos\theta_A + \cos\theta_B)}\right) \tag{12}$$

$$\overline{\tau}_1 = \frac{\int_{\theta_A}^{\pi-\theta_B}\sigma_1{}'(1-K_a)\sin\theta\cos\theta R\sin\theta d\theta}{b} = \sigma_1{}'\frac{(1-K_a)(\sin^3\theta_B - \sin^3\theta_A)}{3(\cos\theta_A + \cos\theta_B)} \tag{13}$$

The horizontal reaction force of the retaining wall in the new coordinate system can be expressed as:

$$\sigma_{xA}{}' = \sigma_1{}'\left(\cos^2\theta_A + K_a\sin^2\theta_A\right) \tag{14}$$

Therefore, in the new coordinate system, the lateral active earth pressure coefficient and interlayer shear stress coefficient are:

$$K_1 = \frac{\sigma_{xA}{}'}{\overline{\sigma}_{z1}{}'} = \frac{3(\cos\theta_A + \cos\theta_B)(\cos^2\theta_A + K_a\sin^2\theta_A)}{3(\cos\theta_A + \cos\theta_B) + (K_a - 1)(\cos^3\theta_A + \cos^3\theta_B)} \tag{15}$$

$$k_1 = \frac{\overline{\tau}_1}{\overline{\sigma}_{z1}{}'} = \frac{(1-K_a)(\sin^3\theta_B - \sin^3\theta_A)}{3(\cos\theta_A + \cos\theta_B) + (K_a - 1)(\cos^3\theta_A + \cos^3\theta_B)} \tag{16}$$

Zone II:

The deflection angle of point C has the same expression as that of point A. Considering the deflection of the principal stress at the sliding surface, θ_C and θ_D is obtained, as follows:

$$\theta_C = \frac{\pi}{2} - \frac{1}{2}\arcsin\left(\frac{\sin\delta_1}{\sin\varphi'}\right) + \frac{\delta_1}{2} \tag{17}$$

$$\theta_D = \frac{\pi}{4} + \beta - \frac{\varphi'}{2} \tag{18}$$

The width of the horizontal element in zone II and the radius of the soil arch conform to the following relationship:

$$b_z = R(\cos\theta_C - \cos\theta_D) \tag{19}$$

Referring to Equations (12)–(14), the average vertical stress, average shear stress and horizontal reaction force of the retaining wall in zone II can be expressed as:

$$\overline{\sigma}_{z2}{}' = \frac{\int_{\theta_C}^{\theta_D}\sigma_1{}'(\sin^2\theta + K_a\cos^2\theta)R\sin\theta d\theta}{b_z} = \sigma_1{}'\left(1 + \frac{(K_a-1)(\cos^3\theta_C - \cos^3\theta_D)}{3(\cos\theta_C - \cos\theta_D)}\right) \tag{20}$$

$$\overline{\tau}_2 = \frac{\int_{\theta_C}^{\theta_D} \sigma_1{}'(1-K_a)\sin\theta\cos\theta R\sin\theta d\theta}{b_z} = \sigma_1{}'\frac{(1-K_a)(\sin^3\theta_D-\sin^3\theta_C)}{3(\cos\theta_C-\cos\theta_D)} \tag{21}$$

$$\sigma_{xC}{}' = \sigma_1{}'\left(\cos^2\theta_C + K_a\sin^2\theta_C\right) \tag{22}$$

Then, the lateral pressure coefficient and interlayer shear stress coefficient of zone II in the new coordinate system can be expressed as:

$$K_2 = \frac{\sigma_{xC}{}'}{\overline{\sigma}_{z2}{}'} = \frac{3(\cos\theta_C-\cos\theta_D)(\cos^2\theta_C+K_a\sin^2\theta_C)}{3(\cos\theta_C-\cos\theta_D)+(K_a-1)(\cos^3\theta_C-\cos^3\theta_D)} \tag{23}$$

$$k_2 = \frac{\overline{\tau}_2}{\overline{\sigma}_{z2}{}'} = \frac{(1-K_a)(\sin^3\theta_D-\sin^3\theta_C)}{3(\cos\theta_C-\cos\theta_D)+(K_a-1)(\cos^3\theta_C-\cos^3\theta_D)} \tag{24}$$

Figure 4 shows the relationship between the interlayer shear stress coefficients (k_1 and k_2) and the interface friction angle of the fixed wall–soil δ_2 under different values of the interface friction angle of the moving wall–soil δ_1. The curves in the figures are obtained from Equations (16) and (24). As shown in the Figure 4a, the increase of δ_1, k_1 increases obviously, but decreases with the increase of δ_2. Moreover, from Figure 4b, k_2 increases with the increase of δ_1, but does not change with the increase of δ_2. This is because the soil arch in zone II is not formed to the fixed wall. Additionally, when δ_1 = δ_2, k_1 = 0. As δ_2 continues to increase, k_1 will take a negative value, and the direction of the interlayer shear stress will change.

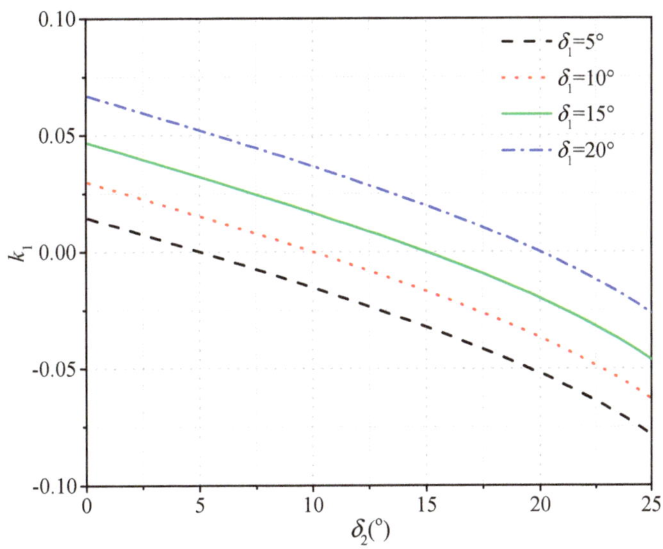

(a)

Figure 4. *Cont.*

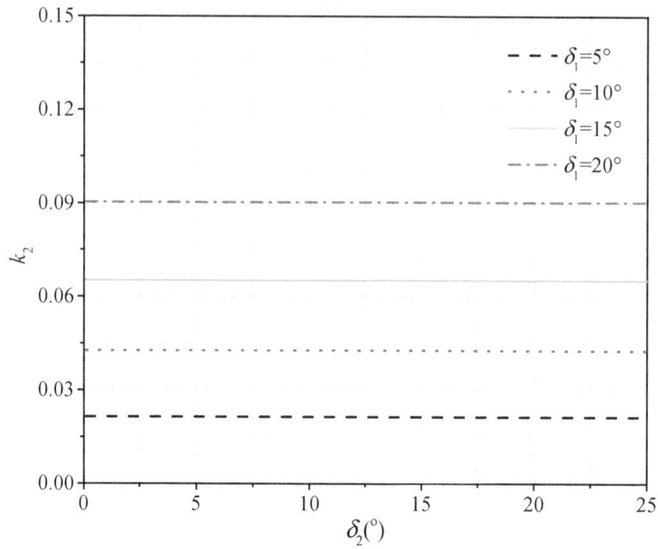

(b)

Figure 4. Relationship between interlayer shear stress coefficients and δ_2: (**a**) k_1; (**b**) k_2.

4. Derivation of Active Earth Pressure Formula

For unsaturated backfills, the tension cracks on the backfill surface cannot be ignored due to matric suction. Combining Equation (9) and the transformation relationship between the two coordinate systems, at the critical depth ($z = h_c$), the following relationship is given:

$$\begin{aligned}\sigma_{xA} &= \sigma_1{}'\left(\cos^2\theta_A + K_a\sin^2\theta_A\right) - c_t\cot\varphi' \\ &= \left(\gamma h_c + q_0 + c_t\cot\varphi'\right)\left(\cos^2\theta_A + K_a\sin^2\theta_A\right) - c_t\cot\varphi' = 0\end{aligned} \tag{25}$$

Rewriting the above equation, the tension crack depth can be expressed as:

$$h_c = \frac{1}{\gamma}\left(\frac{c_t\cot\varphi'}{\cos^2\theta_A + K_a\sin^2\theta_A} - c_t\cot\varphi' - q_0\right) \tag{26}$$

In Equation (26), when $h_c < 0$, take $h_c = 0$.

Zone I:

Carry out force balance analysis for rectangular thin layer element, as shown in Figure 5, in which σ_{x1} and σ_{x2} are the horizontal reaction force of the retaining wall and fixed wall; τ_{w1} and τ_{w2} are the interface shear stress of the retaining wall and fixed wall; $\overline{\sigma}_z$ and $\overline{\sigma}_z + d\overline{\sigma}_z$ are the average vertical stress acting on the upper surface and lower surface; $\overline{\tau}$ and $\overline{\tau} + d\overline{\tau}$ are the average interlayer shear stress acting on the upper surface and lower surface; dW is the self-weight of thin layer element.

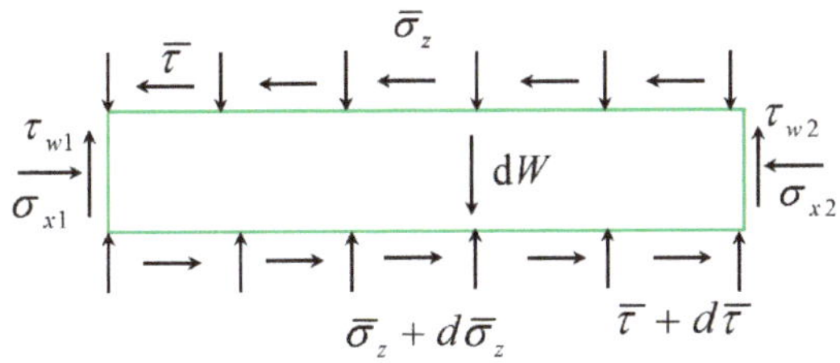

Figure 5. Horizontal thin-layer element model of finite soil of element I.

The force equilibrium equations in the vertical and horizontal directions are established as:

$$\sigma_{x1}dz - \sigma_{x2}dz + (\overline{\tau} + d\overline{\tau})b - \overline{\tau}b = 0 \tag{27}$$

$$\tau_{w1}dz + \tau_{w2}dz - \overline{\sigma}_z b + (\overline{\sigma}_z + d\overline{\sigma}_z)b - dW = 0 \tag{28}$$

where

$$\begin{cases} \tau_{w1} = \sigma_{x1}\tan\delta_1 + c_{wt1} \\ \tau_{w2} = \sigma_{x2}\tan\delta_1 + c_{wt2} \\ \sigma_{x1} = K_1\overline{\sigma}_z + (K_1 - 1)c_t\cot\varphi' \\ \overline{\tau} = k_1\overline{\sigma}_z + k_1 c_t\cot\varphi' \\ dW = \gamma b dz \end{cases} \tag{29}$$

The adhesive force of wall and soil can be considered as the same as the friction force, which is given by:

$$c_{wt1} = c_t\frac{\tan\delta_1}{\tan\varphi'},\ c_{wt2} = c_t\frac{\tan\delta_2}{\tan\varphi'} \tag{30}$$

Neglecting the second-order terms, the first-order differential equation can be obtained by solving Equations (27) and (28):

$$\frac{d\overline{\sigma}_z}{dz} + M_1\overline{\sigma}_z + N_1 = 0 \tag{31}$$

where

$$\begin{cases} M_1 = \frac{K_1(\tan\delta_1+\tan\delta_2)}{b(k_1\tan\delta_2+1)} \\ N_1 = \frac{K_1 c_t\cot\varphi'(\tan\delta_1+\tan\delta_2)-\gamma b}{b(k_1\tan\delta_2+1)} \end{cases} \tag{32}$$

By integrating Equation (30), the average vertical stress can be solved as:

$$\overline{\sigma}_z = C_1 e^{-M_1 z} - \frac{N_1}{M_1} \tag{33}$$

Substituting the boundary conditions ($z = h_c, \overline{\sigma}_z = q_0 + \gamma h_c$) into Equation (33), C_1 is obtained as follows:

$$C_1 = e^{M_1 h_c}\left(q_0 + \gamma h_c + \frac{N_1}{M_1}\right) \tag{34}$$

Zone II:

As shown in Figure 6, according to the equilibrium conditions of stresses acting on the trapezoidal thin layer element, the two equations are expressed as:

$$\sigma_x dz - \sigma_n dz\frac{\sin\beta}{\sin\beta} + \tau_s dz\frac{\cos\beta}{\sin\beta} + (\overline{\tau} + d\overline{\tau})(b_z - dz\cot\beta) - \overline{\tau}b_z = 0 \tag{35}$$

$$\tau_w dz + \tau_s dz\frac{\sin\beta}{\sin\beta} + \sigma_n dz\frac{\cos\beta}{\sin\beta} + (\overline{\sigma}_z + d\overline{\sigma}_z)(b_z - dz\cot\beta) - \overline{\sigma}_z b_z - dW = 0 \tag{36}$$

in which

$$\begin{cases} \tau_w = \sigma_x \tan \delta_1 + c_{wt1} \\ \tau_s = \sigma_n \tan \varphi' + c_t \\ \sigma_x = K_2 \overline{\sigma}_z + (K_2 - 1) c_t \cot \varphi' \\ \overline{\tau} = k_2 \overline{\sigma}_z + k_2 c_t \cot \varphi' \\ b_z = (H - z) \cot \beta \\ dW = \gamma b_z dz - \frac{1}{2} \gamma dz \cot \beta dz \end{cases} \tag{37}$$

where σ_n and τ_s are the normal stress and shear stress on the slip surface.

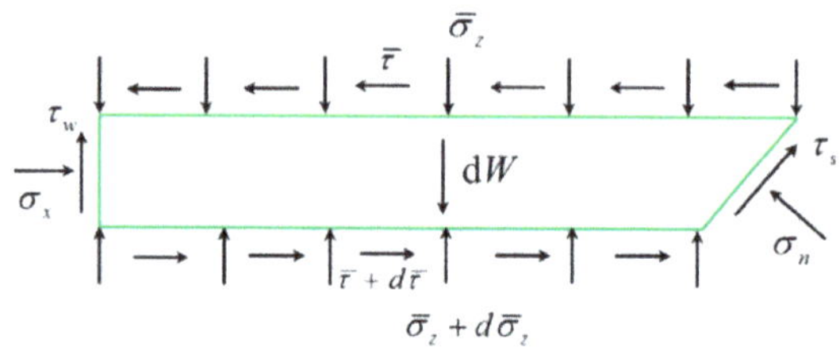

Figure 6. Horizontal thin-layer element model of finite soil of element II.

Combining Equations (35) and (36), a differential equation is obtained as:

$$\frac{d\overline{\sigma}_z}{dz} + \frac{M_2}{H - z} \overline{\sigma}_z + \frac{N_1}{H - z} + P_2 = 0 \tag{38}$$

where

$$\begin{cases} M_2 = \frac{K_2 \left[\tan\left(\frac{\pi}{2} - \beta + \varphi'\right) + \tan \delta_1 \right]}{\cot \beta \left[1 + k_2 \tan\left(\frac{\pi}{2} - \beta + \varphi'\right) \right]} - 1 \\ N_2 = c_t \frac{\tan\left(\frac{\pi}{2} - \beta + \varphi'\right)((K_2 - 1) \cot \varphi' + \cot \beta (1 - k_2 \cot \varphi')) + K_2 \cot \varphi' \tan \delta_1 + 1}{\cot \beta \left[1 + k_2 \tan\left(\frac{\pi}{2} - \beta + \varphi'\right) \right]} \\ P_2 = \frac{-\gamma}{1 + k_2 \tan\left(\frac{\pi}{2} - \beta + \varphi'\right)} \end{cases} \tag{39}$$

Integrating Equation (38) can be solved as

$$\overline{\sigma}_z = C_2 (H - z)^{M_2} + \frac{P_2}{1 - M_2} (H - z) - \frac{N_2}{M_2} \tag{40}$$

By substituting the boundary value $z = h$, $\overline{\sigma}_z = C_1 e^{-M_1 h} - \frac{N_1}{M_1}$, C_2 is obtained, as follows:

$$C_2 = \frac{C_1 e^{-M_1 h} - \frac{N_1}{M_1} - \frac{P_2}{1 - M_2}(H - h) + \frac{N_2}{M_2}}{(H - h)^{M_2}} \tag{41}$$

Therefore, the distribution of the horizontal active earth pressure is obtained as:

$$\sigma_x = \begin{cases} K_1 \left[C_1 e^{-M_1 z} - \frac{N_1}{M_1} \right] + (K_1 - 1) c_t \cot \varphi' \ (h_c \le z \le h) \\ K_2 \left[C_2 (H - z)^{M_2} + \frac{P_2}{1 - M_2} (H - z) - \frac{N_2}{M_2} \right] + (K_2 - 1) c_t \cot \varphi' \ (h \le z \le H) \end{cases} \tag{42}$$

The resultant force of the horizontal active earth pressure and the total overturning moment of the retaining wall can be expressed, respectively, as

$$\begin{aligned} E_x = & \int_{h_c}^{H} \sigma_x dz = \int_{h_c}^{h} K_1 \left[C_1 e^{-M_1 z} - \frac{N_1}{M_1} \right] + (K_1 - 1) c_t \cot \varphi' dz \\ & + \int_{h}^{H} K_2 \left[C_2 (H - z)^{M_2} + \frac{P_2}{1 - M_2} (H - z) - \frac{N_2}{M_2} \right] + (K_2 - 1) c_t \cot \varphi' dz \end{aligned} \tag{43}$$

$$\begin{aligned} M = & \int_{h_c}^{H} \sigma_x (H - z) dz = \int_{h_c}^{h} K_1 \left[C_1 e^{-M_1 z} - \frac{N_1}{M_1} \right] (H - z) + (K_1 - 1) c_t \cot \varphi' (H - z) dz \\ & + \int_{h}^{H} K_2 \left[C_2 (H - z)^{M_2} + \frac{P_2}{1 - M_2} (H - z) - \frac{N_2}{M_2} \right] (H - z) + (K_2 - 1) c_t \cot \varphi' (H - z) dz \end{aligned} \tag{44}$$

It should be noted that a special working condition may also occur. When $h < h_c$, there will be no zone I. In this case, the boundary conditions ($z = h_c, \overline{\sigma}_z = q_0 + \gamma h_c$) are substituted into Equation (40), and C_2 is obtained as follows:

$$C_2 = \frac{q_0 + \gamma h_c - \frac{P_2}{1-M_2}(H - h_c) + \frac{N_2}{M_2}}{(H - h_c)^{M_2}} \tag{45}$$

Then, the horizontal active earth pressure, resultant force of horizontal active earth pressure and the total overturning moment of the retaining wall can be expressed, respectively, as

$$\sigma_x = \begin{cases} 0\ (0 \le z \le h_c) \\ K_2\left[C_2(H-z)^{M_2} + \frac{P_2}{1-M_2}(H-z) - \frac{N_2}{M_2}\right] + (K_2 - 1)c_t \cot\varphi'\ (h_c \le z \le H) \end{cases} \tag{46}$$

$$E_x = \int_{h_c}^{H} \sigma_x dz = \int_{h_c}^{H} K_2\left[C_2(H-z)^{M_2} + \frac{P_2}{1-M_2}(H-z) - \frac{N_2}{M_2}\right] + (K_2 - 1)c_t \cot\varphi' dz \tag{47}$$

$$M = \int_{h_c}^{H} \sigma_x(H-z)dz = \int_{h_c}^{H} \begin{matrix} K_2\left[C_2(H-z)^{M_2} + \frac{P_2}{1-M_2}(H-z) - \frac{N_2}{M_2}\right](H-z) \\ +(K_2 - 1)c_t \cot\varphi'(H-z)dz \end{matrix} \tag{48}$$

The distance from the location of the resultant of the active earth pressure to the heel of the wall, denoted as H_a, can be easily derived as the ratio of M to E_x:

$$H_a = \frac{M}{E_x} \tag{49}$$

5. Comparison and Verification

The novelty of this study is to propose a theoretical framework for calculating the active earth pressure of narrowed unsaturated backfill. To the knowledge of the authors, no similar experimental and theoretical methods have been developed so far, so the proposed method is only compared with the results in the absence of the matric suction. Frydman et al. [6] carried out centrifugal tests of limited sand active earth pressure for vertical retaining walls under different aspect ratio (b/H) conditions. In the test, b/H = 0.235, γ = 15.8 kN/m^3, φ' = 36°, $\delta_1 = \delta_2$ = 25°, and the backfill width b = 2 m. Additionally, Chen et al. [13] used the horizontal differential element method to calculate the $\sigma_x/(\gamma H)$ value of the narrow backfill behind the wall. Figure 7 shows the comparison results of the normalized active earth pressure $\sigma_x/(\gamma H)$ value obtained by the method in this paper and the centrifugal test [6] and Chen et al. [13]. It can be observed that the method in this paper is in good agreement with the results of the centrifugation test, which demonstrates the rationality and accuracy of the method in this paper. In addition, the horizontal differential element method adopted by Chen et al. [13] ignores the shear force between the adjacent elements; thus, the calculation result is larger than that of the method in this study.

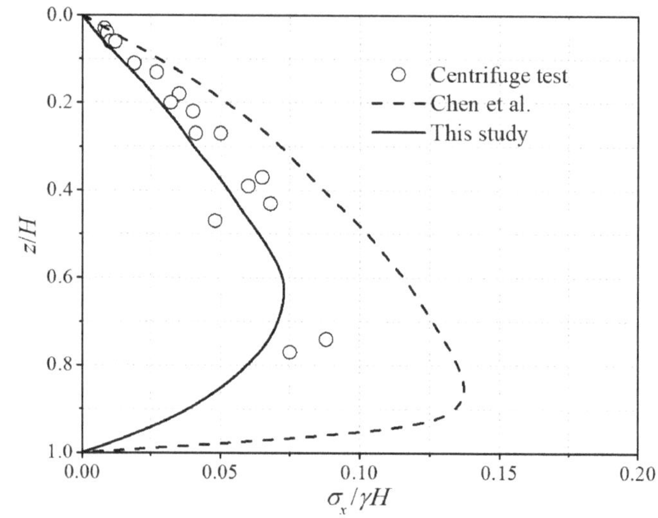

Figure 7. Comparison of centrifugal test and theoretical calculation results [6,13].

6. Parametric Study

In this section, a parametric study was performed to analyze the effective internal friction angle φ', surcharge pressure q_0, interface friction angle δ, effective cohesion c', matric suction $(u_a - u_w)$, and width-height ratio of backfill b/H on the distribution of the horizontal active earth pressure. In the analysis, these parameters were fixed: γ = 18 kN/m^3, α = 0.02 kPa^{-1}, n = 3 and H = 10 m. Other parameters for analysis can be found in the analysis figures. In addition, the interface friction angle δ_1 of the moving wall–soil and the interface friction angle δ_2 of the fixed wall–soil are equal, that is, $\delta_1 = \delta_2 = \delta$.

Figure 8 depicts the normalized horizontal active earth pressure ($\sigma_x/\gamma H$) along the normalized height (z/H) for the various effective internal friction angle φ'. To investigate the effect of the effective internal friction angle on the horizontal active earth pressure against the retaining structure, different values of φ' (i.e., φ' = 20°, 25°, 30°, 35° and 40°) were used. From Figure 8, it is obvious that the horizontal earth pressure first increases and then decreases along the depth, showing a nonlinear "drum" distribution, reaching a peak above the wall bottom. With the effective internal friction angle increases, the active earth pressure acting on the retaining wall at each depth decreases significantly, and the area where the earth pressure is zero gradually increases, indicating that the depth of the tension crack also increases with the increase in the effective internal friction angle.

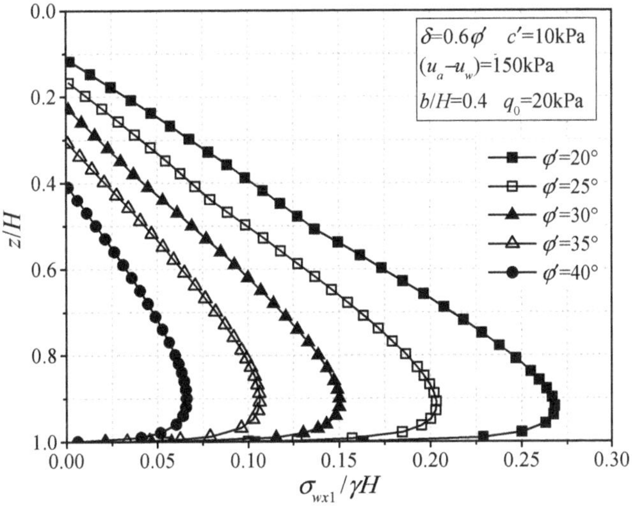

Figure 8. Values of $\sigma_x/\gamma H$ versus z/H for different values of the effective internal friction angle φ'.

Figure 9 shows the normalized horizontal active earth pressure ($\sigma_x/\gamma H$) along the normalized height (z/H) for different values of the surcharge pressure q_0. Herein, the aim is to discuss the effect of the surcharge pressure on the horizontal active earth pressure against the retaining structure, and thus different constant values of q_0 (i.e., q_0 = 10 kPa, 20 kPa, 30 kPa, 40 kPa and 50 kPa) were used. Figure 9 shows that the horizontal active earth pressure increases significantly with the increase in surcharge pressure applied on the backfill surface, whereas the shape of the horizontal active earth pressure distribution remains unchanged. Moreover, the depth of the tension crack decreases with the increase in the surcharge pressure, and when values of q_0 are large enough, the tension crack will not form.

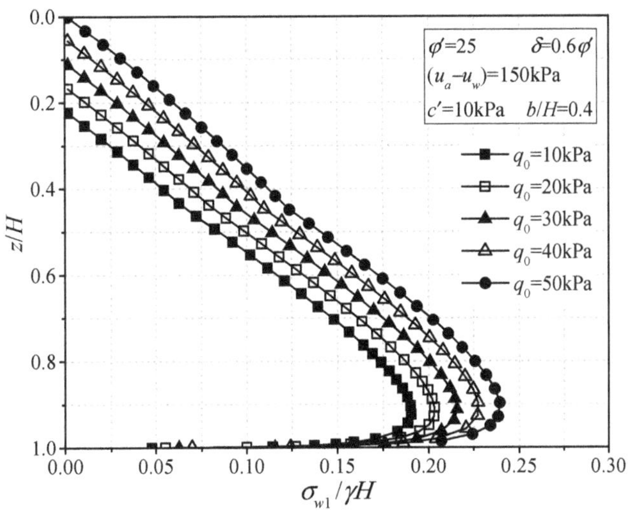

Figure 9. Values of $\sigma_x/\gamma H$ versus z/H for different values of the surcharge pressure q_0.

Figure 10 illustrates the normalized horizontal active pressure ($\sigma_x/\gamma H$) along the normalized height (z/H) for the various interface friction angle δ. To discuss the effect of interface friction angle on the horizontal active earth pressure against the retaining structure, different values of δ varying from $0.2\varphi'$ to $0.8\varphi'$ were used here. As shown in Figure 10, with the increase in the interface friction angle, the horizontal active earth pressure in the upper part of the retaining wall increases slightly, while the horizontal active earth pressure in the lower part decreases obviously. Furthermore, the action point of the horizontal earth pressure resultant force has a tendency to move upward with the increase in interface friction angle.

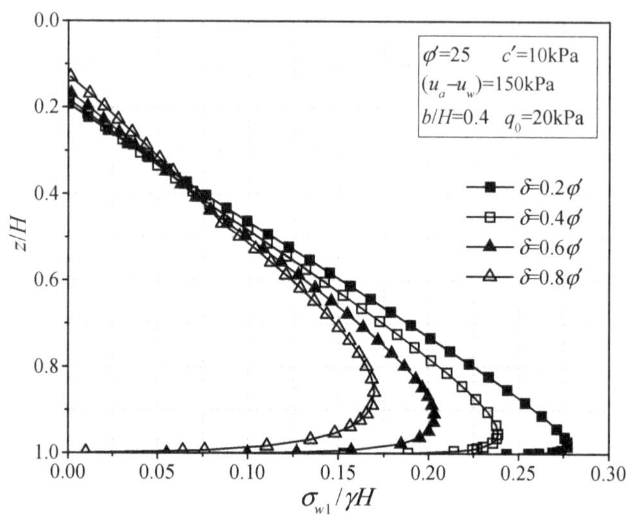

Figure 10. Values of $\sigma_x/\gamma H$ versus z/H for different values of the interface friction angle δ.

Figure 11 shows the normalized horizontal active earth pressure ($\sigma_x/\gamma H$) along the normalized height (z/H) for the various values of soil effective cohesion c'. To analyze the impact of the backfill inclination angle on the horizontal active earth pressure against the retaining wall, different values of c' (i.e., c' = 5 kPa, 10 kPa, 15 kPa, 20 kPa and 25 kPa) were used. Figure 11 shows a pronounced decrease in the horizontal active earth pressure against the rigid retaining wall with the increasing value of c', and the depth of the tension crack in soil increases correspondingly. From the earth pressure distribution curves, with the increase in the effective cohesion of the soil, the earth pressure distribution curves gradually shift to the left axis.

Figure 12 depicts the normalized horizontal active earth pressure distribution ($\sigma_x/\gamma H$) along the normalized height (z/H) for various matric suction ($u_a - u_w$). To study the effect of the matric suction ($u_a - u_w$) on the horizontal active earth pressure against the retaining wall, various values of ($u_a - u_w$) varying from 0 kPa to 200 kPa were used. From Figure 11, with the increase in matric suction, the horizontal active earth pressure first decreases rapidly and then increases gradually, and the depth of the tension crack first increases obviously and then decreases, while the shape of the horizontal active earth pressure distribution remains unchanged. The reason is that the total cohesion increases first and then decreases with the increase in matric suction, and this is consistent with the conclusion given by Song [42]. the horizontal active earth pressure reaches its minimum value at the matric suction equaling 50 kPa; this is because the matric suction takes the air entry pressure (approximately equal to the reciprocal of parameter α), and the apparent cohesion provided by the matric suction is the largest.

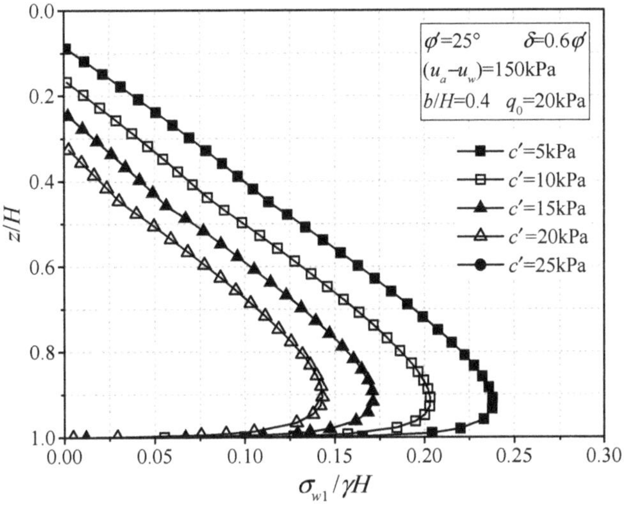

Figure 11. Values of $\sigma_x/\gamma H$ versus z/H for different values of the effective cohesion c'.

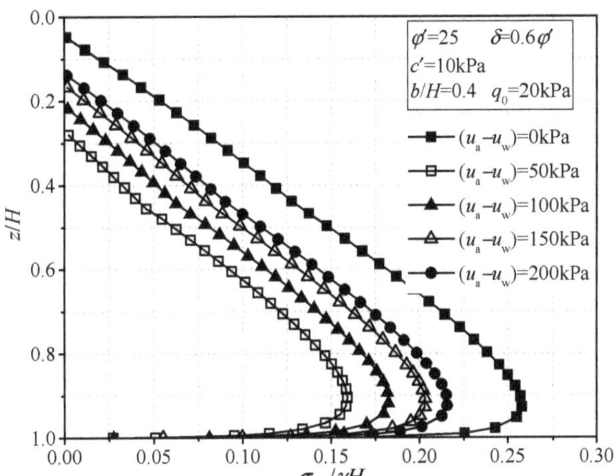

Figure 12. Values of $\sigma_x/\gamma H$ versus z/H for different values of the matric suction $(u_a - u_w)$.

The normalized horizontal active earth pressure distribution ($\sigma_x/\gamma H$) along the normalized height (z/H) for different width-height ratio of backfill *b/H* is shown in Figure 13. In order to study the effect of the width of backfill on the horizontal active earth pressure against the retaining wall, various values of b/H (i.e., b/H = 0.2, 0.3, 0.4, 0.6, and 0.8) were employed. As shown in Figure 13, the magnitude of the horizontal active earth pressure decreases with the decrease in the b/H value, and the attenuation becomes clearer when the b/H value is smaller. When the width of backfill reaches a certain value, the horizontal active earth pressure remains unchanged with the increase in the b/H value. This is because the backfill with a sufficient width allows the slip surface to fully develop.

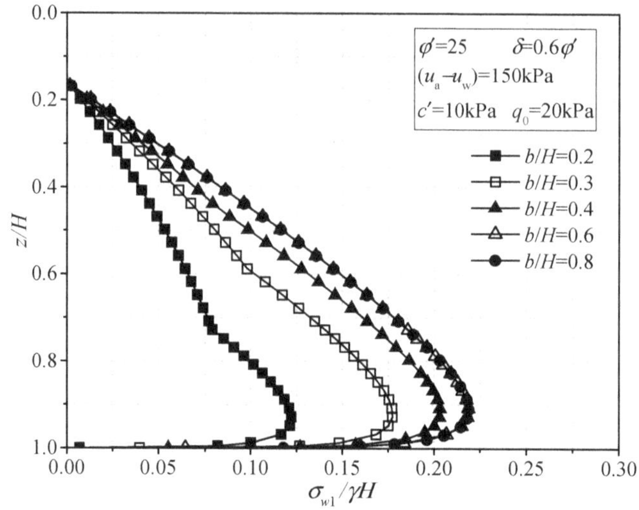

Figure 13. Values of $\sigma_x/\gamma H$ versus z/H for different values of the width–height ratio b/H.

7. Conclusions

The variation of soil suction has a significant effect on the active earth pressure of narrow unsaturated backfill in retaining structures. An analytical framework was developed for evaluating the active earth pressures of narrow unsaturated backfills affected by arching when the interlayer shear stress was considered. By comparing the results of the proposed method with other theoretical and experimental results, the method's effectiveness was verified. A detailed parametric study was carried out to discuss the effect of various factors on active earth pressure. The main conclusions include the following:

(1) With the increase in the interface friction angle of the moving wall–soil, the average shear stress coefficient of zone I and zone II increases gradually, but with the increase in the interface friction angle of the fixed wall–soil, the average shear stress coefficient of zone I decreases, while the average shear stress coefficient of zone II remains unchanged. The horizontal earth pressure shows a nonlinear "drum" distribution along the depth, which increases first and then decreases, and reaches a peak above the wall bottom.

(2) With the increase in effective internal friction angle and effective cohesion, the horizontal active earth pressure decreases significantly, while the tensile depth increases. In addition, the horizontal active earth pressure increases significantly with the increase in surcharge pressure, but the distribution shape of the horizontal active earth pressure remains unchanged. Additionally, when the interface friction angle increases, the horizontal active earth pressure at the upper part of the retaining wall slightly increases, while the horizontal active earth pressure at the lower part obviously decreases. In addition, the action point of the horizontal earth pressure force tends to move up with the increase in the friction angle.

(3) With the increase in matric suction, the horizontal active earth pressure first decreases rapidly and then increases gradually, and the tension crack depth first increases obviously and then decreases, but the distribution pattern of the horizontal active earth pressure remains unchanged.

(4) The horizontal active earth pressure decreases with the decrease in the aspect ratio. When the aspect ratio is smaller, the attenuation is more obvious. When the aspect ratio reaches a certain value, the horizontal active earth pressure is basically unchanged.

Author Contributions: Formal analysis, M.Y. and H.G.; Funding acquisition, M.Y. and B.D.; Investigation, M.Y.; Methodology, H.G.; Writing—original draft, D.F. and B.D. All authors have read and agreed to the published version of the manuscript.

Funding: This research was supported by the Transportation Science and Technology project of Henan province (2019J-2-12, 2021J7) and Science Foundation for Youths of Hunan Province of China (2021JJ40460), which are gratefully acknowledged.

Institutional Review Board Statement: Not applicable.

Informed Consent Statement: Not applicable.

Data Availability Statement: The data used to support the findings of this study are available from the corresponding author upon reasonable request.

Conflicts of Interest: The authors declare no conflict of interest.

References

1. Tsagareli, Z.V. Experimental investigation of the pressure of a loose medium on retaining walls with a vertical back face and horizontal backfill surface. *Soil Mech. Found. Eng.* **1965**, *2*, 197–200. [CrossRef]
2. Fang, Y.S.; Ishibashi, I. Static earth pressures with various wall movements. *J. Geotech. Eng.* **1986**, *112*, 317–333. [CrossRef]
3. Take, W.A.; Valsangkar, A.J. Earth pressures on unyielding retaining walls of narrow backfill width. *Can. Geotech. J.* **2001**, *38*, 1220–1230. [CrossRef]
4. O'Neal, T.S.; Hagerty, D.J. Earth pressures in confined cohesionless backfill against tall rigid walls—A case history. *Can. Geotech. J.* **2011**, *48*, 1188–1197. [CrossRef]
5. Yang, M.; Tang, X. Rigid retaining walls with narrow cohesionless backfills under various wall movement modes. *Int. J. Geomech.* **2017**, *17*, 04017098. [CrossRef]
6. Frydman, S.; Keissar, I. Earth pressure on retaining walls near rock faces. *J. Geotech. Eng.* **1987**, *113*, 586–599. [CrossRef]
7. Tang, Y.; Chen, J. A computational method of active earth pressure from finite soil body. *Math. Probl. Eng.* **2018**, *2018*, 9892376. [CrossRef]
8. Yang, M.; Deng, B. Simplified method for calculating the active earth pressure on retaining walls of narrow backfill width based on DEM analysis. *Adv. Civ. Eng.* **2019**, *2019*, 1507825. [CrossRef]
9. Xie, M.; Zheng, J.; Zhang, R.; Cui, L.; Miao, C. Active earth pressure on rigid retaining walls built near rock faces. *Int. J. Geomech.* **2020**, *20*, 04020061. [CrossRef]
10. Liu, H.; Kong, D. Active earth pressure of finite width soil considering intermediate principal stress and soil arching effects. *Int. J. Geomech.* **2022**, *22*, 04021294. [CrossRef]
11. Wang, Z.; Liu, X.; Wang, W.; Tao, Z.; Li, S. Inclined layer method-based theoretical calculation of active earth pressure of a finite-width soil for a rotating-base retaining wall. *Sustainability* **2022**, *14*, 9772. [CrossRef]
12. Hu, H.; Yang, F.; Tang, H.; Zeng, Y.; Zhou, J.; Gong, X. Field Study on Earth Pressure of Finite Soil Considering Soil Displacement. *Appl. Sci.* **2022**, *12*, 8059. [CrossRef]
13. Chen, J.J.; Li, M.G.; Wang, J.H. Active earth pressure against rigid retaining walls subjected to confined cohesionless soil. *Int. J. Geomech.* **2017**, *17*, 06016041. [CrossRef]
14. Greco, V. Active thrust on retaining walls of narrow backfill width. *Comput. Geotech.* **2013**, *50*, 66–78. [CrossRef]
15. Yang, M.; Tang, X.; Wu, Z. Slip surface and active earth pressure of cohesionless narrow backfill behind rigid retaining walls under translation movement mode. *Int. J. Geomech.* **2020**, *20*, 04020115. [CrossRef]
16. Janssen, H.A. Versuche uber getreidedruck in silozellen. *Z. Ver. Dtsch. Ing.* **1895**, *39*, 1045–1049.
17. Handy, R.L. The arch in soil arching. *J. Geotech. Eng.* **1985**, *111*, 302–318. [CrossRef]
18. Paik, K.H.; Salgado, R. Estimation of active earth pressure against rigid retaining walls considering arching effects. *Geotechnique* **2003**, *53*, 643–653. [CrossRef]
19. Goel, S.; Patra, N.R. Effect of arching on active earth pressure for rigid retaining walls considering translation mode. *Int. J. Geomech.* **2008**, *8*, 123–133. [CrossRef]
20. Nadukuru, S.S.; Michalowski, R.L. Arching in distribution of active load on retaining walls. *J. Geotech. Geoenviron. Eng.* **2012**, *138*, 575–584. [CrossRef]
21. Khosravi, M.H.; Pipatpongsa, T.; Takemura, J. Theoretical analysis of earth pressure against rigid retaining walls under translation mode. *Soils Found.* **2016**, *56*, 664–675. [CrossRef]
22. Patel, S.; Deb, K. Study of active earth pressure behind a vertical retaining wall subjected to rotation about the base. *Int. J. Geomech.* **2020**, *20*, 04020028. [CrossRef]
23. Yoo, C.; Jung, H.Y. Case history of geosynthetic reinforced segmental retaining wall failure. *J. Geotech. Geoenviron. Eng.* **2006**, *132*, 1538–1548. [CrossRef]
24. Vahedifard, F.; Leshchinsky, B.A.; Mortezaei, K.; Lu, N. Active earth pressures for unsaturated retaining structures. *J. Geotech. Geoenviron. Eng.* **2015**, *141*, 04015048. [CrossRef]

25. Tang, H.; Wasowski, J.; Juang, C.H. Geohazards in the three Gorges Reservoir Area, China–Lessons learned from decades of research. *Eng. Geol.* **2019**, *261*, 105267. [CrossRef]
26. Pufahl, D.E.; Fredlund, D.G.; Rahardjo, H. Lateral earth pressures in expansive clay soils. *Can. Geotech. J.* **1983**, *20*, 228–241. [CrossRef]
27. Fredlund, D.G.; Morgenstern, N.R.; Widger, R.A. The shear strength of unsaturated soils. *Can. Geotech. J.* **1978**, *15*, 313–321. [CrossRef]
28. Liang, W.B.; Zhao, J.H.; Li, Y.; Zhang, C.G.; Wang, S. Unified solution of Coulomb's active earth pressure for unsaturated soils without crack. In *Applied Mechanics and Materials*; Trans Tech Publications Ltd: Bäch, Switzerland, 2012; Volume 170, pp. 755–761.
29. Deng, B.; Yang, M. Analysis of passive earth pressure for unsaturated retaining structures. *Int. J. Geomech.* **2019**, *19*, 06019016. [CrossRef]
30. Abdollahi, M.; Vahedifard, F.; Abed, M.; Leshchinsky, B.A. Effect of tension crack formation on active earth pressure encountered in unsaturated retaining wall backfills. *J. Geotech. Geoenviron. Eng.* **2021**, *147*, 06020028. [CrossRef]
31. Vo, T.; Russell, A.R. Slip line theory applied to a retaining wall–unsaturated soil interaction problem. *Comput. Geotech.* **2014**, *55*, 416–428. [CrossRef]
32. Vo, T.; Russell, A.R. Interaction between retaining walls and unsaturated soils in experiments and using slip line theory. *J. Eng. Mech.* **2017**, *143*, 04016120. [CrossRef]
33. Li, Z.W.; Yang, X.L. Active earth pressure for soils with tension cracks under steady unsaturated flow conditions. *Can. Geotech. J.* **2018**, *55*, 1850–1859. [CrossRef]
34. Li, Z.; Yang, X. Three–dimensional active earth pressure for retaining structures in soils subjected to steady unsaturated seepage effects. *Acta Geotech.* **2020**, *15*, 2017–2029. [CrossRef]
35. Chehade, H.A.; Dias, D.; Sadek, M.; Jenck, O.; Chehade, F. Upper bound seismic limit analysis of geosynthetic–reinforced unsaturated soil walls. *Geotext. Geomembr.* **2020**, *48*, 419–430. [CrossRef]
36. Escario, V.; Saez, J. The shear strength of partly saturated soils. *Geotechnique* **1986**, *36*, 453–456. [CrossRef]
37. Zhang, L.L.; Fredlund, D.G.; Fredlund, M.D.; Wilson, G.W. Modeling the unsaturated soil zone in slope stability analysis. *Can. Geotech. J.* **2014**, *51*, 1384–1398. [CrossRef]
38. Vanapalli, S.K.; Fredlund, D.G.; Pufahl, D.E.; Clifton, A.W. Model for the prediction of shear strength with respect to soil suction. *Can. Geotech. J.* **1996**, *33*, 379–392. [CrossRef]
39. Van Genuchten, T.M. A closed–form equation for predicting the hydraulic conductivity of unsaturated soils. *Soil Sci. Soc. Am. J.* **1980**, *44*, 892–898. [CrossRef]
40. Terzaghi, K. *Theoretical Soil Mechanics*; John Wiley and Sons: Hoboken, NJ, USA, 1943.
41. Rao, P.; Chen, Q.; Zhou, Y.; Nimbalkar, S.; Chiaro, G. Determination of active earth pressure on rigid retaining wall considering arching effect in cohesive backfill soil. *Int. J. Geomech.* **2016**, *16*, 04015082. [CrossRef]
42. Song, Y.S. Suction stress in unsaturated sand at different relative densities. *Eng. Geol.* **2014**, *176*, 1–10. [CrossRef]

sustainability

MDPI

Article

A New Discrete Form of Hoek–Brown Criterion and Its Application to Limit Equilibrium Analysis of Rock Slope Stability

Youn-Kyou Lee [1,*] and S. Pietruszczak [2]

[1] Department of Coastal Construction Engineering, Kunsan National University, Gunsan 54150, Korea
[2] Department of Civil Engineering, McMaster University, Hamilton, ON L8S 4L7, Canada
* Correspondence: kyoulee@kunsan.ac.kr; Tel.: +82-63-469-1864

Abstract: The generalized Hoek–Brown criterion (GHB) is recognized as one of the standard failure criteria in rock engineering and its validity extends to a wide range of rock mass quality. A drawback of this criterion is the difficulty of transforming it into an explicit form defining the Mohr failure envelope when its strength parameter a is not equal to 0.5. The information on the functional form of the Mohr envelope for the full range of rock mass conditions enables the implementation of classical engineering approaches, such as the limit equilibrium method and limit analysis, in the framework of the GHB criterion. Knowing that for $a \neq 0.5$ the exact closed-form representation of the Mohr envelope is not feasible, an alternative is to express it in an approximate analytical form. The main purpose of this study is to propose a new improved method to define an approximate Mohr envelope of the GHB criterion that is much more accurate compared with the recently published approximations. The idea behind the formulation is to expand the Balmer's equation, which defines the relationship between the normal stress and minor principal stress at failure, by invoking the finite Taylor series centered at the known solution for $a = 0.5$. The formulation is then completed by substituting this solution into another Balmer's equation, defining the relationship between the shear strength and the minor principal stress. The Taylor polynomial approximations of up to third degree are considered in the formulation. The accuracy of the shear strength prediction is shown to be much better than that of the approximate formula of Lee and Pietruszczak proposed in 2021. An illustrative example of limit equilibrium analysis of rock slope stability, incorporating the new approximate expression for the Mohr envelope, is provided. The analysis incorporates a modified version of the Bishop approach, which is simpler and more rigorous than the original nonlinear expression. The study confirms that the new approximate representation of the Mohr failure envelope can facilitate the application of the GHB criterion to a range of practical rock engineering calculations.

Keywords: Hoek–Brown criterion; Mohr failure envelope; Balmer's equation; limit equilibrium analysis; modified Bishop approach

Citation: Lee, Y.-K.; Pietruszczak, S. A New Discrete Form of Hoek–Brown Criterion and Its Application to Limit Equilibrium Analysis of Rock Slope Stability. *Sustainability* **2022**, *14*, 12113. https://doi.org/10.3390/su141912113

Academic Editors: Mahdi Hasanipanah, Danial Jahed Armaghani and Jian Zhou

Received: 30 July 2022
Accepted: 22 September 2022
Published: 25 September 2022

1. Introduction

The generalized Hoek–Brown (GHB) criterion [1] is a nonlinear failure condition that is commonly used in rock engineering and can be applied to intact rock as well as jointed rock mass. This criterion defines the major principal stress at failure for a given minor principal stress and its strength parameters are identified using the respective empirical formula based on the GSI value [2,3]. A weakness of this criterion is the difficulty of transforming it to the corresponding explicit shear strength–normal stress equation, i.e., the Mohr envelope, which is required for applications of classical rock engineering approaches, such as the limit equilibrium method [4,5] and the upper bound limit analysis [6–10]. In the latter methodology, the energy dissipation along the sliding surface is calculated using the normal and shear stresses. It is noted that in cases when the strength parameter a equals

0.5, the closed form solution for the Mohr envelope is available [11–17]. However, when $a \neq 0.5$, an exact analytical expression of the Mohr envelope relating the shear strength to the normal stress cannot be obtained, which limits the scope of the applications of the GHB criterion.

In order to resolve the difficulty associated with the lack of a closed-form solution, various approximate analytical expressions for the Mohr envelope have been sought. The simplest approach is to obtain the equivalent friction angle and cohesion by approximating the GHB criterion by a linear form in a specified range of minor principal stress [1,18–21]. However, this linear approximation has an evident shortcoming that the strength nonlinearity, which is inherent in the GHB criterion, cannot be accounted for. Moreover, the accuracy of linear approximation depends on the size of the approximation interval. In the original approximation by Hoek et al. [1], the upper limit of minor principal stress was different for deep tunnels and slopes. Therefore, in later works dealing with slope stability [21,22] some empirical equations were employed for a more accurate assessment of this upper bound. At the same time, in order to improve the efficiency of the linear approximation, Wei et al. [23] presented a method of dividing the GHB curve into several sections and then linearly approximating each segmented interval. The methods of approximating the GHB envelope with a simpler form of nonlinear power function have also been attempted [24–26], but they do not retain the original meaning of the strength parameters employed in the GHB criterion.

Several other efforts to formulate the nonlinear Mohr envelope, while preserving the original meaning of strength parameters of the GHB criterion, have been made (e.g., Kumar [27] and Yang and Yin [28,29]). However, these formulations are implicit in the sense that the shear strength and normal stress are expressed as functions of instantaneous friction angle. Recently, Lee and Pietruszczak [14,15] and Lee [16] have formulated an explicit nonlinear expression approximating Mohr envelope equations by converting the power function terms appearing in the implicit form of this envelope to quadratic and/or cubic polynomial equations. Although the accuracy of these GHB envelopes was found to be good overall, these kinds of approximation still have room for further improvement.

As mentioned earlier, most of the existing approaches employ approximations of the GHB criterion in an a priori specified range of minor principal stress. This implies that in the field conditions the shear strength prediction may not be accurate if this range is not appropriately selected. In this study, a new approximate form of the Mohr envelope of the GHB criterion is proposed, which is not affected by the anticipated range of values of minor principal stress. The accuracy of the newly proposed Mohr envelopes is validated by calculating the percentage errors in the shear strength predictions for various rock mass conditions and the results are compared with those obtained using the approximate formula of Lee and Pietruszczak [15]. In the latter part of this paper, an example of limit equilibrium analysis, involving assessment of rock slope stability based on the proposed form of the Mohr envelope, is provided. The analysis employs a modified form of the classical Bishop approach, which is believed to be more rigorous than the original nonlinear expression. The study also includes a scenario in which the loss of stability is triggered by the distributed load acting on the horizontal upper surface.

2. Generalized Hoek–Brown Criterion

2.1. General form in Terms of Minor Principal Stress

In the generalized Hoek–Brown (GHB) failure condition [1], the major principal stress (σ_1) at failure is a nonlinear function of minor principal stress (σ_3) defined as

$$\sigma_1 = \sigma_3 + \sigma_{ci}\left(m_b \frac{\sigma_{ci}}{\sigma_3} + s\right)^a, \tag{1}$$

where σ_{ci} denotes the uniaxial compressive strength of the intact rock, while m_b, s and a are the strength parameters of rock mass defined empirically as follows:

$$m_b = m_i \exp\left(\frac{\text{GSI} - 100}{28 - 14D}\right), \tag{2}$$

$$s = \exp\left(\frac{\text{GSI} - 100}{9 - 3D}\right), \tag{3}$$

$$a = \frac{1}{2} + \frac{1}{6}\left(e^{-\text{GSI}/15} - e^{-20/3}\right). \tag{4}$$

In the equations above, GSI denotes the Geological Strength Index [2] and D is the disturbance factor which varies from 0 to 1 depending on the excavation damage. For undisturbed rock mass, $D = 0$, while for highly disturbed rock mass (e.g., as in an open pit mine slope) there is $D = 1$.

2.2. Normalized Form of the GHB Criterion

In order to facilitate the mathematical treatment of the GHB criterion, Rojat et al. [30] introduced the following normalization rule for normal stress (σ):

$$N = \frac{\sigma}{m_b^{a/(1-a)}\sigma_{ci}} + \frac{s}{m_b^{1/(1-a)}} \tag{5}$$

Applying this rule to σ_1 and σ_3, the GHB criterion takes the simplified nondimensional form as

$$N_1 = N_3 + N_3^a, \tag{6}$$

where

$$N_1 = \frac{\sigma_1}{m_b^{a/(1-a)}\sigma_{ci}} + \frac{s}{m_b^{1/(1-a)}};\ N_3 = \frac{\sigma_3}{m_b^{a/(1-a)}\sigma_{ci}} + \frac{s}{m_b^{1/(1-a)}} \tag{7}$$

In the transformed space ($N_1 - N_3$ space), the normalized GHB criterion, i.e., Equation (6), is completely defined by the strength parameter a and all the GHB curves start from the origin as shown in Figure 1.

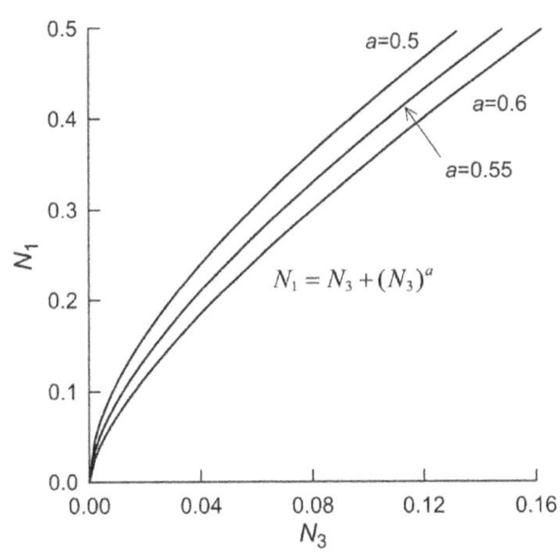

Figure 1. Generalized Hoek–Brown criterion in the transformed nondimensional stress space.

2.3. Shear–Normal Stress Relation of the GHB Criterion

The shear stress (τ) acting on an incipient failure plane is a function of the normal stress σ. In general, this function is nonlinear in $\sigma - \tau$ space and it is referred to as the Mohr failure envelope. Invoking the following normalization rule for τ [14],

$$T = \frac{\tau}{m_b^{a/(1-a)}\sigma_{ci}} \tag{8}$$

it can be shown that the normalized Mohr envelope of the GHB criterion is a concave downward curve always starting from the origin of $N - T$ space as depicted in Figure 2. ϕ_i and c_i in this figure represent the instantaneous friction angle and cohesion, respectively.

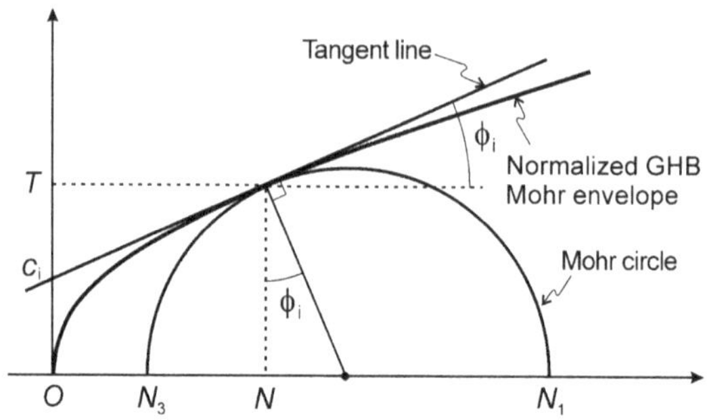

Figure 2. Normalized Mohr envelope and Mohr circle of the generalized Hoek–Brown criterion.

According to Yang and Yin [28] and Lee and Pietruszczak [14], the $\phi_i - c_i$ relationship implied in the GHB criterion is given by

$$\frac{c_i}{\sigma_{ci}} = \frac{\cos\phi_i}{2}\left[\frac{m_b a(1-\sin\phi_i)}{2\sin\phi_i}\right]^{a/(1-a)} - m_b^{a/(1-a)}\left[\frac{a(1-\sin\phi_i)}{2\sin\phi_i}\right]^{1/(1-a)}\left(1+\frac{\sin\phi_i}{a}\right)\tan\phi_i + \frac{s}{m_b}\tan\phi_i. \tag{9}$$

The formulation of the Mohr envelope can be accomplished by invoking Balmer's procedure [31], in which the envelope can be expressed in the form of the following implicit parametric functions of N_3:

$$N = N_3 + \frac{N_1 - N_3}{dN_1/dN_3 + 1}, \tag{10}$$

$$T = (N - N_3)\sqrt{\frac{dN_1}{dN_3}} \tag{11}$$

Noting that, based on Equation (6), there is $dN_1/dN_3 = 1 + aN_3^{a-1}$, the above two parametric equations can be restated as

$$f(N_3) + 2N_3 - 2N = 0, \tag{12}$$

$$T = \frac{N_3^a}{aN_3^{a-1} + 2}\sqrt{aN_3^{a-1} + 1} \tag{13}$$

where

$$f(N_3) = (a+1)N_3^a - aNN_3^{a-1}. \tag{14}$$

Thus, it is evident that the establishment of an explicit analytical equation of the Mohr envelope is equivalent to finding the root N_3 of Equation (12) for a given value of N. Substituting this root into Equation (13) defines the normalized shear strength T, which can then be used to calculate the shear strength τ from Equation (8).

Lee and Pietruszczak [15] have derived the closed form solution of Equation (12) for $a = 0.5$, which takes the form

$$N_3 = \left(\frac{16N+9}{24}\right)\cos\left(\frac{\theta}{3} + \frac{4\pi}{3}\right) + \frac{32N+9}{48}, \tag{15}$$

where

$$\theta = \cos^{-1}\left[\frac{-4096N^3 + 6912N^2 + 3888N + 729}{(16N+9)^3}\right] \tag{16}$$

In general, however, for $a \neq 0.5$, an explicit solution is not available. To overcome this difficulty, Lee and Pietruszczak [14,15] have used explicit analytical functions approximating the exact solution. In this paper, another approach to find an approximate solution of Equation (12) is proposed which leads to a new approximate formulation of the Mohr envelope which is much more accurate. The details of the formulation are described in the following section.

3. New Approximate Formulation of the Mohr Envelope for GHB Criterion

3.1. Approximate Mohr Envelope Based on Taylor Expansion of the Balmer's Equation

The idea of the new formulation of the Mohr envelope is to approximate the power function terms in the first Balmer's equation, i.e., Equation (12), with the Taylor series. Denoting by N_{3^*} the exact solution of Equation (12) for $a = 0.5$ (cf. Equation (15)), i.e.,

$$N_{3^*} = \left(\frac{16N+9}{24}\right)\cos\left(\frac{\theta+4\pi}{3}\right) + \frac{32N+9}{48} \tag{17}$$

and noting that this solution may not be far from that corresponding to $a \neq 0.5$, the above value of N_{3^*} can be selected as the expansion center for the Taylor series. In this case, the approximation of Equation (12) can be expressed as the following polynomial equation of degree n:

$$\sum_{k=0}^{n} \frac{1}{k!} f^{(k)}(N_{3^*})(N_3 - N_{3^*})^k + 2N_3 - 2N = 0, \tag{18}$$

where $f^{(k)}(N_{3^*})$ denotes the kth derivative of the function f with respect to N_3 evaluated at N_{3^*}. In this paper, the polynomials of degree up to three are considered and the corresponding solutions for N_3 are presented below.

(i) Linear approximation ($n = 1$)

If $n = 1$, Equation (18) simplifies to a linear form, and the solution for N_3 is obtained as follows:

$$N_3 = \frac{2N + (a-1)(a+1)N_{3^*}^{a} - a(a-2)N_{3^*}^{a-1}N}{a(a+1)N_{3^*}^{a-1} - a(a-1)N_{3^*}^{a-2}N + 2}. \tag{19}$$

(ii) Quadratic approximation ($n = 2$)

If $n = 2$, Equation (18) becomes a quadratic equation, and its solution for N_3 is given by

$$N_3 = \frac{-K_2\sqrt{K_2^2 - 4K_1K_3}}{2K_1} \tag{20}$$

where

$$\kappa_1 = -\frac{1}{2}a(a-1)(a-2)N_{3^*}^{a-3}N + \frac{1}{2}a(a-1)(a+1)N_{3^*}^{a-2} \tag{21}$$

$$\kappa_2 = a(a-1)(a-3)N_{3^*}^{a-2}N - a(a-2)(a+1)N_{3^*}^{a-1} + 2 \tag{22}$$

$$\kappa_3 = -\frac{1}{2}a(a-2)(a-3)N_{3^*}^{a-1}N + \frac{1}{2}(a-1)(a-2)(a+1)N_{3^*}^{a} - 2N. \tag{23}$$

(iii) Cubic approximation ($n = 3$)

If $n = 3$, Equation (18) reduces to a cubic polynomial equation, which has the following solution

$$N_3 = \frac{2}{3}\sqrt{\eta_1^2 - 3\eta_2}\cos\left(\frac{\theta}{3} + \frac{4\pi}{3}\right) - \frac{1}{3}\eta_1, \tag{24}$$

where

$$\theta = \cos^{-1}\left[\frac{9\eta^1\eta^2 - 27\eta^3 - 2\eta_1^3}{2\left(\eta_1^2 - 3\eta_2\right)^{3/2}}\right] \tag{25}$$

$$\eta_1 = \frac{-3(a-1)(a-3)(a+1)N_{3*}^2 + \left(3a^3 - 18a^2 + 27a - 9\right)N_{3*}N}{(a-1)(a-3)(a+1)N_{3*} + [1 - a(a-2)(a-3)N]} \tag{26}$$

$$\eta_2 = \frac{3a(a-2)(a-3)(a+1)N_{3*}^3 + 3a\left(a^3 - 7a^2 + 14a - 7\right)N_{3*}^2 N + 12N_{3*}^{4-a}}{(a-1)(a-2)(a+1)N_{3*} + [1 - a(a-2)(a-3)N]} \tag{27}$$

$$\eta_3 = \frac{-(a-1)(a-2)(a-3)(a+1)N_{3*}^4 + a\left(a^3 - 8a^2 + 21a - 19\right)N_{3*}^3 N - 12N_{3*}^{4-a}N}{(a-1)(a-2)(a+1)N_{3*} + [1 - a(a-2)(a-3)]N}. \tag{28}$$

Finally, referring to Equations (8) and (13), the equations for the Mohr envelope corresponding to the above three approximate solutions for N_3, i.e., Equations (19), (20) and (24), take the following form:

$$\tau = \sigma_{ci} m_b^{a/1-a} \frac{N_3^a}{aN_3^{a-1} + 2}\sqrt{aN_3^{a-1} + 1} \tag{29}$$

Thus, substitution of Equations (19), (20) and (24) into Equation (29) yields three new analytical expressions of the approximate Mohr envelope of the GHB criterion, which are based on the linear, quadratic and cubic approximations of the Balmer's equation, respectively. It is important to note that N_{3*} appearing in Equations (19), (20) and (24) is different from the actual calculated value of N_3. As such, unlike in the existing approximate expressions for the Mohr envelope, there is no restriction here on the range of values of σ_3. Consequently, the predictive abilities of this representation are significantly enhanced. Due to the nature of the Taylor approximation, it is not difficult to deduce that when the value of GSI approaches 100, that is, when the strength parameter a approaches 0.5, the newly proposed approximate Mohr envelopes converge to the exact solution.

3.2. Discussions on the Accuracy of New Formulations of the Mohr Envelope

In this section, the accuracy of the three new approximate Mohr envelopes adopting Equations (19), (20) and (24), respectively, is investigated. In Figure 3, the approximate Mohr envelopes are compared with the numerically determined 'exact' solutions for the rock mass with $m_i = 20$, $D = 0.0$ and five different GSI values, i.e., 20, 40, 60, 180 and 100. In this figure, the solid lines represent the approximate envelopes, while the dashed lines are the exact solutions. It should be noted here that the term 'exact' refers to the general analytical form of the Mohr envelope constructed with a numerical solution for N_3, viz. Equation (12), which can be obtained by implementing a suitable numerical algorithm such as the Newton–Raphson method [32]. The respective Mohr envelopes are normalized by σ_{ci}. The approximate curves shown in Figure 3a–c are the plots of the analytical form (29) with three different approximations for N_3s, i.e., Equations (19), (20) and (24), respectively.

Figure 3 clearly shows that the accuracy of the approximate envelope is excellent, and even the envelopes corresponding to linear approximation (Figure 3a) are difficult to distinguish from the exact envelopes. On the other hand, in Figure 4 that presents the approximate Mohr envelopes recently proposed by Lee and Pietruszczak [15] for the same rock mass conditions as in Figure 3, slight differences between the exact and the approximate solutions can be seen when GSI $\geq$ 60.

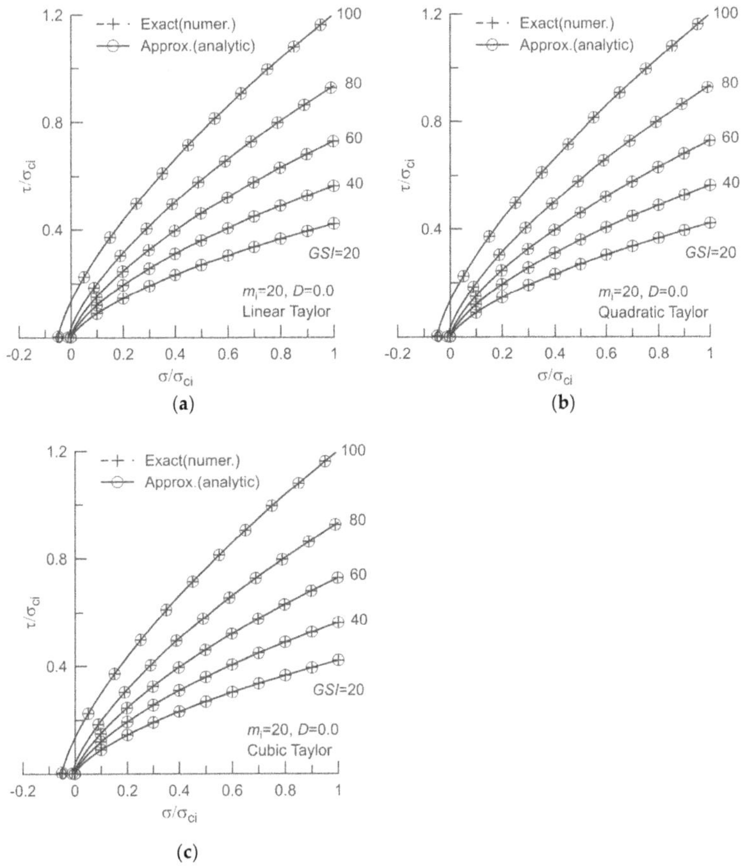

Figure 3. Comparison of the proposed approximate Mohr envelopes with those obtained from the numerical (exact) solution: the approximate Mohr envelopes are calculated by using the proposed analytical shear strength formulas (Mohr envelopes) incorporating the approximate solutions for N_3 corresponding to (**a**) the linear, (**b**) quadratic and (**c**) cubic Taylor polynomial approximations of the Balmer's equation.

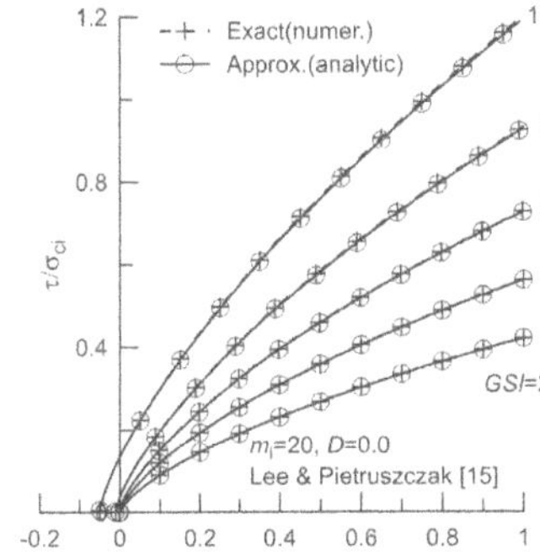

Figure 4. Comparison of the approximate Mohr envelopes from Lee and Pietruszczak's formula [15] with the numerical (exact) solution for the same rock mass conditions as in Figure 3.

In order to understand how the accuracy of the newly proposed approximate Mohr envelopes varies with the selection of the degree of the Taylor approximation, the percentage errors in the shear strength predictions are calculated and the results are compared with the prediction errors from the corresponding Lee and Pietruszczak's approximate Mohr envelopes [15] (Figure 4) in Figure 5. Here, the percentage error is defined as

$$\text{Percentageerror}(\%) = \left| \frac{\tau_{\text{approx}} - \tau_{\text{exact}}}{\tau_{\text{exact}}} \right| \times 100 \quad (30)$$

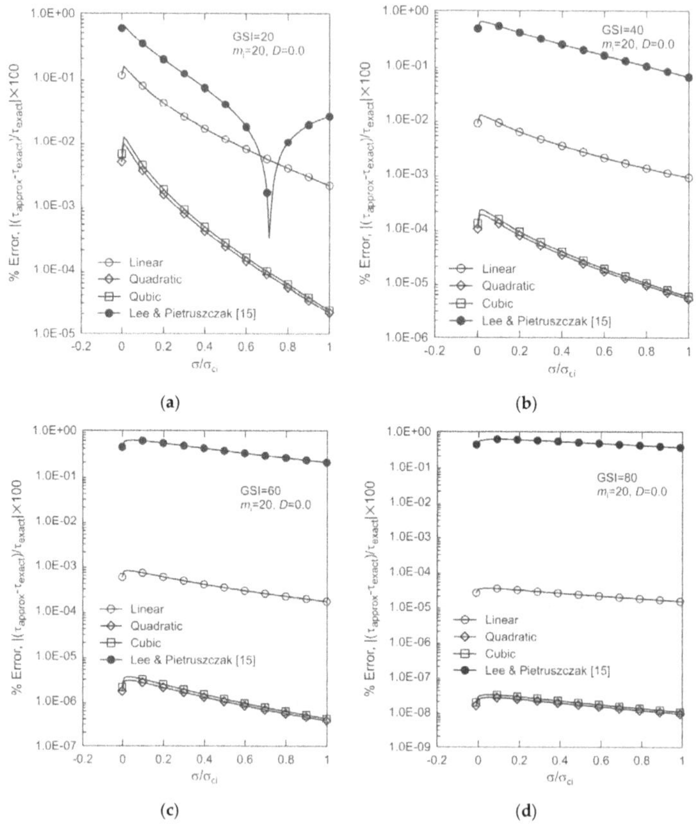

Figure 5. Percentage errors incurred in the shear strength predictions using the newly formulated approximate Mohr envelopes and Lee and Pietruszczak's formula [15] for rock masses with GSI values of (**a**) 20, (**b**) 40, (**c**) 60 and (**d**) 80.

The number of data points in each curve shown in Figure 5 is 101, but only 11 symbols are displayed to indicate the curve.

Figure 5 clearly shows that the errors in the proposed approximations of the Mohr envelope decrease with the increase in the normal stress σ, except for a slight increase in the beginning. Additionally, it is evident that the percentage errors decrease significantly as the GSI value increases. This tendency is because the greater the normal stress and the GSI value, the smaller the curvature of the Mohr envelope. A decrease in the curvature implies that the curve becomes more linear, so that the Taylor polynomial approximation can produce more accurate results. Another interesting fact is that, contrary to the expectation, the accuracy of the Mohr envelope based on the quadratic Taylor approximation is slightly higher than that based on the cubic approximation. This may be due to the fact that the geometric shape of the Balmer function of the GHB criterion is more favorable for the quadratic Taylor approximation. Overall, Figure 5 shows that the accuracy of the three new representations of the Mohr envelope is much better than that for the Lee and Pietruszczak envelope [15]. For example, when GSI = 80, the order of percentage error reduction by the new approach is 7 or more compared with the case of the Lee and Pietruszczak approach [15]. It can also be seen that the quadratic approximation is slightly better than the cubic one, although the difference is rather subtle. Thus, among the three forms, it is evident that the Mohr envelope based on the quadratic approximation of Balmer's equation, i.e., Equation (20), is the best choice when considering both the accuracy and the computational time, as the latter is directly related to the complexity of the formula involved.

The effect of the value of GSI on the percentage error in the shear strength prediction by the newly formulated approximate Mohr envelopes is shown in Figure 6. Here, the assumed rock mass conditions are the same as in Figure 3 and four normal stress intensities, i.e., $\sigma/\sigma_{ci} = 0.2$, 0.4, 0.6 and 0.8, are considered. Evidently, for the assumed values of normal stress, the trends in the variation of the percentage errors with GSI are similar in all approximate Mohr envelopes. When the value of GSI is very small, the percentage errors increase, but after reaching the peak, the errors tend to decrease with increasing GSI. Again, Figure 6 confirms that in the full range of GSI values the new approximate Mohr envelopes based on the quadratic and cubic Taylor approximations of Balmer's equation are more accurate than that based on the linear Taylor approximation.

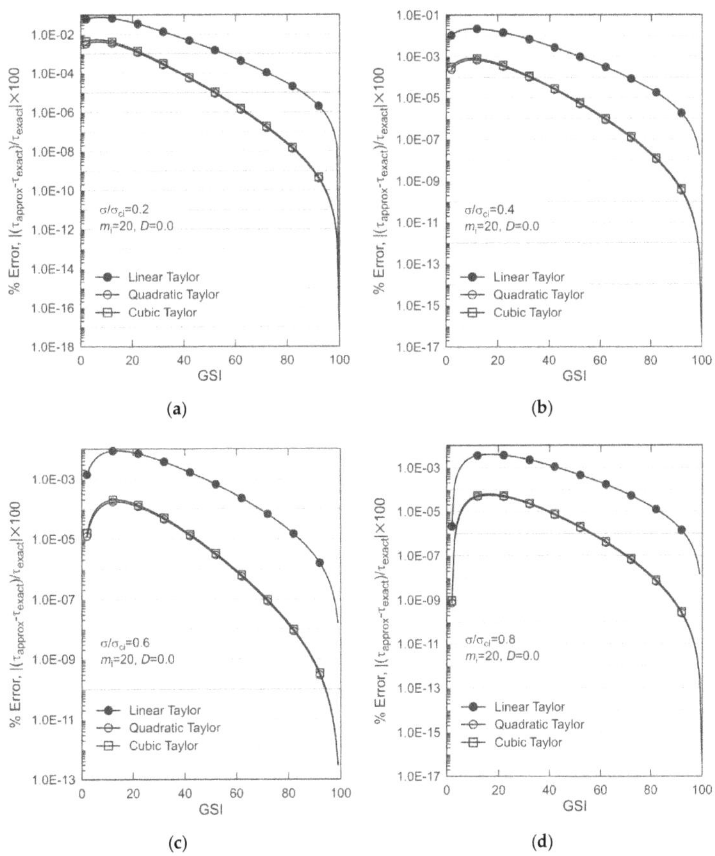

Figure 6. Effect of GSI value on the percentage errors of shear strength prediction based on newly formulated approximate Mohr envelopes; (**a**) σ/σ_{ci} = 0.2, (**b**) 0.4, (**c**) 0.6 and (**d**) 0.8.

4. Limit Equilibrium Analysis of a Slope in GHB Rock Mass

In order to illustrate the proposed formulation, the limit equilibrium analysis of a rock slope is conducted by employing the contact failure criterion based on the quadratic approximation of Balmer's equation, i.e., Equation (29) combined with Equation (20). It should be emphasized here that the original expression of the GHB criterion, i.e., the $\sigma_1 - \sigma_3$ relationship, is not suitable for this type of analysis as the equivalent explicit relation between the shear and the normal stress along the failure surface is not defined. Thus, in order to address the problem, the representation developed in the current study is required.

4.1. Geometry of Rock Slope Models

Figure 7a,b show the geometric configuration together with the assumed failure mechanisms. The slope geometry is defined by its height H and the face angle β_f. In both models analyzed here, it is assumed that a vertical tension crack of depth z is embedded at a distance x_c from the crest and the distributed load p acts on the horizontal upper surface of the slope. Two distinct failure modes are considered involving a planar surface with the inclination angle β_p and a circular failure surface, both passing through the toe of the slope and the tip of the crack. Given that the coordinates of the crack tip are (x_0, y_0), the minimum radius of the circular surface is

$$r_{\min} = \frac{x_0^2 + y_0^2}{2x_0} \tag{31}$$

The analysis employs the method of slices. For a circular slip surface, α_i represents the inclination of the base segment of the ith slice, whereas the corresponding distance from its mid-point to the center of rotation is $\overline{x_i} = r \sin \alpha_i$.

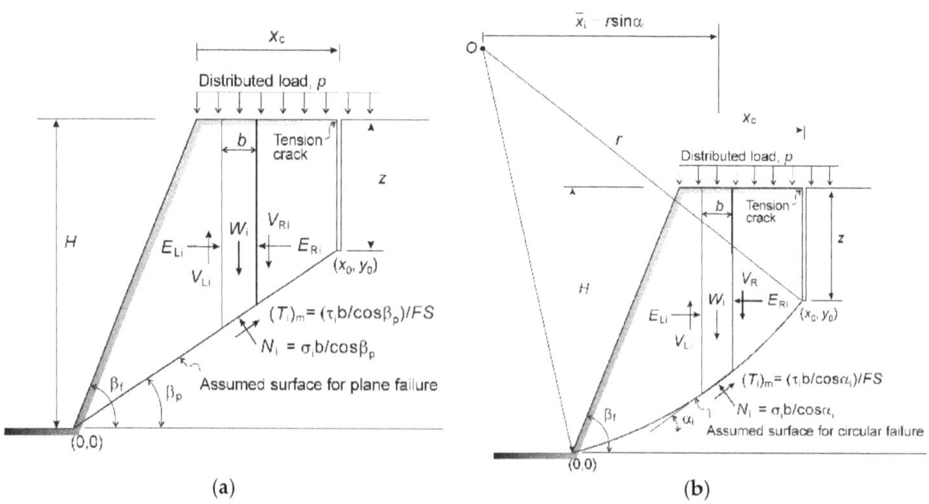

Figure 7. Geometry of slope showing forces of interaction acting on a typical slice: (**a**) model for plane failure; (**b**) model for circular failure.

4.2. Modified Bishop Approach for the Assessment of Safety Factor

The analysis presented here is conducted using a modified version of Bishop's simplified method. This approach is conceptually different from the original one (e.g., [33]). It invokes the classical framework of limit equilibrium analysis whereby an a priori assumption is made regarding the geometry of the failure surface along which the failure criterion is satisfied, and the safety factor is assessed by considering the global conditions of equilibrium. This is unlike the original Bishop approach in which the assessment of global stability is based on the notion of a local safety factor defined to estimate the current/mobilized shear stress. It is noted that the formulation of the simplified Bishop method, which incorporates a nonlinear relation for the safety factor requiring an iterative solver, raises some concerns. First of all, the assessment of the value of mobilized shear stress based on the failure condition that is satisfied *only* at the onset of failure may be questioned. In fact, prior to failure, the shear stress cannot be perceived as a unique function of normal stress. In addition, there is no basis for assuming that the local safety factor is constant within the domain. Given those concerns, the modified approach proposed here is not only computationally more efficient but also appears to be more rigorous.

For a circular failure surface (Figure 7b), the global safety factor is defined as the ratio of the moment resisting sliding to the overturning moment, both taken about the center of rotation. For the entire sliding wedge considered as a free body, the overturning is triggered by the own weight and the external load p, while the resisting moment is due to the shear stress distribution along the failure surface. Thus, summing up the contribution from individual slices, the safety factor (FS) is defined as

$$FS = \frac{\sum \tau_i \, b \sec \alpha_i}{\sum (w_i + pb) \sin \alpha_i} \tag{32}$$

where w_i is the weight of the slice and τ_i is the shear stress which, at the inception of the loss of stability, satisfies the local failure condition $\tau_i = \tau(\sigma_i)$ as defined by Equation (29). It should be noted here that for all slices intercepting the slope there is $p = 0$, as the distributed load acts only along the horizontal boundary.

The expression for the safety factor, viz. Equation (32), is now supplemented by considering the equilibrium of an individual slice. Referring again to Figure 7, V_i and E_i are the shear and normal forces of interaction between the slices. Neglecting now the variation in shear forces, i.e., taking $V_{Li} = V_{Ri}$ as commonly assumed in the Bishop simplified approach [25], and invoking the force equilibrium in the vertical direction, yields

$$W_i + pb - \sigma_i \, b - \tau_i b \tan \alpha_i = 0 \tag{33}$$

Again, since along the rapture surface the failure criterion is said to be satisfied, there is $\tau_i = \tau(\sigma_i)$ as stipulated in Equation (29). Thus, given Equation (33), the value of σ_i for each slice can be determined, which in turn defines the individual terms in the expression for the global safety factor (32).

It should be noted that in the original version of the simplified Bishop method, the equilibrium statement explicitly incorporates a local safety factor, i.e.,

$$W_i + pb - \sigma_i \, b - \frac{1}{FS} \tau_i b \tan \alpha_i = 0 \tag{34}$$

In this case, the problem becomes nonlinear and the simultaneous Equations (32) and (34) are solved in an iterative manner. Apparently, in case of a planar failure mode (Figure 7a), α_i in Equations (32)–(34) is replaced by a constant angle β_p.

For the circular failure mode, the factor of safety varies with the assumed radius r of the sliding block. In this case, the determination of FS represents a unimodal optimization problem for finding the critical radius that minimizes the factor of safety within the interval $r_{min} \leq r < \infty$. Recall that the minimum radius (r_{min}) can be calculated by Equation (31). In this paper, the minimum FS was determined by employing the golden section search algorithm [34]. In the examples that follow, the proposed approach is named as the *modified* Bishop method, while the approach incorporating Equation (34) is referred to by its original name, i.e., the *simplified* Bishop method.

4.3. Comparison of Safety Factors Based on the Simplified and Modified Bishop Methods

The simulations have been carried out assuming $H = 35$ m and $\beta_f = 70°$, while the strength parameters were taken as $\sigma_{ci} = 20$ MPa, $m_i = 10$, $D = 0$. The unit weight of rock was assumed as $\gamma = 26$ kPa/m. In the example given here, no distributed load was considered, i.e., $P = 0$.

Figure 8 shows the variation in the safety factor as a function of GSI for the slope models having a 5 m deep tension crack. Two different horizontal positions of the crack, i.e., (a) $x_c = 5$ m, (b) $x_c = 10$ m, are considered. In the figure, the results obtained by assuming the plane and circular failure surfaces are presented together for the purpose of comparison. It is evident that the simplified Bishop method predicts larger *FS* than the modified approach. When $x_c = 10$ m and GSI = 40, for example, the safety factors for circular failure surface are 1.84 with the simplified method and 1.27 with the modified approach, while the respective factors of safety for a plane failure surface are 1.94 and 1.30. However, as the GSI value decreases, the difference becomes smaller. This is because the global *FS*s from both methods approach unity as the rock mass quality becomes poorer. Here, it should be noted that both methodologies predict the same condition for the onset of the loss of stability, i.e., the case when $FS = 1$. Another interesting feature is that, in the case of the modified method, there is little change in the safety factor when the GSI value is between 10 and 50. Thus, the modified method predicts a more conservative safety factor in a wide range of rock mass quality.

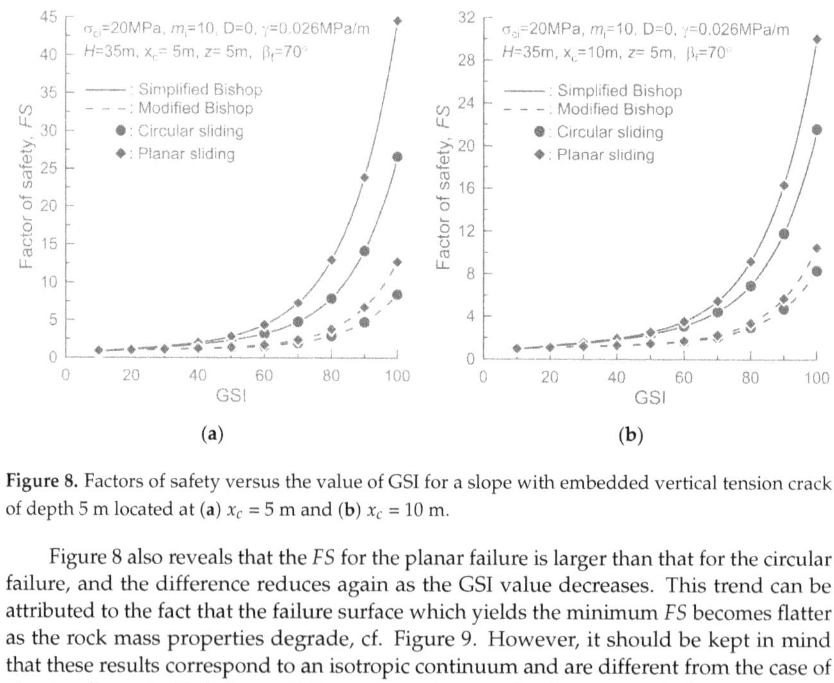

Figure 8. Factors of safety versus the value of GSI for a slope with embedded vertical tension crack of depth 5 m located at (**a**) x_c = 5 m and (**b**) x_c = 10 m.

Figure 8 also reveals that the *FS* for the planar failure is larger than that for the circular failure, and the difference reduces again as the GSI value decreases. This trend can be attributed to the fact that the failure surface which yields the minimum *FS* becomes flatter as the rock mass properties degrade, cf. Figure 9. However, it should be kept in mind that these results correspond to an isotropic continuum and are different from the case of structurally controlled planar sliding commonly occurring in many rock slopes.

Finally, Figure 10 shows the variation in safety factor with the distance (x_c) defining the crack location. The trend is different for both methodologies. It is interesting to note that for large values of GSI, e.g., GSI = 60, the safety factor based on the *simplified method* decreases quite abruptly with increasing value of x_c when the crack is near the slope. The latter is intuitively not correct as it implies that the cracks at locations closer to the crest result in a more stable configuration. Here, the results based on the modified method appear to be more consistent.

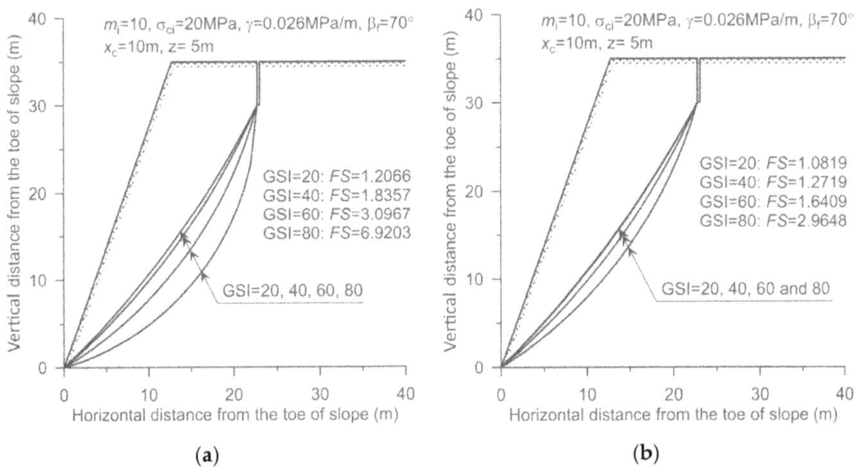

Figure 9. The circular failure surfaces giving the minimum *FS* for four different GSI values; slope model with x_c = 10 m and z = 5 m: (**a**) simplified Bishop method; (**b**) modified Bishop method.

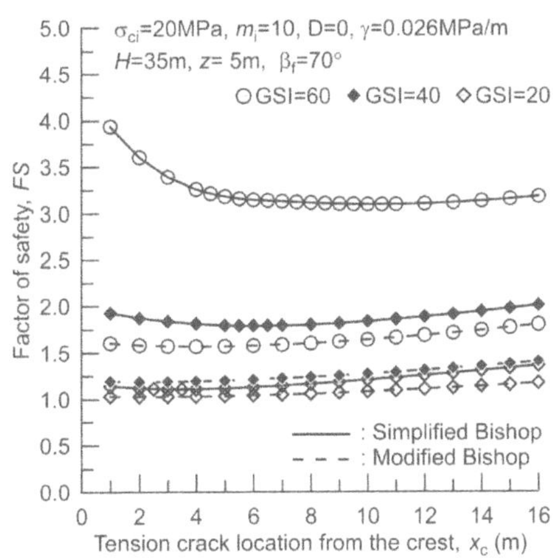

Figure 10. Variation in factor of safety with the location of 5 m deep tension crack for three different values of GSI: simplified vs. modified Bishop approach.

4.4. Assessment of the Critical Value of Surface Load

In this section, the critical value of distributed load p, which results in the loss of stability, was assessed for the same slope geometry as that shown in Figure 8. Since in this case the solution requires $FS = 1$, both methodologies yield the same results, while the *modified* Bishop method is simpler and computationally more efficient. The latter approach was implemented here in an iterative manner by adjusting the value of p until the safety factor became close to 1.

Figure 11 shows the predicted variation in critical load as a function of GSI for two different crack depths, i.e., z = 5 m and 10 m, and two crack locations, i.e., x_c = 5 m and 10 m. It is seen that the critical load increases exponentially as the GSI value increases. It is also evident that the critical load for plane failure is larger than that for circular failure, which is consistent with the *FS* calculation results given in Figure 8. Another observation, which stems from the result shown in Figure 11, is that the critical load increases with increasing crack depth. This may be due to the fact that, as the crack depth increases, the average inclination of the sliding surface decreases, and consequently it becomes more resistant to failure. To examine the effect of crack depth in more detail, the critical loads were calculated for four different depths by assuming a circular failure mode (Figure 12). In this case, it is apparent that the effect of crack depth is less significant when the crack is located close to the slope, but the influence increases as the position of the vertical crack moves further away from the slope.

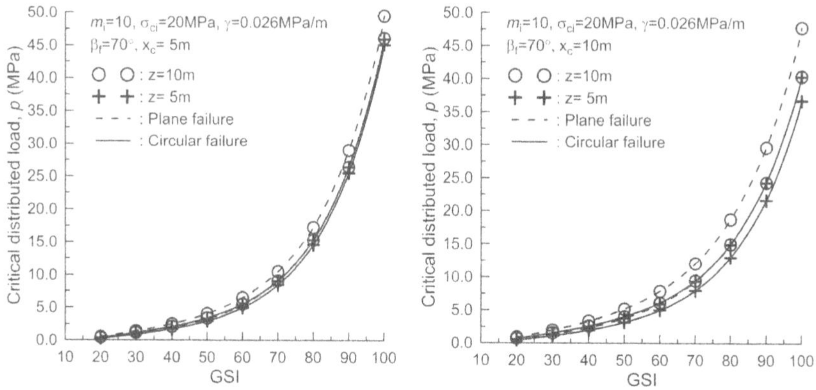

(a) (b)

Figure 11. The critical value of distributed load versus GSI for two different crack depths, i.e., 5 m and 10 m; planar and circular failure modes: (**a**) x_c = 5 m and (**b**) x_c = 10 m.

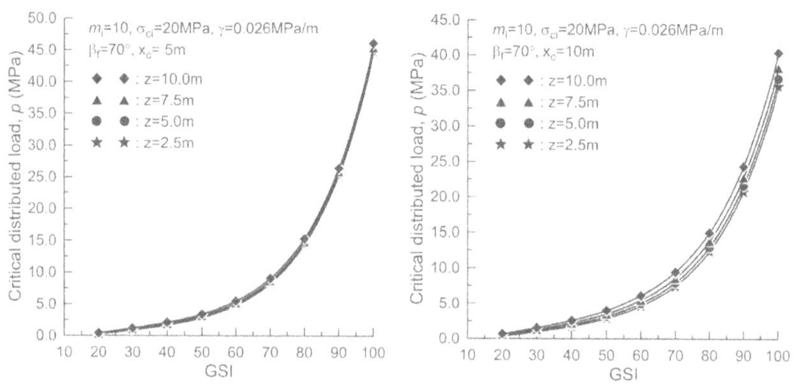

(a) (b)

Figure 12. The critical value of distributed load versus GSI for four different crack depths, i.e., 2.5 m, 5.0 m, 7.5 m and 10.0 m; circular failure mode: (**a**) x_c = 5 m; (**b**) x_c = 10 m.

Finally, Figure 13 shows the geometry of critical failure surfaces associated with the loss of stability for GSI = 20, 40, 60 and 90. In this case, a 10 m deep vertical crack is assumed to be present at two locations, i.e., $x_c = 5$ m and 10 m. The results indicate that, in this case, the influence of GSI value on the shape of the critical surface is not very significant.

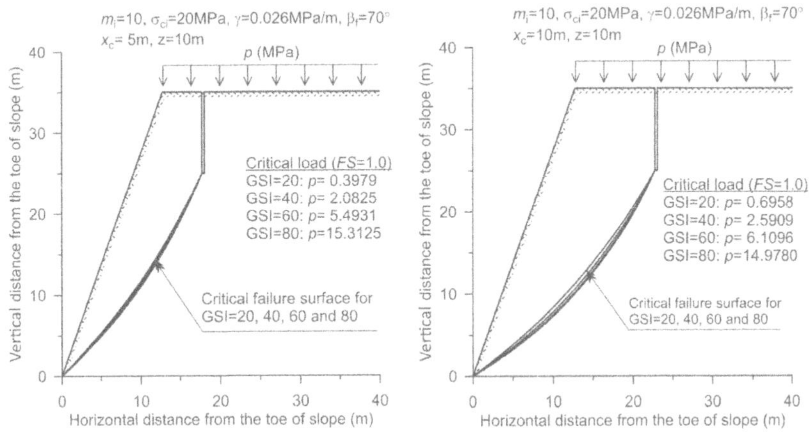

(**a**) (**b**)

Figure 13. The geometry of the critical circular failure surfaces for four different values of GSI, i.e., 20, 40, 60 and 80; slope model with a 10 m deep crack: (**a**) x_c = 5 m; (**b**) x_c = 10 m.

5. Conclusions

The GHB criterion considers the nonlinearity of the rock mass strength and is applicable to a broad spectrum of rock masses, from weak to competent ones. However, a serious disadvantage of this criterion is that when $a \neq 0.5$, its shear strength–normal stress relationship, i.e., the Mohr envelope, is not available in the form of an explicit analytical expression. An alternative to overcome this difficulty is to define the Mohr envelope in an approximate analytical form. In this paper, a new approximate expression of Mohr envelope of the GHB criterion, which has much higher accuracy compared to other existing approaches, was proposed. The idea behind this formulation is to approximate Balmer's equation [31], which defines the relationship between normal stress and minor principal stress at failure, by the Taylor polynomial equations of a finite degree that can be solved analytically in an explicit form.

At a first glance, the proposed formulation looks similar to that of Lee and Pietruszczak [15] in that it starts from the approximation of Balmer's parametric equation which defines the relationship between the normal stress and the minor principal stress at failure. However, in the approach pursued here Balmer's equation is approximated much more accurately by replacing the power function terms with the finite Taylor series centered at the exact root of Balmer's equation for $a = 0.5$. The accuracy of the resulting approximate Mohr envelopes, incorporating the Taylor approximation of degree up to 3, was found to be superior to that of the recently published approximation of Lee and Pietruszczak [15]. Among the three cases considered, the one based on the quadratic Taylor approximation exhibited the best accuracy. Due to the mathematical constraints embedded in the Taylor approximation, as the GSI value approaches 100, the new three approximate Mohr envelopes come close to the exact Mohr envelope. Moreover, it can be shown that the accuracy of the proposed approach can be further improved if the solution is expressed in the form of a Coulomb equation employing the tangential friction angle of the approximate Mohr envelope, although in this case the resulting equations become algebraically more complicated. More importantly, since the proposed formulation of the Mohr envelope does not impose any restrictions on the range of σ_3, the high accuracy of the calculated shear strength can be retained in the whole range of normal stress. Therefore, it is expected that the newly proposed approximate Mohr envelope can find its applications when assessing the stability in the vicinity of a rock excavation surface where a relatively high stress gradient can occur.

As an example of application, the limit equilibrium analysis of a rock slope was carried out by incorporating the newly derived equation of the Mohr envelope based on the quadratic Taylor approximation. Two different approaches were considered, viz. the conventional simplified Bishop method and the modified Bishop method. In the proposed modified method, the factor of safety for slope failure is calculated in a global sense with the assumption that the current stress state satisfies the failure condition. The slope models employed in this study considered a vertical tension crack embedded in the upper horizontal surface, and the factors of safety have been calculated for varying crack position, crack depth and GSI value. In addition, the critical value of distributed load causing the loss of stability has been assessed for different cases. The results of the limit equilibrium analysis have shown that the factor of safety for plane failure is larger than that for circular failure. Furthermore, in a good quality rock mass the safety factor is more sensitive to the value of GSI than in a poor quality rock. Of the two equilibrium methods, the modified approach has resulted in a lower value of *FS*. However, the difference in the calculated *FS*s became smaller with the decrease in the value of GSI. The analysis has also shown that the critical magnitude of distributed load triggering the slope failure increases exponentially with the increase in GSI. However, for a given geometry of vertical crack, the GSI value did not significantly affect the shape of the critical circular surface. As the crack deepened, the critical load showed the tendency to increase, and this trend was more pronounced as the crack moved further away from the slope.

In conclusion, the illustrative examples given here demonstrated that the approximate equation of the Mohr envelope derived in this study can be conveniently used for stability analysis of slopes in the GHB rock mass, which is not feasible with the original form of the GHB criterion. The limit equilibrium analysis incorporated a *modified* version of the Bishop method. The latter is simpler and more rigorous than the original approach. Its implementation is quite straightforward, as it does not involve a nonlinear expression for the safety factor, and the estimates of stability are more conservative than those obtained using the conventional methodology.

Finally, it needs to be emphasized that the current study of slope stability is preliminary and serves mainly as an exploratory example. As mentioned earlier, the analysis employed a simple failure mechanism involving a circular surface passing through the toe of the slope and the tip of a pre-existing tension crack. Certainly, a proper verification of the proposed modified Bishop method requires a more in-depth study incorporating other failure mechanisms. Thus, even though the predicted basic trends are in line with an intuitive assessment, the quantitative verification is still required. In addition, some real case studies involving slope failures in rocks need to be examined to gain more confidence in the proposed methodology. Such studies are planned to be carried out in the future. In particular, the use of commercial software, such as FLAC, will be explored for comparison purposes. In addition, an upper bound limit analysis, incorporating the proposed approximation to the Mohr envelope, will also be pursued for a class of problems dealing with assessment of bearing capacity.

Author Contributions: Conceptualization, Y.-K.L. and S.P.; methodology, Y.-K.L. and S.P.; software, Y.-K.L.; validation, Y.-K.L.; formal analysis, Y.-K.L.; investigation, Y.-K.L. and S.P.; resources, Y.-K.L.; data curation, Y.-K.L.; writing—original draft preparation, Y.-K.L.; writing—review and editing, Y.-K.L. and S.P.; visualization, Y.-K.L.; supervision, Y.-K.L.; project administration, Y.-K.L.; funding acquisition, Y.-K.L. All authors have read and agreed to the published version of the manuscript.

Funding: This research was funded by the National Research Foundation of Korea (NRF) (No. 2021R1F1A1048311).

Institutional Review Board Statement: Not applicable.

Informed Consent Statement: Not applicable.

Data Availability Statement: Not applicable.

Acknowledgments: This work was supported by the National Research Foundation of Korea (NRF) grant funded by the Korean government (MSIT) (No. 2021R1F1A1048311). This financial support is very much appreciated.

Conflicts of Interest: The authors declare no conflict of interest.

References

1. Hoek, E.; Carranza-Torres, C.T.; Corkum, B. Hoek–Brown failure criterion-2002 edition. In Proceedings of the 5th North American Rock Mechanics Symposium, Toronto, ON, Canada, 7–10 July 2002.
2. Marinos, P.; Hoek, E. GSI: A geologically friendly tool for rock mass strength estimation. In Proceedings of the GeoEng2000 at the International Conference on Geotechnical and Geological Engineering, Melbourne, Australia, 19–24 November 2000.
3. Marinos, V.; Marinos, P.; Hoek, E. The geological strength index: Applications and limitations. *Bull. Eng. Geol. Environ.* **2005**, *64*, 55–65. [CrossRef]
4. Wyllie, D.C.; Mah, C.W. *Rock Slope Engineering*, 4th ed.; Spon Press: London, UK, 2004.
5. Deng, D.P.; Li, L.; Wang, J.F.; Zhao, L.H. Limit equilibrium method for rock slope stability analysis by using the generalized Hoek-Brown criterion. *Int. J. Rock Mech. Min. Sci.* **2016**, *89*, 176–184.
6. Merifield, R.S.; Lyamin, A.V.; Sloan, S.W. Limit analysis solutions for the bearing capacity of rock masses using the generalized Hoek–Brown criterion. *Int. J. Rock Mech. Min. Sci.* **2006**, *43*, 920–937. [CrossRef]
7. Yang, X.L.; Zou, J.F. Stability factors for rock slopes subjected to pore water pressure based on the Hoek-Brown failure criterion. *Int. J. Rock Mech. Min. Sci.* **2006**, *43*, 1146–1152. [CrossRef]
8. Chen, W.F. *Limit Analysis and Soil Plasticity*; J. Ross Publishing: Fort Lauderdale, FL, USA, 2007.
9. Michalowski, R.L.; Park, D. Stability assessment of slopes governed by the Hoek-Brown strength criterion. *Int. J. Rock Mech. Min. Sci.* **2020**, *127*, 104217. [CrossRef]
10. Park, D.; Michalowski, R.L. Three-dimensional roof collapse analysis in circular tunnels in rock. *Int. J. Rock Mech. Min. Sci.* **2020**, *128*, 104275. [CrossRef]
11. Hoek, E. Strength of jointed rock masses, 23rd. *Rankine Lecture. Geotechnique* **1983**, *23*, 187–223. [CrossRef]
12. Ucar, R. Determination of shear failure envelope in rock masses. *J. Geotech. Eng. Div. ASCE* **1986**, *112*, 303–315. [CrossRef]
13. Londe, P. Discussion on "Determination of shear failure envelope in rock masses" by R. *Ucar, J. Geotech. Eng. Div. ASCE* **1988**, *114*, 374–376. [CrossRef]
14. Lee, Y.K.; Pietruszczak, S. Analytical representation of Mohr envelope approximating the generalized Hoek-Brown failure criterion. *Int. J. Rock Mech. Min. Sci.* **2017**, *100*, 90–99. [CrossRef]
15. Lee, Y.K.; Pietruszczak, S. Limit equilibrium analysis incorporating generalized Hoek-Brown failure criterion. *Rock Mech. Rock Eng.* **2021**, *54*, 4407–4418. [CrossRef]
16. Lee, Y.K. Approximate shear strength formula implied in the generalized Hoek-Brown failure criterion. *Tunn. Undergr. Space* **2018**, *28*, 426–441.
17. Gao, K.; Bozorgzadeh, N.; Harrison, P. The equivalence of three shear-normal stress forms of the Hoek-Brown criterion. *Rock Mech. Rock Eng.* **2019**, *52*, 3501–3507. [CrossRef]
18. Sofianos, A.I.; Nomikos, P.P. Equivalent Mohr–Coulomb and generalized Hoek–Brown strength parameters for supported axisymmetric tunnels in plastic or brittle rock. *Int. J. Rock Mech. Min. Sci.* **2006**, *43*, 638–704. [CrossRef]
19. Jimenez, R.; Serrano, A.; Olalla, C. Linearization of the Hoek and Brown rock failure criterion for tunneling in elasto-plastic rock masses. *Int. J. Rock Mech. Min. Sci.* **2008**, *45*, 1153–1163. [CrossRef]
20. Lee, Y.K. Analytical Formula for the Equivalent Mohr-Coulomb Strength Parameters Best-fitting the Generalized Hoek-Brown Criterion in an Arbitrary Range of Minor Principal Stress. *Tunn. Undergr. Space* **2019**, *29*, 172–183.
21. Renani, H.R.; Martin, C.D. Slope stability analysis using equivalent Mohr-Coulomb and Hoek-Brown criteria. *Rock Mech. Rock Eng.* **2020**, *53*, 13–21. [CrossRef]
22. Li, A.J.; Merifield, R.S.; Lyamin, A.V. Stability chars for rock slopes based on the Hoek-Brown failure criterion. *Int. J. Rock Mech. Min. Sci.* **2008**, *45*, 689–700. [CrossRef]
23. Wei, Y.; Fu, W.; Ye, F. Estimation of the equivalent Mohr-Coulomb parameters using Hoek-Brown criterion and its application in slope analysis. *Eur. J. Environ. Civ. Eng.* **2021**, *25*, 599–617. [CrossRef]
24. Fraldi, M.; Guarracino, F. Limit analysis of collapse mechanism in cavities and tunnels according to the Hoek-Brown failure criterion. *Int. J. Rock Mech. Min. Sci.* **2009**, *46*, 665–673. [CrossRef]
25. Yang, X.L.; Huang, F. Collapse mechanism of shallow tunnel based on nonlinear Hoek-Brown failure criterion. *Tunn. Undergr. Space Technol.* **2011**, *26*, 686–691. [CrossRef]
26. Suchowerska, A.M.; Merifield, R.S.; Carter, J.P.; Clausen, J. Prediction of underground cavity roof collapse using the Hoek-Brown failure criterion. *Comput. Geotech.* **2012**, *44*, 93–103. [CrossRef]
27. Kumar, P. Shear failure envelope of Hoek-Brown criterion for rockmass. *Tunn. Undergr. Space Technol.* **1998**, *13*, 453–458. [CrossRef]
28. Yang, X.L.; Yin, J.H. Linear Mohr-Coulomb strength parameters from the non-linear Hoek-Brown rock masses. *Int. J. Non-Linear Mech.* **2006**, *41*, 1000–1005. [CrossRef]

29. Yang, X.L.; Yin, J.H. Upper bound solution for ultimate bearing capacity with a modified Hoek-Brown failure criterion. *Int. J. Rock Mech. Min. Sci.* **2005**, *42*, 550–560. [CrossRef]
30. Rojat, F.; Labiouse, V.; Mestat, P. Improved analytical solutions for the response of underground excavations in rock mass satisfying the generalized Hoek-Brown failure criterion. *Int. J. Rock Mech. Min. Sci.* **2015**, *79*, 193–204. [CrossRef]
31. Balmer, G. A general analytical solution for Mohr's envelope. *Proc. ASTM* **1952**, *52*, 1260–1271.
32. Press, W.H.; Teukolsky, S.A.; Vetterling, W.T.; Flannery, B.P. *Numerical Recipes in Fortran*, 2nd ed.; William H. Press: New York, NY, USA, 1992.
33. Das, B.M. *Principles of Geotechnical Engineering*, 4th ed.; PWS Publishing Company: Boston, MA, USA, 1997.
34. Chapra, S.C.; Canale, R.P. *Numerical Methods for Engineers*, 5th ed.; McGraw-Hill: New York, NY, USA, 2005.

MDPI
St. Alban-Anlage 66
4052 Basel
Switzerland
www.mdpi.com

Sustainability Editorial Office
E-mail: sustainability@mdpi.com
www.mdpi.com/journal/sustainability

www.ingramcontent.com/pod-product-compliance
Lightning Source LLC
Chambersburg PA
CBHW061454220926
51850CB00004B/93
9783036597720